Electroweak interactions

Electroweak interactions

an introduction to the physics of quarks and leptons

PETER RENTON
Nuclear Physics Laboratory, University of Oxford

CAMBRIDGE UNIVERSITY PRESS
Cambridge
New York Port Chester
Melbourne Sydney

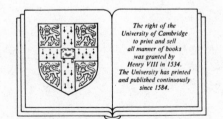

Published by the Press Syndicate of the University of Cambridge
The Pitt Building, Trumpington Street, Cambridge CB2 1RP
40 West 20th Street, New York, NY 10011, USA
10 Stamford Road, Oakleigh, Melbourne 3166, Australia

First published 1990

Printed in Great Britain at the University Press, Cambridge

British Library Cataloguing in Publication Data

Renton, Peter
Electroweak interactions: an introduction
to the physics of quarks and leptons.
1. Leptons & quarks
I. Title
539.7'21

Library of Congress Cataloguing in Publication Data

Renton, Peter.
Electroweak interactions: an introduction to the physics of
quarks and leptons / Peter Renton.
p. cm.
Bibliography: p.
Includes index.
ISBN 0 521 26603 3
1. Electroweak interactions. 2. Quarks. 3. Leptons (Nuclear
physics) I. Title.
QC794.8.E44R46 1990
539.7'21–dc19 88-7354 CIP

ISBN 0 521 26603 3 hard covers
ISBN 0 521 36692 5 paperback

To Diana

'Beauty is truth, truth beauty – that is all ye
know on earth, and all ye need to know'

John Keats 1795–1821

Earliest reference to $SU(2)$ symmetry for heavy flavours?

Contents

Preface

The development of our present understanding of the weak and electromagnetic interactions, and of their unification, has proven to be a fascinating dialogue between profound theoretical insights and remarkable experimental discoveries. The electromagnetic and weak interactions appear to be describable by local gauge theories. Phenomena involving charged leptons and photons can be described, to impressive precision, by the relativistic field theory of Quantum Electrodynamics. The incorporation of weak phenomena led, eventually, to the Glashow–Salam–Weinberg model, the so-called standard model. This theory describes the electroweak properties of leptons and quarks, and successfully predicted the existence of the massive $W^{\pm}$ and Z^0 vector bosons. At present, the standard model is compatible with all well established experimental data, amassed by some tens of thousands of man-years of effort. This is an impressive state of affairs.

This book has its origins in a course of lectures given to first-year postgraduate students (mainly experimentalists) in high energy physics in Oxford. An elementary knowledge of high energy particle physics, together with the basic ideas of quantum mechanics and relativity, is assumed. However, these topics, in as far as they are required for later use, are reviewed in Chapters 1 and 2. An introduction to group theories and symmetries, concepts of great importance in particle physics, is also given in Chapter 2. In Chapter 3, the single particle wave equations for spin 0, spin $\frac{1}{2}$ and spin 1 particles are developed, and an introduction to some of the concepts of field theory is given. An intuitive, rather than a completely rigorous, approach is adopted.

Emphasis is given in the text to practical calculations. In Chapter 4 the cross-sections for a variety of electromagnetic processes, using the Feynman rules of Quantum Electrodynamics, are derived. A brief discussion as to how the higher order divergences in the theory can be rendered finite is also given. Chapter 5 traces the development of the theory of weak interactions from the theory of Fermi to the standard model. This discussion includes the ideas of spontaneous symmetry breaking and of the need for a scalar Higgs particle to give masses to the massive vector bosons and fermions. In Chapter 6 the cross-sections and decay rates for a variety of purely leptonic processes are derived. These constitute the cleanest tests of the theory, and a comparison with the experimental results is given.

Quantum Chromodynamics (QCD) is a theory which describes the strong interactions of quarks, antiquarks and gluons, and is also a local gauge theory. However, perturbative calculations can only be made in the regime where the momentum transferred in the process is large. In this case the effective strong coupling constant is small. In Chapter 7 the properties of quarks and gluons, as they have been established from studies of deep inelastic scattering of leptons off nucleons, are discussed. The ideas of QCD, in as far as they are needed to describe electroweak phenomena, are outlined. In Chapter 8 the weak decays of hadrons are discussed. In this non-perturbative regime the ideas of QCD can be used only in a rather qualitative fashion. Electroweak interference phenomena involving quarks are also described.

In Chapter 9 the two intriguing topics of *CP* violation and the neutrino mass (and potential oscillations) are discussed. In Chapter 10 the discovery and properties of the $W^\pm$ and Z^0 bosons are described, together with a discussion of the precision tests of the standard model which are planned to be carried out at the new generation of e^+e^- colliders (SLC at Stanford in the USA and LEP at CERN in Switzerland). Potential methods for detecting the top quark and the Higgs scalar particle, which are two missing ingredients of the standard model, are also described. Finally, the theoretical shortcomings of the standard model are outlined, and attempts to go beyond it are briefly reviewed.

It is hoped that the text will prove useful in the formulation of graduate and specialist undergraduate courses. No specific exercises are given, but it is suggested that the interested reader performs detailed calculations on each type of the lowest order processes discussed. The book may also serve as a reference text for the more seasoned workers in the field.

A systematic review of all the experimental results on any particular topic is not, in general, attempted. The examples of the experimental results used for the purposes of illustration have been selected in a somewhat arbitrary fashion. Experimental results produced up to the end of 1986 are included, with some selected additional data from 1987.

Discussion of the various experiments is restricted to the results, plus a few additional remarks. Results quoted without reference are taken from the 'Review of Particle Properties' (1986, *Physics Letters*, **170B**). If two errors are quoted with a measurement, then the first is statistical and the second systematic. The lack of experimental detail is not meant in any way to undervalue the role of experiments; indeed, high energy physics is ultimately an experimental subject. However, in order to do justice to the ingenuity of the experimenters, it would require detailed discussion far beyond the scope of this book. The reader is referred to the works by, for example, Kleinknecht (1986) and Williams (1986).

Although the main emphasis is on the understanding of physical processes in terms of lowest order calculations in the standard model, it should be stressed that higher order contributions, and the development of theories beyond the standard model, are, of course, important. These topics are discussed in the more advanced texts of Aitchison (1982), Mandl and Shaw (1984), Cheng and Li (1984), and Ross (1985). Finally, I would like to acknowledge my debt to the many excellent articles and texts used in the formulation of the approaches adopted here. I gratefully acknowledge the constructive comments of my colleagues Robin Devenish, Gerald Myatt, Ray Rook, Graham Ross, John Wheater, and Bill Williams, on various parts of the text. I thank also Irmgard Smith for her careful help in the preparation of the figures.

1

Introduction

1.1 STRONG, ELECTROMAGNETIC AND WEAK INTERACTIONS

One of the main objectives of physics is to find out what, if any, are the basic constituents of matter and to understand the nature of the forces by which they interact. Fundamental particles appear, at present, to be of two distinct types. The first group consists of *quarks* and *leptons*. These are spin $\frac{1}{2}$ particles obeying Fermi–Dirac statistics (*fermions*). The second group consists of the so-called *gauge bosons*. These are integral spin particles obeying Bose–Einstein statistics (*bosons*). The gauge bosons appear to be responsible for mediating the interaction forces between quarks and leptons. Existing results show clear evidence for four types of interactions in nature. These are the strong, electromagnetic, weak and gravitational interactions. Our knowledge of these interactions stems, to a great extent, from our understanding of the underlying symmetries which appear to exist in nature and in the way in which they appear to be broken.

The world is made up of ninety-two naturally occurring chemical elements. The properties of a given isotope of an element do not, as far as we know, depend on its origin. These elements are composed of electrons and nuclei, which are in turn composed of protons and neutrons. The electrons are fermions and obey the Pauli exclusion principle. This leads to an elaborate shell structure and important differences in the chemical properties of the elements. Prior to the development of particle accelerators, studies in particle physics were limited to indirect means. These consisted of using either the low energy particles produced in the radioactive decays of nuclei or the higher energy protons, nuclei and other particles which make up cosmic radiation.

Direct study of the interactions of particles at high energy physics laboratories is made essentially by two methods. In the first, the interactions of beams of high energy particles on a stationary (fixed) target are studied. The target particle is either an atomic electron or, more likely, a nucleus. Hydrogen and deuterium targets are used if nuclear effects, such as Fermi motion, are to be avoided. The beam or projectile particle is obtained by first accelerating to high energy either protons (e.g. CERN 450 GeV, Fermi National Accelerator Lab, FNAL, 1 TeV) or electrons (e.g. Stanford Linear Accelerator Center, SLAC, 20 GeV). In the case of proton accelerators, the extracted protons can also be used to make secondary beams of $\pi^{\pm}$, $K^{\pm}$, K^0, $\bar{K}^0$ (*mesons*) or $\bar{p}$, n, $\bar{n}$, Λ, Σ, Ξ, Ω^- (*baryons*) by suitably collecting the decay products from the primary proton interactions. Beams are restricted to those *hadrons* (a class consisting of mesons and baryons) which are both sufficiently long-lived to give a reasonable flight path and light enough to be produced copiously. Beams of $e^{\pm}$, $\mu^{\pm}$, ν_e, $\bar{\nu}_e$, ν_μ and $\bar{\nu}_\mu$ (*leptons*) are produced either by starting with atomic electrons or allowing mesons to decay.

The second method of study is that of the collisions of two high energy beams of particles. Particles which are essentially stable must be used, so studies with this technique are restricted to pp, $\bar{p}p$ (e.g. CERN Collider, up to 450 + 450 GeV) and e^+e^- (e.g. DESY, 25 + 25 GeV). Beams of heavy ions can also be considered for colliding beam machines. Since the collisions take place in the centre-of-mass (cms) system, high values of the cms energy can be reached. For example, in the CERN Collider the cms energy for 270 GeV beams is 540 GeV, whereas the cms energy of a 270 GeV proton collision with a stationary proton is only 27.5 GeV. However, this gain is at the expense of effective luminosity (luminosity = reaction rate/cross-section). The most recent and planned future accelerators are mainly of this type and include machines at FNAL ($\bar{p}p$ 1 + 1 TeV), TRISTAN in Japan (e^+e^- 30 + 30 GeV), the Stanford Linear Collider (SLC) at SLAC (e^+e^- 50 + 50 GeV), LEP at CERN (e^+e^- phase I 55 + 55 GeV, 1989; phase II 95 + 95 GeV, 1993), HERA at DESY (e^-p 30 + 820 GeV, 1990), UNK in the Soviet Union (pp 3 + 3 TeV, 1993?) and possibly the Superconducting Super Collider (SSC) in the USA (pp 20 + 20 TeV, 1995?). Note that for the HERA Collider the protons are much more energetic than the electrons, and hence the collisions are not in the ep cms.

Hadrons (from the Greek hadros = strong) all have *strong* interactions. The total cross-section for π^+p collisions, as a function of the lab momentum of the π^+ meson (p_{lab}) is shown in Fig. 1.1. For $p_{lab} \sim 10$ GeV, the total cross-section is about 25 mb. The leptons (from Greek meaning light, small) do not feel the strong force and their interaction cross-sections are significantly less than those of hadrons. The total cross-section for the process $e^+e^- \to \mu^+\mu^-$ as a function of $(s)^{1/2}$, the cms energy of the e^+e^-

system, is shown in Fig. 1.2. Photon beams can also be produced at proton accelerators and the γp cross-section, as a function of the laboratory (lab) energy of the photon, is shown in Fig. 1.1. Charged leptons and photons interact by the *electromagnetic* interaction. The main contribution to the cross-section shown in Fig. 1.2 is from this interaction. The cross-sections for neutral leptons (ν and $\bar{\nu}$) are many orders of magnitude less than those for charged leptons. The total cross-section for ν_μ interactions as a function of the lab neutrino energy E_ν is shown in Fig. 1.1. This is an example of a *weak* interaction and the cross-section shows an approximately linear increase with E_ν. For $E_\nu \sim 10$ GeV the cross-section is about 11 orders of magnitude less than for 10 GeV π^+p (strong) interactions.

Fig. 1.1 Sketch of the total cross-sections, as a function of laboratory momentum p_{lab}, for π^+p, γp and ν_μN interactions. For ν_μ the results are the cross-section per nucleon (average of proton and neutron), obtained from heavy nuclear targets (e.g. iron).

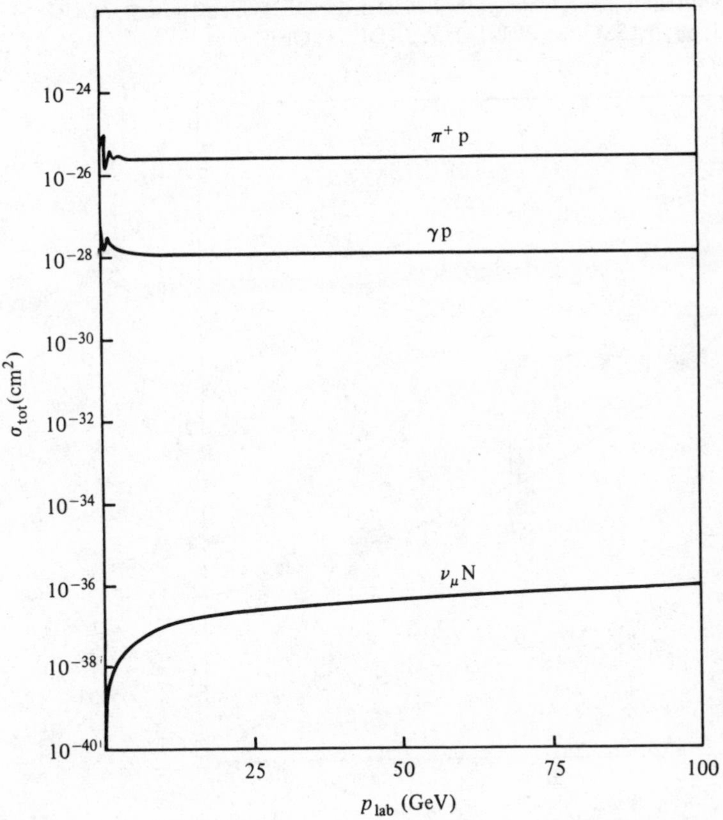

The decay rates, or equivalently the lifetimes, of particles also span an enormous range, as shown in Fig. 1.3. The typical width Γ of a resonance decaying by the strong interaction (e.g. $\rho \to \pi\pi$) is approximately 100 MeV. Use of the uncertainty principle $\Gamma\tau \sim \hbar = 6.6 \times 10^{-22}$ MeV s (see Appendix A) gives a corresponding lifetime $\tau \sim 10^{-23}$ s. The decay of the neutral pion $\pi^0 \to \gamma\gamma$, which is an electromagnetic interaction, has a lifetime $\tau \sim 10^{-16}$ s. The charged pion, however, decays weakly (mainly via the decay $\pi \to \mu\nu_\mu$) with a lifetime $\tau \sim 10^{-8}$ s. These are typical values; however, it can be seen from Fig. 1.3 that the range of lifetimes for a given interaction is large, in particular for the weak interaction. Part of the reason for this large spread in lifetimes is that the density of final states (or phase space) available is often small due to the small energy release in the decay. For example, the expression for the decay rate for neutron beta decay,

Fig. 1.2 Total cross-section for the interaction $e^+e^- \to \mu^+\mu^-$. The data points are from e^+e^- experiments at PETRA and have been corrected for higher order radiative effects and hence are comparable with the full line, which is the lowest order QED calculation. ■, MARK J; ●, TASSO; ▲, PLUTO; ▼, JADE; ◆, CELLO.

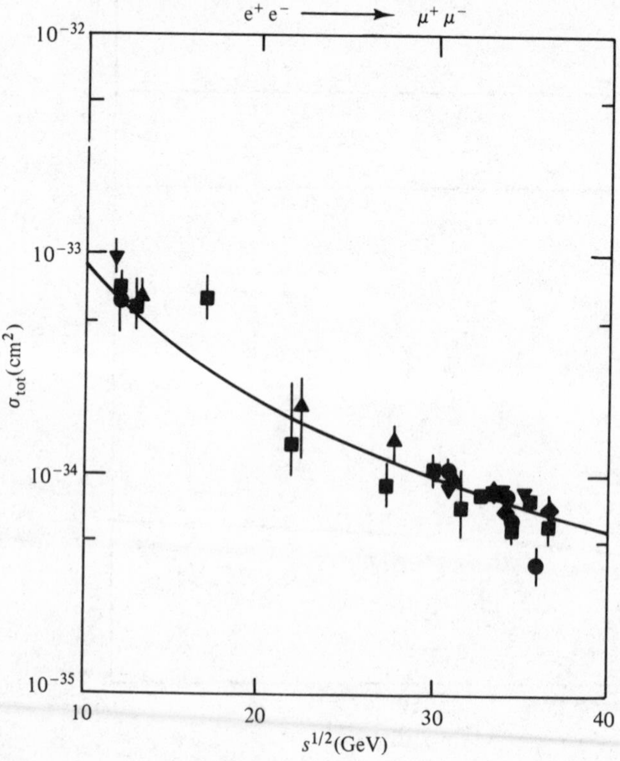

$n \to p + e^- + \bar{\nu}_e$, contains a factor Δ^5, where $\Delta = m_n - m_p = 1.3$ MeV, and this small energy release leads to a rather long lifetime ($\tau_n \simeq 900$ s).

The classification of interactions into a hierarchy of strong, electromagnetic and weak is a convenient framework. However, a particular scattering or decay process can of course have contributions from more than one of these forces. Hence this classification is most meaningful if one of these interactions dominates. For example, the contribution of the electromagnetic and weak interactions to the inelastic π^+p scattering cross-section for $p_{\text{lab}}^{\pi^+} \sim 10$ GeV is negligible. However, for an incident momentum of, say, 10^{15} GeV this may well not be the case. The π^0-meson decays electromagnetically into two photons, there being no competing strong decay because the pion is the lightest hadron. Weak decays become observable when both the strong and electromagnetic decays are suppressed. For example, the weak decay process $K^+ \to \pi^+\pi^0$ is observable because potentially faster decay modes, via the strong or electromagnetic interactions, are forbidden. This is because the quantum number strangeness is conserved in these interactions and the kaon is the lightest strange particle.

As the energy of a particular scattering process increases, the classification in terms of a specific interaction becomes less distinct. For example, the e^+e^- annihilation process at a value of the cms energy squared $s \sim 10$ GeV^2 is predominantly electromagnetic, whereas at $s \sim 2000$ GeV^2 there is a sizeable interference term between the competing electromagnetic and weak processes.

Fig. 1.3 Lifetimes of various particles decaying by the strong, electromagnetic and weak forces.

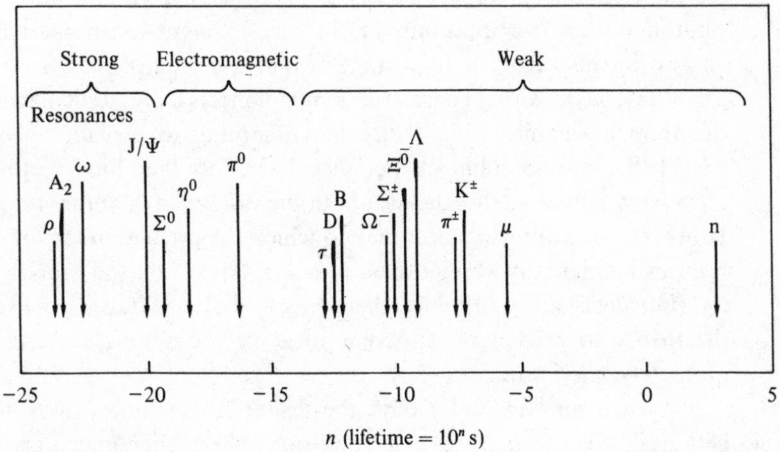

From a theoretical point of view it is clearly desirable to have a single theory which describes all of the fundamental interactions in nature. Some considerable progress in this direction has been made. The *electroweak theory* of Glashow (1961), Weinberg (1967) and Salam (1968) 'unifies' the electromagnetic and weak interactions. This theory, the so-called *standard model*, and the relevant experimental data form the main subject matter of this book.

1.2 THE LIGHT LEPTONS

The discovery and interpretation of both leptons and quarks have depended upon rapid advances in both experimental methods and our theoretical understanding. The starting point was the discovery of the electron by J.J. Thomson in 1897 from a study of the electrical properties of gases at low pressure. Some 30 years later, the use of a cloud chamber in the study of cosmic radiation led to the discovery of the positron, the antiparticle of the electron (Anderson 1932, Blackett and Occhialini 1933). The possible existence of the positron had been predicted by Dirac. Similar experimental techniques led to the discovery (although not initially the correct interpretation) of the muon by Street and Stevenson (1937) and Neddermeyer and Anderson (1937). Further work in cosmic radiation led to the discovery of the π-meson (Lattes *et al.*, 1947), the K-meson (Rochester and Butler, 1947) and the Λ, Σ and Ξ hyperons.

By the early 1950s important advances in our theoretical understanding of matter and the various forces of nature had also been made. The ideas of relativity and quantum mechanics had been combined into the field theory called *quantum electrodynamics* (QED), which successfully accounts for the electromagnetic properties of electrons, muons and photons (the photon is the field quantum of the electromagnetic interaction). The success of this theory is important in that it contains the basic ideas of quantum mechanics. There are many conceptually difficult ideas in quantum mechanics which arise in attempting to explain phenomena outside the range of human experience. Ever since their introduction these ideas have caused intense debate about the validity of quantum mechanics. However, the impressive accuracy to which the predictions of QED have been experimentally verified has, to some extent, allayed worries about the statistical nature of the predictions, i.e. that it is outside the scope of the theory to predict the outcome of a specific interaction and hence, ultimately, the future.

Early attempts to understand the results from studies of radioactive beta decays led to considerable confusion. These phenomena are associated with the weak force, and interpretation of the results in terms of the then known particles (electron, proton and neutron) led to the conclusion that the hallowed laws of energy-momentum and angular-momentum conservation were violated in these decays. Pauli (1933) gave a way out of this dilemma by postulating the existence of an approximately massless neutral particle with half-integral spin, called the neutrino. Fermi (1934), assuming the existence of the neutrino, formulated a theory of beta decays. This was based, to some extent, on the ideas of QED, but with important differences. The Fermi theory postulates the weak interaction to be that of four spin-$\frac{1}{2}$ fermions at the same space-time point (point-like coupling). Thus the theory contains no equivalent of the exchanged boson of QED (i.e. the photon) to mediate the force. Any doubts about the existence of the neutrino were removed by its discovery (or, more precisely, the discovery of the antineutrino) by Reines and Cowan (1953, 1959), who used the intense flux produced by a nuclear reactor.

The advent of new accelerators, with proton energies up to about 30 GeV at Brookhaven National Laboratory and at CERN, allowed the construction of neutrino beams. The neutrinos came mainly from the decays $\pi \to \mu\nu$ and $K \to \mu\nu$. Danby *et al.* (1962) observed that the neutrinos produced in this way gave interactions containing a final state muon, but not containing a final state electron. This showed that there are (at least) two types of neutrino and that the neutrino associated with the muon (ν_μ) is different to that associated with the electron (ν_e).

These and other experimental results on leptons can be explained by assigning two additive quantum numbers defined as follows:

$$\begin{aligned} L_e &= 1\ (e^-, \nu_e), & L_\mu &= 1\ (\mu^-, \nu_\mu), \\ &\quad -1\ (e^+, \bar{\nu}_e), & &\quad -1\ (\mu^+, \bar{\nu}_\mu), \\ &\quad 0\ \text{(other particles)}, & &\quad 0\ \text{(other particles)}. \end{aligned} \tag{1.1}$$

L_e and L_μ are the lepton numbers for the electron and muon families respectively, and $\sum L_e$ and $\sum L_\mu$ appear to be separately conserved in interactions. For example, the principal decay mode of the μ^- is

$$\begin{array}{lcccc} & \mu^- \to & e^- & \bar{\nu}_e & \nu_\mu \\ L_e & 0 & 1 & -1 & 0 \\ L_\mu & 1 & 0 & 0 & 1 \end{array} \tag{1.2}$$

However, the decay $\mu^- \to e^-\gamma$, which would violate these rules, has never been seen, and has an upper limit for its branching ratio of $< 1.7 \times 10^{-10}$.

1.3 THE LIGHT QUARKS

These new proton accelerators also led to many other important discoveries. A rich spectrum of excited states of mesons (e.g. $\rho(770)\to\pi\pi$) and baryons (e.g. $\Delta(1232)\to N\pi$) was found (N stands for nucleon i.e. p or n). These states decay via the strong interaction to the experimentally observed particles.

Studies on the production of the so-called strange particles showed that they are produced in pairs (associated production) by the strong interaction. An additive quantum number *strangeness* (S) can be assigned to all particles and this is conserved by the strong interaction, e.g.

$$\begin{array}{lcccc} & \pi^- + & \mathrm{p} \to & \mathrm{K}^0 + & \Lambda^0 \\ S & 0 & 0 & +1 & -1 \end{array} \tag{1.3}$$

Many excited states of strange mesons (e.g. $\mathrm{K}^*(892)\to\mathrm{K}\pi$) and strange baryons (e.g. $\Sigma(1385)\to\Lambda\pi$) exist, and these have decay widths (or equivalently lifetimes) similar to those of their non-strange counterparts. However, the lifetimes of the K^0 (which decays predominantly to two pions) and the Λ (mainly to $N\pi$) are both about 10^{-10} s. These decays are examples of the weak interaction. In these weak decays strangeness is not conserved.

The baryon quantum number B, like strangeness, is additive. However, it appears that the total baryon number is conserved to good accuracy in the strong, electromagnetic and weak interactions. By definition mesons have $B=0$, whereas baryons and antibaryons have $B=1$ and $B=-1$ respectively. Further, mesons have integral spin (bosons), whereas baryons have half-integral spin (fermions). Baryon states other than the proton are unstable and decay either directly or sequentially to a proton plus one or more particles, e.g.

$$\Delta^+(1232) \xrightarrow{\text{strong}} \pi^+\mathrm{n} \xrightarrow[\text{weak}]{} \mathrm{pe}^-\bar{\nu}_e \tag{1.4}$$

$$\pi^+ \xrightarrow[\text{weak}]{} \mu^+\nu_\mu, \qquad \mu^+ \xrightarrow[\text{weak}]{} \mathrm{e}^+\nu_e\bar{\nu}_\mu$$

The lifetime of the proton is in excess of 10^{30} yr and is thus effectively stable. Thus the particles produced in a particular interaction eventually decay to 'stable' particles, i.e. p, $\bar{\mathrm{p}}$, e^-, e^+, ν, $\bar{\nu}$, γ. The photon is the end product of an electromagnetic decay sequence, e.g. $\eta(548)\to 3\pi^0\to 6\gamma$.

The observed spectrum of hadrons contains both particles and antiparticles. By definition, the charge conjugation operator (C) changes a particle to its antiparticle. This operation changes the charge, magnetic moment, baryon number and lepton number of the particle, leaving the space-time coordinates and momenta unchanged. Hence the spin $\boldsymbol{\sigma}$ is unchanged under C, since one can write $\boldsymbol{\sigma} = \mathbf{r} \times \mathbf{p}$. Some examples of the effects of the operator C are $C(\text{p}) = \bar{\text{p}}$, $C(\text{e}^-) = \text{e}^+$, and $C(\pi^+) = \pi^-$. In these cases the particle and its antiparticle are separate entities. However, neutral non-strange particles which have lepton and baryon number zero, such as the photon and π^0, do not have distinct antiparticles. These particles are eigenstates of C with eigenvalues $C = \pm 1$. The photon has $C = -1$; this follows, since the electromagnetic field produced, say, by an electron changes sign under C. A system of n photons has $C = (-1)^n$. Hence, since the π^0-meson decays predominantly to two photons, $C(\pi^0) = 1$. The other two members of the isospin triplet of pions are not eigenvalues of C, since $C(\pi^+) = \pi^-$. If we define R_2 to be a rotation of 180° around the isospin axis I_2, such that $I_3 \rightarrow -I_3$, then $R_2(\pi^-) = \pi^+$. In general, one can write $R_2 = (-1)^{\text{I}}$. This expression is analogous to that for parity, $P = (-1)^l$, in terms of the orbital angular momentum l. Thus, all three members of the pion isotriplet are eigenstates of the combined operation $G = CR_2$. The eigenvalues of G are ± 1. For the pion G is negative since $C(\pi^0) = 1$ and $R_2 = -1$. For a system of $n\pi$, $G = (-1)^n$. A fermion–antifermion system (e.g. $\text{p}\bar{\text{p}}$, e^+e^-, $\text{q}\bar{\text{q}}$) is also an eigenstate of C with eigenvalues, $C = (-1)^{l+s}$, where l and s are the relative orbital angular momentum and spin of the pair respectively.

The observed spectrum of hadron states could be neatly classified by the quark model of Gell-Mann (1964) and Zweig (1964) in terms of three fractionally charged spin-$\frac{1}{2}$ quarks, called u (up), d (down) and s (strange) – see Table 1.1. The various types of quarks are referred to as *flavours*. In the quark model mesons are bound states of a quark and antiquark ($\text{q}\bar{\text{q}}$) whereas baryons and antibaryons are composed of three quarks (qqq) and three antiquarks ($\bar{\text{q}}\bar{\text{q}}\bar{\text{q}}$) respectively.

The approximate symmetry of the strong interaction with respect to transformations of u and d quarks can be specified in terms of *isospin*. The u and d quarks have $I = \frac{1}{2}$ with third components $I = \frac{1}{2}$ and $-\frac{1}{2}$ respectively. The inclusion of the strange quark necessitates the introduction of a further quantum number, the hypercharge $Y = B + S$. The approximate invariance of the strong interaction under transformations of u, d and s quarks is expressed using the unitary symmetry group SU(3). This group consists of rotations in the three-dimensional coordinate system corresponding to u, d and s quarks with the 'special' condition

Table 1.1. *Quantum numbers of the quarks.* For antiquarks the quantum numbers have the opposite sign.

Fla-vour	Constituent mass (GeV)	Baryon number B	Charge (units of e)	U	D	S	C'	B'	T'
u	0.35	$\frac{1}{3}$	$\frac{2}{3}$	1	0	0	0	0	0
d	0.35	$\frac{1}{3}$	$-\frac{1}{3}$	0	-1	0	0	0	0
s	0.5	$\frac{1}{3}$	$-\frac{1}{3}$	0	0	-1	0	0	0
c	1.5	$\frac{1}{3}$	$\frac{2}{3}$	0	0	0	1	0	0
b	5	$\frac{1}{3}$	$-\frac{1}{3}$	0	0	0	0	-1	0
t	?	$\frac{1}{3}$	$\frac{2}{3}$	0	0	0	0	0	1

The symbols C', B' and T' are used to denote quark flavour numbers, in order to distinguish these from charge conjugation (C), baryon number (B) and time reversal (T).

that the length of a rotated vector is conserved (Chapter 2). This leads to multiplets of particles of a given spin and parity, which would be degenerate in mass if the symmetry were exact and if the electromagnetic effects could be 'switched off'. The quark compositions of the lightest multiplets of mesons and baryons are given in Tables 1.2 and 1.3 respectively. Associated with a particular hadron one can define an intrinsic parity, e.g. for the π-meson $P_\pi = -1$. The various spin-parity states of the multiplets correspond to specific configurations of the quarks' spins and relative orbital angular momenta. This latter separation is, however, non-relativistic and such a simplification is not, in all cases, justifiable.

A reasonable understanding of the mass spectrum of these hadrons can be obtained by assuming that the quarks are relatively light. The mass attributed to the quarks in this context is called the *constituent* mass and is about 0.35 GeV for the u and d quarks. The d quark is slightly heavier than the u quark. The mass difference, $m_d - m_u \sim 3$ MeV, can be determined from the neutron (udd) and proton (uud) mass difference. The mass assigned to the strange quark is somewhat larger (about 0.5 GeV). The mass spectrum of the spin parity $J^P = \frac{1}{2}^+$ baryon octet (p, n, Λ^0, Σ^+, Σ^0, Σ^-, Ξ^0, Ξ^-) gives a relatively clear demonstration of the relative mass assignments. These topics are discussed in detail by Flamm and Schöberl (1982).

A more explicit demonstration that the nucleon is composed of quarks came from the pioneering experiments performed using the high energy electron beam (up to 20 GeV) at the two-mile-long linear accelerator at

Table 1.2. *Quark composition of lightest mesons containing u, d, s and c quarks*

$J^{PC} = 0^{-+}$			$J^{PC} = 1^{--}$		
Meson	Composition	Mass (GeV)	Meson	Composition	Mass (GeV)
π^+	$u\bar{d}$	0.140	ρ^+	$u\bar{d}$	0.770
π^0	$(u\bar{u} - d\bar{d})/2^{1/2}$	0.135	ρ^0	$(u\bar{u} - d\bar{d})/2^{1/2}$	0.770
π^-	$\bar{u}d$	0.140	ρ^-	$\bar{u}d$	0.770
K^+	$u\bar{s}$	0.494	$(K^*)^+$	$u\bar{s}$	0.892
K^0	$d\bar{s}$	0.498	$(K^*)^0$	$d\bar{s}$	0.897
$\bar{K}^0$	$\bar{d}s$	0.498	$\overline{(K^*)}^0$	$\bar{d}s$	0.897
K^-	$\bar{u}s$	0.494	$(K^*)^-$	$\bar{u}s$	0.892
η	$(u\bar{u} + d\bar{d} - 2s\bar{s})/6^{1/2}$	0.549	ω	$(u\bar{u} + d\bar{d})/2^{1/2}$	0.783
η'	$(u\bar{u} + d\bar{d} + s\bar{s})/3^{1/2}$	0.958	ϕ	$s\bar{s}$	1.020
η_c	$c\bar{c}$	2.98	J/Ψ	$c\bar{c}$	3.097
D^+	$c\bar{d}$	1.869	$(D^*)^+$	$c\bar{d}$	2.010
D^0	$c\bar{u}$	1.865	$(D^*)^0$	$c\bar{u}$	2.007
$\bar{D}^0$	$\bar{c}u$	1.865	$\overline{(D^*)}^0$	$\bar{c}u$	2.007
D^-	$\bar{c}d$	1.869	$(D^*)^-$	$\bar{c}d$	2.010
D_s^+ or F^+	$c\bar{s}$	1.97	$(D_s^*)^+$ or $(F^*)^+$	$c\bar{s}$	2.109
D_s^- or F^-	$\bar{c}s$	1.97	$(D_s^*)^-$ or $(F^*)^-$	$\bar{c}s$	2.109

Table 1.3. *Quark composition of lightest* $J^P = \frac{1}{2}^+$ *baryons containing u, d and s quarks*

Baryon	Composition	Mass (GeV)
p	uud	0.938
n	udd	0.940
Σ^+	uus	1.189
Σ^0	uds	1.192
Σ^-	dds	1.197
Ξ^-	dss	1.321
Ξ^0	uss	1.315
Λ	uds	1.116

SLAC. When these electrons were scattered off protons, many more electrons were found to be scattered through large angles than was expected. These observations indicate that the proton has point-like constituents, in a way analogous to that used by Rutherford to demonstrate the existence of the nucleus. Further work in these, and analogous experiments made using high energy neutrino beams, showed that these scattering centres can be identified with the fractionally charged quarks of hadron spectroscopy. This is discussed further in Section 1.7.

1.4 HEAVY QUARKS AND LEPTONS

November 1974 was a month which produced great excitement in the world of high energy physics. Results were produced by Ting and co-workers on a study of the effective mass spectrum in the reaction $p + Be \rightarrow e^+e^- + X$ (anything else) made at the Brookhaven National Laboratory (USA). A narrow peak (called J), with a width of 63 KeV and a mass of about 3.1 GeV, was found (Aubert *et al.*, 1974). At about the same time this state was also observed as a narrow resonance (called Ψ) in e^+–e^- annihilations at SPEAR (Stanford) by Richter and co-workers (Augustin *et al.*, 1974). These discoveries were particularly spectacular because of the large signals observed (Fig. 1.4).

Initially several alternative explanations of this J/Ψ particle were advanced. However, the accepted interpretation is that it is a $J^{PC} = 1^{--}$ bound state of a charmed quark and antiquark, i.e. $c\bar{c}$ (Table 1.1). In fact, as discussed in Chapter 5, the existence of the charmed quark, with approximately the correct mass, had been previously predicted. These discoveries were also important in that the usefulness of colliding beam e^+e^- accelerators became firmly established. Indeed, several excited states

(Ψ', Ψ'', etc.) of the J/Ψ were soon discovered at SPEAR. In addition, charmed mesons and baryons, i.e. containing a single charm quark, have been discovered. The narrow width of the J/Ψ can be explained by the fact that the decay to a pair of charmed particles, which is presumably the most favoured decay mode, is kinematically forbidden. The lightest charmed hadron is the D-meson at 1.865 GeV. Thus, the J/Ψ must decay to non-charmed hadrons and such modes are suppressed by the (empirical) Zweig rule, since they involve unconnected quark lines between the initial and final states (see Section 1.7 and Chapter 10). Since the *G*-parity of the J/Ψ is negative, decay modes to an odd number of pions are favoured. The J/Ψ is interpreted as a rather loosely bound $c\bar{c}$ system (i.e. hidden charm) with the mass of the charmed quark about 1.5 GeV. The Ψ' (3686 MeV) is also relatively narrow with a width of 215 KeV; again, the $D\bar{D}$ mode is forbidden. However, the next excited state, the Ψ'' (3770),

Fig. 1.4 Discovery of the J/ψ particle: (*a*) the e^+e^- spectrum in $p + Be \to e^+e^-X$ (Aubert *et al.*, 1974); (*b*) $\sigma(e^+e^- \to \text{hadrons})$ measured at SPEAR (Augustin *et al.*, 1974), similar peaks were found in $e^+e^- \to e^+e^-$ and $e^+e^- \to \mu^+\mu^-$.

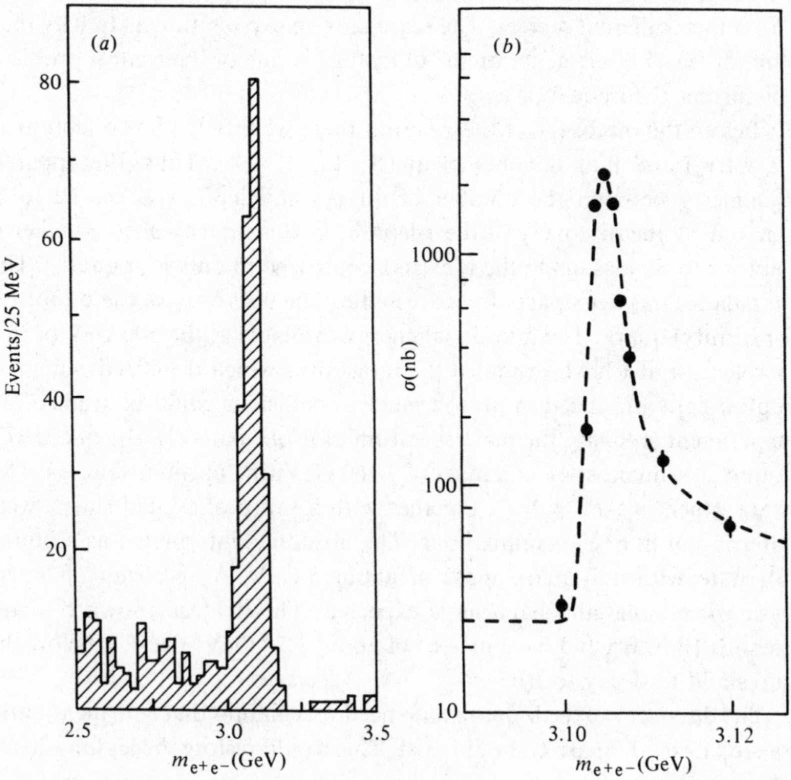

is above $D\bar{D}$ threshold and decays predominantly to this channel. The production of a charmed meson or baryon, e.g. D^+ with quark content $c\bar{d}$, is referred to as open charm production. The charmed quark mass is relatively heavy compared to the equivalent mass values of u, d and s quarks.

The SU(3) symmetry scheme, used to classify the hadronic states of u, d and s quarks, can be extended to SU(4) with the addition of the charmed (c) quark. The lightest charmed mesons expected in this SU(4) scheme are shown in Table 1.2. Only a few of the many expected charmed meson and baryon states are, as yet, well established.

Almost 40 years after the discovery of the muon a new charged lepton (τ) was discovered by Perl and co-workers in the reaction $e^+e^- \to \tau^+\tau^-$ at SPEAR (Perl *et al.*, 1975). The τ-lepton has a mass of 1.78 GeV and spin $\frac{1}{2}$. Although it is assumed that there is a separate neutrino ν_τ associated with the τ-lepton, there is as yet no direct evidence for its existence. The lepton number for the τ family can be defined in a similar way to that for the e and μ families, i.e. $L_\tau = 1$ for τ^- and ν_τ and $L_\tau = -1$ for τ^+ and $\bar{\nu}_\tau$. The known properties of the e-, μ- and τ-leptons are consistent with the hypothesis that the only difference in their properties is that arising from their different masses. The search for an explanation as to why there are (at least) three 'generations' of leptons is one of the central problems of current theoretical physics.

Before the discovery of the τ-lepton there were four known leptons (e, μ, ν_e, ν_μ) and four flavours of quarks (u, d, s, c). Thus, the apparent symmetry between the number of quarks and leptons appeared to be broken by the discovery of the τ-lepton, as this increased the number of leptons to six (assuming the ν_τ exists), compared to only four quarks. This imbalance has been partially redressed by the discovery of the b (bottom or beauty) quark. The additional energy available at the 500 GeV proton accelerator at FNAL extended the range over which the effective mass of lepton pairs produced in proton–nucleus collisions could be studied. An experiment studying the mass spectrum of $\mu^+\mu^-$ pairs (Herb *et al.*, 1977) found a resonant state at a mass of 9.460 GeV (the upsilon state, Υ). This state which has $J^{PC} = 1^{--}$, together with a series of excited states, were later found in e^+e^- annihilations. The upsilon is interpreted as a bound $b\bar{b}$ state with a b-quark mass of about 5 GeV. A spectrum of 'open' b-quark mesons and baryons is expected. The lightest known b-quark mesons ($B^\pm$, B^0, $\bar{B}^0$) have masses of about 5.27 GeV; the Υ is below the threshold to decay to $B\bar{B}$.

The discovery of the b-quark enhanced speculation that a further quark, the top or truth quark t, should exist. This would restore the lepton–quark symmetry. An unsuccessful search for the t-quark in e^+e^- annihilations at the PETRA accelerator (DESY) has set a lower limit for the quark mass of about 22 GeV. Some evidence for the existence of the t-quark has been obtained by the UA1 collaboration at the CERN $p\bar{p}$ Collider (Arnison *et al.*, 1984b). These observations would indicate that the t-quark mass is about 40 GeV; however, a more recent analysis (Albajar *et al.*, 1987c), which includes more data and a more detailed study of the backgrounds, now quotes a lower limit of $m_t \geqslant 44$ GeV (95% c.l.).

It is useful to assign an additive flavour number, defined to be either ± 1, to each type of quark such that the charge of a quark, or an ensemble of quarks, can be calculated in terms of this number. The values of the quark charges, namely $\frac{2}{3}$ and $-\frac{1}{3}$, are asymmetric about zero. However, since all quarks have $B = \frac{1}{3}$, the charge values are symmetric around $B/2$, i.e. $Q = (B \pm 1)/2$. Hence, for an assembly of quarks, one can write

$$Q = (U + D + S + C' + B' + T')/2 + B/2. \tag{1.5}$$

Here, U, C' and T' receive contributions of $+1$ for each u, c and t quark respectively, -1 for each antiquark of these kinds, and zero otherwise. D, S and B' receive -1 for each d, s and b respectively, $+1$ for their antiquarks, and zero otherwise (see Table 1.1). The near mass degeneracy of the u and d quarks leads to the assignment of isospin $\frac{1}{2}$ with $I_3 = +\frac{1}{2}$, $-\frac{1}{2}$ respectively (Section 1.3). Therefore, for an assembly of u and d quarks and antiquarks, $I_3 = (U + D)/2$. Consequently, for a system of u, d and s quarks and antiquarks, we obtain the Gell–Mann–Nishijima relationship

$$Q = (U + D)/2 + (B + S)/2 = I_3 + Y/2. \tag{1.6}$$

1.5 FORCES AND PARTICLE EXCHANGE

Classical field theory gives an adequate description of the properties of the two long-range interactions, namely electromagnetism and gravity, for large scale phenomena. However, it fails at the subatomic scale and, in this domain, quantum field theory must be used. In the description of large-scale phenomena the fluctuations associated with the quantum field, due to the uncertainty principle, do not produce observable effects, so that the quantum field looks like a classical field; however, the underlying physics is markedly different.

Whereas, classically, an interaction at a distance is described in terms of a field or potential, in quantum field theory forces are mediated by the exchange of particles. Fig. 1.5(*a*) shows an example of an electromagnetic interaction in which a photon is exchanged between an electron and a muon. In this diagram, which is called a *Feynman diagram*, time runs horizontally and some space coordinate vertically. A free electron cannot absorb or emit a photon, or any other particle, and conserve energy and momentum. From the uncertainty principle, conservation of energy for such a photon can be violated by an amount ΔE for a time up to $\Delta t \sim \hbar/\Delta E$, during which the photon can travel a distance $r \sim c\hbar/\Delta E$. Since the photon has zero mass, ΔE can be arbitrarily small, and hence the electromagnetic interaction has infinite range. The gravitational interaction also has infinite range, because the exchanged quanta, gravitons, have zero mass. If a particular interaction is mediated by a particle of mass m, then $r \sim \hbar/mc$ (the Compton wavelength). Thus, the more massive the particle to be created, the shorter the range of the interaction. These concepts led Yukawa (1935) to postulate the existence of the pion, to explain the strong short-range nuclear force. We now know that the pion is not a fundamental particle but is composite. The idea of pion exchange, however, gives a reasonable understanding of some aspects of strong interaction phenomena.

A more rigorous and relativistic treatment of the above ideas shows that the cross-section for a diagram, similar to that shown in Fig. 1.5(*a*), involving the exchange of a particle of mass m contains a factor G^2 with

$$G(m) \propto \frac{1}{(q^2 - m^2)}, \tag{1.7}$$

where q is the four-momentum of the exchanged particle. The four-momentum is conserved at each vertex. The value of q^2 in such an exchange process is negative, corresponding to an imaginary value of 'mass'. A particle for which $E^2 - p^2 \neq m^2$ is said to be 'virtual' or 'off its mass shell', in contrast to a free particle, which has $E^2 - p^2 = m^2$, and is said to be real or on-shell. The term G is called the *propagator* term, corresponding to the particle which mediates or propagates the interaction.

Fig. 1.5 Lowest order Feynman diagrams for (*a*) $e^-\mu^- \to e^-\mu^-$ and (*b*) $e^-e^+ \to \mu^-\mu^+$. These diagrams involve the exchange of a single photon, *e* is the coupling constant at each vertex.

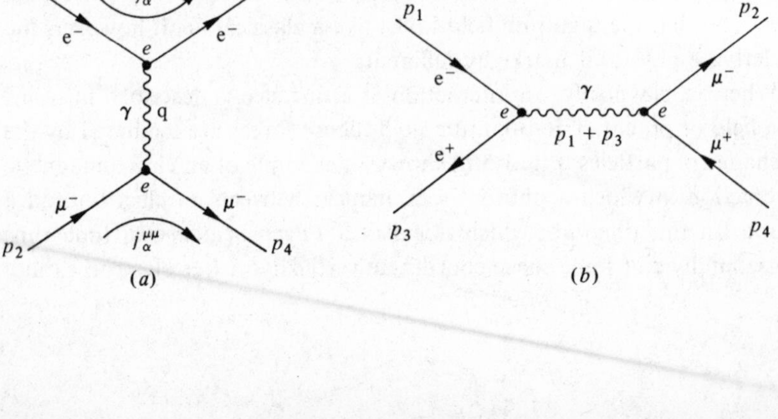

1.6 THE ELECTROWEAK GAUGE BOSONS

The interactions of leptons and photons can be calculated perturbatively in quantum electrodynamics. The results are calculated in an expansion in terms of the fine structure constant $\alpha = e^2/4\pi \sim 1/137$, where e is the electron charge. As discussed later (Chapter 4), these calculations can conveniently be expressed in terms of Feynman diagrams. Fig. 1.5(a) shows the diagram with the lowest order in α for the process $e^-\mu^- \to e^-\mu^-$. The electromagnetic force is mediated by the exchange of the photon between the electron and muon 'currents' j^{ee} and $j^{\mu\mu}$ respectively. The electric charges carried by the electron and muon appear to be identical, whereas the photon has no charge.

In non-relativistic quantum mechanics the transition rate for a particular process is given by (Section 2.2)

$$\lambda = 2\pi |\mathcal{M}|^2 \rho(E), \tag{1.8}$$

where $\mathcal{M}$ is the transition matrix element and $\rho(E)$ the density of the final states (or phase space density), as a function of their energy E. Transition rates calculated using relativistic field theories can be expressed in a similar form. Thus, if x represents some relevant kinematic variable, then the differential cross-section can be written (Chapter 4)

$$\frac{d\sigma}{dx} \propto |\mathcal{M}(x)|^2 \rho(x), \tag{1.9}$$

where $M(x)$ is the Lorentz invariant matrix element and $\rho(x)$ the Lorentz invariant phase space density.

The strength of an interaction enters the calculation as a factor at each vertex. In Fig. 1.5(a) there are two vertices and a factor e is associated with each. The virtual photon propagator gives a factor $G(m=0) = 1/q^2$ to the matrix element, hence (1.9) can be written

$$\frac{d\sigma}{dx} \propto e^4 |\mathcal{M}_{ee}(x) \cdot G(0) \cdot \mathcal{M}_{\mu\mu}(x)|^2 \rho(x), \tag{1.10}$$

where $\mathcal{M}_{ee}$ and $\mathcal{M}_{\mu\mu}$ represent the transition matrix elements at the upper and lower vertices. The basic element in a quantum field theory is the vertex diagram, which involves the transition from a particle to two or more other particles. The basic vertex diagram in QED represents the transitions $e^- \to e^- + \gamma$, $e^+ + e^- \to \gamma$, $\gamma \to e^+ + e^-$ and $e^+ \to e^+ + \gamma$ (an ingoing e^- is equivalent to an outgoing e^+ and *vice versa*). This process of creation and annihilation of particles distinguishes quantum from classical fields, e.g. an electron propagating in free space continuously emits and reabsorbs photons. That is, in order to describe phenomena on distance scales less than the Compton wavelength ($\hbar/mc$), a relativistic quantum field theory is needed.

The total cross-section for a given process is obtained, in principle, by forming all the diagrams in which vertices are connected in all possible ways. The amplitude for each of these diagrams can be written down using the Feynman rules, and the square of the sum of these amplitudes must then be computed. In practice, these computations are lengthy and are restricted to a limited number of higher orders.

In QED, higher order processes have contributions with increasingly larger powers of the electromagnetic coupling constant α. For example, if a real photon is emitted on either the ingoing or outgoing electron lines in Fig. 1.5(a), then there is an additional factor e in the matrix element and hence the contribution to the cross-section is proportional to $\alpha^3(e^6)$. This term is a factor α less than the lowest order term (1.10) and hence is, in general, small.

Calculations of various phenomena in QED show that there are problems with this theory, in that divergent (infinite) results are obtained for physically measurable quantities such as the mass and charge of the electron. These infinities are related to diagrams with closed loops of virtual particles. However, these divergences can be controlled by a recipe called *renormalisation*, in such a way that calculations of higher orders in the perturbation expansion cause no additional problems. The virtual particles emitted and absorbed in these processes cannot be observed directly. However, impressive agreement is obtained between theory and experiment, in particular for the magnetic moment of the electron and the muon, hence there is considerable confidence in these methods.

Fig. 1.5(b) shows the Feynman diagram, with the lowest order in α, for the reaction $e^-e^+ \to \mu^-\mu^+$. This reaction is similar to that for $e^-\mu^- \to e^-\mu^-$, but with the ingoing μ^- replaced by an outgoing μ^+ and the outgoing e^- replaced by an ingoing e^+ respectively. This substitution of an ingoing fermion by an outgoing antifermion, and *vice versa*, is quite general and simplifies many calculations. The virtual photon, γ, in Fig. 1.5(b) is essentially an intermediate state which later (time runs horizontally) 'decays' to $\mu^-\mu^+$. Such a photon is called a time-like photon, in contrast to the photon in Fig. 1.5(a) which is a space-like photon (distance runs vertically).

The lowest order Feynman diagram for the weak decay process $\mu^- \to e^-\bar{\nu}_e\nu_\mu$ is shown in Fig. 1.6(a). This interaction, in the standard electroweak theory, is mediated by the charged boson W. This diagram represents both the emission of a W^- and the absorption of a W^+ by the μ^-. The weak 'current' $\mu^- \to \nu_\mu(j^{\mu\nu_\mu})$ at the upper vertex involves a change in the electric charge and is called a *charged current* interaction. The cross-section can be written in a form similar to (1.10), namely

$$\frac{d\sigma}{dx} \propto g^4 |\mathcal{M}_{\mu\nu_\mu}(x) \cdot G(M_W) \cdot \mathcal{M}_{e\nu_e}(x)|^2 \rho(x). \qquad (1.11)$$

The weak coupling constant g is roughly equal in magnitude to the electromagnetic coupling constant e. Since M_W^2 is large compared to q^2 ($M_W \sim 82$ GeV), the propagator term becomes $G \simeq 1/M_W^2$.

Until 1973 all experimental data on weak interactions, both from studies of the weak decays of strange and non-strange mesons and baryons and from low energy neutrino interactions, were consistent with there being only charged current interactions. For example, the reaction $\nu_\mu p \to \mu^- \Delta^{++}$ can, at the quark level, be interpreted as $\nu_\mu d \to \mu^- u$ (Fig. 1.6(*b*)).

Further work on weak interactions showed that, in addition to the $u \leftrightarrow d$ charged current transition, there are also $u \leftrightarrow s$, $c \leftrightarrow d$ and $c \leftrightarrow s$ transitions. Ignoring for the moment the additional complications arising from b and t quarks, the relative strengths of the matrix elements of these transitions are given by single parameter, θ_C (see Table 1.4). The parameter θ_C is called the Cabibbo angle ($\theta_C \sim 13°$) and was originally introduced to explain the relative strengths of $u \leftrightarrow d$ and $u \leftrightarrow s$ transitions. The significance of the negative sign for the $c \leftrightarrow d$ transitions is discussed later.

Theoretically, all of these observations could be explained within the framework of the Fermi theory of weak interactions, which involves the point-like coupling of four fermions. However, this theory predicted, for example, that the total neutrino–electron cross-section would grow

Fig. 1.6 Lowest order Feynman diagrams for (*a*) $\mu^- \to e^- \bar{\nu}_e \nu_\mu$ and (*b*) $\nu_\mu p \to \mu^- \Delta^{++} (\nu_\mu d \to \mu^- u)$. These diagrams involve the exchange of a single W, $g/2^{1/2}$ is the coupling constant at each vertex.

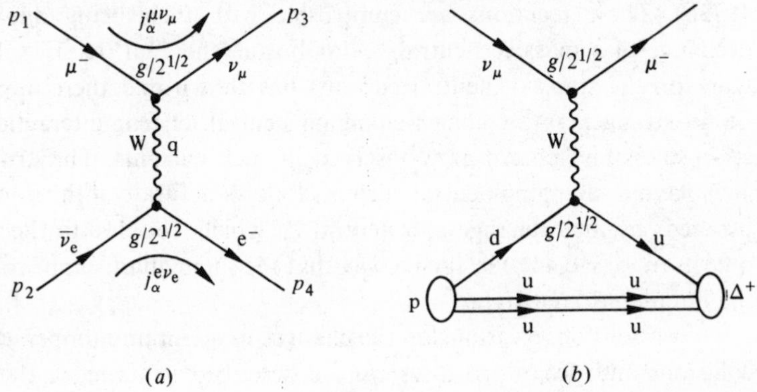

Table 1.4. *Relative strengths of the transition matrix elements between u, d, s and c quarks*

Transition	Coupling constant
u↔d	$g \cos \theta_C$
u↔s	$g \sin \theta_C$
c↔d	$-g \sin \theta_C$
c↔s	$g \cos \theta_C$

without bound, as the neutrino energy was increased. The introduction of the charged vector boson (W) to mediate the weak interaction (in a way analogous to the γ in QED) solved this problem, but only at the expense of introducing uncontrollable divergences in higher order calculations. The theoretical remedy to this problem postulated in the standard model was the introduction of a neutral vector boson (Z^0), whose couplings cancelled the divergences. This postulated that the Z^0 particle mediates *weak neutral currents* and so the total neutral current consists of both γ and Z^0 exchange in a precisely defined mixture; hence the 'unification' of the weak and electromagnetic forces.

In 1973 evidence was found for weak interactions which could not be classified as charged current reactions. A study of ν_μ interactions, with neutrino energies up to about 10 GeV, in the heavy liquid bubble chamber Gargamelle, showed the existence of the reaction $\nu_\mu e^- \to \nu_\mu e^-$ (Hasert *et al.*, 1973a). Furthermore, in about one third of the interactions off nuclei, it was found that there was no muon in the final state (Hasert *et al.*, 1973b). These reactions are compatible with their being due to the exchange of a massive neutral vector boson, the Z^0 (Fig. 1.7). Further work on the study of neutral currents has shown that there appears to be no strangeness- or charm-changing neutral current interactions, i.e. strangeness and charm are conserved in such currents. The absence of such flavour-changing neutral currents follows naturally if the interaction proceeds by the exchange of a neutral Z^0 which 'couples' to the various quarks (u, d, s, c, etc.) in such a way that the same quarks appear in both the initial and final states.

The various observations on the changes in quantum numbers in weak collisions and decay processes can be described in terms of the vertex transitions shown in Fig. 1.8. For the W^+, the possible transitions are (Fig. 1.8(*a*))

$$
\begin{aligned}
W^+ &\to l^+ \nu_l \qquad && l = e, \mu, \tau, \ldots, \\
&\to q_1 \bar{q}_2 && \begin{cases} q_1 = u, c, t, \ldots, & Q = \tfrac{2}{3}, \\ \bar{q}_2 = \bar{d}, \bar{s}, \bar{b}, \ldots, & Q = \tfrac{1}{3}. \end{cases}
\end{aligned}
\tag{1.12}
$$

Fig. 1.7 Lowest order Feynman diagrams for (*a*) $\nu_\mu e^- \to \nu_\mu e^-$ and (*b*) $\nu_\mu N \to \nu_\mu +$ hadrons. The appropriate coupling constants are given in Appendix C.

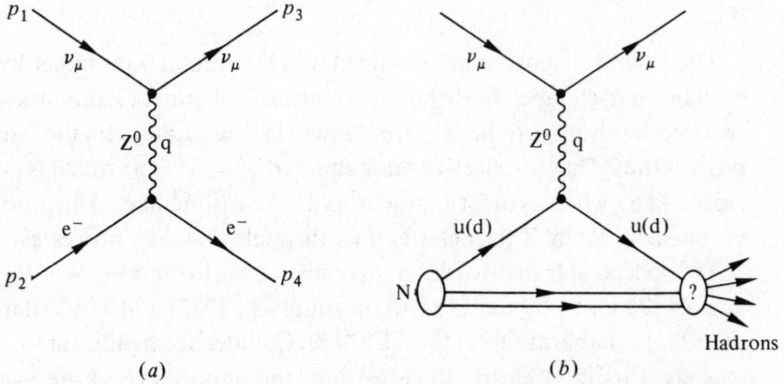

Fig. 1.8 Vertex diagrams (which represent potential decay modes) coupling the heavy bosons, (*a*) W and (*b*) Z^0, to leptons and quarks. Diagram (*c*) shows an example of an exchange and a formation process.

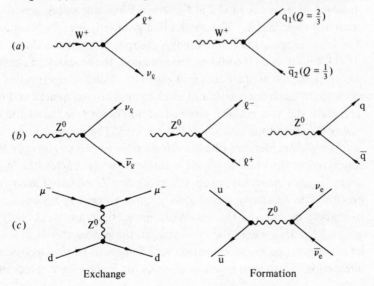

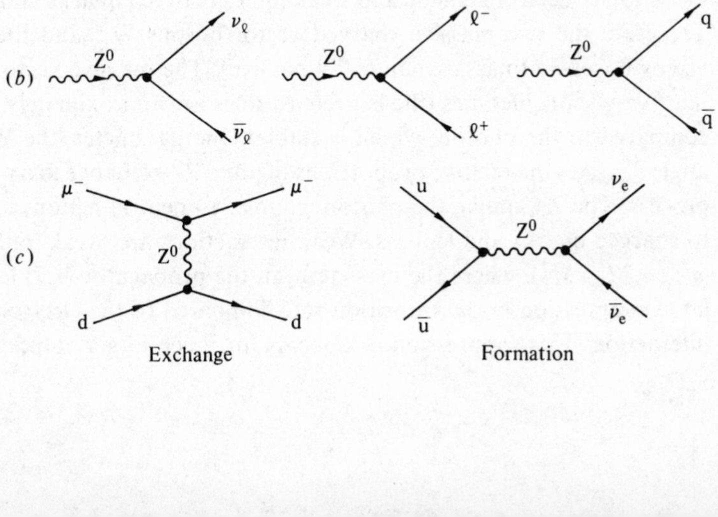

Thus, the possible decay modes of a W^+ are to a charged lepton and its neutrino (total lepton number zero) or to a quark of charge $\frac{2}{3}$ and an antiquark of charge $\frac{1}{3}$. Of course, a decay mode must also be kinematically possible. Similarly, a collision of these pairs of particles can create a W^+. For a W^-, the transitions are $W^- \to l^- \bar{\nu}_l$ and $W^- \to q_2\bar{q}_1$. The vertex transitions for a Z^0 are (Fig. 1.8(*b*)),

$$\begin{aligned} Z^0 &\to \nu_l\bar{\nu}_l \quad \text{or} \quad l^+l^- && l = e, \mu, \tau, \ldots \\ &\to q\bar{q} && q = u, d, s, c, b, t, \ldots. \end{aligned} \tag{1.13}$$

The W and Z^0 bosons are involved in weak interactions either by being exchanged (*t*-channel) or through formation (*s*-channel). Examples of these two mechanisms for the Z^0 are shown in Fig. 1.8(*c*). In the exchange process the Z^0 has an effective mass squared $q^2 = E^2 - p^2$ which is negative (space-like), whereas in formation q^2 is positive (time-like). The production of 'on-shell' W or Z particles is thus through time-like processes.

The spectacular discoveries of the massive vector bosons $W^\pm$ (82 GeV) and Z^0 (93 GeV) by the UA1 (Arnison *et al.*, 1983) and UA2 (Banner *et al.*, 1983) collaborations at the CERN $\bar{p}p$ Collider are significant for several reasons. Firstly, it clearly demonstrates the importance of the ingenious techniques used in the construction and exploitation of the Collider; work for which Rubbia and Van de Meer were awarded the Nobel Prize in 1984. Secondly, it represents a major theoretical triumph, in that the standard electroweak model of Glashow, Weinberg and Salam accurately predicted their masses. This model had gradually become accepted since it was first proposed in 1967. A major theoretical input was that of t'Hooft (1971), who showed it could be renormalised. In the standard electroweak theory (Chapter 5) there are four particles, called gauge bosons, which mediate the electromagnetic and weak forces between quarks and leptons. These are the two massive charged vector bosons $W^\pm$ and the neutral vector bosons γ (massless) and Z^0 (massive). The massive vector bosons have very short lifetimes (the expected values are approximately 10^{-25} s) compared to the photon, which is stable. One parameter, the Weinberg angle θ_W, fixes the relative proportions of γ and Z^0 exchange in a particular process. The Z^0, unlike the photon, couples to neutral leptons as well as to charged quarks and leptons. Weak interactions are 'weak' only at low $|q^2|$ ($\ll M_W^2, M_Z^2$), where the mass term in the propagator (1.7) leads to a large suppression in the transition rate compared to the electromagnetic interaction. This suppression disappears for much larger values of q^2.

1.7 THE STRONG GAUGE BOSONS

Throughout this chapter the existence of quarks and quark currents to interpret the data has been assumed. However, free quarks, if they exist at all, are produced only very rarely in the collisions or decays of hadrons. Hence, if it is assumed that free quarks do not exist, there must be some extremely strong force which confines quarks within hadronic boundaries. The theory of *Quantum Chromodynamics* (QCD) offers a possible explanation for this phenomenon of confinement. In QCD the strong force between quarks and/or antiquarks is transmitted by massless vector particles called *gluons*. A hadron is thus a confined system of quarks, antiquarks and gluons (generically referred to as *partons*). The static properties of hadrons, such as the charge of the proton or neutron, do not directly reveal this substructure. However, the magnetic moments of the proton and neutron are different to those expected for point-like particles carrying charge e and zero respectively. These anomalous magnetic moments suggest that the nucleon is not elementary.

The internal substructure only becomes directly apparent if the nucleon is probed to distances less than the nucleon size (i.e. $\lesssim 1$ fm). From Heisenberg's uncertainty principle ($\Delta p \Delta x \gtrsim 0.197$ GeV fm), it follows that probing to small distances requires the use of large momenta. Note that the simultaneous measurements of the momentum and position of a particle made in an actual detector are far from this limit. For example, if the decay point of a particle can be measured to 1 μm and the momentum to 1 MeV, then $\Delta p \Delta x \sim 10^6$ GeV fm. Small distances can only be resolved by scattering experiments in which the particles themselves act as the probes. To resolve distances on atomic scales (10^{-10} m) requires a probe with a momentum in the electronvolt to kiloelectronvolt range. Significantly higher momenta, typically 100 MeV, are required to probe the internal structure of the nucleus. For example, a series of measurements made at Stanford on the scattering of 100–550 MeV electrons off various nuclei allowed a detailed study of the nuclear charge density distributions (Hofstadter, 1956). Similar measurements made at higher momenta (up to a few GeV) showed that the proton was not point-like, but that its charge was distributed over a radius of about 0.8 fm. The electron, on the other hand, has no apparent internal structure down to distances of less than 10^{-17} m, and hence acts as a point-like probe.

The use of electrons of even higher energies at SLAC (up to 20 GeV) gave results for hydrogen and deuterium targets which could be readily explained if the scattering centres within the nucleon were point-like. These scattering centres or partons can be identified with the quarks that had previously been proposed to explain the spectrum and symmetries of hadrons. However, the total momentum fraction of the nucleon carried by these partons was found to be only about 50%. The remaining fraction is explained in QCD as being carried by the gluons, the strong binding particles holding the quarks together and responsible for their confinement. Various pieces of experimental support for these ideas have gradually been built up.

Experiments using beams of muons, with momenta up to 280 GeV, and of complementary studies using high energy neutrino beams, have shown that the quarks appear to have an apparent substructure when resolved to even smaller distances. What appears to be a quark at one particular distance (momentum) scale shows substructure when probed to smaller distances, in that additional quark–antiquark pairs contribute to the scattering. Such an evolution is indeed predicted by QCD. The transitions $q \rightarrow q + g$, $g \rightarrow q + \bar{q}$, $g \rightarrow g + g$ can occur and, as the nucleon is probed to smaller and smaller distances, more components are revealed, each carrying a progressively smaller fraction of the total momentum. Thus the nucleon consists of 'valence' quarks (e.g. proton = uud), plus a 'sea' of $q\bar{q}$ pairs. Note that the quarks, antiquarks and gluons in this picture have no internal substructure, that is they are not composite. The apparent substructure is explained by the dynamics of QCD. However, internal substructure may exist and become apparent at energies higher than currently accessible.

Further support for the QCD picture comes from measurements of $e^+e^- \rightarrow$ hadrons as a function of the centre-of-mass energy, $(s)^{1/2}$, over the range $5 \lesssim (s)^{1/2} \lesssim 45$ GeV. At the lower values of $(s)^{1/2}$ the hadrons form two back-to-back 'jets'. The distribution of the angle θ between the jet axis and the incident beam direction shows a $(1 + \cos^2\theta)$ dependence. This can be explained if the underlying process is initially $e^+e^- \rightarrow q\bar{q}$, with spin $\frac{1}{2}$ quarks and antiquarks, followed by the formation of *jets of hadrons* along the quark and antiquark directions. Evidence that quarks have spin $\frac{1}{2}$ also comes from the analysis of high energy lepton–nucleon scattering experiments (Chapter 7) and from the classification of hadron multiplets. The jets of hadrons in this QCD picture are produced by the forces confining the quark and antiquark. The momenta of particles transverse to the quark direction are observed to be limited. The mean value of the transverse momentum is typically $\langle p_T \rangle \sim 350$ MeV, which by the uncertainty principle corresponds to a distance of about 1 fm, a typical hadronic size.

At the larger values of $(s)^{1/2}$ ($\gtrsim 20$ GeV), an additional process occurs giving, in about 10% of the events, final states with three discernible jets of hadrons. An example of such an event is shown in Fig. 1.9. These are interpreted as being due to the process $e^+e^- \to q\bar{q}g$; the third jet coming from the gluon *'fragmenting'* to hadrons. The diagrams, at the quark level, corresponding to the two- and three-jet final states are shown in Fig. 1.10. The configurations of these three-jet final states are compatible with a spin 1 assignment for the gluon. Measurement of the fraction of three-jet final states allows a determination of the QCD coupling constant α_S (analogous to the fine structure constant α in QED). The value found for α_S, at $(s)^{1/2} \sim 30$ GeV, is in the range 0.1 to 0.3. The large uncertainty arises due to the, *a priori*, unknown processes by which the quarks and gluons fragment to the observed final state hadrons and also on the size of the higher ($0(\alpha_S^2)$ etc.) processes. There is, as yet, no entirely satisfactory method of reconstructing the underlying parton dynamics from the observed final state hadrons.

Fig. 1.9 Example of an e^+e^- annihilation with three hadronic jets in the final state using the JADE detector at PETRA. The jets reflect the underlying $e^+e^- \to q\bar{q}g$ process.

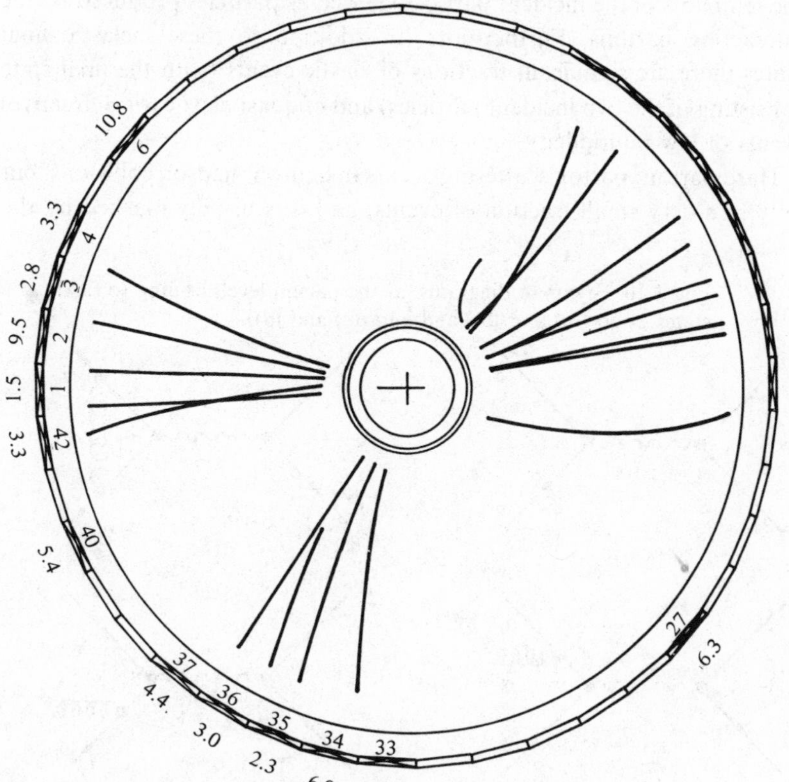

Jets of hadrons are observed in other processes. In lepton–nucleon scattering, at currently accessible energies ($(s)^{1/2} \lesssim 20$ GeV), the bulk of the final states have two jets of hadrons, one along the direction of the quark which has been struck, and the other along the direction of the remaining target remnants. At $(s)^{1/2} \sim 20$ GeV, a small fraction of three-jet final states is also observed, compatible with the expectations of QCD. The interpretation of hadron–hadron interactions in terms of parton collisions is more complicated. This is because both the projectile and the target have complex parton substructures and many possibilities for single and multiple collisions of both quarks and gluons exist. The parton–parton scattering processes are generally *soft*, that is they involve rather small momentum transfers. The major part of the cross-section comes from these soft multiple collisons. This is in contrast to e^+e^- annihilation and deep inelastic lepton–nucleon scattering, which are *hard* processes involving large momentum transfers. The bulk of the hadron–hadron final states contains two jets, corresponding to the projectile and target particles respectively. These jets are more complicated to interpret, as they contain the remnants of the incident particles as well as particles produced by the interacting partons. Furthermore, in addition to these inelastic final states there are significant fractions of elastic events (with the final state consisting of the two incident particles) and of quasi-elastic (or diffractive) events of low multiplicity.

Hard parton–parton scattering occurs in hadron–hadron collisions, but only in a very small fraction of events, and it is usually masked by the

Fig. 1.10 Feynman diagrams, at the parton level, leading to final states of $q\bar{q}$ ((*a*) and (*b*)) and $q\bar{q}g$ ((*c*) and (*d*)).

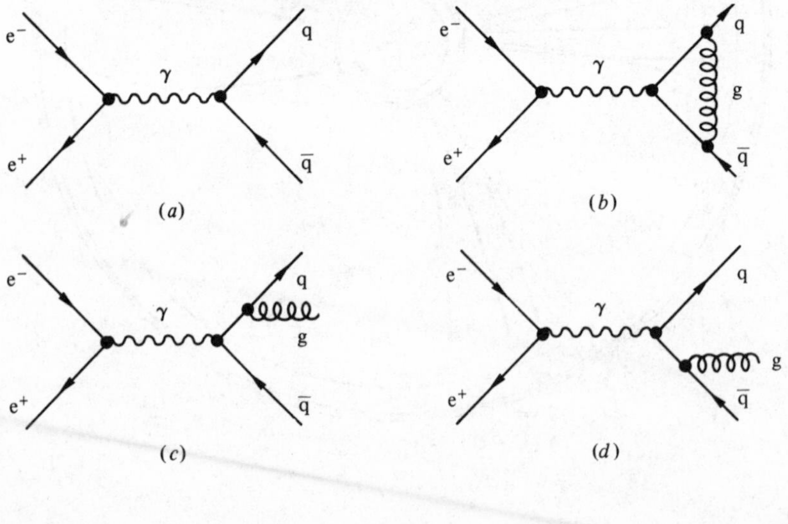

tails of the soft processes. However, at the very large cms energy values of the CERN $\bar{p}p$ Collider ($(s)^{1/2} \sim 600$ GeV) clear hadron jets are observed. Events are selected by a trigger requiring a large amount of energy to be produced in a plane roughly transverse to the incident beam direction. These events show back-to-back jet configurations; an example from the UA2 collaboration is shown in Fig. 1.11. Some of the underlying processes at the parton level are shown in Fig. 1.12. Detailed calculations show that the largest contribution comes from process (c) in Fig. 1.12.

Fig. 1.11 Transverse energy deposited in the UA2 calorimeters for an event with two clear hadronic jets.

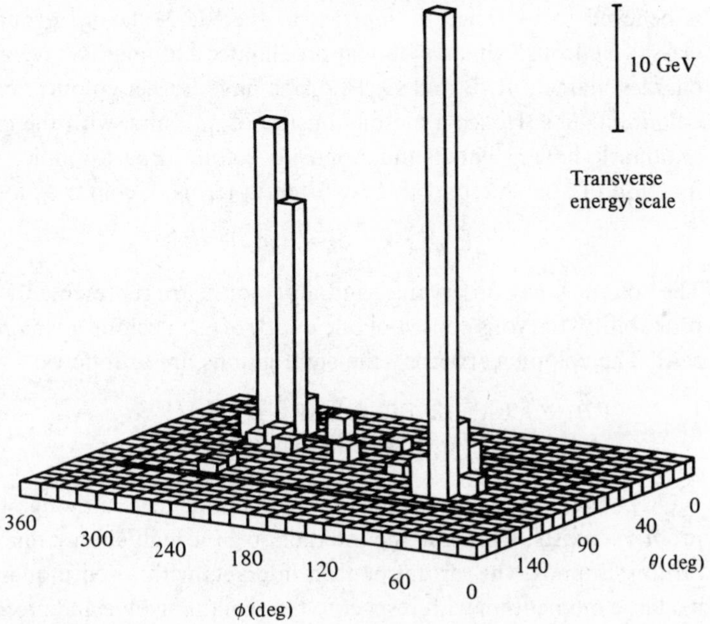

Fig. 1.12 Some lowest order Feynman diagrams involving gluon exchange which contribute to hadron–hadron scattering.

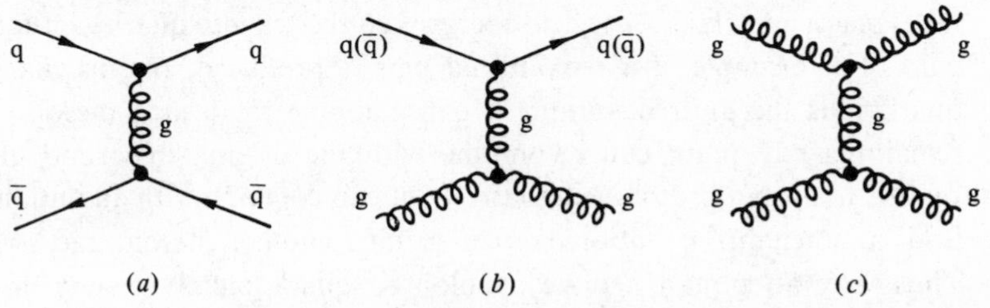

The interactions between quarks and antiquarks in QCD are mediated by eight massless gauge bosons, the gluons. The quarks and antiquarks carry a 'strong charge', the so-called *colour charge* or *colour*. This is analogous to the electric charge in QED; however, there is a very important difference. Whereas the photon carries no electric charge in QED, the gluons themselves have a colour charge. QCD is a non-Abelian field theory, i.e. the field quanta themselves are a field source, in contrast to QED which is an Abelian theory. Thus a vertex coupling three or four gluons is possible in QCD, but there is no equivalent for photons in QED. There are three possible colours for a quark, which we shall call (arbitrarily) red, blue and green, i.e. the additive primary colours. In QCD, the underlying symmetry group is SU(3) of colour. Note that this symmetry is believed to be exact, in contrast to the SU(3) flavour group used to classify hadrons, which is only approximate. Antiquarks have anticolour charges, namely $\bar{\mathrm{R}}$, $\bar{\mathrm{B}}$ and $\bar{\mathrm{G}}$. Hadrons have no net colour, i.e. they are colour singlets. Hence a meson consists of $\mathrm{q}\bar{\mathrm{q}}$ pairs, with the quark and antiquark having equal and opposite colour. For example, the wave function of a π^+-meson can be written in terms of colour as follows

$$|\pi^+\rangle = (\tfrac{1}{3})^{1/2}[\mathrm{u}_R\bar{\mathrm{d}}_R + \mathrm{u}_B\bar{\mathrm{d}}_B + \mathrm{u}_G\bar{\mathrm{d}}_G]. \tag{1.14}$$

The colour is not observable, and all colours are represented with equal probability. Baryons consist of one quark of each colour giving net colour zero. The colours carried by the eight gluons are as follows:

$$\begin{gathered}\mathrm{R\bar{B}, R\bar{G}, B\bar{R}, G\bar{R}, B\bar{G}, G\bar{B}, (R\bar{R} - G\bar{G})/2^{1/2}},\\ \mathrm{(R\bar{R} + G\bar{G} - 2B\bar{B})/6^{1/2}}.\end{gathered} \tag{1.15}$$

QCD offers a possible explanation for confinement, although no formal proof yet exists. Consider the interaction of a high energy muon with a proton. Suppose the virtual photon interacts with a red u-quark, giving it a large momentum with respect to the remaining blue and green quarks. From QCD, the force between the red quark and the other two quarks, which considered together have the colour antired, increases roughly as their separation. Thus, enormous energy would be needed to isolate the quark. However, before this can happen, it appears that another process takes place, namely that the field energy is converted into quark–antiquark pairs. For example, if a red–antired pair is produced, the original red quark plus the antired antiquark can combine to form a meson. The remaining red quark can recombine with the original blue and green quarks to form a proton. Alternatively, it can combine with an antiquark from another quark–antiquark pair to give another meson, and so on. Thus, any attempt to isolate a coloured quark merely results in the production of colourless hadrons. Note that the above illustration has been given in terms of a specific colour, red. The amplitude, from quantum mechanics, must contain corresponding colour terms for blue and green quarks.

If a coloured quark cannot be isolated, then only quark systems which are 'colourless', that is colour singlets, can exist. The simplest colour singlets which can be formed are $q\bar{q}$, qqq and $\bar{q}\bar{q}\bar{q}$; these correspond to mesons, baryons and antibaryons respectively. Thus, these combinations would be expected to exist in nature but not, for example, qq or qqqq systems. More complex systems of multiples of three quarks, e.g. a six-quark system, are also colourless.

An interesting question is to what extent nuclei, which consist of $3n$ quarks with n running from 2 to about 240, can be considered to be made from a collection of nucleons rather than directly from quarks. Many models of the nucleus consider that, in addition to being composed of confined three-quark colourless 'bags', or nucleons, there are significant fractions of 'bags' of six or more quarks. The traditional idea that the nucleus is composed of individual nucleons is, perhaps, not as obvious as it might seem, since the density of (valence) quarks in a nucleon is only a few times that of nucleons in a nucleus. Thus, if the matter density can be increased, for example during the collisions of very high energy nuclei, there exists the possibility that a transition to a new state of matter, the so-called *quark–gluon plasma*, might be achieved. In this state the quarks and gluons would no longer be confined to individual nucleon boundaries.

The colour force increases as the separation between quarks increases. However, as the separation decreases, that is the relative momentum increases, the force becomes weaker. At infinite momentum the quarks would no longer interact; this is called *asymptotic freedom*. The strong coupling constant can be defined as a perturbation series. To first order

$$\alpha_S(Q^2) = \frac{12\pi}{(33 - 2n_f)} \bigg/ \ln(Q^2/\Lambda^2), \tag{1.16}$$

where n_f is the number of active quark flavours, Q^2 is the absolute value of the four-momentum transfer squared and Λ is a parameter to be determined from the data. The value of Λ appears to be in range of 50 to 500 MeV. (This gives, for example, $\alpha_S \sim 0.15$ for $Q^2 = 10^3$ GeV^2, $n_f = 4$ and $\Lambda = 0.2$ GeV.) This rather larger range for Λ stems from uncertainties as to the minimum value of Q^2 which can be used to fit to this asymptotic form. This is because additional contributions to a particular process arise from higher order terms, which often have not been calculated, and from non-asymptotic (higher twist) terms with a $1/Q^2$ fall-off, which are generally uncalculable. A consistent treatment in the extraction of Λ from various processes requires the inclusion of higher order terms in both the expansion of α_S and in the matrix element of the process itself, and these must be calculated using the same renormalisation scheme. The currently favoured value is, for the so-called modified minimal subtraction ($\overline{\text{MS}}$) scheme discussed in Chapter 7, $\Lambda_{\overline{\text{MS}}} \sim 200$ MeV.

For very large Q^2 values α_S becomes asymptotically zero. The decrease of α_S with increasing Q^2 can be understood qualitatively as being due to the colour charge being spread out by gluon branching processes; these processes growing with increasing Q^2 (Fig. 1.13). For small Q^2 ($\sim\Lambda^2$), α_S becomes very large; a hint that quark confinement occurs at a distance $r_c \sim 1/\Lambda$. This is plausible since r_c is roughly the size of a hadron. This latter argument also shows that quark and gluon degrees of freedom are important in a nucleus.

The introduction of the colour degree of freedom to the internal quark structure of baryons in fact solved an old problem in the quark classification of baryons. The Δ^{++} (1232) resonance consists, in the quark model, of uuu and is a member of the lowest lying spin $\frac{3}{2}$ decuplet. Assuming that the quarks are in their ground state, the symmetric $l = 0$ state, then all three u-quarks have the same spin configuration. Hence, the Δ^{++} is a state which is symmetric under the interchange of quarks. Since quarks are spin $\frac{1}{2}$ fermions, this violates the Pauli exclusion principle. The addition of the colour quantum number, in which the quarks are antisymmetric, solves this spin-statistics problem.

There are several reasonably direct pieces of experimental evidence which support this colour hypothesis. The first comes from measurements of R_h, the ratio of the cross-section for the production of hadrons compared to the production of $\mu^+\mu^-$, in e^+e^- annihilations, namely

$$R_h = \sigma(e^+e^- \to \text{hadrons})/\sigma(e^+e^- \to \mu^+\mu^-). \tag{1.17}$$

Fig. 1.13 QCD branching processes leading to an effective weakening of the colour charge of the gluon seen by the quark for different regimes of Q^2.

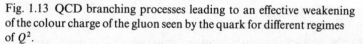

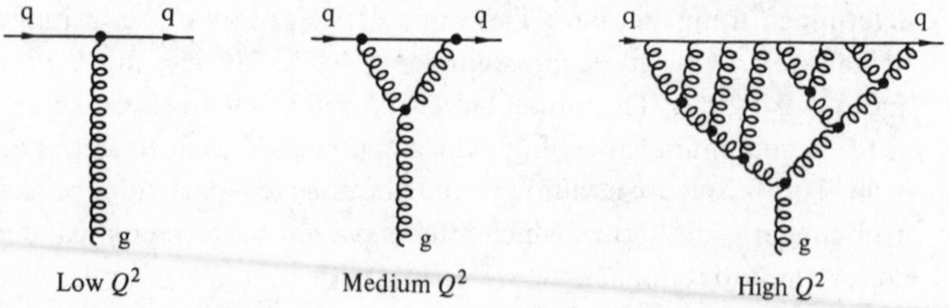

As shown in Chapter 4, the cross-section to produce a pair of point-like fermions, each of charge e_f, at a cms energy $s^{1/2}$ which is large enough that the relevant masses can be neglected, is

$$\sigma(e^+e^- \to f\bar{f}) = \left(\frac{4\pi\alpha^2}{3s}\right)e_f^2. \tag{1.18}$$

Thus, if the process $e^+e^- \to q\bar{q} \to$ hadrons is point-like, then the value expected for R_h is

$$R_h = \sum_i e_i^2, \tag{1.19}$$

where i is the sum over all contributing quark flavours. The point-like condition requires that the hadron production occurs on a timescale which is relatively long compared to that for $q\bar{q}$ production, so that the final state effects do not significantly modify the initial cross-section. Some experimental measurements of R_h as a function of $s^{1/2}$ are shown in Fig. 1.14. The contributing quark flavours at the highest values of $s^{1/2}$ are u, d, s, c and b. For these quarks

$$R_h = \left(\frac{2}{3}\right)^2 + \left(\frac{-1}{3}\right)^2 + \left(\frac{-1}{3}\right)^2 + \left(\frac{2}{3}\right)^2 + \left(\frac{-1}{3}\right)^2 = \frac{11}{9}. \tag{1.20}$$

However, each of these $q\bar{q}$ pairs can be produced in three possible colour states. Hence, with the inclusion of the colour factor, one expects $R_h = 11/3$; in reasonable agreement with the data. Further support for the colour hypothesis comes from the analysis of the π^0 lifetime (as discussed by Cheng and Li (1984)), this is 'wrong' by roughly a factor $N_c^2 = 9$ without the inclusion of colour), the Drell–Yan process, $q\bar{q} \to \gamma^* \to \mu^+\mu^-$ (Chapter 7) and the values of the leptonic branching ratios for the τ-lepton and hadrons containing c and b quarks (Chapter 8).

The decay widths for the $J^{PC} = 1^{--}$ states Ψ and Υ are narrow (63 and 43 keV respectively). They are both below the threshold for the decays to pairs of c- or b-quark hadrons respectively. The simplest decay mode (Fig. 1.15) is via three gluons (the intermediate state must be a colour singlet and C conservation requires an odd number of gluons). Calculations with $\alpha_S \sim 0.2$ can reproduce the observed widths (Chapter 10).

1.8 SUMMARY ON INTERACTIONS AND SYMMETRIES

Throughout this book the most commonly accepted theoretical model will be used as a guide to understanding the experimental data. It should be borne in mind, however, that not all aspects of this theoretical

Fig. 1.14 Measurements of the value of R_h as a function of the centre- of-mass energy $(s)^{1/2}$. ▾, LENA; ◆, DASP II; *, CLEO; △, DHHM; ●, Orsay; ■, Frascati; ◇, Novosibirsk; ×, SLAC-LBL; ○, DASP; □, CELLO; ⧗, JADE; +, MARK J; ▾, PLUTO; ▴, TASSO.

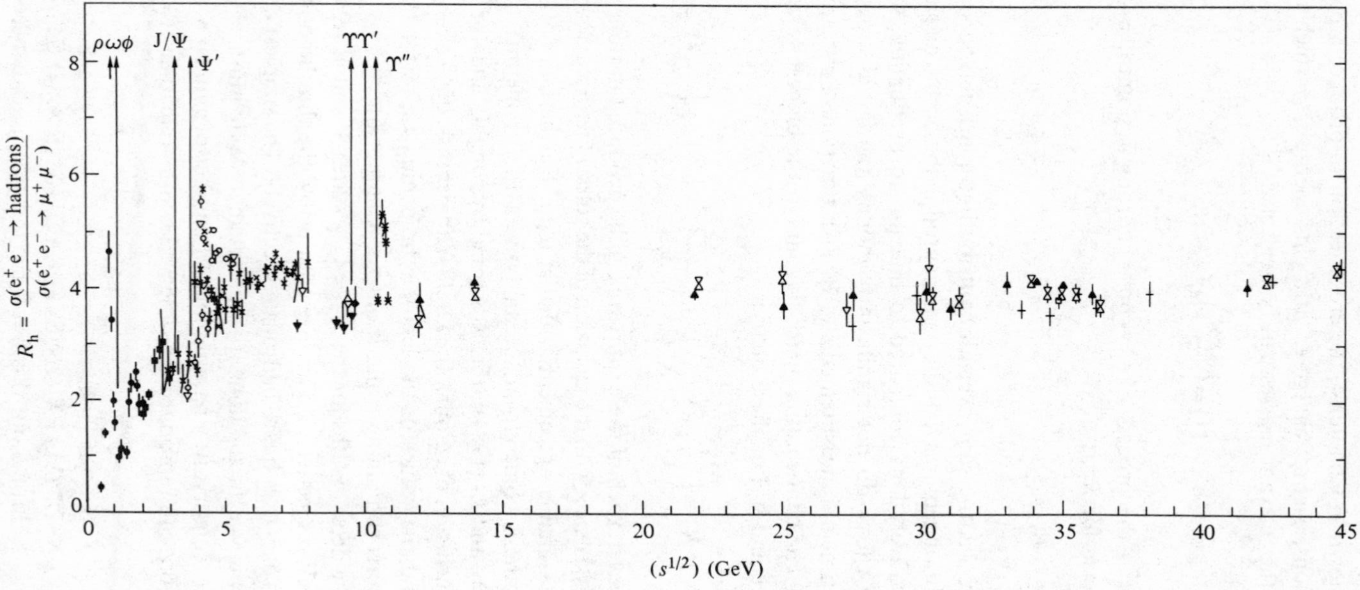

picture are universally accepted, and indeed a critical examination of both the theoretical ideas and of the experimental data should always be made.

In the standard theoretical framework the fundamental constituents of matter, at the distance scales which have so far been experimentally explored, are the spin $\frac{1}{2}$ quarks and leptons and the spin 1 gauge bosons. The quarks come in different flavours or types, of which five are firmly established (u, d, s, c, b). Each of these flavours comes in three colours. The electroweak interaction, that is a unified theory of the weak and electromagnetic interactions, has four gauge bosons. Amongst these, two (γ, Z^0) are electrically neutral and two are charged (W^+, W^-). The photon is massless, whereas the Z^0 (93 GeV) and the $W^\pm$ (82 GeV) are massive. As discussed later, the origin of particle masses is of great interest and may imply the existence of fundamental scalar particles. The strong interaction is mediated, in QCD, by eight massless coloured vector gluons. Since quarks and gluons are not observable directly their 'masses' must be inferred by somewhat indirect means. This relies heavily on theoretical input. The gauge bosons mediating the various interactions are summarised in Table 1.5.

The gravitational force between identical fundamental particles, unlike the other forces, is attractive. This implies the graviton has even spin. Spin 0 is excluded since that would imply light would not bend towards the sun. The simplest assignment for the graviton is thus spin 2.

The possible interactions amongst these particles are represented by the diagrams shown in Fig. 1.16. In this figure, i and j represent different quark flavours and a and b represent colours. The label l represents a lepton, with α and β denoting charged and neutral states (l^- or ν_l). In the minimal standard model there are no reactions which change the lepton generation (e, μ or τ). The quarks interact via the weak, electromagnetic or strong interactions. Thus emission or absorption of a γ or Z^0 (Fig. 1.16(*a*)) does not change the flavour or colour of the quark. For $W^\pm$ (Fig. 1.16(*b*)), there is a change in the quark flavour but no change in the colour (i.e. $W^\pm$ is colour blind), whereas for gluon emission (Fig. 1.16(*c*)) there is a change in colour but not in quark flavour. The charged leptons can emit a γ or Z^0 with no change in lepton type (Fig. 1.16(*d*)).

Fig. 1.15 Possible decay scheme for heavy $Q\bar{Q}$ state with $J^{PC} = 1^{--}$ to decay to three jets of hadrons via a three-gluon intermediate state.

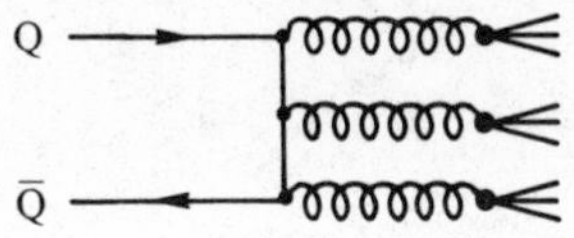

Neutral leptons (i.e. neutrinos) can also emit a Z^0. The emission of a $W^\pm$ (Fig. 1.16(*e*)) causes a change from a charged to neutral lepton. The corresponding diagrams for antiquarks and antileptons can be constructed in a similar manner. These diagrams can also be used to represent the line reversed reactions, e.g. γ, $Z^0 \rightarrow q_i^a \bar{q}_i^a$ (Fig. 1.16(*a*)), etc. In addition there are vertices involving the coupling of either three or four gauge bosons. The trilinear couplings are γW^+W^-, $Z^0W^+W^-$ and ggg. Since quarks are fermions, the total number of quarks minus antiquarks is conserved; however, there is no such restriction on the gauge bosons.

These possible vertex transitions allow a simple understanding of the empirical rules relating to the differences in the quantum numbers between the initial and final states in the various interactions. Strong interactions

Table 1.5. *The fundamental interactions in nature*

	Gauge bosons			
Interaction	Type	Spin	Mass	Coupled fermions
Strong	8 gluons	1	0	q, $\bar{q}$
Electromagnetic	1 photon	1	0	q, $\bar{q}$, l^+, l^-
Weak	2 Ws	1	82 GeV	q, $\bar{q}$, l^+, l^-, ν_l, $\bar{\nu}_l$
	1 Z^0	1	93 GeV	
Gravity	1 graviton	2	0	All matter

Fig. 1.16 Vertex diagrams for the emission of gauge bosons by quarks ((*a*) to (*c*)) and leptons ((*d*) and (*e*)). For the quarks, the indices *i* and *j* denote quark flavours and *a* and *b* colours. For the leptons, α and β denote charged and neutral states.

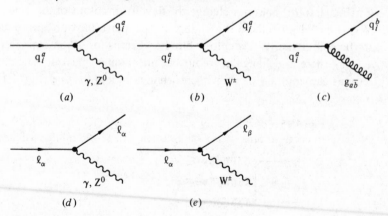

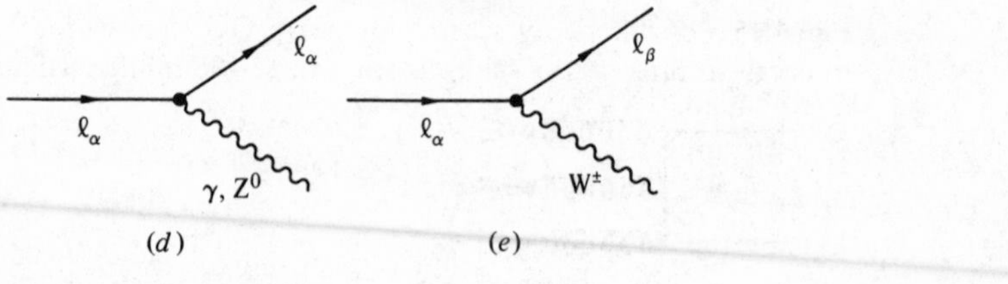

are mediated by gluons, which change only colour; therefore, since colour is not observable, all the quantum numbers referring to quark flavours are conserved. Hence, the total S, C', B' and T' quantum numbers should be equal in the initial and final states, since any particles with non-zero values of these quantum numbers can only be produced in particle–antiparticle pairs. The approximate degeneracy in mass of the u and d quarks means that the quantum numbers for 'u-ness' and 'd-ness' can be combined so that the strong interaction is invariant under isospin transformations. The differences in the strong interactions of hadrons of different quark content (e.g. between π^+p and K^+p collisions) are attributable to the differences in the respective quark masses. Thus, whereas the coupling of a gluon to a quark is independent of its flavour, the probability of creating a quark of a particular flavour is strongly dependent on its mass.

The electromagnetic interaction depends on the charge of the quark, and hence its flavour. Thus the cross-section for e^+p is different to that for e^+n, since the quark content of the proton and neutron are different. The quantum numbers associated with quark flavours (e.g. strangeness) are, however, conserved, since any quark–antiquark pairs created have equal and opposite flavour quantum numbers.

In the weak interactions mediated by $W^\pm$ exchange, the quark flavours change. Any additional quark–antiquark pairs produced in the interaction have zero net flavour, so that only very specific changes in the quark flavours are expected. These arise from transitions between the charge $\frac{2}{3}$ quarks (u, c, t) and the charge $-\frac{1}{3}$ quarks (d, s, b). Weak interactions involving Z^0 exchange do not change the quark flavour.

The conservation or violation of the various interactions under the transformations of spatial inversion or parity (P), charge conjugation (C) and time reversal (T), has given great insight into our understanding of these interactions. If two applications of a discrete transformation restore the original state, then one can define an associated multiplicative quantum number with eigenvalues ± 1, in contrast to the additive quantum numbers discussed above. In strong and electromagnetic interactions, P, C and T appear to be conserved. However, in the weak interaction, P and C are not only violated, but seem to be violated maximally.

The invariance of physical laws under the spatial inversion $\mathbf{r} \to -\mathbf{r}$ had been implicitly assumed until Lee and Yang (1956) suggested that parity was not conserved in weak interactions. This hypothesis gave an explanation to the so-called 'θ–τ puzzle'. Two charged particles, θ and τ, had been found with approximately the same mass but with different decay modes, namely $\theta^+ \to \pi^+\pi^-$ and $\tau^+ \to \pi^+\pi^+\pi^-$. Application of parity conservation for the θ-decay gives the possible spin-parity assignments

$J^{P(\theta)} = 0^+, 1^-$, etc. $(P(\theta) = (P_\pi)^2(-1)^J)$. Study of the τ-decay Dalitz plot showed that the τ had spin 0. Hence, since the pions have no relative angular momentum, $J^P(\tau) = 0^-$. The possibility of parity violation removes this conflict and θ and τ are identified as the $K \to 2\pi$ and $K \to 3\pi$ decay modes of the charged K-meson. Confirmation of parity violation came from an experiment by Wu *et al.* (1957), studying the β-decay of polarised ^{60}Co nuclei. A large decay asymmetry was found, indicating almost maximal parity violation. It appears that the weak interaction creates or annihilates neutrinos that are lefthanded and antineutrinos that are righthanded. Lefthanded (or negative helicity) means that the spin points in the direction opposite to the direction of motion. The production of particles in helicity eigenstates is a consequence of parity violation.

Under the operation of charge conjugation, a lefthanded neutrino becomes a lefthanded antineutrino; contrary to what is observed in nature. Thus, charge conjugation also appears to be violated maximally. The combined operation of C and P (i.e. CP) appears, however, to be almost exactly conserved in weak interactions. Under CP, a lefthanded ν becomes a righthanded $\bar{\nu}$. However, as first observed by Christenson *et al.* (1964) and confirmed by numerous other experiments, CP is clearly violated in the decays of K^0-mesons. This violation is rather small (a few parts per thousand) and its origin is somewhat unclear. The combined operation CPT, whose conservation is fundamental in quantum field theory, appears to be conserved. Thus, the small violation of CP may imply that T is also violated. These topics are discussed in Chapter 9.

The interaction mechanisms depicted in Fig. 1.16 assume that the lepton number for each lepton family is separately conserved. This conservation law implies, for example, that a ν_e beam propagates through space retaining its lepton number identity. The possibility exists, however, that transitions or oscillations between the neutrino types could occur (Pontecorvo, 1968). Such transitions would require the existence of a force which violates lepton number conservation. The experimental data on this subject are discussed in Chapter 9.

Another quantum number, whose conservation would follow from Fig. 1.16, is that of baryon number. Baryons can only be created as baryon–antibaryon pairs, whereas no such restriction applies for mesons. It should be noted that the conservation laws relating to quark flavour or lepton number are empirical and do not come from some deeper physical principle. For example, the existence of a massive X boson, with a $q l X$ vertex coupling, would transform quarks into leptons and *vice versa*. Although this may happen only with a very small probability (the predicted mass for such a boson is typically in the region of 10^{15} GeV), it would represent a violation of both baryon number conservation and of the other laws relating to quark flavour and lepton number conservation. In particular, the proton would no longer be stable. Such transitions between quarks and leptons are predicted in attempts to unify the electroweak interaction with QCD (Chapter 10). On the other hand, conservation of charge is expected to be exact. This is related to the principle of gauge invariance, and it is widely believed to be a fundamental principle.

In the light of the ideas discussed above it is worthwhile to review the experimental facts introduced in Section 1.1. In particular, the relative magnitudes of the cross-sections and decay rates for the strong, electromagnetic and weak interactions can be understood. The total strong interaction cross-section for hadron–hadron collisions is made up mainly from multiple soft (and relatively long-range) collisions. Further, the strong coupling α_S is large ($\alpha_S \sim 0(1)$), particularly for soft collisions. Hence, since both the projectile and target hadrons can be regarded as 'bags' of partons having large interaction cross-sections in the regions of geometrical overlap, the total cross-section is roughly proportional to the geometric cross-section. That is, $\sigma_{\text{tot}} \sim \pi R^2$, where R is a typical hadronic radius giving, for $R \sim 1$ fm, $\sigma_{\text{tot}} \sim 30$ mb (see Fig. 1.1). The time scale for a highly unstable resonance to decay (e.g. $\rho \rightarrow \pi\pi$) will be roughly that required for the colourless decay products to separate completely. To achieve a separation of, say, 2 fm at velocity c takes a time $\tau \sim 6 \times 10^{-24}$ s or equivalently a resonant width of $\Gamma \sim 100$ MeV. Thus, these phenomena, which we classify as strong interactions, are rather indirect manifestations of the underlying, and extremely strong, forces of QCD.

The electromagnetic cross-sections and decay rates are significantly smaller than those for strong processes, since $\alpha \ll \alpha_S$. The interaction of a real photon with a nucleon can be envisaged as a two-stage process. In the first stage, the photon transforms into a quark–antiquark pair with a probability proportional to α. This virtual quark–antiquark pair has the same quantum numbers ($J^{PC} = 1^{--}$) as the photon and hence the same as the vector mesons ρ, ω, ϕ, J/Ψ etc. In the second stage, this virtual vector meson interacts with the nucleon with a typical hadronic strong interaction cross-section, i.e. $\sigma_{\gamma\text{N}} \sim \alpha\sigma_{\text{hN}}$. In the inelastic scattering of high energy electrons or muons, the virtual photon can also interact with the nucleon through a virtual vector meson intermediate state. At low values of the four-momentum squared of the virtual photon (Q^2), this process is important (*vector meson dominance*). However, for values of $Q^2 \gtrsim 2\,\text{GeV}^2$, this process becomes negligible and the scattering occurs mainly between the virtual photon and the quarks inside the nucleon (*deep inelastic scattering*). Large fluctuations of the virtual photon are required to reach large values of Q^2, and this is reflected in the cross-section formula which has a $1/Q^4$ suppression factor for the virtual photon propagator.

The weak interaction has a similar structure to the electromagnetic interaction except for two important differences. Firstly, the virtual particles which are exchanged are massive and, secondly, they can be charged as well as neutral. The suppression factor for the propagator becomes $1/(q^2 - M^2)^2$ for a massive virtual boson. The weak coupling constant is similar in magnitude to the electromagnetic coupling constant α, hence for $Q^2(=-q^2) \ll M^2$, which is the case for the weak decays of leptons and hadrons and for existing neutrino beams, the weak interaction is suppressed by a factor of about $1/M^4$ with respect to equivalent electromagnetic processes. From the magnitude of the νN cross-section (Fig. 1.1) one can calculate that the effective geometric size of a nucleon as seen by a 10 GeV neutrino beam is about 10^{-19} cm. This effective size increases as the neutrino energy increases.

Compared to the other interactions, gravity is extremely weak, with a coupling strength of approximately 10^{-40}. A hydrogen atom bound with the gravitational, rather than the electromagnetic, interaction would be larger than the size of the known universe. Note that any particular interaction receives contributions, through higher order graphs, from all of the interactions at some level.

The coupling constants of the various interactions are not, in fact, constants. They are defined at some value of Q^2 and their evolution with Q^2 can be predicted. The electromagnetic constant α increases with Q^2, whereas the QCD coupling constant α_S decreases with Q^2. The value of Q^2 at which the coupling constants for the electroweak and strong interactions become equal sets the scale of the possible unification of these interactions. Estimates of this scale are of the order of 10^{15} GeV. For the gravitational interaction, the scale is set by considering the distance d at which the gravitational energy becomes equal to the rest mass energy, i.e. $d \sim G_g M$, where G_g is the Newtonian gravitational constant. Quantum gravity effects (e.g. virtual black holes) would be expected to be important for a Compton wavelength $1/M \sim d$. This sets the mass scale $M = (1/G_g)^{1/2} \sim 10^{19}$ GeV (the Planck mass) for a complete unification of forces.

Our understanding of the interactions in nature and, in particular, the enormous range in the energy scales of the interactions is far from complete. However, it appears important to identify the underlying symmetries in nature and to find out the mechanisms by which they are broken. Symmetries have played an important role throughout the history of physics. For example, the observed regularities in the motion of the moon and planets were important in our initial understanding of gravity. A symmetry (i.e. the invariance of the physical equations describing a system under a particular group of transformations) leads, in quantum mechanics, to a conserved quantity. Invariance with respect to Lorentz space-time translations and to rotations lead to the conservation of momentum-energy and angular-momentum respectively. However, the degeneracy in the energy levels of atoms, stemming from rotational invariance, is broken by the application of an external electromagnetic field, which causes the levels to split because the system is no longer rotationally symmetric. These are examples of *global* symmetries, in which the coordinate system is translated or rotated by some amount, but the physical equations in the new system are identical to those in the original system.

A more powerful symmetry is the *local* symmetry, in which the physical equations are invariant under transformations applied independently at each space-time point. A theory with such a property is known as a *gauge theory*, and it is widely believed that all interactions can be described by gauge theories. For example, invariance under local colour transformations leads to the theory of QCD, in which the colour forces are transmitted by the coloured gauge bosons. The group for this local symmetry in colour is SU(3). Note that the SU(3) group applied to quark flavours is a global symmetry. Maxwell's equations also obey a gauge symmetry. Conversely, assuming the existence of the appropriate gauge symmetry in nature, one can derive Maxwell's equations and the fact that the photon appears to be massless. The unified electroweak interaction has, as its underlying symmetry, the gauge group $SU(2)_L \otimes U(1)$, which is, however, manifestly broken. The $SU(2)_L$ group corresponds to the quantum number called *weak isospin*. This is a local symmetry, in contrast to the SU(2) group of the isospin properties of u and d quarks, which is global. Weakly interacting particles seem to appear in doublets, e.g. ν_e and e^-, u and d. These doublets have weak isospin $t = \frac{1}{2}$, with $t_3 = \frac{1}{2}$ for ν_e and $t_3 = -\frac{1}{2}$ for e^- etc. The gauge group U(1) corresponds to *weak hypercharge y*, defined such that $Q = t_3 + y/2$. The subscript L signifies that the weak interaction is lefthanded and it is only the lefthanded component of a particle which participates in the interaction. The righthanded component of particles, e.g. e_R^-, are singlet states of weak isospin with $t = 0$.

A further symmetry, for which there is as yet no convincing experimental evidence, but which is being actively investigated theoretically, is supersymmetry (SUSY). The postulated symmetry supposes the invariance of the laws of nature under transformations turning bosons into fermions and *vice versa*. Each particle has a supersymmetric partner, e.g. the spin $\frac{1}{2}$ electron has a spin 0 scalar electron partner, the spin $\frac{1}{2}$ quark has a spin 0 'squark' partner. The theory of supergravity deals with the consequences of local SUSY transformations; in particular, the spin 2 graviton has a spin $\frac{3}{2}$ SUSY partner, the gravitino. The development of such theories will not solve all the outstanding problems, however. The formalism of field theory is merely a descriptive framework in which the probabilities of particular physical processes can be calculated. Field theory does not explain why certain particles exist and not others. The input to such calculations is a list of elementary particles with their known properties (mass, spin, charge, etc.) and the form of the interaction between such particles. This list, for a particular calculation, could well be incomplete. The assumption than an interaction has some underlying group structure gives, however, an indication of the possible number of gauge particles.

Recently, however, theoretical ideas have been advanced which may (ultimately) provide a solution to the above problems. The fundamental entities in this approach are represented by 'strings', and not by points as in QED. Such strings sweep out (with time) two-dimensional world sheets rather than world lines. The size of the string is very roughly that of the Planck scale. The theory naturally incorporates supersymmetry and may prove to be free of divergences (i.e. finite). Although work on these 'superstring' theories is still in its infancy, the ambitious goal that the various interactions, coupling constants and all the properties of the associated particles may emerge is being actively pursued. These ideas are discussed further in Chapter 10.

2

Towards a quantum field theory

2.1 THE BUILDING BLOCKS OF QUANTUM FIELD THEORY

The current consensus of theoretical opinion is that the forces of nature can be described in terms of local gauge field theories. The main ideas which are the ingredients of a quantum field theory are those of

(i) quantum mechanics,
(ii) special relativity, and
(iii) classical field theory.

In this chapter these, plus other related topics, are briefly reviewed. The emphasis is on the formulation of an adequate description of the scattering process. This is because the bulk of our knowledge in particle physics comes from studies of the collisions of elementary particles. New particles can be formed in these collisions and the ideas relating to the decays of unstable particles are also considered.

2.2 NON-RELATIVISTIC QUANTUM MECHANICS

2.2.1 Classical mechanics

Before discussing quantum mechanics we first recall some of the laws of classical mechanics and electromagnetism. The equations of classical mechanics in the Lagrangian formulation and the canonical equations of Hamilton have essentially the same physical content as Newton's second law of motion. The Lagrangian function is defined in terms of some generalised coordinates q_i $(i=1,2,\ldots,\mathrm{n})$ and their derivatives with respect to time $(\dot{q}_i)$ as follows,

$$\mathscr{L}(q_i, \dot{q}_i) = \mathrm{T}\,(\mathrm{q}_i, \dot{q}_i) - V(q_i), \tag{2.1}$$

where T and V are the kinetic and potential energies respectively. Hamilton's principle of least action states that, for a conservative system, the action integral A does not depend on the actual path of the integration; thus

$$\delta A = \delta \int_{t_1}^{t_2} \mathscr{L}\, dt = 0. \tag{2.2}$$

In Fig. 2.1 this is shown for the simplified case of one coordinate. For this case we have

$$\delta A = \int_{t_1}^{t_2} \left(\frac{\partial \mathscr{L}}{\partial q_i} \delta q_i + \frac{\partial \mathscr{L}}{\partial \dot{q}_i} \delta \dot{q}_i \right) \mathrm{d}t. \tag{2.3}$$

Integrating the second term by parts and imposing the conditions that $\delta q_i(t_1) = \delta q_i(t_2) = 0$, and generalising to n coordinates, we obtain

$$\delta A = \sum_{i=1}^{n} \int_{t_1}^{t_2} \mathrm{d}t \left(\frac{\partial \mathscr{L}}{\partial q_i} - \frac{\mathrm{d}}{\mathrm{d}t} \frac{\partial \mathscr{L}}{\partial \dot{q}_i} \right) \delta q_i = 0. \tag{2.4}$$

This gives the *Euler–Lagrange equation of motion*

$$\frac{\partial \mathscr{L}}{\partial q_i} - \frac{\mathrm{d}}{\mathrm{d}t} \left(\frac{\partial \mathscr{L}}{\partial \dot{q}_i} \right) = 0. \tag{2.5}$$

As an example, consider the one-dimensional motion of a particle of mass m in a potential $V(x)$, that is

$$\mathscr{L} = \tfrac{1}{2} m \dot{x}^2 - V(x). \tag{2.6}$$

Fig. 2.1 Possible integration paths between point 1 $(t_1, (q_i)_1)$ and point 2 $(t_2, (q_i)_2)$.

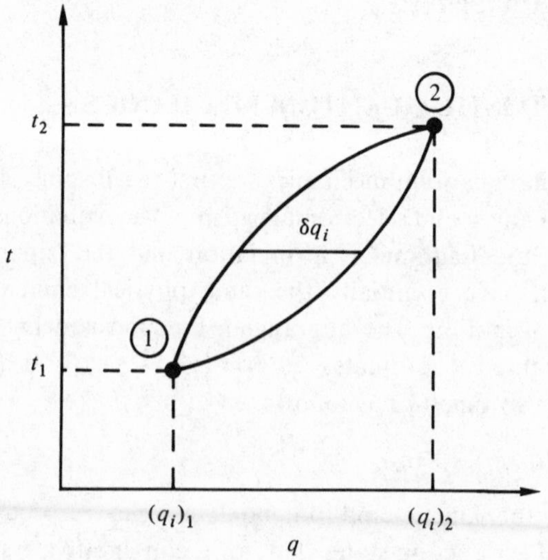

Equation (2.5) gives $m\ddot{x} = -\mathrm{d}V/\mathrm{d}x$, or more generally $m\ddot{\mathbf{r}} = -\nabla V$, that is Newton's equation.

The Euler–Lagrange equations can be written in another form by defining, as new variables, the generalised momenta $p_i = \partial\mathscr{L}/\partial\dot{q}_i$ and a new function, the Hamiltonian

$$H(p_i, q_i) = \sum_{i=1}^{n} p_i\dot{q}_i - \mathscr{L}(q_i, \dot{q}_i). \tag{2.7}$$

Differentiating (2.7) gives

$$\begin{aligned} \mathrm{d}H &= \dot{q}_i\mathrm{d}p_i + p_i\mathrm{d}\dot{q}_i - \frac{\partial\mathscr{L}}{\partial\dot{q}_i}\mathrm{d}\dot{q}_i - \frac{\partial\mathscr{L}}{\partial q_i}\mathrm{d}q_i \\ &= \dot{q}_i\mathrm{d}p_i - \frac{\partial\mathscr{L}}{\partial q_i}\mathrm{d}q_i = \dot{q}_i\mathrm{d}p_i - \dot{p}_i\mathrm{d}q_i. \end{aligned} \tag{2.8}$$

Now, since $H = H(p_i, q_i)$, we can write

$$\mathrm{d}H = \frac{\partial H}{\partial p_i}\mathrm{d}p_i + \frac{\partial H}{\partial q_i}\mathrm{d}q_i, \tag{2.9}$$

Therefore, equating (2.8) and (2.9) gives *Hamilton's equations of motion*

$$\dot{q}_i = \frac{\partial H}{\partial p_i}, \qquad \dot{p}_i = -\frac{\partial H}{\partial q_i}. \tag{2.10}$$

The Hamiltonian is the total energy of the system. This can be seen from (2.1) and (2.7) by putting $T_i = p_i\dot{q}_i/2$, to give

$$H = \left(\sum_{i=1}^{n} 2T_i\right) - T + V = T + V. \tag{2.11}$$

The total derivative of the Hamiltonian with respect to time is

$$\frac{\mathrm{d}H}{\mathrm{d}t} = \frac{\partial H}{\partial q_i}\dot{q}_i + \frac{\partial H}{\partial p_i}\dot{p}_i = 0, \tag{2.12}$$

where use has been made of (2.10). Thus, the total energy is conserved.

2.2.2 Classical electromagnetism

Electromagnetic phenomena are described classically in terms of the magnetic field $\mathbf{B}(t, \mathbf{x})$ and electric field $\mathbf{E}(t, \mathbf{x})$. These are related to the vector potential $\mathbf{A}(t, \mathbf{x})$ and scalar potential $\phi(t, \mathbf{x})$ by Maxwell's equations, which in rationalised Heaviside–Lorentz units are, in vacuo,

(see, e.g, Panofsky and Phillips, 1969)

$$\mathbf{B} = \nabla \times \mathbf{A}, \tag{2.13a}$$

$$\mathbf{E} = -\frac{\partial \mathbf{A}}{\partial t} - \nabla\phi, \tag{2.13b}$$

$$\nabla \times \mathbf{B} = \mathbf{j} + \frac{\partial \mathbf{E}}{\partial t}, \tag{2.13c}$$

$$\nabla \cdot \mathbf{E} = \rho, \tag{2.13d}$$

where $\mathbf{j}(t, \mathbf{x})$ and $\rho(t, \mathbf{x})$ are the electric current and charge density respectively. Taking the divergence of (2.13c), using (2.13d) and the property div curl $\mathbf{B} = 0$, gives

$$\nabla \cdot \mathbf{j} = -\partial\rho/\partial t. \tag{2.14}$$

Thus, charge is conserved in Maxwell's equations. This is a global conservation law and applies to all space-time.

The differential equations for $\mathbf{A}$ and ϕ are determined by eliminating $\mathbf{B}$ and $\mathbf{E}$ from Maxwell's equations. Putting (2.13a) and (2.13b) into (2.13c) gives

$$\nabla \times (\nabla \times \mathbf{A}) = \mathbf{j} - \frac{\partial^2 \mathbf{A}}{\partial t^2} - \nabla \frac{\partial \phi}{\partial t}.$$

Using $\nabla \times (\nabla \times \mathbf{A}) = \nabla(\nabla \cdot \mathbf{A}) - \nabla^2 \mathbf{A}$ and rearranging gives

$$\nabla\left(\nabla \cdot \mathbf{A} + \frac{\partial \phi}{\partial t}\right) - \nabla^2 \mathbf{A} + \frac{\partial^2 \mathbf{A}}{\partial t^2} = \mathbf{j}. \tag{2.15}$$

Putting (2.13b) into (d) gives

$$-\frac{\partial}{\partial t}(\nabla \cdot \mathbf{A}) - \nabla^2 \phi = \rho. \tag{2.16}$$

The solutions to equations (2.15) and (2.16) are not unique; any new potentials $\mathbf{A}'$ and ϕ' such that

$$\begin{aligned} \mathbf{A}' &= \mathbf{A} - \nabla\chi, \\ \phi' &= \phi + \partial\chi/\partial t. \end{aligned} \tag{2.17}$$

where χ is an arbitrary function of $\mathbf{x}$ and t, also satisfy these equations. This can be checked by substitution. Transformation (2.17) is called a *gauge transformation*. If the charge distribution is static, i.e. $\rho(\mathbf{x})$ is independent of t, it is convenient to choose a gauge, the *Coulomb gauge*, such that

$$\nabla \cdot \mathbf{A} = 0. \tag{2.18}$$

In this gauge, equations (2.16) and (2.15) become

$$\nabla^2 \phi = -\rho \tag{2.19}$$

and, since ϕ is independent of time,

$$-\nabla^2 \mathbf{A} + \frac{\partial^2 \mathbf{A}}{\partial t^2} = \mathbf{j}. \tag{2.20}$$

Equation (2.19) is *Poisson's equation*. The scalar potential ϕ is independent of time and, further, both $\mathbf{A}$ and ϕ are real.

In preparation for the field theory description of photons (Chapters 3 and 4) it is useful to consider the case of a free photon, that is $\mathbf{j} = \mathbf{0}$, $\rho = 0$. Hence, ϕ is zero and a possible solution for $\mathbf{A}$ is

$$\mathbf{A} = \boldsymbol{\varepsilon} A_0 \exp[i(\mathbf{k} \cdot \mathbf{x} - \omega t)], \tag{2.21}$$

where A_0 is the amplitude, and $\boldsymbol{\varepsilon}$ is a unit vector (called the *polarisation vector*) which is in the direction of $\mathbf{A}$, and ω and $\mathbf{k}$ are the photon energy and momentum respectively. Now, for the condition $\nabla \cdot \mathbf{A} = 0$ to be satisfied, we require $\boldsymbol{\varepsilon} \cdot \mathbf{k} = 0$, that is the polarisation is transverse to the direction of motion. From (2.13b) and (a) we have $\mathbf{E} = i\omega \mathbf{A}$ and $\mathbf{B} = \nabla \times \mathbf{A} = \mathrm{i}\mathbf{k} \times \mathbf{A}$. Thus, $\mathbf{E}$, $\mathbf{B}$ and $\mathbf{k}$ are mutually orthogonal, and $\boldsymbol{\varepsilon}$ is in the direction of the electric field.

For a particle of mass m and charge e, a possible Lagrangian is

$$\mathscr{L} = T - e\phi + e\dot{\mathbf{x}} \cdot \mathbf{A} = \sum_i \tfrac{1}{2} m \dot{x}_i^2 + e\dot{x}_i A_i - e\phi. \tag{2.22}$$

Inserting this form in (2.5), the Lagrangian equation of motion for the jth coordinate (x_j) is

$$e \sum_i \dot{x}_i \frac{\partial A_i}{\partial x_j} - e \frac{\partial \phi}{\partial x_j} - m\ddot{x}_j - e\dot{A}_j = 0. \tag{2.23}$$

Now,

$$\dot{\mathbf{A}}_j = \frac{\partial A_j}{\partial t} + \sum_i \dot{x}_i \frac{\partial A_j}{\partial x_i}. \tag{2.24}$$

Hence,

$$\begin{aligned} m\ddot{x}_j &= \mathrm{e}\left(-\frac{\partial \phi}{\partial x_j} - \frac{\partial A_j}{\partial t}\right) + e \sum_i \dot{x}_i \left(\frac{\partial A_i}{\partial x_j} - \frac{\partial A_j}{\partial x_i}\right) \\ &= e\left(-\nabla \phi - \frac{\partial \mathbf{A}}{\partial t}\right)_j + e(\mathbf{v} \times \nabla \times \mathbf{A})_j. \end{aligned} \tag{2.25}$$

Since $\mathbf{B} = \nabla \times \mathbf{A}$, (2.25) is thus the equation for the *Lorentz force*

$$m\dot{\mathbf{v}} = e(\mathbf{E} + \mathbf{v} \times \mathbf{B}). \tag{2.26}$$

The momenta p_i can be obtained from (2.22) using $p_i = \partial L/\partial \dot{x}_i$; thus

$$p_i = m\dot{x}_i + eA_i. \tag{2.27}$$

Finally, using (2.7) and (2.27), the Hamiltonian is

$$\begin{aligned} H &= \sum_i p_i \dot{x}_i - \tfrac{1}{2} m \dot{x}_i^2 - e\dot{x}_i A_i + e\phi \\ &= \frac{1}{2m}(\mathbf{p} - e\mathbf{A})^2 + e\phi. \end{aligned} \tag{2.28}$$

Thus, compared to the no-field case, $\mathbf{p}$ is replaced by $\mathbf{p} - e\mathbf{A}$.

2.2.3 The wave equation

In non-relativistic quantum mechanics the state of a particle is represented by a complex *state* or *wave* function ψ. The probability of finding the particle at time t in a volume element d^3x at position $\mathbf{r}$ is $|\psi(t, \mathbf{r})|^2 \mathrm{d}^3x$. If ψ_1 and ψ_2 are wave functions, then it is convenient to use the Dirac notation to define the scalar product,

$$\int \psi_1^* \psi_2 \mathrm{d}^3x = \langle \psi_1 | \psi_2 \rangle. \tag{2.29}$$

A wave function is normalised so that $\langle \psi | \psi \rangle = 1$.

The wave function itself is not 'observable', but it is assumed that a complete knowledge of ψ represents all the information obtainable about the system. Observable quantities are represented by *operators* acting on ψ. For example, the operators for momentum, $\hat{\mathbf{p}}$, kinetic energy, $\hat{T}$, and energy, $\hat{E}$, are

$$\begin{aligned} \hat{\mathbf{p}} &= -\mathrm{i}\,\mathrm{grad} = -\mathrm{i}\nabla, \\ \hat{T} &= -\nabla^2/2m \qquad (\nabla^2 = \partial^2/\partial x^2 + \partial^2/\partial y^2 + \partial^2/\partial z^2), \\ \hat{E} &= \mathrm{i}\partial/\partial t. \end{aligned} \tag{2.30}$$

Operators do not necessarily commute. From the definition of $\hat{p}_x = -\mathrm{i}\partial/\partial x$, it follows that $\hat{x}\hat{p}_x - \hat{p}_x\hat{x} = [\hat{x}, \hat{p}_x] = \mathrm{i}$, where $\hat{x}$ is the position operator.

If an operator $\hat{l}$ acting on a state u_s satisfies $\hat{l}u_s = l_s u_s$, then u_s is said to be an *eigenfunction* of $\hat{l}$ with *eigenvalue* l_s. The wave function ψ can be defined in terms of a complete orthonormal set of eigenfunctions

$$\psi = \sum_{s=1}^{n} c_s u_s, \tag{2.31}$$

with the orthogonality condition $\langle u_s | u_r \rangle = \delta_{sr}$ (Kronecker symbol). The abstract space spanned by such base vectors is called a *Hilbert space*. For the state ψ the probability that $\hat{l}$ is found to have an eigenvalue l_s is $|c_s|^2$, with $c_s = \langle u_s | \psi \rangle$. Consider a space with n linearly independent base vectors, then

$$\sum_{s=1}^{n} |u_s\rangle\langle u_s | u_r\rangle = \sum_{s=1}^{n} |u_s\rangle \delta_{sr} = |u_r\rangle. \tag{2.32}$$

This gives the *completeness* or *closure relationship*

$$\sum_{s=1}^{n} |u_s\rangle\langle u_s| = 1. \tag{2.33}$$

The overall phase of the total wave function cannot be observed. For example, writing $\psi' = \exp(\mathrm{i}\delta)\psi$, then $|\psi'|^2 = |\psi|^2$ for any value of δ.

If two operators do not commute (e.g. $[\hat{x}, \hat{p}_x] = \mathrm{i}$), then the state cannot be, simultaneously, an eigenfunction of both. That is, one cannot make simultaneous measurements of, for example, both x and p_x. If the wave function for a particle is localised around the origin, e.g. $\psi(x) \sim \exp(-x^2/2\Delta_x^2)$, with corresponding probability density $P(x) \sim \exp(-x^2/\Delta_x^2)$, then the probability that the particle has momentum p_x is $|\phi(p_x)|^2$, where

$$\begin{aligned}\phi(p_x) &\sim \int_{-\infty}^{\infty} \exp(-\mathrm{i}p_x x)\psi(x)\mathrm{d}x \\ &\sim \int_{-\infty}^{\infty} \exp\left[-\frac{1}{2}\left(\frac{x}{\Delta_x} + \mathrm{i}p_x\Delta_x\right)^2\right]\exp\left[-\left(\frac{p_x^2\Delta_x^2}{2}\right)\right]\mathrm{d}x \\ &\sim \exp(-p_x^2\Delta_x^2/2).\end{aligned} \tag{2.34}$$

This follows by putting $y = (x/\Delta_x + \mathrm{i}p_x\Delta_x)$ and noting that $\int \exp(-y^2/2)\mathrm{d}y$ just gives a constant. Thus, if we define $\phi(p_x) \sim \exp(-p_x^2/2\Delta_p^2)$, then comparison with (2.34) gives $\Delta_p\Delta_x = \hbar\ (=1)$. The above argument depends on the Gaussian forms assumed for the wave functions. More generally we obtain the *Heisenberg uncertainty principle*

$$(\Delta p)(\Delta x) \gtrsim \hbar. \tag{2.35}$$

Two useful operators in quantum mechanics are the *Hermitian* and *unitary* operators. The eigenvalues of an Hermitian operator are real.

Defining

$$\langle u_s|\hat{l}|u_r\rangle = \langle \hat{l}u_s\,|\,u_r\rangle = \langle u_s\,|\,\hat{l}u_r\rangle, \tag{2.36}$$

then an operator is Hermitian if $\hat{l} = \hat{l}^\dagger$ where

$$\langle u_s|\hat{l}^\dagger|u_r\rangle = (\langle u_r|\hat{l}|u_s\rangle)^*. \tag{2.37}$$

That is, the Hermitian conjugate involves both transposing and complex conjugation. Using these relationships it follows that the eigenvalues are real:

$$\langle u_s|\hat{l}^\dagger|u_s = (\langle u_s|\hat{l}|u_s\rangle)^* = l_s^*\langle u_s|u_s\rangle,$$

$$\langle u_s|\hat{l}|u_s\rangle = l_s\langle u_s\,|\,u_s\rangle, \tag{2.38}$$

thus $l_s = l_s^*$. Hence, observables are represented by linear Hermitian operators. A unitary operator is such that its Hermitian conjugate is equal to its inverse, that is (dropping the ˆ symbol)

$$U^\dagger = U^{-1} \qquad \text{or} \qquad U^\dagger U = 1. \tag{2.39}$$

Note that U is not necessarily Hermitian. Consider an operator l with an eigenvector $|u_s\rangle$ and eigenvalue l_s, then, if under a unitary transformation $l' = UlU^{-1}$ and $|u_s'\rangle = U|u_s\rangle$, it follows that

$$\langle u_s'|\,l'\,|\,u_r'\rangle = \langle u_s|\,U^{-1}l'U\,|u_r\rangle = \langle u_s|\,l\,|u_r\rangle. \tag{2.40}$$

Thus, the physical equations are invariant under unitary transformations. Further, if the operator l is Hermitian ($l^\dagger l = 1$), then l' is also Hermitian.

An infinitesimal unitary tranformation can be written

$$U = 1 - \mathrm{i}\varepsilon F, \tag{2.41}$$

where ε is small and F is called the *generator*. Hence

$$U^\dagger U = (1 + \mathrm{i}\varepsilon F^\dagger)(1 - \mathrm{i}\varepsilon F) = 1 + \mathrm{i}\varepsilon(F^\dagger - F) + O(\varepsilon^2) \tag{2.42}$$

Setting $U^\dagger U$ equal to unity implies $F^\dagger = F$, i.e. F is Hermitian. The usefulness of unitary transformations can be seen by considering the effect of F on an arbitrary operator l. Thus, we have

$$l' = UlU^{-1} = (1 - \mathrm{i}\varepsilon F)l(1 + \mathrm{i}\varepsilon F),$$

i.e.

$$l' = l - \mathrm{i}\varepsilon(Fl - lF) = l - i\varepsilon[F, l]. \tag{2.43}$$

Thus, if F commutes with l then $l' = l$, so l is invariant under the unitary transformation. A finite unitary transformation can be constructed from successive infinitesimal transformations. Putting $\alpha = n\varepsilon$, (2.41) becomes

$$U(\varepsilon) = 1 - \mathrm{i}\frac{\alpha}{n}F. \tag{2.44}$$

After n successive applications we obtain

$$U(\alpha) = U(\varepsilon)^n = \left(1 - \mathrm{i}\frac{\alpha}{n}F\right)^n \overset{n\to\infty}{=} \exp(-\mathrm{i}\alpha F). \tag{2.45}$$

Note that in general if an operator H is Hermitian, then $\exp(\mathrm{i}H)$ is unitary.

The wave function ψ satisfies the *time-dependent Schrödinger equation*. This can be obtained from consideration of the total energy operator or Hamiltonian $\hat{H} = \hat{T} + \hat{V}$ and the energy operator $\hat{E}$. Using equation (2.30) gives

$$\hat{E}\psi(t, \mathbf{r}) = \hat{H}\psi(t, \mathbf{r}), \tag{2.46}$$

so

$$\mathrm{i}\frac{\partial\psi(t, \mathbf{r})}{\partial t} = H\psi(t, \mathbf{r}) = \left[-\frac{1}{2m}\nabla^2 + V(t, \mathbf{r})\right]\psi(t, \mathbf{r}). \tag{2.47}$$

If V is independent of t, then, putting $\psi(t, \mathbf{r}) = u(\mathbf{r})g(t)$, (2.47) can be written in the form

$$\frac{\mathrm{i}}{g(t)}\frac{\partial g(t)}{\partial t} = \frac{1}{u(\mathbf{r})}\left[-\frac{1}{2m}\nabla^2 + V(\mathbf{r})\right]u(\mathbf{r}) = E, \tag{2.48}$$

where E is the energy eigenvalue. Solving for $g(t)$ gives therefore

$$\psi(t, \mathbf{r}) = u(\mathbf{r})\exp(-\mathrm{i}Et). \tag{2.49}$$

The complex conjugate of the Schrödinger equation (2.47) is

$$-\mathrm{i}\frac{\partial\psi^*}{\partial t} = \left[-\frac{1}{2m}\nabla^2 + V(t, \mathbf{r})\right]\psi^*. \tag{2.50}$$

The rate of change of the probability density $\rho = \psi^*\psi$ is

$$\frac{\partial\rho}{\partial t} = \left(\frac{\partial\psi^*}{\partial t}\right)\psi + \psi^*\left(\frac{\partial\psi}{\partial t}\right) = \frac{-\mathrm{i}}{2m}[(\nabla^2\psi^*)\psi - \psi^*(\nabla^2\psi)]. \tag{2.51}$$

The righthand side can be written in terms of a vector current (the probability current)

$$\mathbf{j}(t, \mathbf{r}) = \frac{\mathrm{i}}{2m}[(\boldsymbol{\nabla}\psi^*)\psi - \psi^*(\boldsymbol{\nabla}\psi)], \tag{2.52}$$

since the divergence

$$\operatorname{div}\mathbf{j} = \frac{\mathrm{i}}{2m}[(\nabla^2\psi^*)\psi - \psi^*(\nabla^2\psi)] = -\frac{\partial\rho}{\partial t}. \tag{2.53}$$

Consider an arbitrary volume V enclosed by a surface S, then using

Gauss's theorem

$$\frac{\partial}{\partial t}\int_V \rho \mathrm{d}^3x = -\int_V \operatorname{div}\mathbf{j}\mathrm{d}^3x = -\int_S j_n \mathrm{d}S, \tag{2.54}$$

where n is the component in the direction of the outward normal to surface element $\mathrm{d}S$. Equation (2.54) (and also (2.53)) represent the continuity equation for the current $\mathbf{j}$, since the rate of change of finding the particle in V equals the incoming flux of the current $\mathbf{j}$ crossing the surface.

For a *free particle*, equation (2.47) becomes

$$\mathrm{i}\frac{\partial\psi}{\partial t} = -\frac{1}{2m}\nabla^2\psi, \tag{2.55}$$

which has a solution (which can be checked by substitution)

$$\psi = N\exp[-\mathrm{i}(Et - \mathbf{p}\cdot\mathbf{x})]. \tag{2.56}$$

Thus, for a free particle, this plane wave gives

$$\rho = |N|^2, \qquad \mathbf{j} = |N|^2\mathbf{p}/m. \tag{2.57}$$

The symmetry properties of the Hamiltonian lead to *conserved quantities*. For example, consider the parity operator P, defined as

$$P\psi(\mathbf{x}) = \psi(-\mathbf{x}). \tag{2.58}$$

Further application restores the original state, so $P^2 = 1$ and P has eigenvalues ± 1. If we operate with P on the Schrödinger equation, we obtain

$$\mathrm{i}\frac{\partial}{\partial t}(P\psi(t,\mathbf{x})) = PH\psi(t,\mathbf{x}). \tag{2.59}$$

If $H(\mathbf{x})$, and hence $V(\mathbf{x})$, is an even function of $\mathbf{x}$, then $PH\psi(t,\mathbf{x}) = HP\psi(t,\mathbf{x})$; that is, P commutes with H ($[P,H]=0$) and so $P\psi(t,\mathbf{x})$ is a solution to Schrödinger's equation. In this case the two parity solutions

$$\psi^{\pm}(t,\mathbf{x}) = \tfrac{1}{2}(1 \pm P)\psi(t,\mathbf{x}), \tag{2.60}$$

are both solutions, and do not mix as functions of time. In general an operator that commutes with H and has no explicit time dependence is a constant of the motion.

As discussed later (Section 4.7), there is more than one form in which the equations of motion of quantum mechanics may be expressed. In the *Schrödinger representation* the state vectors depend on time, whereas the operators do not, in contrast to the *Heisenberg representation*, where the converse applies.

Quantum mechanics may also be formulated in a different (but equivalent) way using the Feynman path integral approach. In Section 2.2.1, the classical propagation between two points was considered. In quantum mechanics one must construct an amplitude representing the sum of all possible paths, $\psi = \sum_{\text{paths}} \exp(\mathrm{i}A(x)/\hbar)$, where A is the action. A particular path will have a certain probability. The classical trajectory corresponds to the $\hbar \to 0$ limit. The development of these ideas in the context of field theory is discussed by Cheng and Li (1984).

2.2.4 Perturbation theory

Many problems in quantum mechanics cannot be solved exactly and various approximate methods, such as bound state perturbation theory, time-dependent perturbation theory, variational methods and the WKB approximation are employed. In this section some aspects of the time-dependent perturbation theory of the Schrödinger equation (i.e. $\mathrm{i}\partial\psi/\partial t = H\psi$, equation (2.47)) are considered. The Hamiltonian is written in terms of an unperturbed part H_0 and a time-dependent perturbation H' (e.g. a scattering potential)

$$H = H_0 + H' \qquad \text{with} \quad H_0 u_n = E_n u_n. \tag{2.61}$$

It is assumed that the eigenvalues and functions (E_n and u_n) are known and that H' is small. The procedure is to expand ψ in terms of the eigenfunctions $u_n \exp(-\mathrm{i}E_n t)$ of the unperturbed wave equation. Thus

$$\psi = \sum_n a_n(t) u_n \exp(-\mathrm{i}E_n t), \tag{2.62}$$

where the coefficients a_n depend on t. Substituting (2.62) in (2.47) gives

$$\sum_n (\mathrm{i}\dot{a}_n + a_n E_n) u_n \exp(-\mathrm{i}E_n t) = (H_0 + H') \sum_n a_n u_n \exp(-\mathrm{i}E_n t). \tag{2.63}$$

Use of $H_0 u_n = E_n u_n$ simplifies this equation to

$$\mathrm{i} \sum_n \dot{a}_n u_n \exp(-\mathrm{i}E_n t) = \sum_n a_n H' u_n \exp(-\mathrm{i}E_n t). \tag{2.64}$$

Multiplying from the left by u_f^*, and using the orthonormality relationships for the eigenfunctions, gives

$$\mathrm{i}\, \dot{a}_f \exp(-\mathrm{i}E_f t) = \sum_n a_n H'_{fn} \exp(-\mathrm{i}E_n t),$$

where

$$H'_{fn} = \int u_f^* H' u_n \,\mathrm{d}^3 x. \tag{2.65}$$

Thus,

$$\dot{a}_f(t) = -\mathrm{i}\sum_n a_n(t)H'_{fn}\exp[\mathrm{i}(E_f - E_n)t]. \tag{2.66}$$

To proceed further depends on the nature of the problem to be solved. Certain scattering processes can be considered by assuming that the perturbation acts for a short interval of time at $t \sim 0$. The system is assumed to be in an initial state i at $t = -t'$, such that $a_i(-t') = \delta_{ni}$. Hence, (2.66) becomes

$$\dot{a}_f(t) = -\mathrm{i}H'_{fi}\exp[\mathrm{i}(E_f - E_i)t]. \tag{2.67}$$

At some time t' after the interaction a_f can be obtained by integrating this equation, which remains approximately valid if the perturbation is small. Thus,

$$a_f(t') = -\mathrm{i}\int_{-t'}^{t'} \mathrm{d}tH'_{fi}(t)\exp[\mathrm{i}(E_f - E_i)t]. \tag{2.68}$$

If H' is independent of time during the time it is switched on, then for $t' \to \infty$

$$a_f = -\mathrm{i}H'_{fi}\int_{-\infty}^{\infty} \mathrm{d}t\exp[\mathrm{i}(E_f - E_i)t],$$

so

$$a_f = -2\pi\mathrm{i}H'_{fi}\delta(E_f - E_i), \tag{2.69}$$

where δ is the Dirac δ-function and expresses energy conservation. Exact energy conservation implies, by the uncertainty principle, an infinite time interval between the initial and final states. Hence, a_f is not directly useful in this form.

The transition probability per unit time is defined as

$$\begin{aligned}\lambda &= \lim_{t'\to\infty}\frac{|a_f|^2}{t'}\\ &= \lim_{t'\to\infty}(2\pi)^2\delta(E_f - E_i)\left[\frac{1}{(2\pi)}\int_{-t'/2}^{t'/2}\mathrm{d}t\exp[\mathrm{i}(E_f - \mathrm{E}_i)\mathrm{t}]\right]|H'_{fi}|^2/t'\\ &= \lim_{t'\to\infty}2\pi\delta(E_f - E_i)\left[\int_{-t'/2}^{t'/2}\mathrm{d}t\right]|H'_{fi}|^2/t'\\ &= 2\pi|\mathrm{H}'_{fi}|^2\delta(E_f - E_i).\end{aligned} \tag{2.70}$$

The above derivation makes use of the properties of the δ-function (see Appendix D). A more complete discussion can be found in Muirhead

(1965). In practice, many final states are possible. Let $\rho(E_f)\mathrm{d}E_f$ be the number of final states in the interval $\mathrm{d}E_f$ around E_f. It is assumed that the *density of final states* ρ is a slowly varying function of E_f. The total transition rate is achieved by integrating over E_f. Thus,

$$\begin{aligned}\lambda &= 2\pi \int |H'_{fi}|^2 \delta(E_f - E_i)\rho(E_f)\mathrm{d}E_f \\ &= 2\pi |H'_{fi}|^2 \rho(E_i).\end{aligned} \tag{2.71}$$

This equation, *Fermi's Golden Rule*, can be considered the starting point for many calculations. The above calculation is to first order in the perturbative expansion. Successive corrections can be made by substituting the result for $a_f(t)$ from (2.68) into equation (2.66), and so on.

As an introduction to the relativistic case one can rewrite (2.68) as

$$a_f = -\mathrm{i} \iint \mathrm{d}t\, \mathrm{d}^3x [u_f \exp(-\mathrm{i}E_f t)]^* H'_{fi}(t, \mathbf{x})[u_i \exp(-\mathrm{i}E_i t)].$$

The transition rate T_{fi} between the initial state i and a final state f can be written in covariant form (this is discussed in Section 2.3) in terms of the four-vector $x = (t, \mathbf{x})$, namely

$$T_{fi} = -\int \mathrm{d}^4x \psi_f^*(x) H'_{fi} \psi_i(x). \tag{2.72}$$

The above discussion has been concerned with the scattering of a beam of particles by, for example, a potential. However, the method also allows us to describe the decay of a resonant state. Suppose a system is in an eigenstate of energy E_i at $t = 0$. Then the probability $I(E_f, t)\mathrm{d}E_f$, to find the system at time t in one of the states f in the interval $\mathrm{d}E_f$ around E_f, is given by

$$I(E_f, t)\mathrm{d}E_f = |a_f(t)|^2 \rho(E_f)\mathrm{d}E_f. \tag{2.73}$$

Now, if $H'_{fn} = 0$ for $n \neq i$ (i.e. no final state interactions), then (2.66) becomes

$$\dot{a}_f(t) = -\mathrm{i}a_i(t) H'_{fi} \exp[\mathrm{i}(E_f - E_i)t]. \tag{2.74}$$

The decay of the initial state i can be expressed as $|a_i(t)|^2 = \exp(-\lambda t)$, so $a_i(t) = \exp(-\lambda t/2)$. Putting this in (2.74), setting $\omega = E_f - E_i$ and integrating gives

$$\begin{aligned}a_f(t) &= -\mathrm{i} \int_0^t H'_{fi} \exp[(-\lambda/2 + \mathrm{i}\omega)t']\mathrm{d}t' \\ &= H'_{fi}\{1 - \exp[(-\lambda/2 + \mathrm{i}\omega)t]\}/(\omega + \mathrm{i}\lambda/2).\end{aligned} \tag{2.75}$$

Hence, using (2.71), (2.73) becomes

$$I(E_f, t) = \frac{\lambda[1 + \exp(-\lambda t) - 2\exp(-\lambda t/2)\cos \omega t]}{2\pi(\omega^2 + \lambda^2/4)}, \tag{2.76}$$

so for $t \to \infty$, and putting $\Gamma = \lambda$, the familiar resonance distribution is obtained

$$I(E_f) = \frac{\Gamma/2}{\pi[(E_f - E_i)^2 + \Gamma^2/4]}. \tag{2.77}$$

2.2.5 The Born approximation

The collision processes of an incident particle (e.g. an electron) scattering from a target (e.g. an atom or nucleus) can be classified as being either elastic or inelastic. In *elastic collisions*, the final state particles are the same as those in the initial state. Such collisions are equivalent to the scattering of a single particle by a fixed centre of force, represented by a potential, i.e. effectively an infinitely heavy target, or to the use of the centre of mass system and of the reduced mass for the projectile. *Inelastic collisions* are those in which the final state particles are not the same as those in the initial state.

If the perturbation to the initial free particle momentum eigenstate is provided by a potential $V(\mathbf{r})$, generated by a scattering centre at $\mathbf{r} = 0$, then the resulting elastic scattering cross-section can be evaluated using equation (2.71), and this gives the Born approximation result. It is assumed that the potential is relatively weak and has a short range, such that $rV(\mathbf{r}) \to 0$ as $r \to \infty$. Note that the bare Coulomb potential does not satisfy this condition. Thus, the scattering potential $V(\mathbf{r})$ is a small perturbation which distorts slightly the incident plane wave. The initial and final plane waves are taken to be normalised by a box volume L^3, so that from (2.49) and (2.56) we have

$$u_i(\mathbf{r}) = L^{-3/2} \exp(\mathrm{i}\mathbf{k}_i \cdot \mathbf{r}), \tag{2.78}$$

$$u_f(\mathbf{r}) = L^{-3/2} \exp(\mathrm{i}\mathbf{k}_f \cdot \mathbf{r}), \tag{2.79}$$

with, for elastic scattering, $|\mathbf{k}_i| = |\mathbf{k}_f| = k$, the momentum of the particle. Thus, the matrix element for the perturbation is, from (2.65), and defining $\mathbf{q} = \mathbf{k}_i - \mathbf{k}_f$,

$$\begin{aligned} H'_{fi} &= L^{-3} \int \exp(-\mathrm{i}\mathbf{k}_f \cdot \mathbf{r}) V(\mathbf{r}) \exp(\mathrm{i}\mathbf{k}_i \cdot \mathbf{r}) \mathrm{d}^3 x \\ &= L^{-3} \int V(\mathbf{r}) \exp(\mathrm{i}\mathbf{q} \cdot \mathbf{r}) \mathrm{d}^3 x. \end{aligned} \tag{2.80}$$

The density of states corresponding to the box normalisation must be calculated. Consideration of the one-dimensional free-particle wave function $u(x) = N\exp(ik_x x)$, and the requirement that both u and $\partial u/\partial x$ must be equal at the box boundaries ($x = \pm L/2$), gives the condition that $\exp(ik_x L) = 1$. Hence, the allowed momentum values are $k_x = 2\pi n_x/L$, etc., where n_x is the number of states. So the number of states in d^3k is

$$\mathrm{d}n = \left(\frac{L}{2\pi}\right)^3 \mathrm{d}k_x \mathrm{d}k_y \mathrm{d}k_z = \left(\frac{L}{2\pi}\right)^3 k^2 \mathrm{d}k \mathrm{d}\Omega, \tag{2.81}$$

where $\mathrm{d}\Omega$ is an element of solid angle. Now $\rho(E_f) = \mathrm{d}n/\mathrm{d}E_f$ and, since $E_f = k^2/2m$ where m is the particle mass, so $\mathrm{d}k = (m/k)\mathrm{d}E_f$. Thus,

$$\rho(E_f) = \left(\frac{L}{2\pi}\right)^3 mk\mathrm{d}\Omega. \tag{2.82}$$

The cross-section is, by definition, the transition rate per unit flux, i.e. $\sigma = \lambda/j$ where j is the particle flux or current. Here, the unit flux corresponds to one particle entering the box volume L^3 with velocity $v = k/m$, and so $j = v/L^3 = k/(mL^3)$. This expression for j can also be derived directly from (2.57). Hence the differential cross-section $\mathrm{d}\sigma$ is

$$\begin{aligned} \mathrm{d}\sigma &= \lambda/j = \lambda m L^3/k \\ &= \left(\frac{m}{2\pi}\right)^2 \left| \int V(\mathbf{r}) \exp(i\mathbf{q}\cdot\mathbf{r}) \mathrm{d}^3x \right|^2 \mathrm{d}\Omega, \end{aligned} \tag{2.83}$$

where use has been made of equations (2.71), (2.80) and (2.82). This is the Born approximation result. Note that the range of validity (see, e.g., Schiff, 1955) for a square well potential V_0 extending up to a radius a is, for high energy incident particles with $ka \gg 1$, such that $v \gg V_0 a$.

The asymptotic solution for the wave function can be written, for short range scattering processes, in the form

$$u(\mathbf{r}) \xrightarrow{r\to\infty} \exp(ikz) + f(\theta)\exp(ikr)/r. \tag{2.84}$$

In this formula $\exp(ikz)$ is the incident plane wave travelling along the z-axis, $\exp(ikr)/r$ represents an outgoing spherical wave, and $f(\theta)$ is called the *scattering amplitude*; θ is the scattering angle with respect to the z-axis. The incident flux from (2.57) is $j_{\text{in}} = k/m = v$. For the outgoing spherical wave $j_{\text{out}} = v|f(\theta)|^2/r^2$. The flux of particles scattered into an area $\mathrm{d}S = r^2\mathrm{d}\Omega$ is $j_{\text{out}}\mathrm{d}S$ and, dividing this by the incident flux, gives the differential cross-section

$$\mathrm{d}\sigma/\mathrm{d}\Omega = |f(\theta)|^2, \tag{2.85}$$

with

$$f(\theta) = -\frac{m}{2\pi}\int V(r)\exp(i\mathbf{q}\cdot\mathbf{r})\mathrm{d}^3x. \tag{2.86}$$

Thus, up to a numerical factor, f is just the Fourier transform of the potential evaluated at $\mathbf{q}$. The phase of f can be deduced in a more rigorous derivation. If V is a spherically symmetric potential, then an integration over the polar and azimuthal angles α and β can be performed, giving

$$f(\theta) = -\frac{m}{2\pi}\int_0^{2\pi}\mathrm{d}\beta\int_0^{\pi}\mathrm{d}\alpha\sin\alpha\int_0^{\infty}\mathrm{d}r r^2 V(r)\exp(iqr\cos\alpha), \tag{2.87}$$

where $\mathbf{q} = \mathbf{k}_i - \mathbf{k}_f$ is the polar axis. The magnitude of $\mathbf{q}$ is (Fig. 2.2)

$$q = |\mathbf{k}_i - \mathbf{k}_f| = 2k\sin\theta/2. \tag{2.88}$$

Integration over the polar angles gives

$$f(\theta) = -2m\int_0^{\infty}\mathrm{d}r r^2 V(r)\sin(qr)/qr. \tag{2.89}$$

Thus, the scattering amplitude $f(\theta)$, and hence the cross-section (from (2.85)), depends only on the magnitude of the momentum transfer q.

As an example, consider the scattering of an electron by a neutral atom with atomic number Z, for which the potential can be approximated by

$$V(\mathbf{r}) = -\frac{e^2}{4\pi}\left(\frac{Z}{r} - \int\frac{\rho(\mathbf{r}')}{|\mathbf{r}-\mathbf{r}'|}\mathrm{d}^3x'\right), \tag{2.90}$$

where $\mathbf{r}$ is the position vector of the incident electron. The first term arises from the field of the nucleus (assumed point-like), whereas the second is due to the atomic electrons, represented by an effective density ρ. The normalisation is such that $\int\rho(\mathbf{r})\mathrm{d}^3x = Z$. The potential satisfies Poisson's

Fig. 2.2 Kinematics for the scattering of a particle by a potential.

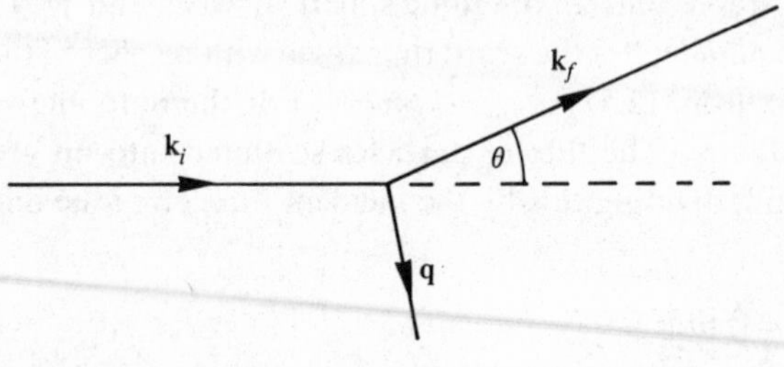

equation

$$\nabla^2 V(r) = e^2[Z\delta(\mathbf{r}) - \rho(\mathbf{r})]. \tag{2.91}$$

The next step is to evaluate the scattering amplitude, f, in (2.86). Consider the integral

$$\begin{aligned}\int V(r)\exp(\mathrm{i}\mathbf{q}\cdot\mathbf{r})\mathrm{d}^3x &= -\frac{1}{q^2}\int V(r)\nabla^2\exp(\mathrm{i}\mathbf{q}\cdot\mathbf{r})\mathrm{d}^3x \\ &= -\frac{1}{q^2}\int \exp(\mathrm{i}\mathbf{q}\cdot\mathbf{r})\nabla^2 V(r)\mathrm{d}^3x,\end{aligned} \tag{2.92}$$

where integration by parts, and the requirement that $V(r)$ is negligible for large distances, have been used. Inserting $\nabla^2 V(r)$ from (2.91) gives

$$\int V(r)\exp(\mathrm{i}\mathbf{q}\cdot\mathbf{r})\mathrm{d}^3x = -\frac{e^2}{q^2}[Z - F(\mathbf{q})], \tag{2.93}$$

where

$$F(\mathbf{q}) = \int \rho(\mathbf{r})\exp(\mathrm{i}\mathbf{q}\cdot\mathbf{r})\mathrm{d}^3x \tag{2.94}$$

is the Fourier transform of ρ and is called the *form factor* of the atom. Hence, using equations (2.85), (2.86) and (2.93) gives

$$\frac{\mathrm{d}\sigma}{\mathrm{d}\Omega} = \left(\frac{me^2}{2\pi q^2}\right)^2 [Z - F(\mathbf{q})]^2. \tag{2.95}$$

For large q, $F(\mathbf{q}) \to 0$ since ρ is effectively dilute, so

$$\frac{\mathrm{d}\sigma}{\mathrm{d}\Omega} = \left(\frac{me^2 Z}{2\pi q^2}\right)^2 = \left(\frac{me^2 Z}{8\pi k^2 \sin^2(\theta/2)}\right)^2 = \left(\frac{e^2 Z}{8\pi pv \sin^2(\theta/2)}\right)^2, \tag{2.96}$$

which is the *Rutherford scattering* formula, obtainable by classical means. Note that, in a more rigorous treatment, the validity of the Born approximation for Coulomb scattering must be considered. The latter form of (2.96), involving pv $(p = k)$, remains valid in the relativistic generalisation of (2.96).

The size of the scattering centre can be obtained from consideration of the small $\mathbf{q}$ behaviour. Expanding $Z - F(\mathbf{q})$ in powers of q gives

$$Z - F(\mathbf{q}) = \left[Z - \int \rho(\mathbf{r})\mathrm{d}^3x\right] + \frac{1}{2}\int \rho(\mathbf{r})(\mathbf{q}\cdot\mathbf{r})^2\mathrm{d}^3x + O(q^4), \tag{2.97}$$

where the coefficients in odd powers of q are zero provided $\rho(\mathbf{r}) = \rho(-\mathbf{r})$.

The zero order term is also zero from the normalisation condition. For spherically symmetric atoms, (2.97) becomes

$$Z - F(\mathbf{q}) = \tfrac{1}{6}q^2\langle r^2\rangle Z. \tag{2.98}$$

Substituting in equation (2.95) gives the result that $d\sigma/d\Omega$ is independent of the scattering angle θ for small θ, i.e. forward scattering. Note that there is no singularity at $\theta = 0$, as in the case of the Rutherford formula.

Thus, measurements of the form factor from low q^2 electron scattering can be used to measure the rms radius:

$$\left.\frac{\partial F}{\partial q^2}\right|_{q^2=0} = -\frac{1}{6}\langle r^2\rangle Z. \tag{2.99}$$

Scattering at large q^2 is sensitive to the internal structure of the target. For example, for electron scattering from hydrogen, for which

$$\rho(r) = \left(\frac{1}{\pi a_0^3}\right)\exp(-2r/a_0), \tag{2.100}$$

where $a_0 = 4\pi/(me^2)$ is the Bohr radius, the form factor is, using (2.94),

$$F(q) = \frac{1}{(1 + a_0^2 q^2/4)^2}. \tag{2.101}$$

An alternative, but roughly equivalent, starting point for calculating the scattering of an electron by a neutral atom is by using a screened Coulomb potential (rather than (2.90)):

$$V(r) = -\left(\frac{Ze^2}{4\pi r}\right)\exp(-r/a). \tag{2.102}$$

For small r this behaves as the nuclear Coulomb potential for atomic number Z, whereas for values of r large compared to a, the radius of the atomic electron cloud, screening is obtained. The Thomas–Fermi statistical model (see e.g. Schiff, 1955) gives $a \sim 4\pi/(m_e e^2 Z^{1/3})$, where m_e is the electron mass. Substituting $V(r)$ from (2.102) in (2.89) and integrating over r gives

$$f(\theta) = \frac{mZe^2}{2\pi(q^2 + 1/a^2)}. \tag{2.103}$$

Thus,

$$\frac{d\sigma}{d\Omega} = \frac{m^2 Z^2 e^4}{4\pi^2(q^2 + 1/a^2)^2}. \tag{2.104}$$

Now, from the uncertainty principle, the momentum corresponding to the atomic size is approximately $1/a$. Hence, for $q^2 \gg 1/a^2$, the scattering probes inside the atomic cloud, and in this limit (2.104) reduces to (2.96), the Rutherford scattering formula.

The Born approximation is applicable for a square well potential provided $v \gg V_0 a$; hence, for the case above, with $V \sim Ze^2/4\pi a$, this becomes $v \gg 2e^2/4\pi$. Thus, putting $e^2/4\pi \sim 1/137$, this means that the electron kinetic energy $T \gg T_{\min} \simeq 15Z^2$ eV. For hydrogen $T_{\min} \sim 15$ ev, whereas for a heavy atom such as xenon ($Z = 54$) this gives approximately 45 keV as the lower limit for validity. The upper limit is that the electrons are not relativistic, e.g. $v < c/2$, so that the kinetic energy is less than or equal to 80 keV.

The examples considered above correspond to the electromagnetic interaction. An example for the strong interaction can be evaluated with little extra work. The Yukawa potential has a form similar to that of (2.102), namely

$$V(r) = \left(\frac{g}{r}\right) \exp(-r/a), \tag{2.105}$$

where g represents the strength of the potential and a the range. If the interaction corresponds to the exchange of a particle of mass M, then from the uncertainty principle, $a \sim 1/M$. The differential cross-section is (see (2.104)), for an incident particle of mass m,

$$\frac{d\sigma}{d\Omega} = \frac{4m^2g^2}{(q^2 + 1/a^2)^2} = \frac{4m^2g^2}{(q^2 + M^2)^2}. \tag{2.106}$$

The short range nature of the strong interaction is equivalent to the exchange of a massive particle. The 'propagator' term $(q^2 + M^2)^{-2}$ dampens the range compared to the q^{-4} behaviour of the electromagnetic interaction. In the forward direction $q \to 0$, and the scattering is energy-independent. In fact this is true for any spherically symmetric potential, as can be seen by considering (2.89) in the limit $q \to 0$.

2.2.6 Phase shift analysis

For many scattering processes it is useful to consider the *partial waves* corresponding to states of the definite orbital angular momentum, l, for the initial particle with respect to the target. Again this is formulated in terms of the scattering of a particle by a potential. This is equivalent, however, to the scattering of two spinless particles when described in their cms. The incident plane wave $\exp(ikz)$, corresponding to a particle propagating along the z-axis and normalised such that the incident flux is the velocity v, can be expressed in terms of Legendre polynomials $P_l(\cos\theta)$, where θ is the angle made with the z-axis. It is assumed that the scattering process has an azimuthal symmetry around the z-axis. For the scattering of spinless particles, far away from the scattering centre, the asymptotic wave function can be written (e.g. Schiff, 1955)

$$\begin{aligned}\psi_i &= \exp(ikz) = \exp(ikr\cos\theta)\\ &\simeq \frac{1}{kr}\sum_{l=0}^{\infty}(2l+1)i^l P_l(\cos\theta)\sin(kr - l\pi/2)\\ &= \frac{i}{2kr}\sum_{l=0}^{\infty}(2l+1)i^l P_l(\cos\theta)\\ &\quad \{\exp[-i(kr - l\pi/2)] - \exp[i(kr - l\pi/2)]\}. \end{aligned} \tag{2.107}$$

That is, ψ_i is expressed as a coherent sum of ingoing spherical waves (first term in brackets) and outgoing waves (second term). The effect of the scattering centre is to modify the outgoing wave by a (complex) *scattering coefficient* η_l. Thus, $\psi = \psi_i + \psi_{sc}$ with

$$\psi = \frac{i}{2kr}\sum_{l=0}^{\infty}(2l+1)i^l P_l(\cos\theta)$$

$$\{\exp[-i(kr - l\pi/2)] - \eta_l \exp[i(kr - l\pi/2)]\} \tag{2.108}$$

$$= \psi_{in} + \psi_{out}. \tag{2.109}$$

Now the outgoing scattered wave can be written $\psi_{sc} = f(\theta)\exp(ikr)/r$; thus, using $\psi_{sc} = \psi - \psi_i$, (2.107) and (2.108), together with the identity $i^l = \exp(il\pi/2)$, gives

$$f(\theta) = \frac{i}{2k}\sum_{l=0}^{\infty}(2l+1)(1-\eta_l)P_l(\cos\theta). \tag{2.110}$$

Hence, the differential scattering cross-section

$$\frac{d\sigma}{d\Omega} = |f(\theta)|^2 = \frac{1}{4k^2}\left|\sum_{l=0}^{\infty}(2l+1)(1-\eta_l)P_l(\cos\theta)\right|^2. \tag{2.111}$$

Using the orthogonality relationship of the Legendre polynomials

$$\int_{-1}^{1} d\Omega P_l(\cos\theta)P_{l'}(\cos\theta) = \frac{4\pi}{2l+1}\delta_{ll'}, \tag{2.112}$$

gives the total cross-section

$$\sigma_{sc} = \sigma_{el} = \int d\Omega |f(\theta)|^2 = \frac{\pi}{k^2}\sum_{l=0}^{\infty}(2l+1)|1-\eta_l|^2. \tag{2.113}$$

This cross-section refers to particles scattered *elastically* in the interactions, i.e. with the same value of k before and after the collision. Particles can also be absorbed from the incident beam by *inelastic* scattering. Using (2.108) and (2.109), the inelastic cross-section is thus

$$\sigma_{\text{inel}} = \int \mathrm{d}\Omega\, r^2(|\psi_{\text{in}}|^2 - |\psi_{\text{out}}|^2) \tag{2.114}$$

$$= \frac{\pi}{k^2} \sum_{l=0}^{\infty} (2l+1)(1 - |\eta_l|^2). \tag{2.115}$$

In general, the scattering process can change both the amplitude and phase of the outgoing wave. Thus, one can write $\eta_l = \rho_l \exp(2\mathrm{i}\delta_l)$, where $\rho_l(k)$ and $\delta_l(k)$ are both real. The phase angle δ_l is called the *phase shift*. Thus, the inelastic cross-section can be written

$$\sigma_{\text{inel}} = \frac{\pi}{k^2} \sum_{l=0}^{\infty} (2l+1)(1 - \rho_l^2). \tag{2.116}$$

Thus, for $\rho_l = 1$, the inelastic term vanishes and the elastic cross-section becomes

$$\sigma_{\text{el}} = \frac{4\pi}{k^2} \sum_{l=0}^{\infty} (2l+1) \sin^2 \delta_l. \tag{2.117}$$

The inelastic cross-section is a maximum, for a given l, when $\eta_l = 0$. For this case, $\sigma_{\text{el}} = \sigma_{\text{inel}}$ and also the scattering amplitude $f(\theta)$ is purely imaginary. The total cross-section is $\sigma_{\text{tot}} = \sigma_{\text{el}} + \sigma_{\text{inel}}$. From (2.113) and (2.115) this can be written

$$\sigma_{\text{tot}} = \frac{2\pi}{k^2} \sum_{l=0}^{\infty} (2l+1)(1 - \operatorname{Re} \eta_l). \tag{2.118}$$

The total cross-section is related to the elastic scattering amplitude (2.110), evaluated at $\theta = 0$. Using the property $P_l(1) = 1$, (2.110) and (2.118) give

$$\sigma_{\text{tot}} = (4\pi/k) \operatorname{Im} f(0). \tag{2.119}$$

This relationship is called the *optical theorem* and, in fact, also holds for particles with non-zero spin and also for potentials which are not azimuthally symmetric.

Inelastic scattering cannot occur without the corresponding elastic shadow scattering. The total cross-section comes about by the removal of a certain amount of incident flux from the beam. This can only occur as a result of destructive interference between the incident plane wave and the coherent scattered wave in the forward direction. It must therefore be a linear function of the forward elastic scattering amplitude (Schiff, 1954). The radial current of particles, corresponding to the asymptotic form of the wave function ψ from (2.84) (with $z = r\cos\theta$), is (from (2.52))

$$j_r = \frac{i}{2m}\left(\psi\frac{\partial\psi^*}{\partial r} - \psi^*\frac{\partial\psi}{\partial r}\right)$$
$$= \frac{k}{m}\left(\cos\theta + \frac{|f(\theta)|^2}{r^2} + \frac{(1+\cos\theta)}{2r}\right.$$
$$\left.\times\{f(\theta)\exp[ikr(1-\cos\theta)] + \text{cc}\}\right), \tag{2.120}$$

where terms in $\exp(ikr)/r^2$ have been neglected as they average to zero, and cc means complex conjugate. The first and second terms arise from the component of the plane wave in the radial direction and from the elastically scattered flux respectively. The third term arises from interference between the incident and scattered amplitudes. The differential cross-section $d\sigma/d\Omega$ is obtained by dividing $j_r r^2$ by the incident flux k/m. Integrating over the solid angle $d\Omega$, the cross-section is,

$$\sigma = \int |f(\theta)|^2\, d\Omega + \frac{r}{2}$$
$$\times\left\{\int (1+\cos\theta)f(\theta)\exp[ikr(1-\cos\theta)]\, d\Omega + \text{cc}\right\}. \tag{2.121}$$

This outward flux, plus the inelastic cross-section, must give zero by flux conservation. Thus,

$$\sigma_{\text{tot}} = -\pi r\left\{\int (1+\cos\theta)f(\theta)\exp[ikr(1-\cos\theta)]\, d\cos\theta + \text{cc}\right\}. \tag{2.122}$$

Integration by parts yields the optical theorem

$$\sigma_{\text{tot}} = \frac{2\pi}{ik}[f(0) - f^*(0)] = \frac{4\pi}{k}\,\text{Im}\, f(0). \tag{2.123}$$

The main contribution to the integral comes from values of $\cos\theta$ such that $kr(1-\cos\theta) \lesssim 1$. For large r, this corresponds to a distribution peaked forward with an angular spread $\theta \lesssim (kr)^{-1/2}$. This is the region in which the destructive interference occurs.

The above formalism does not, of course, specify the actual value of the energy dependence of the phase shifts. The method is most useful if the cross-section is dominated by a few low values of l. From the properties of $P_l(\cos\theta)$ it follows that if values of l up to a maximum L contribute, then the angular distribution contains powers of $\cos\theta$ up to $\cos^{2L}\theta$. Thus s-wave ($L=0$) scattering gives an isotropic distribution. As an example, consider the scattering by a completely absorptive (black) sphere, radius R, such that $\eta_l=0$ for values of $l<kR$, but $\eta_l=1$ for larger l values. This gives, using (2.113), (2.115) and (2.118),

$$\sigma_{\rm el}=\sigma_{\rm inel}=\frac{\pi}{k^2}\sum_{l=0}^{L}(2l+1)=\frac{\pi}{k^2}(L+1)^2\simeq\pi R^2,$$

$$\sigma_{\rm tot}=\sigma_{\rm el}+\sigma_{\rm inel}\simeq 2\pi {\rm R}^2, \qquad (2.124)$$

since $L\simeq kR$. The inelastic cross-section is just the area presented by the black sphere. The elastic component ($=\sigma_{\rm inel}$) is the diffractive (shadow) scattering term. Furthermore from (2.110), it can be seen that $f(\theta)$ is purely imaginary. If a large number of values of l contribute then the summation can be replaced by an integral. For small θ, $P_l(\cos\theta)\simeq J_0(l\sin\theta)$, where J_0 is a Bessel function, and the elastic cross-section can be written (using (2.111))

$$\frac{d\sigma_{\rm el}}{{\rm d}\Omega}\simeq\frac{1}{k^2}\left[\int_0^{kR} lJ_0(l\sin\theta)\,{\rm dl}\right]^2=R^2\left[\frac{J_1(kR\sin\theta)}{\sin\theta}\right]^2, \qquad (2.125a)$$

with a minimum at $\sin\theta=3.8/kR$. The relation $\int_0^1 xJ_0(x)\,{\rm d}x=xJ_1(x)$ is used in the derivation. Alternatively, the cross-section can be expressed as a function of $q=2k\sin\theta/2$ (2.88) as follows

$$\frac{{\rm d}\sigma_{\rm el}}{{\rm d}q^2}\simeq\pi R^4\left[\frac{J_1(Rq)}{Rq}\right]^2\simeq\frac{\pi R^4}{4}\exp\left(-\frac{R^2q^2}{4}\right). \qquad (2.125b)$$

For a typical hadronic radius $R\sim 1$ fm, the latter expression is valid for $q^2\lesssim 0.2\ {\rm GeV}^2$. The slope of the exponential fall-off thus depends on R^2. For higher values of Rq the Bessel function exhibits maxima and minima, characteristic of a diffractive process. Equation (2.125) is the formula for diffractive black sphere scattering, and is appropriate for high energy scattering. More realistically, the absorption is not total, i.e. $\eta_l\neq 0$. Writing $\eta_l=1-\varepsilon$ with ε real and independent of l, the cross-sections become

$$\sigma_{\rm el}=\pi R^2\varepsilon^2, \qquad \sigma_{\rm tot}=2\pi R^2\varepsilon. \qquad (2.126)$$

For example, the elastic and total cross-sections for 10 GeV πp interactions are roughly 5 and 25 mb respectively. From (2.126), this gives $\varepsilon\simeq 0.4$ and $R\simeq 1$ fm. Note that a potentially large number of partial waves are involved since $L\simeq kR\simeq 50$, and that the wavelength corresponding to the incident particle $\lambda\!\!\!^{-}=1/k\ll R$.

For elastic scattering $\eta_l = \exp(2i\delta_l)$, so for the lth partial wave the scattering amplitude is (from (2.110))

$$f_l(\theta) = (2l+1)P_l(\cos\theta)\exp(i\delta_l)\sin\delta_l/k. \tag{2.127}$$

In the low energy limit $\lambda \gg R$, so that $kR \ll 1$, the scattering is dominated by $l = 0$ (s-wave). Hence

$$\frac{d\sigma_{el}}{d\Omega} = \left|\frac{\exp(i\delta_0)\sin\delta_0}{k}\right|^2 = \frac{\sin^2\delta_0}{k^2},$$

and

$$\sigma_{el} = (4\pi/k^2)\sin^2\delta_0. \tag{2.128}$$

Now, for scattering off an impenetrable sphere, ψ must vanish at $r = R$, so from (2.108) this condition yields

$$\eta_0 = \exp(2i\delta_0) = \exp(-2ikR). \tag{2.129}$$

Hence, $\delta_0 = -kR$, and therefore

$$\sigma_{el} \simeq 4\pi R^2. \tag{2.130}$$

The elastic scattering amplitude for the lth partial wave can be written

$$f_l = B_l\frac{\exp(i\delta_l)\sin\delta_l}{k} = B_l\frac{1}{k(\cot\delta_l - i)}. \tag{2.131}$$

The quantity

$$F_l = \frac{1}{\cot\delta_l - i} = \sin\delta_l\cos\delta_l + i\sin^2\delta_l, \tag{2.132}$$

when plotted in terms of the real and imaginary parts as a function of δ_l, gives a circle (the unitary circle), Fig. 2.3(a). This plot is called an *Argand diagram* and shows that $|F_l|$ has a maximum value at $\delta_l = \pi/2$, where F_l is purely imaginary. If the resonance is at $E = E_0$, i.e. $\cot\delta(E_0) = 0$, then for values of E near E_0 a Taylor series expansion gives

$$\begin{aligned}\cot\delta(E) &= (E - E_0)\left(\frac{d}{dE}\cot\delta(E)\right)_{E=E_0} + \ldots \\ &= -(E - E_0)2/\Gamma. \end{aligned} \tag{2.133}$$

Hence,

$$f_l(E) = -\frac{B_l}{k}\frac{\Gamma/2}{(E - E_0) + i\Gamma/2}. \tag{2.134}$$

The quantity Γ in (2.133) can thus be identified with the width of the resonance. The total elastic cross-section for $E \simeq E_0$ is, from (2.117) and (2.133),

$$\sigma_{el}(E) = \frac{4\pi(2l+1)}{k^2} \cdot \frac{\Gamma^2/4}{(E - E_0)^2 + \Gamma^2/4}. \tag{2.135}$$

This is the well-known *Breit–Wigner resonance* formula.

The above formalism has been developed to discuss scattering by a potential. Resonance phenomena are common in the energy dependence of the elastic cross-section of two particles, e.g. compound nuclear states, the Δ^{++} resonance in $\pi^+ p \to \pi^+ p$ scattering. Furthermore, resonance phenomena are also common in inelastic channels, e.g. $\pi N \to N^*(1675) \to N\pi\pi$ (N = nucleon), $e^+e^- \to (J/\psi) \to \mu^+\mu^-$. Inelastic channels correspond to $\rho_l < 1$. The scattering amplitude can, in general, be written

$$f_l(\theta) = (2l+1)\frac{P_l(\cos\theta)}{2k}[\rho_l \sin 2\delta_l + i(1 - \rho_l \cos 2\delta_l)]. \tag{2.136a}$$

A resonance in the elastic channel describes a curve which, for $\rho_l < 1$, is inside the unitarity circle (Fig. 2.3(*b*)). At the resonance, $\delta_l = \pi/2$, the amplitude is purely imaginary

$$f_l(\theta) = \frac{i}{2k}(2l+1)P_l(\cos\theta)(1 + \rho_l). \tag{2.136b}$$

The inelastic cross-section (2.115) is independent of δ_l,

$$\sigma^l_{inel} = \frac{\pi}{k^2}(2l+1)(1 - \rho_l^2) \tag{2.137}$$

Fig. 2.3 Argand diagram for (*a*) elastic and (*b*) inelastic resonance.

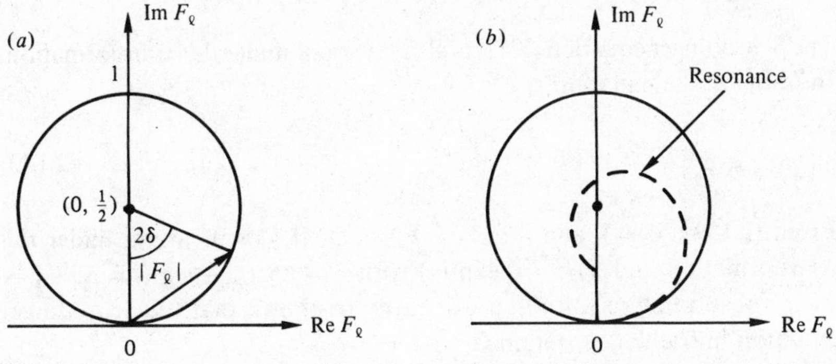

In this case it is ρ_l which exhibits a strong variation around the resonance energy. It is important to note that this partial wave analysis does not predict the existence of resonances. It does provide, however, a convenient formalism in which to discuss resonance phenomena.

Suppose that a resonance is formed at a particular energy in the collision of two particles, a and b. This resonance can, in general, decay either to the initial particles a and b (i.e. elastic scattering) or to a variety of other (inelastic) channels. If the decay is independent of the formation process, then each possible final state f has a probability proportional to its partial width Γ_f (see (2.77)), such that $\Gamma = \sum_f \Gamma_f$. Thus, if Γ_{el} is the width for the elastic channel, then the cross-section to produce a final state f is

$$\sigma(E) = \frac{\pi}{k^2} g \frac{\Gamma_{el}\Gamma_f}{(E - E_0)^2 + \Gamma^2/4}, \tag{2.138}$$

where $g = (2l + 1)$ in the case of the scattering of spinless particles. If the incident particles a and b have spins s_a and s_b and if the resonance has spin J then, for unpolarised particles, $g = (2J + 1)/[(2s_a + 1)(2s_b + 1)]$.

2.3 SPECIAL RELATIVITY

2.3.1 Galilean and special relativity

The special theory of relativity, which was developed by Einstein around 1905, is one of the cornerstones of modern physics. The ideas owed much to the earlier work of Maxwell in unifying the description of electricity and magnetism.

The laws of Newtonian or classical mechanics are invariant under Galilean relativity, and also under spatial transformations of the coordinate system. The Galilean transformation relates the coordinates in some reference frame $S(t, x, y, z)$ to those in a frame $S'(t', x', y', z')$, having a velocity $\boldsymbol{\beta}$ relative to S, by

$$\mathbf{x}' = \mathbf{x} - \boldsymbol{\beta} t, \qquad t' = t. \tag{2.139}$$

The Schrödinger equation (2.47) is also invariant under this transformation. In frame S' the equation is

$$-\frac{1}{2m}\nabla'^2\psi' + V\psi' = \mathrm{i}\frac{\partial\psi'}{\partial t'}. \tag{2.140}$$

From (2.139), $\boldsymbol{\nabla}' = \boldsymbol{\nabla}$ and $\partial/\partial t' = \boldsymbol{\beta}\cdot\boldsymbol{\nabla} + \partial/\partial t$. If V is invariant under the transformation and if $\psi' = \psi\exp(\mathrm{i}\alpha)$ with $\alpha = m\beta^2 t/2 - m\boldsymbol{\beta}\cdot\mathbf{x} = Et - \mathbf{p}\cdot\mathbf{x}$ then, substituting in (2.140), it can easily be shown that the Schrödinger equation in frame S is obtained.

In special relativity space and time are treated on an equal footing. Special relativity is based on two axioms. These deal with the effects on the laws of physics of transformations between inertial frames of reference. An inertial frame is one in which a freely moving particle (i.e. no external forces) has constant velocity. The axioms, which are well tested experimentally, are

(i) the laws of nature are the same in all inertial frames (covariance);
(ii) the velocity of light *in vacuo* is the same in all inertial frames.

Neither of these axioms is true for the Galilean transformation. For example, Maxwell's equations are not invariant under the Galilean transformation. The phase term for the solution of the free photon wave equation (2.21), propagating along the x-axis, can be written $kx - \omega t = k(x - \beta_c t)$, where the phase velocity $\beta_c = \omega/k$. Substituting this expression for $\mathbf{A}$ in (2.20), with the free particle condition that $\mathbf{j} = 0$, gives $\beta_c = 1$, i.e. the propagation travels with the velocity of light. Under a Galilean transformation $k(x - \beta_c t)$ becomes $k'[x' - (\beta_c - \beta)t']$, so that the transformed velocity $\beta_c' = \beta_c - \beta$. This would imply that the velocity of light in S' is different to that in S; contrary to experiment. In relativity, there is no meaningful way to define absolute velocity.

The space-time coordinates $x = (t, x, y, z)$ in some inertial frame S are related to those in S', namely $x' = (t', x', y', z')$, moving with velocity $\boldsymbol{\beta}$ with respect to S, by the *Lorentz transformation*.

$$\begin{aligned} x'_{\parallel} &= \gamma(x_{\parallel} - \beta t), & x_{\parallel} &= \gamma(x'_{\parallel} + \beta t'), \\ \mathbf{x}'_{\perp} &= \mathbf{x}_{\perp}, & \mathbf{x}_{\perp} &= \mathbf{x}'_{\perp}, \\ t' &= \gamma(t - \beta x_{\parallel}), & t &= \gamma(t' + \beta x'_{\parallel}), \end{aligned} \tag{2.141}$$

where $x_{\parallel}$ and $\mathbf{x}_{\perp}$ are the components of the space coordinate along and perpendicular to $\boldsymbol{\beta}$, and $\gamma = (1 - \beta^2)^{-1/2}$, with $\beta = |\boldsymbol{\beta}|$.

An alternative form for the transformation equations (2.141) can be obtained by making the substitution $\beta = \tanh y$, and hence $\gamma = \cosh y$, giving

$$\begin{aligned} t' &= t \cosh y - x_{\parallel} \sinh y, \\ x'_{\parallel} &= -t \sinh y + x_{\parallel} \cosh y. \end{aligned} \tag{2.142}$$

That is, the Lorentz transformation can be regarded as a rotation through an angle iy (since $\cosh y = \cos(\mathrm{i}y)$ and $\sinh y = -\mathrm{i} \sin(\mathrm{i}y)$); y is called the *boost parameter* or *rapidity*. For transformations in the same direction rapidities are additive.

Note that the quantity $x^2 = t^2 - |\mathbf{x}|^2$ is the same in both frames, that is

$$(x')^2 = t'^2 - |\mathbf{x}'|^2 = t^2 - |\mathbf{x}|^2 = x^2. \tag{2.143}$$

The transformation law (2.141) also applies for any four-vector $A=(A^0, A^1, A^2, A^3)$, again with components of the three-vector part $\mathbf{A}$ resolved along and perpendicular to $\boldsymbol{\beta}$. A four-vector may be regarded as a vector whose magnitude is invariant under 'rotations' in space-time, in the same way as the magnitude of a three-vector is invariant under spatial rotations. The scalar product of two four-vectors $A=(A^0, \mathbf{A})$ and $B=(B^0, \mathbf{B})$ is defined as

$$A \cdot B = A^0 B^0 - \mathbf{A} \cdot \mathbf{B} = A^0 B^0 - \sum_{i=1}^{3} A^i B^i. \tag{2.144}$$

Use of (2.141) shows that the scalar product is Lorentz invariant.

The Lorentz invariant interval between two space-time coordinates a and b in S is defined as

$$S_{ab}^2 = (t_a - t_b)^2 - (x_a - x_b)^2 - (y_a - y_b)^2 - (z_a - z_b)^2. \tag{2.145}$$

If $S_{ab}^2 > 0$ the separation is *time-like* because it is always possible to find a frame S' in which $\mathbf{x}'_a = \mathbf{x}'_b$. If $S_{ab}^2 < 0$ the separation is *space-like*, because it is always possible to find a frame S' in which $t'_a = t'_b$. Two events can only be related causally if S_{ab}^2 is time-like. Any event related causally to that at b (taken to be the origin O) must be inside the light cone defined by $t^2 = x^2$ (Fig. 2.4).

Fig. 2.4 Space–time diagram showing the light cones $t^2 = x^2$.

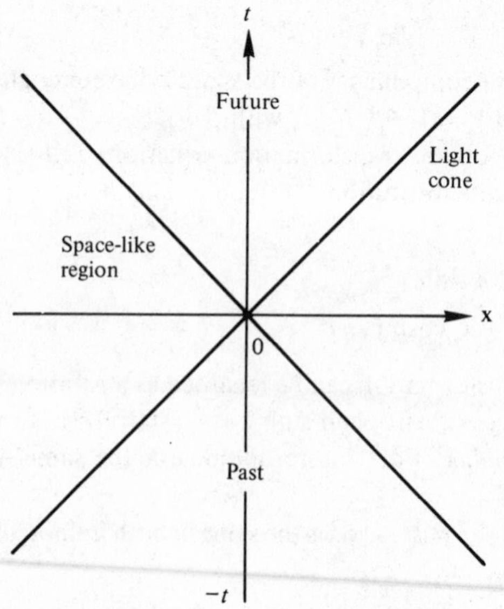

Consider the creation and decay of an unstable particle. The time measured in the rest frame of the particle is called the *proper time*. If τ_0 is the (proper) mean lifetime of the particle, then the value measured in a system in which the particle has velocity β is $\gamma\tau_0$, i.e. it has a *time dilation*. This follows from (2.141) because an observer in the rest (primed) frame has $x'_1 = x'_2$ and $t'_2 - t'_1 = \tau_0$, where the indices 1 and 2 refer to the creation and decays. Thus in S, $t_2 - t_1 = \gamma\tau_0$. The mean distance traversed by the particle in S is $\gamma\beta\tau_0$. For example, a charged pion travels, on average, a distance of 55.88 m $\mathrm{GeV}^{-1}\,\mathrm{c}^{-1}$ whereas a charged kaon travels 7.5 m $\mathrm{GeV}^{-1}\,\mathrm{c}^{-1}$.

The formula for time dilation also holds if the particle undergoes acceleration. In the $g-2$ experiment (Combley and Picasso, 1974), which was performed in CERN, muons were kept in a circular orbit by magnetic fields and had an acceleration of about 4×10^{18} cm s^{-2}. The time-dilated lifetime measured for the muon agrees with that calculated using the rest lifetime to better than 1%. Acceleration does not affect the validity of special relativity. Space-time intervals in different inertial frames depend only on their relative velocity and not on acceleration. Note, however, that during acceleration the rest system of the muon is not an inertial system. Use must be made of instantaneous inertial frames to account correctly for the effects of accelerations. The important property of the Lorentz covariance of physical laws applies even if, for example, a collision process involves large accelerations.

2.3.2 Classification of Lorentz transformations

The scalar product of two four-vectors A and B (2.144) is Lorentz invariant, and it is useful to write this in the form

$$A \cdot B = A^\mu B_\mu = (A^0 B_0 - \mathbf{A} \cdot \mathbf{B}), \tag{2.146}$$

where summation over μ is implied and

$$A^\mu = (A^0, A^1, A^2, A^3) = (A^0, \mathbf{A}), \tag{2.147}$$

$$B_\mu = (B_0, B_1, B_2, B_3) = (B^0, -B^1, -B^2, -B^3) = (B^0, -\mathbf{B}). \tag{2.148}$$

The four-vectors with an upper index are called *contravariant* vectors and transform like x^μ, whereas those with a lower index are called *covariant* vectors.

The invariant space-time interval in differential form is

$$(\mathrm{d}x)^2 = (\mathrm{d}x^0)^2 - (\mathrm{d}x^1)^2 - (\mathrm{d}x^2)^2 - (\mathrm{d}x^3)^2. \tag{2.149}$$

Expressing this as a product of covariant and contravariant elements gives

$$(\mathrm{d}x)^2 = \mathrm{d}x_\alpha\,\mathrm{d}x^\alpha = g_{\alpha\beta}\,\mathrm{d}x^\beta\,\mathrm{d}x^\alpha = g^{\alpha\beta}\,\mathrm{d}x_\alpha\,\mathrm{d}x_\beta, \tag{2.150}$$

where $g_{\alpha\beta}$ is called the *metric tensor* and $g_{\beta\alpha} = g_{\alpha\beta}$. A summation over repeated upper and lower indices is always implied. Comparing (2.149) and (2150), the following properties can be deduced

$$g_{00} = 1, \qquad g_{11} = g_{22} = g_{33} = -1, \tag{2.151}$$

$$g_{\alpha\gamma} g^{\gamma\beta} = \delta^{\beta}_{\alpha}, \qquad g_{\alpha\beta} = g^{\alpha\beta}, \tag{2.152}$$

$$x_{\alpha} = g_{\alpha\beta} x^{\beta}, \qquad x^{\alpha} = g^{\alpha\beta} x_{\beta}. \tag{2.153}$$

Thus, if it is required to change a contravariant index to a covariant one or *vice versa*, it is merely necessary to contract the expression with $g_{\alpha\beta}$ or $g^{\alpha\beta}$ as appropriate. The scalar product is written in terms of the metric tensor as follows

$$A \cdot B = A_{\alpha} B^{\alpha} = A^{\alpha} B_{\alpha} = g_{\alpha\beta} A^{\alpha} B^{\beta} = g^{\alpha\beta} A_{\beta} B_{\alpha}. \tag{2.154}$$

In general, the Lorentz transformation between a space-time four-vector $x = (x^0, x^1, x^2, x^3) = (t, x, y, z)$ in S and x' in S' can be written

$$(x')^{\alpha} = a^{\alpha}_{\beta} x^{\beta} = a^{\alpha\beta} x_{\beta}, \tag{2.155}$$

where a is a 4×4 (real) matrix and summation over index β is implied. The covariant four-vector x_{μ} transforms as

$$(x')_{\alpha} = a^{\beta}_{\alpha} x_{\beta}. \tag{2.156}$$

Now Lorentz invariance gives $(x')^2 = x^2$. Thus,

$$(x')^2 = (x')_{\alpha} (x')^{\alpha} = a^{\beta}_{\alpha} a^{\alpha}_{\gamma} x_{\beta} x^{\gamma} = x_{\gamma} x^{\gamma}. \tag{2.157}$$

Thus,

$$a^{\beta}_{\alpha} a^{\alpha}_{\gamma} = \delta^{\beta}_{\gamma}. \tag{2.158}$$

With the help of (2.155), one can write

$$a^{\beta}_{\alpha} (x')^{\alpha} = a^{\beta}_{\alpha} a^{\alpha}_{\gamma} x^{\gamma}. \tag{2.159}$$

Hence using (2.158),

$$x^{\beta} = a^{\beta}_{\alpha} (x')^{\alpha} = a^{-1} x'. \tag{2.160}$$

Using (2.158) and the well-known matrix properties gives

$$(\det a)^2 = 1. \tag{2.161}$$

Thus, $\det a = \pm 1$, and this property is used in the classification of Lorentz transformations as follows:

(i) *proper Lorentz transformations* which can be achieved by a series of infinitesimal transformations and hence have $\det a = 1$;

(ii) *improper Lorentz transformations*: space inversion ($\det a = -1$), time inversion ($\det a = -1$), space-time inversion ($\det a = 1$).

Observable quantities can be conveniently classifed by their properties under Lorentz transformations. This is discussed further in Section 3.2.5.

Relativistic wave equations contain derivatives with respect to space-time. The transformation of the derivative is as follows:

$$\frac{\partial}{\partial(x')^\alpha} = \frac{\partial x^\beta}{\partial(x')^\alpha}\frac{\partial}{\partial x^\beta}. \tag{2.162}$$

Thus, the components of the derivative are those of a covariant vector operator. It is usual to write

$$\partial^\alpha \equiv \frac{\partial}{\partial x_\alpha} = \left(\frac{\partial}{\partial x^0}, -\frac{\partial}{\partial x^1}, -\frac{\partial}{\partial x^2}, -\frac{\partial}{\partial x^3}\right) = \left(\frac{\partial}{\partial x^0}, -\boldsymbol{\nabla}\right),$$

$$\partial_\alpha \equiv \frac{\partial}{\partial x^\alpha} = \left(\frac{\partial}{\partial x^0}, \boldsymbol{\nabla}\right). \tag{2.163}$$

The divergence of a four-vector A is thus

$$\partial^\alpha A_\alpha = \partial_\alpha A^\alpha = \frac{\partial A^0}{\partial t} + \boldsymbol{\nabla}\cdot\mathbf{A}. \tag{2.164}$$

The second derivative with respect to x^α can be written

$$\Box^2 \equiv \partial_\alpha \partial^\alpha = \frac{\partial^2}{\partial t^2} - \boldsymbol{\nabla}^2. \tag{2.165}$$

$\Box^2$, the D'Alembertian operator, appears in relativistic wave equations. The four-vector operator representing the energy-momentum operator is $p^\alpha = \mathrm{i}\partial^\alpha$.

2.3.3 Relativistic kinematics

If m is the rest mass of a particle, then in a frame in which the particle has velocity $\boldsymbol{\beta}$, the energy and momentum are

$$E = \gamma m, \qquad \mathbf{p} = \gamma m \boldsymbol{\beta}. \tag{2.166}$$

Alternatively, given E and $p = |\mathbf{p}|$, the parameters γ and β are

$$\gamma = E/m, \qquad \beta = |\boldsymbol{\beta}| = p/E. \tag{2.167}$$

The energy-momentum four-vector is written $p^\alpha = (p^0, p^1, p^2, p^3) = (E, \mathbf{p})$ with

$$p^2 = p_\alpha p^\alpha = E^2 - |\mathbf{p}|^2 = m^2. \tag{2.168}$$

The components of p^α transform between different inertial frames with the same equations (2.141) as for space-time with $p^0 = E \equiv t$ and $\mathbf{p} \equiv \mathbf{x}$.

If p_a and p_b are the four-momenta of two particles a and b, then the quantity

$$M_{ab}^2 = (p_a + p_b)^2 = (E_a + E_b)^2 - (\mathbf{p}_a + \mathbf{p}_b)^2, \tag{2.169}$$

is called the effective mass squared of a and b. If a particle R decays to a and b, then M_{ab} is the mass of the parent particle R.

Consider the collision $a + b \to c + d$; energy-momentum conservation gives

$$p_a + p_b = p_c + p_d, \tag{2.170}$$

with $p_a^2 = m_a^2$, etc., and c and d represent either single particles or groups of particles. The *Mandelstam variables* s, t, u are defined as follows:

$$\begin{aligned} s &= (p_a + p_b)^2, \\ t &= (p_a - p_c)^2 = (p_b - p_d)^2, \\ u &= (p_a - p_d)^2 = (p_b - p_c)^2. \end{aligned} \tag{2.171}$$

The sum of s, t and u is a constant

$$\begin{aligned} s + t + u &= 3p_a^2 + p_b^2 + p_c^2 + p_d^2 + 2p_a \cdot (p_b - p_c - p_d) \\ &= m_a^2 + m_b^2 + m_c^2 + m_d^2. \end{aligned} \tag{2.172}$$

As an example consider the process $\pi p \to X$, where X is a system of hadrons (i.e. a and b represent π and p respectively). The physics of the process is probably best described in the *centre-of-mass* or *centre-of-momentum* system (cms), which is defined such that the net sum of the three-momenta is zero. If p_a and p_b are given in the lab system, which in the case of *fixed target* experiments is the rest system of particle b (Fig. 2.5(*a*)), then the velocity $\boldsymbol{\beta}$ of the cms in this system is (using (2.167))

$$\boldsymbol{\beta} = \frac{\mathbf{p}_a + \mathbf{p}_b}{E_a + E_b} = \frac{\mathbf{p}_a}{E_a + m_b}. \tag{2.173}$$

The total energy in the cms system is $s^{1/2} = W$. If $p_a^* = p_b^*$ is the magnitude of the momentum of a or b in the cms, then

$$W = E_a^* + E_b^*, \qquad \text{with } E_b^* = (p_a^{*2} + m_b^2)^{1/2}.$$

Thus,

$$E_a^* = \frac{W^2 + m_a^2 - m_b^2}{2W} \simeq \frac{W}{2}, \tag{2.174}$$

provided W is large compared to m_a and m_b.

If c is one of the final state particles having four-momentum $p_c^* = (E_c^*, \mathbf{p}_c^*)$, and if θ_c^* is the angle between $\mathbf{p}_c^*$ and the direction of the initial pion (Fig. 2.5(*b*)), then the angle of c in the lab system is given by (using (2.141))

$$\tan\theta_c = \frac{(p_T)_c}{(p_L)_c} = \frac{p_c^* \sin\theta_c^*}{\gamma(p_c^* \cos\theta_c^* + \beta E_c^*)}$$
$$= \frac{\sin\theta_c^*}{\gamma(\cos\theta_c^* + \beta/\beta_c^*)}, \qquad (2.175)$$

where $\beta_c^* = p_c^*/E_c^*$. Note that p_T, the momentum transverse to the direction of the incident beam particle, has the same value in all frames, whereas the longitudinal momentum p_L depends on the choice of frame. Particle c can go backwards in the lab system provided that it goes backwards in the cms and also has β_c^* satisfying $(-\beta_c^* \cos\theta_c^*) < \beta$. A proton at rest in the lab system has $p_b^* = -\gamma\beta m_b$ and $E_b^* = \gamma m_b$, and hence $\beta_b^* = \beta$, i.e. goes backwards in the cms with the maximum possible momentum. Thus, no proton can be emitted backwards in the lab system. However, a lighter particle, e.g. a pion, can have $\beta_c^* > \beta$ and hence can be emitted backwards in the lab.

In the frame in which $p'_L = 0$, and hence $E' = (m^2 + p_T^2)^{1/2} = m_T$ (the transverse mass), we have, using (2.142)

$$E = \gamma m_T = m_T \cosh y = m_T[\exp(y) + \exp(-y)]/2,$$
$$p_L = \gamma\beta m_T = m_T \sinh y = m_T[\exp(y) - \exp(-y)]/2. \qquad (2.176)$$

Hence, the rapidity y can be expressed as

$$\exp(y) = (E + p_L)/m_T \qquad \text{or} \qquad y = \ln\left(\frac{E + p_L}{m_T}\right). \qquad (2.177)$$

Some further manipulation gives

$$y = \tfrac{1}{2}\ln\left(\frac{E + p_L}{E - p_L}\right). \qquad (2.178)$$

Fig. 2.5 Kinematics of πp scattering in (*a*) lab and (*b*) cms systems.

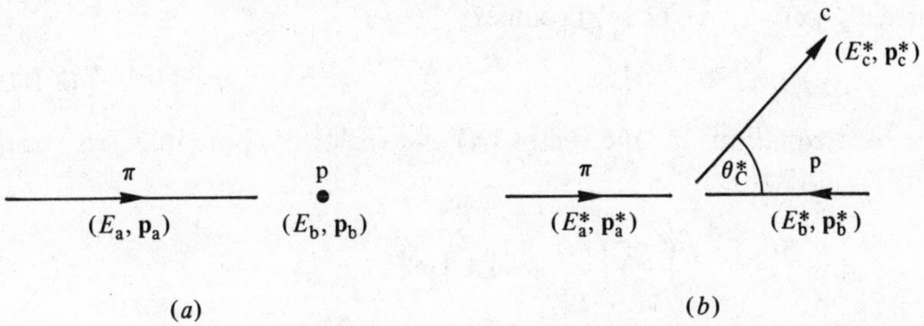

An important property of the rapidity is that a boost along the $\boldsymbol{\beta}$-axis (i.e. affecting the parallel components) changes all rapidities by a constant amount. Rapidity is a useful variable in the description of the hadrons produced in high energy collisions, because it is found that the rapidity density (dN/dy) is roughly constant throughout the accessible rapidity range, and is only a slowly varying function of the cms energy.

Collision processes involve the exchange of particles, and the exchanged particle (with four-momentum p_l) can be either time-like ($p_l^2 > 0$) or space-like ($p_l^2 < 0$). For example, the virtual photon (γ^*) in the process $e^+e^+ \to \gamma^* \to$ hadrons is time-like with $(p_\gamma^*)^2 = (p_a + p_b)^2 = s = W^2$, whereas for $e^-p \to e^- +$ hadrons, the virtual photon exchanged is space-like. In the latter case

$$\begin{aligned} t = q^2 = (p_i - p_f)^2 &= -Q^2 = 2m_e^2 - 2(E_i E_f - \mathbf{p}_i \cdot \mathbf{p}_f) \\ &\simeq -4E_i E_f \sin^2 \theta/2, \end{aligned} \tag{2.179}$$

where $p_i = (E_i, \mathbf{p}_i)$ and $p_f = (E_f, \mathbf{p}_f)$ are the four-momenta of the incident and scattered electrons, and θ is the scattering angle. The approximation holds for the case where m_e is negligible. Note that q^2 is negative, so that Q^2 is positive. The outgoing hadronic system d has an effective mass W given by $W^2 = p_d^2$; that is, W is the total energy in the cms of the outgoing hadrons. For the discussion of the physics of t-channel processes a useful frame, called the *Breit frame*, is that in which the energy in the t-channel is zero. For example, in ep scattering, if the virtual photon has four-momentum $q = (\nu, \mathbf{q})$ in the lab system, then in the Breit system $q = (0, -(Q^2)^{1/2})$. The velocity of the Breit system in the lab is $\beta_B = \nu/|\mathbf{q}| = \nu/(\nu^2 + Q^2)^{1/2}$.

2.3.4 Maxwell's equations in covariant form

Maxwell's equations (2.13) are covariant, i.e. they are the same in all inertial frames. Indeed, the special theory of relativity was developed, to a large extent, to obtain this property as required by axiom 1. The form given in (2.13) is in terms of three vectors, and their derivatives with respect to space and time. Introducing a four-vector for the electromagnetic current $j^\mu(x) = (\rho, \mathbf{j})$, (2.14) becomes

$$\partial_\mu j^\mu(x) = 0. \tag{2.180}$$

The field equations for the vector ($\mathbf{A}$) and scalar (ϕ) potentials are, from (2.15) and (2.16),

$$\left(\frac{\partial^2 \mathbf{A}}{\partial t^2} - \nabla^2 \mathbf{A}\right) + \nabla\left(\frac{\partial \phi}{\partial t} + \nabla \cdot \mathbf{A}\right) = \mathbf{j},$$

$$\left(\frac{\partial^2\phi}{\partial t^2} - \nabla^2\phi\right) - \frac{\partial}{\partial t}\left(\frac{\partial\phi}{\partial t} + \nabla\cdot\mathbf{A}\right) = \rho,$$

which can be written in terms of the four-vector potential $A^\mu = (\phi, \mathbf{A})$

$$\partial_\mu\partial^\mu A^\nu - \partial^\nu(\partial_\mu A^\mu) = j^\nu, \tag{2.181}$$

or

$$\partial_\mu F^{\mu\nu} = j^\nu, \tag{2.182}$$

where $F^{\mu\nu}$ is the *electromagnetic field tensor* defined as

$$F^{\mu\nu} = \partial^\mu A^\nu - \partial^\nu A^\mu. \tag{2.183}$$

The gauge transformation (2.17) can be written

$$(A')^\mu = A^\mu + \partial^\mu\chi, \tag{2.184}$$

and hence χ can be chosen ($\Box^2\chi = 0$) to give the *Lorentz condition* or *Lorentz gauge*#

$$\partial_\mu A^\mu = 0, \tag{2.185}$$

with the field equation (2.181) taking the simple form

$$\partial_\mu\partial^\mu A^\nu = \Box^2 A^\nu = j^\nu. \tag{2.186}$$

A solution to this equation for a free photon ($j^\nu = 0$) is

$$A^\mu = \varepsilon^\mu \exp(-iq\cdot x), \tag{2.187}$$

where q is the four-momentum of the photon and ε^μ the four-vector polarisation. Substitution of (2.187) in $\Box^2 A^\mu = 0$ gives $q^2 = 0$, i.e. the photon is massless. The argument of the exponential (i.e. $-iq\cdot x$) is a Lorentz scalar and hence is the same in all inertial frames.

The physically observable fields in Maxwell's equations are the electric and magnetic fields. The relationships between **E** and **B** and the vector and scalar potentials are given by (2.13). Thus the electromagnetic field tensor $F^{\mu\nu}$ can be written in terms of **E** and **B** as follows

$$F^{\mu\nu} = \begin{pmatrix} 0 & -E_x & -E_y & -E_z \\ E_x & 0 & -B_z & B_y \\ E_y & B_z & 0 & -B_x \\ E_z & -B_y & B_x & 0 \end{pmatrix}. \tag{2.188}$$

The Lorentz condition is covariant. Other gauge fixing conditions, such as the Coulomb ($\nabla\cdot A = 0$), axial ($A^3 = 0$) and temporal ($A^0 = 0$) gauges, are non-covariant.

Consider, for simplicity, a Lorentz transformation along the x-axis. This corresponds to a matrix a^i_j as follows (2.141)

$$a = a^i_j = \begin{pmatrix} \gamma & -\gamma\beta & 0 & 0 \\ -\gamma\beta & \gamma & 0 & 0 \\ 0 & 0 & 1 & 0 \\ 0 & 0 & 0 & 1 \end{pmatrix}, \tag{2.189}$$

such that a four-vector V^j is transformed to $V'^i = a^i_j V^j$. A tensor such as F^{kl} is transformed to $(F')^{ij} = a^i_k a^j_l F^{kl}$. Using this transformation on (2.188) gives the transformed electric and magnetic fields,

$$E'_{\parallel} = E_{\parallel}, \qquad B'_{\parallel} = B_{\parallel},$$

$$E'_{\perp} = \gamma[E_{\perp} + (\mathbf{v} \times \mathbf{B})_{\perp}], \qquad B'_{\perp} = \gamma[B_{\perp} - (\mathbf{v} \times \mathbf{E})_{\perp}], \tag{2.190}$$

where $\parallel$ and $\perp$ correspond to the components along and perpendicular to the direction of motion (e.g. x-axis). Thus, if there is an electric but no magnetic field in some frame, then in the transformed frame there is a magnetic field in the transverse direction.

2.4 ANGULAR MOMENTUM

Classically, angular momentum is defined as the cross product of a position vector $\mathbf{r}$ and a momentum vector $\mathbf{p}$. Putting $\mathbf{p} = -\mathrm{i}\nabla$ (equation (2.30)) we obtain the quantum mechanical angular momentum operator

$$\mathbf{j} = \mathbf{r} \times \mathbf{p} = -\mathrm{i}\mathbf{r} \times \nabla,$$

with components

$$j_x = yp_z - zp_y = -\mathrm{i}\left(y\frac{\partial}{\partial z} - z\frac{\partial}{\partial y}\right),$$

$$j_y = zp_x - xp_z = -\mathrm{i}\left(z\frac{\partial}{\partial x} - x\frac{\partial}{\partial z}\right),$$

$$j_z = xp_y - yp_x = -\mathrm{i}\left(x\frac{\partial}{\partial y} - y\frac{\partial}{\partial x}\right). \tag{2.191}$$

The commutation relations can be evaluated from (2.191). Using the relationships $[x, p_x] = \mathrm{i}$, etc, gives

$$[j_x, j_y] = \mathrm{i}j_z, \qquad [j_y, j_z] = \mathrm{i}j_x, \qquad [j_z, j_x] = \mathrm{i}j_y.$$

In general, writing the indices x, y, z as 1, 2, 3, we have

$$[j_i, j_j] = \mathrm{i}\varepsilon_{ijk} j_k, \tag{2.192}$$

where ε_{ijk} is the totally antisymmetric tensor, which has the value $+1(-1)$ if it is obtained from 1, 2, 3 by an even (odd) number of permutations of the indices, and is equal to zero if any two indices are equal. Hence, $\varepsilon_{123}=\varepsilon_{312}=\varepsilon_{231}=1$ (cyclic) and $\varepsilon_{321}=\varepsilon_{132}=\varepsilon_{213}=-1$ (anticyclic).

The total angular momentum operator $\mathbf{j}^2$ is defined as

$$\mathbf{j}^2=j_x^2+j_y^2+j_z^2. \tag{2.193}$$

Now $\mathbf{j}^2$ commutes with each of j_x, j_y and j_z; hence, $\mathbf{j}^2$ and one component of $\mathbf{j}$ (e.g. j_z) can be simultaneously diagonalised with eigenfunctions Y_{jm} (spherical harmonics) and eigenvalues which can be written

$$\mathbf{j}^2 Y_{jm}=j(j+1)Y_{jm},$$

$$j_z Y_{jm}=mY_{jm}. \tag{2.194}$$

The angular momentum is related to the rotational properties of a system. Consider the rotation of an angle α about the z-axis in a clockwise direction; this can be represented by a rotation matrix R_α such that $\mathbf{r}'=R_\alpha\mathbf{r}$, defined as follows:

$$\begin{pmatrix} x' \\ y' \\ z' \end{pmatrix}=\begin{pmatrix} \cos\alpha & -\sin\alpha & 0 \\ \sin\alpha & \cos\alpha & 0 \\ 0 & 0 & 1 \end{pmatrix}\begin{pmatrix} x \\ y \\ z \end{pmatrix}. \tag{2.195}$$

If the probability amplitude for finding a particle at some point in space is independent of the coordinate system, then

$$\psi'(x',y',z')=\psi(x,y,z).$$

So, equivalently,

$$\psi'(x,y,z)=\psi[R^{-1}(x,y,z)]. \tag{2.196}$$

The relationship between ψ' and ψ can be written in terms of a matrix $U(\alpha)$, which must be unitary in order that the norm $\langle\psi'|\psi'\rangle$ is preserved by the transformation. Thus,

$$\psi'(x,y,z)=U(\alpha)\psi(x,y,z). \tag{2.197}$$

For an infinitesimal rotation angle $\delta\alpha$, (2.196) may be expanded as follows

$$\psi'(x,y,z)=\psi(x+y\,\delta\alpha,y-x\,\delta\alpha,z)=\psi(x,y,z)+y\,\delta\alpha\frac{\partial\psi}{\partial x}-x\,\delta\alpha\frac{\partial\psi}{\partial y}$$

$$=\psi(x,y,z)-\mathrm{i}\,\delta\alpha j_z. \tag{2.198}$$

Now, comparing (2.198) with (2.41) shows that the generator for rotations is j_z (Hermitian and therefore observable). Thus,

$$U(\delta\alpha) = 1 - \mathrm{i}\,\delta\alpha j_z. \tag{2.199}$$

In general, for a rotation ω about an axis $\hat{\mathbf{n}}$, we have (cf. (2.45))

$$U(\omega) = \exp(-\mathrm{i}\omega\hat{\mathbf{n}}\cdot\mathbf{j}). \tag{2.200}$$

Invariance of the Hamiltonian under rotations gives $UH = HU$. Hence, from (2.199), $[\mathbf{j}, H] = 0$; that is, conservation of angular momentum. In particular, note that if U is a unitary operator which is independent of time, then if the Schrödinger equation (2.47) is invariant under the transformation $\psi' = U\psi$ (so that $\mathrm{i}\,\mathrm{d}\psi'/\mathrm{d}t = \mathrm{i}U\,\mathrm{d}\psi/\mathrm{d}t = UH\psi = UHU^\dagger\psi' = H\psi'$), then this implies that $[U, H] = 0$. Thus, for example, if $\mathbf{j}$ is a 'generator' of U (see (2.199)) then $[\mathbf{j}, H] = 0$, so $\mathbf{j}$ is conserved.

In order to discuss the possible eigenvalues of j and m in (2.194), it is useful to introduce the following operators

$$j_\pm = j_x \pm \mathrm{i}j_y. \tag{2.201}$$

The operator $\mathbf{j}^2$ can hence be written

$$\mathbf{j}^2 = j_+j_- + j_z^2 - j_z = j_-j_+ + j_z^2 + j_z. \tag{2.202}$$

Using this and (2.192) the following relations can easily be derived

$$[j_+, j_-] = 2j_z, \tag{2.203}$$

$$j_zj_\pm = j_\pm(j_z \pm 1), \tag{2.204}$$

$$[\mathbf{j}^2, j_\pm] = 0. \tag{2.205}$$

In terms of the eigenfunctions (2.194), (2.205) and (2.204) imply

$$\mathbf{j}^2(j_\pm Y_{jm}) = j_\pm\mathbf{j}^2 Y_{jm} = j(j+1)(j_\pm Y_{jm}), \tag{2.206}$$

and

$$j_z(j_\pm Y_{jm}) = j_\pm(j_z \pm 1)Y_{jm} = (m \pm 1)(j_\pm Y_{jm}). \tag{2.207}$$

Thus, $j_\pm Y_{jm}$ has the same eigenvalue j as Y_{jm} but with eigenvalues of j_z differing by ± 1. Thus, j_+ and j_- act as *raising* and *lowering* operators for m with

$$j_\pm Y_{jm} = C_\pm(j, m)Y_{j,m\pm 1}. \tag{2.208}$$

The coefficients $C_\pm$ can be calculated starting with (2.208),

$$\begin{aligned}|C_\pm(j, m)|^2\langle Y_{j,m\pm 1}|Y_{j,m\pm 1}\rangle &= \langle j_\pm Y_{jm}|j_\pm Y_{jm}\rangle \\ &= \langle Y_{jm}|j_\mp j_\pm|Y_{jm}\rangle = \langle Y_{jm}|\mathbf{j}^2 - j_z^2 \mp j_z|Y_{jm}\rangle \\ &= j(j+1) - m(m \pm 1).\end{aligned}$$

Hence, with a suitable choice of phase,

$$C_{\pm}(j,m)=[j(j+1)-m(m\pm 1)]^{1/2}. \tag{2.209}$$

The operations C_+Y_{jj} and C_-Y_{j-j} both give zero. Thus, for a given value of j, there are $2j+1$ values of m from $-j$ to j. For a system exhibiting spherical symmetry, the energy eigenvalues of the $2j+1$ states are degenerate, but the addition of a symmetry breaking term in the Hamiltonian can separate the degenerate levels (e.g. the application of an external magnetic field to an atom in the Zeeman effect).

The possible eigenvalues of the orbital angular momentum are $l=0,1,2$, etc. This follows, since the operator l_z can be written in spherical coordinates as $l_z=-\mathrm{i}\,\partial/\partial\phi$, where ϕ is the angle of rotation about the z-axis. The solutions for the eigenfunctions thus have the form $\exp(\mathrm{i}m\phi)$. The requirement that a rotation of 2π leaves the system unaltered means that the allowed values of m, which satisfy this periodic boundary condition, are 0, ± 1, ± 2, etc., so that l has integral values only.

In quantum mechanics there is a component of angular momentum, the intrinsic angular momentum or spin **s**, which has no classical analogue. The commutation relations for spin are of the same form as those for orbital angular momentum. Hence, the difference $2s$ between the highest and lowest values of s_z must be integral; thus s can take the values of $0, \frac{1}{2}, 1, \frac{3}{2}$, etc. Note that for particles of half-integral spin the eigenfunction $\exp(\mathrm{i}s_z\phi)$ changes sign (phase) for a rotation of 2π. The wave function itself, however, is not observable, and bilinear combinations of the wave function, which occur in measurable quantities, do not change sign under a 2π rotation.

Particles are either *bosons* (π, K, γ, Z^0, . . .) with integral spin or *fermions* ($e^{\pm}$, p, $\Delta(1232)$, quarks, . . .) with half-integral spin. Particles which are tightly bound states of some more fundamental entities can be characterised by a definite spin, provided that the internal motion and relative spin orientations of the components are not significantly affected by the interactions between the composite particles. A system of bosons obeys *Bose–Einstein statistics* and has a wave function *symmetric* under interchange of two particles, whereas a system of fermions obeys *Fermi–Dirac statistics* and has an *antisymmetric* wave function.

A state with angular momentum zero has only one possible state (singlet), whereas for angular momentum 1 there is a triplet of possible states. The eigenstates for the orbital angular momentum states $l=0$ and 1 are given in Table 2.1 (see, e.g., Schiff, 1955). In the case of spin $\frac{1}{2}$ it is convenient to use a matrix representation for the two possible spin states,

Table 2.1. *Eigenstates* Y_{lm} *for the angular momentum states* $l=0$ *and* $l=1$
For $l=1$ the correspondence to the transformation of a unit vector $\hat{\mathbf{n}}$ along the radius is given

l	$m=-1$	$m=0$	$m=1$
0		$Y_{00}=\left(\frac{1}{4\pi}\right)^{1/2}$	
1	$Y_{1-1}=\left(\frac{3}{4\pi}\right)^{1/2}\sin\theta\,\frac{\exp(-i\phi)}{2^{1/2}}$	$Y_{10}=\left(\frac{3}{4\pi}\right)^{1/2}\cos\theta$	$Y_{11}=-\left(\frac{3}{4\pi}\right)^{1/2}\sin\theta\,\frac{\exp(i\phi)}{2^{1/2}}$
	$(n_x-in_y)/2^{1/2}$	n_z	$-(n_x+in_y)/2^{1/2}$

namely χ_+ for spin up and χ_- for spin down, as follows

$$\chi_+ = \begin{pmatrix} 1 \\ 0 \end{pmatrix}, \qquad \chi_- = \begin{pmatrix} 0 \\ 1 \end{pmatrix}. \tag{2.210}$$

These states are related by the raising or lowering operators. Thus, since $C_\pm(\frac{1}{2}, \mp\frac{1}{2}) = 1$,

$$S_+\begin{pmatrix} 0 \\ 1 \end{pmatrix} = \begin{pmatrix} 1 \\ 0 \end{pmatrix}, \qquad S_-\begin{pmatrix} 1 \\ 0 \end{pmatrix} = \begin{pmatrix} 0 \\ 1 \end{pmatrix}. \tag{2.211}$$

These results, plus the requirement that the eigenvalues of S_z are $\pm\frac{1}{2}$, give the following 2×2 matrix operators

$$S_+ = \begin{pmatrix} 0 & 1 \\ 0 & 0 \end{pmatrix}, \qquad S_- = \begin{pmatrix} 0 & 0 \\ 1 & 0 \end{pmatrix}, \qquad S_z = \frac{1}{2}\begin{pmatrix} 1 & 0 \\ 0 & -1 \end{pmatrix}. \tag{2.212}$$

Thus, one may write $\mathbf{S} = \boldsymbol{\sigma}/2$. The matrices σ_x, σ_y and σ_z are called the *Pauli spin matrices*

$$\sigma_x = \begin{pmatrix} 0 & 1 \\ 1 & 0 \end{pmatrix}, \qquad \sigma_y = \begin{pmatrix} 0 & -\mathrm{i} \\ \mathrm{i} & 0 \end{pmatrix}, \qquad \sigma_z = \begin{pmatrix} 1 & 0 \\ 0 & -1 \end{pmatrix}. \tag{2.213}$$

These matrices are Hermitian ($\boldsymbol{\sigma}^\dagger = \boldsymbol{\sigma}$) and satisfy the commutation relations $[\sigma_x, \sigma_y] = 2\mathrm{i}\sigma_z$, etc. i.e.

$$[\sigma_i, \sigma_j] = 2\mathrm{i}\varepsilon_{ijk}\sigma_k, \qquad \sigma_i\sigma_j = \delta_{ij} + \mathrm{i}\varepsilon_{ijk}\sigma_k. \tag{2.214}$$

All three components have $\sigma_i^2 = I$, the unit matrix. Using these relationships one can show that, for any two vectors **A** and **B**,

$$(\boldsymbol{\sigma}\cdot\mathbf{A})(\boldsymbol{\sigma}\cdot\mathbf{B}) = \mathbf{A}\cdot\mathbf{B}I + \mathrm{i}\boldsymbol{\sigma}\cdot\mathbf{A}\times\mathbf{B}. \tag{2.215}$$

In particular, when $\mathbf{A} = \mathbf{B}$ this becomes $(\boldsymbol{\sigma}\cdot\mathbf{A})^2 = \mathbf{A}^2 I$.

For a spin $\frac{1}{2}$ particle the effect of a rotation through an angle ω about an axis $\hat{\mathbf{n}}$ is given by the unitary matrix

$$U(\omega) = \exp(-\mathrm{i}\omega\hat{\mathbf{n}}\cdot\boldsymbol{\sigma}/2) \tag{2.216}$$

This has a form similar to (2.200) except that the state vector is no longer a scalar but a two-component vector, hence $\mathbf{j} = \boldsymbol{\sigma}/2$.

Let χ represent a general spin $\frac{1}{2}$ state in the form of a column vector, with components a_1 (spin up) and a_2 (spin down), and normalised so that $|a_1|^2 + |a_2|^2 = 1$. Under rotations $\chi^\dagger\chi$ is invariant. In the rotated system $\chi' = U\chi$, with U from (2.216), so that $(\chi')^\dagger\chi' = \chi^\dagger U^\dagger U\chi = \chi^\dagger\chi$. Note, however, that it may easily be shown that the quantity $V = \chi^\dagger\boldsymbol{\sigma}\chi$ transforms as a vector under rotations. Now we can write

$$\chi\chi^\dagger = \begin{pmatrix} a_1 \\ a_2 \end{pmatrix}(a_1^* a_2^*) = \begin{pmatrix} a_1a_1^* & a_1a_2^* \\ a_2a_1^* & a_2a_2^* \end{pmatrix} = \rho. \tag{2.217}$$

The expectation value of some operator **A** (2×2 matrix) is

$$\langle A \rangle = \langle \chi | A | \chi \rangle = a_l^* A_{lm} a_m = \rho_{ml} A_{lm}$$
$$= \mathrm{tr}(\rho \mathrm{A}). \tag{2.218}$$

For the case $A = \sigma_z$

$$\langle \sigma_z \rangle = |a_1|^2 - |a_2|^2 = P_z, \tag{2.219}$$

the polarisation vector in the z-direction. Now ρ is Hermitian, and any 2×2 Hermitian matrix can be written in the form

$$\rho = c(I + \mathbf{d} \cdot \boldsymbol{\sigma}). \tag{2.220}$$

Since $\mathrm{tr}(\rho) = 1$, this gives $c = \frac{1}{2}$. Using (2.220) gives

$$\mathrm{tr}(\rho \boldsymbol{\sigma}) = \tfrac{1}{2} \mathrm{tr}[(\mathbf{d} \cdot \boldsymbol{\sigma}) \boldsymbol{\sigma}] = \mathbf{d}. \tag{2.221}$$

Now, from (2.219), we can identify **d** as the polarisation vector. Hence, we obtain

$$\chi \chi^\dagger = \rho = \tfrac{1}{2}(I + \mathbf{P} \cdot \boldsymbol{\sigma}). \tag{2.222}$$

Multiplying (2.222) by χ from the righthand side, and using $\chi^\dagger \chi = 1$, gives the eigenvalue equation

$$(\mathbf{P} \cdot \boldsymbol{\sigma}) \chi = \chi. \tag{2.223}$$

If a system consists of two particles α and β, each of spin $\frac{1}{2}$, then there are four possible combined spin states. Let J, M be the total and z-components of the angular momentum of the combined system. Since j_z has additive quantum numbers, the four possible values of M are -1, 0, 0 and 1, and the possible J values are 0 and 1. Writing the combined state as ψ_{JM} and the states of α and β as ψ_{jz}^α, ψ_{jz}^β, then

$$\psi_{11} = \psi_{1/2}^\alpha \psi_{1/2}^\beta, \qquad \psi_{1-1} = \psi_{-1/2}^\alpha \psi_{-1/2}^\beta. \tag{2.224}$$

If J_- is the lowering operator for J, then $J_- = j_-^\alpha + j_-^\beta$, hence

$$J_- \psi_{11} = (j_-^\alpha + j_-^\beta) \psi_{1/2}^\alpha \psi_{1/2}^\beta$$
$$= \psi_{-1/2}^\alpha \psi_{1/2}^\beta + \psi_{1/2}^\alpha \psi_{-1/2}^\beta$$
$$= 2^{1/2} \psi_{10}.$$

Hence

$$\psi_{10} = \frac{1}{2^{1/2}} (\psi_{-1/2}^\alpha \psi_{1/2}^\beta + \psi_{1/2}^\alpha \psi_{-1/2}^\beta). \tag{2.225}$$

The $J = 0$ state is orthogonal to ψ_{10} and is thus

$$\psi_{00} = \frac{1}{2^{1/2}}(\psi^{\alpha}_{-1/2}\psi^{\beta}_{1/2} - \psi^{\alpha}_{1/2}\psi^{\beta}_{-1/2}). \tag{2.226}$$

Note that the $J = 1$ state is symmetric, whereas the $J = 0$ state is antisymmetric. In general the combination of states with angular momentum J_1 and J_2 gives $(2J_1 + 1)(2J_2 + 1)$ possible states, with values of J from $|J_1 - J_2|$ to $J_1 + J_2$. The eigenfunctions of the total angular momentum state $\psi(J, M)$ can be described as a superposition of states which are products of the eigenfunctions $\psi(J_1, M_1)\psi(J_2, M_2)$. The coefficients connecting these representations are called *Clebsch–Gordan coefficients*, and can be found in most standard texts on quantum mechanics. The coefficients relating $J_1 = J_2 = \frac{1}{2}$ to $J = 0, 1$ (and *vice versa*) are obtained from equations (2.224) to (2.226).

2.5 GROUP THEORY

2.5.1 Classification of groups

Symmetry transformations on physical systems have the mathematical properties of groups. A group is a set of elements satisfying the following axioms:

(i) The 'product' or combination, ab, of two group members, a and b, is also a member of the group.

(ii) Associativity; $(ab)c = a(bc)$. However, the elements do not necessarily commute.

(iii) One member of the group is the identity element I, which satisfies $aI = Ia = a$.

(iv) For every element a in the group there exists an inverse element a^{-1}, such that $a^{-1}a = aa^{-1} = I$ $(I^{-1} = I)$.

A group is said to be commutative or *Abelian* if all of the elements of the group commute. A group can have either a finite number of elements or, for example, if the group elements are defined in terms of a continuous variable (e.g. $a(x)$), an infinite number of elements. The group of all possible rotations, with each rotation being represented by $\boldsymbol{\alpha}(\alpha_1, \alpha_2, \alpha_3)$ is an example of a continuous group. The rotation group is an example of a *Lie* group, in that each rotation can be expressed as a product of infinitesimal rotations or, more generally, can be represented by a unitary matrix U.

The group of coordinate transformations in one dimension, $x' = x + \varepsilon$, is a one-parameter Lie group. These transformations clearly commute, so that the group is Abelian. The effect of the transformation on the wave function can be written, since $\psi(x) = \psi'(x')$, as follows

$$\psi'(x) = U\psi(x) = \psi(x - \varepsilon). \tag{2.227}$$

For an infinitesimal transformation

$$\begin{aligned}\psi(x-\varepsilon) &= \psi(x) - \varepsilon\frac{\mathrm{d}\psi}{\mathrm{d}x} + \mathrm{O}(\varepsilon^2)\\ &= \psi(x) - \mathrm{i}\varepsilon p_x\psi.\end{aligned} \tag{2.228}$$

Thus, for a finite transformation, $U = \exp(-\mathrm{i}\varepsilon p_x)$. The requirements that the norm $\langle\psi'|\psi'\rangle$ is preserved, and that an infinitesimal transformation can be realised by starting from the identity matrix, imply that U is unitary. Hence, p_x is Hermitian and so is observable. Thus, invariance of the system under translations implies conservation of linear momentum. Similarly, invariance of a system with respect to time leads to conservation of energy.

Discrete transformations, however, cannot be realized by continuous transformations starting with the identity transformation. The parity transformation ($\mathbf{r} \to \mathbf{r}' = -\mathbf{r}$) is an example of a discrete symmetry because there are only two possible frames, the original and the inverted. The parity transformation cannot be produced by the group of proper rotations in three dimensions, R_3, and is thus an improper rotation. Two successive parity transformations restore the original system, thus

$$\psi'(\mathbf{r}') = U_p\psi(\mathbf{r}) = U_p^2\psi'(\mathbf{r}'). \tag{2.229}$$

Hence $U_p^2 = 1$, and therefore U_p is Hermitian ($U_p = U_p^\dagger$) and thus observable. The eigenvalues of parity are $P = \pm 1$ and the eigenvalues are multiplicative. For a single particle, P is the intrinsic parity of the particle. Thus, two particles a and b in a state of relative orbital angular momentum l have

$$P = P_\mathrm{a}P_\mathrm{b}(-1)^l, \tag{2.230}$$

where P_a and P_b are the intrinsic parities. The relative intrinsic parity of particles is only meaningful if they participate in the same interaction.

The charge conjugation operator C changes a particle A to its antiparticle $\bar{\mathrm{A}}$, leaving the four-momentum and spin unchanged. Two successive applications restore the original state, hence the possible eigenvalues of C are ± 1. The time reversal operator T, changing t to $t' = -t$, is again two-valued, but cannot be represented by a unitary operator. This can be seen by taking the complex conjugate of the Schrödinger equation (2.47) from which, for the case of H real and independent of t, it follows that $\psi^*(t)$ is the solution for the time-reversed case. The operator T is thus antilinear in terms of the state function ψ and can be written

$$T = U_T K, \tag{2.231}$$

where K is the complex conjugation operator ($K\psi = \psi^*$) and U_T is unitary; the operator T is antiunitary. The choice that C is unitary rather than antiunitary is dictated by quantum electrodynamics.

2.5.2 SU(2)

Unitary groups are useful, for example, in describing the properties of the various particle multiplets observed in nature. The classification of particles in isospin multiplets stems from the idea of Heinsenberg in 1932 that, if the electromagnetic interaction could be switched off, then there would be no distinction between protons and neutrons. The proton ($I_3 = \frac{1}{2}$) and the neutron ($I_3 = -\frac{1}{2}$) are then states of a single particle, the nucleon, with isospin $\frac{1}{2}$. In terms of quarks, u($I_3 = \frac{1}{2}$) and d($I_3 = -\frac{1}{2}$) are members of an isospin doublet q. Isospin invariance means that, instead of u and d as base states of q, any linear combination of q′ with

$$q' = \begin{pmatrix} \mathrm{u}' \\ \mathrm{d}' \end{pmatrix} = \begin{pmatrix} U_{11} & U_{12} \\ U_{21} & U_{22} \end{pmatrix} \begin{pmatrix} \mathrm{u} \\ \mathrm{d} \end{pmatrix} = U\mathrm{q}, \tag{2.232}$$

is equally as good. The matrix U must be unitary to preserve the norm, and this implies $|\det U| = 1$. Choosing $\det U = 1$, it is easy to show that U can be written

$$U = \begin{pmatrix} U_{11} & U_{12} \\ -U_{12}^* & U_{11}^* \end{pmatrix}. \tag{2.233}$$

Consider an infinitesimal transformation of the form $U = 1 + \mathrm{i}\varepsilon G$, with ε real. From (2.42), G is Hermitian, so G_{11} and G_{22} are real, and hence we can write

$$U = \begin{pmatrix} 1 & 0 \\ 0 & 1 \end{pmatrix} + \mathrm{i}\varepsilon \begin{pmatrix} G_{11} & G_{12} \\ G_{12}^* & G_{22} \end{pmatrix} + \mathrm{O}(\varepsilon^2). \tag{2.234}$$

Using (2.233) gives

$$G_{11} + G_{22} = 0. \tag{2.235}$$

Hence, for a finite transformation, we have

$$U = \exp(\mathrm{i}H) = \exp(\mathrm{i}\alpha G), \tag{2.236}$$

where H is a traceless matrix and $\det U = 1$. In fact this follows in general, since $\det(\exp(\mathrm{i}H)) = \exp(\mathrm{i}\ \mathrm{tr}\ H)$, so $\mathrm{tr}\ H = 0$ implies $\det U = 1$.

The group of all 2×2 unitary matrices, with the special condition that $\det U = 1$, is called SU(2). There are $2^2 - 1 = 3$ independent parameters. It is thus useful to write

$$H = \sum_{j=1}^{3} \alpha_j G_j = \boldsymbol{\alpha} \cdot \mathbf{G}. \tag{2.237}$$

Comparison with (2.216) shows that we can identify $G_j = \sigma_j/2$. Thus we have the commutation relations

$$\left[\frac{\sigma_i}{2}, \frac{\sigma_j}{2}\right] = \mathrm{i}\varepsilon_{ijk}\frac{\sigma_k}{2}. \tag{2.238}$$

This is called the *fundamental* (**2**) *representation* of SU(2). An n-dimensional representation of SU(2) consists of $n \times n$ matrices satisfying the same commutation relation algebra, (2.238).

From the commutation relationship (2.238), by analogy with the results for angular momentum, a state with isospin I will have $2I + 1$ substates with different eigenvalues of I_3. The eigenvalues I_3 are additive quantum numbers. Thus, in the absence of electromagnetic interactions, one would expect degenerate multiplets if isospin were an exact symmetry. Note that no two of the three generators σ_i commute. The *rank* of a group can be defined as the maximum number of mutually commuting generators; thus, SU(2) has rank 1. One can form combinations of generators (Casimir operators) which commute with all the generators of the group. For SU(2) there is one such operator (the number of Casimir operators for a Lie group is equal to the rank of the group), and is $C = \mathbf{J}^2 = (\boldsymbol{\sigma}/2)^2$.

In the above example the isospin properties of u and d quarks have been discussed. Next we consider the corresponding properties of the antiquarks $\bar{\mathrm{u}}$ and $\bar{\mathrm{d}}$. For a rotation θ about an axis $\hat{\mathbf{n}}$ for a spin $\frac{1}{2}$ state we have, from (2.200) and (2.237),

$$U = \exp\left(-\mathrm{i}\frac{\theta}{2}\hat{\mathbf{n}}\cdot\boldsymbol{\sigma}\right) = I\cos\frac{\theta}{2} - \mathrm{i}(\hat{\mathbf{n}}\cdot\boldsymbol{\sigma})\sin\frac{\theta}{2}, \tag{2.239}$$

where use has been made of (2.215). Thus for a rotation of θ about the I_2-axis

$$\mathrm{q}' = \begin{pmatrix} \mathrm{u}' \\ \mathrm{d}' \end{pmatrix} = \begin{pmatrix} \cos\theta/2 & -\sin\theta/2 \\ \sin\theta/2 & \cos\theta/2 \end{pmatrix} \begin{pmatrix} \mathrm{u} \\ \mathrm{d} \end{pmatrix}. \tag{2.240}$$

Under charge conjugation (C), (2.240) becomes

$$\begin{aligned} \bar{\mathrm{u}}' &= \bar{\mathrm{u}}\cos\theta/2 - \bar{\mathrm{d}}\sin\theta/2, \\ \bar{\mathrm{d}}' &= \bar{\mathrm{u}}\sin\theta/2 + \bar{\mathrm{d}}\cos\theta/2. \end{aligned} \tag{2.241}$$

Hence the isospin doublet for antiquarks must have the form

$$\tilde{q}' = \begin{pmatrix} -\bar{d} \\ \bar{u} \end{pmatrix}, \tag{2.242}$$

in order to satisfy (2.241) and transform according to (2.240), together with the requirements that $\bar{d}(\bar{u})$ have I_3 values of $\frac{1}{2}$ $(-\frac{1}{2})$ as necessitated by their charges. This equivalence in the transformation properties of the quark (**2** representation) and antiquark (**2*** representation) is a property of the SU(2) group and does not, for example, hold for SU(3) or SU(4).

A system of two quarks qq has four possible states in terms of isospin $|I, I_3\rangle$ (see (2.224) to (2.226))

$$\begin{aligned} &|1,1\rangle = uu, \\ &|1,0\rangle = (ud + du)/2^{1/2}, \qquad |0,0\rangle = (ud - du)/2^{1/2}, \\ &|1,-1\rangle = dd, \end{aligned} \tag{2.243}$$

consisting of three symmetric and one antisymmetric states. This combination qq is called the *direct product* and is usually written in the form

$$\mathbf{2} \otimes \mathbf{2} = \mathbf{3} \oplus \mathbf{1}. \tag{2.244}$$

Note that this gives a more formal basis for the combination of angular momenta discussed at the end of Section 2.4. The states **3** (triplet with isospin 1) and **1** (singlet with isospin 0) are called *irreducible representations* in that, application of the group operators (the $\boldsymbol{\sigma}$ matrices) on any state and in any order, only results in other states of that representation. The group operators connect states within an irreducible representation, but do not connect different irreducible representations.

Bound states of two quarks do not appear to exist in nature. This can be readily understood in terms of the colour quantum numbers. A qq system has net colour (or, more precisely, anticolour) and hence will strongly attract a further quark. A $q\bar{q}$ system, however, has no net colour and hence will be strongly bound. The π-meson is the lightest symmetric state **3** of $q\bar{q}$. Hence the π has isospin 1 and its isospin properties can be derived in a similar way to that used for spin $\frac{1}{2}$ in Section 2.4. In this case the matrices I_1, I_2 and I_3 are 3×3 matrices satisfying the same commutation relations (2.238), namely

$$[I_i, I_j] = i\varepsilon_{ijk} I_k. \tag{2.245}$$

Specific formulations of these matrices are discussed, for example, in Close (1979). This representation of the SU(2) group has dimension three. In general, a particle of isospin I corresponds to a dimension $2I + 1$.

A representation of SU(n) with dimension n^2-1 is called a *regular representation.*

Under charge conjugation a π^+-meson is transformed into a π^-, and *vice versa.* The π^0-meson is, however, an eigenstate of C. Since the π^0 decays electromagnetically to two photons and (experimentally) the electromagnetic interaction conserves C, we have $C(\pi^0)=C(\gamma)^2=1$. The photon has, in fact, $C(\gamma)=-1$; this is because all the components of the electromagnetic current change sign under C, and the product of the field and current is invariant. If the eigenstates 1, 0 and -1 of I_3 are identified with π^+, π^0 and π^- respectively, then a rotation of 180° around the two-axis ($R_2(\pi)$, cf. (2.240)) will, to within a phase factor, transform π^+ to π^- and *vice versa.* A more detailed analysis (see, e.g., Gibson and Pollard, 1976) shows that the phase factor is -1 for both π^+ and π^-, and that the same factor also applies to π^0, which is an eigenstate of $R_2(\pi)$. The combined operation, which is known as *G-parity,*

$$G=CR_2(\pi), \tag{2.246}$$

has the same eigenvalue for all values of I_3. Thus $G(\pi)=-1$ and, for a system of n pions,

$$G(n\pi)=(-1)^n. \tag{2.247}$$

2.5.3 SU(3)

The extension of the above ideas to include three rather than two quarks is conceptually straightforward, but leads to more complicated mathematics. If any combination q′ of u, d and s quarks is equally as good as the base states u, d and s, then we have

$$\mathrm{q}'=\begin{pmatrix}\mathrm{u}'\\ \mathrm{d}'\\ \mathrm{s}'\end{pmatrix}=U\begin{pmatrix}\mathrm{u}\\ \mathrm{d}\\ \mathrm{s}\end{pmatrix}=U\mathrm{q}, \tag{2.248}$$

where U is unitary and can be written

$$U=\exp(\mathrm{i}H), \qquad \text{with tr } H=0,\ \det U=1. \tag{2.249}$$

For the fundamental (**3**) representation we can express H in terms of eight 3×3 traceless matrices λ_j as follows

$$U=\exp\left(\sum_{j=1}^{8}\mathrm{i}\alpha_j\lambda_j/2\right), \tag{2.250}$$

where α_j are real continuous parameters and λ_j are the generators of SU(3).

The matrices λ_j are the SU(3) equivalents of the Pauli matrices. If the system is invariant under interchange of u, d and s quarks, then it must be invariant separately under the interchange of the pairs of quarks ud, ds and us respectively. Thus there are three corresponding SU(2) subgroups of SU(3). Each of these SU(2) subgroups has two shift operators, analogous to the operators $I_\pm$ (see (2.201) and (2.212)) of isospin. If we define M_{ij} to be a 3×3 matrix whose elements are all zero except for the element i, j which has a value of one, then the isospin raising and lowering operators are $I_+ = M_{12}$ and $I_- = M_{21}$ respectively. The raising and lowering operators corresponding to the symmetry between d and s quarks (U-spin) and between u and s quarks (V-spin) are $U_+ = M_{23}$, $U_- = M_{32}$, $V_+ = M_{13}$ and $V_- = M_{31}$. Thus the values of (I_3, U_3, V_3) for the three quarks are $(\frac{1}{2}, 0, \frac{1}{2})$ for u, $(-\frac{1}{2}, \frac{1}{2}, 0)$ for d and $(0, -\frac{1}{2}, -\frac{1}{2})$ for s.

The eight matrices λ_j can be formed from these six shift operators, together with the two remaining matrices which are diagonal and traceless. Thus the λ matrices can be chosen to be (Gell-Mann, 1962)

$$\lambda_1 = \begin{pmatrix} 0 & 1 & 0 \\ 1 & 0 & 0 \\ 0 & 0 & 0 \end{pmatrix}, \quad \lambda_2 = \begin{pmatrix} 0 & -\mathrm{i} & 0 \\ \mathrm{i} & 0 & 0 \\ 0 & 0 & 0 \end{pmatrix}, \quad \lambda_3 = \begin{pmatrix} 1 & 0 & 0 \\ 0 & -1 & 0 \\ 0 & 0 & 0 \end{pmatrix},$$

$$\lambda_4 = \begin{pmatrix} 0 & 0 & 1 \\ 0 & 0 & 0 \\ 1 & 0 & 0 \end{pmatrix}, \quad \lambda_5 = \begin{pmatrix} 0 & 0 & -\mathrm{i} \\ 0 & 0 & 0 \\ \mathrm{i} & 0 & 0 \end{pmatrix}, \quad \lambda_6 = \begin{pmatrix} 0 & 0 & 0 \\ 0 & 0 & 1 \\ 0 & 1 & 0 \end{pmatrix},$$

$$\lambda_7 = \begin{pmatrix} 0 & 0 & 0 \\ 0 & 0 & -\mathrm{i} \\ 0 & \mathrm{i} & 0 \end{pmatrix}, \quad \lambda_8 = \frac{1}{3^{1/2}} \begin{pmatrix} 1 & 0 & 0 \\ 0 & 1 & 0 \\ 0 & 0 & -2 \end{pmatrix}. \tag{2.251}$$

Writing $F_i = \lambda_i/2$ we can identify

$$I_\pm = I_1 \pm \mathrm{i}I_2 = F_1 \pm \mathrm{i}F_2, \tag{2.252}$$

$$U_\pm = U_1 \pm \mathrm{i}U_2 = F_6 \pm \mathrm{i}F_7, \tag{2.253}$$

$$V_\pm = V_1 \pm \mathrm{i}V_2 = F_4 \pm \mathrm{i}F_5 \tag{2.254}$$

The additional two matrices are the diagonal matrices F_3 and F_8. The matrix F_3 is the operator I_3, with eigenvalues $\frac{1}{2}$, $-\frac{1}{2}$ and 0 for the u, d and s quarks respectively. The SU(3) group has rank 2 and the second mutually commuting generator, F_8, has a form chosen such that its eigenvalues for u and d quarks are equal. From the definitions of I_3, U_3 and V_3 we have $V_3 = I_3 + U_3$. Hence, writing

$$F_8 = aI_3 + bU_3, \tag{2.255}$$

and substituting the values of I_3 and U_3 for the u and d quarks, gives $b=2a$; hence, $F_8(\mathrm{u})=F_8(\mathrm{d})=a/2$, and so $F_8(\mathrm{s})=-\mathrm{a}$. The requirement that the distance in the (F_8, F_3) plane between the two U-spin multiplet members (i.e. d and s) is unity gives, taking a to be positive, $a=1/3^{1/2}$. Hence,

$$F_8=(I_3+2U_3)/3^{1/2}. \tag{2.256}$$

This specifies the form of the matrix F_8, and hence λ_8 in (2.251). Fig. 2.6 shows the values of F_8 and F_3 for the u, d and s quarks and the effect of the operators I_+, U_+ and V_+. The operators I_-, etc., work in the opposite sense. The difference in charge resulting from the operation I_+ is one unit (of e); hence, $Q_\mathrm{u}-Q_\mathrm{d}=1$. We can define a *hypercharge* Y, such that the difference in Y between u (or d) and s is also one unit. Thus

$$Y=\frac{2}{3^{1/2}}F_8. \tag{2.257}$$

Fig. 2.6 Triplet representation of SU(3) in terms of F_8 and F_3. The action of the shift operators is also shown.

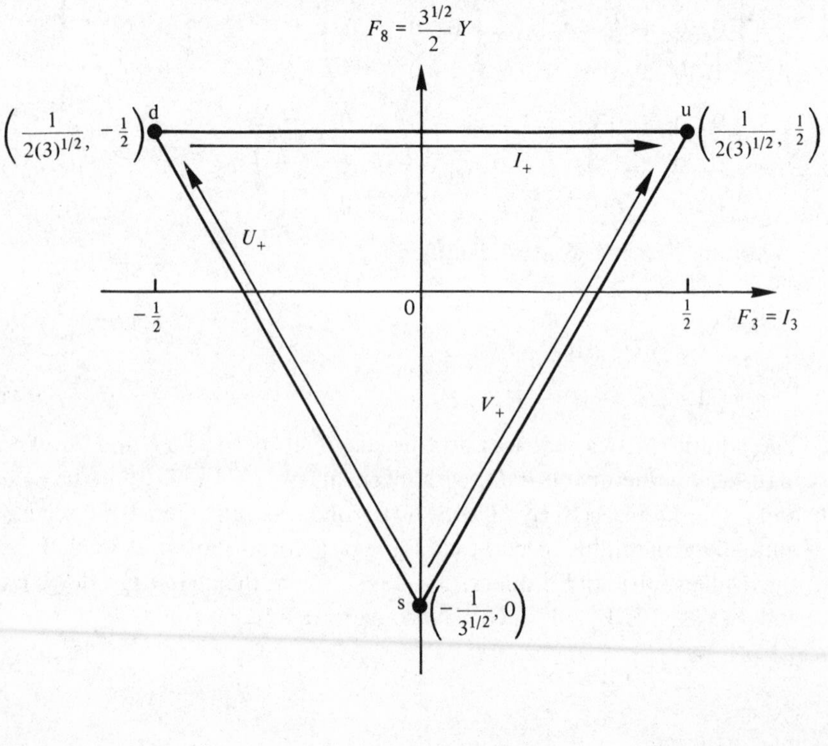

The values of the third components of U-spin and V-spin are thus

$$U_3 = -\frac{F_3}{2} + \frac{3^{1/2}}{2} F_8 = -\frac{I_3}{2} + \tfrac{3}{4}Y, \tag{2.258}$$

$$V_3 = \frac{F_3}{2} + \frac{3^{1/2}}{2} F_8 = \frac{I_3}{2} + \tfrac{3}{4}Y. \tag{2.259}$$

From the above considerations it can be seen that the quark charge is related linearly to I_3 and Y, with the form

$$Q = I_3 + bY + c. \tag{2.260}$$

This implies that the sum of the charges of the u, d and s quarks is $3c$. The *Gell-Mann–Nishijima relationship*

$$Q = I_3 + Y/2, \tag{2.261}$$

corresponds to $b = \frac{1}{2}$ and $c = 0$. In this case the quark charges are $\frac{2}{3}$, $-\frac{1}{3}$ and $-\frac{1}{3}$ for u, d and s respectively, with sum zero. The quark charges are the same for the two members of the U-spin doublet. Use of the definition that $Y = B + S$, with the *strangeness* $S = 0, 0, -1$ for u, d and s quarks respectively, implies that the *baryon number* $B = \frac{1}{3}$, $-\frac{1}{3}$ for quarks and antiquarks respectively. Note that the choice as to which physical quantity $(Y, S, \ldots)$ is used as the coordinate orthogonal to I_3 is arbitrary. The hypercharge is useful in that it centres the multiplets on zero.

The generators of the SU(3) group F_i satisfy the following closed algebra of commutation relations

$$[F_i, F_j] = \mathrm{i} f_{ijk} F_k, \qquad i, j, k = 1, \ldots, 8, \tag{2.262}$$

where f_{ijk} are the SU(3) structure constants. The following anticommutation relations also hold:

$$\{F_i, F_j\} = \tfrac{1}{3}\delta_{ij} + d_{ijk} F_k. \tag{2.263}$$

The f_{ijk} and d_{ijk} are respectively *symmetric* and *antisymmetric* under the interchange of any two indices. The values of f_{ijk} and d_{ijk} can be found from (2.251) by explicit calculation, and the non-zero values are given in Table 2.2. Use of this table shows that

$$[Q, U_\pm] = [Q, U_3] = 0, \tag{2.264}$$

so that charge is a U-spin scalar.

The irreducible representations of SU(3) can be derived by consideration of the action of the shift operators for I, U and V (Fig. 2.6). The resulting multiplets, which transform amongst themselves under the action of the

Table 2.2. *Non-zero values of f_{ijk} and d_{ijk}*

ijk	f_{ijk}	ijk	d_{ijk}
123	1	118	$1/3^{1/2}$
147	$\frac{1}{2}$	146	$\frac{1}{2}$
156	$-\frac{1}{2}$	157	$\frac{1}{2}$
246	$\frac{1}{2}$	228	$1/3^{1/2}$
257	$\frac{1}{2}$	247	$-\frac{1}{2}$
345	$\frac{1}{2}$	256	$\frac{1}{2}$
367	$-\frac{1}{2}$	338	$1/3^{1/2}$
458	$3^{1/2}/2$	344	$\frac{1}{2}$
678	$3^{1/2}/2$	355	$\frac{1}{2}$
		366	$-\frac{1}{2}$
		377	$-\frac{1}{2}$
		448	$-1/(2\times 3^{1/2})$
		558	$-1/(2\times 3^{1/2})$
		668	$-1/(2\times 3^{1/2})$
		778	$-1/(2\times 3^{1/2})$
		888	$-1/3^{1/2}$

shift operators, are specified by the eigenvalues of the two additive constants of the SU(3) group F_3 and F_8. The simplest state is the singlet (0, 0), specified by **1**. The next highest multiplet is the triplet (**3**) representation, which is the one so far considered in this section. The antiquarks are represented by the $\bar{\mathbf{3}}$ representation, as shown in Fig. 2.7. The regular, adjoint or octet (**8**) representation is also shown in Fig. 2.7. Since each site is specified by the additive quantum numbers (F_8, F_3), the possible sites for the direct product of say **3** and $\bar{\mathbf{3}}$ are obtained by adding all possible (F_8, F_3) values of **3** and $\bar{\mathbf{3}}$. This can be achieved by positioning the centre of gravity of the $\bar{\mathbf{3}}$ multiplet on top of each position of the **3** multiplet. This gives an octet (Fig. 2.7), whose states transform amongst themselves, and a singlet. Thus the direct product of the **3** (quark) and $\bar{\mathbf{3}}$ (antiquark) representations gives

$$\mathbf{3}\otimes\bar{\mathbf{3}}=\mathbf{8}\oplus\mathbf{1}. \tag{2.265}$$

That is, the nine possible meson states consist of two irreducible representations, an octet **8** and a singlet **1**. A detailed discussion of how representations can be built up by such graphical methods is given by Gasiorowicz (1966).

The multiplicity of the site at the origin (0, 0) is three, since there are two contributions from the octet and one from the singlet. The resulting states can be represented by a 3×3 matrix M constructed as follows:

$$M = \mathrm{q} \otimes \bar{\mathrm{q}} = \begin{pmatrix} \mathrm{u\bar{u}} & \mathrm{u\bar{d}} & \mathrm{u\bar{s}} \\ \mathrm{d\bar{u}} & \mathrm{d\bar{d}} & \mathrm{d\bar{s}} \\ \mathrm{s\bar{u}} & \mathrm{s\bar{d}} & \mathrm{s\bar{s}} \end{pmatrix} \tag{2.266}$$

$$= \begin{pmatrix} (2\mathrm{u\bar{u}} - \mathrm{d\bar{d}} - \mathrm{s\bar{s}})/3 & \mathrm{u\bar{d}} & \mathrm{u\bar{s}} \\ \mathrm{d\bar{u}} & (2\mathrm{d\bar{d}} - \mathrm{u\bar{u}} - \mathrm{s\bar{s}})/3 & \mathrm{d\bar{s}} \\ \mathrm{s\bar{u}} & \mathrm{s\bar{d}} & (2\mathrm{s\bar{s}} - \mathrm{u\bar{u}} - \mathrm{d\bar{d}})/3 \end{pmatrix} + \frac{1}{3}\begin{pmatrix} T & 0 & 0 \\ 0 & T & 0 \\ 0 & 0 & T \end{pmatrix}, \tag{2.267}$$

where T is the trace of M.

The assignment of the off-diagonal elements in terms of mesons is straightforward. However, there are three states with $F_8 = F_3 = 0$, which must be constructed out of $\mathrm{u\bar{u}}$, $\mathrm{d\bar{d}}$ and $\mathrm{s\bar{s}}$. These can be assigned as follows

$$M_{81} = \frac{(\mathrm{u\bar{u}} - \mathrm{d\bar{d}})}{2^{1/2}}, \qquad M_{10} = \frac{(\mathrm{u\bar{u}} + \mathrm{d\bar{d}} + \mathrm{s\bar{s}})}{3^{1/2}},$$

$$M_{80} = \frac{(\mathrm{u\bar{u}} + \mathrm{d\bar{d}} - 2\mathrm{s\bar{s}})}{6^{1/2}}. \tag{2.268}$$

Fig. 2.7 The **3** (quark), $\bar{\mathbf{3}}$ (antiquark) and **8** (octet) representations of SU(3) in terms of F_8 and F_3.

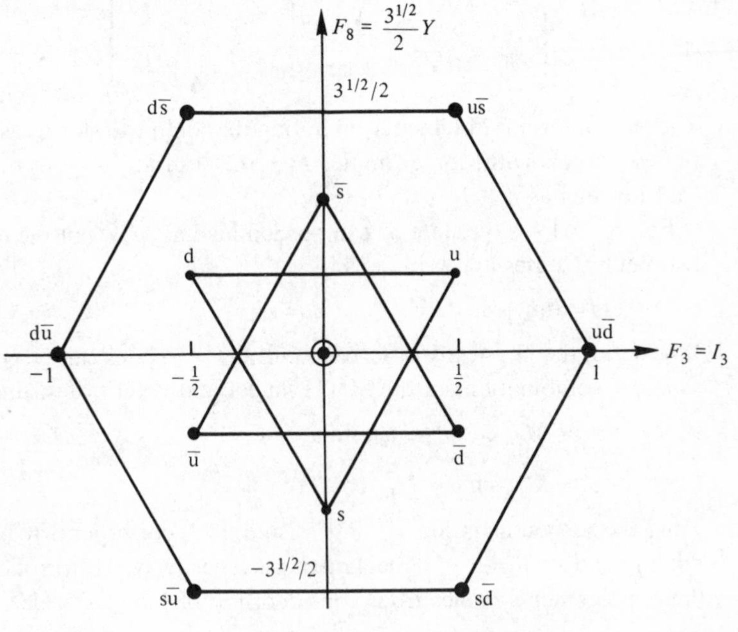

The first index indicates whether the meson beongs to the octet (**8**) or singlet (**1**), and the second index gives the I-value of the state (0 or 1). The state M_{81} is the neutral member of the isovector triplet which contains $u\bar{d}$ and $d\bar{u}$; this state was determined in (2.243). Note that, for a $q\bar{q}$ system, the phases for the $\bar{q}$ given in (2.242) are used; this changes the relative signs of the symmetric and antisymmetric $I=0$ terms in (2.243). The state M_{10} is an SU(3) singlet and M_{80} is orthogonal to both M_{81} and M_{10}, with all the states normalised to unity.

Let us now consider the quark assignment of the $J^{PC}=0^{-+}$ and 1^{--} meson nonets (Table 1.2). For $J^{PC}=0^{-+}$, the off-diagonal elements are $\pi^{\pm}$, $K^{\pm}$, K^0 and $\bar{K}^0$. We can identify π^0, η' and η with M_{81}, M_{10} and M_{80} respectively. Thus, the first matrix (the octet) in (2.267) can be written

$$M=\begin{pmatrix} \dfrac{\pi^0}{2^{1/2}}+\dfrac{\eta}{6^{1/2}} & \pi^+ & K^+ \\ \pi^- & -\dfrac{\pi^0}{2^{1/2}}+\dfrac{\eta}{6^{1/2}} & K^0 \\ K^- & \bar{K}^0 & -\dfrac{2}{6^{1/2}}\eta \end{pmatrix}$$

$$=\begin{pmatrix} \dfrac{F_3}{2^{1/2}}+\dfrac{F_8}{6^{1/2}} & F_1+iF_2 & F_4+iF_5 \\ F_1-iF_2 & -\dfrac{F_3}{2^{1/2}}+\dfrac{F_8}{6^{1/2}} & F_6+iF_7 \\ F_4-iF_5 & F_6-iF_7 & -\dfrac{2}{6^{1/2}}F_8 \end{pmatrix}. \tag{2.269}$$

The second form is obtained from noting that $qF_j\bar{q}$ transforms as the octet representation with, for example, $F_1 \pm iF_2$ transforming as $\pi^{\pm}$ and F_8 transforming as η.

For $J^{PC}=1^{--}$, the state ρ^0 can be identified as M_{81}, but the remaining two neutral states are written as

$$\omega=(u\bar{u}+d\bar{d})/2^{1/2}, \qquad \phi=s\bar{s}. \tag{2.270}$$

In general, the physically observed neutral $I=0$ states can be considered as linear combinations of the SU(3) singlet and octet states, thus

$$\text{`}\omega\text{'}=M_{10}\cos\theta+M_{80}\sin\theta,$$

$$\text{`}\phi\text{'}=M_{10}\sin\theta-M_{80}\cos\theta. \tag{2.271}$$

Thus the assignments for $J^{PC}=0^{-+}$ and 1^{--} correspond to $\theta=0$ (no mixing) and $\sin\theta=1/3^{1/2}$ (ideal mixing) respectively. The justification for these assignments comes from consideration of the meson decay modes and of models for their masses (see, e.g., Close, 1979). However, these assignments should not be taken to be exact.

The nonets of mesons correspond to different values of the quark–antiquark spin (s) and orbital angular momentum (l). For a fermion–antifermion pair $P=(-1)^{l+1}$, and for neutral states $C=(-1)^{l+s}$. Thus for $s=0$, $C=-P$ and so for $l=0$ we have $J^{PC}=0^{-+}$. For $l=0$ and $s=1$ we have $J^{PC}=1^{--}$. The heavier nonets of mesons correspond to both higher values of l and/or radial excitations of the quark—antiquark state. Note that the inclusion of the two possible spin states for the quarks into the symmetry extends the symmetry group to be considered from SU(3) to SU(6).

The combination of two quarks gives (Fig. 2.8)

$$\mathbf{3}\otimes\mathbf{3}=\mathbf{6}\oplus\bar{\mathbf{3}}, \tag{2.272}$$

where the **6** is symmetric (since it contains uu, etc.) and the $\bar{\mathbf{3}}$ is antisymmetric. Addition of a third quark gives the multiplet structure for baryons

$$\begin{aligned}\mathbf{3}\otimes\mathbf{3}\otimes\mathbf{3}&=(\mathbf{6}\otimes\mathbf{3})\oplus(\bar{\mathbf{3}}\otimes\mathbf{3})\\&=\mathbf{10}\oplus\mathbf{8}\oplus\mathbf{8}\oplus\mathbf{1}.\end{aligned} \tag{2.273}$$

Fig. 2.8 The **6** and $\bar{\mathbf{3}}$ representations formed by the product $\mathbf{3}\otimes\mathbf{3}$.

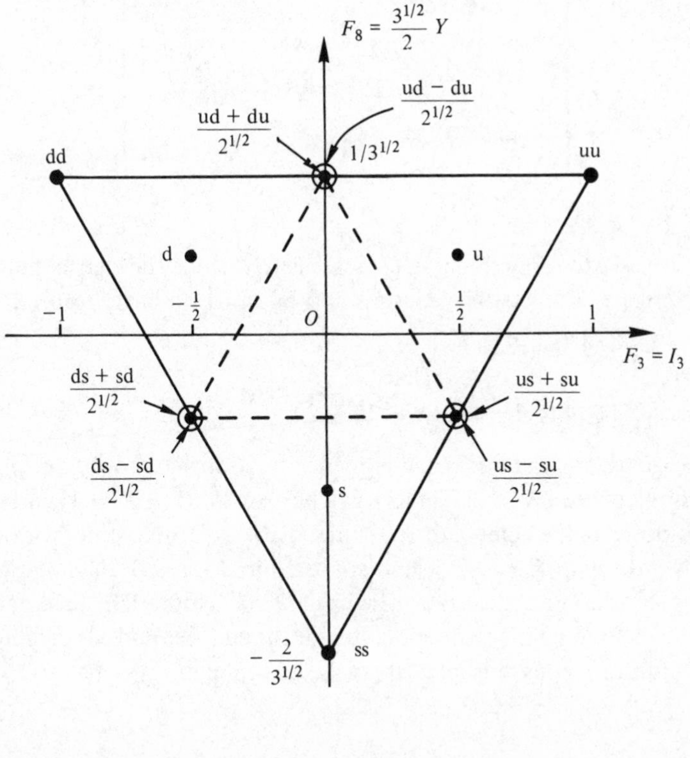

Table 2.3. *Quark content of the* $J^P = \frac{1}{2}^+$ *baryons containing at least one charmed quark*

C'	S	Quark content	Name	Quark content	Name
1	0	cuu	Σ_c^{++}		
		c(ud + du)	Σ_c^+	c(ud − du)	Λ_c^+
		cdd	Σ_0^c		
	−1	c(us + su)	Ξ_c^+	c(us − su)	Ξ_c^+
		c(ds + sd)	Ξ_c^0	c(ds − sd)	Ξ_c^0
	−2	css	Ω_c^0		
2	0	ccu	Ω_{cc}^{++}		
		ccd	Ω_{cc}^+		
	−1	ccs	Ω_{cc}^+		

The decuplet and singlet are symmetric and antisymmetric respectively under the interchange of quarks, whereas the octet states have mixed symmetry. The lightest baryon multiplet is the $J^P = \frac{1}{2}^+$ octet (Table 1.3), which can be written as a 3×3 matrix B, analogous to M in (2.269),

$$B = \begin{pmatrix} \dfrac{\Sigma^0}{2^{1/2}} + \dfrac{\Lambda^0}{6^{1/2}} & \Sigma^+ & p \\ \Sigma^- & -\dfrac{\Sigma^0}{2^{1/2}} + \dfrac{\Lambda^0}{6^{1/2}} & n \\ \Xi^- & \Xi^0 & -\dfrac{2}{6^{1/2}}\Lambda^0 \end{pmatrix}. \tag{2.274}$$

2.5.4 SU(4)

Extension of the SU(3) scheme to include the charm quarks leads to larger multiplets. For mesons and baryons we have, respectively,

$$\mathbf{4} \otimes \bar{\mathbf{4}} = \mathbf{15} \oplus \mathbf{1},$$

and

$$\mathbf{4} \otimes \mathbf{4} \otimes \mathbf{4} = \mathbf{4} \oplus \mathbf{20} \oplus \mathbf{20} \oplus \mathbf{20}. \tag{2.275}$$

The quark content and usual nomenclature for the $J^P = 0^-$ and 1^- meson multiplets are given in Table 1.2. The twenty $J^P = \frac{1}{2}^+$ baryon states are composed of the octet of $C' = 0$ states (Table 1.3), nine states with $C' = +1$ and three with $C' = +2$. These are listed in Table 2.3, the notation being that of the Particle Data Group (1986). Note that there are both symmetric and antisymmetric states for the cud, cus and cds combinations. In addition to this multiplet $\mathbf{20}_\mathrm{M}$ of mixed symmetry, a symmetric multiplet

$\mathbf{20}_S$ with $J^P = \frac{3}{2}^+$ is also expected. This latter multiplet contains the decuplet of non-charmed baryons, which includes the Δ^{++}. Most of these states, however, remain as yet to be discovered. Inclusion of b and t quarks in the unitary symmetry scheme leads to even larger multiplets, which are likely to remain sparsely filled for some time to come.

2.5.5 SU(3) colour

The unitary symmetry group SU(3) was first used in particle physics in order to try and understand the large number of observed hadronic states. Indeed this classification led to the idea of quarks, in that the hadron states could all be built out of the fundamental $\mathbf{3}$ and $\bar{\mathbf{3}}$ representations, and that the basic states in these representations are the u, d and s quarks and antiquarks respectively. The discussion of the group SU(n), $n = 2, 3, \ldots$, in the preceding sections was developed on the hypothesis that, for example for $n = 3$, any combination of u, d and s quarks was equally good. The application of SU(3) in this way is usually referred to as $\mathrm{SU(3)}_{\mathrm{flavour}}$, in that symmetry between different quark flavours is implied. Although these ideas are useful in developing the discussion of the underlying group theory ideas, this symmetry is far from exact. The estimated quark masses range from a few MeV for u- and d-quarks, to about 5 GeV for the b-quark. Despite the fact that $\mathrm{SU}(n)_{\mathrm{flavour}}$ is badly broken, it is useful in the classification of hadronic states, not least in looking for states which cannot be accommodated within this scheme. Furthermore, the weak currents involved in the decays of $J^P = \frac{1}{2}^+$ baryons can also be described by SU(3). In this case the group symmetry properties can be used to derive relationships between these currents. Nonetheless, these applications of SU(3) are not, in general, expected to give quantitative results to better than about 5 to 10%.

The application of SU(3) in the description of the colour properties of quarks and gluons (Section 1.7) is, however, widely believed to be exact. Since each quark comes in three possible colour states (c) there are, for example, nine possible ways to make a charged pion ($\pi^+ = \mathrm{u}_c\bar{\mathrm{d}}_{c'}$, c, $c' = 1, 2, 3$). Each of these states could have a different mass. However, such a proliferation of meson states is not observed, and it is postulated that all interactions satisfy SU(3) colour exactly and that all observed hadrons are colour singlets.

The base states for SU(3) colour are column three-vectors, whose elements correspond to the possible colour states called (arbitrarily) R (red), G (green) and B (blue). The $\mathbf{3}$ and $\bar{\mathbf{3}}$ representations can be combined to give an octet (1.15) and a singlet (see (2.268)), namely

$$(\mathrm{R}\bar{\mathrm{R}} + \mathrm{G}\bar{\mathrm{G}} + \mathrm{B}\bar{\mathrm{B}})/3^{1/2}. \tag{2.276}$$

The colour singlet does not carry any colour, and is postulated not to carry any force between colour charges. Such a singlet would give rise to long range forces between baryons. In QCD the eight members of the octet are the gluons. The various SU(3) properties established in Section 2.5.3 apply equally well to $SU(3)_{colour}$. There are SU(2) subgroups and shift operators equivalent to *I*, *U* and *V*-spin; however, the physical interpretation of these operators is, of course, quite different. The colour content of a gluon can be pictured as that of a coloured quark and antiquark. Some examples of possible vertex diagrams are shown in Fig. 2.9.

Let us now consider the force between two coloured quarks or antiquarks, mediated by the exchange of a gluon. The possible gluons which can be exchanged in the scattering of two red quarks are shown in Fig. 2.10(*a*). The colour factors involved at each vertex are $g/2^{1/2}$ and $g/6^{1/2}$ for the gluons $(R\bar{R} - G\bar{G})/2^{1/2}$ and $(R\bar{R} + G\bar{G} - 2B\bar{B})/6^{1/2}$ respectively, where g is the colour charge. The strength of the interaction, in an analogous way to the electromagnetic case, is proportional to the product of these colour charges. Hence the contributions are $g^2/2$ and $g^2/6$ giving for Fig. 2.10(*a*) a total of $2g^2/3$. The colour factor C_F is conventionally defined

Fig. 2.9 Examples of possible quark and gluon vertex diagrams, together with the corresponding colour flow.

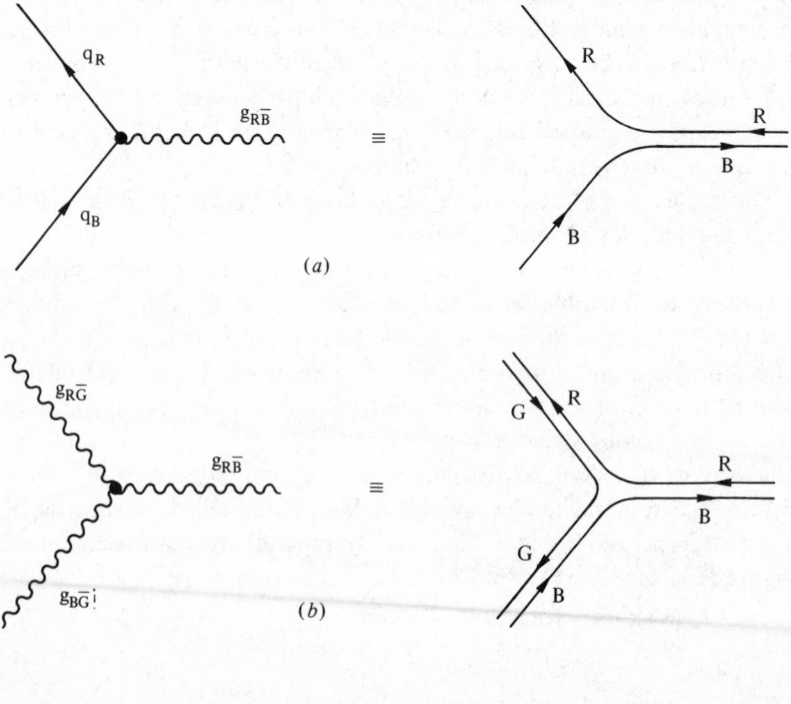

to be one-half of this coefficient; hence for Fig. 2.10(*a*) $C_F = \frac{1}{3}$. For the scattering of a red and a blue quark there is a similar diagram (Fig. 2.10(*b*)) involving the exchange of the gluon $(R\bar{R} + G\bar{G} - 2B\bar{B})/6^{1/2}$, with colour factor $C_F = (1/6^{1/2})(-2/6^{1/2})/2 = -\frac{1}{6}$. There is also an exchange diagram RB → BR (Fig. 2.10(*c*)) involving the gluon $R\bar{B}$; this has $C_F = (1)/2$. The sum of the two terms for RB is thus $\frac{1}{3}$. However, these indistinguishable amplitudes should be added only if the colour wave function is symmetric. For the antisymmetric case the amplitudes should be subtracted, giving $C_F^A = -\frac{2}{3}$. It is easily checked that other colour combinations give the same results, as indeed they must from colour symmetry.

The interaction of a red quark and antired antiquark involves the gluon exchanges shown in Fig. 2.10(*d*) and (*e*) respectively. For the two possible gluons which can be exchanged in Fig. 2.10(*d*), the strengths are $-g^2/2$ and $-g^2/6$, giving a total $C_F = (-\frac{1}{2} - \frac{1}{6})/2 = -\frac{1}{3}$. The negative sign is introduced because the colour of the antiquark has the opposite sign to that of the quark. The diagram for $R\bar{R} \to B\bar{B}$ (or, equivalently, $R\bar{R} \to G\bar{G}$) is shown in Fig. 2.10(*e*), and has $C_F = -(1)/2$. The interaction of a quark and antiquark of different colour (Fig. 2.10(*f*)) gives $C_F = (1/6^{1/2})(2/6^{1/2})/2 = \frac{1}{6}$, which is of opposite sign to that of $R\bar{R}$.

Fig. 2.10 Interactions between quarks of various colours by gluon exchange.

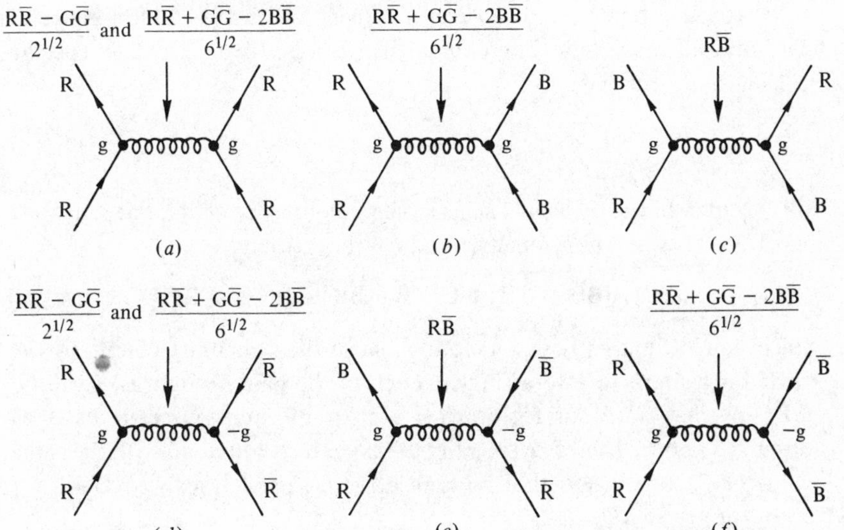

The potential between two quarks or antiquarks can, to a reasonable approximation, be written as a sum of a short range Coulomb-like term plus a long range confining term. Thus

$$V(r) = C_F \frac{g^2}{4\pi r} + kr = C_F \frac{\alpha_S}{r} + kr, \tag{2.277}$$

with $k \simeq 0.2\ \text{GeV}^2$. The strong coupling constant $\alpha_S = g^2/4\pi$. The colour factors for the short range part of the interaction can be derived from the above results. We will consider the following cases:

(i) *$q\bar{q}$ in colour singlet state*. The wave function for the colour singlet (2.276) contains equal contributions from $R\bar{R}$, $G\bar{G}$ and $B\bar{B}$. Hence it is only necessary to evaluate the $R\bar{R}$ contribution and multiply this by three. Thus, we have

$$3 \cdot \frac{1}{3^{1/2}} (R\bar{R}) \cdot \frac{1}{3^{1/2}} (R\bar{R} + G\bar{G} + B\bar{B}).$$

The colour factors for $R\bar{R} \to R\bar{R}$, $R\bar{R} \to G\bar{G}$ and $R\bar{R} \to B\bar{B}$ are $C_F = -\frac{1}{3}$, $-\frac{1}{2}$ and $-\frac{1}{2}$ respectively, giving an overall colour factor which is negative (corresponding to an attractive potential) of

$$C_F(q\bar{q} \text{ in } \mathbf{1}) = -\tfrac{4}{3}. \tag{2.278}$$

(ii) *$q\bar{q}$ in colour octet state*. The scattering, for example, of a red quark and blue antiquark was discussed above and gives $C_F = \frac{1}{6}$, corresponding to a repulsive potential. The octet members are just the gluon states of 1.15 and all these states are equivalent (this can be checked by specific calculation); thus,

$$C_F(q\bar{q} \text{ in } \mathbf{8}) = \tfrac{1}{6}. \tag{2.279}$$

(iii) *qq* in $\bar{\mathbf{3}}$ (*baryon colour singlet*). The colour part of the baryon wave function must be antisymmetric and can be written

$$(1/6^{1/2})[R(BG - GB) + G(RB - BR) + B(GR - RG)]. \tag{2.280}$$

Each pair of quarks is in a colour $\bar{\mathbf{3}}$, since the pair must couple to the remaining quark to give a singlet. Each of the pairs of quarks, e.g. BG, GB, BR, has a colour factor $C_F = -\frac{2}{3}$ for the antisymmetric case, as discussed above. There are six pairs to take into account, all with the same value of C_F, together with a normalisation factor of $\frac{1}{6}$, giving

$$C_F(qq \text{ in } \bar{\mathbf{3}}) = -\tfrac{2}{3}. \tag{2.281}$$

(iv) *qq in sextet*. The **6** representation has the symmetric form

$$\frac{1}{6^{1/2}}\left[\mathrm{RR}+\mathrm{GG}+\mathrm{BB}+\frac{1}{2^{1/2}}(\mathrm{RG}+\mathrm{GR})\right.$$
$$\left.+\frac{1}{2^{1/2}}(\mathrm{RB}+\mathrm{BR})+\frac{1}{2^{1/2}}(GB+\mathrm{BG})\right]. \quad (2.282)$$

In this case we have

$$C_{\mathrm{F}}(\mathrm{qq\ in\ }\mathbf{6})=\tfrac{1}{3}. \quad (2.283)$$

Hence, the states with net colour have a repulsive force, whereas the colour singlets have an attractive force. This is in accord with the observation that only these combinations are stable. However, this is perhaps just a hint and does not constitute a proof, as the latter requires a complete description of the hadrons.

The gluon octet can equivalently be described by the λ_a matrices (2.251), where $a=1,\ldots,8$ represents the gluon label. The matrices $T_a=\lambda_a/2$ are generally used. The commutation relations of T_a are those of (2.262). Note that the gluons are now combinations of the representation used above with, e.g., $\mathrm{R\bar{G}}=T_1+\mathrm{i}T_2$.

The vertex factor for the transition $\mathrm{q}_i\to\mathrm{q}_j+\mathrm{g}_a$ is

$$v_{\mathrm{qg}}=\mathrm{i}g(T_a)_{ij}, \quad (2.284)$$

The force between two quarks 1 and 2 depends on the expectation value of $\mathbf{T}_1\cdot\mathbf{T}_2$ with

$$V(r)=\alpha_{\mathrm{S}}(\mathbf{T}_1\cdot\mathbf{T}_2)/r. \quad (2.285)$$

If $\mathbf{T}=\mathbf{T}_1+\mathbf{T}_2$, then we have

$$\mathbf{T}_1\cdot\mathbf{T}_2=(\mathbf{T}^2-\mathbf{T}_1^2-\mathbf{T}_2^2)/2. \quad (2.286)$$

That is, $\mathbf{T}_1\cdot\mathbf{T}_2$ is expressed in terms of the quadratic Casimir operators, which are invariant under group rotations and depend only on the representations of quarks 1 and 2 and their sum. To calculate $\mathbf{T}^2$ for a specific representation it suffices to find the average value of a *single* generator over the members of the representation, and then multiply by the number of generators (8). This follows since all the generators are equivalent. Choosing, for example, the generator I_3, we must thus calculate the mean value (M) of I_3^2. For the fundamental (**3**) representation (or $\bar{\mathbf{3}}$) this is $M=\frac{1}{6}$, since the states have $I_3=\frac{1}{2}$, $-\frac{1}{2}$ and 0. For the symmetric **6** representation of $\mathbf{3}\otimes\mathbf{3}$ (see Fig. 2.8) the members have $I_3=-1$, 0, 1, $-\frac{1}{2}$, $\frac{1}{2}$ and 0, giving $M=\frac{5}{12}$. For the octet (Fig. 2.7) $M=\frac{3}{8}$. Thus $\mathbf{T}^2$, which is equal to $8M$, is as follows

$$\mathbf{T}^2(\mathbf{3}) = \tfrac{4}{3},$$

$$\mathbf{T}^2(\mathbf{6}) = \tfrac{10}{3},$$

$$\mathbf{T}^2(\mathbf{8}) = 3. \tag{2.287}$$

It is straightforward to reproduce the results (i) to (iv) above for the colour factor $C_{\mathrm{F}} = \mathbf{T}_1 \cdot \mathbf{T}_2$. For example, for $q\bar{q}$ in a colour singlet $\mathbf{T}_1 = \mathbf{T}_2 = \mathbf{3}$ and $\mathbf{T} = \mathbf{1}$ ($\mathbf{T}^2 = 0$ in this case), so using (2.286) we find $\mathbf{T}_1 \cdot \mathbf{T}_2 = -\tfrac{4}{3}$.

The colour factor for the emission of an external gluon by a quark can be calculated starting from the vertex factor (2.284). The contribution to the cross-section for the colour terms is

$$\sum_{a=1}^{8} \sum_{k=1}^{3} (T_a)_{ik}(T_a)_{kj} = C_{\mathrm{F}}\delta_{ij}. \tag{2.288}$$

Summing over possible colour states we obtain

$$3C_{\mathrm{F}} = \sum_{a=1}^{8} \mathrm{tr}(T_a T_b) = 8 \cdot \tfrac{1}{2},$$

Fig. 2.11 Diagrams for the calculation of the colour factors for (*a*) $q \to q + g$; (*b*) $g \to g + g$; and (*c*) fermion loop.

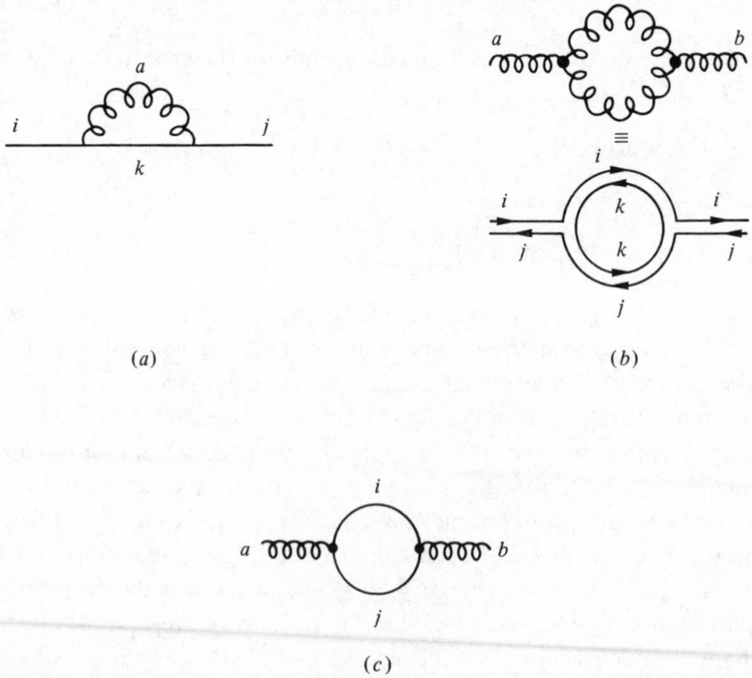

so

$$C_F = \tfrac{4}{3}. \tag{2.289}$$

Here the relationship $\mathrm{tr}(T_a T_b) = \delta_{ab}/2$ (which can easily be derived from (2.262) and (2.263)) has been used. Fig. 2.11(*a*) shows the contribution graphically, the colour factor for $g \to g + g$ (Fig. 2.11(*b*)) can be calculated by consideration of the corresponding colour loops. For gluon states *a* and *b* there are three possible internal colour loops; thus $C_F = 3$. Note that the ratio of the colour factors is

$$R_{col} = \frac{q \to q + g}{g \to g + g} = \tfrac{4}{9}. \tag{2.290}$$

A consequence is that, in a QCD 'cascade' process involving the evolution through many such vertices, gluons are preferentially produced.

Finally, for completeness, we note that the colour factors are often referred to in terms of the corresponding SU(3) representation, i.e. fundamental or adjoint. There are two Casimir terms $C_2(F) = \frac{4}{3}$ (i.e. (2.289)) and $C_2(A)$, which is given by (Fig. 2.11(*b*))

$$C_2(A)\delta_{ab} = \sum_{c,d} f_{acd} f_{bcd} = 3\delta_{ab}. \tag{2.291}$$

That is $C_2(A) = 3$, as deduced above, and corresponds to the $g \to g + g$ vertex. In addition, for fermion loops (Fig. 2.11(*c*)), we obtain a term containing a trace factor, namely

$$T_2(F)\delta_{ab} = \sum_{i,j=1}^{3} T^a_{ji} T^b_{ij} = \delta_{ab}/2. \tag{2.292}$$

In general, for n_f fermion families, we have $T_2(F) = n_f/2$.

3

Wave equations, propagators and fields

The description of scattering processes using quantum mechanics, as discussed in Chapter 2, has two obvious deficiencies. Firstly, it is not formulated in a relativistically invariant manner. The particle spin, which was incorporated by having a multicomponent wave function, must also be treated in a Lorentz invariant way. Secondly, it deals with scattering processes where the incident particle is scattered by a force represented by a fixed potential (or effective potential). However, in high energy scattering processes particles can be created or annihilated. Fermions must be created or destroyed as particle–antiparticle pairs, whereas any number of bosons may be involved. Thus the formalism needed to discuss the interaction of particles must include these points. Indeed, great care must be taken as to the actual definition of a particle and its properties. For example, a 'free' electron can dissociate (see Fig. 3.1) into an electron and a photon, or more complex states involving loops of e^+e^- (or any other charged particles). A 'dressed' particle, with which we deal experimentally, has different properties to the 'bare' particle, the wave equations of which are discussed in this chapter.

The fundamental particles in the standard electroweak model are the spin $\frac{1}{2}$ leptons and quarks, the spin 1 gauge bosons $W^\pm$, γ and Z^0 (which propagate the forces), and the spin 0 Higgs bosons (which are necessary to give the $W^\pm$, Z^0 and the fermions their masses). In addition to the free particle wave equations, which are introduced in the order of increasing spin (and hence complexity), this chapter also deals with the methods used to describe the propagation of particles (and hence forces) from one space-time point to another. The description of a multiparticle system, in terms of the field theory creation and annihilation operators, is also discussed. A rigorous and general field theory approach is not, however, pursued. Some field theory ideas are introduced, however, mainly with the aim of 'underwriting' the more intuitive approach adopted.

3.1 SPIN 0 PARTICLES

3.1.1 Klein–Gordon equation for spin 0 particles

The simplest free-particle relativistic wave equation, the Klein–Gordon equation, can be constructed from the energy ($\hat{E}$) and momentum ($\hat{p}$) operators, in a way analogous to that used for the Schrödinger equation (2.47), but using $\hat{E}^2 = \hat{p}^2 + m^2$ rather than the non-relativistic $\hat{E} = \hat{p}^2/2m$. Thus

$$(\hat{E}^2 - \hat{p}^2)\phi(x) = -\left(\frac{\partial^2}{\partial t^2} - \nabla^2\right)\phi(x) = -\Box^2\phi(x) = m^2\phi(x). \quad (3.1)$$

A free particle solution, which can be checked by substitution, is of the form

$$\phi(x) = N\exp(-\mathrm{i}p\cdot x) = N\exp[-\mathrm{i}(Et - \mathbf{p}\cdot\mathbf{x})], \quad (3.2)$$

where N is a (complex) normalisation constant. The wave function $\phi(x)$ has only one component. In general, the description of a particle of spin s requires $2s + 1$ components. Thus $\phi(x)$ is suitable for the description of a spin 0 particle. The covariance requirements of (3.1) imply that ϕ must be a scalar or pseudoscalar quantity, with thus no preferred direction. Note that, since $E^2 = p^2 + m^2$, this does not fix the sign of E, which can be positive or negative, and thus constitutes a potential problem.

A further problem arises when we consider the probability and current densities. Proceeding in a manner similar to that used for the Schrödinger equation (Section 2.2.3), we calculate [ϕ^*(eqn. 3.1) – ϕ(eqn. 3.1)*] and

Fig. 3.1 Virtual states of a free electron.

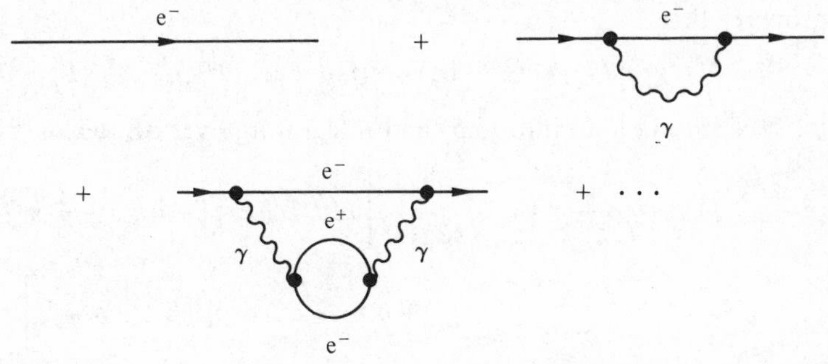

cast this in the form (2.180)

$$\partial_\mu j^\mu = \frac{\partial \rho}{\partial t} + \nabla \cdot \mathbf{j} = 0,$$

with the four-current $j^\mu = (\rho, \mathbf{j})$ given by

$$j^\mu = \mathrm{i}(\phi^* \, \partial^\mu \phi - \phi \, \partial^\mu \phi^*),$$

$$\rho = \mathrm{i}\left(\phi^* \frac{\partial \phi}{\partial t} - \phi \frac{\partial \phi^*}{\partial t}\right), \qquad \mathbf{j} = \mathrm{i}(\phi \nabla \phi^* - \phi^* \nabla \phi). \tag{3.3}$$

For the plane wave solution (3.2), this gives

$$j^\mu = 2|N|^2 p^\mu = 2|N|^2 (E, \mathbf{p}). \tag{3.4}$$

The sign of the probability density ρ is the same as the sign of E and can thus be positive or negative. In general, it can be seen from the form of (3.3) that ρ is not necessarily positive definite, and hence cannot, as such, be interpreted as the probability density.

3.1.2 Creation and annihilation operators

In 1934 Pauli and Weisskopf proposed a solution to the problem of negative values of ρ by suggesting that ϕ should be treated as a field operator rather than a single-particle wave function. As outlined below, ϕ can be used to describe a system containing both positive and negative charges and ρ interpreted as the net charge density function.

The Fourier transform of the free particle solution ($\phi \sim \exp(-\mathrm{i}p \cdot x)$) in momentum space, namely

$$\phi(x) = \frac{1}{(2\pi)^2} \int_{-\infty}^{\infty} \mathrm{d}^4 p (4\pi E_p)^{1/2} a(p) \, \delta(p^2 - m^2) \exp(-\mathrm{i}p \cdot x), \tag{3.5}$$

is also a solution. This can be seen by operating on $\phi(x)$ in (3.5) by $(\Box^2 + m^2)$, and using the property that the result contains a factor $y\delta(y)$ which yields zero on integration. The factor containing $E_p = +(\mathbf{p}^2 + m^2)^{1/2}$ is for later convenience. The Dirac δ-function (see Appendix D) has the property that

$$\delta(p^2 - m^2) = [\delta(E - E_p) + \delta(E + E_p)]/(2E_p), \qquad E_p > 0, \tag{3.6}$$

so, substituting (3.6) into (3.5) and integrating over $\mathrm{d}E$, yields

$$\phi(x) = \frac{1}{(2\pi)^{3/2}} \int_{-\infty}^{\infty} \frac{\mathrm{d}^3 p}{(2E_p)^{1/2}} \Big[a(\mathbf{p}, E_p) \exp[-\mathrm{i}(E_p t - \mathbf{p} \cdot \mathbf{x})] + a(\mathbf{p}, -E_p) \exp[\mathrm{i}(E_p t + \mathbf{p} \cdot \mathbf{x})] \Big]. \tag{3.7}$$

Changing variables from $\mathbf{p}$ to $-\mathbf{p}$ in the second term gives $a(-\mathbf{p}, -E_p) \times \exp(ip \cdot x)$. Thus, writing $a(\mathbf{p}) = a(\mathbf{p}, E_p)$ and $a(-\mathbf{p}) = a(-\mathbf{p}, -E_p)$, (3.7) becomes

$$\phi(x) = \frac{1}{(2\pi)^{3/2}} \int_{-\infty}^{\infty} \frac{d^3p}{(2E_p)^{1/2}} \left[a(\mathbf{p}) \exp(-ip \cdot x) + a(-\mathbf{p}) \exp(ip \cdot x) \right]. \quad (3.8)$$

If $\phi(x)$ is Hermitian, i.e. $\phi^\dagger(x) = \phi(x)$, then from (3.8) we obtain the conditions that $a^\dagger(\mathbf{p}) = a(-\mathbf{p})$ and $a^\dagger(-\mathbf{p}) = a(\mathbf{p})$, hence

$$\begin{aligned} \phi(x) &= \frac{1}{(2\pi)^{3/2}} \int_{-\infty}^{\infty} \frac{d^3p}{(2E_p)^{1/2}} \left[a(\mathbf{p}) \exp(-ip \cdot x) + a^\dagger(\mathbf{p}) \exp(ip \cdot x) \right] \\ &= \phi^{(+)}(x) + \phi^{(-)}(x), \end{aligned} \quad (3.9)$$

where $\phi^{(+)}$ and $\phi^{(-)}$ correspond to the positive and negative frequency components $\exp(-ip \cdot x)$ and $\exp(ip \cdot x)$ respectively.

It is often useful to normalise to a box of side L and volume V, with a periodic boundary condition giving discrete momenta

$$\mathbf{p} = (n_x, n_y, n_z) 2\pi / L. \quad (3.10)$$

The discrete case can be obtained by the replacement

$$\frac{1}{(2\pi)^{3/2}} \int d^3p \to \frac{1}{V^{1/2}} \sum_p, \qquad \delta(\mathbf{p} - \mathbf{p}') \to \delta_{pp'}. \quad (3.11)$$

Thus, (3.9) can be written

$$\phi(x) = \sum_p \frac{1}{(2E_p V)^{1/2}} \left[a(\mathbf{p}) \exp(-ip \cdot x) + a^\dagger(\mathbf{p}) \exp(ip \cdot x) \right]. \quad (3.12)$$

In Chapter 2 the classical Lagrangian and Euler–Lagrange equation for an ensemble of particles with generalised coordinates q_i was discussed. In classical field theory the discrete coordinates q_i are replaced by field functions $\phi(x)$, where x is a continuous variable. This can be accomplished in a covariant way and, as shown below, conserved quantities (e.g. energy, momentum) are associated with the disturbance represented by $\phi(x)$. In quantum field theory $\phi(x)$ no longer represents ordinary functions, but rather linear field operators which act on the state vectors of an ensemble of particles and can change these quantum states. These operators satisfy certain commutation relations and, furthermore, the eigenvalues of Hermitian operators constructed from the field functions (e.g. E, $\mathbf{p}$) represent observable quantities. In the rest of this section this process of *second quantisation* is outlined for the scalar field. The approach adopted here follows that of Muirhead (1965) to which the interested reader is referred for further details. An alternative, and ultimately equivalent method, is to start with the commutation relations and deduce the properties of the operators a and $a^\dagger$; see, e.g. Bjorken and Drell, 1965, Vol. 2.

Consider a normalised state vector $|n_p\rangle$ in an occupation number space containing n bosons with momentum $\mathbf{p}$. In the second quantisation process, the operators $a(\mathbf{p})$ and $a^\dagger(\mathbf{p})$ in (3.9) and (3.12) are defined to be annihilation and creation operators as follows

$$a^\dagger(\mathbf{p})|n_p\rangle = (n+1)_p{}^{1/2}|(n+1)_p\rangle,$$
$$a(\mathbf{p})|n_p\rangle = (n_p)^{1/2}|(n-1)_p\rangle. \tag{3.13}$$

Successive application of these operators gives

$$a(\mathbf{p})a^\dagger(\mathbf{p})|n_p\rangle = (n+1)_p|n_p\rangle,$$
$$a^\dagger(\mathbf{p})a(\mathbf{p})|n_p\rangle = n_p|n_p\rangle. \tag{3.14}$$

However, for $\mathbf{p} \neq \mathbf{p}'$

$$a(\mathbf{p})a^\dagger(\mathbf{p}')|n_p, n_{p'}\rangle = [n_p(n+1)_{p'}]^{1/2}|(n-1)_p, (n+1)_{p'}\rangle,$$
$$a^\dagger(\mathbf{p}')a(\mathbf{p})|n_p, n_{p'}\rangle = [(n+1)_{p'}n_p]^{1/2}|(n-1)_p, (n+1)_{p'}\rangle, \tag{3.15}$$

Thus, a and $a^\dagger$ satisfy the commutation relations

$$[a(\mathbf{p}), a^\dagger(\mathbf{p}')] = \delta_{pp'} \text{ or } \delta^3(\mathbf{p}-\mathbf{p}'),$$
$$[a(\mathbf{p}), a^\dagger(\mathbf{p}')] = [a^\dagger(\mathbf{p}), a^\dagger(\mathbf{p}')] = 0. \tag{3.16}$$

From (3.14) it can be seen that the operator $N(\mathbf{p}) = a^\dagger(\mathbf{p})a(\mathbf{p})$ gives the number of particles in the state with momentum $\mathbf{p}$. This *occupation number operator* $N(\mathbf{p})$ is the product of an operator and its Hermitian conjugate, and so is positive definite with lowest eigenvalue zero. Therefore a state $|0\rangle$, the so-called *vacuum* state, must exist with

$$a(\mathbf{p})|0\rangle = 0, \qquad N(\mathbf{p})|0\rangle = 0. \tag{3.17}$$

States with one, two (etc.) particles can be built from the vacuum by successive application of the creation operator $a^\dagger(\mathbf{p})$, using (3.13), giving

$$|1_p\rangle = a^\dagger(\mathbf{p})|0\rangle,$$

$$|2_p\rangle = \frac{1}{2^{1/2}}a^\dagger(\mathbf{p})|1_p\rangle = \frac{1}{2^{1/2}}a^\dagger(\mathbf{p})a^\dagger(\mathbf{p})|0\rangle,$$

$$|n_p\rangle = \frac{1}{(n_p!)^{1/2}}[a^\dagger(\mathbf{p})]^{n_p}|0\rangle. \tag{3.18}$$

The scalar particles arising from the quantisation obey symmetric, or Bose–Einstein, statistics.

The $\phi^{(+)}(x)$ and $\phi^{(-)}(x)$ parts of (3.9), which are useful for practical calculations, satisfy

$$\phi^{(+)}(x)|0\rangle = 0, \qquad \langle 0|\phi^{(-)}(x) = 0. \tag{3.19}$$

The application $\phi^{(-)}(x)|0\rangle$ can be shown, by consideration of the Fourier transformation of $a^\dagger(\mathbf{p})$, to correspond to the creation of a particle within a distance of approximately $1/m$ (Compton wavelength) of the space point $\mathbf{x}$.

3.1.3 Lagrangian for a scalar field

The Lagrangian and Euler–Lagrange equation for an ensemble of classical particles was discussed in Section 2.2.1. For a relativistically invariant theory, the action integral A must be a Lorentz scalar. Thus, if we define a four-dimensional action integral

$$A = \int_\Omega \mathrm{d}^4x \mathscr{L}(\phi, \partial\phi/\partial x^\mu), \tag{3.20}$$

then $\mathscr{L}$, the Lagrangian density, should be invariant under proper Lorentz transformations (since d^4x is invariant). Furthermore, $\mathscr{L}$ should not contain second or higher-order derivatives, otherwise the theory would have negative metric states. The requirement $\delta A = 0$, for $\phi \to \phi + \delta\phi$ with $\delta\phi$ having the boundary condition that it is equal to zero everywhere on the surface of Ω, gives, writing $\alpha_\mu = \partial_\mu\phi = \partial\phi/\partial x^\mu$,

$$\begin{aligned} \delta A &= \int_\Omega \mathrm{d}^4x\left(\frac{\partial\mathscr{L}}{\partial\phi}\,\delta\phi + \frac{\partial\mathscr{L}}{\partial\alpha_\mu}\,\delta\alpha_\mu\right) \\ &= \int_\Omega \mathrm{d}^4x\left(\frac{\partial\mathscr{L}}{\partial\phi}\,\delta\phi + \frac{\partial\mathscr{L}}{\partial\alpha_\mu}\frac{\partial(\delta\phi)}{\partial x^\mu}\right) = 0. \end{aligned} \tag{3.21}$$

Integrating the second term by parts, this expression gives

$$\int_\Omega \mathrm{d}^4x\left[\frac{\partial\mathscr{L}}{\partial\phi} - \frac{\partial}{\partial x^\mu}\frac{\partial\mathscr{L}}{\partial\alpha_\mu}\right]\delta\phi = 0, \tag{3.22}$$

where the other term in the partial integration, which is a 'surface' term in four-dimensions, is zero because $\delta\phi$ is everywhere zero on the boundary. Thus (3.22) gives the Euler–Lagrange field equation

$$\frac{\partial\mathscr{L}}{\partial\phi} - \frac{\partial}{\partial x^\mu}\left(\frac{\partial\mathscr{L}}{\partial\alpha_\mu}\right) = 0. \tag{3.23}$$

In order to derive the field equations, further conditions on $\mathscr{L}$ are usually required. It is assumed that the field equations are linear in $\phi(x)$, so that $\mathscr{L}$ should contain only bilinear combinations of ϕ and $\partial\phi/\partial x^\mu$. It is further assumed that the state of the field and its interactions at x are completely specified by ϕ and $\partial\phi/\partial x^\mu$ at x; that is, the fields are *local*. For a scalar particle $\mathscr{L}$ can be constructed as follows

$$\mathscr{L} = \frac{1}{2}\left(\frac{\partial\phi}{\partial x^\mu}\frac{\partial\phi}{\partial x_\mu} - m^2\phi^2\right) = \tfrac{1}{2}(\partial_\mu\phi\,\partial^\mu\phi - m^2\phi^2)$$
$$= \tfrac{1}{2}(\alpha_\mu\alpha^\mu - m^2\phi^2). \tag{3.24}$$

Application of (3.23) on $\mathscr{L}$ gives the Klein–Gordon equation.

3.1.4 Invariance properties of the Lagrangian

If the Lagrangian, or more generally the action integral, is invariant under a continuous group of transformations, then there exist certain conserved quantities for the associated fields (*Noether's theorem*). Consider the infinitesimal displacement

$$x_\mu \to x'_\mu = x_\mu + \varepsilon_\mu, \tag{3.25}$$

which changes the Lagrangian $\mathscr{L}(\phi, \partial\alpha^\mu)$ to $\mathscr{L}'$ with

$$\delta\mathscr{L} = \mathscr{L}' - \mathscr{L} = \varepsilon_\mu\,\partial\mathscr{L}/\partial x_\mu$$
$$= \frac{\partial\mathscr{L}}{\partial\phi}\,\delta\phi + \frac{\partial\mathscr{L}}{\partial\alpha^\mu}\,\delta\alpha^\mu. \tag{3.26}$$

In the latter equation it is assumed that $\mathscr{L}$ is invariant under translations, with no explicit space dependence. Noting that

$$\delta\phi = \phi(x+\varepsilon) - \phi(x) = \varepsilon_\nu\frac{\partial\phi}{\partial x_\nu} = \varepsilon_\nu\alpha^\nu, \tag{3.27}$$

and using (3.23) and (3.27) in (3.26), gives

$$\delta\mathscr{L} = \frac{\partial}{\partial x_\mu}\frac{\partial\mathscr{L}}{\partial\alpha^\mu}\,\varepsilon_\nu\alpha^\nu + \frac{\partial\mathscr{L}}{\partial\alpha^\mu}\,\delta\alpha^\mu = \varepsilon_\mu\frac{\partial\mathscr{L}}{\partial x_\mu}. \tag{3.28}$$

This can be rewritten as

$$\frac{\partial}{\partial x_\mu}\left[-\varepsilon_\mu\mathscr{L} + \frac{\partial\mathscr{L}}{\partial\alpha^\mu}\,\varepsilon_\nu\alpha^\nu\right] = 0, \tag{3.29}$$

which is of the form

$$\partial J_{\mu\nu}/\partial x_\mu = 0, \tag{3.30}$$

where, noting that $\varepsilon_\mu = g_{\mu\nu}\varepsilon^\nu$, $\varepsilon_\nu\alpha^\nu = \varepsilon^\nu\alpha_\nu$, and that ε^ν is arbitrary,

$$J_{\mu\nu} = -g_{\mu\nu}\mathscr{L} + \frac{\partial\mathscr{L}}{\partial\alpha^\mu}\,\alpha_\nu. \tag{3.31}$$

The conserved quantities, corresponding to the conservation equation (3.30), are

$$P_\nu = \int d^3x J_{0\nu} = \int d^3x \left(\frac{\partial \mathscr{L}}{\partial \alpha^0} \alpha_\nu - g_{0\nu} \mathscr{L} \right), \tag{3.32}$$

with

$$\frac{\partial P_\nu}{\partial t} = \int d^3x \frac{\partial J_{0\nu}}{\partial x_0} = -\int d^3x \frac{\partial J_{k\nu}}{\partial x_k} = 0, \tag{3.33}$$

where use has been made of (3.30), and the term on the right can be replaced by a surface integral which gives zero contribution at infinity.

Consider, as an example, the conservation of energy and momentum associated with the field. This requires a definition of the Hamiltonian density $\mathscr{H}$ of the field. We can write

$$H = \int d^3x \mathscr{H} = \int d^3x \left(\frac{\partial \mathscr{L}}{\partial \alpha^0} \alpha_0 - \mathscr{L} \right) = \int d^3x J_{00}, \tag{3.34}$$

which has a form similar to that for a mechanical system; $H = p_i \dot{q}_i - L$. The equivalent momentum conjugate of the field is $\partial \mathscr{L}/\partial \alpha^0 = \partial \mathscr{L}/\partial(\partial \phi/\partial t) = \Pi(x)$ (cf $p_i = \partial L/\partial \dot{q}_i$). Thus we may identify P_v as the energy-momentum four-vector associated with the field.

Additional conservation laws, associated with internal symmetries, arise if $\mathscr{L}$ is invariant under local transformations of the type

$$\phi(x) \to \exp(-i\varepsilon)\phi(x) = \phi(x) - i\varepsilon\phi(x), \qquad \text{for } \varepsilon \to 0, \tag{3.35}$$

where ε is real and independent of x. Equation (3.35) thus represents a *global phase transformation*. Proceeding in a similar fashion to the previous example, with $\delta\phi = -i\varepsilon\phi$,

$$\begin{aligned} \delta \mathscr{L} &= \frac{\partial \mathscr{L}}{\partial \phi} \delta\phi + \frac{\partial \mathscr{L}}{\partial \alpha_\mu} \delta\alpha_\mu \\ &= -i\varepsilon \frac{\partial}{\partial x^\mu} \left(\frac{\partial \mathscr{L}}{\partial \alpha_\mu} \phi \right) = 0. \end{aligned} \tag{3.36}$$

Hence, there is again a conserved current

$$\partial J^\mu / \partial x^\mu = 0, \tag{3.37}$$

with

$$J^\mu = -i \frac{\partial \mathscr{L}}{\partial \alpha_\mu} \phi, \tag{3.38}$$

and a corresponding conserved 'charge' for each internal symmetry

$$Q = \int d^3x J^0 = -i \int d^3x \frac{\partial \mathscr{L}}{\partial \alpha_0} \phi, \qquad \frac{\partial Q}{\partial t} = 0. \tag{3.39}$$

The above considerations apply to a relativistically invariant, but classical field, ϕ. In quantum field theory, the field functions $\phi(x)$ become *operators* acting on the state vectors. As an example, let us consider the quantum field operator for the four-momentum. For the scalar field this can be derived by evaluating the effect of the operator P_ν (3.32) on the scalar Lagrangian (3.24), expressed in terms of the field operator (3.12). Thus the *energy operator* P_0 is

$$P_0 = \int d^3x \left(\frac{\partial \mathscr{L}}{\partial \alpha^0} \alpha_0 - \mathscr{L} \right) = \int d^3x \mathscr{H}(\alpha_0, \phi). \tag{3.40}$$

For $\mathscr{L}$ given by (3.24), and noting that $\alpha_0 = \partial\phi/\partial t$, $\mathscr{H}$ can be written,

$$\mathscr{H} = \tfrac{1}{2}(\alpha_0^2 + \nabla^2\phi + m^2\phi^2). \tag{3.41}$$

Using (3.41) and ϕ from (3.12) gives

$$\begin{aligned} P_0 = \frac{1}{2} \sum_{p,p'} \int \frac{d^3x}{2VE_p} \Big\{ & (-E_pE_{p'} - \mathbf{p}\cdot\mathbf{p}')[a(\mathbf{p}) \exp(-ip\cdot x) - a^\dagger(\mathbf{p}) \exp(ip\cdot x)] \\ & \times [a(\mathbf{p}') \exp(-ip'\cdot x) - a^\dagger(\mathbf{p}') \exp(ip'\cdot x)] \\ & + m^2[a(\mathbf{p}) \exp(-ip\cdot x) + a^\dagger(\mathbf{p}) \exp(ip\cdot x)] \\ & \times [a(\mathbf{p}') \exp(-ip'\cdot x) + a^\dagger(\mathbf{p}') \exp(ip'\cdot x)] \Big\}. \end{aligned} \tag{3.42}$$

The term in curly brackets can be expressed as

$$\begin{aligned} \{\ \} = \{ & (-E_pE_{p'} - \mathbf{p}\cdot\mathbf{p}' + m^2) a(\mathbf{p}) a(\mathbf{p}') \exp[-i(p+p')\cdot x] \\ & + (E_pE_{p'} + \mathbf{p}\cdot\mathbf{p}' + m^2) a(\mathbf{p}) a^\dagger(\mathbf{p}') \exp[-i(p-p')\cdot x] \\ & + (E_pE_{p'} + \mathbf{p}\cdot\mathbf{p}' + m^2) a^\dagger(\mathbf{p}) a(\mathbf{p}') \exp[i(p-p')\cdot x] \\ & + (-E_pE_{p'} - \mathbf{p}\cdot\mathbf{p}' + m^2) a^\dagger(\mathbf{p}) a^\dagger(\mathbf{p}') \exp[i(p+p')\cdot x] \}. \end{aligned} \tag{3.43}$$

Integration over d^3x yields zero for the first and fourth of these terms, since they involve contributions of the type

$$\int \frac{d^2x}{V} \exp[i(\mathbf{p}+\mathbf{p}')\cdot\mathbf{x}] = \delta_{\mathbf{p},-\mathbf{p}'},$$

and so the requirement that $\mathbf{p}' = -\mathbf{p}$ gives

$$(-E_pE_{p'} - \mathbf{p}\cdot\mathbf{p}' + m^2) = (-E_p^2 + \mathbf{p}^2 + m^2) = 0.$$

The corresponding integration of the second and third terms involves their evaluation with $\mathbf{p}' = \mathbf{p}(E_{p'} = E_p)$, so that (3.42) simplifies to

$$P_0 = \frac{1}{2}\sum_p \frac{1}{2E_p}(E_p^2 + \mathbf{p}^2 + m^2)[a(\mathbf{p})a^\dagger(\mathbf{p}) + a^\dagger(\mathbf{p})a(\mathbf{p})]$$

$$= \frac{1}{2}\sum_p E_p[2N(\mathbf{p}) + 1] = \sum_p E_p[N(\mathbf{p}) + \tfrac{1}{2}], \tag{3.44}$$

where the operators a and $a^\dagger$ have been expressed in terms of the occupation number operator $N(\mathbf{p})$.

The expectation value of P_0 for a state $|\psi\rangle$, consisting of n_1 particles with energy E_1, n_2 with E_2, etc., is

$$E = \langle H \rangle = \langle \psi | \sum_p E_p[N(\mathbf{p}) + \tfrac{1}{2}] | \psi \rangle = \sum_p E_p(n_p + \tfrac{1}{2}). \tag{3.45}$$

The momentum operator $\mathbf{P}$ can be analysed in similar fashion, giving

$$\mathbf{P} = \sum_p \mathbf{p}[N(\mathbf{p}) + \tfrac{1}{2}] = \sum_p \mathbf{p}N(\mathbf{p}),$$

$$\langle \mathbf{P} \rangle = \sum_p n_p \mathbf{p}. \tag{3.46}$$

Note that, even if there are no particles in the field, the energy operator has a contribution from the term $\sum_p \frac{1}{2}E_p$. This term is called the *zero point energy* and, when summed over all possible states, gives an infinite energy. If we exclude gravity from our discussion, then this infinity, as with others which arise in field theory, does not correspond to a measurable quantity. Only changes in energy, and not the absolute value, are measurable so that the zero point energy can be subtracted off and ignored. The corresponding term in the momentum sum gives zero when summed over all directions. A more detailed analysis shows that the infinite energy term of the vacuum can be removed by ordering all products of operators such that the annihilation operators appear after (i.e. to the right of) creation operators (normal ordering). More generally, however, these infinities are a problem in field theory.

3.1.5 The charged scalar field

The Hermitian field, outlined in Section 3.1.2, has only one component and is thus suitable only for the description of neutral scalar particles, e.g. the π^0 or the neutral Higgs particle H^0. The arguments outlined above indicate that the charge of the particle should be related to an internal symmetry of $\mathscr{L}$. This can be achieved (see (3.35) and subsequent discussion) if $\phi(x)$ is non-Hermitian (i.e. complex). If ϕ_1 and ϕ_2 are both Hermitian (i.e. real) then we can construct

$$\phi = \frac{1}{2^{1/2}}(\phi_1 + i\phi_2), \qquad \phi^\dagger = \frac{1}{2^{1/2}}(\phi_1 - i\phi_2), \tag{3.47}$$

where ϕ and $\phi^\dagger$ each satisfy the Klein–Gordon equation

$$(\Box^2 + m^2)\phi = 0, \qquad (\Box^2 + m^2)\phi^\dagger = 0, \tag{3.48}$$

and the corresponding Lagrangian is

$$\mathscr{L} = \frac{\partial \phi^\dagger}{\partial x_\mu}\frac{\partial \phi}{\partial x^\mu} - m^2\phi^\dagger\phi. \tag{3.49}$$

Equations (3.48) can be derived from (3.49) using the Euler–Lagrange equation.

Each of the solutions ϕ_1 and ϕ_2 has a field operator equation of the form of (3.9), with corresponding operators $a_1(\mathbf{p})$, $a_1^\dagger(\mathbf{p})$, $a_2(\mathbf{p})$ and $a_2^\dagger(\mathbf{p})$, with a_1 and a_2 each satisfying commutation relations of the type (3.16). Introducing the linear combinations

$$a(\mathbf{p}) = \frac{1}{2^{1/2}}[a_1(\mathbf{p}) + ia_2(\mathbf{p})], \qquad b^\dagger(\mathbf{p}) = \frac{1}{2^{1/2}}[a_1^\dagger(\mathbf{p}) + ia_2^\dagger(\mathbf{p})], \tag{3.50}$$

and using (3.47), we can write

$$\begin{aligned}\phi(x) &= \frac{1}{(2\pi)^{3/2}}\int_{-\infty}^{\infty}\frac{d^3p}{(2E_p)^{1/2}}[a(\mathbf{p})\exp(-ip\cdot x) + b^\dagger(\mathbf{p})\exp(ip\cdot x)],\\ \phi^\dagger(x) &= \frac{1}{(2\pi)^{3/2}}\int_{-\infty}^{\infty}\frac{d^3p}{(2E_p)^{1/2}}[a^\dagger(\mathbf{p})\exp(ip\cdot x) + b(\mathbf{p})\exp(-ip\cdot x)].\end{aligned} \tag{3.51}$$

The commutation relations for $a(\mathbf{p})$ and $b(\mathbf{p})$ can be easily derived and are

$$\begin{aligned}&[a(\mathbf{p}), a^\dagger(\mathbf{p}')] = [b(\mathbf{p}), b^\dagger(\mathbf{p}')] = \delta^3(\mathbf{p} - \mathbf{p}'),\\ &[a(\mathbf{p}), b(\mathbf{p}')] = [a^\dagger(\mathbf{p}), b^\dagger(\mathbf{p}')] = [a(\mathbf{p}), b^\dagger(\mathbf{p}')] = [a^\dagger(\mathbf{p}), b(\mathbf{p}')] = 0.\end{aligned} \tag{3.52}$$

Thus $a(\mathbf{p})$ and $b(\mathbf{p})$ satisfy the same algebra as $a_1(\mathbf{p})$ and $a_2(\mathbf{p})$ and hence we can define corresponding occupation number operators which will have the same integer eigenvalues, namely

$$N^{(+)}(\mathbf{p}) = a^\dagger(\mathbf{p})a(\mathbf{p}), \qquad N^{(-)}(\mathbf{p}) = b^\dagger(\mathbf{p})b(\mathbf{p}). \tag{3.53}$$

In terms of these, the four-momentum operator can be written

$$P_\mu = \sum_p p_\mu[N^{(+)}(\mathbf{p}) + N^{(-)}(\mathbf{p})].$$

We will now justify the use of the notation $N^{(+)}$ and $N^{(-)}$ for the occupation numbers. For complex solutions of the Klein–Gordon equation, (3.36) becomes

$$\partial\mathscr{L} = -\mathrm{i}\varepsilon \frac{\partial}{\partial x^\mu}\left(\frac{\partial\mathscr{L}}{\partial\alpha_\mu}\phi - \frac{\partial\mathscr{L}}{\partial\alpha_\mu^\dagger}\phi^\dagger\right) = 0. \tag{3.54}$$

This gives, for $\mathscr{L}$ from (3.49), a conserved current of the form (3.3). Substituting for ϕ and $\phi^\dagger$ from (3.51) it can be shown, with a derivation similar to that above for energy operator, that the total charge

$$\begin{aligned} Q &= \int \mathrm{d}^3x\rho = \mathrm{i}\int \mathrm{d}^3x\left(\phi^\dagger \frac{\partial\phi}{\partial t} - \phi\frac{\partial\phi^\dagger}{\partial t}\right) \\ &= \int \mathrm{d}^3p[N^{(+)}(\mathbf{p}) - N^{(-)}(\mathbf{p})], \qquad \text{or} \qquad \sum_p [N^{(+)}(\mathbf{p}) - N^{(-)}(\mathbf{p})]. \end{aligned} \tag{3.55}$$

That is, the positive and negative frequency states carry quantum numbers with eigenvalues $+1$ and -1 respectively. In order to give a physical significance to this charge Q, the coupling of the current j^μ to the electromagnetic field must be established. We may identify the states with positive eigenvalues of Q as particles (e.g. π^+) and those with negative values of Q as antiparticles (e.g. π^-). Thus, ρ in equation (3.3) may be identified with the net charge density (Pauli and Weisskopf, 1934). In this case the appearance of negative values of ρ is no longer a problem (as it was in trying to interpret it as the probability to find a particle). Charge conservation is thus a consequence of the invariance of the Lagrangian under global phase transformations. Finally we note that the above approach to internal symmetries can be applied to quantum numbers other than electric charge.

3.1.6 Propagator theory

The aim of a quantum field theory is to provide a framework in which the interactions of the basic field particles can be described. In the formal approach to field theory the route to this goal is rather lengthy. The end product is that the interactions of particles can be described in terms of Feynman diagrams, which describe the propagation of the particles involved, and their interactions, in space-time. A non-rigorous, but more intuitive, approach using Feynman diagrams directly is adopted here. The formalism needed to describe the propagation of particles is introduced in this section. After having established the techniques required to make calculations using Feynman diagrams, the relationship to the more formal approach is described briefly in Section 4.7. The reader interested in more details on the formal approach is referred to Itzykson and Zuber (1980), Mandl and Shaw (1984) or to many of the other excellent texts on field theory.

In scattering theory we desire to discover the space–time development of a system.† Suppose that, in non-relativistic quantum mechanics, the solution $\psi(x) = \psi(t, \mathbf{x})$ is known at some time t. The solution at $x' = (t', \mathbf{x}')$, with $t' > t$, can be written, using Huygens' principle, in the form

$$\psi(t', \mathbf{x}') = \mathrm{i} \int \mathrm{d}^3x G(t', \mathbf{x}'; t, \mathbf{x}) \psi(t, \mathbf{x}) \qquad t' > t, \tag{3.56}$$

where the propagator or *Green's function* $G(x'; x)$ gives the strength of the wave propagation from x to x'. Let us denote ψ_0 to be the wave function, H_0 the Hamiltonian and G_0 the Green's function for the free particle case. Suppose a potential $V(x_1)$ $(x_1 = t_1, \mathbf{x}_1)$ is turned on for a brief period $\mathrm{d}t_1$ around t_1. For $t > t_1$, this gives an additional source of waves $\Delta\psi(x_1)$ compared to the free particle case. Now $\psi(x_1)$ satisfies

$$\mathrm{i} \frac{\partial \psi(x_1)}{\partial t} = [H_0 + V(x_1)]\psi(x_1), \tag{3.57}$$

so $\Delta\psi(x_1)$ is obtained by integrating (3.57), which gives

$$\Delta\psi(x_1) = -\mathrm{i} V(x_1) \psi_0(x_1)\, \mathrm{d}t_1. \tag{3.58}$$

The development at later times, using (3.56), is

$$\begin{aligned} \Delta\psi(x') &= \int \mathrm{d}^3x_1 G_0(x'; x_1) V(x_1) \psi_0(x_1)\, \mathrm{d}t_1 \\ &= \mathrm{i} \int \mathrm{d}^3x \int \mathrm{d}^4x_1 G_0(x'; x_1) V(x_1) G_0(x_1; x) \psi_0(x), \end{aligned} \tag{3.59}$$

where the second line is obtained by substituting $\psi_0(x_1)$ from (3.56) and writing $\mathrm{d}^4x_1 = \mathrm{d}t_1 \mathrm{d}^3x_1$. Using this expression for $\Delta\psi(x')$, and comparing $\psi(x') = \psi_0(x') + \Delta\psi(x')$ with (3.56), gives the following relation for the Green's function:

$$G(x'; x) = G_0(x'; x) + \int \mathrm{d}^4x_1 G_0(x'; x_1) V(x_1) G_0(x_1; x). \tag{3.60}$$

That is, it is expressed in terms of the free-particle Green's function G_0, which we assume to be known. The above considerations can be

† The discussion in this section follows closely that in Bjorken and Drell (1964), where more details on some points are given.

represented graphically by space-time diagrams. Fig. 3.2(*a*) represents free propagation from x to x', whereas Fig. 3.2(*b*) corresponds to free propagation from x to x_1, a scattering at x_1, then free propagation to x'.

The above methods describing a single scatter may readily be generalised to multiple scatters. Firstly, let us consider a second scatter at $x_2 = (t_2, \mathbf{x}_2)$, with $t_2 > t_1$, produced by turning on a potential $V(x_2)$ for a time $\mathrm{d}t_2$. For $t' > t_2$, the additional contribution is

$$\begin{aligned}\Delta\psi(x') &= \int \mathrm{d}^3x_2 G_0(x'; x_2)V(x_2)\psi(x_2)\,\mathrm{d}t_2 \\ &= \mathrm{i}\int \mathrm{d}^3x\,\mathrm{d}^4x_2 G_0(x'; x_2)V(x_2)[G_0(x_2; x) \\ &\quad + \int \mathrm{d}^4x_1 G_0(x_2; x_1)V(x_1)G_0(x_1; x)]\psi_0(x). \end{aligned} \tag{3.61}$$

Thus the total wave function is

$$\begin{aligned}\psi(x') = \psi_0(x') &+ \int \mathrm{d}^4x_1 G_0(x'; x_1)V(x_1)\psi_0(x_1) \\ &+ \int \mathrm{d}^4x_2 G_0(x'; x_2)V(x_2)\psi_0(x_2) \\ &+ \int \mathrm{d}^4x_1\,\mathrm{d}^4x_2 G_0(x'; x_2)V(x_2)G_0(x_2; x_1)V(x_1)\psi_0(x_1). \end{aligned} \tag{3.62}$$

That is, the individual terms correspond to free propagation, a single scatter at x_1, a single scatter at x_2, and double scattering respectively. The corresponding space-time diagrams are shown in Fig. 3.2(*a*) to (*d*). Thus we are led to an iterative or perturbative approach. This conclusion does not apply to all problems, and we note that Green's function can

Fig. 3.2 Propagation of a particle from x to x'; (*a*) free propagation; (*b*) scatter at x_1; (*c*) scatter at x_2 and (*d*) double scatter.

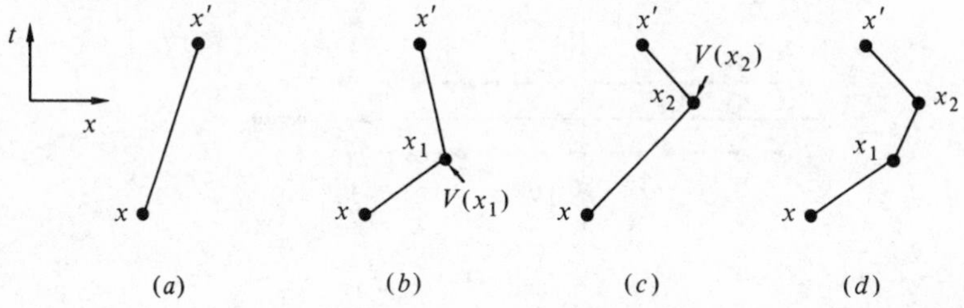

be used to solve Poisson's equation $\nabla^2\phi = -\rho$ (2.19) in electrostatics. In this case, $\phi(\mathbf{x}') = \int G(\mathbf{x}', \mathbf{x})\rho(\mathbf{x})\mathrm{d}^3x$; that is, it can be written in a closed form since the righthand side of (2.19) does not involve ϕ, in contrast to the applications we shall consider.

Next we consider the solution for the free-particle Green's function G_0. Equation (3.56) can be written in the form

$$\theta(t'-t)\psi(x') = \mathrm{i}\int \mathrm{d}^3x G(x'; x)\psi(x), \tag{3.63}$$

where the restriction $t' > t$ is removed and replaced by the step function θ, defined such that $\theta(\tau) = 1$ for $\tau > 0$, else $\theta(\tau) = 0$. Alternatively, $\theta(\tau)$ can be written in terms of a complex variable ω, as

$$\theta(\tau) = \lim_{\varepsilon\to 0} \frac{-1}{2\pi \mathrm{i}} \int_{-\infty}^{\infty} \frac{\mathrm{d}\omega \exp(-\mathrm{i}\omega\tau)}{\omega + \mathrm{i}\varepsilon}. \tag{3.64}$$

The pole at $\omega = -\mathrm{i}\varepsilon$ and the integration contours for positive and negative τ are shown in Fig. 3.3. The lowest contour ($\tau > 0$) encloses the pole, so gives unity by Cauchy's theorem, whereas the upper contour yields zero. Application of $[\mathrm{i}\partial/\partial t' - H(x')]$ to the lefthand side of (3.63) gives

$$\left[\mathrm{i}\frac{\partial}{\partial t'} - H(x')\right]\theta(t'-t)\psi(x') = \mathrm{i}\frac{\partial\theta(t'-t)}{\partial t'}\psi(x') = \mathrm{i}\,\delta(t'-t)\psi(x'), \tag{3.65}$$

since the derivative of θ is a δ-function. Thus, application to both sides of (3.63) gives, using (3.65),

$$\delta(t'-t)\psi(x') = \int \mathrm{d}^3x\left[\mathrm{i}\frac{\partial}{\partial t'} - H(x')\right]G(x'; x)\psi(x). \tag{3.66}$$

The lefthand side of (3.66) can be expressed as an integral over x together

Fig. 3.3 Integration contour for step function $\theta(\tau)$.

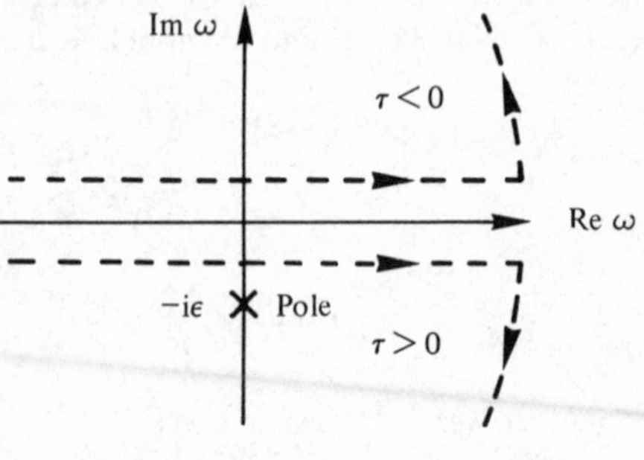

with the δ-function $\delta^3(\mathbf{x}' - \mathbf{x})$. Hence, since the equation is valid for any solution ψ, we obtain the Schrödinger Green's function equation

$$\left[i\frac{\partial}{\partial t'} - H(x')\right]G(x'; x) = \delta^4(x' - x). \tag{3.67}$$

Note that $G(x'; x) = 0$ for $t' < t$, so that G is a retarded Green's function.

For the free particle case $H_0(x') = -(\nabla^2)'/2m$. Since $G_0(x'; x)$ represents the amplitude of a wave at x' originating at x, it follows that G_0 is a function of $x' - x$. It can be written as a Fourier transform

$$G_0(x'; x) = G_0(x' - x) = \int \frac{d^4p}{(2\pi)^4} \exp[-ip\cdot(x' - x)]G_0(E, p). \tag{3.68}$$

Substituting (3.68) in (3.67) gives, using the definition of $\delta^4(x' - x)$,

$$\int \frac{d^4p}{(2\pi)^4}\left(E - \frac{p^2}{2m}\right)G_0(E, \mathbf{p}) \exp[-ip\cdot(x' - x)]$$
$$= \int \frac{d^4p}{(2\pi)^4} \exp[-ip\cdot(x' - x)].$$

Hence

$$G_0(E, \mathbf{p}) = \frac{1}{E - p^2/2m}, \qquad E \neq p^2/2m. \tag{3.69}$$

For the relativistic case, the creation and destruction of particles and antiparticles must be handled, and the whole formalism must be covariant. In addition to space-time diagrams of the type shown in Fig. 3.2, there are also diagrams such as those shown in Fig. 3.4. In Fig. 3.4(*a*) there is one particle (a π^+) present between t and t_1 and between t_2 and t'. However, between t_1 and t_2 there are two particles and one antiparticle (π^-) present. Thus the process corresponds to the creation of a $\pi^+\pi^-$ pair at 1 and the annihilation of the π^- by the initial π^+ at 2. Now the

Fig. 3.4 Space–time diagrams for $\pi^+\pi^-$ pair production.

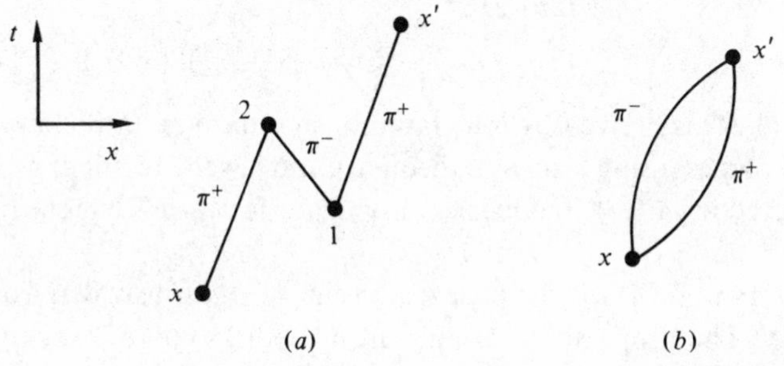

propagation of a positive energy π^+ forwards in time is equivalent to that of a negative energy π^- propagating backwards in time (see Section 3.1.5). Hence Fig. 3.4(*a*) can also be interpreted as the scattering of a π^+ at 2 which travels backwards in time and with negative energy to 1; then, after the scatter at 1, travels forwards in time with positive energy. Closed loops (Fig. 3.4(*b*)) can be interpreted in a similar manner.

The propagator for a scalar particle can be derived from the Klein–Gordon equation, in a similar manner to the non-relativistic case, starting with (3.63). The required Green's function, G_s, is a solution to the equation (the equivalent of (3.67)),

$$[(\square^2)' + m^2]G_s(x', x) = -\delta^4(x' - x) \tag{3.70}$$

Again, G_s is a function of $x' - x$ and has a Fourier transform given by

$$G_s(x', x) = G_s(x' - x) = \int \frac{\mathrm{d}^4 p}{(2\pi)^4} \exp[-\mathrm{i}p\cdot(x' - x)]G_s(p). \tag{3.71}$$

Substituting in (3.70) gives

$$G_s(p) = \frac{1}{p^2 - m^2}, \qquad p^2 \neq m^2. \tag{3.72}$$

The boundary conditions on $G_s(x' - x)$ must be utilised in order to handle the singularity at $p^2 = m^2$ in (3.71). The required condition is that the wave propagating from x to x' into the future or, more precisely, the future light-cone, should only contain positive frequency ($\phi \sim \exp(-\mathrm{i}p\cdot x)$) π^+ and π^- components. The propagator (3.72) has poles at $p_0 = \pm E_p$, where $E_p = +(\mathbf{p}^2 + m^2)^{1/2}$. If we treat p_0 as complex and let $m^2 \rightarrow m^2 - \mathrm{i}\varepsilon$ ($\varepsilon \rightarrow 0^+$), so that the poles are in the lower and upper half-planes, then (3.71) becomes

$$G_s(x', x) = \int \frac{\mathrm{d}^4 p}{(2\pi)^4} \frac{\exp[-\mathrm{i}p\cdot(x' - x)]}{p^2 - m^2 + \mathrm{i}\varepsilon} \tag{3.73}$$

$$= -\mathrm{i} \int \frac{\mathrm{d}^3 p}{(2\pi)^3 2E_p} \{\exp[-\mathrm{i}p\cdot(x' - x)]\theta(t' - t) + \exp[\mathrm{i}p\cdot(x' - x)]\theta(t - t')\}. \tag{3.74}$$

Equation (3.74) is derived by integration over contours in the lower and upper half-planes (using Cauchy's theorem), and has the required property that positive (negative) frequencies propagate forwards (backwards) in time.

The equation (3.74) for the propagator can be rigorously derived from field theory. The propagation of a π^+ from $x = (t, \mathbf{x})$ to $x' = (t', \mathbf{x}')$ (Fig.

3.2(*a*)) corresponds to the creation from the vacuum of a π^+ at x and the destruction at $x'(t'>t)$. A physically equivalent contribution is the creation of a π^- at x' and the destruction of a π^- at x with $t'<t$. The propagator contains the sum of these two amplitudes, namely

$$\begin{aligned} \mathrm{i}G_s(x'-x) &= \langle 0|\phi(x')\phi^\dagger(x)|0\rangle\theta(t'-t) + \langle 0|\phi^\dagger(x)\phi(x')|0\rangle\theta(t-t') \\ &= \langle 0|T\{\phi(x')\phi^\dagger(x)\}|0\rangle, \end{aligned} \tag{3.75}$$

where ϕ and $\phi^\dagger$ are field operators as given in (3.51) and T is the *time-ordering operator* giving $\phi(x')\phi^\dagger(x)$ if $t'>t$ and $\phi^\dagger(x)\phi(x')$ if $t>t'$. Noting that the operators a and b satisfy $a|0\rangle = b|0\rangle = \langle 0|a^\dagger = \langle 0|b^\dagger = 0$, the first term in (3.75) can be written, using (3.51),

$$\begin{aligned} \langle 0|\phi(x')\phi^\dagger(x)|0\rangle\theta(t'-t) = \int & \frac{\mathrm{d}^3p'\,\mathrm{d}^3p}{(2\pi)^3(2E_{p'}2E_p)^{1/2}} \exp(-\mathrm{i}p'\cdot x')\exp(\mathrm{i}p\cdot x) \\ & \times \langle 0|a(p')a^\dagger(p)|0\rangle\theta(t'-t). \end{aligned} \tag{3.76}$$

The second term has a similar form but involves $\langle 0|b(p)b^\dagger(p')|0\rangle$. Noting that $\langle 0|a(p')a^\dagger(p)|0\rangle = \langle 0|[a(p'), a^\dagger(p)]|0\rangle$, and thus contributes $\delta^3(\mathbf{p}'-\mathbf{p})$ from (3.16), then integrating over p' gives (3.74). In summary, the propagation of a scalar particle between x and x' is represented by $\mathrm{i}G_s(x', x)$ in *configuration space*, and by $\mathrm{i}G_s(p) = \mathrm{i}/(p^2 - m^2 + \mathrm{i}\varepsilon)$ in *momentum space*.

Using the expression for $\phi(x)$ from (3.9) and the commutation relations (3.16), it can also be shown that $[\phi(x'), \phi^\dagger(x)] = \mathrm{i}G_s(x', x) = 0$, if $(x'-x)^2<0$. The commutation relation can be shown to be zero by evaluating it for $t'=t$ (equal time commutation relation) and noting that Lorentz invariance of G_s requires that this holds for any space-like separation of x and x'. This is equivalent to the requirement that no effect may propagate with a velocity $\beta > c$ (microcausality). It can also be shown that two observable quantities (which are bilinear in the fields) commute if they have a space-like separation. This property of microcausality would not hold if the scalar field was quantised with Fermi–Dirac, rather than Bose–Einstein statistics.

3.2 SPIN $\frac{1}{2}$ PARTICLES

3.2.1 The Dirac equation

In order to avoid the difficulties of negative probabilities associated with the Klein–Gordon equation, which were discussed in Section 3.1.1, Dirac proposed an equation linear in $\partial/\partial t$, rather than second order. The equation should also be linear in the space coordinates $\partial/\partial x_i$ $(i = 1, 2, 3)$, in order to treat space and time on an equal footing. The Dirac equation is

$$\mathrm{i}\frac{\partial\psi(x)}{\partial t} = H\psi(x) = \left(-\mathrm{i}\alpha^i\frac{\partial}{\partial \mathrm{x}^i} + \beta m\right)\psi(x)$$

$$= (\boldsymbol{\alpha}\cdot\mathbf{p} + \beta m)\psi(x). \tag{3.77}$$

The relativistic energy-momentum relation for a free particle requires that

$$H^2\psi = (\boldsymbol{\alpha}\cdot\mathbf{p} + \beta m)(\boldsymbol{\alpha}\cdot\mathbf{p} + \beta m)\psi$$

$$= \left[\sum_{i,j}\alpha^i\alpha^j p^i p^j + \sum_i(\alpha^i\beta + \beta\alpha^i)mp^i + \beta^2 m^2\right]\psi$$

$$= (\mathbf{p}^2 + m^2)\psi. \tag{3.78}$$

Thus (3.78) implies (where $\{\ \}$ denotes anticommutation)

$$\{\alpha^i, \beta\} = 0, \qquad \{\alpha^i, \alpha^j\} = 2\delta_{ij}, \qquad \beta^2 = 1. \tag{3.79}$$

Since pure numbers will not satisfy (3.79), Dirac proposed that α and β are matrices and that the wavefunction ψ be multicomponent and written as a column vector (Dirac spinor). The requirement that the energy operator in (3.77) has real eigenvalues implies α^i and β are Hermitian. Since $\alpha_i^2 = \beta^2 = 1$, the eigenvalues of α and β are ± 1. Further, α and β are traceless. This follows since, from (3.79),

$$\alpha^i = -\beta\alpha^i\beta \qquad \text{and} \qquad \beta = -\alpha^i\beta\alpha^i. \tag{3.80}$$

So, taking traces, and using the property that $\mathrm{tr}(AB) = \mathrm{tr}(BA)$, gives $\mathrm{tr}(\alpha^i) = \mathrm{tr}(\beta) = 0$. The number of positive and negative eigenvalues must be equal, and since the trace is the sum of the eigenvalues, hence the dimension of the matrices must be even. The simplest representation satisfying (3.79) is 4×4 matrices. Thus ψ is a four-component column vector.

For later use, it is convenient to define

$$\gamma^0 = \beta, \qquad \gamma^i = \beta\alpha^i, \qquad i = 1, 2, 3. \tag{3.81}$$

The relationships (3.79) become ($\mu, \nu = 0, 1, 2, 3$)

$$\gamma^\mu\gamma^\nu + \gamma^\nu\gamma^\mu = \{\gamma^\mu, \gamma^\nu\} = 2g^{\mu\nu}, \tag{3.82}$$

giving, for the squares of the γ-matrices, $(\gamma^\mu)^2 = g^{\mu\nu}$ so

$$(\gamma^0)^2 = 1, \qquad (\gamma^i)^2 = -1 \qquad i = 1, 2, 3. \tag{3.83}$$

Since α^i and β are Hermitian, we further obtain

$$(\gamma^\mu)^\dagger = \gamma^0\gamma^\mu\gamma^0, \qquad (\gamma^0)^\dagger = \gamma^0. \tag{3.84}$$

Equation (3.82) is only a constraint on the possible γ-matrices, however specific representations are often useful. The *Pauli—Dirac representation* has

$$\alpha^i = \begin{pmatrix} 0 & \sigma^i \\ \sigma^i & 0 \end{pmatrix}, \qquad \beta = \gamma^0 = \begin{pmatrix} I & 0 \\ 0 & -I \end{pmatrix}, \qquad \gamma^i = \begin{pmatrix} 0 & \sigma^i \\ -\sigma^i & 0 \end{pmatrix}, \tag{3.85}$$

where σ^i are the 2×2 Pauli spin matrices (2.213) and I the unit 2×2 matrix. A further useful shorthand is the matrix γ^5

$$\gamma^5 = \gamma_5 = i\gamma^0\gamma^1\gamma^2\gamma^3 = \begin{pmatrix} 0 & I \\ I & 0 \end{pmatrix}, \tag{3.86}$$

with

$$\{\gamma^\mu, \gamma^5\} = 0, \qquad (\gamma^5)^2 = 1. \tag{3.87}$$

The Dirac equation can be expressed in a simpler form in terms of the γ-matrices. Using (3.81), and multiplying (3.77) from the left by γ^0, gives

$$\left[i\left(\gamma^0 \frac{\partial}{\partial x^0} + \gamma^i \frac{\partial}{\partial x^i} \right) - m \right] \psi = (i\,\gamma^\mu \partial_\mu - m)\psi = (\gamma^\mu p_\mu - m)\psi$$
$$= (\not{p} - m)\psi = 0. \tag{3.88}$$

where the slash notation $\not{A} = \gamma^\mu A_\mu = \gamma_\mu A^\mu$ has been used.

If we take the Hermitian conjugate of (3.88), multiply from the right by γ^0 and use the property (3.84) for $(\gamma^\mu)^\dagger$, then we obtain the *adjoint* equation

$$\psi^\dagger \gamma^0 (\not{p} + m) = \bar{\psi}(\not{p} + m) = 0, \tag{3.89}$$

where $\bar{\psi} = \psi^\dagger \gamma^0$ and the operator $\not{p}$ acts to the left. A conserved current $(\partial_\mu j^\mu = 0)$ can be constructed for a Dirac particle, in a similar way to that for the Klein–Gordon equation. The method is to multiply (3.89) from the right by ψ and (3.88) from the left by $\bar{\psi}$, and add the results, giving

$$\bar{\psi}(i\,\partial_\mu \gamma^\mu + m)\psi + \bar{\psi}(i\gamma^\mu \partial_\mu - m)\psi = i\partial_\mu(\bar{\psi}\gamma^\mu\psi) = 0. \tag{3.90}$$

Thus the conserved Dirac current is

$$j^\mu_{\mathrm{D}} = \bar{\psi}\gamma^\mu\psi. \tag{3.91}$$

Note that $\rho = j^0 = \bar{\psi}\gamma^0\psi = \psi^\dagger\psi$ is positive definite, in contrast to ρ from the Klein–Gordon equation (3.3). The electromagnetic current for a Dirac particle must be obtained by including the electric charge, giving for the electron,

$$j^\mu = -e\bar{\psi}\gamma^\mu\psi. \tag{3.92}$$

3.2.2 Spin of a Dirac particle

The wave function for a Dirac particle is multicomponent and is thus suitable for describing a particle with non-zero spin. In this section the spin $\frac{1}{2}$ assignment is justified. In order to do this we must find a quantity which represents the total angular momentum of the particle and which commutes with H. First we consider $[\mathbf{L}, H]$, where $\mathbf{L} = \mathbf{r} \times \mathbf{p}$ is the orbital angular momentum operator. Using H from (3.77)

$$\begin{aligned}[\mathbf{L}, H] &= [\mathbf{r} \times \mathbf{p}, \gamma^0 \boldsymbol{\gamma} \cdot \mathbf{p} + \gamma^0 m] \\ &= \gamma^0 [\mathbf{r} \times \mathbf{p}, \boldsymbol{\gamma} \cdot \mathbf{p}] \\ &= \gamma^0 [\mathbf{r}, \boldsymbol{\gamma} \cdot \mathbf{p}] \times \mathbf{p} \\ &= \mathrm{i}\gamma^0 \boldsymbol{\gamma} \times \mathbf{p},\end{aligned} \tag{3.93}$$

where the last line is derived using $[\mathbf{r}, \mathbf{p}] = \mathrm{i}$. Thus $\mathbf{L}$ does not commute with H. Consider the operator $\boldsymbol{\Sigma}$ with components

$$\Sigma^1 = \frac{\mathrm{i}}{2}[\gamma^2, \gamma^3], \qquad \Sigma^2 = \frac{\mathrm{i}}{2}[\gamma^3, \gamma^1], \qquad \Sigma^3 = \frac{\mathrm{i}}{2}[\gamma^1, \gamma^2]$$

i.e. $$\boldsymbol{\Sigma} = \frac{\mathrm{i}}{2}\boldsymbol{\gamma} \times \boldsymbol{\gamma}. \tag{3.94}$$

We can define

$$\sigma^{\mu\nu} = \frac{\mathrm{i}}{2}[\gamma^\mu, \gamma^\nu], \tag{3.95a}$$

which has the property, using the Pauli–Dirac representation and the properties of the 2×2 Pauli spin matrices (2.213)

$$\sigma^{ij} = \begin{pmatrix} \sigma^k & 0 \\ 0 & \sigma^k \end{pmatrix}, \qquad i, j, k = 1, 2, 3 \text{ in cyclic order.} \tag{3.95b}$$

Thus $\boldsymbol{\Sigma}$ can be written in the form

$$\boldsymbol{\Sigma} = \begin{pmatrix} \boldsymbol{\sigma} & 0 \\ 0 & \boldsymbol{\sigma} \end{pmatrix}. \tag{3.96}$$

The commutation of $\boldsymbol{\Sigma}$ with H gives the following result

$$\begin{aligned}[\boldsymbol{\Sigma}, H] &= \frac{\mathrm{i}}{2}[\boldsymbol{\gamma} \times \boldsymbol{\gamma}, \gamma^0 \boldsymbol{\gamma} \cdot \mathbf{p} + \gamma^0 m] \\ &= \frac{\mathrm{i}}{2}\gamma^0 [\boldsymbol{\gamma} \times \boldsymbol{\gamma}, \boldsymbol{\gamma} \cdot \mathbf{p}] = -2\mathrm{i}\gamma^0 \boldsymbol{\gamma} \times \mathbf{p},\end{aligned} \tag{3.97}$$

where the last step can be shown, for example, by explicit evaluation of the individual components. From (3.93) and (3.97) we can construct a total angular momentum operator $\mathbf{J}$, which commutes with H, and in which we identify $\mathbf{\Sigma}$ as the spin operator

$$\mathbf{J}=\mathbf{L}+\mathbf{S}=\mathbf{L}+\tfrac{1}{2}\mathbf{\Sigma}. \tag{3.98}$$

Thus we can conclude that a Dirac particle has spin $\frac{1}{2}$, since the eigenvalues of $\mathbf{\Sigma}$ are ± 1. Furthermore, since it can be shown that

$$[\mathbf{\Sigma}\cdot\mathbf{p}, H]=0 \qquad \text{and} \qquad (\mathbf{\Sigma}\cdot\mathbf{p})^2=\mathbf{p}^2, \tag{3.99}$$

the operator $\mathbf{\Sigma}\cdot\mathbf{p}$ has eigenvalues $\pm|\mathbf{p}|$. Thus the eigenfunctions of $\mathbf{\Sigma}\cdot\mathbf{p}$ represent systems in which the spin is parallel or antiparallel to $\mathbf{p}$. Alternatively, we can define the *helicity operator* $h(\mathbf{p})=(\mathbf{\Sigma}\cdot\mathbf{p})/|\mathbf{p}|$, which has eigenvalues $+1$ and -1 for the parallel and antiparallel configurations respectively.

3.2.3 Free particle solutions

Many problems can be solved without resort to specific solutions of the Dirac equation or to specific representations of the γ-matrices. However, some insight into the properties of a Dirac particle can be gained from consideration of specific solutions. As for the Klein–Gordon equation, we seek plane wave solutions for the free particle case. For the Dirac equation these have the form

$$\psi(x)=\frac{1}{L^{3/2}}u(\mathbf{p})\exp(-ip\cdot x), \qquad \psi(x)=\frac{1}{L^{3/2}}v(\mathbf{p})\exp(ip\cdot x), \tag{3.100}$$

where u and v are four-component Dirac spinors. The spinor u refers to states with four-momentum $p=[+(\mathbf{p}^2+m^2)^{1/2}, \mathbf{p}]$, whereas v refers to states with four-momentum $-p=[-(\mathbf{p}^2+m^2)^{1/2}, -\mathbf{p}]$, that is negative energy and momentum. Negative energy solutions for the Dirac equation arise from the quadratic nature of (3.78). The factor $L^{3/2}$ can be replaced by $(2\pi)^{3/2}$ where appropriate. Using the 2×2 matrix form (3.85), and applying the operators on the plane wave for the u-spinor, we can write (3.88) as

$$(\gamma^0 p_0-\boldsymbol{\gamma}\cdot\mathbf{p}-m)u=\begin{pmatrix} p_0-m & -\boldsymbol{\sigma}\cdot\mathbf{p} \\ \boldsymbol{\sigma}\cdot\mathbf{p} & -p_0-m\end{pmatrix}\begin{pmatrix}\phi \\ \chi\end{pmatrix}=0. \tag{3.101}$$

This gives two coupled equations for the two-component (*Weyl*) spinors ϕ and χ

$$(p_0-m)\phi=(\boldsymbol{\sigma}\cdot\mathbf{p})\chi, \qquad (p_0+m)\chi=(\boldsymbol{\sigma}\cdot\mathbf{p})\phi. \tag{3.102}$$

Thus u can be written in terms of ϕ only as

$$u = N\begin{pmatrix} \phi \\ \dfrac{\boldsymbol{\sigma}\cdot\mathbf{p}}{p_0+m}\phi \end{pmatrix}, \tag{3.103}$$

where N is a normalisation factor. Note that the two-spinor ϕ can be chosen to be an eigenstate of the operator $(\boldsymbol{\sigma}\cdot\mathbf{p})/|\mathbf{p}|$ (see (2.223)) with eigenvalues ± 1. In this form u is then an eigenfunction of the helicity operator $h(\mathbf{p}) = (\boldsymbol{\Sigma}\cdot\mathbf{p})/|\mathbf{p}|$ with eigenvalues ± 1. The choice for ϕ discussed below will be, however, more frequently used.

Similarly for $\psi = L^{-3/2} v \exp(ip\cdot x)$ we obtain, with $p_0 = (\mathbf{p}^2 + m^2)^{1/2}$,

$$v = N\begin{pmatrix} \dfrac{\boldsymbol{\sigma}\cdot\mathbf{p}}{p_0+m}\chi_v \\ \chi_v \end{pmatrix}. \tag{3.104}$$

Thus for a particle at rest we have

$$\psi = \frac{1}{L^{3/2}} u \exp(-imt) = \frac{N}{L^{3/2}}\begin{pmatrix}\phi \\ 0\end{pmatrix}\exp(-imt),$$

$$\psi = \frac{1}{L^{3/2}} v \exp(imt) = \frac{N}{L^{3/2}}\begin{pmatrix}0 \\ \chi\end{pmatrix}\exp(imt), \tag{3.105}$$

Now the rest-frame two-spinors ϕ and χ can be chosen to correspond to the two independent spin states of a spin $\frac{1}{2}$ particle, giving the four rest frame solutions

$$u^1 = N\begin{pmatrix}1\\0\\0\\0\end{pmatrix}, \qquad u^2 = N\begin{pmatrix}0\\1\\0\\0\end{pmatrix},$$

$$v^1 = N\begin{pmatrix}0\\0\\0\\1\end{pmatrix}, \qquad v^2 = N\begin{pmatrix}0\\0\\1\\0\end{pmatrix}. \tag{3.106}$$

The reason for the choice of the third and fourth components is discussed later. Now the third component of the spin operator Σ^3 (equation (3.96)) gives eigenvalues 1, -1, -1 and 1 for the four solutions respectively in (3.106). Thus the four independent solutions represent positive energy spin up, positive energy spin down, negative energy spin down and negative energy spin up respectively. Under proper Lorentz transformations the positive and negative energy solutions do not mix, so that we can choose u to represent a particle (e.g. e^-) with positive energy and v to represent a state with negative energy, corresponding to an antiparticle (e.g. e^+).

The starting point in the solution of many problems is writing down the amplitude for the process in terms of the spinors u and v. The solutions refer always to the particle (e.g. e^-). The antiparticle solutions do not appear directly, but are treated as negative energy particles propagating backwards in time. Note that the arguments of $v(p, s)$ are those of the physical antiparticle. It is useful to write the Dirac equations for the spinors u and v explicitly. Proceeding as above we obtain

$$(\not{p} - m)u = 0, \qquad (\not{p} + m)v = 0,$$
$$\bar{u}(\not{p} - m) = 0, \qquad \bar{v}(\not{p} + m) = 0. \tag{3.107}$$

The normalisation constant N in (3.103) and (3.104) has so far not been considered. From (3.91) the probability density of Dirac particles is $j^0 = \rho = \psi^\dagger\psi$. Integrating over a normalisation volume we obtain

$$\int d^3x\rho = \int d^3x u^\dagger u/L^3 = u^\dagger u$$
$$= |N|^2\left[1 + \frac{(\boldsymbol{\sigma}\cdot\mathbf{p})^2}{(E+m)^2}\right] = |N|^2\frac{2E}{E+m}, \tag{3.108}$$

where use has been made of (2.215) and the solution for u in (3.103). If we normalise this to a flux of $2E$ particles (this is discussed further later) then $N = (E + m)^{1/2}$, and we have the following normalisation relations

$$u^\dagger u = v^\dagger v = 2E,$$
$$\bar{u}u = 2m, \qquad \bar{v}v = -2m. \tag{3.109}$$

The relationship $\bar{u}u = 2m$ can be derived by multiplying $\gamma^\mu p_\mu u = mu$ (i.e. (3.107)) from the left by $\bar{u}\gamma^\nu$ and then adding this to the result of $\bar{u}\gamma^\mu p_\mu = m\bar{u}$ multiplied from the right by $\gamma^\nu u$ and then utilising (3.82), giving $2\bar{u}p^\nu u = 2m\bar{u}\gamma^\nu u$. Thus, for $\nu = 0$, $2E\bar{u}u = 2mu^\dagger u$. The relationship $\bar{v}v = -2m$ can be shown in a similar way.

3.2.4 Projection operators

In the evaluation of matrix elements the combinations $u\bar{u}$ and $v\bar{v}$ frequently appear. Note that these are 4×4 matrices compared to the normalisation relations for $\bar{u}u$ and $\bar{v}v$, which are just numbers. Using u

from (3.103), $u\bar{u}$ can be expressed as

$$u\bar{u} = (E+m)\begin{pmatrix} \phi \\ \dfrac{\boldsymbol{\sigma}\cdot\mathbf{p}}{E+m}\phi \end{pmatrix}\begin{pmatrix} \phi^\dagger & -\dfrac{\boldsymbol{\sigma}\cdot\mathbf{p}}{E+m}\phi^\dagger \end{pmatrix}$$

$$= \begin{pmatrix} E+m & -\boldsymbol{\sigma}\cdot\mathbf{p} \\ \boldsymbol{\sigma}\cdot\mathbf{p} & -E+m \end{pmatrix} = \not{p} + m, \tag{3.110}$$

where the relationship $\phi\phi^\dagger = \phi_1\phi_1^\dagger + \phi_2\phi_2^\dagger = I$ (i.e. summing over the possible spin states) has been used. Similarly, for the v spinor

$$v\bar{v} = \sum_{i=1}^{2} v^i\bar{v}^i = \not{p} - m. \tag{3.111}$$

From (3.110) and (3.111) we obtain the *completeness relation*,

$$\sum_{r=1,2} (u^r\bar{u}^r - v^r\bar{v}^r)_{\alpha\beta} = 2mI_{\alpha\beta} = 2m\delta_{\alpha\beta}. \tag{3.112}$$

It is useful to define projection operators, which project out states of a particular energy or spin. The *energy projection operators* Λ_+ and Λ_- are defined as follows

$$\Lambda_+ = \frac{\not{p}+m}{2m}, \qquad \Lambda_- = \frac{-\not{p}+m}{2m}, \tag{3.113}$$

so that $\Lambda_+ + \Lambda_- = 1$, and

$$\Lambda_+ u = u, \qquad \Lambda_- u = 0, \qquad \Lambda_+ v = 0, \qquad \Lambda_- v = v. \tag{3.114}$$

Hence Λ_+ and Λ_- project out the positive and negative energy states respectively.

A similar operator is required to project out the spin up and spin down (i.e. helicity) states. The operator must commute with $\not{p}$ so that the four solutions have unique energy and spin eigenvalues. In order to treat spin in a covariant way, a four-vector for the spin polarisation s^μ is required. In the rest system of the particle, by classical analogy, there will be only a 'space' component. That is, we can write $s^\mu = (0, \hat{\mathbf{s}})$, where $\hat{\mathbf{s}}$ is a unit vector along the spin direction. The usual Lorentz transformation properties can be used to transform s^μ to other frames in which s^μ acquires a 'time' component, with no classical analogue. Note that $s^2 = s_\mu s^\mu = -1$, and that in the rest system $p^\mu = (m, \mathbf{0})$, so that $s\cdot p = 0$: by Lorentz covariance these relations are true in any frame.

The two spin components of the positive energy spinor (u) are, in the rest frame, eigenfunctions of $\boldsymbol{\Sigma}\cdot\hat{\mathbf{s}}$, with $\boldsymbol{\Sigma}$ as defined in (3.96); i.e.

$$(\boldsymbol{\Sigma}\cdot\hat{\mathbf{s}})\mathrm{u}^1(p,s) = u^1(p,s), \qquad (\boldsymbol{\Sigma}\cdot\hat{\mathbf{s}})u^2(p,s) = -u^2(p,s). \tag{3.115}$$

The required covariant operator should reduce to this form in the rest frame. Consider the operator $\gamma^5 \not{s} = \gamma^5 s^\mu \gamma_\mu$. In the rest system (r) this becomes, using (2.144), (3.85) and (3.86)

$$(\gamma^5 \not{s})_r = \begin{pmatrix} \boldsymbol{\sigma}\cdot\hat{\mathbf{s}} & 0 \\ 0 & -\boldsymbol{\sigma}\cdot\hat{\mathbf{s}} \end{pmatrix}. \tag{3.116}$$

For the rest frame spinors u^1 and u^2 this operator has the required properties of (3.115). For the negative energy spinors v^1 and v^2

$$(\gamma^5 \not{s})_r v^i = \begin{pmatrix} \boldsymbol{\sigma}\cdot\hat{\mathbf{s}} & 0 \\ 0 & -\boldsymbol{\sigma}\cdot\hat{\mathbf{s}} \end{pmatrix}\begin{pmatrix} 0 \\ \chi^i \end{pmatrix} = -\begin{pmatrix} 0 \\ \boldsymbol{\sigma}\cdot\hat{\mathbf{s}}\chi^i \end{pmatrix} = c^i v^i, \tag{3.117}$$

where $c^1 = 1, c^2 = -1$. This follows since $(\boldsymbol{\sigma}\cdot\hat{\mathbf{s}})\chi^1 = -\chi^2$ and $(\boldsymbol{\sigma}\cdot\hat{\mathbf{s}})\chi^2 = \chi^2$. Thus, $\gamma^5 \not{s}$ acting on $v(p, s)$ corresponds to a state with polarisation $-s$. This explains the form adopted in (3.106) for the v spinors. The solution $u(p, s)$ describes a particle with momentum p and polarisation s, whereas $v(p, s)$ describes a particle with momentum $-p$ and polarisation $-s$, i.e. the labels attached to v are for the equivalent antiparticle. The operator $\gamma^5 \not{s}$ has the required properties to label the states, since it may easily be shown that it commutes with $\not{p}$ and has eigenvalues ± 1, i.e.

$$[\gamma^5 \not{s}, \not{p}] = 0, \qquad (\gamma^5 \not{s})^2 = 1. \tag{3.118}$$

Thus we can construct the required *spin projection operator*

$$\Lambda(s) = (1 + \gamma^5 \not{s})/2,$$
$$\Lambda(s) u(s) = u(s), \qquad \Lambda(s) u(-s) = 0, \tag{3.119}$$

The expression for $u\bar{u}$ (3.110), which allows this spinor combination to be replaced by $(\not{p} + m)$ in the evaluation of matrix elements, was derived by summing over the possible spin states. If it is required to calculate the contribution of a specific spin state, then the appropriate replacement is

$$u\bar{u} = (\not{p} + m)(1 + \gamma^5 \not{s})/2. \tag{3.120}$$

This can be derived from the rest frame solutions and using the expression for $\phi\phi^\dagger$ from (2.222). A similar treatment for $v\bar{v}$ gives

$$v\bar{v} = (\not{p} - m)(1 + \gamma^5 \not{s})/2. \tag{3.121}$$

3.2.5 Lorentz covariance

The Dirac equation in some Lorentz frame O′ must yield the same physical content as that in O. The coordinates and the four-momentum operators in O′ and O are related by (from (2.155))

$$(x')^\mu = a^\mu_\nu x^\nu = a^{\mu\nu} x_\nu, \qquad (p')^\mu = a^\mu_\nu p^\nu = a^{\mu\nu} p_\nu. \tag{3.122}$$

Assuming the transformation between the wavefunctions $\psi'(x')$ and $\psi(x)$ is linear and that the γ-matrices remain unaltered*, then the Dirac equation in O′ will be

$$(\gamma^\mu p'_\mu - m)\psi'(x') = 0, \tag{3.123}$$

with

$$\psi'(x') = S\psi(x), \tag{3.124}$$

where S is a 4×4 matrix. If we multiply the Dirac equation for frame O (3.88) from the left by S and insert $S^{-1}S = 1$, we obtain

$$(S\gamma^\mu S^{-1} p_\mu - mSS^{-1})S\psi(x) = 0. \tag{3.125}$$

This is of the required form (3.123), provided that

$$S\gamma^\mu S^{-1} = \gamma^\nu a^\mu_\nu \qquad \text{or} \qquad S^{-1}\gamma^\nu S = a^\nu_\mu \gamma^\mu, \tag{3.126}$$

where use has been made of (2.158) in the latter form.

For the case of a *pure rotation* (R), only the space coordinates are affected and $|\mathbf{p}'| = |\mathbf{p}|$. Equation (3.126) then simplifies to

$$S_R\gamma^0 S_R^{-1} = \gamma^0, \qquad S_R\gamma^k S_R^{-1} = \gamma^j a^k_j \ \text{ or } \ S_R\boldsymbol{\gamma} S_R^{-1}\cdot\mathbf{p} = \boldsymbol{\gamma}\cdot\mathbf{p}'. \tag{3.127}$$

The solution to (3.127) is the unitary matrix (a generalisation of (2.216))

$$S_R = \begin{pmatrix} \exp[-\mathrm{i}(\omega/2)\boldsymbol{\sigma}\cdot\hat{\mathbf{n}}] & 0 \\ 0 & \exp[-\mathrm{i}(\omega/2)\boldsymbol{\sigma}\cdot\hat{\mathbf{n}}] \end{pmatrix}. \tag{3.128}$$

For the case of a *Lorentz boost* (B) of velocity β along the x^1-axis, the required transformation is

$$S_B = \exp(\gamma^0\gamma^1 y/2), \tag{3.129}$$

where $\cosh y = \gamma$ (see (2.142)). This can be demonstrated by showing that

$$\exp(\gamma^0\gamma^1 y/2) = I\cosh(y/2) + (\gamma^0\gamma^1)\sinh(y/2), \tag{3.130}$$

and, using this relation, that

$$\begin{aligned} S_B\gamma^0 S_B^{-1} &= \gamma^0\cosh y - \gamma^1 \sinh y, \\ S_B\gamma^1 S_B^{-1} &= -\gamma^0 \sinh y + \gamma^1\cosh y, \end{aligned} \tag{3.131}$$

which is the required form of (3.126), since the coefficients for this boost are $a^0_0 = \cosh y$ and $a^1_0 = -\sinh y$ etc.

The proof of invariance under a *general proper Lorentz transformation* is as follows. Consider the infinitesimal transformation

$$a^\nu_\mu = g^\nu_\mu + \lambda\varepsilon^\nu_\mu, \qquad \lambda \to 0. \tag{3.132}$$

* The proof that the form of the γ-matrices is invariant (Pauli's fundamental theorem) is given in Sakurai (1967).

Applicaton of (2.158) gives $\varepsilon^{\mu}_{\nu} = -\varepsilon^{\nu}_{\mu}$ or $\varepsilon^{\mu\nu} = -\varepsilon^{\nu\mu}$. If we write the transformation S in the form

$$S = 1 + \lambda T, \qquad S^{-1} = 1 - \lambda T, \tag{3.133}$$

then, inserting (3.132) and (3.133) in (3.126), gives

$$(1 - \lambda T)\gamma^{\nu}(1 + \lambda T) = (g^{\nu}_{\mu} + \lambda\varepsilon^{\nu}_{\mu})\gamma^{\mu},$$

or

$$\gamma^{\nu} + \lambda[\gamma^{\nu}, T] = \gamma^{\nu} + \lambda\varepsilon^{\nu}_{\mu}\gamma^{\mu}, \tag{3.134}$$

where λ^2 terms are neglected. Hence

$$[\gamma^{\nu}, T] = \varepsilon^{\nu}_{\mu}\gamma^{\mu} = \varepsilon^{\nu\mu}\gamma_{\mu}. \tag{3.135}$$

Thus the problem is reduced to finding T which satisfies (3.135), and with tr $T = 0$ (which follows from the normalisation det $S = 1$). A suitable form, which can be checked by substitution, is

$$T = \frac{\varepsilon^{\nu\mu}}{8}[\gamma_{\nu}, \gamma_{\mu}] = -\mathrm{i}\frac{\varepsilon^{\nu\mu}}{4}\sigma_{\nu\mu}. \tag{3.136}$$

Finite transformations gives an exponential for S and, for the specific case ε^{01} treated above, it can be seen that (3.136) is equivalent to (3.129). Thus the Dirac equation is invariant under proper Lorentz transformations.

For the parity (P) transformation, $a^0_0 = 1$, $a^1_1 = a^2_2 = a^3_3 = -1$. Hence (3.126) becomes

$$S_P\gamma^0 S_P^{-1} = \gamma^0,$$
$$S_P\gamma^k S_P^{-1} = -\gamma^k \qquad (k = 1, 2, 3), \tag{3.137}$$

which is satisfied by (ignoring an arbitrary phase factor)

$$S_P = \gamma^0. \tag{3.138}$$

The Hamiltonian H (3.77) commutes with γ^0 only if $\mathbf{p} = 0$. Application of S_P to the spinor solutions (3.106) indicates that the positive and negative energy rest states have opposite intrinsic parity. The assignment of opposite intrinsic parities for fermions and antifermions is supported by the observation that the polarisation vectors of the two photons in the decay of the $^1S_0$ positronium (e^+e^-) state are perpendicular. This implies a negative parity state, and using $P(e^+e^-) = P_{e^+}P_{e^-}(-1)^l$ with $l = 0$, gives $P_{e^+} = -P_{e^-}$. A further example is the pion ($q\bar{q}$ in $l = 0$ state), which has negative intrinsic parity because the q and $\bar{q}$ have opposite intrinsic parities ($P = (-1)^{l+1}$).

The matrix elements encountered in practical calculations involve, in general, the products of γ-matrices. It is possible to construct 16 linearly independent 4×4 matrices Γ_n from these products, namely

$$\Gamma_{\rm S} = I(1), \qquad \Gamma_{\rm V}^{\mu} = \gamma^{\mu}(4), \qquad \Gamma_{\rm T}^{\mu\nu} = \sigma^{\mu\nu}(6),$$
$$\Gamma_{\rm A}^{\mu} = \gamma^5\gamma^{\mu}(4), \qquad \Gamma_{\rm P} = \gamma^5(1). \tag{3.139}$$

The subscripts refer to the properties of the quantity $\bar{\psi}\Gamma_n\psi$ under Lorentz transformation (as discussed below), and the number of components is shown in parentheses. From (3.133) and (3.136) it follows that

$$S^{\dagger} = \gamma^0 S^{-1}\gamma^0. \tag{3.140}$$

Thus, for the transformation of $\bar{\psi}$ we have (this also holds for S_P)

$$\bar{\psi}' = \psi'^{\dagger}\gamma^0 = \psi^{\dagger}S^{\dagger}\gamma^0 = \psi^{\dagger}\gamma^0 S^{-1} = \bar{\psi}S^{-1}. \tag{3.141}$$

Therefore, under proper Lorentz transformations, the S, V and A terms transform as follows

$$\bar{\psi}\Gamma_{\rm S}\psi = \bar{\psi}\psi \to \bar{\psi}'\psi' = \bar{\psi}S^{-1}S\psi = \bar{\psi}\psi, \tag{3.142}$$

$$\bar{\psi}\Gamma_{\rm V}^{\mu}\psi = \bar{\psi}\gamma^{\mu}\psi \to \bar{\psi}'\gamma^{\mu}\psi' = \bar{\psi}S^{-1}\gamma^{\mu}S\psi = a_{\nu}^{\mu}\bar{\psi}\gamma^{\nu}\psi, \tag{3.143}$$

$$\bar{\psi}\Gamma_{\rm A}^{\mu}\psi = \bar{\psi}\gamma^5\gamma^{\mu}\psi \to \bar{\psi}'\gamma^5\gamma^{\mu}\psi' = \bar{\psi}S^{-1}\gamma^5\gamma^{\mu}S\psi = a_{\nu}^{\mu}\bar{\psi}\gamma^5\gamma^{\nu}\psi. \tag{3.144}$$

In deriving (3.144) the property $[S, \gamma^5] = 0$, which follows from $[\sigma^{\mu\nu}, \gamma^5] = 0$, has been used. The transformation properties under parity (S_P) are as follows, where the upper line is for $\mu = 0$ and the lower $\mu = k = 1, 2, 3$.

$$\bar{\psi}\psi \xrightarrow{P} \bar{\psi}\psi, \qquad \bar{\psi}\gamma^{\mu}\psi \xrightarrow{P} \begin{cases} \bar{\psi}\gamma^0\psi \\ -\bar{\psi}\gamma^k\psi \end{cases}, \qquad \bar{\psi}\gamma^5\gamma^{\mu}\psi \xrightarrow{P} \begin{cases} -\bar{\psi}\gamma^5\gamma^0\psi \\ \bar{\psi}\gamma^5\gamma^k\psi \end{cases}. \tag{3.145}$$

Thus $\Gamma_{\rm S}$, $\Gamma_{\rm V}$ and $\Gamma_{\rm A}$ transform as a scalar, vector and axial-vector respectively. Similarly it can be shown that $\Gamma_{\rm T}$ and $\Gamma_{\rm P}$ transform as a rank-two tensor and a pseudoscalar (i.e. changes sign under P) respectively. The $\Gamma_{\rm V}$ properties justify the use of the γ^{μ} vector notation.

3.2.6 Massless spin $\frac{1}{2}$ particles

The family of neutrinos consists of neutral spin $\frac{1}{2}$ particles, one or more of which may be massless. Except when indicated to the contrary, we will assume that $m_v = m_{\bar{v}} = 0$. For the massless case, the spinor solutions (3.107) reduce to the same form, namely

$$\not{p}u = 0, \qquad \not{p}v = 0, \qquad \bar{u}\not{p} = 0, \qquad \bar{v}\not{p} = 0. \tag{3.146}$$

For $m = 0$ there is no βm term in the expression for H in (3.77), and the commutation relations for $\boldsymbol{\alpha}$ (3.79) can be satisfied by the 2×2 Paul spin matrices. Such a two-component spinor was considered by Weyl in 1929, but was initially rejected since it meant that the parity operator ($P = \beta = \gamma^0$) was lost. The charge conjugation C is also lost; however, the combined operation CP is still valid as discussed below.

Returning to the four-component spinors, the helicity eigenstates can be found by expressing the spin operator $\mathbf{\Sigma}$ (3.96) in the following form

$$\mathbf{\Sigma} = \gamma^5\gamma^0\boldsymbol{\gamma}, \tag{3.147}$$

which can be shown using (3.85) and (3.86). Starting with the free particle Dirac equation, and multiplying from the left by $\gamma^5\gamma^0$, we obtain (with $E = |\mathbf{p}|$)

$$\gamma^5\gamma^0\boldsymbol{\gamma}\cdot\mathbf{p}u = \gamma^5\gamma^0 E\gamma^0 u,$$

i.e.
$$\mathbf{\Sigma}\cdot\mathbf{p}u = E\gamma^5 u. \tag{3.148}$$

Now, from (3.99) we know that $\mathbf{\Sigma}\cdot\mathbf{p} = \pm|\mathbf{p}|$, so $\gamma^5 u = \pm u$. The two solutions are thus

$$\gamma^5 u = u \quad : \mathbf{\Sigma}\cdot\mathbf{p}u = Eu, \qquad h = 1 \tag{3.149}$$

$$\gamma^5 u = -u \quad : \mathbf{\Sigma}\cdot\mathbf{p}u = -Eu, \qquad h = -1 \tag{3.150}$$

That is, the solution $\gamma^5 u = u$ has helicity $h = 1$ (spin parallel to $\mathbf{p}$), whereas $\gamma^5 u = -u$ has $h = -1$ (spin antiparallel to $\mathbf{p}$). The wave function can be split into lefthanded ($h = -1$) and righthanded ($h = 1$) parts, as follows

$$u = u_\mathrm{L} + u_\mathrm{R}, \tag{3.151}$$

with the properties $\gamma^5 u_\mathrm{L} = -u_\mathrm{L}$ and $\gamma^5 u_\mathrm{R} = u_\mathrm{R}$, hence

$$u_\mathrm{L} = \frac{(1-\gamma^5)}{2}u, \qquad u_\mathrm{R} = \frac{(1+\gamma^5)}{2}u. \tag{3.152}$$

It is useful to define left- and righthanded projection operators as follows

$$\Lambda_\mathrm{L} = \frac{(1-\gamma^5)}{2}, \qquad \Lambda_\mathrm{R} = \frac{(1+\gamma^5)}{2}, \tag{3.153}$$

which have the properties $\Lambda_\mathrm{L}u_\mathrm{L} = u_\mathrm{L}$, $\Lambda_\mathrm{L}u_\mathrm{R} = 0$, etc.

The spin projection operator Λ_s for a particle with non-zero mass (3.119) reduces to the form of (3.153) in the extreme relativistic limit $E \gg m$. Boosting antiparallel to the spin direction in the rest frame, the positive helicity component becomes $s_+^\mu \to p^\mu/m$, so $\not{s}_+ u = u$. Thus $\Lambda_s = (1+\gamma^5)/2$ for the positive energy spinor. For the negative energy v spinor, $\Lambda_s = (1-\gamma^5)/2$. For a particle of non-zero mass, the spinor can also be written in the form of (3.151) and (3.152). Using (3.107), and the property (3.87), gives

$\not{p}u_R = mu_L$ and $\not{p}u_L = mu_R$, which shows how the solutions decouple in the massless limit. The operator γ^5 is important in our understanding of weak interactions and is called the *chirality* operator. Note that for the Dirac Hamiltonian (3.77) $[H, \gamma^5] = 0$, for $m = 0$.

The normalisation of the spinors for $m = 0$ (cf. (3.109)) is

$$u^\dagger u = v^\dagger v = 2E. \tag{3.154}$$

3.2.7 Dirac particle in an electromagnetic field

In the presence of an electromagnetic potential $A^\mu = (A^0, \mathbf{A})$ the wave equation for a particle of charge e is obtained by making the (minimal) substitution (see Section 2.2.2)

$$p^\mu \to p^\mu - eA^\mu, \qquad \partial^\mu \to D^\mu = \partial^\mu + \mathrm{i}eA^\mu. \tag{3.155}$$

The Dirac equation (3.77) becomes, writing $D^\mu = (D^0, -\mathbf{D})$,

$$\mathrm{i}\frac{\partial \psi}{\partial t} = (-\mathrm{i}\boldsymbol{\alpha}\cdot\mathbf{D} + \beta m + eA^0)\psi. \tag{3.156}$$

This introduces an effective potential $e(A^0 - \boldsymbol{\alpha}\cdot\mathbf{A})$ into $\hat{H}$. The *magnetic moment* of a Dirac particle can be found by considering the non-relativistic limit of (3.156). In this limit $\hat{H} = m + H_{nr}$, where H_{nr} is the non-relativistic Hamiltonian. From (3.156), this gives

$$H_{nr} = -\mathrm{i}\boldsymbol{\alpha}\cdot\mathbf{D} + \beta m + eA^0 I - mI. \tag{3.157}$$

The effect of H_{nr} acting on ψ, in terms of two-component spinors, is

$$H_{nr}\begin{pmatrix}\psi_L\\ \psi_S\end{pmatrix} = -\mathrm{i}\boldsymbol{\sigma}\cdot\mathbf{D}\begin{pmatrix}\psi_S\\ \psi_L\end{pmatrix} - 2m\begin{pmatrix}0\\ \psi_S\end{pmatrix} + eA^0\begin{pmatrix}\psi_L\\ \psi_S\end{pmatrix}, \tag{3.158}$$

where ψ_L and ψ_S still contain the space-time dependence, and use has been made of (3.81) and (3.85). The second of these two coupled equations gives

$$(H_{nr} + 2m - eA^0)\psi_S = -\mathrm{i}\boldsymbol{\sigma}\cdot\mathbf{D}\psi_L. \tag{3.159}$$

For small kinetic and field energies we have

$$\psi_S \simeq -\mathrm{i}(\boldsymbol{\sigma}\cdot\mathbf{D}/2m)\psi_L, \tag{3.160}$$

so ψ_S is 'small' compared to the 'large' component ψ_L. Substituting ψ_S in the first equation of (3.158) gives

$$H_{nr}\psi_L = [-(\boldsymbol{\sigma}\cdot\mathbf{D})^2/2m + eA^0]\psi_L. \tag{3.161}$$

Using (2.215), $(\boldsymbol{\sigma}\cdot\mathbf{D})^2$ can be written as

$$\begin{aligned}(\boldsymbol{\sigma}\cdot\mathbf{D})^2 &= \mathbf{D}^2 + \mathrm{i}\boldsymbol{\sigma}\cdot\mathbf{D}\times\mathbf{D} \\ &= \mathbf{D}^2 + \mathrm{i}e\boldsymbol{\sigma}\cdot(\mathbf{p}\times\mathbf{A}+\mathbf{A}\times\mathbf{p}) \\ &= \mathbf{D}^2 + e\boldsymbol{\sigma}\cdot\mathbf{B},\end{aligned} \tag{3.162}$$

where $\mathbf{B} = \nabla\times\mathbf{A}$, and use has been made of the relation

$$(\mathbf{p}\times\mathbf{A}+\mathbf{A}\times\mathbf{p})\psi = -\mathrm{i}[\nabla\times(\mathbf{A}\psi)+\mathbf{A}\times(\nabla\psi)] = -i(\nabla\times\mathbf{A})\psi.$$

Thus (3.161) can be written in the form of the non-relativistic Pauli equation

$$H_{\mathrm{nr}}\psi_{\mathrm{L}} = \left[\frac{(\mathbf{p}-e\mathbf{A})^2}{2m} - \frac{e}{2m}\boldsymbol{\sigma}\cdot\mathbf{B} + eA^0\right]\psi_{\mathrm{L}}. \tag{3.163}$$

The term $(-e\boldsymbol{\sigma}\cdot\mathbf{B}/2m)$ is of the form $-\boldsymbol{\mu}\cdot\mathbf{B}$, describing the potential energy of a magnetic moment $\boldsymbol{\mu} = e\boldsymbol{\sigma}/2m$. Since the spin $\mathbf{s} = \boldsymbol{\sigma}/2$ (Section 2.4), the gyromagnetic ratio g is

$$g = \boldsymbol{\mu}\bigg/\left(\frac{e}{2m}\mathbf{s}\right) = 2. \tag{3.164}$$

The values of g for the electron and the muon are very close to this prediction. The small deviations in $g-2$ can be accurately predicted in quantum electrodynamics (Section 4.9.4). However, for the proton $g = 5.586$, indicative that the proton is not point-like.

The wave equation in the presence of an electromagnetic field should provide a consistent treatment for both particles and antiparticles. For an electron (charge $-e$) the wave function ψ is the solution of

$$[\gamma^\mu(\mathrm{i}\,\partial_\mu + eA_\mu) - m]\psi = 0. \tag{3.165}$$

A similar equation should exist for the positron (charge $+e$) with wave function ψ_C satisfying

$$[\gamma^\mu(\mathrm{i}\,\partial_\mu - eA_\mu) - m]\psi_C = 0, \tag{3.166}$$

In the specific representation (3.85), $\gamma^2(\gamma^\mu)^* = -\gamma^\mu\gamma^2$. Hence, taking the complex conjugate of (3.165), then multiplying by $\mathrm{i}\gamma^2$ and comparing with (3.166), gives

$$\psi_C = \mathrm{i}\gamma^2\psi^* = \mathrm{i}\gamma^2\gamma^0\bar{\psi}^{\mathrm{t}} = C'\bar{\psi}^{\mathrm{t}}, \tag{3.167}$$

where A_μ is taken to be real, the phase of ψ_C is arbitrary and t denotes transpose. Note that here, and in the rest of this section, equivalent transformations exist in other representations (see Appendix B). Thus

(3.167) links the solutions for the electron and positron. However, this interpretation should be made with care, as the Dirac equation is a single particle theory. In the equation for the electron, the negative energy solutions were interpreted as being equivalent to a positron. Alternatively one could equally well start from (3.166) and identify the negative energy solutions of ψ_C with the electron.

A more complete discussion of this problem involves the use of field theory, which naturally provides a framework for the description of multiparticle (e.g. e^- and e^+) states. Following a prescription similar to that developed for the scalar field (Sections 3.1.2 and 3.1.5), creation ($a_r^\dagger$) and destruction (a_r) operators for e^- and creation ($b_r^\dagger$) and destruction (b_r) operators for e^+ can be defined. The index $r = 1, 2$ refers to the possible components of the spinors u_r and v_r. Occupation numbers for e^- ($N_r^- = a^\dagger a$) and for e^+ ($N_r^+ = b^\dagger b$) can be defined. However, in order to obtain the following form for the energy operator ($E_p = +(\mathbf{p}^2 + m^2)^{1/2}$),

$$H = \sum_{p,r} E_p[N_r^-(\mathbf{p}) + N_r^+(\mathbf{p}) - 2], \tag{3.168}$$

it is necessary to assume that operators a and b *anticommute*, e.g.

$$\{a_r(\mathbf{p}), a_s^\dagger(\mathbf{p}')\} = \delta_{rs}\delta_{pp'}, \qquad \{a_r, a_s\} = 0, \tag{3.169}$$

rather than commute as for the scalar field. For the Dirac field, commuting operators would change the sign of $N_r^+(\mathbf{p})$ in (3.168), leading to unacceptable negative energy values. The occupation number operator N_r^- has, using (3.169), the following property

$$(N_r^-)^2 = a_r^\dagger a_r a_r^\dagger a_r = a_r^\dagger a_r = N_r^-. \tag{3.170}$$

A similar relation holds for N_r^+, so $N_r^\pm$ have eigenvalues 0 and 1 only. This is the *Pauli exclusion principle* for particles obeying Fermi–Dirac statistics. More details of the Dirac field are given in Appendix F.

Charge conjugation C is a unitary operator, which transforms a particle to an antiparticle, changing the signs of the internal quantum numbers (charge, strangeness, baryon number, etc.), but leaving the mechanical properties (e.g. $\mathbf{p}$) unchanged. Note that the operator C' defined in (3.167) reverses the sign of the energy, and so it cannot be identified directly with C. For the Dirac field it can be shown that the total charge operator Q changes sign under C

$$CQC^{-1} = -Q, \qquad Q = -e\sum_{p,r}[N_r^-(\mathbf{p}) - N_r^+(\mathbf{p})]. \tag{3.171}$$

Under C' the wave function transforms to $\psi_C = i\gamma^2\gamma^0\bar{\psi}^t$ (3.167), hence the bilinear covariants (3.139) transform as, using (3.85) etc.,

$$\bar{\psi}_a\Gamma_i\psi_b \xrightarrow{C'} C^i\bar{\psi}_b\Gamma_i\psi_a, \tag{3.172}$$

where $C^i = 1 \qquad i = \mathrm{V, T},$

$\qquad = -1 \qquad i = \mathrm{S, A, P}.$

The electromagnetic current j^μ and the charge Q both change sign under C. Thus C transforms a positive energy e^- current into a positive energy e^+ current. The operator $C' = i\gamma^2\gamma^0$ (3.167) acts on the single particle states. From (3.172) we have $\bar{\psi}_C\gamma^\mu\psi_C = \bar{\psi}\gamma^\mu\psi$, i.e. does not have the required sign change. An additional minus sign must be added for every negative energy e^- state when constructing the matrix element from Feynman diagrams in terms of these single particle states. The operator C transforms positive energy spin up electrons into positive energy spin up positrons. This can be seen by applying C' to the rest frame solutions (3.106).

In a more complete field theory approach, the transformation (3.172) includes the products of the intrinsic charge conjugation values C^{a} and C^{b} of the particles a and b. Furthermore, the bilinear covariant term should be antisymmetrised. This follows from a more general field theory requirement that Lagrangians must be symmetrised in their boson fields and antisymmetrised in their fermion fields. This process has the bonus of removing the infinite vacuum terms appearing in the expressions for the physical variables as well as giving a sign change in j^μ.

The Lagrangian $\mathscr{L} = -j_\mu A^\mu$ for the $e\gamma$ interaction is invariant under C if both j_μ and A^μ change sign. The photon associated with field A^μ is neutral and so self-conjugate (particle not distinguished from its anti-particle), with intrinsic charge conjugation $C_\gamma = -1$. For an n photon state, therefore, $C = (-1)^n$. An example is the classification of the (e^+e^-) positronium system. C-invariance implies that the ${}^1\mathrm{S}_0$ singlet state decays to two photons, whereas the ${}^3\mathrm{S}_1$ triplet state decays to three photons.

The Dirac equation (3.156) is covariant under the *time reversal transformation* $x(t, \mathbf{x}) \rightarrow x'(-t, \mathbf{x})$, if the transformed wave function $\psi_T(x')$ satisfies

$$i\frac{\partial\psi_T(x')}{\partial t'} = [\boldsymbol{\alpha}\cdot(\mathbf{p}' - e\mathbf{A}') + \beta m + e(A^0)']\psi_T(x'),$$

or

$$-i\frac{\partial\psi_T(x')}{\partial t} = [\boldsymbol{\alpha}\cdot(\mathbf{p} + e\mathbf{A}) + \beta m + eA^0]\psi_T(x'), \tag{3.173}$$

where $\mathbf{A}' = -\mathbf{A}$ and $(A^0)' = A^0$, because the former is generated by currents and the latter by charges. In the case of the Schrödinger equation it was found that T was not unitary, and that $\psi_T(x') = \psi^*(x)$ (Section 2.5.1). Under complex conjugation $\alpha_1^* = \alpha_1$, $\alpha_2^* = -\alpha_2$, $\alpha_3^* = \alpha_3$ and $\beta^* = \beta$ (from

(3.85)). Thus, taking the complex conjugate of (3.156), multiplying by $\alpha_1\alpha_3$ and commuting it through using (3.79), gives (choosing a phase factor $\exp(\mathrm{i}\phi) = -\mathrm{i}$)

$$\psi_T(x') = -\mathrm{i}\alpha^1\alpha^3\psi^*(x) = \mathrm{i}\gamma^1\gamma^3\psi^*(x). \tag{3.174}$$

Next, the effect of the combined transformation PCT is considered. Under PCT, (3.156) becomes

$$\mathrm{i}\frac{\partial\psi_{PCT}(x')}{\partial t'} = [\boldsymbol{\alpha}\cdot(\mathbf{p}' + e\mathbf{A}') + \beta m - eA^{0'}]\psi_{PCT}(x'),$$

or

$$-\mathrm{i}\frac{\partial\psi_{PCT}(x')}{\partial t} = [\boldsymbol{\alpha}\cdot(-\mathbf{p} + e\mathbf{A}) + \beta m - eA^0]\psi_{PCT}(x'). \tag{3.175}$$

Multiplying (3.156) by γ^5, commuting it through and comparing with (3.175), shows that the required operator is $\mathrm{i}\gamma^5$ (phase arbitrary). The combined application of (3.138), (3.167) and (3.174) on $\psi(x)$ gives

$$\psi_{PCT}(x') = \mathrm{i}\gamma^5\psi(x). \tag{3.176}$$

Thus there is an equivalence between positrons and negative energy electrons running backwards in space-time (Stückelberg–Feynman approach). Finally we note that similar considerations on P, C and T can be applied to the simpler case of the Klein–Gordon equation for a scalar particle of charge e in the presence of an electromagnetic field A, namely

$$(\mathrm{i}\,\partial^\mu - eA^\mu)(\mathrm{i}\,\partial_\mu - eA_\mu)\phi(x) = m^2\phi(x),$$

i.e.

$$[\Box^2 + m^2 + \mathrm{i}e(\partial^\mu A_\mu + A^\mu\partial_\mu) - e^2A^2]\phi(x) = 0. \tag{3.177}$$

This equation is invariant under P, and it is easy to show that $\phi_C(x) = \phi^*(x)$ and $\phi_T(x') = \phi^*(x)$.

3.2.8 Propagator for a Dirac particle

The Dirac propagator can be derived in a way similar to that for the Klein–Gordon equation. The required Green's function G_D is a solution to the equation

$$[\mathrm{i}\gamma_\mu D^\mu(x') - m]G_\mathrm{D}(x', x) = \delta^4(x' - x). \tag{3.178}$$

The free particle propagator ($D^\mu(x') = \partial'^\mu$, see (3.155)) is obtained by Fourier transforming into momentum space, giving

$$G_\mathrm{D}(x', x) = G_\mathrm{D}(x' - x) = \int \frac{\mathrm{d}^4p}{(2\pi)^4}\exp[-\mathrm{i}p\cdot(x' - x)]G_\mathrm{D}(p). \tag{3.179}$$

Substitution in (3.178) gives $(\not{p} - m)G_D(p) = 1$, or, using $\not{p}\not{p} = p^2$,

$$G_D(p) = \frac{1}{\not{p} - m} = \frac{\not{p} + m}{p^2 - m^2}, \qquad p^2 \neq m^2. \tag{3.180}$$

The poles can be handled by letting $m^2 \to m^2 - i\varepsilon(\varepsilon \to 0^+)$, as for the scalar case. For $t' > t$ the integration contour for (3.179) is over the lower half-plane (since p_0 in $\exp[-ip_0(t'-t)]$ cannot be positive imaginary), which encloses the positive energy pole at $p_0 = E - i\delta$ ($E = +(\mathbf{p}^2 + m^2)^{1/2}$, $\delta = \varepsilon/2E$). For $t > t'$, the contour is over the upper half-plane enclosing the pole $p_0 = -E + i\delta$ (see Fig. 3.5). Thus (3.179) gives

$$G_D(x'-x) = \int \frac{d^3p}{(2\pi)^3} \exp[i\mathbf{p}\cdot(\mathbf{x}'-\mathbf{x})] \int \frac{dp_0}{(2\pi)} \frac{\exp[-ip_0(t'-t)]}{p^2 - m^2 + i\varepsilon} (\not{p} + m)$$

$$= -i \int \frac{d^3p}{(2\pi)^3} \frac{\exp[i\mathbf{p}\cdot(\mathbf{x}'-\mathbf{x})]}{2E}$$

$$\times \{\exp[-iE(t'-t)](E\gamma^0 - \boldsymbol{\gamma}\cdot\mathbf{p} + m)\theta(t'-t)$$

$$\times \exp[iE(t'-t)](-E\gamma^0 - \boldsymbol{\gamma}\cdot\mathbf{p} + m)\theta(t-t')\}, \tag{3.181}$$

where use has been made of Cauchy's theorem. Equation (3.181) can be simplified, using (3.113), to

$$G_D(x'-x) = -i \int \frac{d^3p}{(2\pi)^3} \frac{m}{E} \{\Lambda_+ \exp[-ip\cdot(x'-x)]\theta(t'-t)$$

$$+ \Lambda_- \exp[ip\cdot(x'-x)]\theta(t-t')\}. \tag{3.182}$$

Fig. 3.5 Integration contours for Dirac propagator.

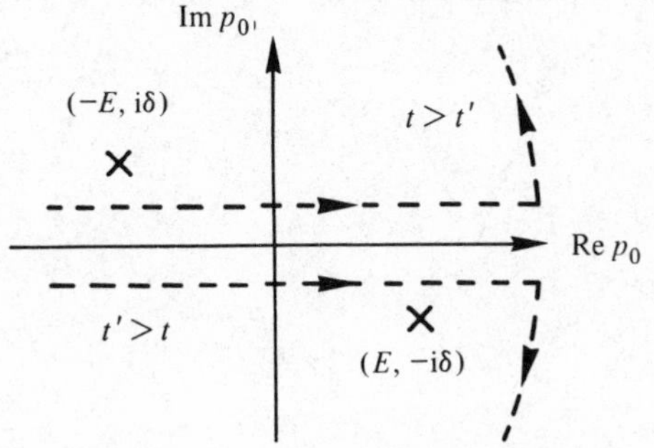

Using the expressions for $\Lambda_\pm$ from (3.113), together with (3.110), (3.111) and (3.100), equation (3.182) can be rearranged as follows

$$G_D(x'-x) = -i\theta(t'-t)\int \frac{d^3p}{2E}\,\psi_0(x')\bar{\psi}_0(x) + i\theta(t-t')\int \frac{d^3p}{2E}\,\psi_0(x')\bar{\psi}_0(x)$$

$$= -i\langle 0|T\{\psi(x')\bar{\psi}(x)\}|0\rangle, \tag{3.183}$$

where the $\theta(t'-t)$ term is for positive energy solutions and the $\theta(t-t')$ term is for negative energy solutions. The last line gives the expression for the propagator in terms of the quantum field operators $\psi(x')$ and $\bar{\psi}(x)$. The electron propagator corresponds to the creation of an electron from the vacuum at x and its annihilation at x'. For $t'>t$, an e^- is created at x, propagates forwards in time, and is annihilated at x' (Fig. 3.6(*a*)); whereas, for $t'<t$, an e^+ is created at x', propagates backwards in time, and is annihilated at x (Fig. 3.6(*b*)). The arrow in both graphs is thus in the same sense. The separation as to whether t or t' is earlier is not Lorentz invariant for space-like separation; however, these two diagrams are topologically equivalent and are subsequently treated together, with no distinction on time-ordering.

The form of (3.183) is similar to that for the scalar field (3.75), except that the time-ordering operator T gives $\psi_\alpha(x')\bar{\psi}_\beta(x)$ for $t>t'$ but $-\bar{\psi}_\beta(x)\psi_\alpha(x')$ for $t'>t$, the negative sign stemming from Fermi–Dirac statistics. The relationship (3.183) can be proved (see, e.g., Mandl and Shaw, 1984) by inserting $\psi(x)$ and $\bar{\psi}(x')$, defined in terms of the field operators $a_r(\mathbf{p}')$ and $a_r^\dagger(\mathbf{p})$, into the righthand side then integrating over p' using the anti-commutation relations $\{a_r(\mathbf{p}'), a_s^\dagger(\mathbf{p})\} = \delta_{rs}\,\delta^3(\mathbf{p}'-\mathbf{p})$ and the properties $a(\mathbf{p}')|0\rangle = 0$, etc. (see Appendix F).

Fig. 3.6 Propagation of an electron from x to x' for (*a*) $t'>t$ and (*b*) $t'<t$.

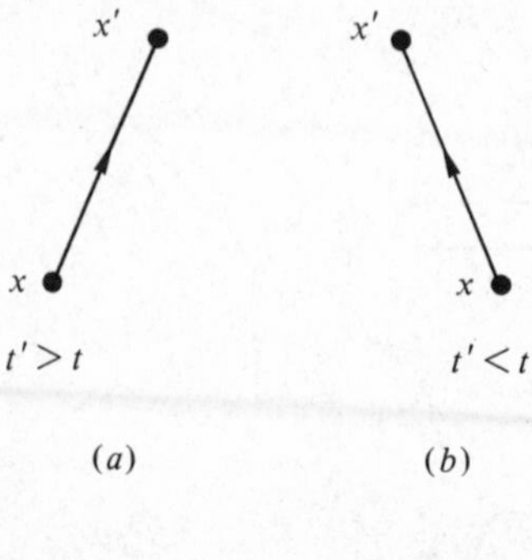

From (3.182), and the normalisation relations (3.109), the following results can be easily derived

$$i\int G_D(x'-x)\gamma_0\psi^{(+)}(x)\,d^3x=\theta(t'-t)\psi^{(+)}(x'),$$

$$-i\int G_D(x'-x)\gamma_0\psi^{(-)}(x)\,d^3x=\theta(t-t')\psi^{(-)}(x'). \tag{3.184}$$

where $\psi^{(+)}$ and $\psi^{(-)}$ refer to the positive and negative energy solutions respectively. Thus the propagator $G_D(x'-x)$, the Feynman propagator, carries the positive energy solutions forward in time (corresponding to say e^-) and the negative energy solutions backwards in time (corresponding to e^+). Equation (3.184) can be compared to its non-relativistic equivalent (3.63). Note that we have shown that the equation for the propagator (3.179), with $G_D(p)$ given by (3.180) and $m^2 \to m^2 - i\varepsilon$, is valid for all times.

In the presence of an electromagnetic field, (3.178) can be written in the form

$$[i\gamma_\mu\partial'^\mu - m]G_D(x',x)=\delta^4(x'-x)+e\gamma_\mu A^\mu(x')G_D(x',x).$$

Comparison with the non-relativistic case shows that $e\gamma_\mu A^\mu$ represents an effective potential V for the electromagnetic interaction. The wave function is given by

$$\psi(x')=\psi_0(x')+e\int d^4x_1 G_D(x'-x_1)\not{A}(x_1)\psi(x_1), \tag{3.185}$$

which can be demonstrated by operating on $\psi(x')$ with $(i\gamma_\mu\partial'^\mu - m)$ and noting that the righthand side reduces to $e\not{A}(x')\psi(x')$, i.e. the Dirac equation.

If we start with some initial state $\psi^i(t' \to -\infty)$, then we require to calculate the probability of obtaining a specific final plane wave state $\psi_0^f(t' \to \infty)$. That is, it is required to calculate the *scattering* or *S-matrix*

$$S_{fi}=\lim_{t'\to\infty}\int \psi_0^f(x')^\dagger\psi^i(x')\,d^3x', \tag{3.186}$$

where we assume, for the moment, that ψ^i has positive energy. Starting with (3.185) and using $G_D(x'-x_1)$ from (3.183), the expression for the S-matrix for the electromagnetic potential becomes, after integration over d^3x' and d^3p,

$$S_{fi}=\delta_{fi}-ie\int d^4x_1\bar{\psi}_0^f(x_1)\not{A}(x_1)\psi^i(x_1). \tag{3.187}$$

In this expression, the term δ_{fi} describes the possibility that the incident wave is propagated without any scatter taking place. Equation (3.187) refers to initial positive energy solutions ψ^{i}, starting in the remote past, and final positive energy solutions ψ^{f} in the remote future. For negative energy solutions, the roles of past and future are exchanged and there is a change of sign in the interaction term. Note that ψ^{i} is not, in general, a plane wave state. To first order in e, we have $\psi^{\mathrm{i}}(x_1)=\psi_0^{\mathrm{i}}(x_1)$; but for higher orders the solution must be calculated perturbatively, in a manner similar to that outlined for the non-relativistic case in Section 3.1.6. Feynman diagrams with the appropriate propagation terms are again used. A somewhat more rigorous treatment of the S-matrix in terms of field theory operators is outlined in Section 4.7. Before discussing the interactions of electrons and positrons, the propagators of vector particles are first considered.

3.3 SPIN 1 PARTICLES

3.3.1 Massless spin 1 particles

The classical electromagnetic field equations for the massless photon were discussed in Section 2.3.5. For a free photon, in the Lorentz gauge, the four-vector potential satisfies

$$\Box^2 A^\mu = 0, \qquad \text{with } A^\mu = \varepsilon^\mu \exp(-\mathrm{i}q\cdot x). \tag{3.188}$$

The Lorentz condition, $\partial_\mu A^\mu = 0$ (2.185), gives

$$\varepsilon\cdot q = \varepsilon^0 q^0 - \boldsymbol{\varepsilon}\cdot\mathbf{q} = 0. \tag{3.189}$$

This constraint leaves three independent polarisation vectors, as required to describe a spin 1 particle. Now the Lorentz condition is satisfied provided $\Box^2\chi = 0$ (from (2.184)), hence χ is of the form $\chi \sim \exp(-\mathrm{i}q\cdot x)$, and $q^2=0$. Therefore, if the solution (3.188) is to satisfy the gauge condition (2.184), then ε^μ must be invariant under

$$(\varepsilon')^\mu = \varepsilon^\mu + \lambda q^\mu, \tag{3.190}$$

where λ is some constant. That is, ε' and ε describe the same photon. Therefore, using (3.190), ε^0 can always be chosen to be zero. Equation (3.189) then reads $\boldsymbol{\varepsilon}\cdot\mathbf{q}=0$, so that only two independent possible polarisation states exist for a free photon. Useful choices are (for $\mathbf{q}$ along the z-axis) $\boldsymbol{\varepsilon}_x=(1,0,0)$ and $\boldsymbol{\varepsilon}_y=(0,1,0)$, describing linearly polarised photons or, for the description of circularly polarised photons, $\boldsymbol{\varepsilon}_{\mathrm{R}}=(\boldsymbol{\varepsilon}_x+\mathrm{i}\boldsymbol{\varepsilon}_y)/2^{1/2}$ and $\boldsymbol{\varepsilon}_{\mathrm{L}}=(\boldsymbol{\varepsilon}_x-\mathrm{i}\boldsymbol{\varepsilon}_y)/2^{1/2}$. The helicity values J_z of the states $\boldsymbol{\varepsilon}_{\mathrm{R}}$ and $\boldsymbol{\varepsilon}_{\mathrm{L}}$ can be found by considering a rotation of θ around the z-axis under which $\boldsymbol{\varepsilon}_{\mathrm{R}}$ and $\boldsymbol{\varepsilon}_{\mathrm{L}}$ transform to $\exp(-\mathrm{i}\theta)\boldsymbol{\varepsilon}_{\mathrm{R}}$ and $\exp(\mathrm{i}\theta)\boldsymbol{\varepsilon}_{\mathrm{L}}$ respectively. Comparison with the corresponding expression (2.200), given in terms of the angular momentum J_z, shows that $\boldsymbol{\varepsilon}_R$ and $\boldsymbol{\varepsilon}_L$ have helicity values 1 and -1 respectively. Note that there is no helicity zero state for a real photon, and that the square of the polarisation four-vector satisfies $\varepsilon \cdot \varepsilon = -1$.

The second quantisation of the electromagnetic field follows a similar recipe to that for scalar particles. This leads to creation and destruction operators which commute, and hence to Bose–Einstein statistics. However, some formal difficulties arise in the quantisation, due to there being more variables than internal degrees of freedom (see, e.g., Bjorken and Drell, 1965).

3.3.2 Massive spin 1 particles

If the substitution $\Box^2 + M^2$ for $\Box^2 (= \partial_\mu \partial^\mu)$ is made in (2.181), then the following equation is obtained

$$(\Box^2 + M^2) W^\nu - \partial^\nu (\partial_\mu W^\mu) = J^\nu, \tag{3.191}$$

where J^ν, in analogy to the photon case, represents a vector current coupled to the vector field W^ν. Multiplying (3.191) by ∂_ν gives

$$(\Box^2 + M^2) \partial_\nu W^\nu - \partial_\nu \partial^\nu (\partial_\mu W^\mu) = \partial_\nu J^\nu,$$

or

$$M^2 \partial_\nu W^\nu = \partial_\nu J^\nu. \tag{3.192}$$

Thus, for free particles ($J^\nu = 0$), $\partial_\nu W^\nu = 0$, so (3.191) becomes

$$(\Box^2 + M^2) W^\nu = 0. \tag{3.193}$$

This has solutions

$$W^\nu = \varepsilon^\nu \exp(-iq \cdot x), \tag{3.194}$$

with $q^2 = M^2$ (from (3.193)) and $\varepsilon^\nu q_\nu = 0$ (from $\partial_\nu W^\nu = 0$).

If a gauge transformation of the type

$$(W')^\nu = W^\nu + \partial^\nu \chi, \tag{3.195}$$

is made in (3.191), then the following equation is obtained

$$(\Box^2 + M^2) W^\nu - \partial^\nu \partial_\mu W^\mu + M^2 \partial^\nu \chi = J^\nu. \tag{3.196}$$

This equation is obviously different to (3.191), and only gives (3.191) for the case $M = 0$. Hence, for massive vector particles (including virtual photons), there is no further constraint and so there are three independent polarisation states, i.e. three independent four-vectors. In the Feynman graphs describing virtual vector bosons there will, in general, be both helicity 0 and 1 contributions. However, in the case of coupling to a conserved current ($\partial_\nu J^\nu = 0$), then $\partial_\nu W^\nu = 0$ from (3.192) (for $M^2 \neq 0$). Hence, $\varepsilon \cdot q = 0$, and so there are only helicity 1 contributions.

3.3.3 Propagators for spin 1 particles

The propagators for both massless and massive vector particles can be found in a similar fashion to that used for the scalar particle (Section 3.1.6). Corresponding to the free photon equation $g^{\mu\nu}\Box^2 A_\nu = 0$, the propagator or Green's function (3.63) is given by (see (3.67))

$$\Box'^2 G_{\rm P}^{\mu\nu}(x' - x) = \delta^4(x' - x) g^{\mu\nu}. \tag{3.197}$$

A Fourier transform to momentum space gives

$$G_{\rm P}^{\mu\nu}(x' - x) = \int \frac{{\rm d}^4 q}{(2\pi)^4} \exp[-iq\cdot(x' - x)] G_{\rm P}^{\mu\nu}(q^2),$$

hence, using (3.197),

$$G_{\rm P}^{\mu\nu}(q^2) = -g^{\mu\nu}/q^2, \qquad q^2 \neq 0. \tag{3.198}$$

Again, as for the scalar case, a prescription for handling the pole in $G_{\rm P}$ at $q^2 = 0$, is needed. This can be achieved by giving the photon a small negative imaginary mass (see (3.73)), so that

$$G_{\rm P}^{\mu\nu}(x' - x) = \int \frac{{\rm d}^4 q}{(2\pi)^4} \exp[-iq\cdot(x' - x)] \left(\frac{-g^{\mu\nu}}{q^2 + {\rm i}\varepsilon}\right). \tag{3.199}$$

Note that the above is for the Lorentz gauge. For the most general case $[g^{\mu\nu}\Box^2 - \partial^\mu \partial^\nu] A_\nu = 0$, there is no inverse possible until some gauge conditions are applied (see Appendix (C.2)).

The free massive vector particle (*Proca*) equation is

$$[g^{\mu\nu}(\Box^2 + M^2) - \partial^\nu \partial^\mu] W_\mu = 0, \tag{3.200}$$

and the propagator is given by

$$[g^{\mu\nu}(\Box'^2 + M^2) - (\partial')^\mu (\partial')^\nu] G_{\mu\nu}^{\rm V}(x' - x) = \delta^4(x' - x). \tag{3.201}$$

Proceeding as before this gives, in momentum space,

$$[g^{\mu\nu}(-q^2 + M^2) + q^\mu q^\nu] G_{\mu\nu}^{\rm V}(q^2) = 1. \tag{3.202}$$

To solve for $G_{\rm V}$ the inverse of the term in square brackets is needed. Lorentz invariance dictates that the inverse must be constructed from $g_{\mu\nu}$ and $q_\mu q_\nu$, and the coefficients A and B of these terms are determined from the matrix identity

$$[g^{\mu\nu}(-q^2 + M^2) + q^\mu q^\nu][A g_{\nu\lambda} + B q_\nu q_\lambda] = \delta^\mu_\lambda,$$

i.e.,

$$A(-q^2+M^2)\delta^\mu_\lambda + q^\mu q_\lambda[A + B(-q^2+M^2) + Bq^2] = \delta^\mu_\lambda, \quad (3.203)$$

where the relationship $g^{\mu\nu}g_{\nu\lambda} = \delta^\mu_\lambda$ has been used. Thus $A = -1/(q^2 - M^2)$ and $B = -A/M^2$. Hence, the propagator for a massive vector particle is

$$G^{\mu\nu}_V(q^2) = (-g^{\mu\nu} + q^\mu q^\nu/M^2)/(q^2 - M^2). \quad (3.204)$$

4

Quantum electrodynamics

In this chapter the propagator ideas which have been introduced are utilised in order to make calculations of processes involving electrons, positrons and photons. The calculations are perturbative and the lowest orders only are initially considered. The theory describing these particles, *quantum electrodynamics*, can be solved (at least in principle) to arbitrary order using the approach of Feynman diagrams. The rules for these diagrams will be introduced in a mainly non-rigorous way, using propagator ideas and intuitive arguments.

4.1 SCATTERING OF AN ELECTRON IN A COULOMB FIELD

The transition matrix element for this process is given by (3.187) and, to lowest order (Fig. 4.1(*a*)), is (noting that the charge of electron is $-e$)

$$S_{\mathrm{fi}} = \int \mathrm{d}^4x\bar{\psi}_{\mathrm{f}}(x)[\mathrm{i}e\not{A}(x)]\psi_i(x), \qquad (\mathrm{f} \neq \mathrm{i}), \tag{4.1}$$

where ψ_{i} and $\bar{\psi}_{\mathrm{f}}$ are plane wave states (3.100) normalised to a volume V, and the Coulomb potential is given by $A^\mu = (Ze/4\pi|\mathbf{x}|, 0, 0, 0)$ for a nucleus charge Ze $(e > 0)$. Inserting these ingredients in (4.1) gives

$$S_{\mathrm{fi}} = \frac{\mathrm{i}Ze^2}{4\pi V}\int \mathrm{d}^4x\,\frac{\bar{u}_{\mathrm{f}}\gamma^0 u_{\mathrm{i}}}{|\mathbf{x}|}\exp[\mathrm{i}(p_{\mathrm{f}} - p_{\mathrm{i}})\cdot x]. \tag{4.2}$$

Integrating over $\mathrm{d}x_0$ (see Appendix D) gives $2\pi\,\delta(E_{\mathrm{f}} - E_{\mathrm{i}})$, that is energy conservation in the static potential. Writing $\mathbf{q} = \mathbf{p}_{\mathrm{i}} - \mathbf{p}_{\mathrm{f}}$, the space integral part is $\int \mathrm{d}^3x \exp(\mathrm{i}\mathbf{q}\cdot\mathbf{x})/|\mathbf{x}| = 4\pi/|\mathbf{q}|^2$, which is the Fourier transform of the

Coulomb potential. Hence

$$S_{\mathrm{fi}} = \frac{\mathrm{i}Ze^2}{V} \bar{u}_{\mathrm{f}}\gamma^0 u_{\mathrm{i}} 2\pi\, \delta(E_{\mathrm{f}} - E_{\mathrm{i}})/|\mathbf{q}|^2. \tag{4.3}$$

In order to calculate the transition probability, the matrix element squared $|S_{\mathrm{fi}}|^2$ must be multiplied by the factor $\Phi = V\,\mathrm{d}^3p_{\mathrm{f}}/(2\pi)^3$, for the density of final states in a volume V (see (2.81)), and by the appropriate volume normalisation factors for the spinors. The expression (4.3) for S_{fi} contains $\delta(E_{\mathrm{f}} - E_{\mathrm{i}})$, which must be squared. Assuming the transition takes place over a time interval $-T/2$ to $T/2$ and in a volume V, then the double δ-function may be written (see, e.g., Gasiorowicz, 1966)

$$\delta^4(p_{\mathrm{f}} - p_{\mathrm{i}}) \lim_{V,T\to\infty} \int_{VT} \frac{\mathrm{d}^4x}{(2\pi)^4} \exp[\mathrm{i}(p_{\mathrm{f}} - p_{\mathrm{i}})\cdot x] = \delta^4(p_{\mathrm{f}} - p_{\mathrm{i}})VT/(2\pi)^4. \tag{4.4}$$

For the energy part alone, $[\delta(E_{\mathrm{f}} - E_{\mathrm{i}})]^2 = \delta(E_{\mathrm{f}} - E_{\mathrm{i}})T/2\pi$. The spinors are normalised by equation (3.109) to $2E$ particles in a volume V. Hence the transition probability per unit time is

$$\begin{aligned} \omega_{\mathrm{fi}} &= |S_{\mathrm{fi}}|^2 N_{\mathrm{i}} N_{\mathrm{f}} \Phi/T \\ &= \frac{Z^2\alpha^2}{VE_{\mathrm{i}}} \frac{|\bar{u}_{\mathrm{f}}\gamma^0 u_{\mathrm{i}}|^2}{|\mathbf{q}|^4} \frac{\mathrm{d}^3p_{\mathrm{f}}}{E_{\mathrm{f}}} \delta(E_{\mathrm{f}} - E_{\mathrm{i}}), \end{aligned} \tag{4.5}$$

where $N_{\mathrm{i}} = (2E_{\mathrm{i}})^{-1}$ and $N_{\mathrm{f}} = (2E_{\mathrm{f}})^{-1}$ are the spinor normalisations and $\alpha = e^2/4\pi$. The cross-section is the transition rate ω_{fi} divided by the flux of incident particles of velocity $\beta_{\mathrm{i}}(=p_{\mathrm{i}}/E_{\mathrm{i}})$, namely $J_{\mathrm{i}} = \beta_{\mathrm{i}}/V$. The differential

Fig. 4.1 Coulomb scattering to lowest order of (*a*) e^- and (*b*) e^+ by a static charge.

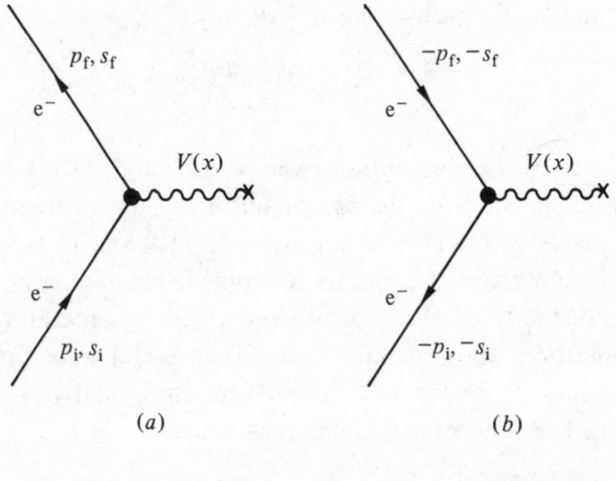

cross-section $d\sigma$ for scattering into a solid angle $d\Omega$ is thus (using $d^3p_f = p_f^2\,dp_f\,d\Omega$)

$$\frac{d\sigma}{d\Omega} = \frac{Z^2\alpha^2|\bar{u}_f\gamma^0u_i|^2}{\beta_iE_i|\mathbf{q}|^4}\,p_f^2\frac{dp_f}{E_f}\,\delta(E_f - E_i)$$

$$= \frac{Z^2\alpha^2}{|\mathbf{q}|^4}\,|\bar{u}_f\gamma^0u_i|^2, \tag{4.6}$$

where use has been made of the relation $p_f\,dp_f = E_f\,dE_f$ and integration over E_f has been performed.

So far the polarisation states of the electron have not been considered. The way in which the calculation proceeds depends on the experimental set-up for which the calculation is being performed. If the incident electron is unpolarised and the polarisation of the scattered electron is not measured, then we must sum over the final state spins and average over the initial spins. The latter gives a factor $1/(2s_e + 1) = 1/2$. A technique which can be used to calculate $|\bar{u}_f\gamma^0u_i|^2$, summed over the final spin states, is one which, in fact, can be applied generally. The method is to write the expression in the form of a trace, enabling various trace theorems to be utilised, and to replace the spinors by appropriate projection operators.

Consider the evaluation of the more general form

$$|\bar{u}_f\Gamma u_i|^2 = (\bar{u}_f\Gamma u_i)(u_i^\dagger\Gamma^\dagger\gamma^0u_f)$$

$$= \bar{u}_f\Gamma u_i\bar{u}_i\bar{\Gamma}u_f, \tag{4.7}$$

where $\bar{\Gamma} = \gamma^0\Gamma^\dagger\gamma^0$. Note that for $\Gamma = \gamma^\mu$ or $\gamma^\mu\gamma^5$ then $\bar{\Gamma} = \Gamma$. Now, inserting specific matrix indices, (4.7) can be rearranged in the form of a trace

$$(\bar{u}_f)_\alpha\Gamma_{\alpha\beta}(u_i\bar{u}_i)_{\beta\gamma}(\bar{\Gamma})_{\gamma\delta}(u_f)_\delta = (u_f\bar{u}_f)_{\delta\alpha}\Gamma_{\alpha\beta}(u_i\bar{u}_i)_{\beta\gamma}(\bar{\Gamma})_{\gamma\delta}$$

$$= \mathrm{tr}[(u_f\bar{u}_f)\Gamma(u_i\bar{u}_i)\bar{\Gamma}]. \tag{4.8}$$

The trace arises from the summation over index δ and, in the last line, the indices are dropped. Now the spinor terms $u\bar{u}$, when summed over spins, can be replaced by $\not{p} + m = 2m\Lambda_+$ (using (3.110) and (3.113)). Thus the use of specific forms for the spinors is avoided. The remaining task is the calculation of the trace of an expression which, in general, contains products of γ-matrices and four-momenta. Some useful trace properties are given in Appendix B. The proofs of these theorems is relatively straightforward and can be found in many standard texts.

Returning to the Coulomb scattering example, we have, summing over final state spins,

$$\begin{aligned}|\bar{u}_{\rm f}\gamma^0 u_{\rm i}|^2 &= \mathrm{tr}[(u_{\rm f}\bar{u}_{\rm f})\gamma^0(u_{\rm i}\bar{u}_{\rm i})\gamma^0] \\ &= \mathrm{tr}[(\not{p}_{\rm f} + m)\gamma^0(\not{p}_{\rm i} + m)\gamma^0] \\ &= \mathrm{tr}[\not{p}_{\rm f}\gamma^0\not{p}_{\rm i}\gamma^0 + m^2\gamma^0\gamma^0],\end{aligned} \tag{4.9}$$

where the property (B.1), that the trace of an odd number of γ-matrices is zero, has been used. Now $\not{p}_{\rm f}\gamma^0\not{p}_{\rm i}\gamma^0$ is of the form of (B.3) with $a_1 = p_{\rm f}$, $a_3 = p_{\rm i}$ and $a_2 = a_4 = (1, 0, 0, 0)$. Using these and $\gamma^0\gamma^0 = 1$ gives

$$|\bar{u}_{\rm f}\gamma^0 u_{\rm i}|^2 = 4(2E_{\rm f}E_{\rm i} - p_{\rm f}\cdot p_{\rm i} + m^2), \tag{4.10}$$

and the differential cross-section (4.6), averaging over initial spin states, becomes

$$\frac{d\sigma}{d\Omega} = \frac{2Z^2\alpha^2}{|\mathbf{q}|^4}(2E_{\rm f}E_{\rm i} - p_{\rm f}\cdot p_{\rm i} + m^2). \tag{4.11}$$

Defining θ to be the angle through which the electron is scattered, and writing $E_{\rm i} = E_{\rm f} = E$ and $|\mathbf{p}_{\rm i}| = |\mathbf{p}_{\rm f}| = p$, then $|\mathbf{q}|^2 = 4p^2\sin^2(\theta/2)$ and $p_{\rm f}\cdot p_{\rm i} = m^2 + 2p^2\sin^2(\theta/2)$, so that (4.11) may be cast in the *Mott scattering* form

$$\frac{d\sigma}{d\Omega} = \frac{Z^2\alpha^2}{4p^2\beta^2\sin^4\theta/2}(1 - \beta^2\sin^2\theta/2), \tag{4.12}$$

which reduces to the Rutherford scattering formula (2.96) in the limit $\beta \to 0$.

The differential cross-section for the scattering of a positron in a Coulomb field is, to lowest order, identical to that for an electron. The Feynman diagram for the process is shown in Fig. 4.1(*b*). The 'incoming' state $\psi_{\rm i}$ is interpreted as an electron with four-momentum $-p_{\rm f}$, propagating backwards in time. Similarly, the 'outgoing' state $\psi_{\rm f}$ corresponds to $-p_{\rm i}$. Thus the plane wave states are

$$\psi_{\rm i} = \frac{1}{V^{1/2}}v_{\rm f}\exp(\mathrm{i}p_{\rm f}\cdot x), \qquad \psi_{\rm f} = \frac{1}{V^{1/2}}v_{\rm i}\exp(\mathrm{i}p_{\rm i}\cdot x), \tag{4.13}$$

and the expression for $S_{\rm fi}$, corresponding to (4.2), is

$$S_{\rm fi} = \frac{-\mathrm{i}Ze^2}{4\pi V}\int d^4x\,\frac{\bar{v}_{\rm i}\gamma^0 v_{\rm f}}{|\mathbf{x}|}\exp[\mathrm{i}(p_{\rm f} - p_{\rm i})\cdot x]. \tag{4.14}$$

The rest of the calculation is similar to that for the electron. The sum over spins for the term involving the spinors (cf. (4.9)) is

$$|\bar{v}_i\gamma^0 v_f|^2 = \mathrm{tr}[(\not{p}_i - m)\gamma^0(\not{p}_f - m)\gamma^0]$$
$$= \mathrm{tr}[\not{p}_i\gamma^0\not{p}_f\gamma^0 + m^2]. \quad (4.15)$$

Hence the lowest order cross-section is the same for e^- and e^+. From charge conjugation, the behaviour of an electron in a potential A^μ is the same as that of a positron in $-A^\mu$. The lowest order cross-section is proportional to e^4, so the sign of A^μ does not enter. For the next order (e^6) this equality does not hold.

4.2 ELECTRON–PROTON SCATTERING

In this section the scattering of an electron by a (hypothetical) point-like proton is considered. This process can be considered as the scattering of an electron in the Coulomb field of a proton or *vice versa*. The solution for the potential $A^\mu(x)$ from $\Box^2 A^\mu = j^\mu$ (Lorentz gauge), is

$$A^\mu(x_1) = \int d^4x_2 G_F^{\mu\nu}(x_1 - x_2) j_\nu(x_2). \quad (4.16)$$

This can easily be checked by operating on $A^\mu(x_1)$ with $\Box^2$ and (using (3.199)) showing that the righthand side reduces to $j^\mu(x_1)$. Introducing $A^\mu(x_1)$ into S_{fi} in (3.187) gives, to lowest order,

$$S_{fi} = \int d^4x_1\, d^4x_2 [ie\bar{\psi}_3(x_1)\gamma_\mu\psi_1(x_1)] G_F^{\mu\nu}(x_1 - x_2) j_\nu(x_2). \quad (4.17)$$

The corresponding Feynman diagram is shown in Fig. 4.2. The initial and final state electrons (protons) are labelled 1(2) and 3(4) respectively.

Fig. 4.2 Lowest order diagram for e^-p scattering.

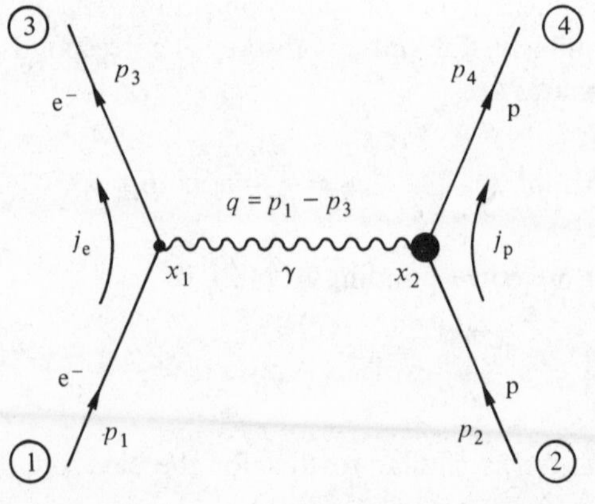

That is, we are considering an elastic scattering process of the type $1+2 \rightarrow 3+4$. Much of the discussion below is valid for a general elastic scattering process of this type. Note that the blob at the proton vertex in Fig. 4.2 indicates that the proton has structure which is (for the moment) neglected. It is plausible to identify the electromagnetic Dirac current for the proton with (see (3.92))

$$j^{\nu}(x_2) = e\bar{\psi}_4(x_2)\gamma^{\nu}\psi_2(x_2). \tag{4.18}$$

Thus the matrix element has the form of an electron current and a proton current, connected by a photon propagator. S_{fi} has the desirable feature of being symmetric in the currents; we could have started with the proton and inserted j^{ν} for the electron. Inserting the expression for the current (4.18) (with the wave functions in plane wave form, normalised to a volume V) and for the photon propagator (3.199) into (4.17), and performing the integrations, gives

$$S_{\mathrm{fi}} = \frac{(2\pi)^4}{V^2}\,\delta^4(p_3+p_4-p_1-p_2)$$
$$\times\left[\bar{u}_3(\mathrm{i}e\gamma^{\mu})u_1\left(\frac{-\mathrm{i}g_{\mu\nu}}{q^2+\mathrm{i}\varepsilon}\right)\bar{u}_4(-\mathrm{i}e\gamma^{\nu})u_2\right]. \tag{4.19}$$

The δ-function, arising from the integration of the exponentials from the plane waves and the propagator, expresses overall four-momentum conservation with $q = p_1 - p_3 = p_4 - p_2$. Referring to Fig. 4.2, the expression for S_{fi} in (4.19) can be written down by assigning (a) the spinors $u_1(\bar{u}_3)$ and $u_2(\bar{u}_4)$ to the ingoing (outgoing) electron and proton respectively; (b) vertex factors $\mathrm{i}e\gamma^{\mu}$ and $-\mathrm{i}e\gamma^{\nu}$ at the electron and proton vertices; (c) a photon propagator term $-\mathrm{i}g_{\mu\nu}/(q^2+\mathrm{i}\varepsilon)$; and (d) the δ-function expressing overall four-momentum conservation. These are some of the *Feynman rules* for QED, which are listed in Appendix C. The factors of i are such that the matrix elements for higher order diagrams can be written down using the same rules. Thus the Feynman diagrams, constructed in *momentum* rather than *configuration space*, together with the appropriate Feynman rules, can be considered as the starting point for practical calculations.

The transition rate ω_{fi} can be calculated in a similar way to that for Coulomb scattering. The spinor volume normalisation N is the product of factors $1/2E$ for each of the incident and final state particles, and the phase space factor Φ is the product of terms $V\,\mathrm{d}^3p_{\mathrm{f}}/(2\pi)^3$ for each of the final state particles (see equation (2.81)). Thus we have

$$\omega_{\text{fi}} = |S_{\text{fi}}|^2 N\Phi/T$$
$$= \frac{(2\pi)^4\, \delta^4(p_3 + p_4 - p_1 - p_2)|\mathcal{M}_{\text{fi}}|^2 V^3\, \mathrm{d}^3p_3\, \mathrm{d}^3p_4}{V^4(2E_1)(2E_2)(2\pi)^3 2E_3(2\pi)^3 2E_4}, \tag{4.20}$$

where the four-momenta for the incident electron and proton and final electron and proton are $p_1 = (E_1, \mathbf{p}_1)$, $p_2 = (E_2, \mathbf{p}_2)$, $p_3 = (E_3, \mathbf{p}_3)$ and $p_4 = (E_4, \mathbf{p}_4)$ respectively. In deriving (4.20), use has been made of (4.4) and $\mathcal{M}_{\text{fi}}$ is the term in square brackets in (4.19), namely

$$\mathcal{M}_{\text{fi}} = \bar{u}_3(\mathrm{i}e\gamma^\mu)u_1\left(\frac{-\mathrm{i}g_{\mu\nu}}{q^2 + \mathrm{i}\varepsilon}\right)\bar{u}_4(-\mathrm{i}e\gamma^\nu)u_2. \tag{4.21}$$

$\mathcal{M}_{\text{fi}}$ is Lorentz invariant and is called the *invariant amplitude*. The limit $\varepsilon \to 0$ for the propagator may be taken here for this process.

The cross-section is the transition rate divided by the incident flux. For a proton at rest $J_{\text{inc}} = \beta_1/V$ and, in this frame, the kinematic terms involving incident particles namely $E_1E_2\beta_1$, can be written

$$E_1E_2\beta_1 = m_2|\mathbf{p}_1| = [(\mathbf{p}_1 \cdot \mathbf{p}_2)^2 - m_1^2m_2^2]^{1/2}, \tag{4.22}$$

where m_1 and m_2 are the electron and proton masses, and the final form is constructed in a Lorentz invariant way. It can easily be checked that this latter form is true in any frame. The differential cross-section, which no longer contains the arbitrary normalisation volume V, thus becomes

$$\mathrm{d}\sigma = \frac{|\mathcal{M}_{\text{fi}}|^2}{4[(p_1 \cdot p_2)^2 - m_1^2m_2^2]^{1/2}} \cdot \mathscr{P}_2, \tag{4.23}$$

with $\mathscr{P}_2$, which is called the two-body Lorentz invariant phase space factor, given by

$$\mathscr{P}_2 = (2\pi)^4\, \delta^4(p_3 + p_4 - p_1 - p_2)\frac{\mathrm{d}^3p_3\, \mathrm{d}^3p_4}{(2\pi)^3 2E_3(2\pi)^3 2E_4}. \tag{4.24}$$

The term $\mathscr{P}_2$ is Lorentz invariant since it is the product of factors d^3p/E, which are separately invariant. This can be seen by considering, for example, a boost along the z-axis, under which $\mathrm{d}p'_z = E'\, \mathrm{d}p_z/E$. Alternatively, using the properties of the δ-function (Appendix D), one can directly show that $\int \mathrm{d}^3p/2E = \int \mathrm{d}^4p\, \delta(p^2 - m^2)$; this latter form being manifestly Lorentz invariant. Note that, for an n-body final state, $\mathscr{P}_n$ contains a factor $\mathrm{d}^3p/[(2\pi)^3 2E]$ for each final state particle. It is often useful to calculate the cross-section expected for $|\mathcal{M}_{\text{fi}}|^2 = \text{constant}$ (i.e. that arising from phase space alone) in order to compare with the full calculation, so that the effects of the 'physics' in $|\mathcal{M}_{\text{fi}}|^2$ can be seen.

The evaluation of $\mathscr{P}_2$ in the centre-of-mass system of the collision is straightforward. First we integrate over d^3p_4 using the δ^3-function giving

$$\mathscr{P}_2(\text{cms}) = \int \frac{\delta(E_3 + E_4 - W)\, d^3p_3}{16\pi^2 E_3 E_4}, \tag{4.25}$$

where $W = E_1 + E_2$ is the cms energy, and it is understood that E_4 is evaluated with p_4 given by momentum conservation from the δ^3 integral. The differential $d^3p_3 = p_3^2\, dp_3\, d\Omega = p_3 E_3\, dE_3\, d\Omega$, since $p_3\, dp_3 = E_3\, dE_3$. Thus

$$\mathscr{P}_2(\text{cms}) = \int \frac{\delta(E_3 + E_4 - W) p_3\, dE_3\, d\Omega}{16\pi^2 E_4}. \tag{4.26}$$

This integral is of the form $\int dE_3\, \delta[g(E_3)] = |dg/dE_3|^{-1}$, with

$$g(E_3) = E_3 + (E_3^2 - m_1^2 + m_2^2)^{1/2} - W,$$
$$dg/dE_3 = 1 + E_3/E_4 = W/E_4. \tag{4.27}$$

Hence, we obtain the simple expression

$$\mathscr{P}_2(\text{cms}) = \frac{p_3\, d\Omega}{16\pi^2 W}. \tag{4.28}$$

That is, the two-body phase space factor is proportional to the momentum p_3 $(=p_4)$ of the final state particle in the cms. In the cms it can be easily shown that $(p_1 \cdot p_2)^2 - m_1^2 m_2^2 = p_1^2 W^2 = p_1^2 s$, thus the differential cross-section (4.23) becomes

$$\frac{d\sigma}{d\Omega}(\text{cms}) = \frac{|\mathscr{M}_{\text{fi}}|^2 p_3}{64\pi^2 W^2 p_1} = \frac{|\mathscr{M}_{\text{fi}}|^2 p_3}{64\pi^2 s p_1}. \tag{4.29}$$

For elastic scattering, where particles of the same mass appear in the initial and final states, $p_3 = p_1$, thus simplifying (4.29). Equation (4.29) is general, and is also valid for bosons provided the wave functions are normalised to give $2E$ particles per unit volume.

To evaluate the cross-section in the system in which the initial proton is at rest, with $p_2 = (m_2, \mathbf{0})$, (which is often called the *lab system*), we can transform (4.29) to this system. Alternatively, we can evaluate $\mathscr{P}_2$ in the lab system giving, after integration over d^3p_4, and writing $d^3p_3 = p_3 E_3\, dp_3\, d\Omega$

$$\mathscr{P}_2(\text{lab}) = \int \frac{\delta(E_3 + E_4 - E_1 - m_2) p_3\, dE_3\, d\Omega}{16\pi^2 E_4}. \tag{4.30}$$

The momentum δ-function integration fixes $\mathbf{p}_4 = \mathbf{p}_1 - \mathbf{p}_3$, and thus E_4 is given by

$$E_4^2 = m_2^2 + p_1^2 + p_3^2 - 2p_1 p_3 \cos\theta, \tag{4.31}$$

where θ is the angle between the incident and scattered electrons. The integration over dE_3 can be carried out in the same way as for the cms. In this case, we define (cf. (4.27))

$$g(E_3) = E_3 + (m_2^2 + p_1^2 + p_3^2 - 2p_1 p_3 \cos\theta)^{1/2} - E_1 - m_2,$$
$$dg/dE_3 = 1 + [2E_3 - 2(p_1 E_3/p_3)\cos\theta]/(2E_4)$$
$$= [E_1 + m_2 - (p_1 E_3/p_3)\cos\theta]/E_4. \tag{4.32}$$

Thus

$$\mathscr{P}_2(\text{lab}) = \frac{p_3\, d\Omega}{16\pi^2[E_1 + m_2 - (p_1 E_3/p_3)\cos\theta]}, \tag{4.33}$$

and hence

$$\frac{d\sigma}{d\Omega}(\text{lab}) = \frac{|\mathscr{M}_{\text{fi}}|^2 p_3}{64\pi^2 m_2 p_1[E_1 + m_2 - (p_1 E_3/p_3)\cos\theta]}. \tag{4.34}$$

The forms for the differential cross-sections, (4.29) and (4.34), are general for any two-body scattering processes Note that $|\mathscr{M}_{\text{fi}}|^2$ must be evaluated in each case with the appropriate energy–momentum constraints. For final states containing three or more particles, this factorisation of $|\mathscr{M}_{\text{fi}}|^2$ from the phase space integration in the calculation of single particle distribution, will not, in general, hold.

The next step depends on the experimental arrangement for which the calculation is to be performed. If the incident electron and proton beams are unpolarised, and the final state polarisations are not measured, then we must sum over final state polarisations, and average over initial spins (i.e. factor $\frac{1}{4}$), giving from (4.21)

$$|\mathscr{M}_{\text{fi}}|^2 = \frac{e^4}{4}\sum_{\text{spins}}\left|\bar{u}_3\gamma^\mu u_1\left(\frac{1}{q^2}\right)\bar{u}_4\gamma_\mu u_2\right|^2$$
$$= \frac{e^4}{4q^4}\sum_{\text{spins}}[(\bar{u}_3\gamma^\mu u_1\bar{u}_1\gamma^\nu u_3)(\bar{u}_4\gamma_\mu u_2\bar{u}_2\gamma_\nu u_4)]$$
$$= \frac{e^4}{4q^4}L^{\mu\nu}W_{\mu\nu}, \tag{4.35}$$

where (4.7) has been used and with

$$L^{\mu\nu} = \text{tr}[(\not{p}_3 + m_1)\gamma^\mu(\not{p}_1 + m_1)\gamma^\nu],$$
$$W_{\mu\nu} = \text{tr}[(\not{p}_4 + m_2)\gamma_\mu(\not{p}_2 + m_2)\gamma_\nu]. \tag{4.36}$$

The separation of the spin sums in (4.35) into a product of a tensor $L^{\mu\nu}$ (for the transition $1 \to 3$) and a tensor $W_{\mu\nu}$ (for $2 \to 4$), arises because of the one-photon exchange nature of the interaction. In deriving (4.36), the projection operators (Section 3.2.4) for $\bar{u}u$, summed over the spin states, have been used, together with (4.7) and (4.8). The traces $L^{\mu\nu}$ and $W_{\mu\nu}$ are of the same form. Using the property (B.1), $L^{\mu\nu}$ can be simplified to

$$L^{\mu\nu} = \mathrm{tr}[\not{p}_3\gamma^\mu\not{p}_1\gamma^\nu + m_1^2\gamma^\mu\gamma^\nu]. \tag{4.37a}$$

The first term in (4.37a) is of the form (B.3), but with the corresponding four-vectors a_2 and a_4 having respectively only μ and ν components The term in m_1^2 involves $\mathrm{tr}(\gamma^\mu\gamma^\nu) = \mathrm{tr}(2g^{\mu\nu} - \gamma^\nu\gamma^\mu) = 4g^{\mu\nu}$. Thus, (4.37a) gives

$$L^{\mu\nu} = 4[p_3^\mu p_1^\nu + p_3^\nu p_1^\mu - g^{\mu\nu}(p_3 \cdot p_1) + m_1^2 g^{\mu\nu}]. \tag{4.37b}$$

The trace $W_{\mu\nu}$ gives a similar result, and so the product $L^{\mu\nu}W_{\mu\nu}$ in (4.35) may be formed. Using the property that $g^{\mu\nu}g_{\mu\nu} = 4$, and grouping together similar terms, gives

$$|\mathscr{M}_{\mathrm{fi}}|^2 = \frac{8e^4}{q^4}[(p_3 \cdot p_4)(p_1 \cdot p_2) + (p_3 \cdot p_2)(p_1 \cdot p_4) - m_1^2(p_4 \cdot p_2) - m_2^2(p_3 \cdot p_1) + 2m_1^2 m_2^2]. \tag{4.38}$$

This is the general and Lorentz invariant form for $|\mathscr{M}_{\mathrm{fi}}|^2$. Two limiting cases in the lab system are interesting to consider:

(i) *Non-relativistic limit*, $E_1 \ll m_2$. In this limit $E_1 \simeq E_3 = E$ and $E_4 \simeq m_2$. Hence, $p_3 \cdot p_4 = p_1 \cdot p_2 = p_3 \cdot p_2 = p_1 \cdot p_4 = Em_2$, $p_4 \cdot p_2 = m_2^2$ and $q^2 = -4|\mathbf{p}|^2 \sin^2(\theta/2)$. Thus

$$|\mathscr{M}_{\mathrm{fi}}|^2 = \frac{8e^4}{q^4}[2E^2m_2^2 + m_1^2m_2^2 - m_2^2(p_3 \cdot p_1)]. \tag{4.39}$$

Inserting this expression in (4.34) yields a form similar to (4.11) (for $Z = 1$)

$$\frac{d\sigma}{d\Omega}(\mathrm{lab}) = \frac{2\alpha^2}{q^4}[2E^2 + m_1^2 - (p_3 \cdot p_1)]. \tag{4.40}$$

(ii) *Extreme relativistic limit*, $E_1 \gg m_1$. Substituting $p_4 = p_1 + p_2 - p_3$ in (4.38), and neglecting terms in m_1^2, gives

$$|\mathscr{M}_{\mathrm{fi}}|^2 = \frac{8e^4}{q^4}[2(p_3 \cdot p_2)(p_1 \cdot p_2) + (p_3 \cdot p_1)(p_1 \cdot p_2 - p_3 \cdot p_2 - m_2^2)]. \tag{4.41}$$

Now $p_3 \cdot p_2 = m_2E_3$ and $p_1 \cdot p_2 = m_2E_1$. For $p_1 = E_1$, $p_3 = E_3$, then $p_3 \cdot p_1 = E_3E_1(1 - \cos\theta) = 2E_3E_1\sin^2(\theta/2)$. Furthermore $q^2 = (p_2 - p_4)^2 = 2m_2^2 -$

$2m_2E_4 = 2m_2(E_3 - E_1)$. With these simplifications (4.41) becomes

$$|\mathcal{M}_{\rm fi}|^2 = \frac{8e^4}{q^4}\left[2m_2^2E_3E_1 + 2E_3E_1\sin^2\frac{\theta}{2}\left(-\frac{q^2}{2} - m_2^2\right)\right]$$

$$= \frac{16e^4m_2^2E_3E_1}{q^4}\left[\cos^2\frac{\theta}{2} - \frac{q^2}{2m_2^2}\sin^2\frac{\theta}{2}\right]. \tag{4.42}$$

In (4.34), the term $[E_1 + m_2 - (p_1E_3/p_3)\cos\theta] = [E_1 + m_2 - E_1\cos\theta] = \mathrm{m_2E_1/E_3}$, thus

$$\frac{d\sigma}{d\Omega}(\text{lab}) = \frac{\alpha^2}{4E_1^2\sin^4\theta/2}\frac{\mathrm{E_3}}{\mathrm{E_1}}\left[\cos^2\frac{\theta}{2} - \frac{q^2}{2m_2^2}\sin^2\frac{\theta}{2}\right], \tag{4.43}$$

where $q^2 = -4E_1E_3\sin^2\theta/2$ has been used.

In the high energy limit it is also interesting to consider the cross-section in the cms. Neglecting the masses and using the invariant Mandelstam variables (2.171) to describe the scattering process, namely

$$\begin{aligned} s &= (p_1 + p_2)^2 = (p_3 + p_4)^2 \simeq 2p_1\cdot p_2 \simeq 2p_3\cdot p_4, \\ t &= (p_1 - p_3)^2 = (p_2 - p_4)^2 \simeq -2p_1\cdot p_3 \simeq -2p_2\cdot p_4, \\ u &= (p_1 - p_4)^2 = (p_2 - p_3)^2 \simeq -2p_1\cdot p_4 \simeq -2p_2\cdot p_3, \\ s &+ t + u \simeq 0, \end{aligned} \tag{4.44}$$

then $|\mathcal{M}_{\rm fi}|^2$ in (4.38) becomes

$$|\mathcal{M}_{\rm fi}|^2 = 32\pi^2\alpha^2(s^2 + u^2)/t^2, \tag{4.45}$$

and the differential cross-section (4.29) is (with $p_1 = p_3$)

$$\frac{d\sigma}{d\Omega}(\text{cms}) = \frac{\alpha^2}{2s}\left(\frac{s^2 + u^2}{t^2}\right). \tag{4.46}$$

If θ is the angle made by the scattered electron with respect to the incident direction, then $u = -2E^2(1 + \cos\theta) = -s(1 + \cos\theta)/2$. Thus

$$\frac{d\sigma}{d\Omega}(\text{cms}) = \frac{\alpha^2 s}{2t^2}\left[1 + \left(\frac{1 + \cos\theta}{2}\right)^2\right] = \frac{\alpha^2 s}{2t^2}[1 + (1 - y)^2], \tag{4.47}$$

where $y = (1 - \cos\theta)/2$. Alternatively, $y = (s + u)/s = p_2\cdot(p_1 - p_3)/p_1\cdot p_2 = p_2\cdot q/p_1\cdot p_2$ in invariant form. Further, $t = -s(1 - \cos\theta)/2$ so $d\Omega = 4\pi dt/s$.

Note that in the above discussion it has been assumed that both the particles are point-like Dirac particles. However, the proton is not point-like, and has a quark substructure (Chapter 7). The above formula applies for the electromagnetic elastic scattering of any two non-identical point-like spin $\frac{1}{2}$ particles (e.g. $e^-\mu^- \to e^-\mu^-$). For the scattering of quarks by electrons the appropriate quark charges ($2e/3$ or $-e/3$) must be used, giving additional factors of 4/9 or 1/9 respectively. Note that, as discussed in Chapter 7, equation (4.47) is extremely useful in the description of electron (or muon) quark scattering. To lowest order (that considered above), the cross-section does not depend on the actual sign of the charge (i.e. positive or negative) of the particle. However, this is not, in general, the case for higher order terms.

4.3 ELECTRON–ELECTRON AND ELECTRON–POSITRON INTERACTIONS

4.3.1 $e^- + e^- \to e^- + e^-$

For the electromagnetic scattering of two electrons (*Møller scattering*), there is an additional graph arising because of the identity of the electrons. The two graphs and the corresponding four-vectors are shown in Fig. 4.3. The invariant amplitude has two corresponding terms (cf. (4.21)), which can be constructed from the Feynman rules (Appendix C) as follows

$$\mathcal{M}_{\mathrm{fi}} = \bar{u}_3(\mathrm{i}e\gamma^\mu)u_1 \frac{(-\mathrm{i}g_{\mu\nu})}{(p_1 - p_3)^2} \bar{u}_4(\mathrm{i}e\gamma^\nu)u_2 - \bar{u}_4(\mathrm{i}e\gamma^\mu)u_1 \frac{(-\mathrm{i}g_{\mu\nu})}{(p_1 - p_4)^2} \bar{u}_3(\mathrm{i}e\gamma^\nu)u_2$$

$$= \mathrm{i}e^2 \left[\frac{\bar{u}_3\gamma^\mu u_1 \bar{u}_4 \gamma_\mu u_2}{(p_1 - p_3)^2} - \frac{\bar{u}_4\gamma^\mu u_1 \bar{u}_3 \gamma_\mu u_2}{(p_1 - p_4)^2} \right]. \tag{4.48}$$

The relative negative sign between the *direct* (Fig. 4.3(*a*)) and the *exchange* (Fig. 4.3(*b*)) terms ensures that the overall amplitude is antisymmetric under the interchange of the two initial or final state electrons, as required by Fermi–Dirac statistics. This relative factor can, in fact, be deduced from

Fig. 4.3 Lowest order diagrams for $e^-e^- \to e^-e^-$ (Møller scattering)

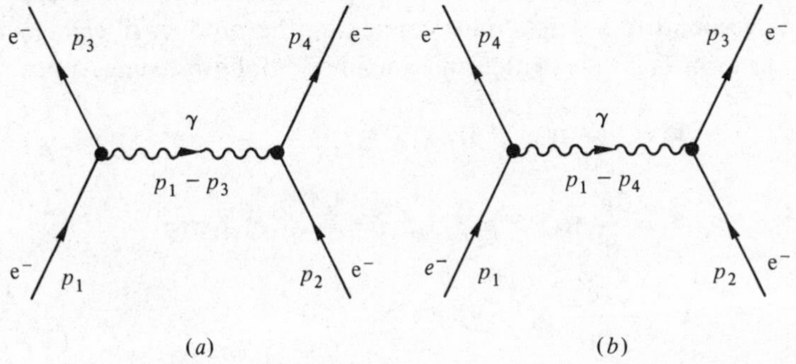

the properties of the annihilation and creation operators (a and $a^\dagger$) for fermions. Graph (a) contains the combination $a_3^\dagger a_1 a_4^\dagger a_2$, whereas for ($b$) it is $a_4^\dagger a_1 a_3^\dagger a_2 = -a_3^\dagger a_1 a_4^\dagger a_2$, with the latter form derived using the anti-commutation relations for the Dirac field (Section 3.2.7). The calculation of the differential cross-section proceeds in a similar way to that for e^-p scattering. First we compute

$$\begin{aligned}|\mathcal{M}_{\text{fi}}|^2 &= e^4\left[\frac{\bar{u}_3\gamma^\mu u_1\bar{u}_4\gamma_\mu u_2}{t} - \frac{\bar{u}_4\gamma^\mu u_1\bar{u}_3\gamma_\mu u_2}{u}\right]\left[\frac{\bar{u}_2\gamma_\nu u_4\bar{u}_1\gamma^\nu u_3}{t} - \frac{\bar{u}_2\gamma_\nu u_3\bar{u}_1\gamma^\nu u_4}{u}\right]\\ &= e^4\left[\frac{(\bar{u}_3\gamma^\mu u_1\bar{u}_1\gamma^\nu u_3)(\bar{u}_4\gamma_\mu u_2\bar{u}_2\gamma_\nu u_4)}{t^2} + \frac{(\bar{u}_4\gamma^\mu u_1\bar{u}_1\gamma^\nu u_4)(\bar{u}_3\gamma_\mu u_2\bar{u}_2\gamma_\nu u_3)}{u^2}\right.\\ &\quad\left. - \frac{(\bar{u}_3\gamma^\mu u_1\bar{u}_1\gamma^\nu u_4\bar{u}_4\gamma_\mu u_2\bar{u}_2\gamma_\nu u_3)}{tu} - \frac{(\bar{u}_4\gamma^\mu u_1\bar{u}_1\gamma^\nu u_3\bar{u}_3\gamma_\mu u_2\bar{u}_2\gamma_\nu u_4)}{tu}\right],\end{aligned} \tag{4.49}$$

where the terms in round brackets have been grouped together so that they can be written as a trace. Note that $\bar{u}_i\gamma^\mu u_j$ is a 1×1 matrix, and so such terms can be written in any order. Next we evaluate each of the four terms in (4.49), which we will call T_1 to T_4 respectively. For simplicity, mass terms will be neglected. Thus, inserting the projection operators (summed over final state spins) we have

$$\begin{aligned}T_1 &= \frac{e^4}{t^2}\,\text{tr}(\not{p}_3\gamma^\mu\not{p}_1\gamma^\nu)\,\text{tr}(\not{p}_4\gamma_\mu\not{p}_2\gamma_\nu)\\ &= \frac{16e^4}{t^2}[p_3^\mu p_1^\nu + p_3^\nu p_1^\mu - g^{\mu\nu}(p_1\cdot p_3)][p_{4\mu}p_{2\nu} + p_{4\nu}p_{2\mu} - g_{\mu\nu}(p_4\cdot p_2)]\\ &= \frac{32e^4}{t^2}[(p_1\cdot p_2)(p_3\cdot p_4) + (p_1\cdot p_4)(p_2\cdot p_3)],\end{aligned} \tag{4.50}$$

where use has been made of (B.3). Similarly,

$$T_2 = \frac{32e^4}{u^2}[(p_1\cdot p_2)(p_3\cdot p_4) + (p_1\cdot p_3)(p_2\cdot p_4)]. \tag{4.51}$$

Inspection of (4.49) shows that the interference terms, T_3 and T_4, each correspond to a single trace containing the product of eight γ-matrices. The term T_3 is thus written and simplified as follows (using Appendix B)

$$\begin{aligned}T_3 &= -\frac{e^4}{tu}\,\text{tr}(\not{p}_3\gamma^\mu\not{p}_1\gamma^\nu\not{p}_4\gamma_\mu\not{p}_2\gamma_\nu)\\ &= \frac{2e^4}{tu}\,\text{tr}(\not{p}_3\gamma^\mu\not{p}_1\not{p}_2\gamma_\mu\not{p}_4) = \frac{8e^4(p_1\cdot p_2)}{tu}\,\text{tr}(\not{p}_3\not{p}_4)\\ &= \frac{32e^4}{tu}(p_1\cdot p_2)(p_3\cdot p_4).\end{aligned} \tag{4.52}$$

The final term $T_4 = T_3$, since T_4 differs from T_3 only by the interchange of indices 3 and 4. Gathering these terms together and using the relations (4.44), $|\mathscr{M}_{\text{fi}}|^2$ becomes

$$|\mathscr{M}_{\text{fi}}|^2 = 32\pi^2\alpha^2\left[\frac{(s^2+u^2)}{t^2}+\frac{(s^2+t^2)}{u^2}+\frac{(2s^2)}{tu}\right], \tag{4.53}$$

where a factor $\frac{1}{4}$ for the average over the initial spin states has been introduced. Comparison with (4.45) shows that the extra contributions arise from the exchange and interference terms. The differential cross-section, using (4.29), is

$$\frac{d\sigma}{d\Omega}(\text{cms}) = \frac{\alpha^2}{2s}\left[\frac{(s^2+u^2)}{t^2}+\frac{(s^2+t^2)}{u^2}+\frac{2s^2}{tu}\right]. \tag{4.54}$$

In terms of the cms scattering angle θ, made by one of the electrons with respect to the incident direction, we have, putting $u/s = -(1+\cos\theta)/2$ and $t/s = -(1-\cos\theta)/2$,

$$\begin{aligned}\frac{d\sigma}{d\Omega}(\text{cms}) &= \frac{\alpha^2}{2s}\left[\frac{(1+\cos^4\theta/2)}{\sin^4\theta/2}+\frac{(1+\sin^4\theta/2)}{\cos^4\theta/2}+\frac{2}{\sin^2\theta/2\cos^2\theta/2}\right]\\ &= \frac{4\alpha^2}{s}\left[\frac{4}{\sin^4\theta}-\frac{2}{\sin^2\theta}+\frac{1}{4}\right].\end{aligned} \tag{4.55}$$

For small scattering angles, only the direct term is important and the Coulomb form is obtained, and this is independent of the statistics obeyed by the particle.

4.3.2 $e^- + e^+ \to e^- + e^+$

The two graphs which contribute to electron–positron elastic scattering (*Bhabha scattering*) are shown in Fig. 4.4. Using the Feynman rules (Appendix C), the invariant amplitude is

$$\mathscr{M}_{\text{fi}} = -ie^2\left[\frac{\bar{u}_3\gamma^\mu u_1\bar{v}_2\gamma_\mu v_4}{(p_1-p_3)^2}-\frac{\bar{v}_2\gamma^\mu u_1\bar{u}_3\gamma_\mu v_4}{(p_1+p_2)^2}\right]. \tag{4.56}$$

Note that the positrons (i.e. electrons running backwards in time) have spinors $\bar{v}$ and v for ingoing and outgoing lines respectively. There is again a relative minus sign between the two terms. The calculation of $d\sigma/d\Omega$ is carried out in a similar way to that for e^-e^- and gives

$$\frac{d\sigma}{d\Omega}(\text{cms}) = \frac{\alpha^2}{2s}\left[\frac{(s^2+u^2)}{t^2}+\frac{(t^2+u^2)}{s^2}+\frac{(2u^2)}{st}\right], \tag{4.57}$$

or, in terms of the cms scattering angle of the electron (noting that $u/s = -\cos^2\theta/2$ and $t/s = -\sin^2\theta/2$),

$$\frac{d\sigma}{d\Omega}(\text{cms}) = \frac{\alpha^2}{2s}\left[\frac{(1+\cos^4\theta/2)}{\sin^4\theta/2} + \frac{(1+\cos^2\theta)}{2} - \frac{(2\cos^4\theta/2)}{\sin^2\theta/2}\right] \tag{4.58a}$$

$$= \frac{\alpha^2}{4s}\left(\frac{3+\cos^2\theta}{1-\cos\theta}\right)^2. \tag{4.58b}$$

Fig. 4.4 (*a*) and (*b*) lowest order diagrams for $e^-e^+ \to e^-e^+$ (Bhabha scattering), (*c*) comparison of QED formula with data from MARK-J for $e^+e^- \to e^+e^-$ $s^{1/2} = 34.6$ GeV.

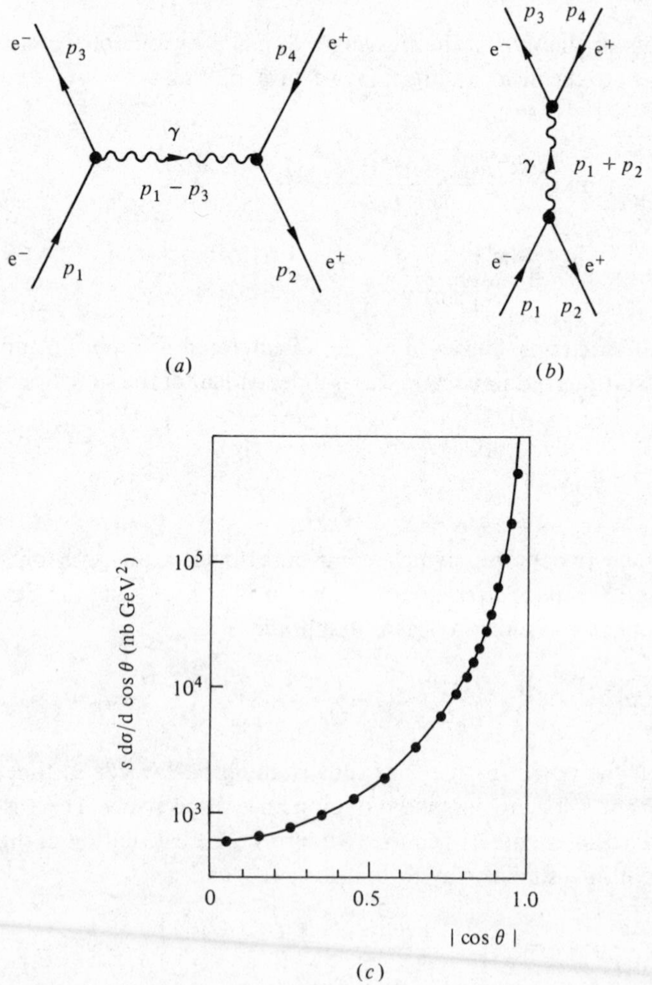

The order of the terms in (4.57) and (4.58a) corresponds to the diagrams for the direct (Fig. 4.4(*a*)), annihilation (Fig. 4.4(*b*)) and interference terms respectively. The direct and annihilation interactions are often referred to as *t*- and *s*-channel processes respectively. A comparison of equation (4.58) with some data is shown in Fig. 4.4(*c*).

4.3.3 $e^- + e^+ \to f + \bar{f}$

In Fig. 4.4(*b*) the final state e^+e^- particles are formed from the virtual photon which is itself created by the annihilation of the incident e^+e^-. The virtual photon, which is of course time-like, i.e. $s > 0$, can also produce any charged particle–antiparticle pair which is kinematically allowed ($s \geqslant 4m^2$). In particular, we next consider a final state of a spin $\frac{1}{2}$ fermion–antifermion pair, with charges $Q_f e$ and $-Q_f e$ respectively ($e > 0$), as shown in Fig. 4.5. Proceeding as before, we can write down

$$\mathscr{M}_{fi} = -ie^2 Q_f(\bar{v}_2\gamma^\mu u_1 \bar{u}_3\gamma_\mu v_4)/s$$

$$\begin{aligned}|\mathscr{M}_{fi}|^2 &= \frac{e^4 Q_f^2}{s^2}[(\bar{v}_2\gamma^\mu u_1\bar{u}_1\gamma^\nu v_2)(\bar{u}_3\gamma_\mu v_4\bar{v}_4\gamma_\nu u_3)] \\ &= \frac{e^4 Q_f^2}{s^2}\,\mathrm{tr}[\not{p}_2\gamma^\mu\not{p}_1\gamma^\nu]\,\mathrm{tr}[(\not{p}_3 + m_f)\gamma_\mu(\not{p}_4 - m_f)\gamma_\nu],\end{aligned} \tag{4.59}$$

where the electron mass has been neglected, and use has been made of (3.110) and (3.111) (i.e. summing over final state spins). Introducing a factor $\frac{1}{4}$ for the average over initial spins (i.e. assuming unpolarised beams), and carrying out the evaluation of the traces as before, gives

$$\begin{aligned}|\mathscr{M}_{fi}|^2 &= \frac{8e^4 Q_f^2}{s^2}[(p_1\cdot p_4)(p_2\cdot p_3) + (p_1\cdot p_3)(p_2\cdot p_4) + m_f^2(p_1\cdot p_2)] \\ &= \frac{2e^4 Q_f^2}{s}[(m_f^2 - u)^2 + (m_f^2 - t)^2 + 2m_f^2 s],\end{aligned} \tag{4.60}$$

Fig. 4.5 Lowest order diagram for $e^+e^- \to f\bar{f}$.

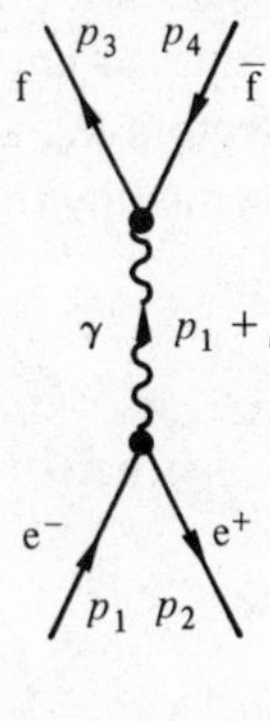

where the latter form uses $s = 2p_1 \cdot p_2$, $t = (m_\mathrm{f}^2 - 2p_1 \cdot p_3) = (m_\mathrm{f}^2 - 2p_2 \cdot p_4)$ and $u = (m_\mathrm{f}^2 - 2p_1 \cdot p_4) = (m_\mathrm{f}^2 - 2p_2 \cdot p_3)$. Note that $E_1 = E_2 = E_3 = E_4 = E$, so $s = 4E^2$. If θ is the cms angle made by f with respect to the incident e^- direction, then we can write the four-vectors as

$$p_1 = (E,\ 0, 0, E), \qquad p_2 = (E, 0, 0, -E),$$
$$p_3 = (E, 0, p_3 \sin\theta, p_3 \cos\theta), \qquad p_4 = (E, 0, -p_3 \sin\theta, -p_3 \cos\theta),$$

where the axes have been chosen (for convenience) such that the e^- is along z and f is in the y–z plane. Thus $p_1 \cdot p_2 = 2E^2$, $p_1 \cdot p_4 = p_2 \cdot p_3 = E^2(1 + \beta\cos\theta)$ and $p_1 \cdot p_3 = p_2 \cdot p_4 = E^2(1 - \beta\cos\theta)$, where $\beta = p_3/E$ is the cms velocity of f and $\bar{\mathrm{f}}$. Inserting these forms in (4.60), and using (4.29), we obtain

$$\begin{aligned}\frac{\mathrm{d}\sigma}{\mathrm{d}\Omega}(\mathrm{cms}) &= \frac{Q_\mathrm{f}^2 \alpha^2 \beta}{4s}\left(1 + \beta^2 \cos^2\theta + \frac{m_\mathrm{f}^2}{E^2}\right) \\ &= \frac{Q_\mathrm{f}^2 \alpha^2 \beta}{4s}(2 - \beta^2 + \beta^2 \cos^2\theta),\end{aligned} \tag{4.61}$$

where the factor in (4.29) $p_3/p_1 = p_3/E = \beta$. Note that, in the relativistic limit $\beta \to 1$, the angular distribution is of the form $(1 + \cos^2\theta)$. To obtain the total cross-section we must integrate over $\mathrm{d}\Omega = \mathrm{d}\cos\theta\, \mathrm{d}\phi$, with $-1 \leqslant \cos\theta \leqslant 1$ and $0 \leqslant \phi \leqslant 2\pi$, giving

$$\sigma_\mathrm{tot} = \frac{4\pi\alpha^2 Q_\mathrm{f}^2 \beta}{3s}\left(\frac{3 - \beta^2}{2}\right) \xrightarrow{\beta \to 1} \frac{4\pi\alpha^2 Q_\mathrm{f}^2}{3s}. \tag{4.62}$$

This formula is very useful, and is discussed further in Chapter 6.

4.4 COMPTON SCATTERING, $\gamma + \mathrm{e}^- \to \gamma + \mathrm{e}^-$

In the reactions discussed in the previous sections the photon has acted as a propagator ($q^2 \neq 0$), and is represented by an internal line on a Feynman diagram. However, real photons ($q^2 = 0$) can also be produced in electromagnetic processes, and these will escape the interaction region. A real photon is described by two polarisation four-vectors (ε_λ, $\lambda = 1, 2$) (see Section 3.3.1). A photon of polarisation λ is represented by the plane wave

$$A^\mu(x, q) = \frac{1}{V^{1/2}} \varepsilon_\lambda^\mu \exp(-\mathrm{i}q \cdot x), \qquad \text{ingoing;}$$

$$A^\mu(x, q) = \frac{1}{V^{1/2}} \varepsilon_\lambda^{*\mu} \exp(\mathrm{i}q \cdot x), \qquad \text{outgoing,} \tag{4.63}$$

with the polarisation vectors ε_λ satisfying

$$\varepsilon_\lambda \cdot q = 0, \qquad \varepsilon_\lambda \cdot \varepsilon_\lambda = -1, \qquad \varepsilon_\lambda^* \cdot \varepsilon_\rho = -\delta_{\lambda\rho}. \tag{4.64}$$

The normalisation factor $1/V^{1/2}$ (or $1/(2\pi)^{3/2}$) is the same as that for scalar particles (Section 4.8) and corresponds to $2k$ photons in the volume V, where k is the photon energy.

As an example of the evaluation of a process involving a real photon, we next find the differential cross-section for the *Compton scattering* process $e^-(p) + \gamma(k) \to e^-(p') + \gamma(k')$. It is assumed that the photons are high enough in energy that atomic effects can be neglected. There are two diagrams in the lowest order, and these are shown in Fig. 4.6. The S-matrix element for this second-order process can be constructed using the methods outlined in the discussion of the first order process (3.187) and from the propagator theory discussion of Section 3.1.6. There are two terms in S_{fi}, corresponding to the two graphs of Fig. 4.6(*a*) and (*b*) respectively:

$$S_{\mathrm{fi}} = e^2 \int \mathrm{d}^4x_1\, \mathrm{d}^4x_2 \bar{\psi}_0^{\mathrm{f}}(x_1)\{[\mathrm{i}\not{A}(x_1, k')][\mathrm{i}G_{\mathrm{D}}(x_1 - x_2)][\mathrm{i}\not{A}(x_2, k)]$$
$$+ [\mathrm{i}\not{A}(x_1, k)][\mathrm{i}G_{\mathrm{D}}(x_1 - x_2)][\mathrm{i}\not{A}(x_2, k')]\}\psi^{\mathrm{i}}(x_2). \tag{4.65}$$

Note that (as for the first order case) we have a factor $\mathrm{i}e\not{A}(y)$ at the $e\gamma$ vertex, and the virtual propagator in each graph is an electron. In order to make the required transformation from configuration to momentum space, we must first insert into (4.65) the plane wave solutions for $\bar{\psi}^{\mathrm{f}}(x_1)$ and $\psi^{\mathrm{i}}(x_2)$ (from (3.100)), and for the ingoing and outgoing photons (A^μ

Fig. 4.6 Lowest order diagrams for $e^-\gamma \to e^-\gamma$ (Compton scattering).

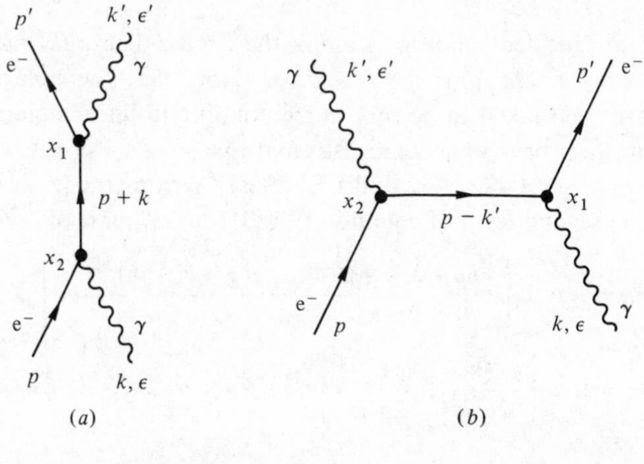

from (4.63)), together with the Dirac propagator $G_D(x_1 - x_2)$ from (3.179) (with the appropriate four-momentum for each graph in $G_D(p)$ in (3.180)). Integration over the space coordinates gives

$$S_{fi} = \frac{e^2}{V^2}(2\pi)^4\,\delta^4(p' + k' - p - k)\bar{u}_f\left\{[i\not{\varepsilon}^*(k')]\frac{i}{\not{p} + \not{k} - m}[i\not{\varepsilon}(k)]\right.$$
$$\left. + [\not{\varepsilon}(k)]\frac{i}{\not{p} - \not{k}' - m}[i\not{\varepsilon}(k')^*]\right\}u_i. \tag{4.66}$$

Writing $\varepsilon = \varepsilon(k)$ and $\varepsilon' = \varepsilon(k')$, it can be seen that (4.66) is invariant under the interchange $k \leftrightarrow -k'$, $\varepsilon \leftrightarrow \varepsilon'^*$ of the two photons (an example of '*crossing*' *symmetry*). Note that the complete matrix element is gauge invariant, but the individual terms are not (see Gasiorowicz (1966) for a detailed discussion). Equation (4.66) is again of the form $S_{fi} = (2\pi)^4\,\delta^4(p_f - p_i)\mathcal{M}_{fi}$, (ignoring the terms in V which always cancel out). The expression for $\mathcal{M}_{fi}$ can be written down directly from the Feynman rules (Appendix C).

In the lab system we can choose the axes such that $k = (k, 0, 0, 0, k)$, $p = (m, 0, 0, 0)$, $k' = (k', 0, k' \sin\theta, k' \cos\theta)$, $p' = (E', 0, p' \sin\alpha, p' \cos\alpha)$. From (4.34), the differential cross-section for the photon to be scattered into $d\Omega$ about θ is

$$\frac{d\sigma}{d\Omega}(\text{lab}) = \frac{|\mathcal{M}_{fi}|^2 k'}{64\pi^2 mk(k + m - k\cos\theta)} = \frac{|\mathcal{M}_{fi}|^2}{64\pi^2 m^2}\left(\frac{k'}{k}\right)^2, \tag{4.67}$$

where use has been made of the kinematic relation $k' = km/(k + m - k\cos\theta)$.

The expression for $\mathcal{M}_{fi}$ can be simplified by choosing the gauge in which the initial and final photons are transversely polarised in the lab system, i.e.

$$\begin{aligned} &\varepsilon(k) = (0, \boldsymbol{\varepsilon}), && \varepsilon\cdot k = \boldsymbol{\varepsilon}\cdot\mathbf{k} = 0, \\ &\varepsilon'(k') = (0, \boldsymbol{\varepsilon}'), && \varepsilon'\cdot k' = \boldsymbol{\varepsilon}'\cdot\mathbf{k}' = 0. \end{aligned} \tag{4.68}$$

That is, if the incident photon is along the z-axis, then $\varepsilon^\mu(k)$ has only x and y components, thus $\varepsilon\cdot p = \varepsilon'\cdot p = 0$. Note that the polarisation vectors have been taken to be real, corresponding to linear polarisation states. Using the property (Appendix B.5) that $\not{a}\not{b} = -\not{b}\not{a}$ if $a\cdot b = 0$, we have $\not{p}\not{\varepsilon} = -\not{\varepsilon}\not{p}$, $\not{p}\not{\varepsilon}' = -\not{\varepsilon}'\not{p}$, $\not{k}\not{\varepsilon} = -\not{\varepsilon}\not{k}$ and $\not{k}'\not{\varepsilon}' = -\not{\varepsilon}'\not{k}'$. Writing the Dirac propagators in the second form of equation (3.180), the $\mathcal{M}_{fi}$ part of (4.66) is

$$\mathcal{M}_{fi} = -ie^2\bar{u}_f\left[\frac{\not{\varepsilon}'(\not{p} + \not{k} + m)\not{\varepsilon}}{2p\cdot k} + \frac{\not{\varepsilon}(\not{p} - \not{k}' + m)\not{\varepsilon}'}{-2p\cdot k'}\right]u_i$$
$$= ie^2\bar{u}_f\left[\frac{\not{\varepsilon}'\not{\varepsilon}\not{k}}{2p\cdot k} + \frac{\not{\varepsilon}\not{\varepsilon}'\not{k}'}{2p\cdot k'}\right]u_i = \mathcal{M}_1 + \mathcal{M}_2 \tag{4.69}$$

where use has been made of $(\not p + m)\not\varepsilon u_i = -\not\varepsilon(\not p - m)u_i = 0$ (from the Dirac equation) and similarly of $(\not p + m)\not\varepsilon' u_i = 0$.

Summing over the final and averaging over the initial electron spins, we can write $|\mathcal{M}_{fi}|^2$ as usual in terms of a trace as follows

$$|\mathcal{M}_{fi}|^2 = \frac{e^4}{2}\,\mathrm{tr}\left[(\not p' + m)\left(\frac{\not\varepsilon'\not\varepsilon\not k}{2p\cdot k} + \frac{\not\varepsilon\not\varepsilon'\not k'}{2p\cdot k'}\right)\right.$$
$$\left.(\not p + m)\left(\frac{\not k\not\varepsilon\not\varepsilon'}{2p\cdot k} + \frac{\not k'\not\varepsilon'\not\varepsilon}{2p\cdot k'}\right)\right]. \tag{4.70}$$

This trace has up to eight γ-matrices, but the evaluation can be considerably simplified using the γ-matrix properties. Splitting the trace into four terms for convenience, we have

$$\begin{aligned}
T_1 &= \mathrm{tr}[(\not p' + m)\not\varepsilon'\not\varepsilon\not k(\not p + m)\not k\not\varepsilon\not\varepsilon'] \\
&= \mathrm{tr}[\not p'\not\varepsilon'\not\varepsilon\not k\not p\not k\not\varepsilon\not\varepsilon'] && (m^2 \text{ term contains } \not k\not k = k^2 = 0) \\
&= 2(k\cdot p)\,\mathrm{tr}[\not p'\not\varepsilon'\not\varepsilon\not k\not\varepsilon\not\varepsilon'] && (\not k\not p\not k = (2k\cdot p - \not p\not k)\not k = 2(k\cdot p)\not k) \\
&= 2(k\cdot p)\,\mathrm{tr}[\not p'\not\varepsilon'\not k\not\varepsilon'] && (\not\varepsilon\not k\not\varepsilon = -\not k\not\varepsilon\not\varepsilon = -\not k\varepsilon^2 = \not k) \\
&= 8(k\cdot p)[2(p'\cdot\varepsilon')(k\cdot\varepsilon') + (p'\cdot k)] && (\text{using (B.3)}) \\
&= 8(k\cdot p)[2(k\cdot\varepsilon')^2 + (k'\cdot p)] && (\text{using } \varepsilon'\cdot p' = \varepsilon'\cdot(k + p - k') = \\
& && \varepsilon'\cdot k,\ p'\cdot k = k'\cdot p).
\end{aligned} \tag{4.71}$$

Similarly, we can deduce

$$T_2 = \mathrm{tr}[(\not p' + m)\not\varepsilon\not\varepsilon'\not k'(\not p + m)\not k'\not\varepsilon'\not\varepsilon] = 8(k'\cdot p)[-2(k'\cdot\varepsilon)^2 + (k\cdot p)]. \tag{4.72}$$

Next we evaluate the cross terms with denominator factors $4(p\cdot k)(p\cdot k')$. Firstly,

$$\begin{aligned}
T_3 &= \mathrm{tr}[(\not p' + m)\not\varepsilon'\not\varepsilon\not k(\not p + m)\not k'\not\varepsilon'\not\varepsilon] \\
&= \mathrm{tr}[(\not p + m)\not\varepsilon'\not\varepsilon\not k(\not p + m)\not k'\not\varepsilon'\not\varepsilon] + \mathrm{tr}[(\not k - \not k')\not\varepsilon'\not\varepsilon\not k(\not p + m)\not k'\not\varepsilon'\not\varepsilon],
\end{aligned} \tag{4.73}$$

since $p' = p + k - k'$. The calculation of T_3 is straightforward, but somewhat tedious. In general, for a product of n γ-matrices, we can use property B.3 successively to reduce the product to expressions containing four γ-matrices, which can then be written directly as vector products. The algebra can often be simplified by using the cyclic properties of the traces and by commuting the terms using $\not a\not b = -\not b\not a + 2a\cdot b$ (especially useful in this example, since $\varepsilon\cdot k = \varepsilon'\cdot k' = \varepsilon\cdot p = \varepsilon'\cdot p = 0$) and $\not a\not a = a^2$, which removes two γ-matrices). Using these properties, T_3 reduces to (recommended as a useful exercise)

$$T_3 = 8(k\cdot p)(k'\cdot p)[2(\varepsilon\cdot\varepsilon')^2 - 1] - 8(k\cdot\varepsilon')^2(k'\cdot p) + 8(k'\cdot\varepsilon)^2(k\cdot p), \tag{4.74}$$

Similarly,

$$T_4 = \mathrm{tr}[(\not{p}' + m)\not{\varepsilon}\not{\varepsilon}'\not{k}'(\not{p} + m)\not{k}\not{\varepsilon}\not{\varepsilon}'] = T_3. \tag{4.75}$$

This identity follows since, writing $\mathcal{M}_{\mathrm{fi}} = \mathcal{M}_1 + \mathcal{M}_2$ in (4.69), then T_3 and T_4 are proportional to $\mathcal{M}_1\mathcal{M}_2^*$ and $\mathcal{M}_1^*\mathcal{M}_2$ respectively, and these are equal, since T_3 is real.

Collecting these terms together gives

$$|\mathcal{M}_{\mathrm{fi}}|^2 = e^4\left[\frac{k'\cdot p}{k\cdot p} + \frac{k\cdot p}{k'\cdot p} + 4(\varepsilon\cdot\varepsilon')^2 - 2\right], \tag{4.76}$$

and inserting this expression, evaluated in the lab system, in (4.67) gives the *Klein–Nishina* formula

$$\frac{\mathrm{d}\sigma}{\mathrm{d}\Omega}(\mathrm{lab}) = \frac{\alpha^2}{4m^2}\left(\frac{k'}{k}\right)^2\left[\frac{k'}{k} + \frac{k}{k'} + 4(\varepsilon\cdot\varepsilon')^2 - 2\right]. \tag{4.77}$$

If the incident photon beam is unpolarised, and the final state photon polarisation is not measured, then we must average over the initial and sum over the final polarisation states. The two possible initial and final polarisation states are perpendicular to $\mathbf{k}$ and $\mathbf{k}'$ respectively. In the coordinate system used to define k and k' above, these states can be chosen (without loss of generality) to be

$$\begin{aligned} \varepsilon_1 &= (0, 1, 0, 0), & \varepsilon_1' &= (0, 1, 0, 0), \\ \varepsilon_2 &= (0, 0, 1, 0), & \varepsilon_2' &= (0, 0, \cos\theta, -\sin\theta). \end{aligned} \tag{4.78}$$

The $(\varepsilon\cdot\varepsilon')^2$ term in the cross-section has thus four contributions

$$(\varepsilon_1\cdot\varepsilon_1')^2 + (\varepsilon_1\cdot\varepsilon_2')^2 + (\varepsilon_2\cdot\varepsilon_1')^2 + (\varepsilon_2\cdot\varepsilon_2')^2 = 1 + \cos^2\theta. \tag{4.79}$$

Inserting this summation, and averaging over the two initial photon spin states (factor $\frac{1}{2}$), (4.77) becomes

$$\frac{\mathrm{d}\sigma}{\mathrm{d}\Omega}(\mathrm{lab}) = \frac{\alpha^2}{2m^2}\left(\frac{k'}{k}\right)^2\left[\frac{k'}{k} + \frac{k}{k'} - \sin^2\theta\right]. \tag{4.80}$$

For the scattering of soft photons ($k \simeq k' \ll m$), the total cross-section is obtained by integrating (4.80) over $\mathrm{d}\Omega = \mathrm{d}\cos\theta\,\mathrm{d}\phi$, giving the classical *Thomson scattering* formula

$$\sigma_{\mathrm{tot}} = \frac{8\pi\alpha^2}{3m^2} = 0.665\times 10^{-24}\ \mathrm{cm}^2. \tag{4.81}$$

The numerical value is obtained by using the classical electron radius $r_0 = \alpha/m = 2.82 \times 10^{-13}$ cm. For very high energy photons, inserting $k'/k = m/(k + m - k\cos\theta)$ into (4.80) and integrating, yields

$$\sigma \simeq \frac{\pi\alpha^2}{km}\left[\ln\left(\frac{2k}{m}\right) + \tfrac{1}{2} + O\left(\frac{m}{k}\ln\frac{k}{m}\right)\right], \tag{4.82}$$

where the main contribution is from the k/k' term (Fig. 4.6(*b*)) in (4.80). Thus $\sigma \sim (\ln k)/k$ and becomes negligible for high energy photons, where pair production is dominant.

The calculation of Compton scattering above made use of a specific photon gauge and Lorentz frame. It is useful to consider a covariant description of the polarisation of a real photon: that is, extending the formalism from the two allowed transverse states to four polarisation states. Consider firstly the sum over polarisation states of a single external photon. We can write $\mathcal{M}_{\mathrm{fi}} = \varepsilon^{\alpha}_{\lambda}\mathcal{M}_{\alpha}$, where $\mathcal{M}_{\alpha}$ is independent of the polarisation (as can be seen from the Feynman rules). Choosing the Lorentz frame in which $\varepsilon_1 = (0, 1, 0, 0)$, $\varepsilon_2 = (0, 0, 1, 0)$ and $k = (\omega, 0, 0, |\mathbf{k}|)$, then the polarisation sum for the two transverse states is

$$\begin{aligned}\sum_{\lambda=1}^{2} |\mathcal{M}_{\mathrm{fi}}|^2 &= \mathcal{M}_{\alpha}\mathcal{M}^*_{\beta} \sum_{\lambda=1}^{2} \varepsilon^{\alpha}_{\lambda}\varepsilon^{\beta}_{\lambda} \\ &= |\mathcal{M}_1|^2 + |\mathcal{M}_2|^2 = |\mathcal{M}_0|^2 - |\mathcal{M}_3|^2 - g^{\alpha\beta}\mathcal{M}_{\alpha}\mathcal{M}^*_{\beta}.\end{aligned} \tag{4.83}$$

Now the gauge invariance property (3.190) means that $k^{\alpha}\mathcal{M}_{\alpha} = 0$, and hence $\omega\mathcal{M}_0 - |\mathbf{k}|\mathcal{M}_3 = 0$. For real photons $\omega = |\mathbf{k}|$, so $\mathcal{M}_0 = \mathcal{M}_3$. Thus (4.83) can be simplified to extract the general properties

$$\sum_{\lambda=1}^{2} |\mathcal{M}_{\mathrm{fi}}|^2 = -g^{\alpha\beta}\mathcal{M}_{\alpha}\mathcal{M}^*_{\beta} \qquad \text{and} \qquad \sum_{\lambda=1}^{2} \varepsilon^{\alpha}_{\lambda}\varepsilon^{\beta}_{\lambda} = -g^{\alpha\beta}. \tag{4.84}$$

This can readily be extended to two photons (i.e. Compton scattering), in which case the photon polarisation sum is

$$\sum_{\lambda} |\mathcal{M}_{\mathrm{fi}}|^2 = \mathcal{M}^{\alpha\beta}\mathcal{M}^*_{\alpha\beta}. \tag{4.85}$$

In the above discussion, real photons are taken to have $k^2 = 0$ and hence have only two allowed (transverse) polarisation states. However, in some sense, all photons are virtual, since a photon line will eventually terminate (e.g. in the detector) and thus has a non-zero value of k^2 (albeit small) and have four possible polarisation states. These two descriptions can, however, be reconciled. For transverse photons we have

$$\sum_{\mathrm{T}} \varepsilon^i_{\mathrm{T}}\varepsilon^j_{\mathrm{T}} = \sum_{\lambda=1}^{2} \varepsilon^i_{\lambda}\varepsilon^j_{\lambda} = \delta_{ij} - \hat{k}_i\hat{k}_j \qquad i, j = 1, 2, 3, \tag{4.86}$$

which can easily be checked using the explicit forms of ε_1, ε_2 and k given above. Now a virtual photon line represents a sum over the four possible polarisation states, and we can identify the term $-g^{\alpha\beta}$ of the photon propagator with this summation. The polarisation sum can be split into components as follows (following Halzen and Martin, 1984)

$$\sum_{\lambda=1}^{4} \varepsilon_\lambda^\alpha \varepsilon_\lambda^\beta = -g^{\alpha\beta} = \sum_{\mathrm{T}} \varepsilon_{\mathrm{T}}^\alpha \varepsilon_{\mathrm{T}}^\beta + \varepsilon_{\mathrm{L}}^\alpha \varepsilon_{\mathrm{L}}^\beta + \varepsilon_{\mathrm{S}}^\alpha \varepsilon_{\mathrm{S}}^\beta$$

$$= (\delta_{ij} - \hat{k}_i \hat{k}_j) + (\hat{k}_i \hat{k}_j) + (-g^{0\alpha} g^{0\beta}), \tag{4.87}$$

where ε_{L} stands for a *longitudinal photon* (i.e along the three-axis in this example) and ε_{S} for a *time-like* or *scalar photon*.

The matrix element for the exchange of a virtual photon between two particles X and Y can be written in the form

$$\mathscr{M} \propto \mathscr{M}_\alpha^{\mathrm{X}}(-g^{\alpha\beta})\mathscr{M}_\beta^{\mathrm{Y}}$$

$$= (\mathscr{M}_1^{\mathrm{X}}\mathscr{M}_1^{\mathrm{Y}} + \mathscr{M}_2^{\mathrm{X}}\mathscr{M}_2^{\mathrm{Y}}) + (\mathscr{M}_3^{\mathrm{X}}\mathscr{M}_3^{\mathrm{Y}} - \mathscr{M}_0^{\mathrm{X}}\mathscr{M}_0^{\mathrm{Y}}), \tag{4.88}$$

where the first term is the transverse contribution. In the second term, the contributions from longitudinal and scalar photons cancel in the limit $k^2 \to 0$ since $\mathscr{M}_3 = \mathscr{M}_0 |\mathbf{k}|/\omega$ (as discussed above); this leaves only the transverse term.

The general form of the Compton matrix element (first line of (4.69)) can be written (putting $m = 0$ for simplicity)

$$\mathscr{M}_{\mathrm{fi}} = -\mathrm{i}e^2 \varepsilon_\alpha \varepsilon'_\beta \bar{u}_{\mathrm{f}} \left[\frac{\gamma^\beta(\not{p} + \not{k})\gamma^\alpha}{2p\cdot k} + \frac{\gamma^\alpha(\not{p} - \not{k}')\gamma^\beta}{-2p\cdot k'} \right] u_{\mathrm{i}}. \tag{4.89}$$

Summing over photon polarisations (using $\sum \varepsilon^\alpha \varepsilon^\mu = -g^{\alpha\mu}$, etc.), the trace terms become, using the standard methods (in particular using (B.3) and (B.5)),

$$T_1 = \mathrm{tr}[\not{p}'\gamma^\beta(\not{p} + \not{k})\gamma^\alpha \not{p}\gamma_\alpha(\not{p} + \not{k})\gamma_\beta] = 32(p\cdot k)(p\cdot k')$$

$$T_2 = \mathrm{tr}[\not{p}'\gamma^\alpha(\not{p} - \not{k}')\gamma^\beta \not{p}\gamma_\beta(\not{p} - \not{k}')\gamma_\alpha] = 32(p\cdot k)(p\cdot k')$$

$$T_3 = \mathrm{tr}[\not{p}'\gamma^\beta(\not{p} + \not{k})\gamma^\alpha \not{p}\gamma_\beta(\not{p} - \not{k}')\gamma_\alpha] = 0$$

$$T_4 = \mathrm{tr}[\not{p}'\gamma^\alpha(\not{p} - \not{k}')\gamma^\beta \not{p}\gamma_\alpha(\not{p} + \not{k})\gamma_\beta] = 0.$$

From these results, the cross-section in the cms (4.29) can be written as follows

$$\frac{\mathrm{d}\sigma}{\mathrm{d}\Omega}(\mathrm{cms}) = \frac{\alpha^2}{2s}\left(-\frac{u}{s} - \frac{s}{u} \right),$$

or

$$\frac{d\sigma}{dt} = \frac{2\pi\alpha^2}{s^2}\left(-\frac{u}{s}-\frac{s}{u}\right), \tag{4.90}$$

where $s=(p+k)^2$, $t=(p-p')^2$ and $u=(p-k')^2$, and where the latter form uses the relation $t=-2k^2(1-\cos\theta)$, so $d\Omega = 2\pi\, d\cos\theta = \pi\, dt/k^2 = 4\pi\, dt/s$.

The cross-section for $e^-(p)+e^+(p') \to \gamma(k)+\gamma(k')$ can be found directly from that for $e^-(p)+\gamma(k) \to e^-(p')+\gamma(k')$, by noting that $\gamma(k)_{in} \to \gamma(k)_{out}$ and $e^-(p')_{out} \to e^+(p')_{in}$. This leads to the 'exchange' of s and t in the matrix element giving, for the polarisation summed and spin averaged result and working in the limit $m_e = 0$, $|\mathcal{M}|^2 = 2e^4(u/t + t/u)$; where $s=(p+p')^2$, $t=(p-k)^2$ and $u=(p-k')^2$. The resulting cross-section (which can be checked by the usual calculation starting with the two Feynman diagrams) is

$$\frac{d\sigma}{d\Omega} = \frac{\alpha^2}{2s}\left[\frac{u}{t}+\frac{t}{u}\right] = \frac{\alpha^2}{s}\left[\frac{2}{\sin^2\theta}-1\right], \tag{4.91}$$

where θ is the cms scattering angle of the photon. The singularity at $\theta=0$ arises from neglect of the electron mass, and a full calculation is finite and gives a total cross-section $\sigma_{tot} = 4\pi\alpha^2/s[\ln(2E/m_e)-\frac{1}{2}]$.

4.5 BREMSSTRAHLUNG AND PAIR PRODUCTION

An electron, or any other charged particles, on passing through a nuclear Coulomb field will emit photons through its interaction with the Coulomb field. This process is called *bremsstrahlung*. The lowest order diagrams (Fig. 4.7) are similar to those for Compton scattering (Fig. 4.6),

Fig. 4.7 Lowest order diagrams for electron bremsstrahlung in a Coulomb field.

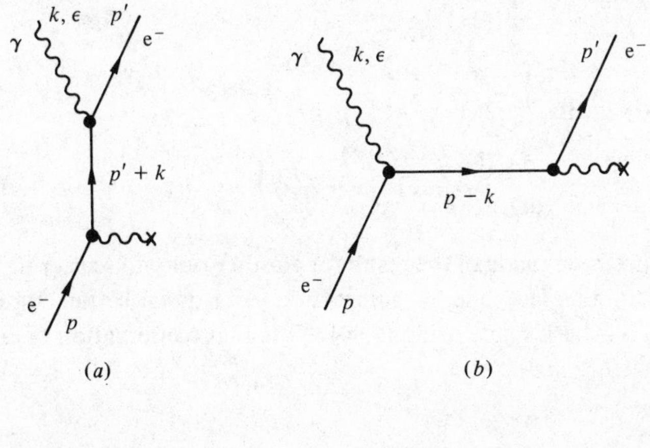

except that the incident photon is that from a static Coulomb field of a charge Ze. As discussed in Section 4.1, the vector potential for the Coulomb field has only one component, namely $A_0 = Ze/4\pi|\mathbf{x}|$, with Fourier transform $Ze/|q|^2$ where $\mathbf{q}$ is the three-momentum of the Coulomb photon. Following similar arguments to those used in deriving (4.66), we obtain

$$\begin{aligned}
\mathcal{M}_{\mathrm{fi}} &= \frac{Ze^3}{|\mathbf{q}|^2}\bar{u}_{\mathrm{f}}\left[(\mathrm{i}\not{\varepsilon})\frac{\mathrm{i}}{\not{p}'+\not{k}-m}(\mathrm{i}\gamma^0)+(\mathrm{i}\gamma^0)\frac{\mathrm{i}}{\not{p}-\not{k}-m}(\mathrm{i}\not{\varepsilon})\right]u_{\mathrm{i}} \\
&= -\mathrm{i}\frac{Ze^3}{|\mathbf{q}|^2}\bar{u}_{\mathrm{f}}\left[\frac{\not{\varepsilon}(\not{p}'+\not{k}+m)\gamma^0}{2p'\cdot k}+\frac{\gamma^0(\not{p}-\not{k}+m)\not{\varepsilon}}{-2p\cdot k}\right]u_{\mathrm{i}} \\
&= -\mathrm{i}\frac{Ze^3}{|\mathbf{q}|^2}\bar{u}_{\mathrm{f}}\left\{\frac{[2\varepsilon\cdot p'-(\not{p}'-m)\not{\varepsilon}+\not{\varepsilon}\not{k}]\gamma^0}{2p'\cdot k}-\frac{\gamma^0[2\varepsilon\cdot p-\not{\varepsilon}(\not{p}-m)-\not{k}\not{\varepsilon}]}{2p\cdot k}\right\}u_{\mathrm{i}} \\
&\overset{k\to 0}{=} -\mathrm{i}\frac{Ze^3}{|\mathbf{q}|^2}\bar{u}_{\mathrm{f}}\gamma^0 u_{\mathrm{i}}\left(\frac{\varepsilon\cdot p'}{p'\cdot k}-\frac{\varepsilon\cdot p}{p\cdot k}\right).
\end{aligned} \tag{4.92}$$

The simplification to the last line makes use of the Dirac equation, and is valid in the soft photon limit $k \to 0$. Comparison of (4.92) with $\mathcal{M}_{\mathrm{fi}}$ for elastic Coulomb scattering ($\mathcal{M}^0_{\mathrm{fi}}$) (Section 4.1) gives

$$\mathcal{M}_{\mathrm{fi}} = -\mathcal{M}^0_{\mathrm{fi}}eF, \qquad F = \left(\frac{\varepsilon\cdot p'}{p'\cdot k}-\frac{\varepsilon\cdot p}{p\cdot k}\right). \tag{4.93}$$

The cross-section can be derived in a similar way to the elastic Coulomb case, giving (Appendix (C.2))

$$\begin{aligned}
\frac{\mathrm{d}\sigma}{\mathrm{d}\Omega_{\mathrm{e}}} &= \frac{(2\pi)\,\delta(E'+k-E)Z^2e^4|\bar{u}_{\mathrm{f}}\gamma^0 u_{\mathrm{i}}|^2e^2F^2p'^2\,\mathrm{d}p'\,\mathrm{d}^3k}{\beta 2E|\mathbf{q}|^4(2\pi)^3 2E'(2\pi)^3 2k} \\
&= \frac{Z^2\alpha^2|\bar{u}_{\mathrm{f}}\gamma^0 u_{\mathrm{i}}|^2\,\mathrm{d}^3k\cdot e^2F^2}{|\mathbf{q}|^4(2\pi)^3 2k} \\
&= \left(\frac{\mathrm{d}\sigma}{\mathrm{d}\Omega_{\mathrm{e}}}\right)_0\frac{\alpha}{(2\pi)^2}\left(\frac{\varepsilon\cdot p'}{p'\cdot k}-\frac{\varepsilon\cdot p}{p\cdot k}\right)^2\frac{\mathrm{d}^3k}{k} \\
&\overset{k\to 0}{=} \left(\frac{\mathrm{d}\sigma}{\mathrm{d}\Omega_{\mathrm{e}}}\right)_0\frac{(-\alpha)}{(2\pi)^2}\left(\frac{p'}{p'\cdot k}-\frac{p}{p\cdot k}\right)^2\frac{\mathrm{d}^3k}{k},
\end{aligned} \tag{4.94}$$

where use has been made of the result for elastic Coulomb scattering from (4.6) and, in the last line, a summation over polarisation states of the photon has been made using (4.84). Note that conservation of energy requires that $E \geqslant m + k$.

The differential cross-section in the soft photon limit (4.94) can thus be factorised into the product of the elastic term (i.e. the bare cross-section without any external photons), which is proportional to α^2, and a term describing the production of the photon. This latter term has an additional factor α, and is thus a first order term compared to the leading term. However, the differential cross-section diverges as $1/k$ as $k \to 0$. This is known as the *infrared divergence* and corresponds to the electron propagator going 'on-shell'. This apparent problem stems from two omissions in the above discussion. Firstly, an experiment, even an idealised one, has a finite resolution ΔE in measuring the energy of the scattered electron. Thus, experimentally, there is no distinction between elastic scattering and inelastic scattering with a photon with energy $0 \leqslant k \leqslant \Delta E$. It is only physically meaningful, therefore, to calculate the sum of these two terms. Indeed pure elastic scattering does not exist, since it is impossible to scatter an electron with the emission of no photons (Feynman, 1961). The second omission is that the bremsstrahlung correction to the elastic scattering process is only one of the possible corrections to order α, and they should all be considered to achieve a consistent result.

The additional graphs which contribute to the order α correction do not involve external photons. One of these graphs, the vertex correction graph, is shown in Fig. 4.8. A virtual photon is emitted by the incident electron and reabsorbed by the scattered electron. The other graphs involve an internal photon starting and ending on the same electron and the Coulomb photon forming a virtual e^+e^- pair ($\gamma \to e^+e^- \to \gamma$).

The cross-section for the emission of a bremsstrahlung photon with energy $0 \leqslant k \leqslant \Delta E$ is

$$\frac{d\sigma}{d\Omega_e} = \left(\frac{d\sigma}{d\Omega_e}\right)_0 \frac{-\alpha}{(2\pi)^2} \int \frac{d^3k}{k} \left(\frac{p'}{p' \cdot k} - \frac{p}{p \cdot k}\right)^2. \tag{4.95}$$

Fig. 4.8 Vertex diagram correction to Coulomb scattering.

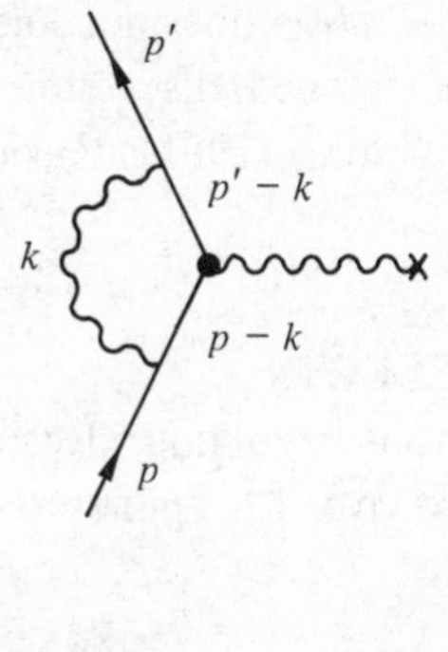

This is, however, divergent in the soft photon limit. This problem can be tackled in several ways. One, due to Feynman, is to give the photon a small non-zero mass λ, and take the limit $\lambda \to 0$ only at the end of the calculation. Repeating the bremsstrahlung calculation (equations (4.92) to (4.94)) for the case $k^2 = \lambda^2$, we obtain

$$\frac{d\sigma}{d\Omega_e} = \left(\frac{d\sigma}{d\Omega_e}\right)_0 \frac{\alpha}{(2\pi)^2} \int \frac{d^3k}{k} \left(\frac{2\varepsilon \cdot p'}{2p' \cdot k + \lambda^2} + \frac{2\varepsilon \cdot p}{-2p \cdot k + \lambda^2}\right)^2. \tag{4.96}$$

Now a photon of finite mass also has a longitudinal polarisation state. The summation over the three possble polarisation states becomes (and this can be inferred from the form of (3.204) and is valid for any massive vector boson)

$$\sum_{\lambda=1}^{3} \varepsilon_\lambda^\alpha \varepsilon_\lambda^\beta = -g^{\alpha\beta} + \frac{k^\alpha k^\beta}{\lambda^2}. \tag{4.97}$$

Note that the longitudinal term does not contribute to the unpolarised cross-section in the limit $\lambda \to 0$. Writing $k = (\omega, \mathbf{k})$, the cross-section for the emission of a photon with energy $\lambda \leqslant \omega \leqslant \Delta E$ is

$$\begin{aligned}\frac{d\sigma}{d\Omega_e} &= \left(\frac{d\sigma}{d\Omega_e}\right)_0 \frac{(-\alpha)}{(2\pi)^2} \int \frac{d^3k}{\omega} \left(\frac{2p'}{2p' \cdot k + \lambda^2} - \frac{2p}{2p \cdot k - \lambda^2}\right)^2 \\ &= \left(\frac{d\sigma}{d\Omega_e}\right)_0 \alpha B(\lambda),\end{aligned} \tag{4.98}$$

The $O(\alpha)$ correction from the internal photon radiative correction graphs arises from the interference term with the lowest order matrix element. The resultant cross-section is that for elastic scattering, times a factor $\alpha V(\lambda)$. Thus the observable cross-section is of the form

$$\left(\frac{d\sigma}{d\Omega_e}\right)_{obs} = \left(\frac{d\sigma}{d\Omega_e}\right)_0 \{1 + \alpha[B(\lambda) + V(\lambda)] + O(\alpha^2)\}. \tag{4.99}$$

The factors $B(\lambda)$, from Fig. 4.7, and $V(\lambda)$, from Fig. 4.8, are separately divergent in the limit $\lambda \to 0$. However, in their sum, the divergences are of opposite sign, and cancel leaving a finite result independent of λ. Thus, a calculation asking a physically meaningful question does not suffer from the so-called infrared catastrophe. A more detailed discussion of this problem can be found, for example, in Feynman (1961), Gasiorowicz (1966) and Mandl and Shaw (1984).

4.6 ELECTRON POLARISATION EFFECTS

In the derivation of the cross-sections involving electrons (or positrons) described above, we have summed over the spin states of the electron. In Section 4.2 we considered the elastic scattering of unpolarised electrons by (point-like) protons. If the incident electron (but not the proton) beam is polarised, then the projection operator (3.120), rather than (3.110), should be used for $u_1\bar{u}_1$. If we sum over the spin states of the other particles, then the trace $W_{\mu\nu}$ (equation (4.36)) remains unaltered, but $L^{\mu\nu}$ becomes

$$L^{\mu\nu} = \mathrm{tr}\left[(\not{p}_3 + m_1)\gamma^\mu(\not{p}_1 + m_1)\frac{(1+\gamma^5\not{s})}{2}\gamma^\nu\right]$$
$$= 2[p_3^\mu p_1^\nu + p_3^\nu p_1^\mu - g^{\mu\nu}(p_3\cdot p_1) + m_1^2 g^{\mu\nu}$$
$$+ im_1 p_{3\alpha}s_\beta\varepsilon^{\alpha\mu\beta\nu} + im_1 p_{1\alpha}s_\beta\varepsilon^{\mu\alpha\beta\nu}]. \quad (4.100)$$

In evaluating the product $L^{\mu\nu}W_{\mu\nu}$, the terms in s_β drop out. This can easily be seen because $g_{\mu\nu}\varepsilon^{\alpha\mu\beta\nu} = 0$, and the remaining terms cancel because of the antisymmetric properties of ε. The expression obtained for $|\mathcal{M}_{\mathrm{fi}}|^2$ is the same as that given in (4.39); the $L^{\mu\nu}W_{\mu\nu}$ term is a factor two less in the polarised case, but the factor for the spin average over the initial states is $\frac{1}{2}$ rather than $\frac{1}{4}$. Hence the differential cross-section for ep scattering is the same, to lowest order in α, for polarised and unpolarised electrons. This is not necessarily the case, however, for higher order terms or if the target proton is polarised (Section 7.4).

Let us now consider the polarisation of the scattered electron for the case were the incident electron is righthanded. The polarisation is

$$P = (N_{\mathrm{R}} - N_{\mathrm{L}})/(N_{\mathrm{R}} + N_{\mathrm{L}}), \quad (4.101)$$

where $N_{\mathrm{R}}(N_{\mathrm{L}})$ is the probability that the scattered electron is right(left)-handed. This is found by inserting the projection operator (3.120), $u_3\bar{u}_3 = (\not{p}_3 + m_3)(1+\gamma^5\not{s}_3)/2$, for the scattering of a righthanded electron of spin s_3 to calculate N_{R}, and reversing the sign of $s_3$to calculate N_{L}. The evaluation of P is somewhat lengthy but gives, in the relativistic limit, that the scattered electron has $P \to 1$, i.e., the helicity is conserved.

A somewhat simpler example is that of the Coulomb scattering of a righthanded electron. Neglecting factors which cancel in the determination of P using (4.101), we have

$$N_{\mathrm{R}} = \tfrac{1}{4}\,\mathrm{tr}[(\not{p}_{\mathrm{f}} + m)(1+\gamma^5\not{s}_{\mathrm{f}})\gamma^0(\not{p}_{\mathrm{i}} + m)(1+\gamma^5\not{s}_{\mathrm{i}})\gamma^0],$$
$$N_{\mathrm{L}} = \tfrac{1}{4}\,\mathrm{tr}[(\not{p}_f + m)(1-\gamma^5\not{s}_{\mathrm{f}})\gamma^0(\not{p}_{\mathrm{i}} + m)(1+\gamma^5\not{s}_{\mathrm{i}})\gamma^0], \quad (4.102)$$

where the appropriate projection operators for $u_{\mathrm{i}}\bar{u}_{\mathrm{i}}$ and $u_{\mathrm{f}}\bar{u}_{\mathrm{f}}$ have been inserted in (4.9). From (4.102), the required combinations to calculate P

are, using (B.1) and (B.4) and noting that $\varepsilon^{\alpha\mu\beta\nu}=0$ if two indices are equal,

$$\begin{aligned}N_{\mathrm{R}}-N_{\mathrm{L}}&=\tfrac{1}{2}\,\mathrm{tr}[(\not{p}_{\mathrm{f}}+m)\gamma^5\not{s}_{\mathrm{f}}\gamma^0(\not{p}_{\mathrm{i}}+m)(1+\gamma^5\not{s}_{\mathrm{i}})\gamma^0]\\&=\tfrac{1}{2}\,\mathrm{tr}[(\not{p}_{\mathrm{f}}+m)\gamma^5\not{s}_{\mathrm{f}}\gamma^0(\not{p}_{\mathrm{i}}+m)\gamma^5\not{s}_{\mathrm{i}}\gamma^0]\\&=\tfrac{1}{2}\,\mathrm{tr}[-\not{p}_{\mathrm{f}}\not{s}_{\mathrm{f}}\gamma^0\not{p}_{\mathrm{i}}\not{s}_{\mathrm{i}}\gamma^0+m^2\not{s}_{\mathrm{f}}\gamma^0\not{s}_{\mathrm{i}}\gamma^0],\end{aligned}$$

and

$$N_{\mathrm{R}}+N_{\mathrm{L}}=\tfrac{1}{2}\,\mathrm{tr}[(\not{p}_{\mathrm{f}}+m)\gamma^0(\not{p}_{\mathrm{i}}+m)\gamma^0]. \tag{4.103}$$

The expression for $N_{\mathrm{R}}-N_{\mathrm{L}}$ is readily reduced using property (B.3) of Appendix B and by using the coordinate system in which (with $\beta=|\mathbf{p}|/E$)

$$p_{\mathrm{i}}=E(1,0,0,\beta),\qquad s_{\mathrm{i}}=\frac{E}{m}(\beta,0,0,1),$$

$$p_{\mathrm{f}}=E(1,\beta\sin\theta,0,\beta\cos\theta),\qquad s_{\mathrm{f}}=\frac{E}{m}(\beta,\sin\theta,0,\cos\theta), \tag{4.104}$$

where $s_{\mathrm{i}}^2=s_{\mathrm{f}}^2=-1$, and $p_{\mathrm{i}}\cdot s_{\mathrm{i}}=p_{\mathrm{f}}\cdot s_{\mathrm{f}}=0$, as required for the spin four-vector. The resulting calculation gives

$$P=\left(1-\frac{2m^2\sin^2\theta/2}{E^2\cos^2\theta/2+m^2\sin^2\theta/2}\right)\xrightarrow{m\to 0}1. \tag{4.105}$$

The property that helicity is conserved in a vector (or axial-vector) interaction in the relativistic limit, can be shown by the following general argument. The matrix element for a spin $\frac{1}{2}$ particle to emit a virtual boson is of the form $\mathcal{M}\propto\bar{u}\gamma^\mu u$. In the limit $m\to 0$, the two helicity states (see (3.152)) are $u_{\mathrm{R}}=(1+\gamma^5)u/2$ and $u_{\mathrm{L}}=(1-\gamma^5)u/2$. For the adjoint spinors we have $\bar{u}_{\mathrm{R}}=\bar{u}(1-\gamma^5)/2$ and $\bar{u}_{\mathrm{L}}=\bar{u}(1+\gamma^5)/2$. There are four possible helicity combinations of which $\bar{u}_{\mathrm{R}}\gamma^\mu u_{\mathrm{R}}$ and $\bar{u}_{\mathrm{L}}\gamma^\mu u_{\mathrm{L}}$ are non-zero, whereas $\bar{u}_{\mathrm{L}}\gamma^\mu u_{\mathrm{R}}=\bar{u}_{\mathrm{R}}\gamma^\mu u_{\mathrm{L}}=0$ (since $(1+\gamma^5)(1-\gamma^5)=0$). A similar conclusion holds for an axial-vector interaction $\bar{u}\gamma^\mu\gamma^5 u$. In the annihilation reaction (Fig. 4.5)$\mathrm{e}^+\mathrm{e}^-\to\mathrm{f}\bar{\mathrm{f}}$, where f is a spin $\frac{1}{2}$ fermion, helicity conservation requires that the ingoing e^+ and e^- and also the outgoing f and $\bar{\mathrm{f}}$ have opposite respective helicities. Although the idea of helicity is a useful one, it should be noted, however, that for a particle with non-zero mass it is not Lorentz invariant. This can be seen by considering the rest-frame spin and four-momenta, $s^\mu=(0,0,0,1)$ and $p^\mu=(m,0,0,0)$ respectively, and computing s'_z/p'_z for boosts in both directions along the z-axis. A frame can always be found in which the helicity is reversed.

4.7 LAGRANGIAN FOR ELECTROMAGNETIC INTERACTIONS

In this section we consider the Lagrangian density for electromagnetic interactions and the connection with Feynmann diagrams. For a free spin $\frac{1}{2}$ Dirac particle the Lagrangian field density can be written

$$\mathscr{L}_{\mathrm{D}} = \bar{\psi}(\mathrm{i}\not{\partial} - m)\psi. \tag{4.106}$$

This has the desired property that, application of the Euler–Lagrange equation (3.23) with respect to ψ, yields the adjoint Dirac equation (3.89) and, with respect to $\bar{\psi}$, the Dirac equation (3.88). In Section 3.1.4 we considered the invariance of the Lagrangian under a *global* phase transformation $\psi'(x) = \exp(-\mathrm{i}\varepsilon)\psi(x)$, where ε is independent of x. If we write this transformation as $\psi'(x) = \exp(-\mathrm{i}\varepsilon e)\psi(x)$, where e is the magnitude of the electron charge then, in the limit $\varepsilon \to 0$, the Lagrangian (4.106) gives a conserved current (3.38). This current is the electromagnetic current for a Dirac particle, (3.92).

Invariance under a global phase transformation means that the phase ε is unobservable and can be freely chosen by observers anywhere in the universe, provided they choose the same value of ε. A more powerful invariance property is that the observer is free to choose the phase independently at *each* space–time point, leaving the Lagrangian and hence the 'physics' unchanged. With respect to a *local* phase transformation, defined to be

$$\psi(x) \to \psi'(x) = \exp[-\mathrm{i}e\theta(x)]\psi(x) = U\psi(x), \tag{4.107}$$

with $\theta(x)$ real, the Lagrangian (4.106) is not invariant but changes by

$$\delta\mathscr{L}_{\mathrm{D}} = e\bar{\psi}\gamma_\mu\psi\,\partial^\mu\theta. \tag{4.108}$$

Invariance of the Lagrangian can be restored by replacing ∂^μ by the covariant derivative $D^\mu = \partial^\mu + \mathrm{i}eA^\mu$ (3.155) and, at the same time, transforming $A^\mu \to A^\mu + \delta A^\mu$. The Lagrangian (4.106) becomes

$$\mathscr{L} = \mathscr{L}_{\mathrm{D}} - e\bar{\psi}\gamma_\mu\psi A^\mu = \mathscr{L}_{\mathrm{D}} + \mathscr{L}_{\mathrm{I}}, \tag{4.109}$$

with the invariance requirement implying that $\delta A^\mu = \partial^\mu\theta(x)$, that is $A^\mu \to A^\mu + \partial^\mu\theta(x)$. This is, of course, the gauge transformation (2.184) (and justifies the factor e in (4.107)). Hence, invariance under local phase transformations requires the introduction of an interaction term $\mathscr{L}_{\mathrm{I}} = -j_\mu A^\mu$ (where j_μ is the electron current (3.92)) into the Lagrangian.

For a free massless spin 1 particle (photon) a suitable Lagrangian is

$$\mathscr{L}_\gamma = -\tfrac{1}{4}F_{\mu\nu}F^{\mu\nu}, \tag{4.110}$$

where $F_{\mu\nu}$ is given by (2.183). Application of the Euler–Lagrange equation gives the Maxwell equation (2.181), for the case of no currents. Note that $F_{\mu\nu}$ is invariant under the gauge transformation (2.184). This term for the free 'gauge field' must be added to the Lagrangian, so that $\mathscr{L} = \mathscr{L}_e + \mathscr{L}_\gamma - j_\mu A^\mu$ for an eγ system. The field A^μ must correspond to a massless particle, because a massive particle would give an additional term $\frac{1}{2}m^2A^2$ in (4.110), which is not gauge invariant.

For a free scalar particle, the Lagrangian $\mathscr{L}_S$ is given by (3.49). Similar arguments to those given for the Dirac field show that the electromagnetic interaction can be introduced by the so-called minimal coupling (3.155). The Lagrangian becomes

$$\begin{aligned}\mathscr{L} &= (\partial^\mu - \mathrm{i}eA^\mu)\phi^*(\partial_\mu + \mathrm{i}eA_\mu)\phi - m^2\phi^*\phi \\ &= \mathscr{L}_S - \mathrm{i}e(\phi^*\,\partial_\mu\phi - \phi\,\partial_\mu\phi^*)A^\mu + e^2A^2\phi^*\phi.\end{aligned} \tag{4.111}$$

Neglecting (for the moment) the term in A^2, this is again of the form $\mathscr{L} = \mathscr{L}_S + \mathscr{L}_I$, with $\mathscr{L}_I = -j_\mu A^\mu$ where j_μ is the current for the scalar field (Section 3.1). Local phase invariance is thus potentially a very powerful symmetry, and we return to it in Chapter 5.

The S-matrix elements for electromagnetic interactions in the previous sections have been derived without direct use of the Lagrangian. The connection between the rather intuitive approach adopted above and the more formal approach based on the interaction Lagrangian is now briefly discussed. This can conveniently be developed in the so-called *interaction representation* of quantum mechanics. In the *Schrödinger representation* SR (Chapter 2) the operators, including the Hamiltonian H^S, do not depend on time and the entire time dependence is carried by the state vector $\psi^S(t)$. From the Schrödinger equation $\mathrm{i}\,\partial\psi^S/\partial t = H^S\psi^S$, the time dependence is

$$\psi^S(t) = \exp(-\mathrm{i}H^S t)\psi(0). \tag{4.112}$$

In the *Heisenberg representation* HR, the state vector $\psi^H = \psi(0)$ has no time dependence. The expectation value of an operator O in the two representations must be equal. Thus

$$\begin{aligned}\langle\psi^H|O^H(t)|\psi^H\rangle &= \langle\psi^S(t)|O^S|\psi^S(t)\rangle \\ &= \langle\psi^H|\exp(\mathrm{i}Ht)O^S\exp(-\mathrm{i}Ht)|\psi^H\rangle,\end{aligned}$$

i.e.

$$O^H(t) = \exp(\mathrm{i}Ht)O^S\exp(-\mathrm{i}Ht), \tag{4.113}$$

where we have written $H^S = H$. In fact $\mathrm{H}^H = H^S = H$, which can easily be verified by putting $O = H$ in (4.113) and evaluating H^{H} at some small interval δt after zero. Expanding the righthand side gives $H^{\mathrm{H}}(\delta t) = H^{\mathrm{S}}(0)$, and hence $H^{\mathrm{H}} = H^{\mathrm{S}}$ for all times. Differentiation of (4.113) with respect to time gives the Heisenberg equation of motion

$$\frac{\mathrm{d}O(t)}{\mathrm{d}t} = \mathrm{i}HO^{\mathrm{H}}(t) - \mathrm{i}O^{\mathrm{H}}(t)H = \mathrm{i}[H, O^{\mathrm{H}}(t)]. \tag{4.114}$$

In this form it is clear that operators, which are constants of the motion, commute with H.

Suppose that, in the SR, $H^{\mathrm{S}} = H_0^{\mathrm{S}} + H_{\mathrm{I}}^{\mathrm{S}}$, where H_0 is the free particle Hamiltonian and H_1 the part due to the interaction. In the *interaction representation* IR, which is intermediate between the SR and HR, we consider first the time development of the H_0 part of the Hamiltonian. That is, we define

$$\begin{aligned} \psi^{\mathrm{I}}(t) &= \exp(\mathrm{i}H_0^{\mathrm{S}}t)\psi^{\mathrm{S}}(t), \\ O^{\mathrm{I}}(t) &= \exp(\mathrm{i}H_0^{\mathrm{S}}t)O^{\mathrm{S}}\exp(-\mathrm{i}H_0^{\mathrm{S}}t). \end{aligned} \tag{4.115}$$

Thus the temporal development of ψ^{I} is given by

$$\begin{aligned} \mathrm{i}\frac{\mathrm{d}\psi^{\mathrm{I}}(t)}{\mathrm{d}t} &= \mathrm{i}\left[\mathrm{i}H_0^{\mathrm{S}}\exp(\mathrm{i}H_0^{\mathrm{S}}t)\psi^{\mathrm{S}} + \exp(\mathrm{i}H_0^{\mathrm{S}}t)\frac{\mathrm{d}\psi^{\mathrm{S}}}{\mathrm{d}t}\right] \\ &= -H_0^{\mathrm{S}}\exp(\mathrm{i}H_0^{\mathrm{S}}t)\psi^{\mathrm{S}} + \exp(\mathrm{i}H_0^{\mathrm{S}}t)(H_0^{\mathrm{S}} + H_{\mathrm{I}}^{\mathrm{S}})\psi^{\mathrm{S}} \\ &= \exp(\mathrm{i}H_0^{\mathrm{S}}t)H_{\mathrm{I}}^{\mathrm{S}}\exp(-\mathrm{i}H_0^{\mathrm{S}}t)\psi^{\mathrm{I}}(t) = H_{\mathrm{I}}\psi^{\mathrm{I}}(t). \end{aligned} \tag{4.116}$$

In order to determine the time development of O^{I} we note that $O^{\mathrm{I}}(t=0) = O^{\mathrm{S}}$, and hence

$$O^{\mathrm{I}}(t) = \exp(\mathrm{i}H_0^{\mathrm{S}}t)O^{\mathrm{I}}(0)\exp(-\mathrm{i}H_0^{\mathrm{S}}t). \tag{4.117}$$

Thus we have

$$\frac{\mathrm{d}O^{\mathrm{I}}(t)}{\mathrm{d}t} = \mathrm{i}[H_0^{\mathrm{S}}, O^{\mathrm{I}}(t)]. \tag{4.118}$$

Hence the operators in the IR have the same time dependence as for the non-interacting case. Any change in the state vectors as a function of time arises from the interaction. Thus we can use, for example, plane wave expansions of the free fields.

The time development of $\psi^{\mathrm{I}}(t)$, with respect to the initial time (t_0 say), can be written (using (4.115) and (4.112))

$$\psi^{\mathrm{I}}(t) = U(t, t_0)\psi(t_0), \tag{4.119}$$

with $U(t_0, t_0) = 1$. Insertion in (4.116) gives

$$\mathrm{i}\,\frac{\mathrm{d}U(t, t_0)}{\mathrm{d}t} = H_\mathrm{I}(t)U(t, t_0), \tag{4.120}$$

so that

$$U(t, t_0) = 1 - \mathrm{i}\int_{t_0}^{t} \mathrm{d}t_1 H_\mathrm{I}(t_1)U(t_1, t_0). \tag{4.121}$$

Assuming H_I is relatively weak, this equation can be solved iteratively, giving

$$\begin{aligned} U(t, t_0) &= 1 - \mathrm{i}\int_{t_0}^{t} \mathrm{d}t_1 H_\mathrm{I}(t_1)\left[1 - \mathrm{i}\int_{t_0}^{t_1} \mathrm{d}t_2 H_\mathrm{I}(t_2)U(t_2, t_0)\right] \\ &= 1 + (-\mathrm{i})\int_{t_0}^{t} \mathrm{d}t_1 H_\mathrm{I}(t_1) + (-\mathrm{i})^2\int_{t_0}^{t} \mathrm{d}t_1 \int_{t_0}^{t_1} \mathrm{d}t_2 H_\mathrm{I}(t_1)H_\mathrm{I}(t_2) + \cdots \\ &\quad + (-\mathrm{i})^n \int_{t_0}^{t} \mathrm{d}t_1 \cdots \int_{t_0}^{t_{n-1}} \mathrm{d}t_n H_\mathrm{I}(t_1)H_\mathrm{I}(t_2)\cdots H_\mathrm{I}(t_n) \\ &= 1 + \sum_{n=1}^{\infty} \frac{(-\mathrm{i})^n}{n!}\int_{t_0}^{t} \mathrm{d}t_1 \cdots \int_{t_0}^{t} \mathrm{d}t_n T[H_\mathrm{I}(t_1)H_\mathrm{I}(t_2)\cdots H_\mathrm{I}(t_n)], \end{aligned} \tag{4.122}$$

where T is the *time-ordering operator* of Dyson, defined as

$$T[H_\mathrm{I}(t_1)H_\mathrm{I}(t_2)\cdots H_\mathrm{I}(t_n)] = H_\mathrm{I}(t_i)H_\mathrm{I}(t_j)\cdots H_\mathrm{I}(t_k), \qquad t_i > t_j \cdots > t_k. \tag{4.123}$$

The factor $n!$ arises from the relative areas of the integration; e.g. for $n = 2$ the integration is over a triangle rather than a square, and thus has half the area.

If we consider the limits $t_0 \to -\infty$ and $t \to \infty$, then comparison with (3.186) shows that we can relate the S-matrix and $U(t, t_0)$ by $S = U(\infty, -\infty)$. Thus the S-matrix is given by the perturbative expansion

$$\begin{aligned} S &= 1 + \sum_{n=1}^{\infty} \frac{(-\mathrm{i})^n}{n!}\int_{-\infty}^{\infty} \mathrm{d}t_1 \cdots \int_{-\infty}^{\infty} \mathrm{d}t_n T[H_\mathrm{I}(t_1)\cdots H_\mathrm{I}(t_n)] \\ &= 1 + \sum_{n=1}^{\infty} \frac{(-\mathrm{i})^n}{n!}\int \mathrm{d}^4x_1 \cdots \int \mathrm{d}^4x_n T[\mathscr{H}_\mathrm{I}(x_1)\cdots \mathscr{H}_\mathrm{I}(x_n)]. \end{aligned} \tag{4.124}$$

The Hamiltonian density $\mathscr{H}_\mathrm{I}$ needed to evaluate (4.124) can be found from the Lagrangian density. From (3.34) we have $\mathscr{H} = (\partial\mathscr{L}/\partial\dot{\phi})\dot{\phi} - \mathscr{L}$. For the case that $\mathscr{L}_\mathrm{I}$ does not include any time derivatives, then $\mathscr{H}_\mathrm{I} = -\mathscr{L}_\mathrm{I}$.

Hence, for the electromagnetic interaction (i.e. QED) we have $\mathscr{H}_{\mathrm{I}} = j_\mu A^\mu$ with, for example, $j_\mu = -e\bar{\psi}\gamma_\mu\psi$ for an electron. Thus the lowest order term ($n = 1$) is exactly that derived (from propagator considerations) in equation (4.1).

The S-matrix is thus built up perturbatively from a series of terms containing $\mathscr{H}_{\mathrm{I}}$ and therefore ψ and A, and these are linear in the creation and annihilation operators or the fields. For example, the $n = 2$ term for the eγ interaction is

$$S_2 = \frac{(-\mathrm{i})^2}{2} e^2 \int \mathrm{d}^4x_1 \int \mathrm{d}^4x_2 T[\bar{\psi}(x_1)\gamma_\mu A^\mu(x_1)\psi(x_1)\bar{\psi}(x_2)\gamma_\nu A^\nu(x_2)\psi(x_2)]. \tag{4.125}$$

If we call the operator in square brackets O, then O contains all possible processes involving e^- and γ to this order (i.e. e^2). Note that the $n = 1$ term corresponds to $\mathrm{e}^- \to \mathrm{e}^- + \gamma$ and is thus unphysical, so $n = 2$ is the lowest order possible. For a particular initial and final state $|\mathrm{i}\rangle$ and $|\mathrm{f}\rangle$ we must pick out from O the annihilation operators needed to destroy the initial state particles and the creation operators to create the desired final state particles. This is facilitated by the so-called *normal ordering* of O, in which all annihilation operators appear to the right of the creation operators, so that they directly correspond to the desired process. The fields consist of negative (positive) frequency parts $\bar{\psi}^{(-)}$, $\psi^{(-)}$, $A^{(-)}$, ($\psi^{(+)}$, $\bar{\psi}^{(+)}$, $A^{(+)}$), which are linear in the creation (annihilation) operators for electrons, positrons and photons respectively.

The normal ordering is simplified by some general properties. For example, a multiplicative factor $(-1)^P$ arises if P permutations of electron and positron operators are required to go from the original order to the normal order. This stems from the anticommutation relations for fermion fields. The time ordered product in (4.125) can be replaced by the normal ordering, which in turn can be expressed as a sum of 'contractions' of fields. This yields terms of the form $A(x_1)B(x_2) = \langle 0|T\{A(x_1)B(x_2)\}|0\rangle$; that is, propagator terms. A detailed discussion of these methods and of Wick's theorem, which is used to simplify the normal ordering and identify the terms with Feynman diagrams, is beyond the scope of this work, but can be found, for example, in Mandl and Shaw (1984). For the case of Compton scattering, $\mathrm{e}^- + \gamma \to e^- + \gamma$, which is one of the possible $n = 2$ processes, we require $\psi^{(+)}(x_2)$ to absorb the initial electron and $\bar{\psi}^{(-)}(x_1)$ to create the final electron. However, either $A^{(+)}(x_1)$ or $A^{(+)}(x_2)$ can annihilate the initial photon, and either $A^{(-)}(x_2)$ or $A^{(-)}(x_1)$ create the final photon. Thus there are two terms in S_2. The contraction involving

$\psi(x_1)\bar{\psi}(x_2)$ gives an internal electron propagator. Thus the S-matrix is

$$S_2 = -e^2 \int d^4x_1 \int d^4x_2 \bar{\psi}^{(-)}(x_1)\gamma^\mu iG_D(x_1 - x_2)\gamma^\nu A_\mu^{(-)}(x_1)A_\nu^{(+)}(x_2)\psi^{(+)}(x_2)$$

$$-e^2 \int d^4x_1 \int d^4x_2 \bar{\psi}^{(-)}(x_1)\gamma^\mu iG_D(x_1 - x_2)\gamma^\nu A_\nu^{(-)}(x_2)A_\mu^{(+)}(x_1)\psi^{(+)}(x_2). \quad (4.126)$$

The corresponding Feynman graphs are shown in Fig. 4.6. Note that there is no explicit time ordering in such a graph, so that it can be drawn in a variety of equivalent ways. Other combinations of the field operators give the various $n = 2$ processes, e.g. $\gamma + \gamma \to e^+ + e^-$. The next step is to evaluate $\langle f|S_2|i\rangle = S_{fi}$. Considering, for the moment, just the incident electron, for which the initial state can be written $|i\rangle = |e^-(\mathbf{p})\ldots\rangle = a^\dagger(\mathbf{p})|0\rangle$. The electron operator $\psi^{(+)}(x_2)$ acting on $|i\rangle$ contains $a(\mathbf{p})$, and thus gives the combination $a(\mathbf{p})a^\dagger(\mathbf{p})|0\rangle = (1 - a^\dagger(\mathbf{p})a(\mathbf{p}))|0\rangle = |0\rangle$; where, in the last step, the anticommutation relation has been used. Similar considerations for the other initial and final state particles allows S_{fi} to be written in the form of (4.65), i.e. independent of the field operators. This can be Fourier-transformed from configuration space into momentum space, as discussed in Section 4.4. This gives the same form for S_{fi} as the more intuitive approach, and the calculation of the cross-section then proceeds in the same way. Thus, the field theory approach, the ideas of which are sketched above, can be considered to underwrite this approach.

4.8 INTERACTIONS OF SCALAR PARTICLES

In this section we consider the electromagnetic interactions of point-like charged scalar particles. These particles will be referred to as π^+ and π^-; however, it should be stressed that the actual π-meson is not point-like and has strong as well as electromagnetic interactions. The form of the $\pi^\pm\gamma$ interaction (Fig. 4.9) can be found from the interaction

Fig. 4.9 Interaction of π^+ with photon field A^μ.

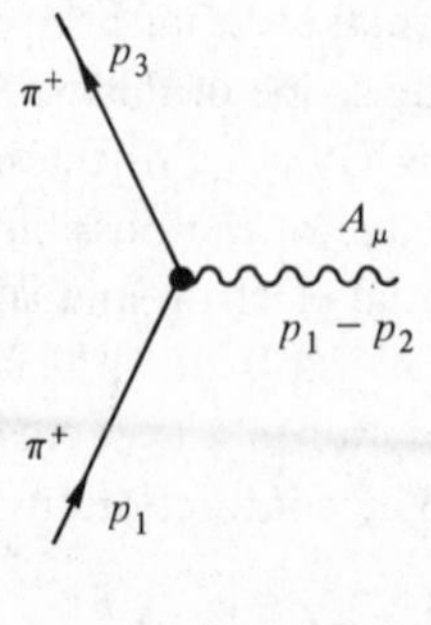

Lagrangian (4.111). Using plane waves for the ingoing (ϕ_1) and outgoing (ϕ_3) π^+ states (of the form (3.2)), then the interaction term is

$$\mathscr{L}_{\mathrm{I}} = -eN_1N_3(p_1 + p_3)_\mu \exp[-\mathrm{i}(p_1 - p_3)\cdot x]A^\mu. \tag{4.127}$$

Now for the eγ interaction $\mathscr{L}_{\mathrm{I}} = -e\bar{\psi}\gamma_\mu\psi A^\mu$ (4.109), and this leads to the Feynman rule that the eγ vertex factor is $-\mathrm{i}e\gamma_\mu$ (Appendix C). Comparison with $\mathscr{L}_{\mathrm{I}}$ from (4.127) for the $\pi^+\gamma$ case thus suggests a vertex factor $-\mathrm{i}e(p_1 + p_3)_\mu$. In terms of field theory, the scalar interaction is rather more complicated than that of the electron, because the Lagrangian contains derivatives. The term in A^2 in (4.111), which has so far been neglected, corresponds to the so-called *contact interaction* of two photons and two scalar particles (Fig. 4.10(*c*)). The corresponding vertex factor is $2\mathrm{i}e^2 g_{\mu\nu}$, as discussed below.

The calculation of the cross-section for π^+ *Coulomb scattering* is similar to that for e$^-$ (Section 4.1), except that the term $\bar{u}_{\mathrm{f}}\gamma^0 u_{\mathrm{i}}$ is replaced by $(p_{\mathrm{i}} + p_{\mathrm{f}})^0 = 2E$. The differential cross-section in the lab frame (see (4.6)) is thus

$$\frac{\mathrm{d}\sigma}{\mathrm{d}\Omega}(\text{lab}) = \frac{4Z^2\alpha^2E^2}{|\mathbf{q}|^4} = \frac{Z^2\alpha^2}{4p^2\beta^2\sin^4\theta/2}, \tag{4.128}$$

i.e. the Rutherford scattering formula (2.96), to be compared to the e$^-$ scattering result of (4.12).

The simplest interaction of two scalar particles is the *scattering of two non-identical charged particles*, which we will call π^+ and K$^+$ for convenience. The Feynman diagram is of the same topology as that for e$^-$p scattering (Fig. 4.2), but with the e$^-$ and p replaced by π^+ and K$^+$ respectively. Thus the invariant matrix element for $\pi^+(p_1) + \mathrm{K}^+(p_2) \to$

Fig. 4.10 Lowest order diagrams for $\pi^+\gamma \to \pi^+\gamma$.

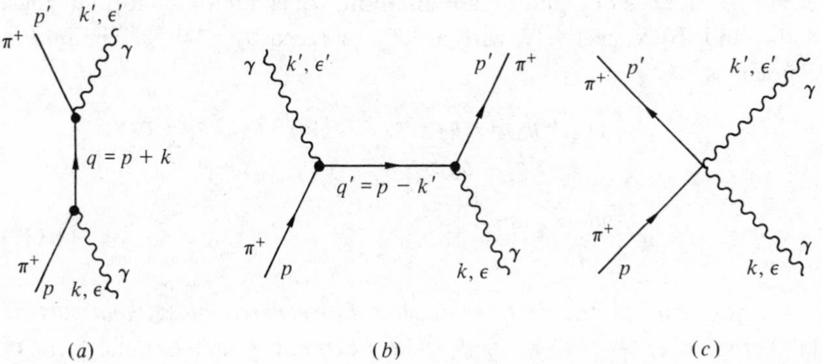

$\pi^+(p_3) + K^+(p_4)$ is (from Appendix C or comparison with (4.21))

$$\mathcal{M}_{fi} = (-ie)(p_1 + p_3)_\mu \frac{(-ig^{\mu\nu})}{(p_1 - p_3)^2}(-ie)(p_2 + p_4)_\nu = ie^2 \frac{(s-u)}{t}, \tag{4.129}$$

where $s = (p_1 + p_2)^2$, $t = (p_1 - p_3)^2$ and $u = (p_1 - p_4)^2$.

The probability density for a scalar particle is $\rho = 2|N|^2 E$ (3.4). The normalisation of $2E$ particles in the volume V is again used so that $N = 1/V^{1/2} = 1/(2\pi)^{3/2}$. Thus we can utilise the formulae established in Section 4.2 and hence the differential cross-section in the cms (using (4.29)) is

$$\begin{aligned}\frac{d\sigma}{d\Omega}(\text{cms}) &= \frac{e^4}{64\pi^2 s}\frac{[(p_1 + p_3)\cdot(p_2 + p_4)]^2}{t^2} \\ &= \frac{\alpha^2}{4s}\left(\frac{s-u}{t}\right)^2.\end{aligned} \tag{4.130}$$

For the *scattering of two identical bosons* (e.g. $\pi^+\pi^+$), $\mathcal{M}_{fi}$ must be symmetric under the exchange of the initial or final particles. There are thus two diagrams, which are of the form shown in Fig. 4.3, but with e^- replaced by π^+, giving

$$\begin{aligned}\mathcal{M}_{fi} &= ie^2\left[\frac{(p_1 + p_3)_\mu(p_2 + p_4)^\mu}{(p_1 - p_3)^2} + \frac{(p_1 + p_4)_\mu(p_2 + p_3)^\mu}{(p_1 - p_4)^2}\right] \\ &= ie^2\left[\frac{s-u}{t} + \frac{s-t}{u}\right].\end{aligned} \tag{4.131}$$

In the case of $\pi^+\pi^-$ *scattering*, the π^- is treated in a similar way to the e^+, namely an ingoing (outgoing) π^- is treated as an outgoing (ingoing) π^+ with negative momenta. There are two graphs: the one-photon exchange and the one-photon annihilation, which are of the form of Figs. 4.4(*a*) and (b) respectively, with $e^-(e^+)$ replaced by $\pi^+(\pi^-)$. The matrix element is

$$\begin{aligned}\mathcal{M}_{fi} &= ie^2\left[\frac{(p_1 + p_3)_\mu(-p_2 - p_4)^\mu}{(p_1 - p_3)^2} + \frac{(p_1 - p_2)_\mu(p_3 - p_4)^\mu}{(p_1 + p_2)^2}\right] \\ &= ie^2\left[\frac{u-s}{t} + \frac{u-t}{s}\right].\end{aligned} \tag{4.132}$$

We next consider the *elastic scattering of an electron and a scalar particle* (π^+), that is $e^- + \pi^+ \to e^- + \pi^+$. The Feynman graph has the form of

Fig. 4.2 (with p replaced by π^+) and, using the Feynman rules, we obtain

$$\mathcal{M}_{\rm fi} = (\mathrm{i}e)\bar{u}_3\gamma_\mu u_1 \frac{(-\mathrm{i}g^{\mu\nu})}{(p_1 - p_3)^2}(-\mathrm{i}e)(p_2 + p_4)_\nu. \tag{4.133}$$

Thus, following the methods of Section 4.2, with $q^2 = t = (p_1 - p_3)^2 = (p_2 - p_4)^2$, we have

$$|\mathcal{M}_{\rm fi}|^2 = \frac{e^4}{q^4}[\bar{u}_3\gamma^\mu u_1 \bar{u}_1\gamma^\nu u_3][(p_2 + p_4)_\mu(p_2 + p_4)_\nu]$$

$$= \frac{e^4}{q^4} L^{\mu\nu} W_{\mu\nu}. \tag{4.134}$$

The lepton tensor $L^{\mu\nu}$ can be evaluated in the usual manner by putting $u_1\bar{u}_1 = (\not{p}_1 + m_1)$, $u_3\bar{u}_3 = (\not{p}_3 + m_1)$, i.e. summing over electron spins, and averaging over initial spins (giving a factor $\frac{1}{2}$). Hence we obtain

$$L^{\mu\nu} = 2(p_3^\mu p_1^\nu + p_1^\mu p_3^\nu + g^{\mu\nu}q^2/2). \tag{4.135}$$

The tensor $L^{\mu\nu}$ has the property that $q_\mu L^{\mu\nu} = q_\nu L^{\mu\nu} = 0$, which can be checked by multiplying (4.135) by $q_\mu = (p_1 - p_3)_\mu$, etc. This property is useful since the hadronic tensor $W_{\mu\nu} = (2p_2 + q)_\mu(2p_2 + q)_\nu$, so that only the p_2 term contributes. Hence we have

$$|\mathcal{M}_{\rm fi}|^2 = \frac{4e^4}{q^4}[4(p_1 \cdot p_2)(p_3 \cdot p_2) + m_2^2 q^2], \tag{4.136}$$

and, using (4.29),

$$\frac{\mathrm{d}\sigma}{\mathrm{d}\Omega}(\mathrm{cms}) = \frac{\alpha^2}{sq^4}[4(p_1 \cdot p_2)(p_3 \cdot p_2) + m_2^2 q^2]. \tag{4.137}$$

In the limit $m_1^2 = m_2^2 = 0$, this takes the simple form $\mathrm{d}\sigma/\mathrm{d}\Omega = \alpha^2(-u)/t^2$, which in terms of the centre-of-mass scattering angle of the $\mathrm{e}^-(\theta)$ or, alternatively, the variable $y = (1 - \cos\theta)/2$, is thus

$$\frac{\mathrm{d}\sigma}{\mathrm{d}\Omega}(\mathrm{cms}) = \frac{\alpha^2 s}{2t^2}(1 + \cos\theta) = \frac{\alpha^2 s}{t^2}(1 - y). \tag{4.138}$$

That is, the interaction of a scalar point-like particle with a virtual photon gives a $(1 - y)$ dependence compared to the $[1 + (1 - y)^2]$ dependence for a spin $\frac{1}{2}$ particle (4.47). The $(1 - y)$ form comes from the electron helicity, which we have seen is conserved in relativistic electromagnetic interactions. For electrons of helicity $+1$ (or -1), the configuration $\theta = \pi$ is forbidden, whereas $\theta = 0$ is fully allowed.

In the lab system of the π^+ we have $p_2 = (m_2, \mathbf{0})$, hence from (4.34) and (4.136) we have

$$\frac{d\sigma}{d\Omega}(\text{lab}) = \frac{\alpha^2}{4E_1^2 \sin^4\theta/2} \frac{E_3}{E_1} \cos^2\theta/2, \tag{4.139}$$

where θ is the lab scattering angle of the e^-. This has the same form as the first term in the corresponding result for spin $\frac{1}{2}$ scattering (4.43). The second term in (4.43) is due to the electron magnetic moment.

In Section 4.3, the annihilation of e^+e^- to form a pair of spin $\frac{1}{2}$ particles of charge Q_f was discussed. It is therefore useful to calculate the *cross-section for* $e^+e^- \to b^+b^-$ (where $b^\pm$ is a spin 0 boson of charge Q_b), in order to compare with the spin $\frac{1}{2}$ result. The diagram is that of Fig. 4.5, with $f(\bar{f})$ replaced by $b^+(b^-)$. Using the Feynman rules, the matrix element is

$$\mathcal{M}_{fi} = -ie^2 Q_b (\bar{v}_2 \gamma^\mu u_1)(p_3 - p_4)_\mu / s. \tag{4.140}$$

Using the standard techniques, the differential cross-section in the cms, averaged over the spin states of the assumed unpolarised e^+ and e^- (i.e. factor $\frac{1}{4}$), is

$$\frac{d\sigma}{d\Omega}(\text{cms}) = \frac{\alpha^2 Q_b^2 \beta}{2s^3}[(m_b^2 - u)(m_b^2 - t) - m_b^2 s], \tag{4.141}$$

where the electron mass has been neglected and β is the velocity of b. In terms of the angle θ, made by the b-particle with respect to the incident e^- direction, the differential cross-section is

$$\frac{d\sigma}{d\Omega}(\text{cms}) = \frac{\alpha^2 Q_b^2 \beta}{8s}\left(1 - \beta^2 \cos^2\theta - \frac{m_b^2}{E^2}\right) = \frac{\alpha^2 Q_b^2 \beta^3}{8s}(1 - \cos^2\theta). \tag{4.142}$$

Thus, scalar particles are produced with a $(1 - \cos^2\theta)$ distribution, in contrast to the $(1 + \cos^2\theta)$ for spin $\frac{1}{2}$. The total cross-section for the production of b^+b^- is

$$\sigma_{\text{tot}} = \frac{\pi\alpha^2 Q_b^2 \beta^3}{3s} \xrightarrow{\beta \to 1} \frac{\pi\alpha^2 Q_b^2}{3s}, \tag{4.143}$$

which is a factor four less than for the production of a spin $\frac{1}{2}$ particle (4.62). Fig. 4.11 shows some measurements of the cross-section of the process $e^+e^- \to \tau^+\tau^-$, normalised with respect to $e^+e^- \to \mu^+\mu^-$, and the expectations for the spin assignments 0 and $\frac{1}{2}$ for the τ-lepton. Spin $\frac{1}{2}$ is clearly preferred.

The final example that we will consider involving scalar particles is the *interaction of a* π^+ *with a real photon*, $\gamma + \pi^+ \to \gamma + \pi^+$, i.e. Compton scattering. The lowest order (e^2) diagrams for $\gamma\pi^+$ are shown in Fig. 4.10. The diagrams 4.10(*a*) and (*b*) have the same form as those for electron Compton scattering (Fig. 4.6). However, for the pion case there is, in addition, the contact diagram (Fig. 4.10(*c*)), which also contributes to order e^2. Using the Feynman rules, the matrix element for the three diagrams of Fig. 4.10 is

$$\mathcal{M}_{\mathrm{fi}} = -\mathrm{i}e^2\left[\frac{(p+q)^\mu\varepsilon_\mu(q+p')^\nu\varepsilon'_\nu}{s-m^2} + \frac{(p+q')^\mu\varepsilon'_\mu(q'+p')^\nu\varepsilon_\nu}{u-m^2} - 2g^{\mu\nu}\varepsilon_\mu\varepsilon'_\nu\right]$$

$$= -2\mathrm{i}e^2\left[\frac{2(p\cdot\varepsilon)(p'\cdot\varepsilon')}{s-m^2} + \frac{2(p\cdot\varepsilon')(p'\cdot\varepsilon)}{u-m^2} - (\varepsilon\cdot\varepsilon')\right], \qquad (4.144)$$

where $q = p + k$, $q' = p - k'$ and m is the pion mass. In the second line the Lorentz gauge condition, $k\cdot\varepsilon = k'\cdot\varepsilon' = 0$, has been used.

In the case of electron Compton scattering (Section 4.4), it was shown that the gauge freedom of the photon polarisation vector (3.190) leads to the condition $k^\alpha\mathcal{M}_\alpha = 0$, where $\mathcal{M}_{\mathrm{fi}} = \varepsilon^\alpha_\lambda\mathcal{M}_\alpha$. Writing $\mathcal{M}_{\mathrm{fi}}$ in (4.144) as $\mathcal{M}_{\mathrm{fi}} = \varepsilon_\mu\mathcal{M}^\mu$, then this gauge condition (i.e. substituting k_μ for ε_μ) gives

$$k_\mu\mathcal{M}^\mu = -2\mathrm{i}e^2\left[\frac{2(p\cdot k)(p'\cdot\varepsilon')}{s-m^2} + \frac{2(p\cdot\varepsilon')(p'\cdot k)}{u-m^2} - (k\cdot\varepsilon')\right]$$

$$= -2\mathrm{i}e^2[(p'\cdot\varepsilon') - (p\cdot\varepsilon') - (k\cdot\varepsilon')] = 0, \qquad (4.145)$$

where the relations $2p\cdot k = s - m^2$ and $2p'\cdot k = m^2 - u$ have been used. Similarly $k'_\nu\mathcal{M}^\nu = 0$. Note that the matrix element for the sum of diagrams

Fig. 4.11 Measurements of $\sigma(e^+e^- \to \tau^+\tau^-)$, in arbitrary units, as a function of $s^{1/2}$; from Bacino *et al.* (1978). The curves correspond to $m_\tau = 1.782$ GeV.

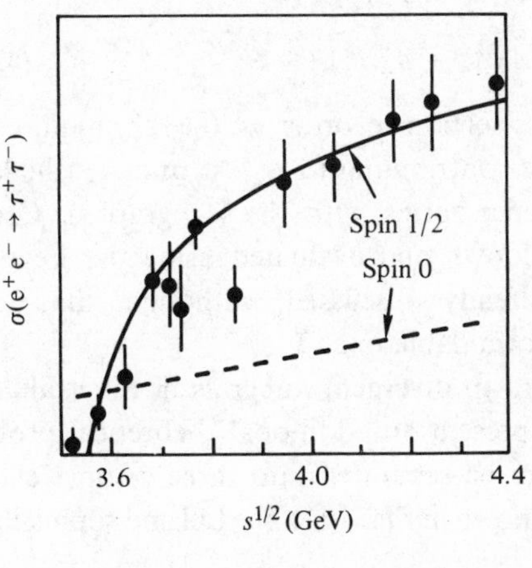

(*a*) and (*b*) is not by itself gauge invariant, and that the contact diagram is needed to achieve gauge invariance. This justifies the choice of the vertex term, $2ie^2 g_{\mu\nu}$, for the Feynman rules in Appendix C.

The matrix element can be simplified by requiring $\varepsilon \cdot p = \varepsilon' \cdot p = 0$ (see (4.68) and subsequent discussion), thus giving $|\mathcal{M}_{\rm fi}|^2 = 4e^4(\varepsilon \cdot \varepsilon')^2$. Hence the differential cross-section for the scattering of a photon into $\mathrm{d}\Omega$ is

$$\frac{\mathrm{d}\sigma}{\mathrm{d}\Omega}(\text{lab}) = \frac{\alpha^2}{m^2}\left(\frac{k'}{k}\right)^2 (\varepsilon \cdot \varepsilon')^2, \tag{4.146}$$

to be compared with the Klein–Nishina formula (4.77).

4.9 HIGHER ORDERS, DIVERGENCES AND RENORMALISATION

In the previous sections the calculation of the lowest order diagrams has been considered, together with the precise formalism needed to handle the potentially infrared divergent cross-sections arising from the emission of soft photons. The lowest order graphs are called *Born diagrams* and involve only *tree diagrams*; that is, diagrams which can be constructed without closed loops. If we wish to calculate the cross-section for a particular physical process to order n in the perturbation series expansion, then all possible diagrams which can give a contribution to this order must be calculated. If A_m represents the sum of all the amplitudes of diagrams with coefficient e^m (i.e. with m vertices), then the total matrix element for, say, Møller scattering is

$$\mathcal{M}_{\rm fi} = e^2 A_2 + e^3 A_3 + \cdots e^m A_m. \tag{4.147}$$

Hence

$$|\mathcal{M}_{\rm fi}|^2 = e^4 A_2 A_2^* + e^5(A_2 A_3^* + A_3 A_2^*) + e^6(A_3 A_3^* + A_2 A_4^* + A_4 A_2^*) + \cdots. \tag{4.148}$$

Thus a calculation of the cross-section to order α^3 (i.e. e^6) requires the evaluation of individual graphs contributing to A_2, A_3 and A_4 (the latter entering through the interference terms with the A_2 graphs). Graphs without loops can be written down and evaluated using the Feynman diagrams and techniques already discussed, although the actual computational task can be a formidable one.

Graphs containing loops lead to divergent integrals in the evaluation of the cross-section, and thus present an additional theoretical problem in QED. These divergences can be separated into three distinct classes, corresponding to the diagrams shown in Fig. 4.12. The full and sophisticated treatment of these divergences is beyond the scope of this book, and the discussion below is restricted to a brief survey of how finite (and indeed remarkably accurate) results are obtained from these potentially divergent diagrams. Each of the three graphs shown in Fig. 4.12 is discussed in turn below.

4.9.1 Self-energy of an electron

The diagram of Fig. 4.12(*a*) corresponds to the emission and subsequent reabsorption of a virtual photon by an electron during the passage of the electron between x_1 and x_2. Using the Feynman rules (Appendix C) this loop gives

$$\mathcal{M}_{(a)} = \mathrm{i}e_0^2\Sigma(p) = \int \frac{\mathrm{d}^4\mathrm{k}}{(2\pi)^4}[\mathrm{i}G_\mathrm{P}^{\mu\nu}(k)][\mathrm{i}e_0\gamma_\mu][\mathrm{i}G_\mathrm{D}(p-k)][\mathrm{i}e_0\gamma_\nu]$$

$$= -e_0^2\int \frac{\mathrm{d}^4\mathrm{k}}{(2\pi)^4}\frac{1}{\mathrm{k}^2+\mathrm{i}\varepsilon}\gamma^\nu\frac{(\not p-\not k+m_0)}{(p-k)^2-m_0^2+\mathrm{i}\varepsilon}\gamma_\nu$$

$$= e_0^2\int \frac{\mathrm{d}^4\mathrm{k}}{(2\pi)^4}\frac{1}{\mathrm{k}^2+\mathrm{i}\varepsilon}\frac{2\not p-2\not k-4m_0}{(p-k)^2-m_0^2+\mathrm{i}\varepsilon}, \tag{4.149}$$

where the expressions for the propagators of the photon (G_P) and electron (G_D) have been inserted, and the simplification in the last line is achieved by using the γ-matrix properties (B.5). The integral $(2\pi)^{-4}\int \mathrm{d}^4k$ arises because the loop momentum k can take any value from 0 to ∞. The propagator terms for the internal lines, together with the δ-functions arising from four-momentum conservation at each vertex, give rise to the factor $(2\pi)^{-4}$. The reasons for writing the electron charge and mass as e_0 and m_0 are explained below. Inspection of the powers in k in the integral (4.149) shows that it has the form $\int \mathrm{d}^4k/k^3$, and is thus divergent as $k \to \infty$ (*ultraviolet divergence*). In fact $\mathcal{M}_{(a)}$ in (4.149) also diverges in the limit $k \to 0$ (*infrared divergence*).

Fig. 4.12 Divergent graphs in QED, (*a*) electron self-energy, (*b*) photon self-energy and (*c*) vertex diagram.

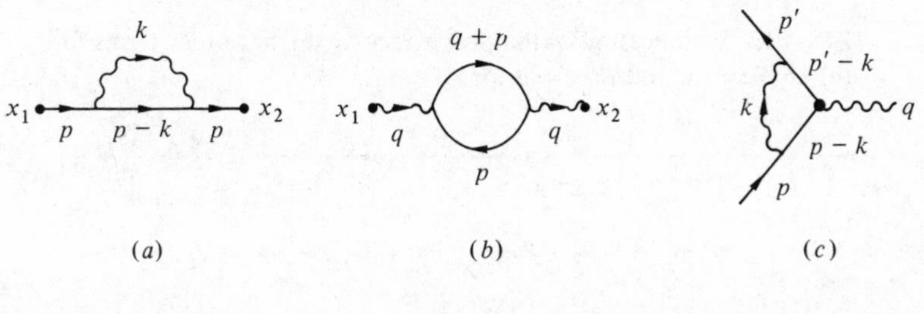

This graph, plus other higher order graphs, will lead to a modification of the form of the electron propagator as illustrated in Fig. 4.13(*a*). Hence, the original propagator is modified as follows

$$\frac{\mathrm{i}}{\not{p}-m_0} \to \frac{\mathrm{i}}{\not{p}-m_0} + \frac{\mathrm{i}}{\not{p}-m_0}\,\mathrm{i}e_0^2\Sigma(p)\,\frac{\mathrm{i}}{\not{p}-m_0} + O(e_0^4)$$

$$\equiv \frac{\mathrm{i}}{\not{p}-m}. \tag{4.150}$$

The original propagator, corresponding to a free non-interacting electron, has a pole at $\not{p} = m_0$, whereas the modified propagator has a pole at some modified mass $m = m_0 + \delta m$. The mass m corresponds to the effect of all orders in the perturbation expansion and thus should give the quantity actually measured in an experiment; that is, the real or '*renormalised*' mass. The quantity m_0, which corresponds to the unrealistic case of an electron without interactions, is called the *bare mass* and is unobservable experimentally.

To explore further the modification of the propagator it is useful to expand $\Sigma(p)$ in terms of $\not{p} - m$ as follows

$$\Sigma(p) = A + (\not{p} - m)B + (\not{p} - m)\Sigma_{\mathrm{f}}(p). \tag{4.151}$$

This form follows from Lorentz invariance, since $\Sigma(p)$ can depend only on $\not{p}$ and $\not{p}\not{p} = p^2$. The constants A and B are independent of p and are divergent. $\Sigma_{\mathrm{f}}(p)$ is finite, and can be expanded as $\Sigma_{\mathrm{f}}(p) \propto (\not{p} - m)$ at $\not{p} = m$. Now, for any two operators X and Y (not necessarily commuting), the following identity holds (Feynman 1949)

$$\frac{1}{X-Y} = \frac{1}{X} + \frac{1}{X}\cdot Y \cdot \frac{1}{X} + \cdots. \tag{4.152}$$

This can be proved by multiplying by $(X - Y)$ from the right. Using this identity, equation (4.150) can be cast in the following form

Fig. 4.13 Modification of the propagator by higher order terms for (*a*) an electron and (*b*) a photon.

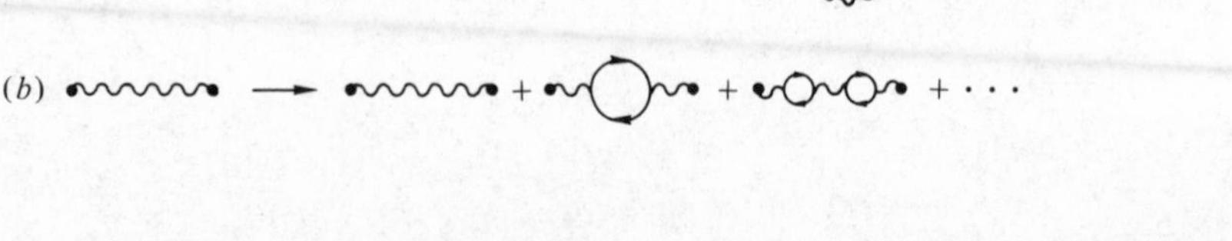

$$\frac{\mathrm{i}}{\not{p}-m_0}\rightarrow\frac{\mathrm{i}}{\not{p}-m_0+e_0^2\Sigma(p)}+O(e_0^4)$$
$$=\frac{\mathrm{i}}{\not{p}-(m_0-e_0^2A)+e_0^2(\not{p}-m)[B+\Sigma_{\mathrm{f}}(p)]}+O(e_0^4)$$
$$=\frac{\mathrm{i}}{(\not{p}-m)[1+e_0^2(B+\Sigma_{\mathrm{f}}(p))]}+O(e_0^4)$$
$$=\frac{\mathrm{i}}{\not{p}-m}[1-e_0^2(B+\Sigma_{\mathrm{f}}(p))]+O(e_0^4)=G_{\mathrm{D}}'(p), \tag{4.153}$$

where we have identified in the second line that $\delta m=-e_0^2A$.

In order to compute the mass shift δm we must evaluate the integral for $\Sigma(p)$ in equation (4.149), computed with the condition $\not{p}=m$, so that from (4.151) only the term A (and hence δm) contributes. The divergent integral (4.149) must be '*regulated*' in some way. This can be done in a variety of ways, but for the following purposes it is sufficient to modify the photon propagator by a regulator function $C(k^2)$, which cuts off the high frequency components of k, these being the cause of the divergence. Following essentially the treatment of Feynman (1961),

$$\frac{1}{k^2}\rightarrow\frac{1}{k^2}C(k^2)=\frac{1}{k^2}\left(\frac{-\Lambda^2}{k^2-\Lambda^2}\right). \tag{4.154}$$

The $\mathrm{i}\varepsilon$ notation is implicit in the subsequent discussion. With this modification, the integral $\Sigma(p)$ is no longer divergent for finite (but large) values of the cut-off constant Λ. Thus, from equations (4.149), (4.151) and (4.154) we obtain (noting that $A=\Sigma(p)|_{\not{p}=m}$)

$$\delta m=-\mathrm{i}e_0^2\int\frac{\mathrm{d}^4k}{(2\pi)^4}\frac{(2\not{p}-2\not{k}-4m)}{[(p-k)^2-m^2]}\frac{\Lambda^2}{k^2(k^2-\Lambda^2)}\bigg|_{\not{p}=m}$$
$$=\mathrm{i}e_0^2\int\frac{\mathrm{d}^4k}{(2\pi)^4}\frac{(2\not{k}+2m)}{[k^2-2p\cdot k]}\frac{\Lambda^2}{k^2(k^2-\Lambda^2)}\bigg|_{\not{p}=m}, \tag{4.155}$$

where the condition $\not{p}=m$ has been used in the simplification.

The evaluation of δm in (4.155) can be carried out by using the following integral relationships (Feynman, 1949; Sakurai, 1967)

$$\frac{\Lambda^2}{k^2(k^2-\Lambda^2)}=\int_0^{\Lambda^2}\frac{\mathrm{d}t}{(k^2-t)^2}, \tag{4.156}$$

$$\frac{1}{a^2b}=\int_0^1\frac{2y\,\mathrm{d}y}{[ay+b(1-y)]^3}, \tag{4.157}$$

$$\int\frac{\mathrm{d}^4k}{(2\pi)^4}\frac{1(k_\mu)}{(k^2-2k\cdot q+s)^3}=\frac{\mathrm{i}\cdot 1(q_\mu)}{32\pi^2(s-q^2)}. \tag{4.158}$$

Using these, the expression for δm can be evaluated as follows

$$\delta m = \mathrm{i}e_0^2 \int \frac{\mathrm{d}^4k}{(2\pi)^4} \int_0^{\Lambda^2} \frac{\mathrm{d}t(2\not{k}+2m)}{(k^2-2p\cdot k)(k^2-t)^2}\bigg|_{\not{p}=m}$$

$$= \mathrm{i}e_0^2 \int_0^1 \mathrm{d}y \int_0^{\Lambda^2} \mathrm{d}t \int \frac{\mathrm{d}^4k}{(2\pi)^4} \frac{(4y\not{k}+4ym)}{[(k^2-t)y+(k^2-2p\cdot k)(1-y)]^3}\bigg|_{\not{p}=m}$$

$$= -\mathrm{i}e_0^2 \int_0^1 \mathrm{d}y \int_0^{\Lambda^2} \mathrm{d}t \frac{\mathrm{i}[4y\not{p}(1-y)+4ym]}{32\pi^2[ty+p^2(1-y)^2]}\bigg|_{\not{p}=m}$$

$$= \frac{e_0^2 m}{8\pi^2} \int_0^1 \mathrm{d}y \int_0^{\Lambda^2} \frac{\mathrm{d}ty(2-y)}{[ty+m^2(1-y)^2]}. \tag{4.159}$$

Integrating over t, with $\Lambda^2 \gg m^2$, we have

$$\int_0^{\Lambda^2} \frac{\mathrm{d}t}{ty+m^2(1-y)^2} = \frac{1}{y}\ln\left[\frac{\Lambda^2 y+m^2(1-y)^2}{m^2(1-y)^2}\right]$$

$$\simeq \frac{1}{y}\left\{\ln\left[\frac{\Lambda^2}{m^2}\right]+\ln\left[\frac{y}{(1-y)^2}\right]\right\}. \tag{4.160}$$

Inserting this in (4.159) gives the final result

$$\delta m = \left(\frac{3\alpha m}{2\pi}\right)\ln\left(\frac{\Lambda}{m}\right) + \text{const. term} + \cdots. \tag{4.161}$$

That is, the mass correction is logarithmically divergent in the cut-off parameter Λ (this divergence is, in fact, slower than that obtained from simple power counting). The potential problem with infrared photons in the above derivation can be circumvented by giving the photon a small but finite mass λ, i.e. $k^2 \to k^2 - \lambda^2$, then letting $\lambda \to 0$ at the end of the calculation. The result obtained for δm is, however, the same. The coefficient B (like A) is logarithmically divergent in Λ; however, $\Sigma_{\mathrm{f}}(p)$ gives a finite contribution.

For a free electron, the above considerations show that the renormalised or observable mass m is equal to the (unobservable) bare mass m_0 plus a correction δm, which is divergent in Λ. If we wish to use the real mass m in the Hamiltonian density, then the free Dirac particle term, i.e. $-\bar{\psi}(\not{p}-m_0)\psi$ will give an addition term $-\delta m\bar{\psi}\psi$, which has the form of an effective interaction. Thus for each graph containing an electron self-energy loop, one must add a corresponding *mass counterterm* $-\delta m\bar{\psi}\psi$ (usually denoted by —×—). The divergence in this latter term exactly cancels that due to the term A in (4.151).

4.9.2 Vacuum polarisation

The photon propagator is modified, to first order, by the graph shown in Fig. 4.12(*b*) (*vacuum polarisation*). Using the Feynman rules this loop contributes

$$\mathscr{M}_{(b)} = ie_0^2\Pi^{\mu\nu}(q) = (-1)\int \frac{d^4p}{(2\pi)^4}[ie_0\gamma^\mu]_{\alpha\beta}[iG_D(p+q)]_{\beta\gamma}[ie_0\gamma^\nu]_{\gamma\delta}[iG_D(p)]_{\delta\alpha}$$

$$= -e_0^2\int \frac{d^4p}{(2\pi)^4}\frac{\mathrm{tr}[\gamma^\mu(\not{p}+\not{q}+m)\gamma^\nu(\not{p}+m)]}{[(p+q)^2-m^2][p^2-m^2]}, \qquad (4.162)$$

where the overall factor (-1) is a consequence of Fermi–Dirac statistics for the fermion loop. The explicit use of the indices of the γ-matrices in the vertex factors $(ie_0\gamma^\mu)$ and $(ie_0\gamma^\nu)$ and the propagators for the electron loop ($iG_D(p+q)$ and $iG_D(p)$ respectively), shows that this term may be written as a trace. As will be shown below, this graph leads to a redefinition of the charge of the electron, and e_0 represents the *bare charge*; that which would be measured if there were no higher order diagrams.

The integral (4.162) is again divergent. The method used to circumvent this divergence is the same as that used for the electron propagator above; namely, the integral must first be regularised by modifying the original expression in such a way that it gives a finite integral. Next, the potentially divergent parts are renormalised in terms of physically measurable quantities (mass, charge). The final result should not depend on the regularisation method. A substitution of the type (4.154) does not work for the modification to the photon propagator (4.162), because it does not give a gauge invariant result. Two methods of regularisation, which are gauge invariant, are commonly used:

(i) *The method of Pauli and Villars* (1949), which consists of subtracting from the integrand (4.162) a similar contribution, but with the electron mass m replaced by some large mass M. The result is logarithmically divergent in M.

(ii) *The method of dimensional regularisation* ('t Hooft and Veltman, 1972), which consists of generalising the metric tensor $g_{\mu\nu}$ from four to, say, four $-\,\eta$ dimensions, together with appropriate redefinitions of four-vectors and γ-matrices. In the new space the integrals are not divergent, and the limit $\eta \to 0$ is taken at the end of the calculation. Details of these methods can be found, for example, in Mandl and Shaw (1984) and Itzykson and Zuber (1980).

The e^+e^- loop diagram (plus higher order diagrams) modify the effective photon propagator (Fig. 4.13(*b*)). To first order

$$\frac{-ig_{\mu\nu}}{q^2} \to \frac{-ig_{\mu\nu}}{q^2} + \left(\frac{-ig_{\mu\alpha}}{q^2}\right)(ie_0^2\Pi^{\alpha\beta}(q))\left(\frac{-ig_{\beta\nu}}{q^2}\right)$$

$$= \frac{-ig_{\mu\nu}}{q^2}[1 - e_0^2A(q^2)/q^2]$$

$$= \frac{-ig_{\mu\nu}}{q^2 + e_0^2A(q^2)} + O(e_0^4), \tag{4.163}$$

where the form $\Pi^{\alpha\beta}(q) = -g^{\alpha\beta}A(q^2)$ has been used. This is the most general form satisfying both Lorentz invariance and the requirement that the photon propagator is coupled to a conserved current. (This latter requirement means that any term in $q^\alpha q^\beta$ may be omitted – see Mandl and Shaw, 1984.) In the last line it is assumed that $A(q^2)$ is finite (regularised), so that an expansion in e_0^2 can be made. This final form must correspond to the propagator of a physical photon, and thus have a pole at $q^2 = 0$; i.e. $A(0) = 0$. (The experimental upper limit is, in fact, $m_\gamma < 3 \times 10^{-33}$ MeV.)

Evaluation of the integral (4.162) can be made using the Pauli–Villars regularisation. This method (see, e.g., Bjorken and Drell, 1964) consists of subtracting from the integrand, evaluated at the electron mass, the same quantity evaluated at some large cut-off mass M (ghost fermion loop), giving

$$\Pi^R_{\alpha\beta} = \Pi_{\alpha\beta}(m^2) - \Pi_{\alpha\beta}(M^2). \tag{4.164}$$

The modified photon propagator, to order e_0^2, is

$$iG'_P(q^2) = \frac{-ig_{\mu\nu}}{q^2}\left[1 - \frac{e_0^2}{12\pi^2}\ln\left(\frac{M^2}{m^2}\right) + e_0^2\Pi_f(q^2)\right]. \tag{4.165}$$

That is, there is a term logarithmically divergent on the cut-off mass M and a finite term

$$\Pi_f(q^2) = \frac{1}{2\pi^2}\int_0^1 dz\, z(1-z)\ln\left[1 - \frac{q^2z(1-z)}{m^2}\right]. \tag{4.166}$$

For small q^2, the log term in $\Pi_f(q^2)$ can be expanded, giving

$$\Pi_f(q^2) = \frac{-q^2}{2\pi^2m^2}\int_0^1 dz\, z^2(1-z)^2 = \frac{-q^2}{60\pi^2m^2}, \tag{4.167}$$

which results in an expression for the photon propagator compatible with the form of that given by (4.163) and the subsequent discussion.

In order to understand the significance of equation (4.165), let us consider the effect of the modified photon propagator on, for example, Coulomb scattering. This process was discussed in lowest order in Section 4.1 The matrix element is modified as follows

$$\frac{\mathrm{i}e_0^2\bar{u}_\mathrm{f}\gamma^0 u_\mathrm{i}}{q^2} \to \frac{\mathrm{i}e_0^2\bar{u}_\mathrm{f}\gamma^0 u_\mathrm{i}}{q^2}\left[1 - \frac{e_0^2}{12\pi^2}\ln\left(\frac{M^2}{m^2}\right) + e_0^2\Pi_\mathrm{f}(q^2)\right]. \tag{4.168}$$

In the limit $q^2 \to 0$, the $\Pi_\mathrm{f}(q^2)$ term does not contribute, in which case the matrix element takes the same form as the unmodified matrix element, but with e_0^2 replaced by e^2, where

$$e^2 = e_0^2\left[1 - \frac{e_0^2}{12\pi^2}\ln\left(\frac{M^2}{m^2}\right) + O(e_0^4)\right] = Z_3 e_0^2. \tag{4.169}$$

The measured cross-section for a process in the $q^2 \to 0$ (i.e. long range) limit is related to the physical charge of the particle; that is, e rather than e_0. The infinite contribution (as $M \to \infty$) from the one-loop diagram (Fig. 4.12(*b*)) can be absorbed into a redefinition (renormalisation) that the physical charge is e rather than e_0 ('on-shell' renormalisation scheme). The matrix element can then be re-expanded in terms of e^2 by inverting (4.169), giving

$$e_0^2 = e^2\left[1 + \frac{e^2}{12\pi^2}\ln\left(\frac{M^2}{m^2}\right) + O(e^4)\right]. \tag{4.170}$$

Inserting this in (4.168) gives, to order e^2, a matrix element independent of the cut-off mass M, and dependent only on the physical charge e, namely

$$\mathrm{i}e^2\frac{\bar{u}_\mathrm{f}\gamma^0 u_\mathrm{i}}{q^2}\left[1 - \frac{e^2q^2}{60\pi^2m^2} + O(e^4)\right]. \tag{4.171}$$

Similar considerations apply to other scattering processes, e.g. Møller scattering, in the $q^2 \to 0$ limit. This procedure, of being able to replace the divergent quantities in the expansion in e_0^2 with finite ones in the expansion in e^2, is quite general and works to all orders. The divergent quantities appear only in intermediate steps, and not in the final result.

4.9.3 Vertex correction graph

The third type of divergent graph is the vertex correction graph (Fig. 4.12(*c*)), which has already been discussed in Section 4.5 in connection with infrared divergences. We are concerned here, however, with the high energy (ultraviolet) behaviour of this graph. The effect of this graph, plus the original vertex graph (i.e. without the connecting photon), is to modify the vertex factor as follows

$$ie_0\gamma^\mu \to ie_0\Gamma^\mu(p', p) = ie_0[\gamma^\mu + e_0^2\Lambda^\mu(p', p)], \tag{4.172}$$

where

$$ie_0^2\Lambda^\mu(p', p) = e_0^2 \int \frac{d^4k}{(2\pi)^4k^2}\, \gamma^\nu \frac{1}{\not p' - \not k - m}\, \gamma^\mu \frac{1}{\not p - \not k - m}\, \gamma_\nu. \tag{4.173}$$

The infrared divergence can be dealt with by the usual $k^2 \to k^2 - \lambda^2$ substitution. In treating the divergence in the electron self-energy, it was useful to consider the $\not p = m$ free particle limit. If we consider Λ^μ in the limit $q = p' - p \to 0$, and for initial and final free particle momenta ($\not p = m$, $\not p' = m$), this isolates the infinite part

$$D(p) = \bar u(p)\Lambda^\mu(p, p)u(p) = a_1\bar u(p)\gamma^\mu u(p) + a_2 p^\mu \bar u(p)u(p). \tag{4.174}$$

The form of the second part follows from Lorentz invariance, and a_1 and a_2 are constants. Now, starting with the Dirac equation, $mu(p) = p_\mu\gamma^\mu u(p)$, and multiplying from the left by $p^\mu\bar u(p)$, we obtain

$$p^\mu \bar u(p)u(p) = m\bar u(p)\gamma^\mu u(p). \tag{4.175}$$

Hence, (4.174) becomes

$$D(p) = \bar u(p)\Lambda^\mu(p, p)u(p) = L\bar u(p)\gamma^\mu u(p), \tag{4.176}$$

where L is a constant. In general we have

$$\Lambda^\mu(p', p) = L\gamma^\mu + \Lambda_f^\mu(p', p), \tag{4.177}$$

where the part $\Lambda_f^\mu(p', p)$ is finite. For a free particle, (4.176) gives

$$\bar u(p)\Lambda_f^\mu(p, p)u(p) = 0. \tag{4.178}$$

Consideration of the form (4.177) in the expression (4.173) shows that L diverges when the cut-off in the regularisation becomes infinite. Inserting (4.177) into (4.172), the vertex modification factor becomes

$$\begin{aligned} ie_0\gamma^\mu \to ie_0\Gamma^\mu(p', p) &= ie_0[\gamma^\mu(1 + e_0^2 L) + e_0^2\Lambda_f^\mu(p', p)] \\ &= ie[\gamma^\mu + e^2\Lambda_f^\mu(p', p) + O(e^4)], \end{aligned} \tag{4.179}$$

where, in the second line, the expression is given in terms of a renormalised charge e, where

$$e = e_0[1 + e_0^2 L + O(e_0^4)] = e_0/Z_1. \tag{4.180}$$

Inspection of the equation (4.153) for the electron propagator, when modified by self-energy effects, shows that the graph in Fig. 4.13(a) also leads to a charge renormalisation. Since the propagator is sandwiched between two vertex factors (i.e. e_0^2), the renormalised charge in this case is

$$e^2 = e_0^2[1 - e_0^2 B + O(e_0^4)] = Z_2 e_0^2. \tag{4.181}$$

We have shown above that (at least to first order) the loop divergences in the photon and electron propagators, corresponding to internal photon and electron lines, can be handled. The same graphs will, however, also lead to a modification of *external* photons and electrons. An external photon can be considered to propagate to or from some distant charge, hence the charge at the interaction vertex is renormalised by $Z_3^{1/2} e_0$ (from (4.169)). Likewise, an external electron leads to a renormalised charge of $Z_2^{1/2} e_0$. The square root arises because, whereas the propagator (factor $e_0^2 Z_2$) is bilinear in the field operators, the external line is linear. The term in $(\not{p} - m)\Sigma_f(p)$, when acting on $u(p)$ and in the limit $\not{p} = m$, gives zero contribution by virtue of the Dirac equation; so there are no further finite higher order terms. (A more rigorous discussion of these points is given in Mandl and Shaw, 1984.) Putting all these factors together, the vertex graph of Fig. 4.12(*c*) (which contains two external electron lines and one external photon line) has a charge renormalisation

$$e = e_0 Z_1^{-1} Z_2 Z_3^{1/2}. \tag{4.182}$$

The vertex correction factor $\Lambda^\mu(p', p)$ in equation (4.173) and the electron self-energy factor $\Sigma(p)$ in (4.149) are related. Comparison of $\Sigma(p)$ and $\Lambda^\mu(p, p)$, together with the identity

$$\frac{\partial}{\partial p_\mu} \frac{1}{\gamma^\mu p_\mu - m} = -\frac{1}{\not{p} - m} \gamma^\mu \frac{1}{\not{p} - m}, \tag{4.183}$$

gives the relation (*Ward identity*)

$$\Lambda^\mu(p, p) = \partial \Sigma(p)/\partial p_\mu. \tag{4.184}$$

Using equations (4.177), (4.178), (4.151) and the property that $\Sigma_f(p)u(p) = 0$, equation (4.184) can be written and reduced as follows

$$\bar{u}(p)\Lambda^\mu(p, p)u(p) = \bar{u}(p)\frac{\partial \Sigma(p)}{\partial p_\mu} u(p),$$

$$L\bar{u}(p)\gamma^\mu u(p) = B\bar{u}(p)\gamma^\mu u(p),$$

i.e.

$$L = B. \tag{4.185}$$

Hence, comparing (4.180) and (4.181), we have (to order e^2), $Z_1 = Z_2$. The charge renormalisation relation (4.182) then takes the simple form

$$e = e_0 Z_3^{1/2} \tag{4.186}$$

That is, the vertex term and the electron self-energy term exactly cancel, and the charge renormalisation comes only from the vacuum polarisation correction to the photon propagator.

The expressions for Z_1 and Z_2 are in fact gauge dependent, whereas Z_3 (and δm) are gauge independent. However, the equality $Z_1 = Z_2$ removes this gauge dependence from (4.186). The above results hold in first order; however, generalised Ward identities and the result (4.186), that the charge is only modified by the photon propagator, can be shown to hold to all orders (see e.g. Itzykson and Zuber, 1980). This is important, since the renormalised charge of point-like particles other than the electron will get contributions from exactly the same diagrams as the electron. In general any charged particle–antiparticle loop can occur in the photon propagator.

4.9.4 Lamb shift and anomalous magnetic moment

Finite results in QED can thus be obtained if charge and mass renormalisation (together with the additional mass counterterm diagrams) are carried out, so that the calculations are performed using the physical mass and charge of the electron. The remaining finite correction terms, such as $\Pi_f(q^2)$ in (4.165), should give experimentally testable predictions. The presence of this additional term $(-e^2q^2/60\pi^2m^2)$ effectively modifies the Coulomb law between point charges. At large distances (small q^2), this term is negligible and the Fourier transform of the $1/q^2$ term gives the usual Coulomb law (Section 4.1). However, at smaller distances, there is an additional effective interaction. This term, for example, leads to a -27 MHz correction to the 1058 MHz Lamb shift of the otherwise degenerate $2s_{1/2}$ and $2p_{1/2}$ levels in the hydrogen atom; and this has been well verified experimentally.

The vacuum polarisation can alternatively be visualised by considering the effective nuclear charge seen by an electron. The effective charge becomes less at large distances (small q^2), due to the creation of e^+e^- pairs by the virtual photons which screen the nuclear charge. At small distances (large q^2) this screening cloud is penetrated and the effective charge increases.

In Section 3.2.7 the magnetic moment of a Dirac particle was found by consideration of its interaction with a magnetic field. Higher order graphs will change the magnetic moment from the free particle value

$(g = 2)$. This difference of $g - 2$, which is usually expressed as $a = (g - 2)/2$, is called the *anomalous magnetic moment*. The value of $g - 2$ for the electron from graphs up to order e^3 is discussed below. The method is to construct the matrix element from the sum of all possible graphs for the interaction of an electron in a static external field, and then to identify the part of the matrix element which gives rise to the magnetic moment. The complete set of graphs to order e^3 for the scattering of an electron by a static external field is shown in Fig. 4.14. The graphs (*b*) and (*c*) contribute only to the renormalisation of the charge and need not be considered further. After renormalisation the matrix element is

$$\mathcal{M} = \mathcal{M}_{(a)} + \mathcal{M}_{(d)} + \mathcal{M}_{(e)}$$
$$= \mathrm{i}e\bar{u}\gamma^{\mu}uA_{\mu} + \mathrm{i}e\bar{u}\gamma^{\mu}u[e^2\Pi_{\mathrm{f}}(q)]A_{\mu} + \mathrm{i}e\bar{u}[e^2\Lambda^{\mu}_{\mathrm{f}}(p', p)]uA_{\mu}, \quad (4.187)$$

where the spinors in full are $\bar{u}(p')$ and $u(p)$, with $q = p' - p$, and the photon field is $A_{\mu}(q)$. We note in passing that this set of graphs, when treated in the context of the hydrogen atom (which means interpreting the electron lines in terms of hydrogen states), explains the entire Lamb shift provided detailed atomic effects are taken into account.

Before discussing the results of the evaluation of the diagrams of Fig. 4.14, we first consider the most general form for the effective current of a Dirac particle. For a point-like Dirac particle we have seen in Section 3.2.1 that the current is proportional to $\bar{u}\gamma^{\mu}u$. The most general form

Fig. 4.14 Order e and e^3 graphs for the interaction of an electron with a static external field. Graph (*a*) is the lowest order, (*b*) and (*c*) are electron self-energy graphs, with corresponding mass counterpart diagrams, (*d*) is the vacuum polarisation and (*e*) is the vertex correction.

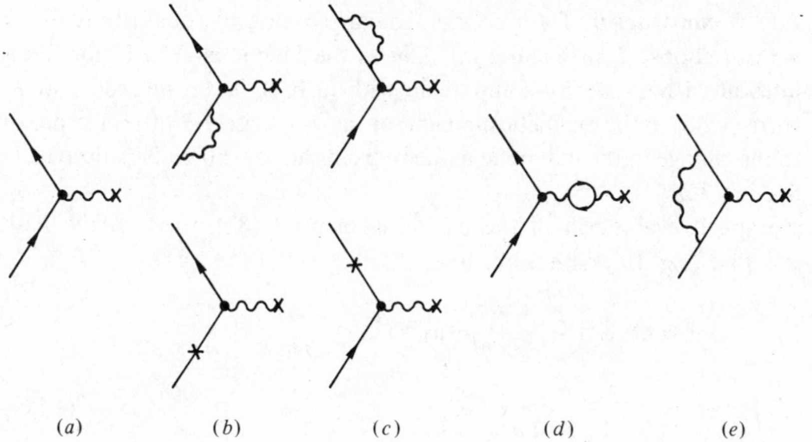

consistent with current conservation is as follows

$$\bar{u}(p')[\gamma^\mu F_1(q^2) + \frac{i}{2m}\sigma^{\mu\nu}q_\nu F_2(q^2)]u(p). \tag{4.188}$$

To see this, we first of all prove the *Gordon identity*, namely

$$2m\bar{u}(p')\gamma^\mu u(p) = \bar{u}(p')[(p'+p)^\mu + i\sigma^{\mu\nu}(p'-p)_\nu]u(p). \tag{4.189}$$

This can be proven by starting with the Dirac equation, together with an arbitrary four-vector a_μ, and manipulating these as follows

$$\begin{aligned} 0 &= \bar{u}(p')[\not{a}(\not{p}-m) + (\not{p}'-m)\not{a}]u(p) \\ &= -2m\bar{u}(p')\not{a}u(p) + \bar{u}(p')(\not{a}\not{p} + \not{p}'\not{a})u(p) \\ &= -2m\bar{u}(p')\not{a}u(p) + \bar{u}(p')\left(\left\{\frac{\not{p}'+\not{p}}{2}, \not{a}\right\} + \left[\frac{\not{p}'-\not{p}}{2}, \not{a}\right]\right)u(p). \end{aligned}$$

Next, this expression is differentiated with respect to a_μ, giving

$$\begin{aligned} 2m\bar{u}(p')\gamma^\mu u(p) &= \frac{\bar{u}(p')}{2}[(p'+p)_\nu(\gamma^\nu\gamma^\mu + \gamma^\mu\gamma^\nu) + (p'-p)_\nu(\gamma^\nu\gamma^\mu - \gamma^\mu\gamma^\nu)]u(p) \\ &= \bar{u}(p')[(p'+p)^\mu + i\sigma^{\mu\nu}q_\nu]u(p), \end{aligned}$$

where $q_\nu = (p'-p)_\nu$, and the commutation properties of the γ-matrices have been used.

Thus the Gordon decomposition of a Dirac vector current (4.189) contains two parts. The first part is a coupling of the form $(p'+p)^\mu$, which is the same as that of a scalar particle (Section 4.8), i.e. an *electric term.* The second part (*magnetic term*) is related to the spin through $\sigma^{\mu\nu}$, and gives rise to the magnetic moment. Each of these terms can be multiplied by an arbitrary function, which can depend on any scalar quantities which can be constructed. The only non-trivial (i.e. not m^2) quantity is q^2, so we can choose $F_1(q^2)$ and $F_2(q^2)/2m$ as the functions. The factor $2m$ is introduced because, by comparison with (3.163), it can be seen that F_2 corresponds to the magnetic moment; or, more precisely, $F_2(0)$ corresponds to the change in the magnetic moment from the value $(-e/2m)$, derived in Section 3.2.7.

Explicit evaluation of the matrix element (4.187) gives, in the limit $|q^2| \ll m^2$ (e.g. Itzykson and Zuber, 1980),

$$\begin{aligned} \mathcal{M} = ie\bar{u}\Bigg\{\gamma^\mu\left[1 + \frac{\alpha q^2}{3\pi m^2}(c_{(d)} + c_{(e)}) + \frac{\alpha q^2}{3\pi m^2}\ln\left(\frac{m}{\lambda}\right)\right] \\ + \frac{\alpha}{2\pi}\frac{i}{2m}\sigma^{\mu\nu}q_\nu\Bigg\}u, \end{aligned} \tag{4.190}$$

Table 4.1. *Experimental and theoretical values of* $g-2$ *for* $e^{\pm}$ *and* $\mu^{\pm}$. The numbers in parentheses are the estimated errors

	$a_e \times 10^{12}$	$a_\mu \times 10^{9}$
Experiment	1 159 652 209 (31)	1 165 923 (9)
Theory	1 159 652 411 (166)	1 165 920 (2)

where the constants $c_{(d)} = -1/5$ and $c_{(e)} = -3/8$ arise from graphs (*d*) and (*e*) respectively. The infrared term in the photon mass λ again causes no problems in physically meaningful calculations. The term in $\sigma^{\mu\nu}$ gives an additional contribution to the magnetic moment, which becomes

$$\mu_e = -\frac{e}{2m}\left(1 + \frac{\alpha}{2\pi}\right), \tag{4.191}$$

giving

$$a_e = (g-2)/2 = \alpha/2\pi. \tag{4.192}$$

To this order in α the correction does not depend on the particle mass; hence $a_\mu = a_e$. However, higher order terms are mass-dependent. The theoretical and experimental values for the electron and muon anomalous magnetic moments are shown in Table 4.1. The prediction from *CPT* invariance that $a_{e^+} = a_{e^-}$ has been checked to 1 part in 10^8. The value of $g-2$ for the muon, for example, has been calculated theoretically to $O(\alpha^4)$ giving (Kinoshita *et al.*, 1984)

$$\begin{aligned} a_\mu(\text{QED}) &= \frac{\alpha}{2\pi} + 0.76585810(10)\left(\frac{\alpha}{\pi}\right)^2 + 24.073(11)\left(\frac{\alpha}{\pi}\right)^3 + 140(6)\left(\frac{\alpha}{\pi}\right)^4 \\ &= 1\,165\,848.0(0.3) \times 10^{-9}, \\ a_\mu(\text{had}) &= 70.2(1.9) \times 10^{-9}, \\ a_\mu(\text{weak}) &= 1.95(0.01) \times 10^{-9}, \end{aligned} \tag{4.193}$$

where the figures in parentheses are the estimated theoretical or experimental errors. The value of α used ($\alpha^{-1} = 137.035963(15)$) is measured from the Josephson effect (Williams and Olsen, 1979). Note that $g-2$ for the muon has been calculated to such a precision that hadronic and $SU(2) \otimes U(1)$ weak correction terms contribute.

The remarkable agreement between theory and experiment for the $g-2$ calculations show that, even if the renormalisation procedure may not be understood at the deepest level, it can nonetheless be readily used. The energy scales at which the divergent terms become problematic are, in fact, enormous. For example, for a cut-off value $\Lambda = 100$ GeV (roughly the highest scale which can currently be explored experimentally), the changes in $\delta e/e$ and $\delta m/m$ are 0.015 and 0.067 respectively. To get to $\delta m \sim m$ needs a value of $\Lambda \sim 10^{120}$ GeV, corresponding to a distance scale of the order of 10^{-134} cm. New particles, new forces or manifestations of the unification of forces, may well come into play at a much lower energy scale than this!

4.9.5 Running coupling constant in QED

The renormalisation of the charge (4.169), arising from the vacuum polarisation term in the photon propagator, was considered above in the low q^2 limit. For large q^2, the term $\Pi_f(q^2)$ in (4.165) has the form (from (4.166))

$$\Pi_f(q^2) \simeq \frac{1}{2\pi^2}\int_0^1 dz\, z(1-z)\ln\left(\frac{Q^2}{m^2}\right) = \frac{1}{12\pi^2}\ln\left(\frac{Q^2}{m^2}\right), \tag{4.194}$$

where $Q^2 = -q^2$ is a positive quantity. In this limit the effective charge becomes, using (4.165),

$$e^2(Q^2) = e_0^2\left[1 + \frac{e_0^2}{12\pi^2}\ln\left(\frac{Q^2}{M^2}\right)\right]. \tag{4.195}$$

In practice, the electron charge is measured at some reference value, $Q^2 = Q_0^2$, say. The dependence on the arbitrary cut-off parameter M can be removed by subtracting $e^2(Q_0^2)$ from $e^2(Q^2)$. Expressing the result in terms of $\alpha = e^2/4\pi$, and noting that the loop diagrams (Fig. 4.13b) lead to a geometric progression $1 - \delta + \delta^2 \cdots = 1/(1-\delta)$, this gives

$$\alpha(Q^2) = \alpha(Q_0^2)\Bigg/\left[1 - \frac{\alpha(Q_0^2)}{3\pi}\ln\left(\frac{Q^2}{Q_0^2}\right)\right]. \tag{4.196}$$

Thus the effective coupling constant 'runs' with Q^2, and increases as Q^2 increases. That is, the effective electron charge seen by the photon at smaller and smaller distances increases. As mentioned earlier, at high enough Q^2, there will be loop contributions from all accessible charged particle–antiparticle pairs. These points are discussed further in Chapter 7.

Fig. 4.15 Examples of graphs with no external lines.

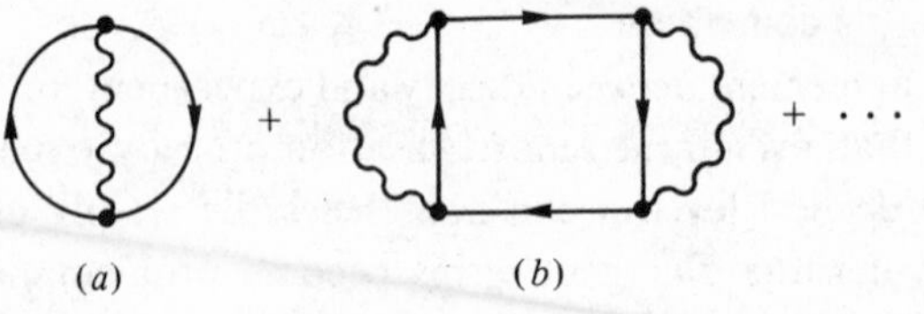

Two final topics are now discussed on the subject of higher order QED effects. The first topic is the effect of graphs which contain no external lines, and are thus *fluctuations in the vacuum*. Examples of such graphs are shown in Fig. 4.15 and these, plus an infinite number of higher order graphs, can occur in conjunction with any QED process containing external lines. The matrix elements for these graphs are again highly divergent. However, the same set of such divergent graphs will occur in conjunction with any real diagram, and their net effect is to modify the S-matrix from S (no vacuum diagrams) to $S' = CS$, where the multiplicative factor C is the sum of the matrix elements of the vacuum diagrams. Now, from conservation of momentum, there is no matrix element connecting the vacuum to another state; hence this factor can be divided out of S (it can contribute only an overall unobservable phase) and all disconnected graphs subsequently neglected.

The second topic is that of *photon–photon elastic scattering*. Since QED is an Abelian theory, the photon couples only to charged particles and so direct $\gamma\gamma$ interactions are not allowed. The simplest allowed possibility is the *box diagram* shown in Fig. 4.16 (plus equivalent graphs). Calculation (e.g. Itzykson and Zuber, 1980) shows that the cross-section for this process (which is, in fact, convergent due to a Ward identity) peaks at an energy value $E_\gamma \sim m$, with a cross-section of $10^{-31}\ \text{cm}^2$.

Fig. 4.16 Lowest order diagram for $\gamma\gamma$ scattering.

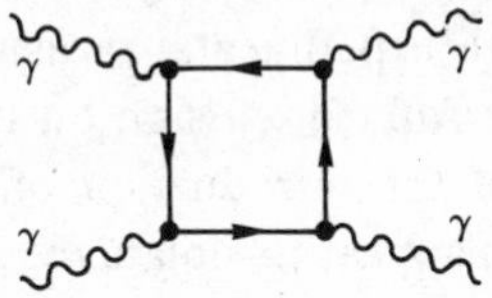

5

From Fermi theory to the standard model

Nuclear beta decay has played a very important historical role in the development of our present understanding of weak interactions. For more than 20 years these reactions were the primary source of information on the weak force. In the first part of this chapter, the theoretical ideas developed to account for the experimental measurements on beta decays (and later on the weak decays of hadrons) are reviewed, leading to the V − A theory of weak interactions. This theory, however, gives badly divergent cross-sections at high energy and, moreover, does not account for the weak neutral current interactions which are observed in nature. The standard model of Glashow, Salam and Weinberg, which is based on the principle of local gauge symmetry, predicted such currents and, indeed, successfully explains the bulk of all experimental measurements on weak and electromagnetic phenomena. After discussion on the basic framework of this model, the problem of the introduction of particle masses is reviewed. Non-zero masses would appear to violate the required gauge invariance of the theory. However, if the underlying group symmetry is spontaneously broken, then the particle masses can be generated at the price of introducing one or more fundamental scalars (Higgs particles) into the theory. The predictions of the standard model, for a wide class of reactions, together with a comparison of the experimental results, are discussed in subsequent chapters.

5.1 BETA DECAY AND THE V − A INTERACTION

5.1.1 Fermi theory

The first attempt to formulate a theory of weak interactions was due to Fermi in 1934. In order to explain the then known properties of nuclear beta decay, Fermi proposed a Lagrangian which, for $\mathrm{n}(p_1) \rightarrow$

$p(p_2) + e^-(p_3) + \bar{\nu}_e(p_4)$, see Fig. 5.1(*a*), has the form

$$\mathscr{L} = -\mathscr{H} = -C_V(\bar{\psi}_p \gamma_\mu \psi_n)(\bar{\psi}_e \gamma^\mu \psi_\nu). \tag{5.1}$$

The fields in (5.1) are all evaluated at the same space-time point x, so that the Lagrangian represents the product of two Dirac currents, for $n \to p$ and $\bar{\nu}_e \to e^-$ respectively. The processes of e^+ emission and e^- capture can easily be included by adding to $\mathscr{L}$ the Hermitian conjugate of (5.1).

If we compare the *current* × *current* form of the Fermi Lagrangian with the QED Lagrangian $\mathscr{L} = -j_\mu A^\mu$, describing say γp scattering, then we see that the form of $j_\mu(\bar{\psi}_p \gamma_\mu \psi_p)$ is rather similar, but the electromagnetic field A^μ is replaced by $\bar{\psi}_e \gamma^\mu \psi_\nu$. Another important difference, as discussed below, is that whereas the electromagnetic coupling constant is dimensionless, the constant C_V in (5.1) has dimension $(\text{mass})^{-2}$.

In the original Fermi Lagrangian (5.1), the currents have the Lorentz transformation properties of polar vectors, i.e. $V \times V$. A more general form of the Lagrangian was proposed by Gamow and Teller, by including the possibility of scalar, tensor, axial vector and pseudoscalar terms. The Lorentz transformation properties of these terms were discussed in Section 3.2.5. The most general Lagrangian which can be constructed, without containing any derivatives of the fields, can thus be written as

$$\mathscr{L} = -\sum_i C_i(\bar{\psi}_p \Gamma_i \psi_n)(\bar{\psi}_e \Gamma^i \psi_\nu), \tag{5.2}$$

where i = S, V, T, A and P (see (3.139)), and the coefficients C_i are complex. (Note that different forms of the tensor term are used in the literature, and care must be taken with the summation convention.)

The above discussion has ignored the fact that beta decay occurs in a nucleus in which the energy release is only a few MeV, so that nuclear

Fig. 5.1 Lowest order diagram for $n \to pe^-\bar{\nu}_e$ in (*a*) Fermi theory and (*b*) in standard model at the quark level.

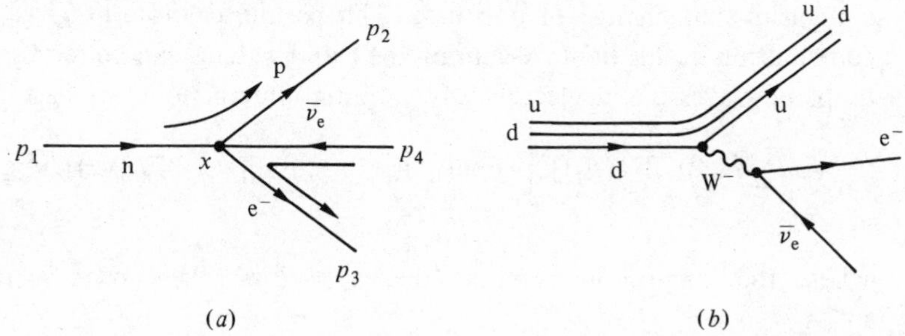

Table 5.1. *Non-relativistic limit of nuclear matrix elements*
The notation $\psi^\dagger\psi = \phi^\dagger\phi = \langle 1\rangle$ and $\phi^\dagger\sigma^k\phi = \langle\sigma^k\rangle$ is used

Coupling	Relativistic expression	Non-relativistic limit
S	$\bar{\psi}\psi$	$\langle 1\rangle$
V	$\bar{\psi}\gamma^\mu\psi$	$\begin{cases}\langle 1\rangle & \mu=0\\ 0 & \mu=1,2,3\end{cases}$
T	$\bar{\psi}\sigma^{\mu\nu}\psi$	$\langle\sigma^k\rangle \quad k=1,2,3$
A	$\bar{\psi}\gamma^5\gamma^\mu\psi$	$\begin{cases}0 & \mu=0\\ -\langle\sigma^k\rangle & \mu=k=1,2,3\end{cases}$
P	$\bar{\psi}\gamma^5\psi$	0

effects are important. The nuclear recoils have energies of only O (keV), therefore a non-relativistic approximation may be used. In the limit $p \to 0$ we have, from (3.103) and neglecting the overal normalisation,

$$\psi = \begin{pmatrix} \phi \\ \dfrac{\boldsymbol{\sigma}\cdot\mathbf{p}}{2M}\phi \end{pmatrix} = \begin{pmatrix}\psi_{\rm L}\\ \psi_{\rm S}\end{pmatrix} \simeq \begin{pmatrix}\phi\\ 0\end{pmatrix}. \tag{5.3}$$

In the limit where the large and small components are $\psi_{\rm L} = \phi$, $\psi_{\rm S} \sim 0$, the vector current becomes

$$\bar{\psi}\gamma^\mu\psi = \psi^\dagger\gamma^0\gamma^\mu\psi = \begin{cases} \psi^\dagger\psi & \mu = 0, \\ \psi^\dagger\begin{pmatrix}0 & \sigma_\mu\\ \sigma_\mu & 0\end{pmatrix}\psi = 0 & \mu = 1, 2, 3. \end{cases} \tag{5.4}$$

Hence, the vector current cannot induce a nuclear spin change, i.e. $\Delta J = |\mathbf{J}_{\rm i} - \mathbf{J}_{\rm f}| = 0$, where $\mathbf{J}_{\rm i}$ and $\mathbf{J}_{\rm f}$ are the initial and final nuclear spins. The non-relativistic limit for the other couplings can be calculated in a similar way, and the results are given in Table 5.1. It can be seen that the scalar term has the form $\psi^\dagger\psi = \langle 1\rangle$, whereas the tensor ($\sigma_{\mu\nu}$) and axial vector ($\gamma_5\gamma_\mu$) terms involve the Pauli spin matrices $\boldsymbol{\sigma}$, and can thus induce a nuclear spin change of one unit. The pseudoscalar term gives zero contribution in this limit. Assuming the e and ν states can be represented by plane waves (i.e. neglecting any Coulomb interactions) we can write

$$\mathscr{L} = -\sum_i [\bar{u}_{\rm e}(p_3)\Gamma^i v(p_4)] \int_V \exp[-{\rm i}(\mathbf{p}_3 + \mathbf{p}_4)\cdot\mathbf{x}][\bar{\psi}_{\rm p}(x)\Gamma_i\psi_{\rm n}(x)]\,{\rm d}^3x, \tag{5.5}$$

where the integral is over the nuclear volume, i.e. up to a radius $r \sim A^{1/3}/m_\pi$.

Table 5.2. *Classification of nuclear beta-decays*

Transition	Lepton spin	Nuclear spin	Example
Fermi	S = 0	$\lvert\Delta \mathbf{J}\rvert = 0$	$^{14}O \to {}^{14}Ne^{+}\nu_e$ $0 \to 0$
Gamow–Teller	S = 1	$\lvert\Delta \mathbf{J}\rvert = 0, 1$ $0 \nrightarrow 0$	$^{12}B \to {}^{12}Ce^{-}\bar{\nu}_e$ $1 \to 0$

The momentum of the lepton pair, $\mathbf{q} = \mathbf{p}_3 + \mathbf{p}_4$, is small ($\lesssim$few MeV), so that the argument of the exponential is small ($\sim 1/100$). Thus expanding $\exp(-i\mathbf{q}\cdot\mathbf{x})$ in terms of spherical harmonics, the first term in the expansion, the so-called *allowed term* $l = 0$, will dominate. In certain cases the selection rules (e.g. parity change) forbid the $l = 0$ transition, and the first ($l = 1$) forbidden transition can be observed, and so on. For allowed transitions, the nuclear spin change follows from the spin of the lepton pair. The classification of decays in this way is summarised in Table 5.2. *Fermi* transitions have no nuclear spin change and thus involve SS or VV terms in the Lagrangian, whereas the nuclear spin changes in the *Gamow–Teller* transitions correspond to TT or AA terms. The free particle decay of a neutron is a mixed transition in this scheme.

5.1.2 General matrix element for beta decay

The experimental situation in 1956 was confused, with different experiments leading to apparently conflicting conclusions. There was, however, good support for S and T contributions. This conclusion was soon to be radically changed. The starting point was the θ–τ puzzle which led Lee and Yang (1956) to the profound suggestion that parity conservation was violated in weak interactions (Section 1.8). The matrix element in this case, which can be written down directly from the Lagrangian, becomes

$$\mathcal{M} = \sum_i [\bar{\psi}_p \Gamma_i \psi_n][\bar{\psi}_e \Gamma^i (C_i + C_i' \gamma^5)\psi_\nu], \tag{5.6}$$

where the fields are evaluated at the same space-time point x. Note that the coefficients C_i and C_i' (in (5.2) and (5.6)) are, in general, functions of q^2. However, q^2 is small in these processes, so that C_i and C_i' can be taken as constants.

The transformation properties of the various bilinear forms under parity (P) were discussed in Section 3.2.5. Application of these results to (5.6) shows that under P, $\mathcal{M}$ transforms to $\mathcal{M}_P$, which has exactly the same form as (5.6), but with $C_i' \to -C_i'$. That is, for each coupling, the term in γ^5 (coefficient C_i') has the opposite parity to the term in C_i, and thus non-zero values of C_i' would imply parity violation.

In the non-relativistic limit, the nuclear part of (5.6), and consequently also the leptonic part, can be considerably simplified. Using the results of Table 5.1, equation (5.6) can be simplified to

$$\mathcal{M} = \mathcal{M}_F[\bar{u}_e(C_S + C_S'\gamma^5)v_\nu + \bar{u}_e\gamma^0(C_V + C_V'\gamma^5)v_\nu] \\ + \mathcal{M}_{GT}[\bar{u}_e\boldsymbol{\sigma}(C_T + C_T'\gamma^5)v_\nu + \bar{u}_e\boldsymbol{\sigma}\gamma^0(C_A + C_A'\gamma^5)v_\nu], \qquad (5.7)$$

where $\mathcal{M}_F$ and $\mathcal{M}_{GT}$ refer to Fermi and Gamow–Teller reactions respectively. They consist of the matrix elements $\langle 1 \rangle$ (for F) and $\langle \boldsymbol{\sigma} \rangle$ (for GT), together with the isospin raising operator (τ^+ for $n \to p$) and are summed over all possible contributing nucleons in the nucleus. It is assumed that the matrix element is a sum of contributions for individual nucleons (impulse approximation). The calculation of the nuclear wave function requires a detailed knowledge of the nucleus in question; however, this is not needed for the discussion below and is not considered further. The form of the leptonic current for the vector term, for example, arises because the only non-zero contribution from the nuclear part is for $\mu = 0$. Note that all the, *a priori*, unknown constants C_i, C_i' depend only on the leptonic part of the matrix element.

The experimental measurements on nuclear beta decay and on free neutron decay can be separated into two distinct classes. The first class is on measurements of the electron spectrum from unpolarised nuclei (or neutrons), whereas the second (and more difficult) class of measurements is on the polarisation of the e^- (or e^+) or on studies of the decay products from polarised nuclei (or neutrons).

5.1.2.1 Electron spectrum from unpolarised nuclei

The techniques used to calculate the differential decay rate closely follow those used for electromagnetic interactions. The derivation of the results for pure couplings is outlined below, and the general result allowing all possible couplings is then discussed. For an unpolarised initial nucleus, then, summing over the spin states of the electron and averaging over the initial nuclear spin states, we obtain for:

(i) *Pure S*

$$\mathcal{M} = \langle 1 \rangle \bar{u}_e(C_S + C_S'\gamma^5)v_\nu, \qquad \mathcal{M}^* = \langle 1 \rangle^* \bar{v}_\nu(C_S^* - C_S'^*\gamma^5)u_e$$

$$|\mathcal{M}|^2 = |\langle 1 \rangle|^2 \operatorname{tr}[(\not{p}_3 + m_e)(C_S + C_S'\gamma^5)\not{p}_4(C_S^* - C_S'^*\gamma^5)] \\ = |\langle 1 \rangle|^2 4(|C_S|^2 + C_S'|^2)(E_3E_4 - \mathbf{p}_3 \cdot \mathbf{p}_4). \qquad (5.8)$$

(ii) *Pure V*

$$\mathscr{M} = \langle 1 \rangle \bar{u}_e \gamma^0 (C_V + C'_V \gamma^5) v_\nu$$

$$|\mathscr{M}|^2 = |\langle 1 \rangle|^2 \operatorname{tr}[(\not{p}_3 + m_e)\gamma^0 (C_V + C'_V\gamma^5)\not{p}_4 (C_V^* - C'^*_V \gamma^5)\gamma^0]$$

$$= |\langle 1 \rangle|^2 4(|C_V|^2 + |C'_V|^2)(E_3 E_4 + \mathbf{p}_3 \cdot \mathbf{p}_4). \tag{5.9}$$

(iii) *Pure T*

$$\mathscr{M} = \langle \sigma_k \rangle \bar{u}_e \sigma^k (C_T + C'_T \gamma^5) v_\nu$$

$$|\mathscr{M}|^2 = \tfrac{1}{3}|\langle \sigma_k \rangle|^2 \operatorname{tr}[(\not{p}_3 + m_e)\sigma^k (C_T + C'_T\gamma^5)\not{p}_4 (C_T^* - C'^*_T \gamma^5)\sigma^k]$$

$$= |\langle \boldsymbol{\sigma} \rangle|^2 4(|C_T|^2 + |C'_T|^2)(E_3 E_4 + \tfrac{1}{3}\mathbf{p}_3 \cdot \mathbf{p}_4). \tag{5.10}$$

(iv) *Pure A*

$$\mathscr{M} = \langle \sigma_k \rangle \bar{u}_e \sigma^k \gamma^0 (C_A + C'_A \gamma^5) v_\nu$$

$$|\mathscr{M}|^2 = \tfrac{1}{3}|\langle \sigma_k \rangle|^2 \operatorname{tr}[(\not{p}_3 + m_e)\sigma^k \gamma^0 (C_A + C'_A\gamma^5)\not{p}_4 (C_A^* - C'^*_A \gamma^5)\gamma^0 \sigma^k]$$

$$= |\langle \boldsymbol{\sigma} \rangle|^2 4(|C_A|^2 + |C'_A|^2)(E_3 E_4 + \tfrac{1}{3}\mathbf{p}_3 \cdot \mathbf{p}_4). \tag{5.11}$$

In deriving these formulae the properties of the γ-matrices (Section 3.2.1), the Pauli spin matrices (Section 2.4) and the projection operators (Section 3.2.4) have been used. Furthermore, we have taken $\mathscr{M}_F = \langle 1 \rangle$ and $\mathscr{M}_{GT} = \langle \boldsymbol{\sigma} \rangle$ for simplicity. The factor $\frac{1}{3}$ for the T and A terms is the average over the spin states of the initial nucleus. We have $\overline{\langle \sigma_k \rangle \langle \sigma_j \rangle} = \delta_{kj} |\langle \boldsymbol{\sigma} \rangle|^2/3$. Note that all of the couplings give a similar form for $|\mathscr{M}|^2$, namely

$$|\mathscr{M}_i|^2 = 4K_i E_3 E_4 (1 + a_i \beta \cos\theta), \tag{5.12}$$

with $K_i = |\langle r \rangle|^2 (|C_i|^2 + |C'_i|^2)$, where $r = 1$ or $\boldsymbol{\sigma}$ as appropriate, and with

$$a_S = -1, \qquad a_V = 1, \qquad a_T = \tfrac{1}{3}, \qquad a_A = -\tfrac{1}{3}, \tag{5.13}$$

where θ is the angle between the e^- and $\bar{\nu}_e$, and β is the electron velocity.

The decay rate for the beta-decay $A \to A' e^- \bar{\nu}_e$ is calculated in a similar way to that used to compute the interaction probability in Section 4.2. The probability per unit time for a transition is

$$d\lambda = |S_{fi}|^2 N \Phi / T$$

$$= \frac{(2\pi)^4}{V^3} \sum_i |\mathscr{M}_{fi}|^2 N \Phi \, \delta^4(p_1 - p_2 - p_3 - p_4), \tag{5.14}$$

where N is the normalisation term consisting of factors $(2E_3)^{-1}$ and $(2E_4)^{-1}$ for the e^- and $\bar{\nu}_e$ spinors and factors of unity for the nuclear wave functions, which are confined to the nuclear volume V. The term Φ is the phase space factor (Section 4.2) for the final state particles. Inserting these factors in (5.14), and integrating over d^3p_2, gives

$$d\lambda = \frac{1}{(2\pi)^5} \sum_i |\mathcal{M}_{fi}|^2 \frac{d^3p_3\, d^3p_4\, \delta(\Delta - E_3 - E_4)}{4E_3E_4}, \tag{5.15}$$

where $\Delta = E_1 - E_2$ is the energy release in the decay. Substituting $|\mathcal{M}|^2$ from (5.12), putting $d^3p_3 = p_3E_3\, dE_3\, d\Omega_3$, $d^3p_4 = E_4^2\, dE_4\, d\Omega_4$ and integrating over the $\bar{\nu}_e$ variables dE_4 and $d\Omega_4$, gives

$$d\lambda = \frac{1}{8\pi^4} \sum_i K_i(1 + a_i\beta \cos\theta)p_3E_3(\Delta - E_3)^2\, dE_3\, d\Omega_3. \tag{5.16}$$

The energy spectrum of the e^- is obtained by integrating over $d\Omega_3$. Writing $\xi = \sum_i K_i$, this gives

$$\frac{d\lambda}{dE_3} = \frac{1}{2\pi^3} \xi p_3E_3(\Delta - E_3)^2. \tag{5.17}$$

A plot of the measured e^- energy spectrum in the form $[N(E_3)/p_3E_3]^{1/2}$ *versus* E_3 (*Kurie plot*) should thus be linear, with intercept $E_3 = \Delta$. The above derivation assumes $m_\nu = 0$; small deviations from the form of (5.17) near the electron end point are expected if $m_\nu \neq 0$ (see Chapter 9). The total decay rate is

$$\lambda = \frac{1}{2\pi^3} \xi \int_{m_e}^{\Delta} p_eE_3(\Delta - E_3)^2\, dE_3 = \frac{1}{2\pi^3} \xi f_0. \tag{5.18}$$

The integral f_0 is called the *Fermi integral* and, in a realistic calculation, includes a further multiplication factor $F(Z, E_3)$ describing the distortion of the e^- (or e^+) wave function in the Coulomb field of the nucleus (see, e.g., Schopper, 1966). The mean lifetime τ is the inverse of (5.18), and the commonly used *ft value* is defined as

$$ft = 2\pi^3 \ln 2/(\xi f_0). \tag{5.19}$$

A more general calculation, allowing for all possible couplings simultaneously gives, for an unpolarised nucleus,

$$\frac{d\lambda}{dE_3\, d\Omega_3} = \frac{p_3E_3(\Delta - E_3)^2}{8\pi^4} \xi(1 + a\beta \cos\theta + bm_e/E_3), \tag{5.20}$$

where

$$\begin{aligned} \xi &= |\mathcal{M}_F|^2(E_{SS} + E_{VV}) + |\mathcal{M}_{GT}|^2(E_{TT} + E_{AA}), \\ \xi a &= |\mathcal{M}_F|^2(-E_{SS} + E_{VV}) + |\mathcal{M}_{GT}|^2(E_{TT} - E_{AA})/3, \\ \xi b &= |\mathcal{M}_F|^2 2\mathrm{Re}(E_{SV}) + |\mathcal{M}_{GT}|^2 2\mathrm{Re}(E_{TA}). \end{aligned} \tag{5.21}$$

In these, and subsequent, formulae it is useful to define the following combinations of coupling constants

$$E_{ij} = C_i C_j^* + C_i' C_j'^*, \qquad F_{ij} = C_i C_j'^* + C_i' C_j^*. \tag{5.22}$$

The additional term compared to the previous calculation is the *Fierz interference* term b. For e^+ emission, b changes sign. If Coulomb effects in the nucleus are taken into account, then the righthand side of the expression for ξb should be multiplied by 2γ, with $\gamma = (1 - \alpha^2 Z^2)^{1/2}$, Z being the atomic number of the daughter nucleus. Experimentally, the value of b is consistent with zero for both F and GT transitions. For example, typical experimental limits are $b = -0.02 \pm 0.09$ for Fermi transitions, and $b = 0.0008 \pm 0.0020$ for Gamow–Teller transitions (see, e.g., Daniel, 1968; Paul, 1970).

The coefficient a in (5.13) and (5.20) is also sensitive to the type of coupling. Experimentally, the values found are consistent with V and A couplings only. For example, Allen *et al.* (1959) measure $a = 0.97 \pm 0.14$ for $^{35}\text{Ar} \to {}^{35}\text{Cl}\, e^+ \nu_e$ (pure F), whereas, for the pure GT decay $^6\text{He} \to {}^6\text{Li}\, e^- \bar{\nu}_e$, Johnson *et al.* (1963) give $a = -0.3343 \pm 0.0030$. Note that for pure F or GT transitions, the predicted values of a do not depend on $\mathscr{M}_F$ or $\mathscr{M}_{GT}$, and hence on any uncertainties in the nuclear matrix elements. Experimentally the neutrino is not detected, and so its direction must be inferred from the electron and nuclear recoil measurements. Fig. 5.2 (see Schopper,

Fig 5.2 Recoil momentum spectrum for the decay $^6\text{He} \to \text{Li}\, e^- \bar{\nu}_e$, together with the predictions for pure A and pure T couplings.

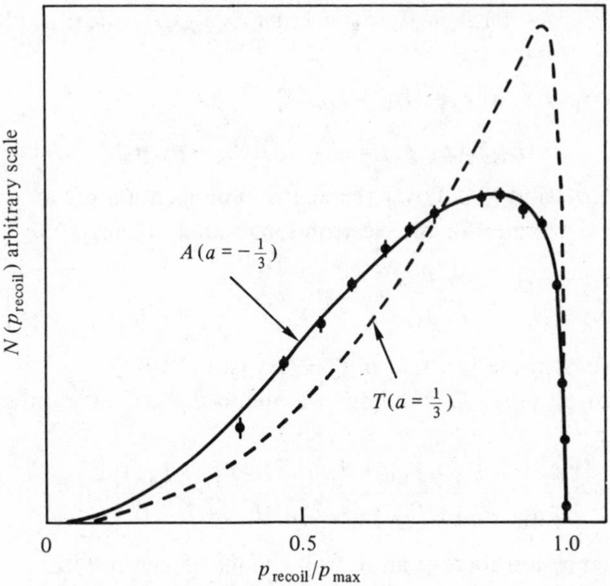

1966) illustrates a measurement of the recoil momentum distribution, and clearly favours pure A coupling.

5.1.2.2 Polarisation measurements

The information which can be extracted from a study of the electron spectrum from unpolarised nuclei is limited. The results show that the contributions of S and T are small relative to V and A. However, the relative magnitudes and signs of C_V, C'_V, C_A and C'_A are not determined. Measurements of the longitudinal polarisation of the decay e^- (or e^+), and of various quantities in the decays of polarised nuclei, give additional information on the coupling constants.

The longitudinal polarisation of an electron in nuclear beta decay can be calculated in a similar way to that used for electromagnetic interactions in Section 4.6. The required quantities are the relative numbers of events where the decay electron is righthanded (N_R) and lefthanded (N_L). For example, for a pure scalar interaction from an unpolarised nucleus

$$N\binom{R}{L} = X_S \operatorname{tr}\left[(\not{p}_3 + m_e)\frac{(1 \pm \gamma^5 \not{s}_e)}{2}(C_S + C'^S\gamma^5)\not{p}_4(C_S^* - C_S'^*\gamma^5)\right]$$

$$= \frac{X_S}{2}\operatorname{tr}[(\not{p}_3 \pm m_e\gamma^5\not{s}_e)\not{p}_4(E_{SS} - F_{SS}\gamma^5)], \tag{5.23}$$

where X_S represents the remaining common terms in the cross-section. Thus

$$N_R - N_L = m_e X_S \operatorname{tr}[\gamma^5 \not{s}_e \not{p}_4 (E_{SS} - F_{SS}\gamma^5)]$$

$$= -4m_e F_{SS} X_S (s_e \cdot p_4) = -4m_e F_{SS} X_S (s_e^0 E_4 - \hat{\mathbf{s}}_e \cdot \mathbf{p}_4), \tag{5.24}$$

and

$$N_R + N_L = X_S \operatorname{tr}[\not{p}_3 \not{p}_4 (E_{SS} - F_{SS}\gamma^5)]$$

$$= 4E_{SS} X_S (p_3 \cdot p_4) = 4E_{SS} X_S (E_3 E_4 - \mathbf{p}_3 \cdot \mathbf{p}_4). \tag{5.25}$$

Integrating (5.24) and (5.25) over the neutrino momentum $\mathbf{p}_4$, we obtain the following expression for the electron longitudinal polarisation

$$P = \frac{N_R - N_L}{N_R + N_L} = \frac{-m_e F_{SS} s_e^0}{E_{SS} E_3} = -\beta \frac{F_{SS}}{E_{SS}}, \tag{5.26}$$

where we have used the fact that $m_e s_e^0 = \beta E_3$ (see (4.104)).

Allowing for all possible couplings we obtain the general expressions (taking $b = 0$)

$$P = \beta \frac{[|\mathcal{M}_F|^2(-F_{SS} + F_{VV}) + |\mathcal{M}_{GT}|^2(-F_{TT} + F_{AA})]}{[|\mathcal{M}_F|^2(E_{SS} + E_{VV}) + |\mathcal{M}_{GT}|^2(E_{TT} + E_{AA})]}. \tag{5.27}$$

For positron emission there is an overall change of sign in (5.27).

Experimentally, the decay electron has a longitudinal polarisation consistent with $P = -\beta$. For example, Brosi *et al.* (1962) measure $P = -(0.990 \pm 0.009)\beta$ for the GT decay of ^{32}P. This dependence has been checked over a wide range of β values (Koks and Van Klinken, 1976). The e^+ results are consistent with $P = +\beta$, but with larger errors. Assuming $P = -\beta$ in (5.27), then, for pure couplings, we can deduce the following

$$\begin{aligned}
&\text{S} \quad F_{SS} = E_{SS} \quad \Rightarrow |C_S - C'_S|^2 = 0 \Rightarrow C'_S = C_S,\\
&\text{V} \quad F_{VV} = -E_{VV} \Rightarrow |C_V + C'_V|^2 = 0 \Rightarrow C'_V = -C_V,\\
&\text{T} \quad F_{TT} = E_{TT} \quad \Rightarrow |C_T - C'_T|^2 = 0 \Rightarrow C'_T = C_T,\\
&\text{A} \quad F_{AA} = -E_{AA} \Rightarrow |C_A + C'_A|^2 = 0 \Rightarrow C'_A = -C_A.
\end{aligned} \tag{5.28}$$

These results place severe restrictions on the possible forms of the interaction. From (5.7) it can be seen that they imply that the $\bar{\nu}_e$ enters $\mathcal{M}$ with the form $(1 + \gamma^5)v_\nu$ for S and T, but $(1 - \gamma^5)v_\nu$ for V and A. This means that for S and T the $\nu_e(\bar{\nu}_e)$ is right(left)handed, whereas for V and A the $\nu_e(\bar{\nu}_e)$ is left(right)handed. Note that the electron always enters $\mathcal{M}$ in the form $\bar{u}_e(1 + \gamma^5)$, irrespective of the type of coupling, i.e. the $e^-(e^+)$ emitted in beta decay is left(right)handed.

The remaining piece in this 'jigsaw puzzle' is to determine the helicity of the neutrino, which is not fixed by the above considerations. The ν_e helicity was measured by the ingenious experiment of Goldhaber *et al.* (1958), which used the following reaction

$$\begin{aligned}
&{}^{152}\text{Eu}(J^P = 0^-) + e^- \rightarrow \nu_e + {}^{152}\text{Sm}^*(J^P = 1^-)\\
&\qquad\qquad\qquad\qquad\qquad\qquad \hookrightarrow {}^{152}\text{Sm}(J^P = 0^+) + \gamma.
\end{aligned} \tag{5.29}$$

The first process is a GT transition. The experiment selected γ-rays from the excited state which were in the opposite direction to the ν_e, and thus in the direction of the recoil nucleus. The γ-rays were found to have negative helicity (using resonance scattering off a target of ^{152}Sm). Since the initial state has $J = 0$, the angular momentum of the ^{152}Sm* state (and hence the γ-ray) is opposite to that of the ν_e. Thus the ν_e helicity is (predominantly) negative.

The beta decays of polarised nuclei are also sensitive to the relative signs of the coupling constants. We will consider one example, namely the decay angular distribution of the e^- (or e^+) with respect to the direction of polarisation. For the case of V and A couplings only, the matrix element is

$$\mathcal{M} = \langle 1 \rangle \bar{u}_e \gamma^0 (C_V + C'_V \gamma^5) v_\nu + \langle \sigma_i \rangle \bar{u}_e \sigma^i \gamma^0 (C_A + C'_A \gamma^5) v_\nu, \tag{5.30}$$

and so

$$\mathcal{M}^* = \langle 1\rangle^* \bar{v}_\nu (C_V^* - C_V'^* \gamma^5)\gamma^0 u_e + \langle \sigma_j \rangle^* \bar{v}_\nu (C_A^* - C_A'^* \gamma^5)\sigma^j \gamma^0 u_e. \tag{5.31}$$

In order to compute the angular distribution of the decay electron with respect to the nuclear polarisation vector $\mathbf{P}_i$, the projection operators $(\not{p}_3 + m_e)$ (for the electron, summing over electron spins) and $E_4\gamma^0$ (for the $\bar{v}_e$, integrating over $\mathbf{p}_4$) are inserted into $|\mathcal{M}|^2$ giving

$$\begin{aligned}|\mathcal{M}|^2 = {} & |\langle 1\rangle|^2 E_4 \operatorname{tr}[(\not{p}_3 + m_e)\gamma^0(E_{VV} + F_{VV}\gamma^5)] \\ & + \langle\sigma_i\rangle\langle\sigma_j\rangle^* E_4 \operatorname{tr}[(\not{p}_3 + m_e)\sigma^i\sigma^j\gamma^0(E_{AA} + F_{AA}\gamma^5)] \\ & + \langle 1\rangle\langle\sigma_j\rangle^* E_4 \operatorname{tr}[(\not{p}_3 + m_e)\sigma^j\gamma^0(E_{VA} + F_{VA}\gamma^5)] \\ & + \langle 1\rangle^*\langle\sigma_i\rangle E_4 \operatorname{tr}[(\not{p}_3 + m_e)\sigma^i\gamma^0(E_{AV} + F_{AV}\gamma^5)].\end{aligned} \tag{5.32}$$

Evaluation of $|\mathcal{M}|^2$ is carried out using the standard trace properties; however, care must be taken in the evaluation of the nucleon spin states, which enter through the Pauli spin matrices. This is considered in detail by Källen (1964), for both nuclear beta decay and free neutron decay. The matrix elements $\mathcal{M}_F$ and $\mathcal{M}_{GT}$ can be assumed real without loss of generality. In the case of a spin transition $J_i \to J_f$, for a nucleus with mean polarisation $\langle \mathbf{P}_i \rangle / P_i$, the decay angular distribution takes the form

$$d\lambda \propto \xi\left[1 + A\frac{\langle \mathbf{P}_i\rangle}{P_i}\cdot\frac{\mathbf{p}_3}{E_3}\right], \tag{5.33}$$

where

$$\xi = |\mathcal{M}_F|^2 E_{VV} + |\mathcal{M}_{GT}|^2 E_{AA},$$

$$A\xi = \pm\Lambda_{J_iJ_f}|\mathcal{M}_{GT}|^2 F_{AA} - 2\delta_{J_iJ_f}|\mathcal{M}_F||\mathcal{M}_{GT}|\left(\frac{J_i}{J_i+1}\right)^{1/2} \operatorname{Re} F_{VA},$$

with

$$\Lambda_{J_iJ_f} = \begin{cases} -J_i/(J_i+1) & \text{if } J_f = J_i + 1 \\ 1/(J_i+1) & \text{if } J_f = J_i \\ 1 & \text{if } J_f = J_i - 1. \end{cases} \tag{5.34}$$

The upper sign refers to e^- emission and the lower to e^+.

The decay $^{60}\text{Co} \to {}^{60}\text{Ni}\, e^- \bar{v}_e$ is a GT transition with spin states $J_i = 5$, $J_f = 4$; hence (5.34) takes the simple form $A = F_{AA}/E_{AA}$. In the important experiment of Wu *et al.* (1957), a value of $A \simeq -1$ was found. That is, the e^- is produced preferentially opposite to the direction of the nuclear spin.

For the case $A = -1$, then $C'_A = -C_A$. Thus parity is violated in these decays (since $C'_A \neq 0$) and, furthermore, the results of many subsequent experiments indicate that it appears to be violated maximally. A value of $C'_A \neq 0$ also shows that charge conjugation invariance is violated in these decays. This follows from equation (3.172), where it can be seen that the C'_A term has opposite C-parity to the C_A term.

The decay asymmetry of electrons from free polarised neutrons is also described by (5.34). The spin state configurations imply $\mathcal{M}_F = 1$ and $\mathcal{M}_{GT} = 3^{1/2}$; hence

$$A = \frac{2F_{AA} - 2\,\mathrm{Re}\,F_{VA}}{E_{VV} + 3E_{AA}}. \tag{5.35}$$

If the direction of the decay neutrino can be deduced, then the decay rate has the form (Jackson *et al.*, 1957)

$$\mathrm{d}\lambda \propto \xi\left[1 + \frac{\langle \mathbf{P}_i \rangle}{P_i}\cdot\left(A\frac{\mathbf{p}_3}{E_3} + B\frac{\mathbf{p}_4}{E_4} + D\frac{\mathbf{p}_3 \times \mathbf{p}_4}{E_3 E_4}\right)\right], \tag{5.36}$$

where

$$B\xi = \mp\Lambda_{J_iJ_f}|\mathcal{M}_{GT}|^2 F_{AA} - 2\delta_{J_iJ_f}|\mathcal{M}_F||\mathcal{M}_{GT}|\left(\frac{J_i}{J_i + 1}\right)^{1/2}\mathrm{Re}\,F_{VA},$$

$$D\xi = 2\delta_{J_iJ_f}|\mathcal{M}_F||\mathcal{M}_{GT}|\left(\frac{J_i}{J_i + 1}\right)^{1/2}\mathrm{Im}\,E_{VA}. \tag{5.37}$$

The term with coefficient D is a triple scalar product. If time reversal (T) invariance holds for weak decays, then $D = 0$. This follows since under T each of the three components of the product changes sign, so that a non-zero value of D can only be produced if T is violated. This argument is only strictly true for the primary interaction and any later secondary interaction effects can give $D \neq 0$, and thus masquerade as T violation. This is important in beta decay, where the decay electron must initially pass through dense nuclear matter.

The requirement of time reversal invariance (in the absence of final state interaction effects) implies that the form factors (e.g. D in (5.37)) are relatively real. The application of T changes the signs of momenta and spins, and changes the initial and final states. The matrix element is Hermitian, i.e. $\mathcal{M}^\dagger = \mathcal{M}$ (Section 2.2.3); hence the operations of Hermitian conjugation and T give

$$\langle \mathrm{f}|\mathcal{M}|\mathrm{i}\rangle \xrightarrow{*} \langle \mathrm{i}|\mathcal{M}|\mathrm{f}\rangle^* = \mathcal{M}^*,$$

$$\langle \mathrm{f}|\mathcal{M}|\mathrm{i}\rangle \xrightarrow{T} \langle \mathrm{i}|\mathcal{M}|\mathrm{f}\rangle = \mathcal{M}^T. \tag{5.38}$$

These considerations lead to the requirement that $|\mathcal{M}^*| = |M^T|$ if T invariance holds. In Section 3.2.7 it was shown that, for spin $\frac{1}{2}$ particles, the T operation corresponds to the transformation (3.174) $\psi \xrightarrow{T} i\gamma^1\gamma^3\psi^*$, and hence $\psi^\dagger \xrightarrow{T} i\psi^t\gamma^3\gamma^1$. Thus under T the bilinear covariants Γ transform as

$$\begin{aligned}\bar{\psi}_f\Gamma\psi_i &\xrightarrow{T} \psi_f^t\gamma^{0t}\gamma^3\gamma^1\Gamma\gamma^1\gamma^3\psi_i^* \\ &= \bar{\psi}_i\gamma^0\gamma^3\gamma^1\Gamma^t\gamma^1\gamma^3\gamma^0\psi_f,\end{aligned} \tag{5.39}$$

where the superscript t means transposed. This follows by using the matrix relation $(AB)^t = B^tA^t$ and noting that $(\gamma^0)^t = \gamma^0$ and $(\gamma^i)^t = -(\gamma^i)$, $i = 1, 2, 3$. Under Hermitian conjugation (h.c.) the following transformation is obtained

$$\bar{\psi}_f\Gamma\psi_i \xrightarrow{\text{h.c.}} \psi_i^\dagger\Gamma^\dagger\gamma^0\psi_f = \bar{\psi}_i\gamma^0\Gamma^\dagger\gamma^0\psi_f. \tag{5.40}$$

Using these transformations, and the specific forms for the bilinear covariants Γ, then the requirement that $\mathcal{M}^* = \mathcal{M}^T$ for the matrix element (5.6) gives $C_i^* = C_i$ ($C_i'^* = C_i'$), i.e. relatively real. Note that if T is conserved then, from the general field theory CPT theorem, CP is also conserved. This is discussed further in Chapter 9.

For polarised neutron decay, some typical experimental results for the coefficients A, B and D are

$A = -0.115 \pm 0.006$	(Erozolimskii *et al.*, 1976),
$B = \quad 1.00 \pm 0.05$	(Christensen *et al.*, 1970),
$D = -0.0011 \pm 0.0017$	(Steinberg *et al.*, 1976).

Assuming that $C_V' = -C_V$ and $C_A' = -C_A$ (these relations being compatible with the e^- polarisation results (5.28)), then

$$D\xi = -2i(C_VC_A^* - C_V^*C_A). \tag{5.41}$$

Thus the value $D = 0$, which is compatible with the experimental results, implies C_V and C_A are relatively real. Assuming this, then the formulae for $A(A\xi = 2F_{AA} - 2F_{VA})$ and $B(B\xi = -2F_{AA} - 2F_{VA})$ from (5.35) and (5.37) respectively can be cast in the form

$$\begin{aligned}(B + A)\xi &= -4F_{VA} = 8C_VC_A, \\ (B - A)\xi &= -4F_{AA} = 8C_A^2.\end{aligned} \tag{5.42}$$

Thus

$$C_A/C_V = (B - A)/(B + A). \tag{5.43}$$

The experimental values quoted above give $C_A/C_V = 1.26 \pm 0.02$. A value for C_A/C_V can be extracted from a measurement of A alone using (5.35). The recent measurement of $A = -0.1146 \pm 0.0019$ by Bopp *et al.* (1986) gives a value $C_A/C_V = 1.262 \pm 0.005$.

A fit to all the data on nuclear beta decay and neutron decay available at the time by Paul (1970) gives the results

(i) $C_A/C_V = 1.240 \pm 0.007$,
(ii) $C_A/C_V = 1.262 \pm 0.008$.

The value (i) is obtained using the neutron and ^{14}O ft-values, whereas the value (ii) is from 12 selected data points. The difference can be taken as an estimate of the error inherent in extracting weak interaction results from hadronic decays in the absence of a full treatment of higher order electroweak and QCD corrections. A global fit to the data is compatible with the absence of S and T terms, although a small admixture cannot be excluded. Assuming that there are only V and A contributions, then the above considerations simplify the general matrix element (5.6) to

$$\mathcal{M} = \frac{G_\beta}{2^{1/2}}[\bar{\psi}_p\gamma_\mu(1 - 1.26\gamma^5)\psi_n][\bar{\psi}_e\gamma^\mu(1 - \gamma^5)\psi_\nu], \tag{5.44}$$

where we have written $C_V = G_\beta/2^{1/2}$. The beta decay coupling constant $G_\beta = 1.1 \times 10^{-5}\ \text{GeV}^{-2} \simeq 10^{-5} M_N^{-2}$ (with the nucleon mass M_N in GeV); this being measured from the total decay rates of, say, $n \to pe^-\bar{\nu}_e$ or the ft-value of ^{14}O. The factor of $2^{1/2}$ is introduced to keep the original numerical value of G unchanged. The leptonic current thus has the form V − A (i.e. equal strengths for V and A, but opposite sign), whereas the hadronic current is of the form V − 1.26A. Note that pure V − A for neutron decay would give the parameter $A = 0$. The difference of the axial part of the hadronic current from V − A can be understood as being due to corrections at the hadron level to a current which is of the form V − A at the quark level. This is discussed further in Chapter 8.

Thus the results of nucleon beta decay, which emerged over a period of a few years after 1956 and were later refined, show that the interaction involved is predominantly V − A in nature. Results from beta decay have played an important role in the development of the theory of weak interactions. The reason for this is partly historical, in that until the advent of sufficiently high energy accelerators beta decay was the easiest means of studying weak interactions. However, in addition, the large variety of measurements which can be performed, together with the selection of angular momentum states which can be made from knowledge of the nuclear transitions involved, have meant that precise values of the coupling constants can be extracted despite the inherent uncertainties of calculations in the multibody nuclear environment.

5.1.3 The V – A theory

The framework of the phenomenological V – A theory was first published by Feynman and Gell-Mann (1958). The scope of the theory was later extended and the resulting Lagrangian successfully describes a wide class of weak interaction phenomena, provided that the four-momentum transfers involved are reasonably small ($Q^2 \lesssim$ few GeV^2). The general V – A Lagrangian is written as the product of two V – A currents, namely

$$\mathscr{L}_{\text{V}-\text{A}} = -\frac{G_\text{F}}{2^{1/2}} J_1^\mu J_{2\mu}^\dagger + \text{herm. conj.}, \tag{5.45}$$

where G_F is the Fermi coupling constant ($G_\text{F} = 1.166 \times 10^{-5}\ \text{GeV}^{-2}$). The currents J_1 and J_2 describe either leptonic or quark transitions. These currents are defined as

$$J^\mu = \begin{cases} J_l^\mu = \bar{\psi}_{\nu_l}\gamma^\mu(1-\gamma^5)\psi_l, & l = \text{e}^-, \mu^-, \tau^- \ldots, \\ J_\text{q}^\mu = \bar{\psi}_{\text{q}_2}\gamma^\mu(1-\gamma^5)\psi_{\text{q}_1}, & \text{q}_1 = \text{d, s, b}, \ldots, \text{q}_2 = \text{u, c, t} \ldots. \end{cases} \tag{5.46}$$

The leptonic current J_l describes, for example, the transition $\text{e}^- \to \nu_\text{e}$ and the Hermitian conjugate $J_l^\dagger$ describes the inverse process $\nu_\text{e} \to \text{e}^-$. Transitions involving antiparticles (e.g. $\text{e}^+ \to \bar{\nu}_\text{e}$, $\bar{\nu}_\text{e} \to \text{e}^+$) are introduced, in a similar way to that used for QED, by using the appropriate field operators (or spinors) in the current. Purely leptonic processes are described by the product of two such currents; for example, the decay $\mu^- \to \text{e}^-\bar{\nu}_\text{e}\nu_\mu$ involves the product of the currents ($\mu^- \to \nu_\mu$) and ($\nu_\text{e} \to \text{e}^-$). The resulting Lagrangian gives a good description of muon decay (Chapter 6) as well as other low Q^2 purely leptonic interactions.

The V – A current (5.46) can be written as $J^\mu(\text{V}-\text{A}) = \bar{\psi}\gamma^\mu(1-\gamma^5)\psi = 2\bar{\psi}_\text{L}\gamma^\mu\psi_\text{L}$, where $\psi_\text{L} = \frac{1}{2}(1-\gamma^5)\psi$ (see (3.152)). Thus this current involves only lefthanded fermions and, correspondingly, righthanded antifermions. Conversely, a V + A model with $J^\mu(\text{V}+\text{A}) = \bar{\psi}\gamma^\mu(1+\gamma^5)\psi = 2\bar{\psi}_\text{R}\gamma^\mu\psi_\text{R}$, with $\psi_\text{R} = \frac{1}{2}(1+\gamma^5)\psi$, involves only righthanded fermions and lefthanded antifermions. Note that the other combinations, namely $\bar{\psi}_\text{L}\gamma^\mu\psi_\text{R}$ and $\bar{\psi}_\text{R}\gamma^\mu\psi_\text{L}$ are both zero because $(1+\gamma^5)(1-\gamma^5) = 0$.

Built into the V – A Lagrangian (5.45) is the assumption of conservation of lepton number. As discussed in Section 1.2, an additive quantum number is defined such that $L_l = +1$ for the lefthanded leptons l^-, ν_l and $L_l = -1$ for the righthanded l^+, $\bar{\nu}_l$ (separately for $l = \text{e}, \mu, \tau$), and the sum of L_l is conserved in any interaction. Some experimental limits on decay branching ratios involving a violation of this rule are, at the 90% c.l.,

$$\mu^- \to e^-\gamma < 1.7 \times 10^{-10}, \qquad \tau^- \to e^-\gamma < 6.4 \times 10^{-4},$$
$$\to e^-e^+e^- < 2.4 \times 10^{-12}, \qquad \to \mu^-\gamma < 5.5 \times 10^{-4},$$
$$\to (3l)^- < 5 \times 10^{-4}.$$

An alternative lepton number scheme (Feinberg and Weinberg, 1961) assigns an additive lepton number $L = -1$ to all leptons and $+1$ to all antileptons, together with a multiplicative muon parity P_μ, which is defined to be $P_\mu = -1$ for μ^-, μ^+, ν_μ, $\bar{\nu}_\mu$ and $+1$ for e^-, e^+, ν_e, $\bar{\nu}_e$. This scheme (proposed before the discovery of the τ-lepton) forbids $\mu \to e\gamma$ etc., but would allow other reactions for which there are now reasonably good experimental limits, e.g. at the 90% c.l.

$$\sigma(\bar{\nu}_\mu e^- \to \mu^- \bar{\nu}_e)/\sigma(\nu_\mu e^- \to \mu^- \nu_e) < 0.05 \qquad \text{(Bergsma } et\ al.\text{, 1983)},$$

$$\Gamma(\mu^+ \to e^+\bar{\nu}_e\nu_\mu)/\Gamma(\mu^+ \to e^+\nu_e\bar{\nu}_\mu) < 0.07 \qquad \text{(Willis } et\ al.\text{, 1980)}.$$

Thus the additive scheme is clearly preferred. It should be stressed that lepton number conservation is empirical, and not based on any deeper principle. Indeed, in models attempting to unify leptons and quarks, there can be transformations between the various leptons and quarks. Lepton number conservation is discussed further in the context of neutrino oscillations and double beta decay in Chapter 9.

The quark current describes transitions between the quarks of charge $-\frac{1}{3}(q_1)$ and those of charge $+\frac{2}{3}(q_2)$. The inverse transitions and the currents for antiquarks are included in the same way as for leptons. The quarks involved in the weak interaction processes are, of course, contained in hadrons. Hence the experimentally measurable quantities correspond to an hadronic current rather than a quark current. Cabibbo (1963) proposed the following form for the hadronic current

$$J_h^\mu = \cos\theta_C J_h^\mu(\Delta S = 0) + \sin\theta_C J_h^\mu(|\Delta S| = 1), \tag{5.47}$$

where ΔS is the difference in strangeness of the initial and final state hadrons, and θ_C is the Cabibbo angle. This description was proposed before the discovery of charm and successfully described the observations that $|\Delta S| > 1$ transitions are not observed, and that strangeness-changing decays (e.g. $K \to \mu\nu$) are suppressed relative to those with no change in strangeness (e.g. $\pi \to \mu\nu$). The value of the Cabibbo angle needed to reproduce the data is $\theta_C \sim 13°$. At the quark level this means that $u \to s$ transitions are suppressed relative to $u \to d$.

The coupling constant G_β extracted from nuclear beta and neutron decay is found to be smaller than that determined from the purely leptonic muon decay (G_F in (5.45)). The experimental results are consistent with the relation $G_\beta = G_F \cos\theta_C$, which is expected from Cabibbo's hypothesis for the non-strangeness changing $d \to u$ transition responsible for beta decay. However, the extraction of θ_C from nuclear beta decay is sensitive to higher order electroweak corrections, and from the uncertainties inherent in a nuclear environment. From the raw (uncorrected) rates a value of $\sin\theta_C = 0.18$ is extracted, compared to a value of $\sin\theta_C = 0.22$ obtained from $K^+ \to \pi^0 e^+ \nu_e$ decays. Agreement is obtained after electroweak corrections have been applied, giving $\sin\theta_C = 0.224 \pm 0.005$, or $\cos\theta_C = 0.9747 \pm 0.0011$ (Sirlin and Zucchini, 1986). Other semi-leptonic interactions (i.e. those involving one lepton current and one quark current), such as $B \to B' l \nu_l$, where B and B′ are baryons, are also well described by (5.47). Purely hadronic weak decays (e.g. $\Lambda \to p\pi^-$) involve the product of two quark currents, and thus the underlying structure is obscured. Nonetheless, the experimental results are compatible with the selection rules implied by (5.47). Thus the Lagrangian (5.45) is *universal* in the sense that all these interactions can be described by the one coupling constant G_F.

The structure of the current in (5.46) requires that all fermions are lefthanded and all antifermions are righthanded. We can project out the left- and righthanded parts of a field ψ by the projection operators (3.153), giving $\psi_L = \Lambda_L \psi$ and $\psi_R = \Lambda_R \psi$ respectively. If we replace ψ in a bilinear covariant term $\bar{\psi}_2 \Gamma_i \psi_1$ by ψ_L then (since $(1+\gamma^5)(1-\gamma^5) = 0$) the resultant term is zero for $i = S, T, P$ and equal to $\frac{1}{2}\bar{\psi}_2 \gamma^\mu (1-\gamma^5)\psi_1$ for $i = V, A$. In the case of massless neutrinos, this form corresponds to neutrinos (or antineutrinos) of definite helicity ($h_\nu = -1$). Only two of the four components of the Dirac field are needed for $m_\nu = 0$ (two-component neutrino theory). The question of the neutrino mass is discussed further in Chapter 9.

The form of the V − A Lagrangian ((5.45) and (5.46)) requires that P and C are both violated maximally, but that the combined operation CP (and hence T via CTP) is a symmetry of the Lagrangian. The operation of C (3.172) on $\mathscr{L}_{V-A}$ changes the $(1-\gamma^5)$ terms to $(1+\gamma^5)$, i.e. corresponding to righthanded fermions. Subsequent application of P (3.145) restores the original form of $(1-\gamma^5)$.

As an example of a typical calculation in the V − A theory we will now calculate the cross-section for the purely leptonic process $\nu_e(p_1) + e^-(p_2) \to e^-(p_3) + \nu_e(p_4)$, see Fig. 5.3(*a*). The interaction is assumed to be a local point-like interaction of the current $J_e^\dagger(\nu_e \to e^-)$ with that of $J_e(e^- \to \nu_e)$. Using the expressions for the currents from (5.46), the S-matrix element can be found in a way similar to that used for electromagnetic interactions in Section 4.2. The S-matrix element contains a Dirac δ-function, representing four-momentum conservation, and a Lorentz invariant matrix element $\mathscr{M}_{\mathrm{fi}}$, written in terms of spinors and γ-matrices. The weak matrix element can be written

$$\mathscr{M}_{\mathrm{fi}} = \frac{G_{\mathrm{F}}}{2^{1/2}} [\bar{u}_3 \gamma^{\mu}(1-\gamma^5)u_1][\bar{u}_4 \gamma_{\mu}(1-\gamma^5)u_2]. \tag{5.48}$$

Proceeding as for the electromagnetic case, we have

$$\mathscr{M}^*_{\mathrm{fi}} = \frac{G_{\mathrm{F}}}{2^{1/2}} [\bar{u}_1 \gamma^{\nu}(1-\gamma^5)u_3][\bar{u}_2 \gamma_{\nu}(1-\gamma^5)u_4], \tag{5.49}$$

and thus

$$\begin{aligned} |\mathscr{M}_{\mathrm{fi}}|^2 &= \frac{G_{\mathrm{F}}^2}{2} [\bar{u}_3 \gamma^{\mu}(1-\gamma^5)u_1 \bar{u}_1 \gamma^{\nu}(1-\gamma^5)u_3] \\ &\quad \times [\bar{u}_4 \gamma_{\mu}(1-\gamma^5)u_2 \bar{u}_2 \gamma_{\nu}(1-\gamma^5)u_4] \\ &= \frac{G_{\mathrm{F}}^2}{2} L^{\mu\nu} M_{\mu\nu}. \end{aligned} \tag{5.50}$$

Taking $m_{\nu_{\mathrm{e}}} = 0$, and summing over the spin states of the outgoing electron, then the projection operators required are $u_3\bar{u}_3 = (\not{p}_3 + m_{\mathrm{e}})$, $u_1\bar{u}_1 = \not{p}$, etc. Hence, using the properties of the γ-matrices from Appendix B and that $(1-\gamma^5)^2 = 2(1-\gamma^5)$, we have

$$\begin{aligned} L^{\mu\nu} &= 2\,\mathrm{tr}[\not{p}_3 \gamma^{\mu} \not{p}_1 \gamma^{\nu}(1-\gamma^5)], \\ M_{\mu\nu} &= 2\,\mathrm{tr}[\not{p}_4 \gamma_{\mu} \not{p}_2 \gamma_{\nu}(1-\gamma^5)]. \end{aligned} \tag{5.51}$$

Writing

$$\not{p}_3 = p_{3\alpha}\gamma^{\alpha}, \qquad \not{p}_1 = p_{1\beta}\gamma^{\beta}, \qquad \not{p}_4 = p_4^{\theta}\gamma_{\theta} \quad \text{and} \quad \not{p}_2 = p_2^{\phi}\gamma_{\phi},$$

Fig. 5.3 Lowest order V − A diagrams for (*a*) $\nu_{\mathrm{e}} + \mathrm{e}^- \to \mathrm{e}^- + \nu_{\mathrm{e}}$ and (*b*) $\bar{\nu}_{\mathrm{e}} + \mathrm{e}^- \to \mathrm{e}^- + \bar{\nu}_{\mathrm{e}}$.

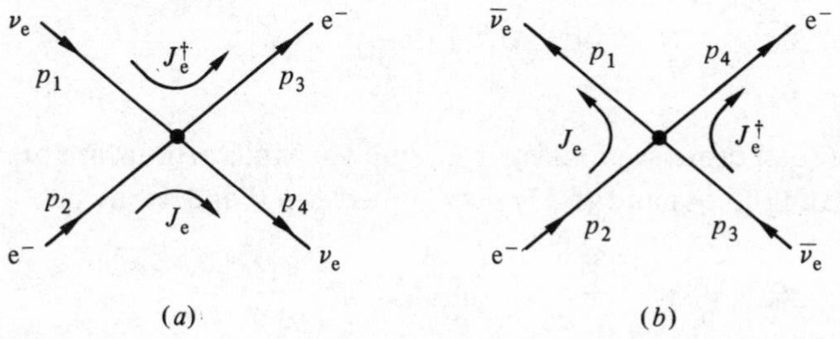

then the product $L^{\mu\nu}M_{\mu\nu}$ is of the form of expression (B.6) of Appendix B, with $C_1 = C_2 = C_3 = C_4 = 1$. Thus, averaging over initial e^- spin states (factor $\frac{1}{2}$), (5.50) becomes

$$\begin{aligned} |\mathcal{M}_{fi}|^2 &= 64G_F^2(p_1 \cdot p_2)(p_3 \cdot p_4) \\ &= 16G_F^2(s - m_e^2)^2. \end{aligned} \tag{5.52}$$

Note that for $m_\nu = 0$ the neutrino has a unique helicity state, hence no summation or averaging of the neutrino spin states is required.

The differential cross-section in the cms is obtained using (4.29), giving

$$\frac{d\sigma}{d\Omega_3}(\text{cms}) = \frac{G_F^2(s - m_e^2)^2}{4\pi^2 s}. \tag{5.53}$$

It is useful to express this result in terms of the invariant quantity $y = (p_1 - p_3) \cdot p_2/(p_1 \cdot p_2) \simeq (1 - \cos\theta)/2$, where θ is the angle between the outgoing e^- and incident ν_e and the limit $m_e^2 \to 0$ has been used. Writing $d\Omega_3 = 2\pi\, d(\cos\theta)$, (5.53) becomes

$$\frac{d\sigma}{dy}(\nu_e e^-) = \frac{G_F^2 s}{\pi}. \tag{5.54}$$

The interaction of antineutrinos with electrons, $\bar{\nu}_e(p_1) + e^-(p_2) \to e^-(p_4) + \bar{\nu}_e(p_3)$, Fig. 5.3(*b*), is obtained from $\nu_e + e^- \to e^- + \nu_e$ by the transformation $\nu_e(\text{in}) \to \bar{\nu}_e(\text{out})$ and $\nu_e(\text{out}) \to \bar{\nu}_e(\text{in})$. Using the Feynman rules for spinors (Appendix C), the corresponding matrix element is

$$\mathcal{M}_{fi} = \frac{G_F}{2^{1/2}}[\bar{v}_1\gamma^\mu(1 - \gamma^5)u_2][\bar{u}_4\gamma_\mu(1 - \gamma^5)v_3]. \tag{5.55}$$

Proceeding as for the $\nu_e e^-$ interaction, we obtain

$$\begin{aligned} |\mathcal{M}_{fi}|^2 &= 64G_F^2(p_1 \cdot p_4)(p_2 \cdot p_3) \\ &= 16G_F^2(u - m_e^2)^2. \end{aligned} \tag{5.56}$$

Now $u = (p_1 - p_4)^2 \simeq -s(1 + \cos\theta)/2$, hence $u = -s(1 - y)$ for $m_e = 0$. Hence the differential cross-section becomes

$$\frac{d\sigma}{dy}(\bar{\nu}_e e^-) = \frac{G_F^2 u^2}{\pi s} = \frac{G_F^2 s}{\pi}(1 - y)^2. \tag{5.57}$$

The total cross-sections for $\nu_e e^-$ and $\bar{\nu}_e e^-$ interactions are obtained by integrating (5.54) and (5.57) over y (between 0 and 1) giving

$$\sigma_{tot}(\nu_e e^-) = \frac{G_F^2 s}{\pi}, \qquad \sigma_{tot}(\bar{\nu}_e e^-) = \frac{G_F^2 s}{3\pi}. \tag{5.58}$$

That is, in the high energy limit, the $\bar{\nu}_e$ cross-section is one third that of ν_e. The helicity configurations of the ingoing and outgoing particles for $\nu_e e^-$ and $\bar{\nu}_e e^-$ are shown in Figs. 5.4(*a*) and (*b*) respectively, for the configuration $y = 1$ (i.e. $\theta = \pi$). It can be seen that for $\bar{\nu}_e e^-$ the configuration $y = 1$ is forbidden by angular momentum conservation. Thus the factor $\frac{1}{3}$ in the total cross-section arises because the helicities of $\bar{\nu}_e$ and e^- are opposite.

The cross-sections (5.54) and (5.57) can be expressed in terms of the lab energy of the neutrino E_ν. In the lab system $p_1 = (E_\nu, 0, 0, E_\nu)$ and $p_2 = (m_e, 0, 0, 0)$; hence $s \simeq 2m_e E_\nu$ for large E_ν. For example, a 5 MeV $\bar{\nu}_e$ (e.g. from a reactor) has a cross-section of 3×10^{-44} cm^2; an experimental challenge!

5.1.4 Problems with the V − A theory

One of the initial problems facing Feynman and Gell-Mann when they formulated the V − A theory was that it appeared to be incompatible with some experimental results. In particular, some recoil measurements on ^{6}He indicated that T coupling was much more likely than A and, further, the branching ratio $\pi \to e\nu/\pi \to \mu\nu$, for which there was an experimental upper limit of 10^{-5}, was incompatible with the V − A prediction of 1.4×10^{-4} (Section 8.1.1). Subsequent experiments clarified the confusion in the favour of the V − A theory. Another apparent problem was that the weak coupling constants found from muon and beta decays differed; this was resolved by the work of Cabibbo (1963).

A somewhat deeper problem arose from the actual current × current form of the Lagrangian. In the Fermi theory, which is based on analogy with QED, the photon is replaced by a νe pair, which interacts with the other two fields (e.g. n and p) at the same space-time point. In QED, the coupling constant α is dimensionless, and the photon acts as a propagator giving a $1/q^2$ term in the matrix element. In the Fermi theory, the νe pair acts as the quantum of the weak force. However, the interaction is assumed

Fig. 5.4 Helicity configurations for the ingoing (in) and outgoing (out) particles in (*a*) $\nu_e e^- \to \nu_e e^-$ and (*b*) $\bar{\nu}_e e^- \to \bar{\nu}_e e^-$ at $y = 1$.

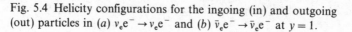

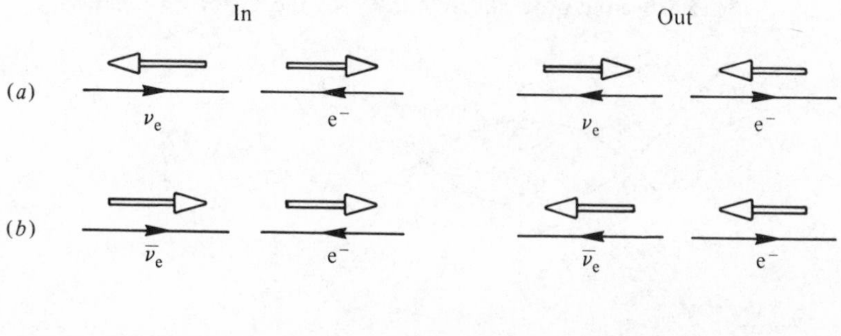

to be local, and thus there is no propagator to give a $1/q^2$ term in the matrix element. The resulting effective weak coupling constant G_F has therefore a dimension of $(\text{mass})^{-2}$. Although the value of $G_F \simeq 10^{-5}/m_p^2$ is numerically small, suggesting that first order perturbation theory is appropriate, the scale m_p^2 is arbitrary.

A specific problem arises in the computation of $\nu_e e^-$ scattering (or similar processes) at ultrahigh energies. In terms of partial waves (see Section 2.2.6), only the $J = 0$ term contributes for the point-like interaction, so that the differential cross-section may be written

$$\frac{d\sigma}{d\Omega} = |f(\theta)|^2 = \frac{1}{s}|\mathcal{M}_0|^2. \tag{5.59}$$

Now unitarity requires that $|\mathcal{M}_J|^2 \leqslant 1$ for all J, limiting the growth of the cross-section with increasing s. On the other hand, the V – A theory cross-section (5.53) grows linearly with s without bound. Hence the theory eventually breaks down when $s \simeq 2\pi/G_F$, i.e. when the cross-sections become equal, which occurs at a cms energy of approximately 750 GeV. The corresponding total cross-section is $\sigma \sim 2G_F \sim 10^{-32}\ \text{cm}^2$.

A possible remedy to this problem might be the neglect of higher order processes such as that shown in Fig. 5.5. The matrix element for this process is

$$\mathcal{M}_{fi} = \frac{G_F^2}{2}\int \frac{d^4p}{(2\pi)^4}\bar{u}_4\gamma_\mu(1-\gamma^5)\frac{1}{\not{p}_1 + \not{p} - m}\gamma_\nu(1-\gamma^5)u_1\bar{u}_3\gamma^\mu(1-\gamma^5)$$
$$\times \frac{1}{\not{p}_2 - \not{p}}\gamma^\nu(1-\gamma^5)u_2. \tag{5.60}$$

The evaluation of such a diagram entails the integration over the four-momentum p, carried by the two fermions in the closed loop, each section of which gives a propagator term of power $1/p$. Hence the form of the loop integral is

$$\mathcal{M}_{fi} \sim \int d^4p/p^2, \tag{5.61}$$

which is clearly divergent for large p. Imposing an upper cut-off parameter

Fig. 5.5 Second order diagram in V – A theory for $\nu_e e^-$ scattering.

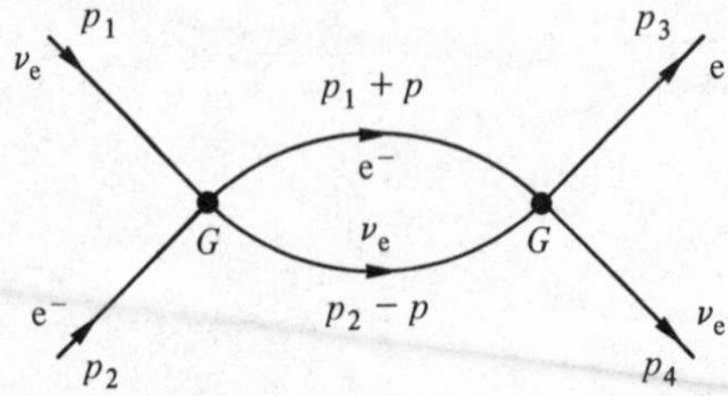

Λ on the value of p gives $\mathcal{M} \propto G_F^2\Lambda^2$ (this form is required on dimensional grounds). To order G^n, the matrix element will diverge as $\mathcal{M} \propto (\Lambda^2)^{n-1}$. In QED divergences also arise in higher orders; however, as discussed in Section 4.9, they behave as $\ln(\Lambda/m)$. That is, they diverge logarithmically rather than quadratically.

QED has the important property that it can be *renormalised*. The divergences which appear in calculating the bare mass and charge of a particle are effectively removed by replacing these infinite (and experimentally unmeasurable) quantities by their 'real' mass and charge. Once this renormalisation has been made, no further divergences appear in higher order calculations. In contrast, the V − A theory cannot be renormalised; a fresh set of divergences appear in each order.

In order to circumvent the problem that there is no scale in the point-like interaction theory, an intermediate weakly interacting boson might provide a possible remedy. This boson (W) must be massive in order to explain the observed short-range nature of the interaction. Further, it must be vector in nature and exist in two charge states ($W^\pm$) in order to reproduce the observed properties of the V − A Lagrangian (5.45). Using the expression (3.204) for the massive vector particle propagator, the matrix element for $\nu_e + e^- \to \nu_e + e^-$ (Fig. 5.6(*a*)) can be written

$$\mathcal{M} = \frac{-ig^2}{2}\left[\bar{u}_3\gamma_\mu \frac{(1-\gamma^5)}{2} u_1\right]\left[\frac{-g^{\mu\nu} + q^\mu q^\nu/M_W^2}{q^2 - M_W^2}\right] \times \left[\bar{u}_4\gamma_\nu \frac{(1-\gamma^5)}{2} u_2\right], \tag{5.62}$$

where $q = p_1 - p_3$. The specific form used for the ν_eWe vertex is taken, for later convenience, as that of the standard model (Appendix C). The relationship between the coupling constants is

$$\frac{G_F}{2^{1/2}} = \frac{g^2}{8M_W^2}, \tag{5.63}$$

Fig. 5.6 ν_e–e^- scattering by (*a*) single W exchange and (*b*) double W exchange.

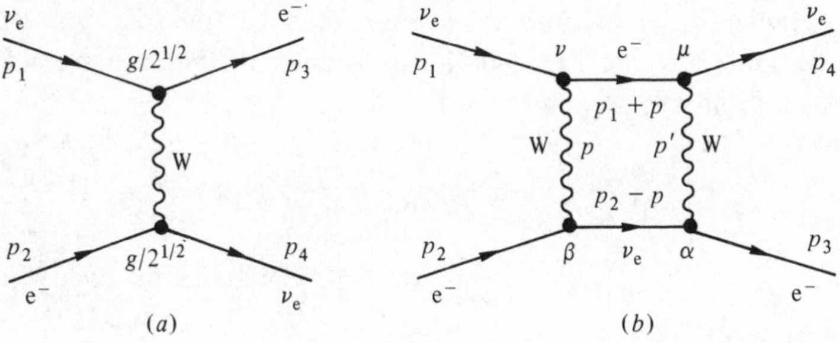

which can be seen by equating (5.62) (in the limit $|q^2| \ll M_W^2$) to (5.48). Note that g (the coupling at a vertex) is dimensionless; and numerically, at low values of Q^2 and for $M_W = 82$ GeV, $g = 0.64$, to be compared with $e = 0.30$ in QED. In the same way as for the electromagnetic case, the diagram (Fig. 5.6(a)) represents the sum of the two possible time orderings for the emission and absorption of the W respectively (i.e. emission from the upper or lower vertex first), and so includes both W^+ and W^- exchange.

The term in $q^\mu q^\nu/M_W^2$ can be simplified by use of the Dirac equation. Writing $q^\nu = p_4^\nu - p_2^\nu$, then

$$\begin{aligned} q^\nu \bar{u}_4 \gamma_\nu (1-\gamma^5) u_2 &= \bar{u}_4 \not{p}_4 (1-\gamma^5) u_2 - \bar{u}_4 \not{p}_2 (1-\gamma^5) u_2 \\ &= -\bar{u}_4 (1+\gamma^5) \not{p}_2 u_2 = -m_e \bar{u}_4 (1+\gamma^5) u_2. \end{aligned} \tag{5.64}$$

A similar exercise can be carried out with $q^\mu = p_1^\mu - p_3^\mu$ on the lefthand side of expression (5.62). It can be seen that the $q^\mu q^\nu/M_W^2$ term in the propagator contributes as $(m_e/M_W)^2$, and is thus negligible. With this simplification, (5.62) becomes

$$\mathcal{M}_{fi} = \frac{ig^2}{8} [\bar{u}_3 \gamma^\mu (1-\gamma^5) u_1][\bar{u}_4 \gamma_\mu (1-\gamma^5) u_2]/(q^2 - M_W^2), \tag{5.65}$$

which has the same form as the V – A expression (5.48), except for the additional $(q^2 - M_W^2)^{-1}$ propagator term. The resulting differential cross-section is thus

$$\frac{d\sigma}{dy}(\nu_e e^-) = \frac{g^4}{32\pi} \frac{s}{(q^2 - M_W^2)^2}. \tag{5.66}$$

Hence for large $q^2 (\gg M_W^2)$ the cross-section falls as $1/q^4$, in the same way as for the electromagnetic interaction. As already remarked, at low $q^2 (\ll M_W^2)$ the V – A form is obtained. Introduction of a massive W boson, for this process, avoids the unitarity problem of the V – A theory in the lowest order diagram (i.e. in the Born or tree diagram). This diagram does not contain any loops.

Higher order diagrams, however, lead to unitarity problems. The diagram shown in Fig. 5.6(b) corresponds to the exchange of two W bosons and has an amplitude

$$\begin{aligned} \mathcal{M}_{fi} = \frac{g^4}{64} \int \frac{d^4 p}{(2\pi)^4} \bar{u}_4 \gamma^\mu (1-\gamma^5) \frac{1}{\not{p}_1 + \not{p} - m} \gamma^\nu (1-\gamma^5) u_1 \bar{u}_3 \gamma^\alpha (1-\gamma^5) \\ \times \frac{1}{\not{p}_2 - \not{p}} \gamma^\beta (1-\gamma^5) u_2 \left[\frac{g_{\mu\alpha} - p'_\mu p'_\alpha/M_W^2}{p'^2 - M_W^2}\right]\left[\frac{g_{\nu\beta} - p_\nu p_\beta/M_W^2}{p^2 - M_W^2}\right], \end{aligned} \tag{5.67}$$

where $p' = p_1 - p_4 + p$. In this case the terms $p'_\mu p'_\alpha/M_W^2$ and $p_\nu p_\beta/M_W^2$, which correspond to the longitudinal polarisation states of the W propagator (Section 3.3.2), are not negligible. For large values of the loop momentum p, the loop integral part behaves as

$$\mathscr{M}_{fi} \sim \int d^4p \frac{1}{p} \cdot \frac{1}{p} \frac{p_\mu p_\alpha}{p^2} \cdot \frac{p_\nu p_\beta}{p^2} \sim \int \frac{d^4 p}{p^2} \sim \Lambda^2, \tag{5.68}$$

i.e. the matrix element is quadratically divergent.

The problem with the longitudinal states of the W arises in the Born term for the reaction $\nu_l + \bar{\nu}_l \to W^+ + W^-$ (Fig. 5.7(*a*)). This reaction, which involves a lepton propagator, has a matrix element (ignoring the lepton mass m_l)

$$\mathscr{M}_{fi} = \frac{-ig^2}{2} \varepsilon_\mu^*(p_4) \varepsilon_\nu^*(p_3) \bar{v}_2 \gamma^\mu \frac{(1-\gamma^5)}{2} \frac{(\not{p}_1 - \not{p}_3)}{(p_1 - p_3)^2} \gamma^\nu \frac{(1-\gamma^5)}{2} u_1. \tag{5.69}$$

Computing $\mathscr{M}_{fi}^*$, then summing over polarisation states of the Ws, we obtain, using (4.97)

$$\begin{aligned} |\mathscr{M}_{fi}|^2 = \frac{g^4}{64(p_1 - p_3)^4} & \left[-g_{\mu\beta} + \frac{p_{4\mu} p_{4\beta}}{M_W^2} \right] \left[-g_{\nu\alpha} + \frac{p_{3\nu} p_{3\alpha}}{M_W^2} \right] \\ & \times \text{tr}[\not{p}_2 \gamma^\mu (1-\gamma^5)(\not{p}_1 - \not{p}_3) \gamma^\nu (1-\gamma^5) \not{p}_1 \gamma^\alpha \\ & \times (1-\gamma^5)(\not{p}_1 - \not{p}_3) \gamma^\beta (1-\gamma^5)]. \end{aligned} \tag{5.70}$$

In the high energy limit ($M_W^2 \to 0$), the longitudinal terms will dominate, and these give

$$|\mathscr{M}_{fi}|^2 = \frac{g^4}{8M_W^4 (p_1 - p_3)^4} \text{tr}[\not{p}_2 \not{p}_4 (\not{p}_1 - \not{p}_3) \not{p}_3 \not{p}_1 \not{p}_3 (\not{p}_1 - \not{p}_3) \not{p}_4 (1-\gamma^5)]. \tag{5.71}$$

Fig 5.7 Possible lowest order diagrams for $\nu_l + \bar{\nu}_l \to W^+ + W^-$, by the exchange of (*a*) a charged lepton $l^\pm$, (*b*) a possible heavy lepton $E_l^\pm$ and (*c*) via the production of Z^0 with a trilinear coupling to W^+W^-

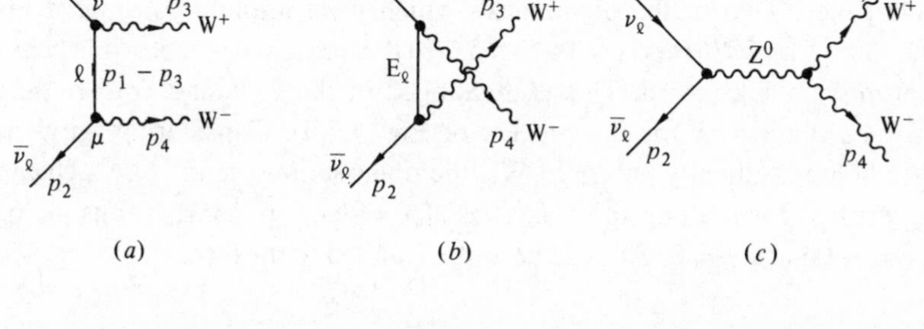

This can be simplified by noting that $(\not{p}_1-\not{p}_3)\not{p}_3=\not{p}_1\not{p}_3-\not{p}_3\not{p}_3=\not{p}_1\not{p}_3$ since $\not{p}_3\not{p}_3=M_W^2$, and this is taken as small. Further $(p_1-p_3)^4=4(p_1\cdot p_3)^2$, so (5.71) becomes

$$|\mathscr{M}_{fi}|^2=\frac{g^4}{32M_W^4(p_1\cdot p_3)^2}\operatorname{tr}[\not{p}_2\not{p}_4\not{p}_1\not{p}_3\not{p}_1\not{p}_3\not{p}_1\not{p}_4(1-\gamma^5)]. \quad (5.72)$$

Now $\not{p}_1\not{p}_3=-\not{p}_3\not{p}_1+2p_1\cdot p_3$ and, using this property twice, gives

$$|\mathscr{M}_{fi}|^2=\frac{g^4}{8M_W^4}\operatorname{tr}[\not{p}_2\not{p}_4\not{p}_1\not{p}_4(1-\gamma^5)]$$

$$=\frac{g^4}{M_W^4}(p_2\cdot p_4)(p_1\cdot p_4). \quad (5.73)$$

If θ is the cms scattering angle of the W^+ with respect to the incident ν_e, then $p_2\cdot p_4=(1-\cos\theta)s/4$ and $p_1\cdot p_4=(1+\cos\theta)s/4$; hence, the differential cross-section (4.29) is

$$\frac{d\sigma}{d\Omega}(\nu_l\bar{\nu}_l\to W^+W^-)=\frac{g^4 s\sin^2\theta}{1024\pi^2M_W^4}=\frac{G_F^2 s\sin^2\theta}{32\pi^2}. \quad (5.74)$$

Now the production of longitudinally polarised Ws is a $J=1$ process, with a corresponding partial wave amplitude $\mathscr{M}_1=G_F s/24\pi$ (Section 2.2.6). Hence, the unitary bound $|\mathscr{M}_1|\leqslant 1$ is violated for $s\geqslant 24\pi/G_F\sim(2\,500\text{ GeV})^2$. Thus again we have a badly divergent high energy behaviour and, in fact, the theory with just $W^\pm$ bosons is not renormalisable. In QED, the process analogous to $\nu\bar{\nu}\to W^+W^-$ is $e^+e^-\to 2\gamma$, and this is not divergent. The essential difference is that in the evaluation of $|\mathscr{M}_{fi}|^2$ above, the polarisation sum (4.97) is used rather than the corresponding sum for the massless boson, (4.84). Note that (4.84) is not just the limit of (4.97) as $M\to 0$. For real (massless) photons the longitudinal polarisation states are absent and, further, this is a consequence of the gauge invariance of QED.

One possible solution to the divergence problem is to include some additional process which has an amplitude which exactly cancels the divergence. Two of the possibilities which were initially considered are shown in Fig. 5.7(*b*) and (*c*). In Fig. 5.7(*b*) the exchange of a possible heavy lepton E_l is considered. This is exchanged in the *u*-channel, rather than in the *t*-channel as for the process of Fig. 5.7(*a*). Considering only the case of longitudinally polarised Ws, the polarisation vectors of which can be written (for $\mathbf{p}$ along the *z*-axis) as $\varepsilon(p)=(|\mathbf{p}|,0,0,E)/M_W$; thus giving $\varepsilon\cdot\varepsilon=-1$ and $\varepsilon\cdot p=0$. This can be approximated to the form $\varepsilon_\mu(p)=p_\mu/M_W$

for high energies. Hence (5.69) (Fig. 5.7(*a*)) can be easily simplified to

$$\mathcal{M}_{\rm fi} = \frac{ig^2}{8M_{\rm W}^2 p_1 \cdot p_3} \bar{v}_2 \not{p}_4(\not{p}_1 - \not{p}_3)\not{p}_3(1-\gamma^5)u_1$$

$$= \frac{ig^2}{4M_{\rm W}^2} \bar{v}_2 \not{p}_4(1-\gamma^5)u_1. \tag{5.75}$$

To derive the latter expression, the properties $\not{p}_3\not{p}_3 = p_3^2 = 0$ and $\not{p}_1\not{p}_3 = -\not{p}_3\not{p}_1 + 2p_1 \cdot p_3$, together with $\not{p}_1 u_1 = 0$, have been used.

For the diagram in Fig. 5.7(*b*) the matrix element, in the high energy limit and noting that the lepton propagator has momentum $p_1 - p_4$ and that a vertex coupling constant g' rather than g is used, is

$$\mathcal{M}_{\rm fi} = \frac{ig'^2}{8M_{\rm W}^2 p_1 \cdot p_4} \bar{v}_2 \not{p}_3\not{p}_1\not{p}_4(1-\gamma^5)u_1$$

$$= \frac{ig'^2}{4M_{\rm W}^2} \bar{v}_2 \not{p}_3(1-\gamma^5)u_1 = \frac{-ig'^2}{4M_{\rm W}^2} \bar{v}_2 \not{p}_4(1-\gamma^5)u_1. \tag{5.76}$$

Thus cancellation of the unitarity violating amplitudes is achieved by choosing $g'^2 = g^2$. For this scheme to work, a heavy lepton $\mathrm{E}_l^{\pm}$ is needed for each neutrino generation ν_l. More details and possible difficulties with this scheme are discussed by Gastmans (1975).

An alternative way of cancelling the divergences, which seems to have been chosen by nature, is to have a neutral vector boson Z^0 coupling to $\nu_l\bar{\nu}_l$ and $\mathrm{W}^+\mathrm{W}^-$ with appropriate constants. The diagram for this process is shown in Fig. 5.7(*c*), and the couplings are discussed later in Section 5.3. The first experimental evidence in favour of this scheme came with the discovery by the Gargamelle neutrino collaboration of the following reactions (Hasert *et al.*, 1973a,b)

$$\text{(i)}\ \bar{\nu}_\mu + \mathrm{e}^- \to \bar{\nu}_\mu + \mathrm{e}^-,$$
$$\text{(ii)}\ \nu_\mu/\bar{\nu}_\mu + \mathrm{N} \to \nu_\mu/\bar{\nu}_\mu + \text{hadrons}. \tag{5.77}$$

These reactions involve neutral currents and cannot be accommodated in the V – A Lagrangian (5.45), which requires the currents to have charges ± 1. Subsequent analysis of these and other reactions has shown (as will be discussed later) that there are both left(V – A) and right(V + A) handed components in neutral current reactions. The spectacular discovery of the $\mathrm{W}^\pm$ and Z^0 bosons at the CERN $\bar{\mathrm{p}}\mathrm{p}$ Collider in 1983 by the UA1 (Arnison *et al.*, 1983) and UA2 (Banner *et al.*, 1983) collaborations finally demonstrated that the weak interactions were indeed mediated by these massive particles.

5.2 CHARM AND THE GLASHOW, ILIOPOULOS AND MAIANI (GIM) MECHANISM

The weak transitions in charged current interactions involve a pair of leptons (e.g. $\mu^- \to \nu_\mu$) or quarks (e.g. $s \to u$). Such pairs can be regarded as *weak doublets*, with the weak interaction inducing transitions between the two components. An analogy from strong interactions is the isospin (p, n) doublet, the nucleon. For leptons these '*weak isospin*' doublets are (ν_l, l^-) with $l = e$, μ and τ. For quarks, the identification of the doublet is less straightforward because, as we have seen, both $u \leftrightarrow d$ and $u \leftrightarrow s$ transitions exist. However, using Cabibbo's hypothesis (5.47), the doublet can be written as

$$q = \begin{pmatrix} u \\ d_c \end{pmatrix} = \begin{pmatrix} u \\ d \cos\theta_C + s \sin\theta_C \end{pmatrix}. \tag{5.78}$$

That is, it is a 'Cabibbo-rotated' d-quark which forms the lower part of the weak doublet.

The transitions between doublet members correspond to SU(2) raising (τ^+) and lowering (τ^-) operators (2.212), giving the charge raising and lowering currents

$$J^+ \sim g(\bar{u}d_c) = g(\bar{u}\ \bar{d}_c)\begin{pmatrix} 0 & 1 \\ 0 & 0 \end{pmatrix}\begin{pmatrix} u \\ d_c \end{pmatrix} = g(\bar{q}\tau^+ q),$$

$$J^- \sim g(\bar{d}_c u) = g(\bar{u}\ \bar{d}_c)\begin{pmatrix} 0 & 0 \\ 1 & 0 \end{pmatrix}\begin{pmatrix} u \\ d_c \end{pmatrix} = g(\bar{q}\tau^- q), \tag{5.79}$$

where overall numerical factors have been omitted. If there exists an appropriate symmetry, based on some underlying gauge theory, then a current involving τ_3 is also expected, since these operators are related via the commutation relation $[\tau^+, \tau^-] = 2\tau_3$. Hence, with such a gauge theory symmetry, one would expect the existence of a neutral current of the form

$$J^0 \sim 2g(\bar{q}\tau_3 q) = g(\bar{u}u - \bar{d}_c d_c)$$
$$= g[\bar{u}u - \bar{d}d\cos^2\theta_C - \bar{s}s\sin^2\theta_C - (\bar{d}s + \bar{s}d)\cos\theta_C \sin\theta_C]. \tag{5.80}$$

The terms $\bar{d}s$ and $\bar{s}d$ correspond to strangeness-changing neutral currents, which do not appear to exist in nature. For example, the decay branching ratio $K^+ \to \mu^+\nu_\mu$ is 63.5%, whereas that for $K_L^0 \to \mu^+\mu^-$ is $(9.1 \pm 1.9) \times 10^{-9}$. That is, the process $d\bar{s} \to Z^0 \to \mu^+\mu^-$ is heavily suppressed relative to $u\bar{s} \to W^+ \to \mu^+\nu_\mu$. A mechanism to suppress these unwanted strangeness-changing neutral currents was suggested in 1970 by Glashow, Iliopoulos and Maiani (GIM). They proposed the existence of a second orthogonal doublet, containing a new quark c (charm) with charge $\frac{2}{3}$, as follows

$$q' = \begin{pmatrix} c \\ s_c \end{pmatrix} = \begin{pmatrix} c \\ -d\sin\theta_C + s\cos\theta_C \end{pmatrix}. \tag{5.81}$$

Adding this term gives the total neutral current

$$J^0 \sim 2g[\bar{q}\tau_3 q + \bar{q}'\tau_3 q'] = g[\bar{u}u + \bar{c}c - \bar{d}_c d_c - \bar{s}_c s_c]$$
$$= g[\bar{u}u + \bar{c}c - \bar{d}d - \bar{s}s], \tag{5.82}$$

That is, the unwanted terms cancel, leaving a flavour diagonal result.

The GIM mechanism also gave an estimate of the mass of the charmed quark in advance of its discovery. The estimate was based on the value of the $K_L^0 \to \mu^+\mu^-$ branching ratio. In terms of the then known quarks, this decay could proceed through the graph shown in Fig. 5.8(*a*). The calculated rate is larger than that observed experimentally. However, with the addition of Fig. 5.8(*b*), the total amplitude is

$$\mathcal{M} = \mathcal{M}_{(a)} + \mathcal{M}_{(b)} \sim f(m_u)g^4 \cos\theta_C \sin\theta_C - f(m_c)g^4 \cos\theta_C \sin\theta_C \tag{5.83}$$

Thus, the c-quark introduces a cancellation. This gives a branching ratio compatible with the experimental value; but only if the c-quark is not too massive in which case the first term again dominates. These arguments led to the prediction (Gaillard and Lee, 1974) that m_c was in the range 1 to 3 GeV; as indeed proved to be the case.

The GIM hypothesis (5.81) represents a generalisation of Cabibbo's hypothesis from three to four quarks. The addition of a fourth quark (c) restored symmetry in the (then known) numbers of quarks and leptons. These ideas were extended by Kobayashi and Maskawa (1973), who introduced a framework to describe six quarks. The additional quarks are the charge $\frac{2}{3}$ t-quark and the charge $-\frac{1}{3}$ b-quark. The 3×3 matrix

Fig. 5.8 The decay $K_L^0 \to \mu^+\mu^-$ via (*a*) u-quark exchange and (*b*) c-quark exchange.

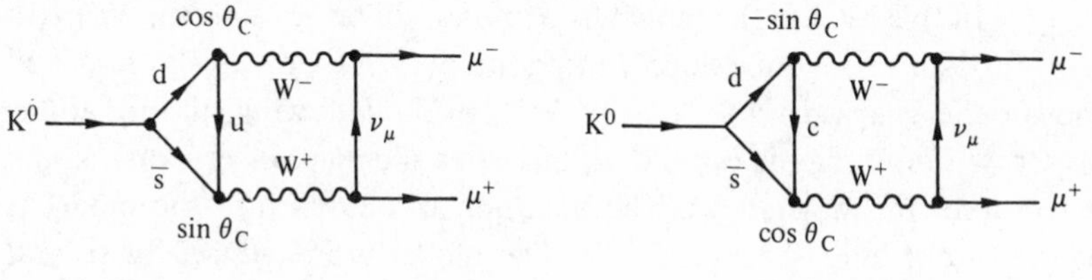

(*a*) (*b*)

needed has three Cabibbo-like angles and one (CP-violating) phase. This is discussed further in Section 5.5.3.

Before describing the standard electroweak model, which unifies the electromagnetic and weak interactions, we list the main facts which must be explained by such a theory.

(i) The leptons and quarks all have spin $\frac{1}{2}$ and, as far as the charged current ($W^{\pm}$ exchange) interaction is concerned, come in weak isospin doublets.

(ii) These CC (charged current) interactions appear to be purely lefthanded (V – A) and to violate P and C maximally, while (almost) conserving CP.

(iii) There seem to be three such generations for both leptons and quarks.

(iv) The neutrinos are very light (possibly massless), whereas the charged lepton and quark masses increase substantially from each generation to the next.

(v) In addition to the charged currents, there exist two types of neutral current, one coupling to all leptons and quarks (Z^0 exchange) and the other to charged particles only (γ exchange). Values of the neutral current couplings have been determined in a wide variety of interactions.

(vi) The electromagnetic and weak forces are mediated by the massless γ and the massive $W^{\pm}$ (82 GeV) and Z^0 (93 GeV).

(vii) A further requirement is that the theory describing all these phenomena must be renormalisable.

When the so-called *standard electroweak* (or *GSW*) *model* (Glashow, 1961; Weinberg, 1967; Salam, 1968) was proposed, however, only the V – A properties of the then known leptons (e^-, ν_e, μ^-, ν_μ) and quarks (u, d, s) were established, together with the successful gauge theory of QED. A comprehensive collection of the original papers on the standard model can be found in the book of Lai (1981).

5.3 INCORPORATING NEUTRAL CURRENTS

In this section the model for leptons will be considered. An extra neutral boson Z^0 is introduced, in addition to the previously discussed photon and charged vector bosons $W^{\pm}$. In the following, the equations are written only for the e^- and ν_e, but extra identical pieces for μ and τ generations are understood. The inclusion of quarks into the model is deferred until later (Section 5.5.3). The model is introduced in several stages. First of all, we consider the extension to the theory of the electromagnetic and charged currents needed to include neutral currents, and try to write this in a 'unified' form. The local gauge invariance which 'generates' these interactions is next discussed (Section 5.4), and finally we consider a mechanism for generating the masses of particles (Section 5.5).

The weak interaction Lagrangian for charged currents (CC), interacting by charged intermediate vector bosons (IVB), is as follows

$$\mathscr{L}_{\mathrm{CC}} = \frac{-g}{2(2)^{1/2}} [\bar{\nu}\gamma_\mu(1-\gamma^5)eW^\mu_+ + \bar{e}\gamma_\mu(1-\gamma^5)\nu W^\mu_-]. \tag{5.84}$$

This Lagrangian has a form similar to that for the strong interaction of a nucleon $\mathrm{N} = (\mathrm{p}, \mathrm{n})$ with a pion $\pi = (\pi_1, \pi_2, \pi_3)$, with $\pi^\pm = (\pi_1 \mp \pi_2)/2^{1/2}$, namely

$$\begin{aligned} \mathscr{L}_{\pi\mathrm{N}} &\sim -[\bar{p}\gamma^5 n\pi^+ + \bar{n}\gamma^5 p\pi^-] \\ &= -\bar{N}\boldsymbol{\tau}\gamma^5 N \cdot \boldsymbol{\pi}. \end{aligned} \tag{5.85}$$

In the last line of (5.85) the Lagrangian is written in an SU(2) invariant way, using the Pauli spin matrices (2.213) (referred to here as τ_1, τ_2, τ_3). There is one overall coupling constant, and the various Clebsch–Gordan coefficients give the relative strengths of specific transitions.

Since the weak CC interaction is lefthanded, one can define a *lefthanded doublet*

$$L = \begin{pmatrix} \nu_{\mathrm{L}} \\ e_{\mathrm{L}} \end{pmatrix} = \frac{(1-\gamma^5)}{2}\begin{pmatrix} \nu \\ e \end{pmatrix}; \qquad \bar{L} = (\bar{\nu}_{\mathrm{L}}\bar{e}_{\mathrm{L}}) = (\bar{\nu}\ \bar{e})\frac{(1+\gamma^5)}{2}. \tag{5.86}$$

The charged lepton current is defined in terms of L as follows

$$\begin{aligned} \bar{\nu}\gamma^\mu(1-\gamma^5)e = 2\bar{\nu}_{\mathrm{L}}\gamma^\mu e_{\mathrm{L}} &= 2(\bar{\nu}_{\mathrm{L}}\bar{e}_{\mathrm{L}})\gamma^\mu\begin{pmatrix} e_{\mathrm{L}} \\ 0 \end{pmatrix} \\ &= 2\bar{L}\gamma^\mu\tau_+ L, \end{aligned} \tag{5.87}$$

where $\tau^\pm = (\tau_1 \pm \mathrm{i}\tau_2)/2$. Following the analogy with the πN strong interaction example, we define

$$W^\mu_\pm = (W^\mu_1 \mp \mathrm{i}W^\mu_2)/2^{1/2}, \tag{5.88}$$

in terms of which (5.84), with the use of (2.213) and (5.86), becomes

$$\begin{aligned} \mathscr{L}_{\mathrm{CC}} &= \frac{-g}{2}[\bar{L}\gamma_\mu\tau_1 L \cdot W^\mu_1 + \bar{L}\gamma_\mu\tau_2 L \cdot W^\mu_2] \\ &= \frac{-g}{2}\bar{L}[\tau_1 \not{W}_1 + \tau_2 \not{W}_2]L. \end{aligned} \tag{5.89}$$

Thus the idea of *weak lefthanded isospin* (t) leads to three currents

$$(j_i^t)^\mu = \bar{L}\gamma^\mu \frac{\tau_i}{2} L \qquad (i = 1, 2, 3). \tag{5.90}$$

The third component is

$$(j_3^t)^\mu = \bar{L}\gamma^\mu \frac{\tau_3}{2} L = [t_3^{\nu_L}\bar{\nu}_L\gamma^\mu\nu_L + t_3^{e_L}\bar{e}_L\gamma^\mu e_L] = \tfrac{1}{2}[\bar{\nu}_L\gamma^\mu\nu_L - \bar{e}_L\gamma^\mu e_L], \tag{5.91}$$

and is clearly a *neutral current* since it couples particles of the same charge. Such a current is inevitable if there is an $SU(2)_L$ symmetry. The quantum number t_3 has the values $\frac{1}{2}$ and $-\frac{1}{2}$ for the lefthanded neutrino and electron respectively.

Next we try and incorporate the electromagnetic neutral currents. We postpone, until Section 5.5, discussion of one obvious problem in trying to unify the weak and electromagnetic interactions, namely that, whereas the photon is massless, the weak vector bosons are massive. The electromagnetic interaction, unlike that for charged currents, has both left- and righthanded parts. Assuming that there are only lefthanded neutrinos, then only the electron has a righthanded component, so we have the *singlet* ($t_3 = 0$)

$$R = e_R = \frac{(1+\gamma^5)}{2} e; \qquad \bar{R} = \bar{e}_R = \bar{e}\,\frac{(1-\gamma^5)}{2}. \tag{5.92}$$

The electromagnetic current for the electron (3.92) can be split into left- and righthanded parts, as follows

$$\begin{aligned} j_e^\mu/e = -\bar{e}\gamma^\mu e &= -\left[\bar{e}\gamma^\mu \frac{(1-\gamma^5)}{2} e + \bar{e}\gamma^\mu \frac{(1+\gamma^5)}{2} e\right] \\ &= -[\bar{e}_L\gamma^\mu e_L + \bar{e}_R\gamma^\mu e_R] \\ &= (j_3^t)^\mu + \tfrac{1}{2}(j^y)^\mu, \end{aligned} \tag{5.93}$$

where, using (5.91),

$$\begin{aligned} (j^y)^\mu &= -(\bar{\nu}_L\gamma^\mu\nu_L + \bar{e}_L\gamma^\mu e_L) - 2\bar{e}_R\gamma^\mu e_R \\ &= -(\bar{L}\,\gamma^\mu L) - 2(\bar{R}\gamma^\mu R). \end{aligned} \tag{5.94}$$

Hence, in addition to the three SU(2) currents $(j_i^t)^\mu$, there is a further current $(j^y)^\mu$, which is called the *weak hypercharge* (y) current. The analogy with strong isospin and hypercharge can be seen from the form of (5.93). The value of the weak hypercharge is chosen to obtain the correct electric charge for each weak SU(2) multiplet. Hence, from (5.94), or using $Q/e = t_3 + y/2$, we have $y = -1$ for the doublet L, and $y = -2$ for the singlet R. The simplest assumption for electroweak unification is that the current $(j^y)^\mu$ has a U(1) gauge symmetry (i.e. as for QED), with a corresponding gauge boson B^μ, coupled with some constant g'. Collecting the $SU(2)_L$ and $U(1)_y$ terms together ($SU(2)_L \otimes U(1)_y$), suggests an interaction Lagrangian of the form

$$\mathscr{L}' = -\left[g\left(\bar{L}\gamma_\mu \frac{\tau_i}{2} L\right)W_i^\mu + \frac{g'}{2}(j^y)_\mu B^\mu\right]$$

$$= \mathscr{L}_{CC} + \mathscr{L}_{NC}, \tag{5.95}$$

where $\mathscr{L}_{CC}$ is given by (5.89) and

$$\mathscr{L}_{NC} = -\left[g\left(\bar{L}\gamma_\mu \frac{\tau_3}{2} L\right)W_3^\mu + \frac{g'}{2}(j^y)_\mu B^\mu\right]$$

$$= -\left[g(j_3^t)_\mu W_3^\mu \qquad + \frac{g'}{2}(j^y)_\mu B^\mu\right]. \tag{5.96}$$

If we demand that the fields W_3^μ and B^μ are orthogonal linear combinations of the fields corresponding to the physical photon (A^μ) and $Z^0(Z^\mu)$, then we can write

$$W_3^\mu = \cos\theta_W Z^\mu + \sin\theta_W A^\mu,$$

$$B^\mu = -\sin\theta_W Z^\mu + \cos\theta_W A^\mu, \tag{5.97}$$

where θ_W is known as the *Weinberg* or *weak mixing angle*. Substituting (5.97) into (5.96) gives

$$\mathscr{L}_{NC} = -\left[g\sin\theta_W (j_3^t)^\mu + g'\cos\theta_W \frac{(j^y)^\mu}{2}\right]A_\mu$$

$$-\left[g\cos\theta_W (j_3^t)^\mu - g'\sin\theta_W \frac{(j^y)^\mu}{2}\right]Z_\mu. \tag{5.98}$$

The coupling of A_μ to the electromagnetic current (5.93) then leads to the identification

$$g\sin\theta_W = g'\cos\theta_W = e, \tag{5.99}$$

and the total neutral current Lagrangian (5.98) becomes

$$\mathscr{L}_{NC} = -j_e^\mu A_\mu - \frac{g}{\cos\theta_W}\left[(j_3^t)^\mu - \sin^2\theta_W \left(\frac{j_e}{e}\right)^\mu\right]Z_\mu.$$

$$= -j_e^\mu A_\mu - j_Z^\mu Z_\mu, \tag{5.100}$$

with

$$j_Z^\mu = \frac{g}{2\cos\theta_W}\bar{f}\gamma^\mu[(t_3^{f_L} - Q_f\sin^2\theta_W)(1-\gamma^5) - Q_f\sin^2\theta_W(1+\gamma^5)]f$$

$$= \frac{g}{2\cos\theta_W}\bar{f}\gamma^\mu[C_L^f(1-\gamma^5) + C_R^f(1+\gamma^5)]f,$$

where the left and righthanded couplings of the fermion $f(\nu, e)$ are given by $C_L^f = t_3^{f_L} - Q_f\sin^2\theta_W$ and $C_R^f = t_3^{f_R} - Q_f\sin^2\theta_W = -Q_f\sin^2\theta_W$ respectively. It is often convenient to specify j_Z in terms of its vector and axial vector components. For the neutrino these are $g_V^\nu = C_L^\nu + C_R^\nu = \frac{1}{2}$ and $g_A^\nu = C_L^\nu - C_R^\nu = \frac{1}{2}$ respectively, whereas for the electron (if unpolarised) $g_V^e = C_L^e + C_R^e = -\frac{1}{2} + 2\sin^2\theta_W$ and $g_A^e = C_L^e - C_R^e = -\frac{1}{2}$ respectively.

Note that, due to (5.93), we have ensured that the electromagnetic interaction is parity conserving, even though left and righthanded leptons have different properties. The additional Feynman rules for leptonic neutral currents can be constructed from (5.100). An attractive feature of the above scheme, which is due to Glashow (1961), is that the above couplings lead to cancellations of the divergences which are the cause of the bad high-energy behaviour of the IVB theory.

5.4 LOCAL GAUGE SYMMETRIES

In Section 4.7 the Lagrangian for a free Dirac particle was discussed. For the following purposes we can write this in the form

$$\mathscr{L}_0 = i\bar{\psi}(x)\gamma_\mu\,\partial^\mu\psi(x) - m\bar{\psi}(x)\psi(x). \tag{5.101}$$

This is not invariant under the local U(1) phase transformation (4.107). However, if the derivative ∂^μ is replaced by $D^\mu = \partial^\mu + ieA^\mu$ (the so-called minimal coupling), then the resulting Lagrangian

$$\mathscr{L} = i\bar{\psi}(x)\gamma_\mu D^\mu\psi(x) - m\bar{\psi}(x)\psi(x), \tag{5.102}$$

is invariant under (4.107), provided that the vector field A^μ is simultaneously 'stretched' from $A^\mu \to A^\mu + \delta A^\mu$, where $\delta A^\mu = \partial^\mu\theta(x)$, with $\theta(x)$ real. Thus, invariance under this change in A^μ (the gauge transformation) is obtained if an interaction term $(-e\bar{\psi}\gamma_\mu A^\mu\psi)$ is included in $\mathscr{L}$. That is, the requirement that $\mathscr{L}$ is invariant under a local phase transformation (i.e. this phase is unobservable) suggests that the fermion f cannot exist on its own, but must be accompanied by a vector particle with which it interacts. Further, this vector particle (the photon in QED) must be massless, since the mass term

$$\mathscr{L}_{\text{mass}} = \tfrac{1}{2}m_\gamma^2 A_\mu(x)A^\mu(x), \tag{5.103}$$

is not gauge invariant. ($\mathscr{L} = \mathscr{L}_{\text{mass}} - \frac{1}{4}F_{\mu\nu}F^{\mu\nu}$ gives (3.193) by application of the Euler–Lagrange equation). The total QED Lagrangian is obtained by adding the photon 'kinetic' term $\mathscr{L}_\gamma$ (4.110) to (5.102).

The photon couples only to charged particles and, since it is electrically neutral, it has no self-coupling. In this respect the photon appears to be unique amongst the gauge bosons. The photon is universal in the sense that there appears to be only one photon; however, the fundamental fermions have different charges (± 1 for $e^\pm$, $\pm\frac{1}{3}$ and $\pm\frac{2}{3}$ for q and $\bar{q}$). For a fermion f of charge $Q_f e$, the QED Lagrangian is

$$\mathscr{L} = i\bar{\psi}_f\gamma_\mu(\partial^\mu + iQ_f e A^\mu)\psi_f - m_f\bar{\psi}\psi - \tfrac{1}{4}F_{\mu\nu}F^{\mu\nu}. \tag{5.104}$$

Other fermions can be included by summing over f. Note that each extra fermion introduces two new parameters (Q_f and m_f) into the Lagrangian. The form of the local phase transformation needed to ensure the invariance of (5.104) is that of (4.107), but with e replaced by $Q_f e$. That is, the phases must be separately tuned, and so QED does not explain the quantisation of charge or the actual values observed in nature. A possible explanation to this problem is offered by grand unified theories, as discussed in Chapter 10.

In the previous section, the $SU(2)_L$ properties of the weak interaction were explored. We will now show that the desired interactions are 'generated' by the requirement of invariance under local phase transformations for the $SU(2)_L$ and $U(1)_y$ groups. Let us consider a doublet of fermions f and f′, having some underlying symmetry such that f_L and f'_L form a doublet (L), but f_R and f'_R are singlets. The free Lagrangian for f and f′ (assumed massless) is, following (5.101),

$$\begin{aligned}\mathscr{L}_0 &= i\bar{f}\gamma_\mu\,\partial^\mu f + i\bar{f}'\gamma_\mu\,\partial^\mu f' \\ &= i\bar{f}_L\gamma_\mu\,\partial^\mu f_L + i\bar{f}_R\gamma_\mu\,\partial^\mu f_R + i\bar{f}'_L\gamma_\mu\,\partial^\mu f'_L + i\bar{f}'_R\gamma_\mu\,\partial^\mu f'_R \\ &= i\bar{L}\gamma_\mu\,\partial^\mu L + i\bar{f}_R\gamma_\mu\,\partial^\mu f_R + i\bar{f}'_R\gamma_\mu\,\partial^\mu f'_R,\end{aligned} \tag{5.105}$$

where each of the terms has been separated into left and righthanded parts (see (5.86) and (5.92)), and the lefthanded fermions are combined into a doublet L. Note that the form of (5.105) results in helicity conservation, because only parts having the same handedness are connected. The grouping of the terms into a doublet and singlets in (5.105) is a consequence of the $SU(2)_L \otimes U(1)_y$ symmetry requirement. If we had also put f_R and f'_R into a doublet, then we would obtain a model with

$SU(2)_L \otimes SU(2)_R \otimes U(1)$ symmetry and with seven gauge bosons. Such a model exhibits *chiral* symmetry, that is symmetry between left- and righthandedness.

In Section 5.3 the symmetry $SU(2)_L \otimes U(1)_y$ was discussed, and the leptons classified into lefthanded doublets (5.86) and righthanded singlets (5.92); i.e. omitting the possibility of a righthanded term ν_R. In this case the Lagrangian, including the free particle term (5.105) and interaction term (5.95), is

$$\mathscr{L}_e = i\bar{L}\gamma_\mu\left[\partial^\mu + ig\frac{\tau_i}{2}W_i^\mu + i\frac{g'}{2}y_L B^\mu\right]L$$

$$+ i\bar{R}\gamma_\mu\left[\partial^\mu \qquad\qquad + i\frac{g'}{2}y_R B^\mu\right]R. \tag{5.106}$$

The local phase transformations corresponding to the weak hypercharge (U_1) and weak isospin (U_2) are

$$U_1 = \exp\left[-i\frac{g'}{2}y\theta(x)\right], \qquad U_2 = \exp\left[-i\frac{g}{2}\boldsymbol{\alpha}(x)\cdot\boldsymbol{\tau}\right], \tag{5.107}$$

where U_1 and U_2 correspond to the U(1) group (cf. (4.107)) and the SU(2) group (Section 2.5.2) respectively. Note that the invariance of the Lagrangian $\mathscr{L}_0$ (5.105) under global SU(2) transformations of the form of U_2 in (5.107) (but with $\boldsymbol{\alpha}$ a constant independent of x) leads, using (3.38) and considering the limit $\boldsymbol{\alpha} \to 0$, to the three conserved weak isospin currents (5.90). Returning to the local case, the combined $SU(2) \otimes U(1)$ transformation is

$$U = U_2U_1 = \exp\left\{-i\left[\frac{g}{2}\boldsymbol{\alpha}(x)\cdot\boldsymbol{\tau} + \frac{g'}{2}y\theta(x)I\right]\right\}, \tag{5.108}$$

under which the doublet L and singlet R are transformed to

$$L' = UL, \qquad R' = UR. \tag{5.109}$$

Before considering this particular example, we first derive the more general invariance properties of a particle interacting with a field E^μ, with the corresponding Lagrangian

$$\mathscr{L} = i\bar{\psi}\gamma_\mu(\partial^\mu + E^\mu)\psi, \tag{5.110}$$

under the local unitary transformation

$$\psi' = U\psi, \qquad \bar{\psi}' = \bar{\psi}U^{-1}. \tag{5.111}$$

If the field E^μ is transformed to $(E')^\mu$, then invariance of the Lagrangian under these transformations leads to the relation

$$(E')^\mu = UE^\mu U^{-1} - (\partial^\mu U)U^{-1}, \tag{5.112}$$

which can be easily shown by substituting (5.111) in (5.110). For the U(1) group, i.e. QED, we have $E^\mu = \mathrm{i}eA^\mu$ and $U = \exp(-\mathrm{i}e\theta(x))$, so that (5.112) gives the usual gauge transformation $A^\mu \to A^\mu + \partial^\mu\theta$.

Next we consider the effect of local SU(2) transformations, corresponding to U_2 in (5.107), on the Lagrangian

$$\mathscr{L}'_\mathrm{e} = \mathrm{i}\bar{L}\gamma_\mu\left(\partial^\mu + \mathrm{i}\frac{g}{2}\boldsymbol{\tau}\cdot\mathbf{W}^\mu\right)L. \tag{5.113}$$

This is the SU(2) part of (5.106) and is of the form (5.110), with $E^\mu = \mathrm{i}(g/2)\boldsymbol{\tau}\cdot\mathbf{W}^\mu$. The transformation properties (5.112) of the field can be found most easily by considering the infinitesimal transformation of U_2 in (5.107), namely

$$U_2 \simeq 1 - \mathrm{i}\frac{g}{2}\boldsymbol{\alpha}(x)\cdot\boldsymbol{\tau}, \qquad \boldsymbol{\alpha}(x) \to 0. \tag{5.114}$$

Using this expression in (5.112), and retaining only first order terms in $\boldsymbol{\alpha}$, gives

$$\begin{aligned}\boldsymbol{\tau}\cdot\mathbf{W}'^\mu &= \left(1 - \mathrm{i}\frac{g}{2}\boldsymbol{\alpha}\cdot\boldsymbol{\tau}\right)\boldsymbol{\tau}\cdot\mathbf{W}^\mu\left(1 + \mathrm{i}\frac{g}{2}\boldsymbol{\alpha}\cdot\boldsymbol{\tau}\right) + \boldsymbol{\tau}\cdot(\partial^\mu\boldsymbol{\alpha})\left(1 + \mathrm{i}\frac{g}{2}\boldsymbol{\alpha}\cdot\boldsymbol{\tau}\right)\\ &= \boldsymbol{\tau}\cdot\mathbf{W}^\mu + \mathrm{i}\frac{g}{2}(\boldsymbol{\tau}\cdot\mathbf{W}^\mu\boldsymbol{\alpha}\cdot\boldsymbol{\tau} - \boldsymbol{\alpha}\cdot\boldsymbol{\tau}\boldsymbol{\tau}\cdot\mathbf{W}^\mu) + \boldsymbol{\tau}\cdot\partial^\mu\boldsymbol{\alpha}\\ &= \boldsymbol{\tau}\cdot(\mathbf{W}^\mu + g\boldsymbol{\alpha}\times\mathbf{W}^\mu + \partial^\mu\boldsymbol{\alpha}),\end{aligned} \tag{5.115}$$

where (2.215) has been used. Thus

$$\mathbf{W}'^\mu = \mathbf{W}^\mu + g\boldsymbol{\alpha}\times\mathbf{W}^\mu + \partial^\mu\boldsymbol{\alpha}. \tag{5.116}$$

Invariance under U_1 (in 5.107), corresponding to the U(1) weak hypercharge group, gives from (5.106) and (5.112)

$$B'^\mu = B^\mu + \partial^\mu\theta. \tag{5.117}$$

Thus, (5.116) and (5.117) are the required gauge transformations giving invariance of the Lagrangian (5.106) under $\mathrm{SU}(2)_\mathrm{L}\otimes\mathrm{U}(1)_y$.

The full Lagrangian for $\mathrm{SU}(2)_\mathrm{L}\otimes\mathrm{U}(1)_y$ requires the addition of terms for the free fields B^μ and W^μ respectively. For B^μ, this term is of the same form as that for the photon in QED, namely, defining $B^{\mu\nu} = \partial^\mu B^\nu - \partial^\nu B^\mu$,

$$\mathscr{L}_B = -\tfrac{1}{4} B_{\mu\nu} B^{\mu\nu}. \tag{5.118}$$

This term is invariant under gauge transformations of both U(1) and, since (5.117) does not involve $\boldsymbol{\alpha}$, SU(2). For the isovector field W_i^μ $(i = 1, 2, 3)$ a first guess for the free particle Lagrangian might be

$$-\tfrac{1}{4}(\partial_\mu W_{i\nu} - \partial_\nu W_{i\mu})(\partial^\mu W_i^\nu - \partial^\nu W_i^\mu). \tag{5.119}$$

However, this expression is not invariant under the SU(2) gauge transformation (5.116). Specifically, the term $g\boldsymbol{\alpha} \times \mathbf{W}^\mu$ violates the symmetry.

A clue to the possible form for W_i^μ fields can be found by re-examining the QED term for $\mathscr{L}_\gamma$ (4.110). The quantity $F^{\mu\nu}$ can be expressed as

$$F^{\mu\nu} = -\frac{\mathrm{i}}{e}[D^\mu, D^\nu]. \tag{5.120}$$

That is, in terms of the commutation relation of the covariant derivative $D^\mu = \partial^\mu + \mathrm{i}eA^\mu$. For the SU(2) ⊗ U(1) case, the covariant derivative is (from (5.106))

$$D^\mu = \partial^\mu + \mathrm{i}\frac{g}{2}\boldsymbol{\tau}\cdot\mathbf{W}^\mu + \mathrm{i}\frac{g'}{2} y B^\mu. \tag{5.121}$$

The commutation relation for the part of the covariant derivative corresponding to the $\mathbf{W}^\mu$ field, is

$$\begin{aligned}
[D^\mu, D^\nu] &= \left[\partial^\mu + \mathrm{i}\frac{g}{2}\boldsymbol{\tau}\cdot\mathbf{W}^\mu, \partial^\nu + \mathrm{i}\frac{g}{2}\boldsymbol{\tau}\cdot\mathbf{W}^\nu\right] \\
&= \mathrm{i}\frac{g}{2}\boldsymbol{\tau}\cdot(\partial^\mu\mathbf{W}^\nu - \partial^\nu\mathbf{W}^\mu) - \frac{g^2}{4}(\boldsymbol{\tau}\cdot\mathbf{W}^\mu\boldsymbol{\tau}\cdot\mathbf{W}^\nu - \boldsymbol{\tau}\cdot\mathbf{W}^\nu\boldsymbol{\tau}\cdot\mathbf{W}^\mu) \\
&= \mathrm{i}\frac{g}{2}\boldsymbol{\tau}\cdot(\partial^\mu\mathbf{W}^\nu - \partial^\nu\mathbf{W}^\mu) - \frac{g^2}{4} 2\mathrm{i}\boldsymbol{\tau}\cdot\mathbf{W}^\mu \times \mathbf{W}^\nu \\
&= \mathrm{i}\frac{g}{2}\boldsymbol{\tau}\cdot\mathbf{W}^{\mu\nu} = \mathrm{i}\frac{g}{2} X^{\mu\nu},
\end{aligned} \tag{5.122}$$

where

$$\mathbf{W}^{\mu\nu} = \partial^\mu\mathbf{W}^\nu - \partial^\nu\mathbf{W}^\mu - g\mathbf{W}^\mu \times \mathbf{W}^\nu. \tag{5.123}$$

Comparison with QED suggests that the required term for the Lagrangian contains the product $X_{\mu\nu}X^{\mu\nu}$, but in a form which is invariant under (5.108). Denoting gauge transformed quantities with a prime, one can see that the following trace satisfies the above requirements

$$\mathrm{tr}(X'_{\mu\nu}X'^{\mu\nu}) = \mathrm{tr}(U^{-1}X_{\mu\nu}UU^{-1}X^{\mu\nu}U) = \mathrm{tr}(X_{\mu\nu}X^{\mu\nu}), \tag{5.124}$$

where the cyclic property that $\mathrm{tr}(AB) = \mathrm{tr}(BA)$ has been used. Substituting for $X^{\mu\nu}$ from (5.122), and using the property (2.215) of the $\boldsymbol{\tau}$-matrices, gives

$$\mathrm{tr}(X_{\mu\nu}X^{\mu\nu}) = \mathrm{tr}(\boldsymbol{\tau}\cdot\mathbf{W}_{\mu\nu}\boldsymbol{\tau}\cdot\mathbf{W}^{\mu\nu}) = \mathrm{tr}(\mathbf{W}_{\mu\nu}\cdot\mathbf{W}^{\mu\nu})$$
$$= 2\mathbf{W}_{\mu\nu}\cdot\mathbf{W}^{\mu\nu}, \tag{5.125}$$

so a suitable SU(2) invariant contribution for the fields $\mathbf{W}^{\mu}$ to the Lagrangian is

$$\mathscr{L}_{\mathrm{W}} = -\tfrac{1}{4}\mathbf{W}_{\mu\nu}\cdot\mathbf{W}^{\mu\nu} \tag{5.126}$$

The numerical factor ensures that $\mathscr{L}_{\mathrm{W}}$ contains the term (5.119). Furthermore, $\mathscr{L}_{\mathrm{W}}$ is invariant under U(1); this follows from the invariance of the fields W_i^{μ}.

The additional term in (5.126), compared to the U(1) Abelian case, arises because the SU(2) generators do not commute. The SU(2) group is *non-Abelian* (non-commuting) and the associated fields $\mathbf{W}^{\mu}$ are called *Yang–Mills fields* (Yang and Mills, 1954). The term $g\mathbf{W}\times\mathbf{W}$ in (5.123) arises from the non-commuting properties. Substitution of (5.123) in (5.126) shows that $\mathscr{L}_{\mathrm{W}}$ contains, in addition to the self-interaction term (5.119), the products of both three and four W_i^{μ} fields. In perturbation theory these terms give rise to vertices connecting three and four field lines. Some examples are given in Fig. 5.9, where the boson fields shown are $\mathrm{W}^{\pm}$, Z^0 and γ (i.e. the physical particles), the being linear combinations of the fields W_i^{μ} and B^{μ}. The existence of such vertices is an important consequence (and thus test) of the theory, and they arise because of the non-Abelian nature of the theory.

The weak gauge bosons carry the weak charge, in contrast to QED where the photon is electrically neutral. In order to illustrate further the

Fig. 5.9 Examples of vertices involving three and four fields arising from the self-interaction terms of the boson fields.

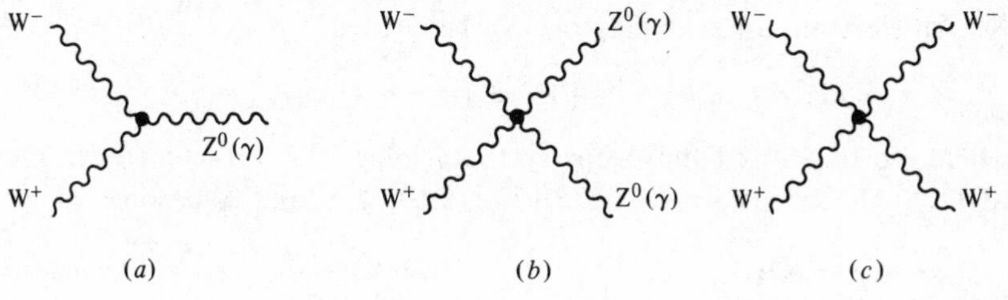

differences in the U(1) and SU(2) properties, let us consider the interactions of two particles a and b with a field E^μ. Let us suppose that the Lagrangian for particle a is that given in (5.110), whereas that for particle b has the same form as (5.110), but with a term QE^μ rather than E^μ. That is, the 'charge' of b is Q times that of a. Under the local transformations

$$\psi'_{\mathrm{a}}(x) = U_{\mathrm{a}}\psi_{\mathrm{a}}(x); \qquad \psi'_{\mathrm{b}}(x) = U_{\mathrm{b}}\psi_{\mathrm{b}}(x), \tag{5.127}$$

invariance of the Lagrangian gives (from (5.112))

$$\begin{aligned} (E')^\mu &= U_{\mathrm{a}}E^\mu U_{\mathrm{a}}^{-1} - (\partial^\mu U_{\mathrm{a}})U_{\mathrm{a}}^{-1}, \\ Q(E')^\mu &= QU_{\mathrm{b}}E^\mu U_{\mathrm{b}}^{-1} - (\partial^\mu U_{\mathrm{b}})U_{\mathrm{b}}^{-1}. \end{aligned} \tag{5.128}$$

For the U(1) group we have $U_{\mathrm{a}} = \exp(-\mathrm{i}e\theta)$ and $U_{\mathrm{b}} = \exp(-\mathrm{i}eQ\theta)$. In both cases the transformation (5.128) gives

$$(E')^\mu = E^\mu + \mathrm{i}e\,\partial^\mu\theta. \tag{5.129}$$

For the SU(2) case U_{a} and U_{b} have the form of (5.114), but with 'charges' g and Qg respectively. Retaining only first order terms in $\boldsymbol{\alpha}$, equations (5.128) become

$$\begin{aligned} (E')^\mu &= E^\mu + \mathrm{i}\frac{g}{2}[E^\mu, \boldsymbol{\alpha}\cdot\boldsymbol{\tau}] + \mathrm{i}\frac{g}{2}\partial^\mu\boldsymbol{\alpha}\cdot\boldsymbol{\tau}, \\ Q(E')^\mu &= QE^\mu + \mathrm{i}Q^2\frac{g}{2}[E^\mu, \boldsymbol{\alpha}\cdot\boldsymbol{\tau}] + \mathrm{i}Q\frac{g}{2}\partial^\mu\boldsymbol{\alpha}\cdot\boldsymbol{\tau}. \end{aligned} \tag{5.130}$$

Hence a common gauge transformation of the field E^μ can only be achieved if $Q^2 = Q$, i.e. $Q = 1$. Thus there is only one constant g associated with the SU(2) group. The SU(2) commutation relations give an extra term over the U(1) case, and the lack of such an equivalent term means that the U(1) coupling of each particle is arbitrary.

5.5 SPONTANEOUS SYMMETRY BREAKING

So far we have not considered the masses of the various particles, except to note that U(1) gauge invariance requires the photon to be massless. The mass term for a fermion in the Lagrangian is given in (5.101). For an electron, for example, this can be written

$$-m_{\mathrm{e}}\bar{e}e = -m_{\mathrm{e}}(\bar{e}_{\mathrm{L}} + \bar{e}_{\mathrm{R}})(e_{\mathrm{L}} + e_{\mathrm{R}}) = -m_{\mathrm{e}}(\bar{e}_{\mathrm{L}}e_{\mathrm{R}} + \bar{e}_{\mathrm{R}}e_{\mathrm{L}}), \tag{5.131}$$

where we have used the property that $(1+\gamma^5)(1-\gamma^5) = 0$. Under the SU(2) ⊗ U(1) transformation ((5.107), (5.108)), e_{L} and e_{R} become

$$e'_{\mathrm{L}} = (U_2)_{21}\nu_{\mathrm{L}} + (U_2)_{22}e_{\mathrm{L}}; \qquad e'_{\mathrm{R}} = U_1 e_{\mathrm{R}}. \tag{5.132}$$

Thus, the mass term is not SU(2) ⊗ U(1) invariant. Furthermore, this symmetry also requires that the bosons B^μ and W^μ are massless. A similar argument to that used for QED shows that m_B is zero, whereas the mass term for the W_i^μ field, namely

$$\tfrac{1}{2}M_W^2 \mathbf{W}_\mu \cdot \mathbf{W}^\mu, \tag{5.133}$$

is not invariant under (5.116). The observed non-zero values for the charged leptons, $W^\pm$ and Z^0 masses clearly violate the assumed SU(2) ⊗ U(1) symmetry. Furthermore, this symmetry must be badly broken, since the particle multiplets are far from being degenerate.

In the standard electroweak model it is assumed that the underlying gauge symmetry is '*spontaneously*' broken, as discussed below. The symmetry breaking mechanism must not only generate the particle masses, but also lead to a renormalisable theory. For example, adding mass terms 'by hand' to the Lagrangian discussed above does not lead to a renormalisable theory.

5.5.1 Higgs model for a U(1) gauge field

In this section we consider the interaction of a complex scalar field ϕ, both with itself and with a U(1) gauge field A^μ. Consider the Lagrangian

$$\mathscr{L} = [(\partial_\mu - \mathrm{i}eA_\mu)\phi^*][(\partial^\mu + \mathrm{i}eA^\mu)\phi] - \tfrac{1}{4}F_{\mu\nu}F^{\mu\nu} - \mu^2\phi^*\phi - \lambda(\phi^*\phi)^2. \tag{5.134}$$

This is invariant under the combined transformations

$$\phi'(x) = U\phi(x) = \exp[-\mathrm{i}e\theta(x)]\phi(x); \qquad (A')^\mu = A^\mu + \partial^\mu\theta. \tag{5.135}$$

Thus the field A^μ corresponds to a massless gauge boson. The terms $\mu^2\phi^*\phi$ and $\lambda(\phi^*\phi)^2$ in (5.134), according to previous arguments, correspond to the mass (see (3.49)) and self-interactions of the field ϕ. Recalling that, classically, the Lagrangian is $L = T - V$, we can define from (5.134) a potential

$$V(\phi) = \mu^2|\phi|^2 + \lambda|\phi|^4, \tag{5.136}$$

such that the ground state (vacuum) of the system corresponds to a minimum in $V(\phi)$. We assume $\lambda > 0$, so that a minium (i.e. vacuum state) exists. If $\mu^2 > 0$ (i.e. if μ is a real particle mass), then V has a minimum at $|\phi| = 0$. This is shown in Fig. 5.10(*a*), which displays $V(\phi)$ as a function of two orthogonal real components ϕ_1, and ϕ_2 defined by

$$\phi(x) = \frac{1}{2^{1/2}}[\phi_1(x) + \mathrm{i}\phi_2(x)]. \tag{5.137}$$

The minimum energy state of the whole system requires $A^\mu = 0$. Thus the lowest energy (vacuum) state is unique (i.e. non-degenerate), and shares the symmetry properties of $\mathscr{L}$. The self-interaction term $\lambda|\phi|^4$ can be treated perturbatively.

If $\mu^2 < 0$, then the minimum of $V(\phi)$ is no longer at $|\phi| = 0$ (which is in fact a local maximum), but rather at a value

$$|\phi| = \phi_0 = (-\mu^2/2\lambda)^{1/2} = v/2^{1/2}. \tag{5.138}$$

This minimum (see Fig. 5.10(*b*)) is degenerate, and any point in the (ϕ_1, ϕ_2) plane satisfying

$$\phi_1^2 + \phi_2^2 = v^2, \tag{5.139}$$

gives the same result. The degeneracy corresponds to invariance under transformations $\phi'(x) = U\phi(x)$ in (5.135). Let us choose the minimum to be at some particular point $(v, 0)$ in the (ϕ_1, ϕ_2) plane. Further, let us introduce two real fields $\eta_1(x)$ and $\eta_2(x)$, such that

$$\phi(x) = \frac{1}{2^{1/2}}[v + \eta_1(x) + i\eta_2(x)]. \tag{5.140}$$

The fields η_1 and η_2 vanish at the minimum, and thus give a measure of deviations from the minimum. Expressing (5.136) in terms of η_1 and η_2,

Fig. 5.10 Higgs potential $V(\phi) = \mu^2|\phi|^2 + \lambda^2|\phi|^4$, with $\lambda > 0$, for (*a*) $\mu^2 > 0$ and (*b*) $\mu^2 < 0$.

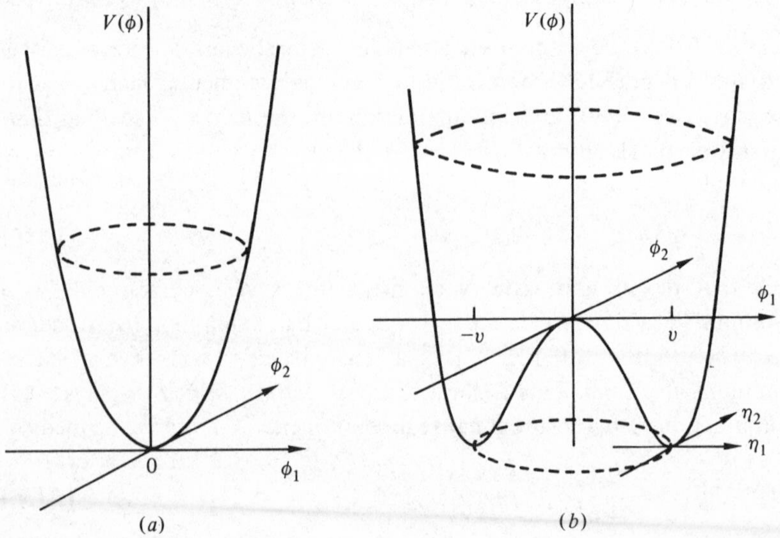

we obtain

$$V = \frac{-\mu^4}{4\lambda} + \lambda v^2 \eta_1^2 + \lambda v \eta_1 (\eta_1^2 + \eta_2^2) + \frac{\lambda}{4} (\eta_1^2 + \eta_2^2)^2. \tag{5.141}$$

If, for the moment, we ignore the gauge field A^μ (i.e. $A^\mu = 0$), then the Lagrangian (5.134), expressed in terms of $\eta_1(x)$ and $\eta_2(x)$ and omitting the constant term, has the following form

$$\mathscr{L} = \tfrac{1}{2}(\partial_\mu \eta_1)(\partial^\mu \eta_1) - \tfrac{1}{2}(2\lambda v^2)\eta_1^2 + \tfrac{1}{2}(\partial_\mu \eta_2)(\partial^\mu \eta_2) - \lambda v \eta_1 (\eta_1^2 + \eta_2^2) - \frac{\lambda}{4} (\eta_1^2 + \eta_2^2)^2. \tag{5.142}$$

Now, for $\mu^2 < 0$, the interpretation that the $\lambda|\phi|^4$ term can be treated as a perturbation around the value $|\phi| = 0$ does not work, since $|\phi| = 0$ is not the stable minimum. Further, this interpretation would correspond to a particle having an imaginary mass, and is thus unphysical. However, inspection of the Lagrangian (5.142) shows that it contains two free particle Klein–Gordon Lagrangians of the type (3.24). That is, it contains two scalar fields, one of (positive) mass $(2\lambda v^2)^{1/2}$ and the other of mass zero; corresponding to η_1 and η_2 respectively.

If one moves away from the chosen minimum (i.e. $(\phi_1, \phi_2) = (v, 0)$ in Fig. 5.10(*b*)) in the η_1 direction, the potential energy increases as η_1^2. In the η_2 direction, however, the potential energy is constant; corresponding to a massless η_2 boson excitation. The remaining cubic and quartic terms in η_1 and η_2 in (5.142) correspond to interactions. Note that the Lagrangian (5.142) has the same physical content as (5.134) (with $A^\mu = 0$). The difference in the interpretation arises when perturbations around the minimum are considered. On quantisation, the requirement that the vacuum contains no particles gives (from (5.140) and (5.138)) the property that the expectation value of the vacuum has the non-zero value

$$\langle 0|\phi(x)|0\rangle = \frac{v}{2^{1/2}} = \phi_0. \tag{5.143}$$

Note that, for a scalar field, this quantity is Lorentz invariant. The vacuum expectation value for a Dirac or vector particle must, however, be zero if it is to be Lorentz invariant.

The method of arbitrarily choosing one of the set of degenerate states as the ground state is called *spontaneous symmetry breaking*. Once the specific state is chosen, the underlying symmetry is broken and thus becomes '*hidden*'. The symmetry is not broken by the addition of a symmetry breaking piece in the Lagrangian, but by the selection of one of the degenerate minima. An example from outside the realm of high energy physics is that of a ferromagnet. The forces between the electron spins, and hence the Hamiltonian, are rotationally invariant. However, in its ground state, the material is magnetised in a specific direction, corresponding to the alignment of the electron spins. Returning to the above example (which, for $A^\mu = 0$, corresponds to a global phase symmetry), we have seen that a massless boson appears in the spontaneous symmetry breaking. Such a boson is called a *Goldstone boson* and, since no such particle has been observed, constitutes a potential theoretical problem.

We next consider the Lagrangian (5.134), including the effect of the U(1) gauge field $A^\mu(x)$. From Lorentz invariance the vacuum expectation value of A^μ must be zero, so the minimum (5.138) is again obtained. Substituting (5.140) into (5.134) gives the Lagrangian, expressed in the fields η_1 and η_2, as follows:

$$\mathscr{L} = \tfrac{1}{2}(\partial_\mu \eta_1)(\partial^\mu \eta_1) - \tfrac{1}{2}(2\lambda v^2)\eta_1^2 + \tfrac{1}{2}(\partial_\mu \eta_2)(\partial^\mu \eta_2)$$
$$- \tfrac{1}{4}\mathrm{F}_{\mu\nu}\mathrm{F}^{\mu\nu} + \tfrac{1}{2}(ev)^2 A_\mu A^\mu + evA^\mu\, \partial_\mu \eta_2 + \cdots. \qquad (5.144)$$

The remaining terms represent interactions involving three or four fields. Written in this form the Lagrangian seems to contain a massless scalar η_2, a massive scalar η_1 and, the new feature, a massive vector field A^μ, with mass ev. However, the term in $A^\mu\, \partial_\mu \eta_2$ shows that A^μ and η_2 do not enter independently. This problem can also be seen if we perform some 'bookkeeping' on the number of degrees of freedom in the two forms of the Lagrangian (5.134) and (5.144). In (5.134), there is a massless vector particle and a complex scalar particle, each having two degrees of freedom. In (5.144), the massive vector particle has three degrees of freedom (the third arising from the longitudinal polarisation state) and the real scalars η_1 and η_2 both have one degree of freedom. Thus we seem to have gained one extra degree of freedom.

This problem can be 'cured' by exploiting the local gauge freedom of ϕ (5.135). We can choose $\theta(x)$ in (5.135) so that $\phi'(x)$ is real, and hence the field η_2' does not appear. Hence we can write (dropping the primes)

$$\phi(x) = \frac{1}{2^{1/2}}\,[v + \eta_1(x)]. \qquad (5.145)$$

The Lagrangian (5.144) then becomes

$$\mathscr{L} = \tfrac{1}{2}(\partial_\mu \eta_1)(\partial^\mu \eta_1) - \tfrac{1}{2}(2\lambda v^2)\eta_1^2 - \tfrac{1}{4}F_{\mu\nu}F^{\mu\nu} + \tfrac{1}{2}(ev)^2 A_\mu A^\mu + \cdots. \qquad (5.146)$$

This has one degree of freedom for the real scalar field and three for the massive vector field, giving the required number. Thus the troublesome massless Goldstone boson has been removed ('eaten') by the gauge transformation. The chosen gauge is called the *unitary gauge*. Starting with the Lagrangian (5.134), describing a complex scalar field (with an apparently imaginary mass) and a massless gauge field, we have obtained, through spontaneous symmetry breaking, the Lagrangian (5.146) describing a real massive scalar field and a massive vector field. The massless Goldstone boson becomes the longitudinal polarisation state of the massive vector boson. This mechanism is called the *Higgs mechanism*, after Higgs (1964, 1966).

5.5.2 The Weinberg–Salam model

Ideas, similar to those used to produce massive U(1) gauge bosons, can be applied to the weak SU(2) gauge bosons. We require to break spontaneously the $\mathrm{SU}(2)\otimes\mathrm{U}(1)$ symmetry of the Lagrangian discussed in Sections 5.3 and 5.4, in such a way that the photon remains massless; that is, there must remain a residual U(1) symmetry for the electromagnetic interaction.

The remaining three gauge bosons must acquire masses in this process. This can be achieved with four independent scalar fields; the simplest choice being a weak isospin doublet of complex scalar fields

$$\phi(x)=\begin{pmatrix}\phi^{\dagger}(x)\\ \phi^{0}(x)\end{pmatrix}. \tag{5.147}$$

The electric charges, positive for $\phi^{\dagger}$ and neutral for ϕ^{0}, follow if the Higgs doublet is assigned the value $y=1$. Under U(1) and SU(2), the doublet will transform as (see (5.108))

$$\phi(x)\to\phi'(x)=\exp\left\{-\mathrm{i}\left[\frac{g}{2}\boldsymbol{\alpha}(x)\cdot\boldsymbol{\tau}+\frac{g'}{2}y\theta(x)I\right]\right\}\phi(x). \tag{5.148}$$

Generalising the $\mathrm{SU}(2)\otimes\mathrm{U}(1)$ Lagrangian density to include the Higgs doublet ϕ, gives the following additional terms (see (5.134))

$$\mathscr{L}_{\mathrm{H}}=[D_{\mu}\phi(x)]^{\dagger}[D^{\mu}\phi(x)]-\mu^{2}\phi^{\dagger}(x)\phi(x)-\lambda[\phi^{\dagger}(x)\phi(x)]^{2}, \tag{5.149}$$

where D_{μ} is given by (5.121). Following the U(1) example of spontaneous symmetry breaking, we take λ to be positive and μ^2 negative. The minimum potential occurs at a value

$$|\phi|^{2}=-\mu^{2}/2\lambda=v^{2}/2. \tag{5.150}$$

Let us choose the ground (vacuum) state to be

$$\phi_{0}=\begin{pmatrix}0\\ v/2^{1/2}\end{pmatrix}. \tag{5.151}$$

Other choices of ϕ_0 can be obtained from (5.151) by a global phase transformation. Now the assignment $y=1$ means that the $t_3=-\frac{1}{2}$ component of (5.147) is neutral, so that the choice (5.151) leads to spontaneous symmetry breaking for the neutral component only. Thus the vacuum is invariant under the electromagnetic U(1) gauge transformation (4.107) of the Higgs field. Charge conservation is not affected by the spontaneous symmetry breaking.

Proceeding in a similar way to the U(1) case, let us introduce four real fields all with zero vacuum expectation values, such that

$$\phi(x)=\frac{1}{2^{1/2}}\begin{pmatrix}\sigma_1(x)+i\sigma_2(x)\\ v+\eta_1(x)+i\eta_2(x)\end{pmatrix}. \tag{5.152}$$

The 'potential' term in (5.149) is

$$\begin{aligned}V&=\mu^2\phi^\dagger\phi+\lambda(\phi^\dagger\phi)^2\\ &=\frac{\mu^2}{2}[\sigma_1^2+\sigma_2^2+\eta_2^2+(v+\eta_1)^2]+\frac{\lambda}{4}[\sigma_1^2+\sigma_2^2+\eta_2^2+(v+\eta_1)^2]^2\\ &=\frac{-\mu^4}{4\lambda}+\lambda v^2\eta_1^2+\lambda v\eta_1(\sigma_1^2+\sigma_2^2+\eta_1^2+\eta_2^2)\\ &\quad+\frac{\lambda}{4}(\sigma_1^2+\sigma_2^2+\eta_1^2+\eta_2^2)^2.\end{aligned} \tag{5.153}$$

Using this expression, and $\phi(x)$ from (5.152), it can easily be seen that the Lagrangian (5.149) corresponds to three massless Goldstone bosons $(\sigma_1,\sigma_2,\eta_2)$, together with a massive scalar particle η_1, with mass $m_H=(2\lambda v^2)^{1/2}$. Hereafter we will refer to the real field $\eta_1(x)$ as the Higgs scalar $H(x)$.

The next step is to 'gauge' away the three unwanted Goldstone bosons. This will ensure that the resulting Lagrangian has the correct number of degrees of freedom. We require a transformation U such that

$$\phi'(x)=\frac{1}{2^{1/2}}\begin{pmatrix}0\\ v+H(x)\end{pmatrix}=U\phi(x)=U\frac{1}{2^{1/2}}\begin{pmatrix}\sigma_1(x)+i\sigma_2(x)\\ v+\eta_1(x)+i\eta_2(x)\end{pmatrix}, \tag{5.154}$$

i.e. leaving only the one real field $\eta_1(x)=H(x)$. A suitable form for U is

$$U=\exp\left[-\frac{i}{v}(\boldsymbol{\Lambda}\cdot\boldsymbol{\tau}-\Lambda_3 I)\right]. \tag{5.155}$$

Comparison with (5.108) shows that

$$\boldsymbol{\alpha}(x)=\frac{2}{gv}\boldsymbol{\Lambda}(x),\qquad \theta(x)=-\frac{2}{g'yv}\Lambda_3(x). \tag{5.156}$$

If we expand (5.155) for infinitesimal values of Λ and Λ_3, and substitute for U in (5.154), then, retaining only first order terms in the fields, we obtain the relationships

$$\sigma_1 = \Lambda_2, \qquad \sigma_2 = \Lambda_1, \qquad \eta_1 = H, \qquad \eta_2 = -2\Lambda_3. \quad (5.157)$$

Using $\phi(x)$ from (5.154) expressed in terms of v and H, then the potential V (5.153) becomes

$$V = -\frac{\mu^4}{4\lambda} - \mu^2 H^2 + \lambda v H^3 + \frac{\lambda}{4} H^4. \quad (5.158)$$

In (5.155) the U(1) term was chosen to be the same function as the third component of the SU(2) part. However, this is still general, since we also have the residual U(1) invariance. Consider the transformation

$$\phi'(x) = \frac{1}{2^{1/2}}\begin{pmatrix} 0 \\ v+H \end{pmatrix} = \exp\left[-\mathrm{i}\frac{\Lambda'}{v}(I+\tau_3)\right]\frac{1}{2^{1/2}}\begin{pmatrix} 0 \\ v+H \end{pmatrix}. \quad (5.159)$$

Comparison with (5.108) for this transformation shows that

$$\alpha_1(x) = \alpha_2(x) = 0, \qquad \alpha_3(x) = \frac{2}{gv}\Lambda'(x), \qquad \theta(x) = \frac{2}{g'vy}\Lambda'(x). \quad (5.160)$$

From (5.116) and (5.117), the corresponding transformation properties of the neutral fields W_3^μ and B^μ are

$$(W'_3)^\mu = W_3^\mu + \frac{2}{gv}\partial^\mu\Lambda',$$

$$(B')^\mu = B^\mu + \frac{2}{g'v}\partial^\mu\Lambda'. \quad (5.161)$$

Hence, the physical fields Z^μ and A^μ, given by (5.97), transform as (using (5.99))

$$(Z')^\mu = Z^\mu,$$

$$(A')^\mu = A^\mu + \frac{2}{ve}\partial^\mu\Lambda'. \quad (5.162)$$

Similar considerations for the fields W_1^μ and W_2^μ show that the combinations $(\mathrm{W}^\pm)^\mu$, given by (5.88), transform as for charges ± 1, whereas Z^μ (5.162) has zero charge. In summary, we are left with four gauge bosons having the desired properties and one neutral scalar Higgs particle.

The next step is to consider the possibility of interactions between the scalar doublet ϕ and the lepton doublet L and singlet R. It is assumed that this electron–Higgs interaction has the so-called *Yukawa coupling* form, with the $\mathrm{SU}(2)_L \otimes \mathrm{U}(1)_y$ invariant Lagrangian

$$\begin{aligned}\mathscr{L}_{eH} &= -g_e[(\bar{L}\phi)R + \bar{R}(\phi^\dagger L)] \\ &= -\frac{g_e}{2^{1/2}}[v(\bar{e}_L e_R + \bar{e}_R e_L) + \bar{e}_L H e_R + \bar{e}_R H e_L]. \qquad (5.163)\end{aligned}$$

Here, g_e is a dimensionless constant for the electron coupling. Similar terms for $l = \mu$ and τ are again understood. If we examine one of the terms in (5.163), for example $\bar{e}_R\phi^\dagger L$, we can see that $\bar{e}_R$ creates a $t = 0$, $y = -2$ state, L represents the destruction of a weak isodoublet $y = -1$ state and $\phi^\dagger$ the creation of a weak isodoublet with $y = +1$. Thus this coupling gives a Lagrangian which transforms as a Lorentz scalar and has the desired $\mathrm{SU}(2)_L$ and $\mathrm{U}(1)_y$ invariance properties. In the second line of (5.163) the spontaneously broken form for ϕ (5.159) has been used. Comparison with (5.131) shows that the electron has acquired a mass $m_e = g_e v/2^{1/2}$. Note that we have assumed that the neutrinos have no righthanded part and remain massless; however, (5.163) can be generalised to include a neutrino term.

Next we consider the term containing the covariant derivative D^μ (5.121) in $\mathscr{L}_H$ (5.149). Writing D^μ as a 2×2 matrix, using the explicit forms (2.213) of the τ-matrices, we obtain

$$D^\mu = \begin{pmatrix} \partial^\mu + \frac{i}{2}(gW_3^\mu + g'B^\mu) & i\frac{g}{2^{1/2}}W_+^\mu \\ i\frac{g}{2^{1/2}}W_-^\mu & \partial^\mu - \frac{i}{2}(gW_3^\mu - g'B^\mu) \end{pmatrix}. \qquad (5.164)$$

From (5.97) and (5.99) we can write

$$\begin{aligned}Z^\mu &= (gW_3^\mu - g'B^\mu)/(g^2 + g'^2)^{1/2}, \\ A^\mu &= (gB^\mu + g'W_3^\mu)/(g^2 + g'^2)^{1/2}. \qquad (5.165)\end{aligned}$$

Using the above expressions and also (5.158), we obtain, after a short calculation,

$$\begin{aligned}\mathscr{L}_H + \mathscr{L}_{eH} = {} & \tfrac{1}{2}(\partial_\mu H\, \partial^\mu H + 2\mu^2 H^2) + \frac{g^2}{4}(v^2 + 2vH + H^2)W_{-\mu}W_+^\mu \\ & + \frac{(g^2 + g'^2)}{8}(v^2 + 2vH + H^2)Z_\mu Z^\mu - \frac{\lambda}{4}(4vH^3 + H^4) \\ & - \frac{g_e}{2^{1/2}}(v + H)(\bar{e}_L e_R + \bar{e}_R e_L). \qquad (5.166)\end{aligned}$$

The $W^{\pm}$ and Z^0 bosons have thus acquired a mass, while the photon has remained massless, and, from (5.166), we have

$$M_W = \frac{gv}{2}, \qquad M_Z = (g^2 + g'^2)^{1/2}\frac{v}{2}. \tag{5.167}$$

Eliminating v and using (5.99), we obtain a simple relationship between the $W^{\pm}$ and Z^0 masses, namely

$$M_W = M_Z \cos\theta_W. \tag{5.168}$$

The full SU(2) ⊗ U(1) Lagrangian in the unitary gauge is obtained by collecting together the component terms $\mathscr{L}_e$ (5.106), $\mathscr{L}_B$ (5.118), $\mathscr{L}_W$ (5.126) and $\mathscr{L}_H + \mathscr{L}_{eH}$ (5.166). Writing these expressions in terms of the fields $W^{\pm}_{\mu}$, Z_{μ} and A_{μ}, we obtain, after some algebra

$$\begin{aligned}
\mathscr{L} = &-\tfrac{1}{4}F_{\mu\nu}F^{\mu\nu} - \tfrac{1}{4}Z_{\mu\nu}Z^{\mu\nu} - \tfrac{1}{2}(F^{\dagger}_W)_{\mu\nu}(F_W)^{\mu\nu}\\
&+ \tfrac{1}{2}\,\partial_{\mu}H\,\partial^{\mu}H + \tfrac{1}{2}M_Z^2 Z_{\mu}Z^{\mu} + M_W^2(W_-)_{\mu}(W_+)^{\mu} - \tfrac{1}{2}m_H^2 H^2\\
&+ \bar{e}(\mathrm{i}\,\not{\partial} - m_e)e + \bar{\nu}\mathrm{i}\,\not{\partial}\nu\\
&+ \mathrm{i}g(\partial_{\mu}W_{+\nu} - \partial_{\nu}W_{+\mu})W^{\nu}_{-}(\cos\theta_W Z^{\mu} + \sin\theta_W A^{\mu})\\
&+ \mathrm{i}g(\partial_{\mu}W_{-\nu} - \partial_{\nu}W_{-\mu})W^{\mu}_{+}(\cos\theta_W Z^{\nu} + \sin\theta_W A^{\nu})\\
&+ \mathrm{i}g(W^{\mu}_{-}W^{\nu}_{+} - W^{\mu}_{+}W^{\nu}_{-})\,\partial_{\mu}(\cos\theta_W Z_{\nu} + \sin\theta_W A_{\nu})\\
&- g^2 W_{+\mu}W^{\mu}_{-}(\cos\theta_W Z_{\nu} + \sin\theta_W A_{\nu})(\cos\theta_W Z^{\nu} + \sin\theta_W A^{\nu})\\
&+ g^2 W^{\nu}_{+}W^{\mu}_{-}(\cos\theta_W Z_{\mu} + \sin\theta_W A_{\mu})(\cos\theta_W Z_{\nu} + \sin\theta_W A_{\nu})\\
&+ \frac{g^2}{2}W_{-\nu}W_{+\mu}(W^{\nu}_{-}W^{\mu}_{+} - W^{\mu}_{-}W^{\nu}_{+}) + e\bar{e}\gamma_{\mu}eA^{\mu}\\
&- \frac{g}{2\cos\theta_W}[\tfrac{1}{2}\bar{\nu}\gamma_{\mu}(1-\gamma^5)\nu + \bar{e}\gamma_{\mu}(g_V - g_A\gamma^5)e]Z^{\mu}\\
&- \frac{g}{2(2)^{1/2}}[\bar{\nu}\gamma_{\mu}(1-\gamma^5)eW^{\mu}_{+} + \bar{e}\gamma_{\mu}(1-\gamma^5)\nu W^{\mu}_{-}]\\
&+ \frac{g^2}{4}(2vH + H^2)W_{-\mu}W^{\mu}_{+} + \frac{(g^2+g'^2)}{8}(2vH + H^2)Z_{\mu}Z^{\mu}\\
&- \frac{\lambda}{4}(4vH^3 + H^4) - \frac{m_e}{v}\bar{e}eH,
\end{aligned} \tag{5.169}$$

where

$$g_V = 2\sin^2\theta_W - \tfrac{1}{2}, \qquad g_A = -\tfrac{1}{2}, \tag{5.170}$$

and

$$(F_W)^{\mu\nu} = \partial^\mu W_+^\nu - \partial^\nu W_+^\mu. \tag{5.171}$$

The Lagrangian (5.169) contains some *a priori* unknown parameters, namely e, g, $\sin\theta_W$, M_W, M_Z, m_H and v. In addition, for each lepton generation l, there are the masses m_l (and m_{ν_l}, if this is taken to be non-zero). (The possibility of 'mixing' between generations is discussed in Section 5.5.3 and Chapter 9.) However, there are relationships ((5.99), (5.167) and (5.168)) between the parameters. Furthermore, the measured value of G_F from muon decay, gives the ratio g/M_W from (5.63). Thus, we find that $v = (G_F(2)^{1/2})^{-1/2} = 246$ GeV, and also that we can express M_W in terms of e, G_F and $\sin\theta_W$ as follows

$$M_W = e/(2^{5/4}G_F^{1/2}\sin\theta_W) \simeq 37/\sin\theta_W \text{ GeV}, \tag{5.172}$$

Indeed, the masses of the $W^\pm$ (~ 80 GeV) and the Z^0 (~ 90 GeV) were accurately predicted, several years before their discovery, from the relationships (5.172) and (5.168). The determination of the parameters in the standard model is discussed in the following chapters. Note, however, that values of the masses of the leptons and of the Higgs particle cannot be predicted in terms of other measured quantities.

The Feynman rules for the standard electroweak theory can be obtained in an analogous way to those in QED, starting with the interaction part of the Lagrangian (5.169). The perturbations from the free particle solutions, found from $\mathcal{L}_0$, are calculated using $\mathcal{H}_I = -\mathcal{L}_I$, as discussed in Section 4.7. Inspection of (5.169) shows that it contains the interaction vertices W_+W_-Z, $W_+W_-\gamma$, W_+W_-ZZ, $W_+W_-Z\gamma$, $W_+W_-\gamma\gamma$, $W_+W_-W_+W_-$, $\bar{l}l\gamma$, $\bar{\nu}_L\nu_L Z$, $\bar{l}lZ$, $\bar{\nu}_L l_L W$, W_+W_-H, W_+W_-HH, ZZH, ZZHH, HHH, HHHH and $\bar{l}lH$. The Feynman rules are given in Appendix C and are discussed further in Section 5.5.4. The asymptotic form of the W/Z propagator (3.204) is $q^\mu q^\nu \ll (M^2q^2)$, hence there is no suppression factor for large loop momenta, and so it would appear that it is not renormalisable. However, the standard GSW model is renormalisable, so that meaningful perturbative calculations can be carried out. The important proof that the theory is renormalisable (using the general R_ξ rather than the unitary gauge) was given by 't Hooft (1971). For more details on this topic, see, for example, Taylor (1978).

5.5.3 Incorporating generations of leptons and quarks in the GSW model

We have seen in Section 5.3 that the charged current interactions of u, d, s and c quarks can be described by two quark doublets q (5.78) and q' (5.81). The weak isospin values for the doublets are $t_3 = \frac{1}{2}$ for the upper (charge $\frac{2}{3}$) components and $t_3 = -\frac{1}{2}$ for the lower (charge $-\frac{1}{3}$) components. The $t_3 = -\frac{1}{2}$ states, which are the eigenstates of the weak charged current interaction, are 'Cabibbo-rotated' mixtures of the d and s quarks. The classification is made in the same way as for the leptons, with the lefthanded components of the quarks being assigned to the doublets and the righthanded parts to singlets.

Following closely the method used for leptons in Section 5.3, the third component of the weak isospin current and the electromagnetic current can be written ((cf. (5.91) and (5.93))

$$(j_3^t)^\mu = \bar{q}\gamma^\mu \frac{\tau_3}{2} q = \tfrac{1}{2}(\bar{u}_L\gamma^\mu u_L - \bar{d}_L^c\gamma^\mu d_L^c), \tag{5.173}$$

$$j_q^\mu/e = \tfrac{2}{3}\bar{u}\gamma^\mu u - \tfrac{1}{3}\bar{d}^c\gamma^\mu d^c, \tag{5.174}$$

and thus we can define a weak hypercharge current

$$(j^y)^\mu = 2[j_q^\mu/e - (j_3^t)^\mu] = \tfrac{1}{3}\bar{q}\gamma^\mu q + \tfrac{4}{3}\bar{u}_R\gamma^\mu u_R - \tfrac{2}{3}\bar{d}_R^c\gamma^\mu d_R^c. \tag{5.175}$$

The weak hypercharges of u_L, d_L^c, u_R and d_R^c are therefore $\frac{1}{3}$, $\frac{1}{3}$, $\frac{4}{3}$ and $-\frac{2}{3}$ respectively. Similar results hold for the second doublet q'. The equation $Q/e = t_3 + y/2$ is again valid, and the weak isospin and hypercharge assignments are summarised in Table 5.3. Additional doublets (e.g. t, b) can, of course, be added to the scheme.

The Lagrangian for the charged current part of the interaction is (see (5.89))

$$\mathscr{L}_{CC}^q = -\frac{g}{2}\bar{q}[\tau_1 W\!\!\!/_1 + \tau_2 W\!\!\!/_2]q - \frac{g}{2}\bar{q}'[\tau_1 W\!\!\!/_1 + \tau_2 W\!\!\!/_2]q' - \cdots \tag{5.176}$$

The neutral current part has the form of (5.100) and, substituting $(j_3^t)^\mu$ and j_q^μ from (5.173) and (5.174) respectively, we obtain

$$\mathscr{L}_{NC}^q = -e\sum_i e_i\bar{q}_i\gamma^\mu q_i A_\mu - \frac{g}{2\cos\theta_W}\sum_i \bar{q}_i\gamma^\mu(g_V^i - g_A^i\gamma^5)q_i Z_\mu, \tag{5.177}$$

where the sum i runs over the quarks u, d, c, s, etc., and the values of the couplings for the various quarks (and also for leptons) are given in Table 5.4. Note that the GIM mechanism gives a weak neutral current contribution in (5.177) which is diagonal in flavour.

In the formulation of the theory for leptons above, the possibility of mixing between generations of leptons was not considered. However, as we have seen, mixing between generations occurs for quarks. A further complication is that, for the quarks, both members of the doublet (e.g.

Table 5.3. *Weak isospin and hypercharge values of left and righthanded leptons and quarks*

L	t_3	y	R	t_3	y
ν_L	$\frac{1}{2}$	-1	ν_R	0	0
e_L	$-\frac{1}{2}$	-1	e_R	0	-2
u_L	$\frac{1}{2}$	$\frac{1}{3}$	u_R	0	$\frac{4}{3}$
d_L^c	$-\frac{1}{2}$	$\frac{1}{3}$	d_R^c	0	$-\frac{2}{3}$
c_L	$\frac{1}{2}$	$\frac{1}{3}$	c_R	0	$\frac{4}{3}$
s_L^c	$-\frac{1}{2}$	$\frac{1}{3}$	s_R^c	0	$-\frac{2}{3}$
t_L	$\frac{1}{2}$	$\frac{1}{3}$	t_R	0	$\frac{4}{3}$
b_L^c	$-\frac{1}{2}$	$\frac{1}{3}$	b_R^c	0	$-\frac{2}{3}$

u, d^c) must acquire a mass through their couplings to the Higgs field, whereas for the leptons the neutrino was assumed massless. If we assume the existence of a righthanded field ν_R, then we can introduce a term for $\bar{\nu}\nu$ into $\mathscr{L}_{eH}$ in (5.163). This can be effected by defining an isodoublet

$$\tilde{\phi} = i\tau^2\phi^* = \frac{1}{2^{1/2}}\begin{pmatrix} v+H \\ 0 \end{pmatrix}, \tag{5.178}$$

which has $y = -1$, and writing $\mathscr{L}_{eH}$ in the form

$$\begin{aligned}\mathscr{L}_{eH} &= -g_e[\bar{L}\phi e_R + \bar{e}_R\phi^\dagger L] - g_\nu[\bar{L}\tilde{\phi}\nu_R + \bar{\nu}_R\tilde{\phi}^\dagger L] \\ &= -\frac{g_e}{2^{1/2}}(v+H)[\bar{e}_L e_R + \bar{e}_R e_L] - \frac{g_\nu}{2^{1/2}}(v+H)[\bar{\nu}_L\nu_R + \bar{\nu}_R\nu_L].\end{aligned} \tag{5.179}$$

The neutrino mass is thus $m_\nu = g_\nu v/2^{1/2}$. Similar Lagrangians for the μ and τ generations can also be defined.

Restricting the discussion, for simplicity, to two lepton generations, then a more general Lagrangian to that of (5.179) is as follows

$$\begin{aligned}\mathscr{L}_{LH} = &-g_{ee}[\bar{E}_L\phi e_R + \bar{e}_R\phi^\dagger E_L] - h_{ee}[\bar{E}_L\tilde{\phi}\nu_{eR} + \bar{\nu}_{eR}\tilde{\phi}^\dagger E_L] \\ &- g_{e\mu}[\bar{E}_L\phi\mu_R + \bar{\mu}_R\phi^\dagger E_L] - h_{e\mu}[\bar{E}_L\tilde{\phi}\nu_{\mu R} + \bar{\nu}_{\mu R}\tilde{\phi}^\dagger E_L] \\ &- g_{\mu e}[\bar{M}_L\phi e_R + \bar{e}_R\phi^\dagger M_L] - h_{\mu e}[\bar{M}_L\tilde{\phi}\nu_{eR} + \bar{\nu}_{eR}\tilde{\phi}^\dagger M_L] \\ &- g_{\mu\mu}[\bar{M}_L\phi\mu_R + \bar{\mu}_R\phi^\dagger M_L] - h_{\mu\mu}[\bar{M}_L\tilde{\phi}\nu_{\mu R} + \bar{\nu}_{\mu R}\tilde{\phi}^\dagger M_L],\end{aligned} \tag{5.180a}$$

where E_L and M_L are lefthanded doublets for (ν_e, e) and (ν_μ, μ) respectively, both of the form (5.86). The 2×2 matrices g_{ij} and h_{ij} encompass possible transitions between the lepton generations, if the off-diagonal elements

Table 5.4. *Vector and axial-vector couplings of leptons and quarks to the weak neutral current*

Particle	g_V	g_A
ν	$\frac{1}{2}$	$\frac{1}{2}$
e	$-\frac{1}{2}+2\sin^2\theta_W$	$-\frac{1}{2}$
q (u-type)	$\frac{1}{2}-\frac{4}{3}\sin^2\theta_W$	$\frac{1}{2}$
q (d-type)	$-\frac{1}{2}+\frac{2}{3}\sin^2\theta_W$	$-\frac{1}{2}$

are non-zero. Now, taking $g_{e\mu}=g_{\mu e}$ and $h_{e\mu}=h_{\mu e}$, (5.180a) simplifies to

$$\mathscr{L}_{LH} = -\frac{(v+H)}{2^{1/2}}[g_{ee}\bar{e}e + g_{\mu\mu}\bar{\mu}\mu + g_{e\mu}(\bar{e}\mu+\bar{\mu}e) + h_{ee}\bar{\nu}_e\nu_e + h_{\mu\mu}\bar{\nu}_\mu\nu_\mu + h_{e\mu}(\bar{\nu}_e\nu_\mu+\bar{\nu}_\mu\nu_e)], \quad (5.180b)$$

Hence, the lepton mass term is (writing the weak interaction eigenstates as primed quantities)

$$\mathscr{L}_{mass} = -\bar{l}'M_l'l' - \bar{\nu}'M_\nu'\nu'; \qquad l'=\begin{pmatrix}e'\\ \mu'\end{pmatrix},\ \nu'=\begin{pmatrix}\nu_e'\\ \nu_\mu'\end{pmatrix}, \quad (5.181)$$

where $(M_l')_{ij}=(v/2^{1/2})g_{ij}$ and $(M_\nu')_{ij}=(v/2^{1/2})h_{ij}$ are 2×2 mass matrices. Now, as shown below in the discussion on quarks, M_l' can be diagonalised by the bi-unitary transformation $U_L^l M_l'(U_R^l)^\dagger=(M_l)_d$ (diagonal), and similarly for M_ν'. That is,

$$\begin{aligned}\mathscr{L}_{mass} &= -\bar{l}_L'(U_L^l)^\dagger(M_l)_d U_R^l l_R' - \bar{\nu}_L'(U_L^\nu)^\dagger(M_\nu)_d U_R^\nu\nu_R' + \text{h.c.}\\ &= -\bar{l}_L(M_l)_d l_R - \bar{\nu}_L(M_\nu)_d\nu_R,\end{aligned} \quad (5.182)$$

where $l_L=U_L^l l_L'$, $l_R=U_R^l l_R'$, etc. Thus, the weak eigenstates (e_L', μ_L') etc., giving currents diagonal in generations, are not necessarily the same as the mass eigenstates (e_L, μ_L), etc. Since the interactions must be expressed in terms of the observed (mass) eigenstates, non-diagonal intergeneration mixing terms can exist. This applies for the charged current case; for neutral currents, the interaction remains diagonal in the generations. However, for the standard model assignments $m_{\nu_e}=m_{\nu_\mu}=0$, the matrix U_L^ν (which is a rotation matrix specified by some angle θ_L^ν) can always be arranged so that the weak and mass eigenstates are the same, since the latter are degenerate. In this case we have diagonal mass matrices, and there are no transformations between lepton generations. Note that it is the nature of the gauge couplings which ensures conservation of the lepton numbers L_e and L_μ (and also L_τ).

A similar treatment to that for leptons can be used to generate the quark masses. Quarks appear to occur in doublets, (u, d), (c, s), (t, b), with the members having charges $\frac{2}{3}$ and $-\frac{1}{3}$ respectively. However these mass eigenstates are not the weak eigenstates, which are, as we have seen, linear combinations of the mass eigenstates. Let us define the weak eigenstates to be, for each generation j, a doublet ($t_3 = \pm\frac{1}{2}$) and two singlets ($t_3 = 0$), as follows

$$Q_{\mathrm{L}}^{j\prime} = \begin{pmatrix} u_{\mathrm{L}}^{j\prime} \\ d_{\mathrm{L}}^{j\prime} \end{pmatrix}, \qquad u_{\mathrm{R}}^{j\prime}, \qquad d_{\mathrm{R}}^{j\prime}. \tag{5.183}$$

Following (5.180), the quark–Higgs interaction has the form

$$\begin{aligned} \mathscr{L}_{\mathrm{QH}} &= -\sum_{i,j} g_{ij}(\bar{Q}_{\mathrm{L}}^{i\prime}\phi d_{\mathrm{R}}^{j\prime} + \bar{d}_{\mathrm{R}}^{j\prime}\phi^{\dagger}Q_{\mathrm{L}}^{i\prime}) - \sum_{i,j} h_{ij}(\bar{Q}_{\mathrm{L}}^{i\prime}\tilde{\phi}u_{\mathrm{R}}^{j\prime} + \bar{u}_{\mathrm{R}}^{j\prime}\tilde{\phi}^{\dagger}Q_{\mathrm{L}}^{i\prime}) \\ &= -\frac{(v+H)}{2^{1/2}}\left[\sum_{i,j} g_{ij}(\bar{d}_{\mathrm{L}}^{i\prime}d_{\mathrm{R}}^{j\prime} + \bar{d}_{\mathrm{R}}^{j\prime}d_{\mathrm{L}}^{i\prime}) + \sum_{i,j} h_{ij}(\bar{u}_{\mathrm{L}}^{i\prime}u_{\mathrm{R}}^{j\prime} + \bar{u}_{\mathrm{R}}^{j\prime}u_{\mathrm{L}}^{i\prime})\right] \\ &= -\left(1 + \frac{H}{v}\right)\sum_{i,j}(\bar{d}_{\mathrm{L}}^{i\prime}M_{\mathrm{d}}^{\prime ij}d_{\mathrm{R}}^{j\prime} + \bar{u}_{\mathrm{L}}^{i\prime}M_{\mathrm{u}}^{\prime ij}u_{\mathrm{R}}^{j\prime} + \mathrm{h.c.}), \end{aligned} \tag{5.184}$$

where

$$M_{\mathrm{d}}^{\prime ij} = \frac{v}{2^{1/2}}g_{ij}, \qquad M_{\mathrm{u}}^{\prime ij} = \frac{v}{2^{1/2}}h_{ij},$$

and, in the second line of (5.184), the spontaneously broken forms for the Higgs doublet ϕ (5.159) and $\tilde{\phi}$ (5.178) have been used.

If there was no mixing between quark generations, then the mass matrices M_{d}' and M_{u}' would be diagonal, giving quark masses $m(\mathrm{d}^j) = M_{\mathrm{d}}^{\prime jj}$ and $m(\mathrm{u}^j) = M_{\mathrm{u}}^{\prime jj}$ respectively. In general, however, the physical and weak quark bases will be related by a unitary transformation of the following type

$$u_h^j = (U_h^u)_{jk}u_h^{k\prime}, \qquad d_h^j = (U_h^d)_{jk}d_h^{k\prime}, \tag{5.185}$$

where the subscript h specifies the helicity (L or R), and the quark states u^j and d^j represent (u, c, t, . . .,) and (d, s, b, . . .,) respectively. The unitary matrices U_h^u and U_h^d are of dimension $n \times n$, where n is the number of quark generations. In the physical basis we require a Lagrangian term of the type, for three generations,

$$\begin{aligned} \mathscr{L}_{\mathrm{QH}} &= -\left(1 + \frac{H}{v}\right)[m_{\mathrm{d}}\bar{d}d + m_{\mathrm{s}}\bar{s}s + m_{\mathrm{b}}\bar{b}b + m_{\mathrm{u}}\bar{u}u + m_{\mathrm{c}}\bar{c}c + m_{\mathrm{t}}\bar{t}t] \\ &= -\left(1 + \frac{H}{v}\right)\left[\sum_j (m_{\mathrm{d}}^j\bar{d}^jd^j + m_{\mathrm{u}}^j\bar{u}^ju^j)\right]. \end{aligned} \tag{5.186}$$

This form can be obtained from (5.184), provided

$$U^{\mathrm{d}}_{\mathrm{L}} M'_{\mathrm{d}} (U^{\mathrm{d}}_{\mathrm{R}})^{\dagger} = \sum_{i,j} \delta_{ij} m^j_{\mathrm{d}}, \qquad U^{\mathrm{u}}_{\mathrm{L}} M'_{\mathrm{u}} (U^{\mathrm{u}}_{\mathrm{R}})^{\dagger} = \sum_{i,j} \delta_{ij} m^j_{\mathrm{u}}, \tag{5.187}$$

that is, the Us diagonalise the mass matrix.

To find the total Lagrangian for the standard model ((cf. (5.169)) the various terms involving quarks must be included. The interaction term for the quarks in the weak interaction basis is, following (5.106),

$$\mathscr{L}_{\mathrm{Q}} = \sum_j \mathrm{i} \bar{Q}^{j\prime}_{\mathrm{L}} \gamma_\mu \left[\partial^\mu + \mathrm{i} g \frac{\tau^{\mathrm{i}}}{2} W^\mu_{\mathrm{i}} + \mathrm{i} \frac{g'}{2} y_{\mathrm{L}} B^\mu \right] Q^{j\prime}_{\mathrm{L}}$$
$$+ \sum_j \mathrm{i} \bar{u}^{j\prime}_{\mathrm{R}} \gamma_\mu \left[\partial^\mu + \mathrm{i} \frac{g'}{2} y_{\mathrm{R}} B^\mu \right] u^{j\prime}_{\mathrm{R}} + \sum_j \mathrm{i} \bar{d}^{j\prime}_{\mathrm{R}} \gamma_\mu \left[\partial^\mu + \mathrm{i} \frac{g'}{2} y_{\mathrm{R}} B^\mu \right] d^{j\prime}_{\mathrm{R}}. \tag{5.188}$$

This can be transformed into the physical basis using (5.185) and the properties $(U^{\mathrm{u}}_{\mathrm{L}})^{\dagger} U^{\mathrm{u}}_{\mathrm{L}} = (U^{\mathrm{d}}_{\mathrm{R}})^{\dagger} U^{\mathrm{d}}_{\mathrm{R}} = 1$, etc. One can easily see that all the terms in (5.188), except the $\tau^1 W^\mu_1 + \tau^2 W^\mu_2$ contribution, take the same form in the physical basis. In particular, the neutral current contribution is diagonal in quark flavour. This is an extension of the GIM mechanism (Section 5.2), and is a consequence of the unitary nature of the transformation (5.185). The classification scheme for quarks is thus compatible with the experimental results that, if they exist, flavour changing neutral currents are heavily suppressed.

The charged current term in (5.188) is not flavour diagonal, and so is not invariant in form under (5.185). This is because $U^{\mathrm{u}}_{\mathrm{L}} (U^{\mathrm{d}}_{\mathrm{L}})^{\dagger}$ is not necessarily equal to unity. The transformation to the physical basis is as follows:

$$\mathscr{L}_{\mathrm{CC}} = -\frac{g}{2^{1/2}} \sum_j \bar{u}^{j\prime}_{\mathrm{L}} \gamma_\mu d^{j\prime}_{\mathrm{L}} W^\mu_+ + \mathrm{h.c.}$$
$$= -\frac{g}{2^{1/2}} \sum_j [\bar{u}^k_{\mathrm{L}} \gamma_\mu (U^{\mathrm{u}}_{\mathrm{L}})_{kj} (U^{\mathrm{d}}_{\mathrm{L}})^{\dagger}_{jm} d^m_{\mathrm{L}}] W^\mu_+ + \mathrm{h.c.}$$
$$= -\frac{g}{2^{1/2}} \sum_j [\bar{u}^k_{\mathrm{L}} \gamma_\mu V_{km} d^m_{\mathrm{L}}] W^\mu_+ + \mathrm{h.c.}, \tag{5.189}$$

where V is the unitary matrix

$$V = U^{\mathrm{u}}_{\mathrm{L}} (U^{\mathrm{d}}_{\mathrm{L}})^{\dagger}. \tag{5.190}$$

The matrix V is specified by n^2 parameters. However, we have freedom in choosing the phases of the $2n$ quark fields, and we can thus remove

$2n-1$ parameters by suitable choice of their relative phases. This leaves $(n-1)^2$ parameters to be determined empirically. For $n=2$ we have one parameter, the Cabibbo angle, giving

$$\mathscr{L}_{\mathrm{CC}} = -\frac{g}{2^{1/2}} W_{+}^{\mu} (\bar{u}_{\mathrm{L}} \bar{c}_{\mathrm{L}}) \gamma_{\mu} \begin{pmatrix} \cos\theta_{\mathrm{C}} & \sin\theta_{\mathrm{C}} \\ -\sin\theta_{\mathrm{C}} & \cos\theta_{\mathrm{C}} \end{pmatrix} \begin{pmatrix} d_{\mathrm{L}} \\ s_{\mathrm{L}} \end{pmatrix}. \tag{5.191}$$

The coupling constants are all real, and the Lagrangian is CP conserving. For three generations the charged current (CC) part of the Lagrangian is

$$\mathscr{L}_{\mathrm{CC}} = -\frac{g}{2^{1/2}} W_{+}^{\mu} (\bar{u}_{\mathrm{L}}\, \bar{c}_{\mathrm{L}}\, \bar{t}_{\mathrm{L}}) \gamma_{\mu} \begin{pmatrix} V_{\mathrm{ud}} & V_{\mathrm{us}} & V_{\mathrm{ub}} \\ V_{\mathrm{cd}} & V_{\mathrm{cs}} & V_{\mathrm{cb}} \\ V_{\mathrm{td}} & V_{\mathrm{ts}} & V_{\mathrm{tb}} \end{pmatrix} \begin{pmatrix} d_{\mathrm{L}} \\ s_{\mathrm{L}} \\ b_{\mathrm{L}} \end{pmatrix} + \text{h.c.} \tag{5.192}$$

The matrix V has four parameters; three rotation angles θ_i ($i=1, 2, 3$), and one phase δ. This matrix can be written in the form (Kobayashi and Maskawa (KM), 1973)

$$V = \begin{pmatrix} c_1 & s_1 c_3 & s_1 s_3 \\ -s_1 c_2 & c_1 c_2 c_3 - s_2 s_3 \exp(\mathrm{i}\delta) & c_1 c_2 s_3 + s_2 c_3 \exp(\mathrm{i}\delta) \\ s_1 s_2 & -c_1 s_2 c_3 - c_2 s_3 \exp(\mathrm{i}\delta) & -c_1 s_2 s_3 + c_2 c_3 \exp(\mathrm{i}\delta) \end{pmatrix}, \tag{5.193}$$

where $c_i = \cos\theta_i$ and $s_i = \sin\theta_i$. CP invariance implies $V = V^*$, thus the phase δ is CP violating. Hence CP violation can be described by the standard model with three generations (see Chapter 9). Note that there is no CP violation in either the electromagnetic or weak neutral currents.

The KM matrix (5.193) can be achieved by the following product of three rotational and one phase matrices (see Jarlskog, 1979)

$$V = \begin{pmatrix} 1 & 0 & 0 \\ 0 & c_2 & s_2 \\ 0 & -s_2 & c_2 \end{pmatrix} \begin{pmatrix} c_1 & s_1 & 0 \\ -s_1 & c_1 & 0 \\ 0 & 0 & 1 \end{pmatrix} \begin{pmatrix} 1 & 0 & 0 \\ 0 & 1 & 0 \\ 0 & 0 & \exp(\mathrm{i}\delta) \end{pmatrix} \begin{pmatrix} 1 & 0 & 0 \\ 0 & c_3 & s_3 \\ 0 & -s_3 & c_3 \end{pmatrix}. \tag{5.194}$$

The determinations of the values of the KM matrix coefficients are discussed in Chapter 8. Other parameterisations of V exist; see, for example, Maiani (1977), Wolfenstein (1983) and Harari and Nir (1987).

The Lagrangian $\mathscr{L}_{\mathrm{LH}}$ for three generations of leptons can be formulated in a similar way to that for quarks; allowing, in general, for transitions between generations. The resulting consequences, including the possibility of neutrino oscillations, are discussed in Chapter 9. We assume here that the weak and physical bases of the leptons are the same (this freedom exists provided the neutrinos are massless, as discussed earlier), so that $\mathscr{L}_{LH}$ becomes (see (5.180b))

$$\mathscr{L}_{LH} = -\left(1+\frac{H}{v}\right)[m_e\bar{e}e + m_\mu\bar{\mu}\mu + m_\tau\bar{\tau}\tau + m_{\nu_e}\bar{\nu}_e\nu_e + m_{\nu_\mu}\bar{\nu}_\mu\nu_\mu + m_{\nu_\tau}\bar{\nu}_\tau\nu_\tau]. \tag{5.195}$$

Note that each mass term in (5.195) (and also in (5.180b)) is arbitrary, and thus represents an additional parameter in the standard model. These mass terms, and also those of the $W^\pm$ and Z^0, all stem from the interactions with the Higgs field. Since the Higgs coupling to a fermion pair is m_f/v ($v = 246$ GeV), the Higgs will decay preferentially to the heaviest $f\bar{f}$ pair with $2m_f < m_H$. The Higgs also couples to W^+W^- and Z^0Z^0, so that these decay modes are also expected, provided m_H is large enough (see Chapter 10).

The standard model contains a considerable number of parameters. These are the coupling constants of the SU(2) and U(1) groups g and g' (or, alternatively, e and $\sin^2\theta_W$), the mass of the Higgs scalar m_H and the constant v related to the vacuum expectation value of the Higgs field. The gauge boson masses can be calculated from these parameters. Of course, the exact choice as to which of these quantities constitute the parameters of the model is somewhat arbitrary. In addition to these four parameters there are, for n quark generations, $2n$ quark masses and $(n-1)^2$ mixing angles and phases; i.e. n^2+1 parameters in all. A similar number appear for n lepton generations if the neutrino masses are taken to be non-zero. Thus we have, in total, $2(n^2+1)+4 = 2n^2+6$ parameters. For $n = 3$, this means that we have a total of 24 free parameters (or 17 if we assume massless neutrinos). Of these only two (g and g') are not associated with the Higgs field. Hence the introduction of a fundamental scalar solves the mass generation problem only at the expense of introducing many arbitrary parameters.

5.5.4 Feynman rules for the GSW model

To lowest order in the perturbation theory, the S-matrix element is related to the interaction Lagrangian through equation (4.124). The first term is

$$S_{fi} = i\int d^4x\,\mathscr{L}_I(x). \tag{5.196}$$

The interaction Lagrangian for the GSW model is contained in (5.169), together with the similar additional terms for the quark transitions as described in the previous section. The QED part of the Lagrangian leads to the Feynman rules 'derived' in Chapter 4 and summarised in Appendix (C.2). The approach sketched below gives results which can be obtained by more formal and rigorous field theory arguments.

The Feynman rules for the standard electroweak theory are given in Appendix (C.3). The factors for external lines and propagators have already been discussed. We will restrict the discussion to a few comments on the remaining terms, the vertex factors. For the charged current (f_1f_2W) and neutral current (ffZ) vertices, the rules can essentially be 'read off' from the Lagrangian (5.169) (and including the factor i from equation (5.196)). The trilinear couplings $W_+W_-\gamma$ and W_+W_-Z both involve derivatives of the W_+ and W_- fields (see (5.169)). Both these terms have the same form except for the overall factor C, which is e for $W_+W_-\gamma$ and $g\cos\theta_W = e\cot\theta_W$ for W_+W_-Z. The derivative terms give contributions of the form $(k_1)_\mu(W_+)_\nu$, etc. Using these factors, and writing the overall contribution in the form $W_+^\nu W_-^\lambda A^\mu$, the vertex factors given in (C.3.3) can be easily obtained. The terms quadrilinear in the gauge bosons in (5.169) (e.g. $W_{+\mu}W_-^\mu A_\nu A^\nu$) can be written as a sum of terms of the type $A^\alpha A^\beta W_+^\mu W_-^\nu$ using the metric tensor, and, including the overall numerical factors, the vertex factors (C.3.4) are obtained. Note that, in deriving these factors, all possible permutations of the fields should be considered. The trilinear and quadrilinear terms containing the Higgs field are treated in a similar way, and yield the vertex factors given in (C.3.5) and (C.3.6) respectively. Again, note that all permutations of the boson fields must be considered. Further we note that we have chosen the unitary gauge in order to make the Higgs sector simple. In other gauges there appear Faddeev–Popov 'ghosts', corresponding to the remaining unphysical scalar particles.

Finally, we note that the discussion in this chapter has been confined to leading order in the perturbation expansion. The role of higher orders, in particular their effects on the constants of the theory, is discussed in Chapter 10.

6

Purely leptonic interactions

The simplest tests of the standard electroweak model come from processes involving only leptons. This is because the leptons appear, at least at present energy scales, to be point-like objects, so that precise theoretical calculations can be made. Experimentally, purely leptonic processes can very often be extracted cleanly. On the other hand, processes involving quarks are more difficult to calculate and, hence, to use to test the theory. This stems from the necessarily indirect way in which the underlying quark properties must be inferred from those of the observed hadrons. Hence, we start with a discussion of muon decay and related processes (Section 6.1), neutrino and antineutrino scattering off electrons (Section 6.2) and, finally, $e^+e^- \to l^+l^-$ $(l = e, \mu)$ in Section 6.3.

6.1 MUON DECAY AND RELATED PROCESSES

The general form for the three-body purely leptonic decay of a spin $\frac{1}{2}$ lepton l_a is

$$l_a^-(p_1, s_a) \to \bar{\nu}_b(p_2) + \nu_a(p_3) + l_b^-(p_4, s_b), \tag{6.1}$$

where (l_a, ν_a) and (l_b, ν_b) belong to lepton generations a and b respectively. The quantities s_a and s_b are the four-polarisation (spin) vectors for l_a and l_b respectively and four-momentum conservation for the decay requires that $p_1 = p_2 + p_3 + p_4$. We assume, for the purpose of this calculation, that ν_a and ν_b are massless and are hence in definite helicity states. In the standard electroweak model this decay takes place, to lowest order, by the diagram shown in Fig. 6.1. Using the Feynman rules given in Appendix (C.3) we obtain the matrix element

$$\mathcal{M}_{fi} = \frac{-ig^2}{8}[\bar{u}_3\gamma_\mu(1-\gamma^5)u_1]\left[\frac{-g^{\mu\nu} + q^\mu q^\nu/M_W^2}{q^2 - M_W^2}\right][\bar{u}_4\gamma_\nu(1-\gamma^5)v_2] \tag{6.2}$$

where $q = p_1 - p_3 = p_2 + p_4$ is the four-momentum of the $W^\pm$ propagator.

Comparison of this matrix element with a typical electromagnetic matrix element (e.g. (4.21)) shows that in the weak case there are axial vector as well as vector current terms and that the propagator is massive, with a term in $q^\mu q^\nu$ arising from the longitudinal polarisation modes of the $W^\pm$. The Dirac equation (3.107) can be used to simplify the second term in the propagator in (6.2), giving

$$\mathscr{M}_{\rm fi} = \frac{ig^2}{8(q^2 - M_{\rm W}^2)}\left[\bar{u}_3\gamma_\mu(1-\gamma^5)u_1\bar{u}_4\gamma^\mu(1-\gamma^5)v_2 - \frac{m_a m_b}{M_{\rm W}^2}\bar{u}_3(1+\gamma^5)u_1\bar{u}_4(1-\gamma^5)v_2\right], \tag{6.3}$$

where $m_{\rm a}$ and $m_{\rm b}$ are the masses of leptons $l_{\rm a}$ and $l_{\rm b}$ respectively. For the decays of the known leptons (i.e. μ, τ), the maximum value of $q^2(\sim m_{\rm a}^2)$ is small compared to $M_{\rm W}^2$. In the following we neglect the q^2 term and also the term in $m_{\rm a}m_{\rm b}/M_{\rm W}^2$. With these simplifications the matrix element $\mathscr{M}_{\rm fi}$, and its conjugate $\mathscr{M}_{\rm fi}^*$, become

$$\mathscr{M}_{\rm fi} = \frac{-ig^2}{8M_{\rm W}^2}[\bar{u}_3\gamma_\mu(1-\gamma^5)u_1][\bar{u}_4\gamma^\mu(1-\gamma^5)v_2], \tag{6.4a}$$

$$\mathscr{M}_{\rm fi}^* = \frac{ig^2}{8M_{\rm W}^2}[\bar{u}_1\gamma_\nu(1-\gamma^5)u_3][\bar{v}_2\gamma^\nu(1-\gamma^5)u_4], \tag{6.4b}$$

where the properties $(ab)^\dagger = b^\dagger a^\dagger$, $(\gamma^\mu)^\dagger = \gamma^0\gamma^\mu\gamma^0$ and $(\gamma^5)^\dagger = \gamma^5$ have been used to obtain $\mathscr{M}_{\rm fi}^*$.

The differential decay rate $d\omega$ is given by

$$d\omega = |S_{\rm fi}|^2 N\Phi/T$$

$$= (2\pi)^4 \frac{\delta^4(p_{\rm i} - p_{\rm f})|\mathscr{M}_{\rm fi}|^2\, {\rm d}^3p_2\, {\rm d}^3p_3\, {\rm d}^3p_4}{2E_1(2\pi)^9 8E_2E_3E_4}, \tag{6.5}$$

Fig. 6.1 Lowest order diagram for the purely leptonic process $l_{\rm a}^- \to l_{\rm b}^- \nu_{\rm a}\bar{\nu}_{\rm b}$.

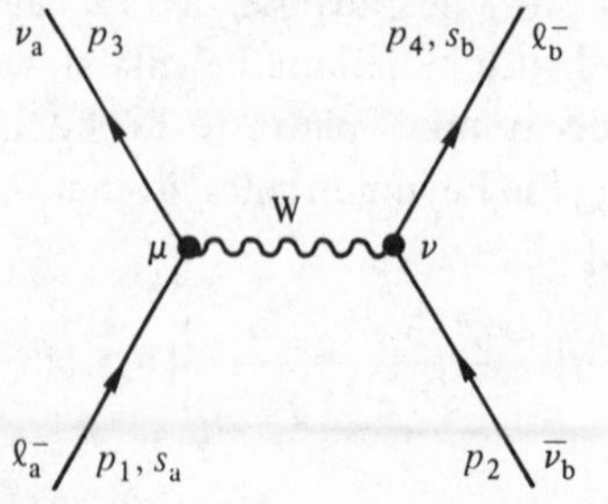

where N is a product of normalisation factors $1/(2E)$ for each particle, Φ is a product of phase space factors $V\,\mathrm{d}^3p/(2\pi)^3$ for each final state particle and the $\delta^4(p_\mathrm{i}-p_\mathrm{f})$ function arises from $|S_{\mathrm{fi}}|^2$, as discussed in Section 4.2. The generalised form of (6.5) for n final state particles is given by equation (3) in Appendix C.

Multiplying $\mathcal{M}_{\mathrm{fi}}$ and $\mathcal{M}^*_{\mathrm{fi}}$ in equation (6.4) gives

$$\begin{aligned}|\mathcal{M}_{\mathrm{fi}}|^2 &= \frac{g^4}{64M_\mathrm{W}^4}[\bar{u}_3\gamma_\mu(1-\gamma^5)u_1\bar{u}_1\gamma_\nu(1-\gamma^5)u_3] \\ &\quad\times[\bar{u}_4\gamma^\mu(1-\gamma^5)v_2\bar{v}_2\gamma^\nu(1-\gamma^5)u_4] \\ &= \frac{g^4}{64M_\mathrm{W}^4}L_{\mu\nu}M^{\mu\nu}. \end{aligned} \tag{6.6}$$

Note that quantities such as $\bar{u}_1\gamma_\nu(1-\gamma^5)u_3$ are 1×1 matrices (i.e. scalars), so that these terms may be written in any order. The evaluation of $L_{\mu\nu}$ and $M^{\mu\nu}$ is made using the same techniques as those used for electromagnetic interactions. We can write $L_{\mu\nu}$ as a trace as follows

$$\begin{aligned}L_{\mu\nu} &= \mathrm{tr}[(u_3\bar{u}_3)\gamma_\mu(1-\gamma^5)(u_1\bar{u}_1)\gamma_\nu(1-\gamma^5)] \\ &= \tfrac{1}{2}\,\mathrm{tr}[\not{p}_3\gamma_\mu(1-\gamma^5)(\not{p}_1+m_\mathrm{a})(1+\gamma^5\not{s}_\mathrm{a})\gamma_\nu(1-\gamma^5)],\end{aligned} \tag{6.7}$$

where, in the second line, the projection operators for the neutrino ν_a and the charged lepton l_a (see (3.120)) have been used.

Using the properties of the γ-matrices (Appendix B), we can evaluate $L_{\mu\nu}$ by the following steps

$$\begin{aligned}L_{\mu\nu} &= \tfrac{1}{2}\,\mathrm{tr}[\not{p}_3\gamma_\mu(1-\gamma^5)(\not{p}_1+m_\mathrm{a}\gamma^5\not{s}_\mathrm{a})\gamma_\nu(1-\gamma^5)] && \text{(from B.1)} \\ &= \mathrm{tr}[\not{p}_3\gamma_\mu(\not{p}_1+m_\mathrm{a}\gamma^5\not{s}_\mathrm{a})\gamma_\nu(1-\gamma^5)] && \text{(moving } (1-\gamma^5) \text{ to right)} \\ &= \mathrm{tr}[\not{p}_3\gamma_\mu(\not{p}_1-m_\mathrm{a}\not{s}_\mathrm{a})\gamma_\nu(1-\gamma^5)] && \text{(moving } \gamma^5 \text{ to right)} \\ &= p_3^\alpha(p_1-m_\mathrm{a}s_\mathrm{a})^\beta\,\mathrm{tr}[\gamma_\alpha\gamma_\mu\gamma_\beta\gamma_\nu(1-\gamma^5)].\end{aligned} \tag{6.8}$$

Similarly, we obtain

$$M^{\mu\nu} = (p_4-m_\mathrm{b}s_\mathrm{b})_\theta(p_2)_\phi\,\mathrm{tr}[\gamma^\theta\gamma^\mu\gamma^\phi\gamma^\nu(1-\gamma^5)], \tag{6.9}$$

and this result, together with (6.8) and using the property (B.6), gives

$$|\mathcal{M}_{\mathrm{fi}}|^2 = \frac{g^4}{M_\mathrm{W}^4}[p_3\cdot(p_4-m_\mathrm{b}s_\mathrm{b})p_2\cdot(p_1-m_\mathrm{a}s_\mathrm{a})], \tag{6.10}$$

with the four-vector dot products arising from the δ^α_θ and δ^β_ϕ factors. Substituting $|\mathcal{M}_{\mathrm{fi}}|^2$ from (6.10) into (6.5) gives the differential decay rate in terms of the three decay particle momenta.

The next step depends on the experimental configuration with which it is desired to compare the theoretical predictions. For example, if we desire to study the properties of the decay lepton l_b, then we must integrate over all possible values of the neutrino momenta compatible with four-momentum conservation. The variables involving the two neutrinos can be factored out of the differential decay rate as follows

$$\mathrm{d}\omega = \frac{g^4(p_4 - m_b s_b)^\alpha (p_1 - m_a s_a)^\beta}{16(2\pi)^5 M_W^4 E_1 E_4}\,\mathrm{d}^3 p_4 I_{\alpha\beta}, \tag{6.11}$$

where the covariant integral $I_{\alpha\beta}$ is defined as

$$I_{\alpha\beta} = \int \frac{\mathrm{d}^3 p_2\, \mathrm{d}^2 p_3\, p_{3\alpha} p_{2\beta}\, \delta^4(p - p_2 - p_3)}{E_2 E_3}. \tag{6.12}$$

The variable p in (6.12) is defined as $p = p_1 - p_4 = p_2 + p_3$.

In order to evaluate $I_{\alpha\beta}$ we note that it must, in general, have the form

$$I_{\alpha\beta} = g_{\alpha\beta} A(p^2) + p_\alpha p_\beta B(p^2). \tag{6.13}$$

Defining the integral I as follows

$$I = \int \frac{\mathrm{d}^3 p_2\, \mathrm{d}^3 p_3\, \delta^4(p - p_2 - p_3)}{E_2 E_3}, \tag{6.14}$$

then multiplying (6.13) by $g^{\alpha\beta}$ and $p^\alpha p^\beta$ successively gives

$$\begin{aligned} g^{\alpha\beta} I_{\alpha\beta} &= 4A + p^2 B = (p^2/2) I, \\ p^\alpha p^\beta I_{\alpha\beta} &= p^2 A + p^4 B = (p^4/4) I, \end{aligned} \tag{6.15}$$

where the relationship $p^2 = 2p_2 \cdot p_3$ has been used. Since the integral I is covariant, we choose to evaluate it in the rest frame of the two neutrinos, in which $|\mathbf{p}_2| = |\mathbf{p}_3| = E_2 = E_3$. Thus

$$I = \int \frac{\mathrm{d}^3 p_2\, \delta(E - 2E_2)}{E_2^2} = 4\pi \int \mathrm{d}E_2\, \delta(E - 2E_2) = 2\pi. \tag{6.16}$$

Hence, solving for A and B in (6.15) gives the result

$$I_{\alpha\beta} = \frac{\pi}{6}(g_{\alpha\beta} p^2 + 2p_\alpha p_\beta). \tag{6.17}$$

Using this result in (6.11) gives

$$\mathrm{d}\omega = \frac{g^4[(p_4 - m_b s_b)\cdot(p_1 - m_a s_a) p^2 + 2p\cdot(p_4 - m_b s_b) p\cdot(p_1 - m_a s_a)]\,\mathrm{d}^3 p_4}{192(2\pi)^4 M_W^4 E_1 E_4}. \tag{6.18}$$

For further evaluation we will use the rest frame of the decaying lepton l_a. In this frame, the four-momenta and polarisation vectors are as follows

$$p_1 = (m_a, \mathbf{0}), \qquad s_a = (0, \hat{\mathbf{s}}_a),$$
$$p_4 = (E_4, \mathbf{p}_4), \qquad s_b = \left(\frac{\mathbf{p}_4 \cdot \hat{\mathbf{s}}_b}{m_b}, \hat{\mathbf{s}}_b + \frac{(\mathbf{p}_4 \cdot \hat{\mathbf{s}}_b)\mathbf{p}_4}{m_b(E_4 + m_b)}\right), \tag{6.19}$$
$$p = p_1 - p_4 = (m_a - E_4, -\mathbf{p}_4), \qquad p^2 = m_a^2 + m_b^2 - 2m_a E_4.$$

The polarisation four-vector for l_b, in the particle's rest frame, is $(0, \hat{\mathbf{s}}_b)$. The form for s_b in (6.19) is obtained by a Lorentz boost to the frame in which the particle has momentum $\mathbf{p}_4$.

Using these forms, and writing $d^3p_4 = p_4^2\, dp_4\, d\Omega$, gives

$$d\omega = \frac{g^4 p_4\, dE_4\, d\Omega}{192(2\pi)^4 M_W^4 m_a} \Bigg\{ [m_a^2 + m_b^2 - 2m_a E_4] \Bigg[m_a(E_4 - \mathbf{p}_4 \cdot \hat{\mathbf{s}}_b) + m_a \left(\mathbf{p}_4 - m_b \hat{\mathbf{s}}_b - \frac{(\mathbf{p}_4 \cdot \hat{\mathbf{s}}_b)\mathbf{p}_4}{E_4 + m_b} \right) \cdot \hat{\mathbf{s}}_a \Bigg] + 2[m_a^2 - m_a E_4 - m_a \mathbf{p}_4 \cdot \hat{\mathbf{s}}_a] \times \Bigg[(m_a - E_4)(E_4 - \mathbf{p}_4 \cdot \hat{\mathbf{s}}_b) + \mathbf{p}_4 \cdot \left(\mathbf{p}_4 - m_b \hat{\mathbf{s}}_b - \frac{(\mathbf{p}_4 \cdot \hat{\mathbf{s}}_b)\mathbf{p}_4}{E_4 + m_b} \right) \Bigg] \Bigg\}. \tag{6.20}$$

This rather lengthy expression can be considerably simplified by neglecting the mass m_b, defining a unit vector $\hat{\mathbf{n}}$ along the direction of $\mathbf{p}_4$ and defining θ to be the angle between the spin direction $\hat{\mathbf{s}}_a$ and $\hat{\mathbf{n}}$ (i.e. $\cos\theta = \hat{\mathbf{s}}_a \cdot \hat{\mathbf{n}}$). Further, if $x = E_4/E_4^{max}$,where E_4^{max} is the maximum allowed value of $E_4(= m_a/2)$, then (6.20) simplifies to

$$d\omega = \left(\frac{g^4}{32 M_W^4}\right) \frac{m_a^5}{192\pi^3} [n(x)][1 + \alpha(x) \cos\theta] \left[\frac{1 - \hat{\mathbf{n}} \cdot \hat{\mathbf{s}}_b}{2}\right] \frac{dx\, d\cos\theta\, d\phi}{4\pi}, \tag{6.21}$$

where

$$n(x) = 2x^2(3 - 2x), \qquad \alpha(x) = (1 - 2x)/(3 - 2x). \tag{6.22}$$

The form of (6.21) is chosen so that each contributing factor, namely the l_b^- energy spectrum $n(x)$, the decay angular asymmetry $\alpha(x)$ of l_b with respect to $\hat{\mathbf{s}}_a$ and the helicity factor $[1 - \hat{\mathbf{n}} \cdot \hat{\mathbf{s}}_b]$ for l_b, gives unity when integrated or summed over. Note that the helicity factor corresponds to that expected for a lefthanded particle and, in particular, gives zero contribution when the spin of l_b is parallel to its direction of motion.

Figure 6.2 shows plots of $n(x)$ and $\alpha(x)$ as a function of x. It can be seen that the most probable value of the energy is the maximum value $m_a/2$, at which point $\alpha = -1$.

The decay $l_a^+ \to \nu_b \bar{\nu}_a l_b^+$ can be calculated in a similar way to that for l_a^-. The resulting expression for $|\mathscr{M}_{fi}|^2$ is the same as that for l_a^- (6.10), except that the signs of the m_a and m_b terms change, reflecting the changes in the helicity values of $h = -1$ for the l_a^- and l_b^- to $h = +1$ for l_a^+ and l_b^+. These lead to changes in the signs of the $\hat{\mathbf{n}} \cdot \hat{\mathbf{s}}_b$ and $\alpha(x)$ terms in the equivalent expression to (6.21) for the differential decay rate. In the configuration $x \simeq 1$, the two neutrinos are emitted, in the rest system of l_a, in the direction opposite to that of l_b. The spins of the two neutrinos (ν_a, $\bar{\nu}_b$ or $\bar{\nu}_a$, ν_b) cancel, so that the spin of l_b must be in the same direction as that of l_a. Since, in the limit $m_b \to 0$, the lepton l_b^- (l_b^+) has $h = -1(1)$, it will be emitted in the opposite (same) direction to the spin direction of l_a^- (l_a^+). This is shown in Fig. 6.3(a). Application of charge conjugation (C) to this configuration leads to that of 6.3(b). Note that the spins and momenta are unaltered by this transformation. Experimentally, it is found that the e^+ in $\mu^+ \to \nu_e \bar{\nu}_\mu e^+$ decay is emitted along the direction of the μ^+ spin, so that Fig. 6.3(b) does not correspond to reality. Now, application of a parity transformation (P) about a vertical axis in Fig. 6.3(b) leads to

Fig. 6.2 Decay distributions for $l_a^- \to \bar{\nu}_b \nu_a l_b^-$ in the standard model; $n(x)$, $n'(x)$ and $\alpha(x)$ are the energy distributions of l_b^- and $\bar{\nu}_b$ and the decay asymmetry of l_b^- respectively.

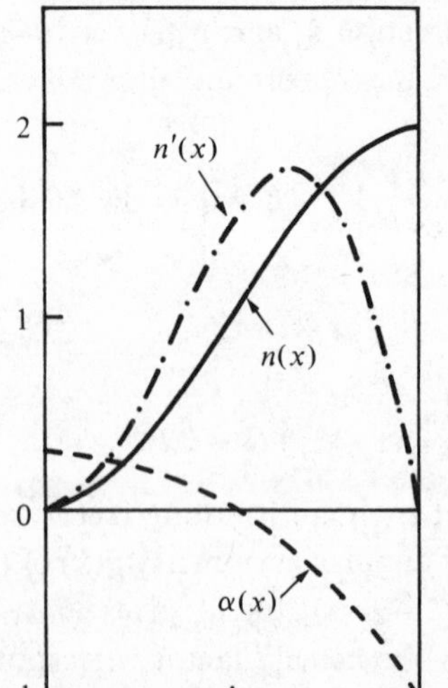

Fig. 6.3(c). The momenta change sign under P, but angular momentum vectors (i.e. $\mathbf{r} \times \mathbf{p}$) are invariant. Thus the combined application of CP gives the experimentally observed configuration. Both C and P are separately violated maximally.

In order to calculate the *total decay rate* we must sum over the spin states of l_b^-, average over the initial spin states s_a (factor $\frac{1}{2}$) and integrate over the solid angle $d\Omega$. This gives the differential energy spectrum for l_b^-, namely

$$\frac{d\omega}{dx} = \left(\frac{g^4}{32M_W^4}\right)\frac{m_a^5 n(x)}{192\pi^3}. \tag{6.23}$$

Finally, integrating over x, gives the total decay rate

$$\Gamma = \left(\frac{g^4}{32M_W^4}\right)\frac{m_a^5}{192\pi^3} = \frac{G_F^2 m_a^5}{192\pi^3}, \tag{6.24}$$

where the relationship (5.63) between G_F, g and M_W has been used.

If the terms in $\varepsilon = m_b/m_a$ are retained, then the total decay rate is found by integrating (6.20). The integration range is $m_b \leqslant E_4 \leqslant m_a(1+\varepsilon^2)/2$ and, after a short calculation in which a change of variables to hyperbolic functions is useful, it is found that

$$\Gamma = \frac{G_F^2 m_a^5}{192\pi^3} F(\varepsilon), \qquad F(\varepsilon) = 1 - 8\varepsilon^2 - 24\varepsilon^4 \ln \varepsilon + 8\varepsilon^6 - \varepsilon^8. \tag{6.25}$$

The decay distributions for ν_a and $\bar{\nu}_b$ can be found by going back to the expression for $d\omega$ (i.e. the equivalent of (6.11)) and integrating over the other two particles in each case, respectively. In the limit $m_b = 0$, we can see from (6.10) that the matrix element is the same for particles 3 (ν_a) and 4(l_b^-). Hence, the decay distributions of the neutrino ν_a are the same as those of the charged lepton l_b^-. Inspection of (6.10) also shows that the decay distributions of particle 2 ($\bar{\nu}_b$) will be different to the others. In evaluating the decay distributions for $\bar{\nu}_b$ we need to integrate over particles

Fig. 6.3 Decay configurations at $x = 1$ for (a) $l_a^- \to \bar{\nu}_b \nu_a l_b^-$, ($b$) after the application of charge conjugation to (a) and in (c) after a parity transformation of (b).

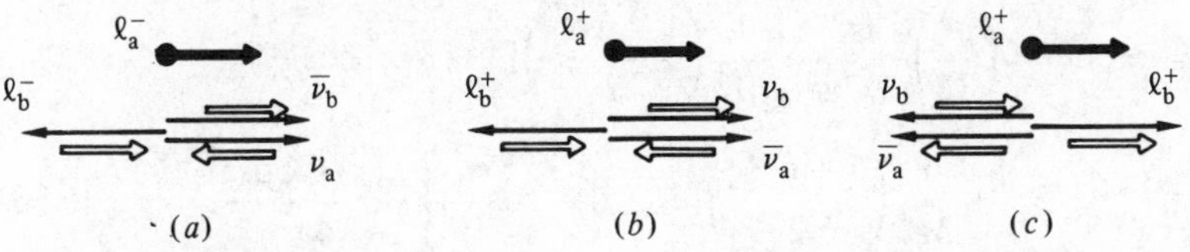

3 and 4, and this gives an integral of the type I in (6.16). The result is

$$d\omega = \left(\frac{g^4}{32M_W^4}\right)\frac{m_a^5}{192\pi^3}[n'(x)][1+\cos\theta]\frac{dx\, d\cos\theta\, d\phi}{4\pi}, \tag{6.26}$$

where

$$n'(x) = 12x^2(1-x). \tag{6.27}$$

The form of $n'(x)$ is shown in Fig. 6.2; the decay asymmetry $\alpha'(x)=1$. Note that, for the massless neutrinos, the projection operator used is that summed over spin states. In the limit $m_b = 0$, the lepton l_b^- can, of course, be treated in the same way. Conversely, a mass can be given to one or both of the neutrinos, if so desired, in the calculation.

6.1.1 More general form of the matrix element

An obvious question about the standard model is why nature has selected lefthanded (V – A) interactions, rather than (V + A) interactions. One possibility is that parity is conserved at small distance scales, say $(10^3\ \text{GeV})^{-1}$, but the vector bosons mediating the (V + A) interaction are more massive than their (V – A) counterparts. Hence, at low values of q^2 the effective Lagrangian appears to be lefthanded; however, the left–right symmetry becomes apparent at large q^2. The relevant part of the charged current Lagrangian for this model is

$$\mathscr{L} = \frac{-g}{2(2)^{1/2}}[(\text{V}-\text{A})_\mu W_\text{L}^\mu + (\text{V}+\text{A})_\mu W_\text{R}^\mu] + \text{hc}. \tag{6.28}$$

The corresponding matrix element for the decay $l_a^- \to \bar{\nu}_b \nu_a l_b^-$ (cf. (6.4a)) is

$$\mathscr{M}_{fi} = \frac{-ig^2}{8M_L^2}[\bar{u}_3\gamma_\mu(1-\gamma^5)u_1][\bar{u}_4\gamma^\mu(1-\gamma^5)v_2]$$
$$-\frac{ig^2}{8M_R^2}[\bar{u}_3\gamma_\mu(1+\gamma^5)u_1][\bar{u}_4\gamma^\mu(1+\gamma^5)v_2]. \tag{6.29}$$

The first term arises from the V – A interaction and the second from V + A.

The evaluation of the l_b^- differential decay rate gives, using the methods described above, and in the limit $m_b = 0$,

$$d\omega = \left(\frac{g^4}{32}\right)\frac{m_a^5}{192\pi^3}\left\{\frac{1}{M_L^4}\left(\frac{1-\hat{\mathbf{n}}\cdot\hat{\mathbf{s}}_b}{2}\right)[1+\alpha(x)\cos\theta]\right.$$
$$\left.+\frac{1}{M_R^4}\left(\frac{1+\hat{\mathbf{n}}\cdot\hat{\mathbf{s}}_b}{2}\right)[1-\alpha(x)\cos\theta]\right\}\frac{n(x)\, dx\, d\Omega}{4\pi}, \tag{6.30}$$

where $n(x)$ and $\alpha(x)$ are given by (6.22). Note that, as expected, the elicity of the lepton l_b^- is $h = -1(+1)$ for the left- and righthanded W exchange respectively. The interference term $(\propto M_L^{-2} M_R^{-2})$ is zero because $(1-\gamma^5)(1+\gamma^5) = 0$.

In general, couplings other than V and A can exist. Allowing for $i =$ S, V, T, A and P couplings (see Section 5.1.2), the matrix element for the decay $l_a^- \to \bar{\nu}_b \nu_a l_b^-$ can be written (Bouchiat and Michel, 1957; Kinoshita and Sirlin, 1957; Sachs and Sirlin, 1975)

$$\mathcal{M}_{fi} = \sum_i [\bar{u}_3 \Gamma_i u_1][\bar{u}_4 \Gamma_i (C_i + C_i' \gamma^5) v_2], \tag{6.31}$$

where a current × current form for the Lagrangian is assumed (i.e. no account is made of the possible corresponding propagating particles) and Γ_i are given by (3.139). The currents $1 \to 3$ and $\bar{2} \to 4$, described by (6.31), both change the charge by one unit and this form is known as the *charge exchange order*. Practical calculations are somewhat simpler if the matrix element is rewritten in the so-called *charge retention order* $1 \to 4$, $\bar{2} \to 3$

$$\mathcal{M}_{fi} = \sum_j [\bar{u}_4 \Gamma_j u_1][\bar{u}_3 \Gamma_j (g_j + g_j' \gamma^5) v_2]. \tag{6.32}$$

These coefficients are linearly related to those of (6.31) by a *Fierz transformation* (Fierz, 1937); see Appendix E. Pure V − A or V + A interactions have the same form in both bases.

The differential decay rate, for $m_b = 0$, has the general form

$$d\omega = \frac{A}{4} \frac{m_a^5 x^2 \, dx}{192\pi^3} \frac{d\Omega}{4\pi}$$

$$\times \{6(1-x) + 4\rho(\tfrac{4}{3}x - 1) - \xi \cos\theta [2(1-x) + 4\delta(\tfrac{4}{3}x - 1)]\}. \tag{6.33}$$

The parameters ρ, δ and ξ are called *Michel parameters* and are defined as follows

$$A\rho = 3b + 6c,$$

$$A\xi = -3a' - 4b' + 14c',$$

$$\delta = (3b' - 6c')/(3a' + 4b' - 14c'), \tag{6.34}$$

where

$$A = a + 4b + 6c,$$

$$a = |g_S|^2 + |g_S'|^2 + |g_P|^2 + |g_P'|^2, \qquad a' = 2\,\mathrm{Re}(g_S g_P'^* + g_P g_S'^*),$$

$$b = |g_V|^2 + |g_V'|^2 + |g_A|^2 + |g_A'|^2, \qquad b' = -2\,\mathrm{Re}(g_V g_A'^* + g_A g_V'^*),$$

$$c = |g_T|^2 + |g_T'|^2, \qquad c' = 2\,\mathrm{Re}(g_T g_T'^*). \tag{6.35}$$

The helicity of the lepton l_b^- is given by

$$h = (a' + 4b' + 6c')/A. \tag{6.36}$$

The above formulae are for the decay of a completely polarised lepton l_a^-. If l_a^- has polarisation P_a, then ξ is replaced by ξP_a in (6.33). The coupling constants are relatively real if time invariance holds. Note that $\rho = \frac{3}{4}$ for V, A, $\rho = 1$ for T, and $\rho = 0$ for S, P interactions. Further, the prediction $\rho = \frac{3}{4}$, and also $\delta = \frac{3}{4}$, is independent of the relative admixture of V and A. However, the parameter $\xi (= -b'/b)$ is sensitive to the relative admixture of V and A, and has the value $\xi = 1(-1)$ for pure V – A(V + A).

If we consider V and A terms only in the charge exchange order (6.31), then the differential decay rate as a function of the energy fraction x of l_b^-, and in the limit $m_b = 0$, is

$$d\omega = \frac{(X+Y)m_a^5 x^2\,dx}{192\pi^3}\left[\frac{X}{(X+Y)}(3-2x) + \frac{6Y}{(X+Y)}(1-x)\right], \tag{6.37}$$

where

$$X = |C_V|^2 + |C'_V|^2 + |C_A|^2 + |C'_A|^2 + 2\,\mathrm{Re}(C_V C_A^* + C'_V C_A'^*),$$
$$Y = |C_V|^2 + |C'_V|^2 + |C_A|^2 + |C'_A|^2 - 2\,\mathrm{Re}(C_V C_A^* + C'_V C_A'^*). \tag{6.38}$$

Comparison with (6.33) shows that we can identify

$$\rho = \frac{3X}{4(X+Y)}. \tag{6.39}$$

Note that pure V or A coupling ($X = Y$) gives $\rho = \frac{3}{8}$, whereas pure V – A ($C_V = C_A = 1$, $C'_V = C'_A = -1$) and pure V + A ($C_V = C_A = C'_V = C'_A = 1$) both give $\rho = \frac{3}{4}$.

If the more restrictive assumption is made that the $\bar{\nu}_b - l_b^-$ current has a pure V – A form, but that the $l_a^- - \nu_a$ current has an, *a priori*, unknown mixture of vector and axial vector terms, then the matrix element can be written

$$\mathscr{M}_{fi} = \bar{u}_3 \gamma_\mu (f_V + f_A \gamma^5) u_1 \bar{u}_4 \gamma^\mu (1 - \gamma^5) v_2. \tag{6.40}$$

The equivalence with the form (6.31) is that $C_V = -C'_V = f_V$ and $C'_A = -C_A = f_A$. For pure V – A ($f_A = -f_V$) this gives the result already derived above that $\rho = \frac{3}{4}$. However, a pure V + A current for $l_a^- - \nu_a$ ($f_A = f_V$) leads to $\rho = 0$.

Comparison of the general result (6.33) with that derived for the manifestly left–right symmetric model (6.30) shows that these forms are equivalent provided $\rho = \delta = \frac{3}{4}$. The parameter ξ is given by $\xi \simeq 1 -$

$2(M_L/M_R)^4$, assuming M_R is more massive than M_L. The left–right model considered above, however, is a somewhat simplified version. In general (Beg *et al.*, 1977), the mass eigenstates W_1 and W_2 are linear combinations of the weak eigenstates as follows

$$W_1 = W_L \cos\chi - W_R \sin\chi,$$
$$W_2 = W_L \sin\chi + W_R \cos\chi. \qquad (6.41)$$

These authors note that the Higgs potential can give rise to masses for both Ws as well as giving an antisymmetric vacuum.

The effective interaction Lagrangian at low q^2 can be found by inserting W_L and W_R from (6.41) into the Lagrangian (6.28). This gives

$$\mathscr{L}_{\text{eff}} = -\frac{G_{LR}}{2^{1/2}}[VV^\dagger + \eta_{AA}AA^\dagger + \eta_{AV}(VA^\dagger + AV^\dagger)], \qquad (6.42)$$

where

$$\frac{G_{LR}}{2^{1/2}} = \frac{g^2}{8M_1^2}(\cos\chi - \sin\chi)^2 + \frac{g^2}{8M_2^2}(\cos\chi + \sin\chi)^2,$$
$$\eta_{AA} = (\varepsilon^2 M_2^2 + M_1^2)/(\varepsilon^2 M_1^2 + M_2^2),$$
$$\eta_{AV} = -\varepsilon(M_2^2 - M_1^2)/(\varepsilon^2 M_1^2 + M_2^2)$$
$$\varepsilon = (1 + \tan\chi)/(1 - \tan\chi). \qquad (6.43)$$

The Michel parameters are $\delta = \frac{3}{4}$ and

$$\rho = \frac{3}{8}\frac{[(1+\eta_{AA})^2 + 4\eta_{AV}^2]}{1 + \eta_{AA}^2 + 2\eta_{AV}^2}, \qquad \xi = \frac{-2\eta_{AV}(1+\eta_{AA})}{1 + \eta_{AA}^2 + 2\eta_{AV}^2}. \qquad (6.44)$$

The $V - A$ limit is $M_2^2/M_1^2 \to \infty$ and $\chi \to 0$.

6.1.2 Results on muon decays

The total muon decay rate gives an accurate determination of the weak coupling constant G_μ, and hence a relationship between g and M_W from (5.63). In the standard model, taking into account first-order radiative corrections, the decay rate is (Sirlin, 1975)

$$\Gamma_\mu = \frac{1}{\tau_\mu} = \frac{G_\mu^2 m_\mu^5}{192\pi^3}\left[1 - \frac{8m_e^2}{m_\mu^2}\right]\left[1 + \frac{\alpha}{2\pi}\left(\frac{25}{4} - \pi^2\right) + \frac{3}{5}\frac{m_\mu^2}{M_W^2}\right]. \qquad (6.45)$$

The correction term to the muon lifetime from the W-boson, and also that to the Michel ρ parameter ($\rho = \frac{3}{4} + (m_\mu/M_W)^2/3$), are very small ($\lesssim 10^{-6}$) and are beyond the present experimental precision.

The μ^+ lifetime has been accurately measured (see Particle Data Group)

Table 6.1. *Experimental values for the Michel parameters of the muon*

Parameter	Value	Reference
ρ	0.7503 ± 0.0026	Peoples, 1966
δ	0.752 ± 0.009	Fryberger, 1968
ξ	$\geqslant 0.9959$ (90% c.l.)	Carr *et al.*, 1983

and, including $O(\alpha^2)$ corrections, Sirlin (1984) gives

$$G_F = G_\mu = (1.16634 \pm 0.00002) \times 10^{-5}\ \mathrm{GeV}^{-2}. \tag{6.46}$$

The ratio of the μ^+ and μ^- lifetimes is 1.000029 ± 0.000078, compatible with the value of unity expected from the *CPT* theorem.

The experimental values for the Michel parameters (6.34) are shown in Table 6.1. The e^+ energy spectrum from the experiment of Bardon *et al.* (1965) is shown in Fig. 6.4. The current value for ρ is very close to the V − A/standard model prediction of $\rho = \frac{3}{4}$. This has not always been the case, as can be seen from Fig. 6.5, which shows the time dependence of the measurements. Each result is essentially compatible with the previous one, but a clear trend towards the now expected result is seen!

The measured values of ρ and δ clearly exclude pure T or pure S/P coupling; however, any admixture of V and A fits the results. This can be seen by comparing the V − A and V + A parts of (6.30) (or from the Beg *et al.* (1977) model, provided $\chi = 0$). The terms $n(x)$ and $\alpha(x)$ are the same in both cases; the differences arise in the sign of the $\cos\theta$ term (corresponding to ξ), and of the helicity term. Thus, an accurate measurement of ξ (or h_e) is sensitive to possible V + A components. Such a measurement has been performed by Carr *et al.* (1983), who measured the ratio of the e^+ spectrum at the end-point from polarised and unpolarised μ^+ decays. At the end-point ($x \simeq 1$) and at $\theta \simeq \pi$, the ratio of the decay rates for muons with polarisation P_μ to unpolarised muons is, from (6.33), $R \simeq 1 + P_\mu \cos\theta\xi\,\delta/\rho \simeq 1 - P_\mu\xi$, if $\delta = \rho$. The muons are produced by pion decay, and the polarisation of the muon will also depend on the presence of any righthanded currents in this decay. For the model of Beg *et al.*, $P_\mu \simeq 1 - 2(\alpha + \chi)^2$, where $\alpha = M_1^2/M_2^2$. This results in a ratio $R \simeq 2(2\alpha^2 + 2\alpha\chi + \chi^2)$. No evidence for righthanded currents is found experimentally, and Carr *et al.* give a limit

$$\xi P_\mu\,\delta/\rho > 0.9959 \qquad 90\%\ \text{c.l.}, \tag{6.47}$$

which can be converted into limits on both α and χ. For no mixing ($\chi = 0$),

$4\alpha^2 < 0.0041$ (90% c.l.), hence $M_2 > 450$ GeV (90% c.l.). Allowing χ to be non-zero gives the limit $M_2 > 380$ GeV and $|\chi| < 0.07$ (90% c.l.).

In the above analysis it was assumed that the mass of the righthanded neutrino was zero (or $\lesssim$ few MeV). Clearly, the decay is kinematically forbidden if the neutrino mass $\gtrsim m_\mu/2$. Hence the limit on $M_2(M_R)$ is only a valid test of those models in which the righthanded neutrino is light (e.g. that of Beg *et al.*, 1977). However, in more recent left–right symmetric models (Mohapatra and Senjanovic, 1981), ν_R is taken to have a mass larger than that of W_R.

The measurement of Carr *et al.* also allows a rather stringent limit to be placed on the value of the helicity of the muon (h_μ), and hence the neutrino (h_{ν_μ}), from the parent pion decay. In order that the muon decay rate be positive, one can easily show that $\xi\delta/\rho \leqslant 1$. Using this limit, Fetscher (1984) has pointed out that (6.47) implies that $P_\mu > 0.9959$

Fig. 6.4 Positron momentum spectrum from unpolarised μ^+ decay, the curve is for $\rho = \frac{3}{4}$.

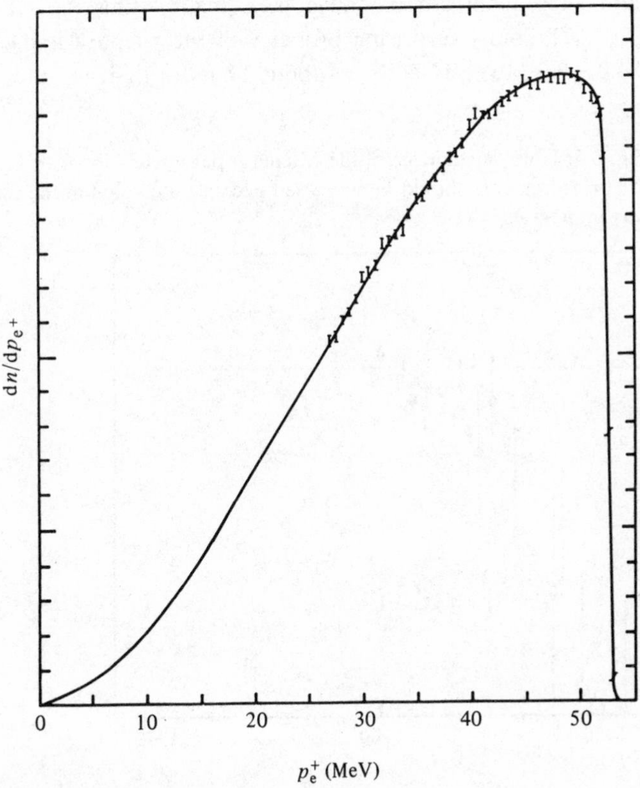

(90% c.l.) and hence that $|h_\mu| = |h_{\nu_\mu}| > 0.9959$ (90% c.l.). Furthermore these conclusions are supported by a measurement of the directional distribution of positrons from polarised muon decays giving, for the integral decay asymmetry, $P_\mu \xi = 1.0027 \pm 0.0084$ (Beltrami *et al.*, 1987). These results, combined with the determinations of the helicity of the $e^\pm$ in $\mu^\pm$ decay ($h(e^+) = 1.010 \pm 0.064$, Corriveau *et al.*, 1981; $h(e^+) = 0.998 \pm 0.045$, Burkard *et al.*, 1985; $h(e^-) = -0.89 \pm 0.28$, Schwartz, 1967), shows that the $l^-(\mu^- e^-)$ is, to a very high degree, a lefthanded particle as far as the charged current weak interaction is concerned. From (6.36), the measurement $h = -1$ implies $g'_i = -g_i$ ($i = 1$ to 5). In summary, therefore, the various results from muon decay are in excellent agreement with the predictions of the V – A theory/standard model. For a detailed review see Engfer and Walter (1986).

6.1.3 Results on leptonic τ-decays

The short lifetime of the τ-lepton ($\sim 3 \times 10^{-13}$ s) makes experiments attempting to isolate and study the τ-decays much more difficult than those for muon decay. The experimental discovery (see Perl (1980) for a review), and subsequent studies, have been made using the reaction $e^+e^- \to \tau^+\tau^-$. The most surprising property of the τ-lepton is its large mass (1784.2 ± 3.2 MeV), a factor of about 17 more than m_μ and 3 500

Fig. 6.5 Time dependence of the Michel ρ parameter. Prior to 1952 the measurements should be increased because the value of m_μ used was imprecise.

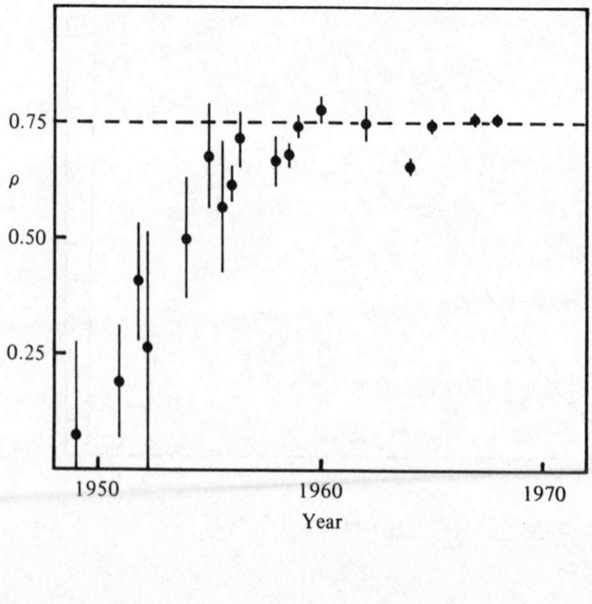

Table 6.2. *Main τ^- lepton decay branching ratios (Particle Data Group,* 1986)

Mode	%
$e^-\bar{\nu}_e\nu_\tau$	17.4 ± 0.5
$\mu^-\bar{\nu}_\mu\nu_\tau$	17.6 ± 0.6
$\pi^-\nu_\tau$	10.1 ± 1.1
$\rho^-\nu_\tau$	21.8 ± 2.0
$\pi^-\rho^0\nu_\tau$	5.4 ± 1.7
$K^-\nu_\tau$	0.67 ± 0.17
$K^{*-}(892)\nu_\tau$	1.7 ± 0.7

more than m_e. The large mass value means that there are many possible decay channels. The decay branching ratios are given in Table 6.2. The purely leptonic decays account for roughly one third of the total, the remaining decay modes contain hadrons. The hadronic decays (see Section 8.7) $\tau^- \to (nh)^-\nu_\tau$ have branching ratios of about 52%, 13%, <0.2% and $<2 \times 10^{-4}$, for $n = 1$, 3, 5 and 7 respectively. Thus, experimentally, the $\tau^+\tau^-$ signature is missing energy (carried off by the neutrinos) plus back-to-back topologies of low multiplicity.

The spin of the τ has been found from a study of its threshold production cross-section. As discussed in Section 4.8, the dependence of $\sigma_{\tau\tau}$ on the τ-velocity β is different for different spin assignments. The data (see Fig. 4.11) show that the τ has spin $\frac{1}{2}$. The threshold behaviour has also been used to obtain the τ-mass. The question of whether the τ is point-like (i.e. no internal structure) has been studied by parameterising the cross-section as follows (see (4.62))

$$\sigma_{\tau\tau} = \frac{4\pi\alpha^2}{3s}\frac{\beta(3-\beta^2)}{2}|F(s)|^2, \qquad F(s) = 1 \mp s/(s - \Lambda_\pm^2). \tag{6.48}$$

The data (e.g. Bartel *et al.*, 1985b) are compatible with a point-like object ($F(s) = 1$) to within the 95% c.l. limits $\Lambda_+ > 285$ GeV, $\Lambda_- > 210$ GeV, or down to a radius of less than or equal to 10^{-16} cm. These measurements also constitute a stringent test of QED.

Assuming the existence of a separate ν_τ, then the observation of the decay mode $\tau^- \to \pi^-\nu_\tau$ shows that the ν_τ has half-integral spin, with the simplest assignment being $\frac{1}{2}$. Taking the above assignments (for more details see Perl, 1980), then the purely leptonic decay modes of the τ^- are (cf. (6.1))

$$\tau^- \to \bar{\nu}_l + \nu_\tau + l^-, \qquad l = e, \mu. \tag{6.49}$$

In the expression for the energy spectrum of l^- from unpolarised τ^-, the mass of l^- enters as $(m_e/m_\tau)^2$. This can be seen by starting with (6.20) and summing over τ and l spin states. Since $(m_\mu/m_\tau)^2 = 3.5 \times 10^{-3}$, the effect of the muon mass is negligible over most of the lepton energy spectrum; thus, the decay spectra for electrons and muons are expected to be essentially the same.

Compared to studies of muon decays, the τ-event samples are small and the possibilities in measuring polarisations limited. Hence the results are much less precise than for muon decay.

The electron energy spectrum has been analysed, for example, by Bacino *et al.* (1979), who find (assuming $m_{\nu_\tau} = 0$) $\rho = 0.72 \pm 0.15$. This is compatible with the expected value $\rho = \frac{3}{4}$ from the V – A theory but, as discussed above, this value is obtained for any admixture of V and A couplings. However, if the more restrictive assumption is made that the couplings at the $Wl\bar{\nu}_l$ vertex (see Fig. 6.1) are known, and have the V – A form, then the transition amplitude (6.40) is obtained. That is, the $W\tau^-\nu_\tau$ coupling is assumed to be, *a priori*, unknown, but restricted to V and A terms. The experimental data and the curves for pure V – A ($\rho = \frac{3}{4}$) and pure V + A ($\rho = 0$) are shown in Fig. 6.6. No significant deviations from the expectations

Fig. 6.6 Electron energy spectrum $x = E_e/E_e^{max}$ from $\tau \to e\nu\bar{\nu}$ decays; the curves are for pure V – A (full line) and V + A (broken line) couplings at the $W\tau\nu_\tau$ vertex.

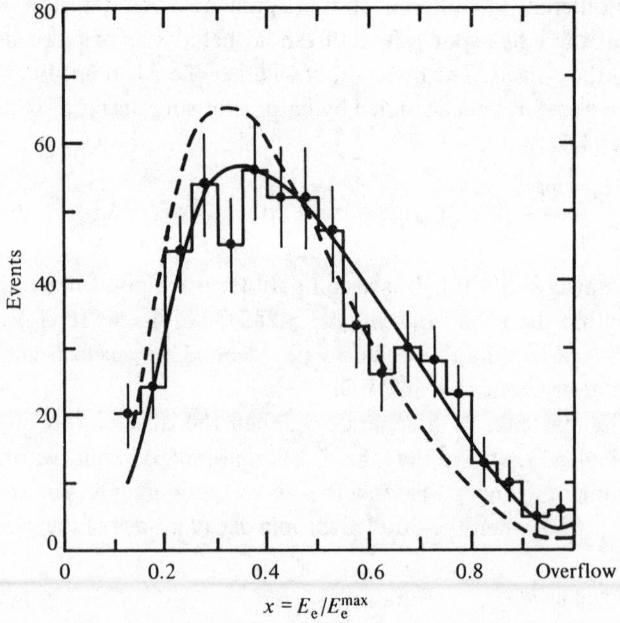

of the standard model are apparent. If the V – A form is assumed, then the data can be used to set an upper limit on the mass of ν_τ, namely $m_{\nu_\tau} < 250$ MeV (95% c.l.). A more precise limit is discussed in Chapter 9. The world average value for the ρ parameter is $\rho = 0.71 \pm 0.08$.

The decay flight paths of the $\tau^\pm$ in $e^+e^- \to \tau^+\tau^-$ at $s^{1/2} \sim 30$ GeV are typically less than or equal to 1 mm. Thus, a reliable measurement of the lifetime is a considerable experimental challenge. Reasonably accurate measurements have, however, been made. The average measured lifetime of the τ-lepton is $(3.3 \pm 0.4) \times 10^{-13}$ s. The expected lifetime can be computed using equation (6.24), together with the value of G_μ from (6.46) and the mean leptonic branching ratio B_l from Table 6.2. The result is $\tau_\tau = (2.79 \pm 0.06) \times 10^{-13}$ s, in excellent agreement with the measured value. Alternatively, if g_τ is the coupling at the $W\tau\nu_\tau$ vertex, and g that for the $W\mu\nu_\mu$ or $We\nu_e$ vertices, then (6.24) gives

$$\frac{g_\tau}{g} = \left[B_l \frac{\tau_\mu}{\tau_\tau} \left(\frac{m_\mu}{m_\tau} \right)^5 \right]^{1/2} = 0.92 \pm 0.06, \qquad (6.50)$$

in agreement with the hypothesis of τ–μ–e universality.

Although there is good evidence that a neutrino is emitted in τ-decay, there is, as yet, no direct observation of the subsequent interactions of the ν_τ. Assuming that the τ^- and ν_τ carry a unique additive lepton number $L_\tau(=1)$, then the charged current interaction $\nu_\tau + N \to \tau^- +$ hadrons is expected. The most fruitful place to search for such events is, perhaps, in the study of neutrinos (or antineutrinos) produced by the interactions of the highest available energy protons in a dense nuclear target. The production of the charmed $D_S(F)$ meson and its subsequent decay $D_S \to \tau\nu_\tau$ is a possible source of $\nu_\tau/\bar{\nu}_\tau$. A so-called *beam-dump* set-up is used for such a study, in which protons interact in a very dense target ($L_{\text{int}} \lesssim 20$ cm), so that the $\pi^\pm$ and $K^\pm$ mesons (whose decays give the neutrinos in a conventional beam) interact predominantly before they have a chance to decay. In this way the conventional neutrino flux is reduced by more than a factor of 1000. However, short-lived particles (lifetimes $\lesssim 10^{-11}$ s) are largely unaffected by the dense material, and so the resulting neutrino beam is rich in these prompt decays. A series of such experiments has been carried out at CERN and FNAL (see, for example, Wachsmuth 1984), and these prompt neutrinos seem to arise from the decays of charmed particles, whose decays give roughly equal fluxes of ν_e and ν_μ. No direct observation of ν_τ interactions has yet been made.

The possibility that the τ^- and ν_τ may have the same lepton number as that of a lighter lepton has been considered by Llewellyn Smith (1977). In this scheme a heavy lepton L^- is called an *ortho-lepton* (*para-lepton*) if it has the same lepton number as a lighter lepton $l^-(l^+)$. If the τ^+ was an ortho-lepton of the e^- type, then the decay $\tau^- \rightarrow e^-\gamma$ would be possible. However, this decay cannot proceed directly, since conservation of the current $\bar{\psi}_e\gamma^\mu\psi_\tau$ forbids the simplest decay matrix element. The decay must then proceed through suppressed higher order diagrams. The experimental limit is less than 6.4×10^{-4} (90% c.l.). The weak decay $\tau^- \rightarrow e^-e^+e^-$ would also be possible; experimentally this branching ratio is less than 4×10^{-4} (90% c.l.). A further limit on the possibility that has the τ^- has the same lepton number as the ν_e has been placed from a study of neutrino interactions in a beam-dump by the BEBC group, Fritze *et al.* (1980), and by CHARM (Winter, 1983). As discussed above, the beam from the dump is rich in ν_e, so that, in addition to the reactions $\nu_e + N \rightarrow e^- + H$, $\nu_e + N \rightarrow \nu_e + H$, the reaction $\nu_e + N \rightarrow \tau^- + H$ can occur (H here stands for final state hadrons). Since the τ branching ratio $B_e \sim 17\%$, this will lead to an apparent increase in the neutral to charged current ratio (NC/CC) for ν_e. The experimental values found are NC/CC $(\nu_e + \bar{\nu}_e) = 0.1 \pm 0.5$ for $E_H > 10$ GeV for BEBC and $0.44^{+0.08}_{-0.10}$ for CHARM, from which it may be concluded that $\sigma(\nu_e N \rightarrow \tau H)/\sigma(\nu_e N \rightarrow eH) < 0.35$ (90% c.l.).

If the τ^- was a para-lepton of the electron type, then $\tau^- \rightarrow e^-\gamma$ is forbidden. However, the ratio of the leptonic decay rates $R_{e/\mu} = \Gamma(\tau^- \rightarrow e^-\bar{\nu}_e\nu_\tau)/\Gamma(\tau^- \rightarrow \mu^-\bar{\nu}_\mu\nu_\tau) \sim 2$ if $\nu_\tau \equiv \bar{\nu}_e$, since there are then two identical particles. This possibility is incompatible with the data.

The possibility that the τ^- is an ortho-(para-)lepton of the muon type can also be excluded. In this case the reaction $\nu_\mu + N \rightarrow \tau^- + H(\nu_\mu + N \rightarrow \tau^+ + H)$ is allowed, leading to anomalous electron events. An analysis of events in a neon bubble chamber by Cnops *et al.* (1978), produced by a ν_μ beam gave no candidates, giving a limit $(g_\tau/g)^2 < 0.025$ (90% c.l.). This is incompatible with (6.50). Furthermore, the decay branching ratio $\tau^- \rightarrow \mu^-\mu^+\mu^-$ ($<4.9 \times 10^{-4}$, 90% c.l.) would be allowed if $\tau^- \equiv \mu^-$, and the hypothesis that $\nu_\tau \equiv \bar{\nu}_\mu$ would give $R_{e/\mu} \sim 0.5$.

Hence, since the other possibilities are excluded, the τ appears to be a sequential lepton with its own unique lepton number, which is also carried by the ν_τ. A direct observation of ν_τ interactions would clearly, however, be more satisfactory. The above discussion is further complicated if the possibility of mixing between neutrino types is admitted. Neutrino mixing is discussed in Chapter 9.

6.2 NEUTRINO AND ANTINEUTRINO ELECTRON SCATTERING

The following reactions are accessible to experimental measurements:

$$\nu_\mu e^- \to \nu_\mu e^- \qquad \text{(NC)}, \qquad (6.51)$$

$$\bar{\nu}_\mu e^- \to \bar{\nu}_\mu e^- \qquad \text{(NC)}, \qquad (6.52)$$

$$\nu_e e^- \to \nu_e e^- \qquad \text{(NC + CC)}, \qquad (6.53)$$

$$\bar{\nu}_e e^- \to \bar{\nu}_e e^- \qquad \text{(NC + CC)}, \qquad (6.54)$$

$$\nu_\mu e^- \to \mu^- \nu_e \qquad \text{(CC)}, \qquad (6.55)$$

where NC and CC refer to neutral current (Z^0 exchange) and charged current ($W^\pm$ exchange) respectively. To lowest order, all these reactions involve one (or both) of the diagrams shown in Fig. 6.7.

For $\nu_\mu e^- \to \nu_\mu e^-$ (6.51), the relevant diagram is Fig. 6.7(*a*), with $a = c = \nu_\mu$, $d = e^-$ and $V = Z^0$. Using the Feynman rules (Appendix C), the matrix element is

$$\mathcal{M}_{fi} = \frac{-ig^2}{2\cos^2\theta_W}\bar{u}_3\gamma_\mu\frac{(1-\gamma^5)}{2}u_1\left[\frac{-g^{\mu\nu}+q^\mu q^\nu/M_Z^2}{q^2-M_Z^2}\right]\bar{u}_4\gamma_\nu\frac{(g_V-g_A\gamma^5)}{2}u_2, \qquad (6.56)$$

where, following the notation of Commins and Bucksbaum (1983),

$$g_V = (C_L + C_R) = 2\sin^2\theta_W - \tfrac{1}{2}, \qquad g_A = C_L - C_R = -\tfrac{1}{2}. \qquad (6.57)$$

Using similar arguments to those following equation (5.62), we can simplify (6.56) to

$$\mathcal{M}_{fi} = \frac{ig^2}{8\cos^2\theta_W(q^2-M_Z^2)}[\bar{u}_3\gamma_\mu(1-\gamma^5)u_1][\bar{u}_4\gamma^\mu(g_V-g_A\gamma^5)u_2]. \qquad (6.58)$$

Summing over final e^- spin states and averaging over the initial spin

Fig. 6.7 Lowest order graphs for $a + e^- \to c + d$ via (*a*) *t*-channel and (*b*) *s*-channel exchange of a vector boson V.

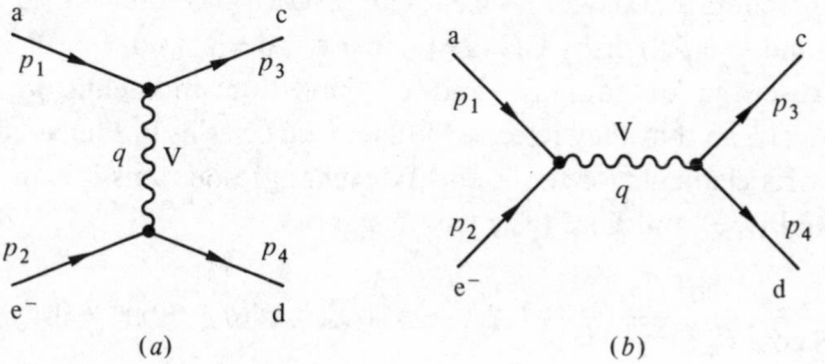

states, we obtain, by the usual methods,

$$|\mathcal{M}_{\rm fi}|^2 = \frac{g^4}{2\cos^4\theta_{\rm W}(q^2 - M_{\rm Z}^2)^2}[(g_{\rm V}+g_{\rm A})^2(p_1\cdot p_2)(p_3\cdot p_4) + (g_{\rm V}-g_{\rm A})^2(p_1\cdot p_4)(p_2\cdot p_3) - m_{\rm e}^2(g_{\rm V}^2 - g_{\rm A}^2)(p_1\cdot p_3)], \tag{6.59}$$

where $g_{\rm V}$ and $g_{\rm A}$ are assumed real. Defining

$$y = (q\cdot p_2)/(p_1\cdot p_2) = (p_1\cdot p_3)/(p_1\cdot p_2),$$

the cms differential cross-section (4.29) is

$$\frac{{\rm d}\sigma}{{\rm d}\Omega}({\rm cms}) = \frac{G_{\rm F}^2(s-m_{\rm e}^2)^2}{16\pi^2 s(1-q^2/M_{\rm Z}^2)^2}[(g_{\rm V}+g_{\rm A})^2 + (g_{\rm V}-g_{\rm A})^2(1-y)^2 - m_{\rm e}^2(g_{\rm V}^2-g_{\rm A}^2)2y/(s-m_{\rm e}^2)], \tag{6.60}$$

where relations (5.63) and (5.168) have been used. If θ is the angle made by the outgoing $e^-(\mathbf{p}_4)$ with respect to the incident ν_μ direction $(\mathbf{p}_1)$, then ${\rm d}\Omega = 2\pi\,{\rm d}\cos\theta = 2\pi\,{\rm d}y(1+\beta)/\beta$, where β is the cms velocity of the e^-. Noting also that $\beta = (s-m_{\rm e}^2)/(s+m_{\rm e}^2)$, (6.60) may be expressed as

$$\frac{{\rm d}\sigma}{{\rm d}y} = \frac{G_{\rm F}^2(s-m_{\rm e}^2)}{4\pi(1-q^2/M_{\rm Z}^2)^2}\left[A + B(1-y)^2 - C\frac{2m_{\rm e}^2 y}{s-m_{\rm e}^2}\right], \tag{6.61}$$

where $A = (g_{\rm V}+g_{\rm A})^2$, $B = (g_{\rm V}-g_{\rm A})^2$ and $C = (g_{\rm V}^2 - g_{\rm A}^2)$.

Equation (6.61) is in a Lorentz invariant form. To obtain the differential cross-section in the lab system, in which the ν_μ has energy E_ν, the substitution $s = m_{\rm e}^2 + 2E_\nu m_{\rm e}$ is required. For high energy beams ($E_\nu \gg m_{\rm e}$), the C term is negligible.

The reaction (6.52), $\bar{\nu}_\mu e^- \to \bar{\nu}_\mu e^-$ (Fig. 6.7(*a*) with $a = c = \bar{\nu}_\mu$, $d = e^-$ and $V = Z^0$), is calculated in a similar way. Constructing $\mathcal{M}_{\rm fi}$ from the Feynman rules one can see that, with respect to (6.58), the roles of particles 1 and 3 are interchanged. Thus, $|\mathcal{M}_{\rm fi}|^2$ is given by (6.59), with p_1 and p_3 interchanged. The resulting coefficients A, B and C are given in Table 6.3.

The reaction (6.53), $\nu_e e^- \to \nu_e e^-$, can proceed by either NC ($a = c = \nu_e$, $d = e^-$ and $V = Z^0$) or by CC ($a = \nu_e$, $c = e^-$, $d = \nu_e$ and $V = W^\pm$). In the latter case the outgoing ν_e and e^- have four-momenta p_3 and p_4 respectively, so that they refer to the same particles as in the NC diagram. The matrix elements for the Z and W exchange diagrams are, in terms of the initial (ν, e) and final (ν', e$'$) state spinors,

$$\mathcal{M}_{\rm Z} = \frac{{\rm i}g^2}{8\cos^2\theta_{\rm W}(q^2 - M_{\rm Z}^2)}[\bar{u}_{\nu'}\gamma_\mu(1-\gamma^5)u_\nu][\bar{u}_{\rm e'}\gamma^\mu(g_{\rm V} - g_{\rm A}\gamma^5)u_{\rm e}], \tag{6.62}$$

Table 6.3. *Coefficients A, B and C (equation 6.61) for different $\nu/\bar{\nu}$ electron scattering reactions*
In the standard model $g_V = 2\sin^2\theta_W - \frac{1}{2}$, $g_A = -\frac{1}{2}$, $g'_V = g_V + 1$, and $g'_A = g_A + 1$

Reaction	A	B	C
$\nu_\mu e^- \to \nu_\mu e^-$	$(g_V + g_A)^2$	$(g_V - g_A)^2$	$g_V^2 - g_A^2$
$\bar{\nu}_\mu e^- \to \bar{\nu}_\mu e^-$	$(g_V - g_A)^2$	$(g_V + g_A)^2$	$g_V^2 - g_A^2$
$\nu_e e^- \to \nu_e e^-$	$(g'_V + g'_A)^2$	$(g'_V - g'_A)^2$	$g_V'^2 - g_A'^2$
$\bar{\nu}_e e^- \to \bar{\nu}_e e^-$	$(g'_V - g'_A)^2$	$(g'_V + g'_A)^2$	$g_V'^2 - g_A'^2$

and

$$\mathscr{M}_W = \exp(i\phi)\frac{ig^2}{8(q^2 - M_W^2)}[\bar{u}_{e'}\gamma_\mu(1-\gamma^5)u_\nu][\bar{u}_{\nu'}\gamma^\mu(1-\gamma^5)u_e], \tag{6.63}$$

where ϕ is the relative phase of the two diagrams.

In the limit $q^2 \ll M_W^2(M_Z^2)$, the matrix elements (6.62) and (6.63) become

$$\mathscr{M}_Z = \frac{-iG_F}{2^{1/2}}[\bar{u}_{\nu'}\gamma_\mu(1-\gamma^5)u_\nu][\bar{u}_{e'}\gamma^\mu(g_V - g_A\gamma^5)u_e], \tag{6.64}$$

$$\mathscr{M}_W = \frac{-iG_F}{2^{1/2}}[\bar{u}_{\nu'}\gamma_\mu(1-\gamma^5)u_\nu][\bar{u}_{e'}\gamma^\mu(1-\gamma^5)u_e], \tag{6.65}$$

where equation (5.168) has been used. The term in $\mathscr{M}_W$ has been reordered by a Fierz transformation (Appendix E) which introduces a sign -1 in addition to the sign $\exp(i\phi) = -1$, which arises from the anticommutation relations of the Dirac creation operators due to Fermi statistics. Thus the total matrix element for $\nu_e e^- \to \nu_e e^-$ is

$$\mathscr{M} = \mathscr{M}_Z + \mathscr{M}_W = \frac{-iG_F}{2^{1/2}}[\bar{u}_{\nu'}\gamma_\mu(1-\gamma^5)u_\nu][\bar{u}_{e'}\gamma^\mu(g'_V - g'_A\gamma^5)u_e], \tag{6.66}$$

where $g'_V = g_V + 1$ and $g'_A = g_A + 1$. We now have a form similar to (6.58), and the resulting coefficients A, B and C in the cross-section are given in Table 6.3.

The calculation of reaction (6.54), $\bar{\nu}_e e^- \to \bar{\nu}_e e^-$, is very similar to that for (6.53). However, for (6.54) the W diagram is in the *s*-channel (Fig. 6.7(*b*)), with $a = c = \bar{\nu}_e$, $d = e^-$. The relative factor between $\mathscr{M}_Z$ and $\mathscr{M}_W$ is again -1 from Fermi statistics, and a Fierz transformation introduces a further factor of -1, leading to, for $q^2 \ll M_W^2(M_Z^2)$,

$$\mathscr{M} = \mathscr{M}_Z + \mathscr{M}_W = \frac{-iG_F}{2^{1/2}}[\bar{v}_\nu\gamma_\mu(1-\gamma^5)v_{\nu'}][\bar{u}_{e'}\gamma^\mu(g'_V - g'_A\gamma^5)u_e]. \tag{6.67}$$

The cross-section calculation follows that for $\bar{\nu}_\mu e^- \to \bar{\nu}_\mu e^-$, resulting in the coefficients given in Table 6.3.

The total cross-sections for the above interactions are obtained by integrating (6.61) over y, which gives, neglecting the m_e^2 term and the q^2 dependence of the propagator,

$$\sigma_{tot} = \frac{G_F^2 m_e E_\nu}{2\pi}\left(A + \frac{B}{3}\right), \tag{6.68}$$

where E_ν is the lab energy of the ν (or $\bar{\nu}$). The constant $G_F^2 m_e/(2\pi)$ has the value 4.31×10^{-42} cm^2 GeV^{-1}. Hence, unless E_ν is very large, experimental observation of these reactions is a formidable problem. Nonetheless, reactions (6.51) to (6.55) have all been observed. Reactions (6.51) and (6.52) have been studied using $\nu_\mu(\bar{\nu}_\mu)$ beams at high energy accelerators, these being produced by π and K meson decays. The energy and angle of the outgoing electron satisfy $E_e^2\theta_e^2 < 2m_e$, giving a characteristic signal in a detector of a single electron produced at a small angle with respect to the incident beam. Barbiellini and Santoni (1985) calculate world average values of $\sigma/E_\nu = 1.55 \pm 0.20$ for (6.51), and 1.26 ± 0.21 for (6.52), both in units of 10^{-42} cm^2 GeV^{-1}. From Table 6.3, we can see that the near equality of these results shows that $g_V g_A \sim 0$. From equation (6.68) we can see that from each reaction we get an elliptic relation between g_V and g_A. The regions of the g_V, g_A plane allowed by these reactions are shown in Fig. 6.8. Taking the reactions together, only the common regions of the ellipses are allowed. These are roughly the (g_V, g_A) points $(\pm\frac{1}{2}, 0)$, $(0, \pm\frac{1}{2})$. The ambiguity arises because the equations are symmetric in g_V and g_A. In terms of the standard model, the measured cross-sections can be used to determine $\sin^2\theta_W$ (using (6.57)) giving 0.23 ± 0.03 from (6.51) and 0.22 ± 0.05 from (6.52). The ratio of the ν_μ and $\bar{\nu}_\mu$ cross-sections is independent of the assumption that the appropriate coupling constant is G_F, and gives $\sin^2\theta_W = 0.22 \pm 0.03$.

Experiments on ν_e or $\bar{\nu}_e$ are even more difficult. Reines *et al.* (1976) used the intense $\bar{\nu}_e$ beam from a reactor to study reaction (6.54). The result can be converted into a contour in the (g_V, g_A) plane as shown in Fig. 6.8. Note that the data are taken with an energy range $1.5 < E_{\bar{\nu}_e} <$ 4.5 MeV, so that the term in C in (6.61) is not negligible. Alternatively, in terms of the standard model, the experiment gives $\sin^2\theta_W = 0.35 \pm 0.06$. Results from the study of reaction (6.53) have also been obtained (Allen *et al.*, 1985); the (g_V, g_A) contour is shown in Fig. 6.8, and the resulting value of $\sin^2\theta_W$ is 0.21, with an error of about 0.12. Further, the possibility of constructive interference between the NC and CC parts (rather than the destructive interference prediction of the standard model) can be ruled out. Examination of Fig. 6.8 shows that there are two possible remaining solutions for (g_V, g_A); namely, around $(0, -\frac{1}{2})$ and $(-\frac{1}{2}, 0)$. The solution around $(0, -\frac{1}{2})$ corresponds to $\sin^2\theta_W \simeq 0.25$.

The final reaction in the list, $\nu_\mu e^- \to \mu^- \nu_e$ (6.55), can only occur through W exchange. This process is related to $\mu^- \to e^- \nu_\mu \bar{\nu}_e$ by the replacement of an outgoing $\bar{\nu}_e$ by an ingoing ν_e, and is thus known as *inverse muon decay*. Reaction (6.55) is of the type of Fig. 6.7(*a*), with $a = \nu_\mu$, $c = \mu^-$, $d = \nu_e$, and $V = W^\pm$. In the standard model, the matrix element is

$$\mathcal{M}_{fi} = \frac{ig^2}{8(q^2 - M_W^2)} [\bar{u}_3\gamma_\mu(1-\gamma^5)u_1][\bar{u}_4\gamma^\mu(1-\gamma^5)u_2], \tag{6.69}$$

which gives, summing over final spins and averaging over initial spins,

$$|\mathcal{M}_{fi}|^2 = \frac{2g^4}{(q^2 - M_W^2)^2}(p_1\cdot p_2)(p_3\cdot p_4). \tag{6.70}$$

Fig. 6.8 Allowed contours in the (g_A, g_V) plane from studies of $\nu(\bar{\nu})$–electron neutral current interactions.

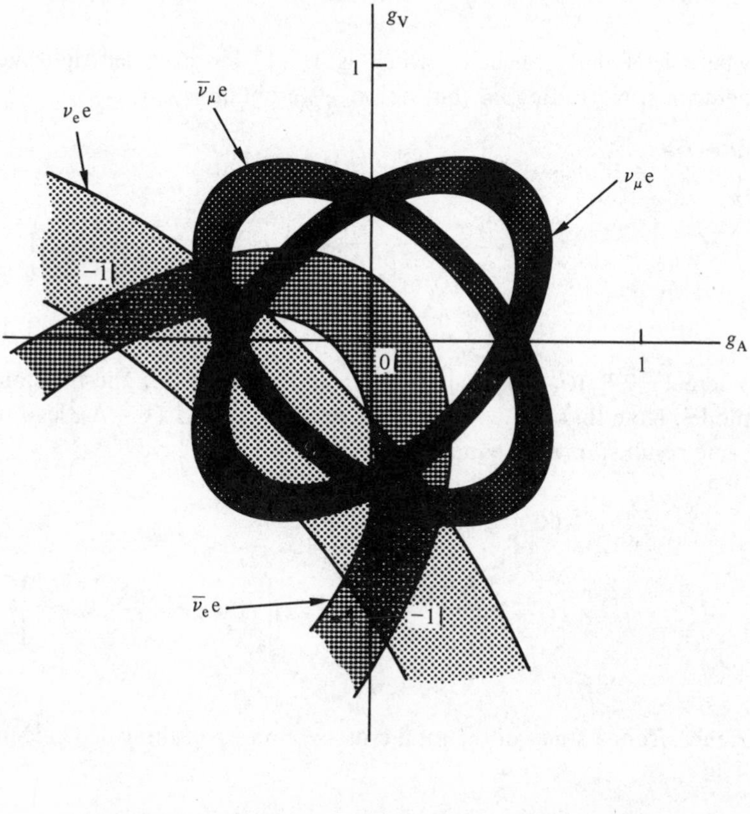

Some care must be taken in the kinematics of this process, since the final and initial particles are different. The differential cross-section is

$$\frac{\mathrm{d}\sigma\,(\mathrm{cms})}{\mathrm{d}\cos\theta}=\frac{G_{\mathrm{F}}^2}{2\pi s}\frac{(s-m_\mu^2)^2}{(1-q^2/M_{\mathrm{W}}^2)^2}, \tag{6.71}$$

where θ is the angle made by the μ^- with respect to the incident ν_μ. Defining $y=(p_1\cdot p_3)/(p_1\cdot p_2)$, as above, this becomes

$$\frac{\mathrm{d}\sigma}{\mathrm{d}y}=\frac{G_{\mathrm{F}}^2(s-m_\mu^2)}{\pi(1-q^2/M_{\mathrm{W}}^2)^2}. \tag{6.72}$$

The presence of the final state muon means that the reaction has a threshold of $E_\nu \sim m_\mu^2/(2m_\mathrm{e}) \sim 11$ GeV.

The experimental beams of ν_μ are formed mainly from π and K decay, and the neutrinos in these decays are predominantly lefthanded (see Section 6.1.2). However, a small ν_R component could exist, and the reaction (6.55) could have some V + A component. Thus, we could have a more general V, A matrix element (in the charge exchange order) of the form

$$\mathscr{M}_{\mathrm{fi}}=\frac{-\mathrm{i}G_{\mathrm{F}}}{2^{1/2}}[\bar{u}_3\gamma_\mu(1-h\gamma^5)u_1][\bar{u}_4\gamma^\mu(C_\mathrm{V}-C_\mathrm{A}\gamma^5)u_2], \tag{6.73}$$

where h is the ν_μ helicity which is $1(-1)$ for pure left(right)handed neutrinos. Neglecting m_e (but not m_μ), we obtain

$$\frac{\mathrm{d}\sigma}{\mathrm{d}y}=\frac{G_{\mathrm{F}}^2}{8\pi}(|C_\mathrm{V}|^2+|C_\mathrm{A}|^2)s\left\{(1+h^2+2h\lambda)(1-m_\mu^2/s)\right.$$
$$\left.+(1+h^2-2h\lambda)\left[(1-y)^2-\frac{m_\mu^2}{s}(1-y)\right]\right\}, \tag{6.74}$$

where $\lambda=2\,\mathrm{Re}(C_\mathrm{V}C_\mathrm{A}^*)/(|C_\mathrm{V}|^2+|C_\mathrm{A}|^2)$, and has the value 1 in the standard model. Note that $h=\lambda=1$ (V − A), and $h=\lambda=-1$ (V + A), lead to the same result. For a lefthanded ν_μ beam ($h=1$), we have

$$\frac{\mathrm{d}\sigma}{\mathrm{d}y}=\frac{G_{\mathrm{F}}^2}{4\pi}(|C_\mathrm{V}|^2+|C_\mathrm{A}|^2)s$$
$$\times\left\{(1+\lambda)\left(1-\frac{m_\mu^2}{s}\right)+(1-\lambda)\left[(1-y)^2-\frac{m_\mu^2}{s}(1-y)\right]\right\}. \tag{6.75}$$

Hence, from a study of the total cross-section, or of the y dependence (or the angular distributions, since $y \simeq (1 - \cos\theta)/2)$, a value of λ can be found. Hence the relative admixture of V and A couplings can be extracted.

The problem can also be analysed using the charge retention order. If we assume the charged leptons are purely lefthanded, then we can write

$$\mathscr{M}_{\mathrm{fi}} = \frac{\mathrm{i}G_{\mathrm{F}}}{2^{1/2}} [\bar{u}_3\gamma_\mu(1-\gamma^5)u_2][\bar{u}_4\gamma^\mu(C'_{\mathrm{V}} - C'_{\mathrm{A}}\gamma^5)u_1]. \tag{6.76}$$

For a righthanded ν_μ, we must replace u_1 by $[(1+\gamma^5)/2]u_1$ and then calculate $\mathrm{d}\sigma_{\mathrm{R}}$. Likewise, for a lefthanded ν_μ, we must replace u_1 by $[(1-\gamma^5)/2]u_1$ giving $\mathrm{d}\sigma_{\mathrm{L}}$. Note that a right(left)handed neutrino has zero cross-section if the current is pure $\mathrm{V-A(V+A)}$. If P is the polarisation of the ν_μ beam, that is $P = [N(\nu_{\mathrm{R}}) - N(\nu_{\mathrm{L}})]/[N(\nu_{\mathrm{R}}) + N(\nu_{\mathrm{L}})]$, then the differential cross-section is, retaining terms in m_μ,

$$\begin{aligned}\frac{\mathrm{d}\sigma\ (\mathrm{cms})}{\mathrm{d}\cos\theta} &= \left(\frac{1+P}{2}\right)\frac{\mathrm{d}\sigma_{\mathrm{R}}}{\mathrm{d}\cos\theta} + \left(\frac{1-P}{2}\right)\frac{\mathrm{d}\sigma_{\mathrm{L}}}{\mathrm{d}\cos\theta}\\ &= \frac{G_{\mathrm{F}}^2(|C'_{\mathrm{V}}|^2 + |C'_{\mathrm{A}}|^2)}{64\pi s}(s - m_\mu^2)^2\\ &\quad\times[(1+P)(1-\lambda')(1-\cos\theta)(a - b\cos\theta) + 4(1-P)(1+\lambda')],\end{aligned} \tag{6.77}$$

where $a = 1 + m_\mu^2/s$, $b = 1 - m_\mu^2/s$, $\lambda' = 2\,\mathrm{Re}(C'_{\mathrm{V}}C'^*_{\mathrm{A}})/(|C'_{\mathrm{V}}|^2 + |C'_{\mathrm{A}}|^2)$ and m_{e} terms have been neglected. The $\mathrm{V-A}$ form (6.71) is obtained with $P = -1$ and $\lambda' = 1$.

The form (6.77) has been used by Bergsma *et al.* (1983) to compare with measurements on $\nu_\mu \mathrm{e}^- \to \mu^- \nu_{\mathrm{e}}$, made in the CERN wide band beam (mostly ν_μ with $E_\nu \sim 20$ GeV). In the lab system the muons are produced at small angles with respect to the ν_μ direction ($\theta_\mu < 10$ mrad is required experimentally). The ratio of the observed number of events to that expected for pure $\mathrm{V-A}$ ($P = -1$, $\lambda' = 1$) is 0.98 ± 0.12, whereas the corresponding ratio for pure $\mathrm{V+A}$ ($P = 1$, $\lambda' = -1$) is 2.63 ± 0.33. The resulting limits on λ', as a function of P, are displayed in Fig. 6.9. For the charge exchange form with $h = 1$ (6.75), a value of λ compatible with pure $\mathrm{V-A}$ can again be deduced. Limits on the contribution of $\mathrm{W_R}$ in the left–right symmetric model of Beg *et al.* (1977) can also be found, giving $M_2 \geqslant 155$ GeV, $|\chi| < 0.26$, (90% c.l.). Note that, as discussed for muon and tau decays, it is important to understand the assumptions used in obtaining the matrix element when considering the implications of an experimental result.

6.3 THE ELECTROWEAK REACTION $e^+e^- \to f\bar{f}$

In Section 4.3 the cross-section was derived for the reaction $e^+e^- \to f\bar{f}$, via single photon exchange and for unpolarised incident beams. The symbol f represents any spin $\frac{1}{2}$ fermion of charge Q_f (in units of e), other than the electron. In addition to single photon exchange (Fig. 6.10(a)) there is, in the standard model, a similar diagram with a single Z^0 exchange (Fig. 6.10(b)) . Before considering the cross-section arising from the sum of these processes, we will first examine the cross-section for single photon exchange in the case where the incident beams are longitudinally polarised. If the incident e^- and e^+ are completely polarised with helicities $h = -1$ and $+1$ respectively, then only the left and righthanded parts of the e^- and e^+ wave functions respectively contribute to the interaction. Hence, the matrix element is (cf. (4.59))

$$\mathcal{M}_\gamma = \frac{ie^2 Q_e Q_f}{s} \bar{v}_2 \frac{(1+\gamma^5)}{2} \gamma_\mu \frac{(1-\gamma^5)}{2} u_1 \bar{u}_3 \gamma^\mu v_4, \tag{6.78}$$

where Q_e is the electron charge in units of e (i.e. $Q_e = -1$). It is assumed that the electron is massless (although it could not, in fact, be transformed into an arbitrary polarisation state if that were the case!), so that the projection operators $(1+\gamma^5)/2$ and $(1-\gamma^5)/2$ for the right and lefthanded states can be used. If the m_e terms are retained, then the projection operator $(\not{p}_e + m_e)(1+\gamma^5\not{s}_e)/2$ must be used. For a longitudinally polarised e^-, the spin appears in the combination $m_e s_e$ which (in order to satisfy $s_e^2 = -1$,

Fig. 6.9 Allowed region (hatched area) for the beam polarisation P and parameter λ' obtained from a study of inverse muon decay.

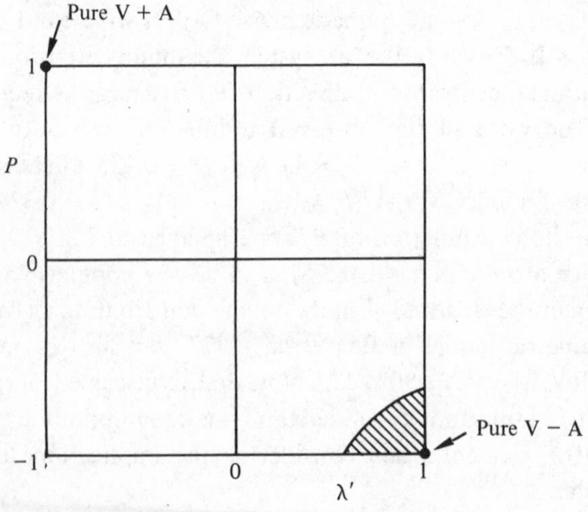

$p_e \cdot s_e = 0$) is, in the limit $m_e \to 0$, equal to $+(-)p_e$ for $h = +1(-1)$ respectively. In this limit the result obtained is the same as that using (6.78). Using the usual methods, we obtain

$$|\mathcal{M}_\gamma|^2 = \frac{16e^4Q_e^2Q_f^2}{s^2}[(p_1 \cdot p_4)(p_2 \cdot p_3) + (p_1 \cdot p_3)(p_2 \cdot p_4) + m_f^2(p_1 \cdot p_2)]$$
$$= 2e^4Q_e^2Q_f^2(1 + \cos^2\theta), \tag{6.79}$$

where m_f is the mass of the outgoing fermion. In the second line of (6.79), m_f is taken as zero, and θ is defined as the angle made by particle 3(f), in the cms system, with respect to the incident direction of particle 1(e^-).

If the incident e^- and e^+ have helicities $h = +1$ and -1 respectively, then the result (6.79) is also obtained. For the two possible remaining cases, namely $h(e^-) = 1$, $h(e^+) = 1$, or $h(e^-) = -1$, $h(e^+) = -1$, the cross-section is zero. This can be seen directly from the matrix element which will, in each case, contain the product $(1 + \gamma^5)(1 - \gamma^5) = 0$. Summing the results for $|\mathcal{M}_\gamma|^2$ for the four possible longitudinal spin configurations, and dividing by four, gives the spin-averaged result (4.60), as required. The resulting differential cross-section (using (4.29)) is

$$\frac{d\sigma}{d\Omega} = \frac{\alpha^2Q_e^2Q_f^2}{4s}(1 + \cos^2\theta). \tag{6.80}$$

It might appear, at first sight, that there is no further information to be gained in studying the polarisation of f (or $\bar{f}$) in the parity conserving electromagnetic interaction. However, this is not the case. For simplicity, we will consider the limit $m_f \to 0$, so that the outgoing f (or $\bar{f}$) is completely longitudinally polarised. The calculation for $m_f \neq 0$ can be made as outlined above, but leads to more cumbersome expressions.

Fig.. 6.10 Lowest order diagrams for the reaction $e^+e^- \to f\bar{f}$ via (*a*) γ^0 and (*b*) Z^0.

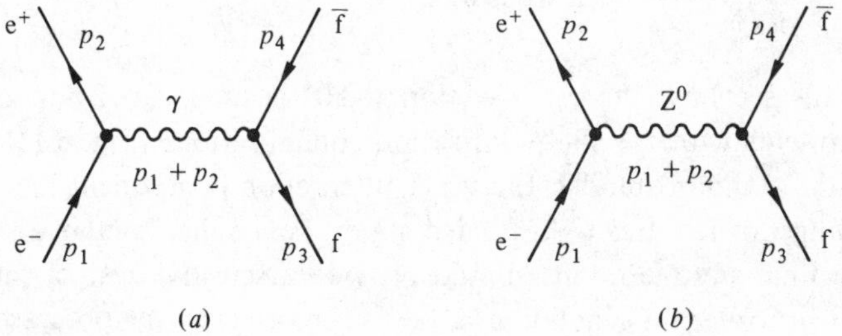

If we consider the possible initial and final state helicities, then we obtain the following matrix element

$$\mathcal{M}_\gamma = \frac{ie^2 Q_e Q_f}{4s} \bar{v}_2 \gamma_\mu (1 + h_i \gamma^5) u_1 \bar{u}_3 \gamma^\mu (1 + h_f \gamma^5) v_4, \tag{6.81}$$

where $h_i = 1$ corresponds to $h(e^-) = 1, h(e^+) = -1$, and $h_i = -1$ corresponds to $h(e^-) = -1, h(e^+) = 1$. Likewise, $h_f = 1$ corresponds to $h(f) = 1, h(\bar{f}) = -1$ and $h_f = -1$ to $h(f) = -1$, $h(\bar{f}) = 1$. The remaining terms, for which the helicities of either both of the initial or both of the final state particles are the same, give zero contribution in the massless limit. The resulting differential cross-section is (with $h_i^2 = h_f^2 = 1$)

$$\begin{aligned}\frac{d\sigma}{d\Omega} &= \frac{2\alpha^2 Q_e^2 Q_f^2}{s^3} [(1 + h_i h_f)(p_2 \cdot p_3)(p_1 \cdot p_4) + (1 - h_i h_f)(p_2 \cdot p_4)(p_1 \cdot p_3)] \\ &= \frac{\alpha^2 Q_e^2 Q_f^2}{8s} [(1 + h_i h_f)(1 + \cos\theta)^2 + (1 - h_i h_f)(1 - \cos\theta)^2] \\ &= \frac{\alpha^2 Q_e^2 Q_f^2}{4s} [(1 + \cos^2\theta) + 2h_i h_f \cos\theta].\end{aligned} \tag{6.82}$$

If we sum over the four possible helicity configurations, the term in $\cos\theta$ disappears. This sum, divided by four, gives the spin-averaged cross-section for unpolarised beams, namely (6.80). If the incident beams are, however, polarised (h_i either $+1$ or -1), then the angular distributions for righthanded ($h_f = 1$) and lefthanded ($h_f = -1$) fermions f are different. For a righthanded incident e^- ($h_i = 1$) and a righthanded f ($h_f = 1$), we have

$$\frac{d\sigma}{d\Omega} = \frac{\alpha^2 Q_e^2 Q_f^2}{4s} (1 + \cos\theta)^2, \tag{6.83}$$

The same result is obtained for the configuration $h_i = h_f = -1$. However, for $h_i = 1$, $h_f = -1$ (or $h_i = -1$, $h_f = 1$), we obtain

$$\frac{d\sigma}{d\Omega} = \frac{\alpha^2 Q_e^2 Q_f^2}{4s} (1 - \cos\theta)^2. \tag{6.84}$$

We thus see how the cross-section (6.80) is made up from individual components with specific polarisation configurations (Fig. 6.11).

In the standard model, the weak interaction component has different couplings for left and righthanded states, hence the angular distribution of the final state fermions should provide a sensitive test of the model. To lowest order the reaction $e^+e^- \to f\bar{f}$ can occur by the processes shown in Fig. 6.10, which lead to the matrix element (using Appendix C)

$$\mathcal{M} = \mathcal{M}_\gamma + \mathcal{M}_Z$$
$$= \frac{ie^2 Q_e Q_f}{s} \bar{v}_2 \gamma_\mu u_1 \bar{u}_3 \gamma^\mu v_4 + \frac{ig^2}{\cos^2\theta_W (s - M_Z^2)}$$
$$\times \bar{v}_2 \gamma_\mu \frac{(g_V^e - g_A^e \gamma^5)}{2} u_1 \bar{u}_3 \gamma^\mu \frac{(g_V^f - g_A^f \gamma^5)}{2} v_4, \tag{6.85}$$

where $g_V = C_L + C_R$ and $g_A = C_L - C_R$. Thus $|\mathcal{M}|^2$ is composed of the following three terms

$$|\mathcal{M}|^2 = |\mathcal{M}_\gamma|^2 + |\mathcal{M}_Z|^2 + (\mathcal{M}_\gamma \mathcal{M}_Z^* + \mathcal{M}_Z \mathcal{M}_\gamma^*). \tag{6.86}$$

The evaluation of these contributions is made by the usual methods. For the collision of unpolarised incident beams, the resulting differential cross-section is

$$\frac{d\sigma}{d\Omega} = \frac{\alpha^2}{4s}[C_1(1 + \cos^2\theta) + C_2 \cos\theta], \tag{6.87}$$

where θ is the angle subtended by f with respect to the incident e^-, and

$$C_1 = Q_e^2 Q_f^2 + \frac{F^2}{16}[(g_V^e)^2 + (g_A^e)^2][(g_V^f)^2 + (g_A^f)^2] + \frac{F}{2} Q_e Q_f g_V^e g_V^f,$$
$$C_2 = \frac{F^2}{2} g_V^e g_A^e g_V^f g_A^f + F Q_e Q_f g_A^e g_A^f, \tag{6.88}$$

Fig. 6.11 Angular distribution of f in $e^+e^- \to \gamma^* \to f\bar{f}$. The contributions from $h_i = h_f$ and $h_i = -h_f$ are shown as broken lines and their sum as a full line.

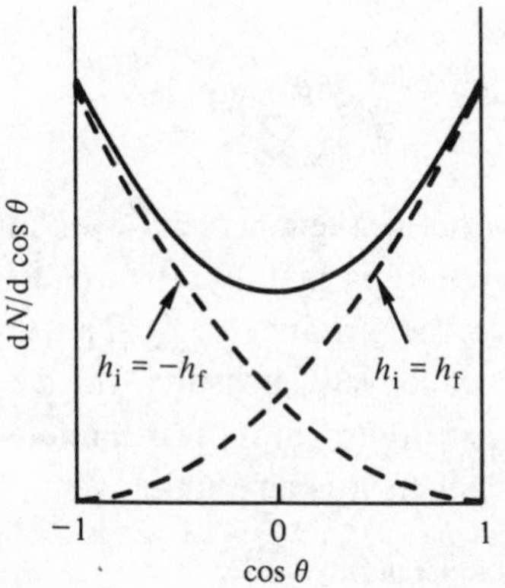

with

$$F = \frac{g^2}{e^2 \cos^2 \theta_W} \frac{s}{(s - M_Z^2)} = \frac{1}{\sin^2 \theta_W \cos^2 \theta_W} \frac{s}{(s - M_Z^2)}. \tag{6.89}$$

Using (5.63) and (5.168), F can be written in the alternative form

$$F = \frac{2^{1/2} G_F}{\pi\alpha} \frac{s M_Z^2}{(s - M_Z^2)}. \tag{6.90}$$

The definitions (6.89) and (6.90) for F will, however, differ when higher order terms are taken into account. This point has been discussed, for example, by Cashmore *et al.* (1986). For the purposes of discussion here, however, higher order electroweak effects are neglected.

At small values of $s^{1/2}$ (say, $s^{1/2} \lesssim 10$ GeV), the effects of the Z^0 are negligible and the pure electromagnetic cross-section (6.80) dominates. This is, of course, symmetric in $\cos\theta$. For intermediate energies (up to say $s^{1/2} \sim 50$ GeV), there is a small, but measurable, effect from the electromagnetic-weak interference term. At, or near, the Z^0 pole ($s^{1/2} \sim M_Z$) the cross-section (6.87) for $e^+e^- \to Z^0 \to f\bar{f}$ becomes large. This process should be considered as the production and subsequent decay of the Z^0 boson, which can be treated as an unstable resonance. This is discussed further in Chapter 10. Extensive measurements on the reactions $e^+e^- \to \mu^+\mu^-$ and $e^+e^- \to \tau^+\tau^-$ have been carried out at PETRA and PEP. The results can be conveniently expressed in terms of $A_{ff} = (N_F - N_B)/(N_F + N_B)$, the forward–backward asymmetry[#], and the ratio R_{ff} of the total cross-section for $f\bar{f}$ to the point-like electromagnetic cross-section $\sigma_0 = 4\pi\alpha^2/3s$, for $Q_e^2 Q_f^2 = 1$. From (6.87) we obtain

$$R_{ff} = \frac{\sigma_{ff}}{\sigma_0} = C_1 \simeq Q_e^2 Q_f^2 \left(1 + \frac{F g_V^e g_V^f}{2 Q_e Q_f}\right), \tag{6.91}$$

and

$$A_{ff} = \frac{3C_2}{8C_1} = \frac{3F g_A^e g_A^f (Q_e Q_f + F g_V^e g_V^f / 2)}{8 R_{ff}} \simeq \frac{3F g_A^e g_A^f}{8 Q_e Q_f}, \tag{6.92}$$

where the approximate form for A_{ff} is good to a few percent, even at the highest PETRA energy. The asymmetry is thus mainly sensitive to the axial coupling, and effectively measures the product $g_A^e g_A^f$. The cross-section ratio R_{ff}, on the other hand, effectively measures the vector coupling product $g_V^e g_V^f$. More generally, if no assumption is made as to the form of the neutral current coupling, then F is also unknown.

[#] The notation A_{FB} for this quantity is also frequently used.

Some measured distributions for $e^+e^- \to f\bar{f}$ are shown in Fig. 6.12. These are from the MAC experiment at PEP for $e^+e^- \to \mu^+\mu^-$ (Ash *et al.*, 1985) and from the JADE experiment at PETRA for $e^+e^- \to \tau^+\tau^-$ (Marshall, 1985). In Fig. 6.13 the results of the measurements of the asymmetry parameters $A_{\mu\mu}$ and $A_{\tau\tau}$ are shown. At the highest values of $s^{1/2}$, the data for $A_{\mu\mu}$ tend to favour a finite value of M_Z.

The reaction $e^+e^- \to e^+e^-$ can proceed, in addition to the *s*-channel and Z^0 graphs shown in Fig. 6.10, by the *t*-channel exchange of a γ or Z^0. The *t*-channel γ exchange graph is shown in Fig. 4.4(*a*); a similar

Fig. 6.12 Angular distribution of l^- in (*a*) $e^+e^- \to \mu^+\mu^-$ (MAC) $s^{1/2} = 29$ GeV and (*b*) $e^+e^- \to \tau^+\tau^-$ (JADE) $s^{1/2} = 42.9$ GeV. The full line is the distribution expected in the standard model and the broken line that from QED alone.

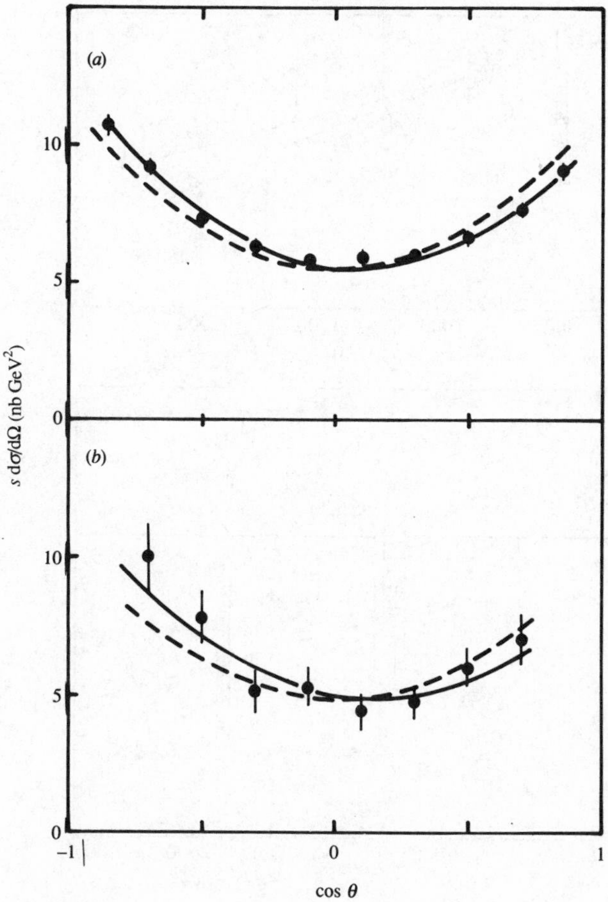

graph can be drawn for Z^0 exchange. In the PETRA range of energies the reaction $e^+e^- \to e^+e^-$ is dominated by the t-channel γ exchange, leading to a sharp peaking of electrons in the forward direction. Electroweak effects are rather small at the level of current experimental precision (Marshall, 1985).

Fig. 6.13 Measurements of the asymmetries (*a*) $A_{\mu\mu}$ ($e^+e^- \to \mu^+\mu^-$) and (*b*) $A_{\tau\tau}$ ($e^+e^- \to \tau^+\tau^-$), as a function of $s^{1/2}$, from PEP and PETRA. The curves are those expected in the standard model, for different values of M_Z. ▾, CELLO; ●, JADE; ■, MARK J; ▴, TASSO; ◆, PLUTO; □, HRS; ×, MARK I; +, MARK II.

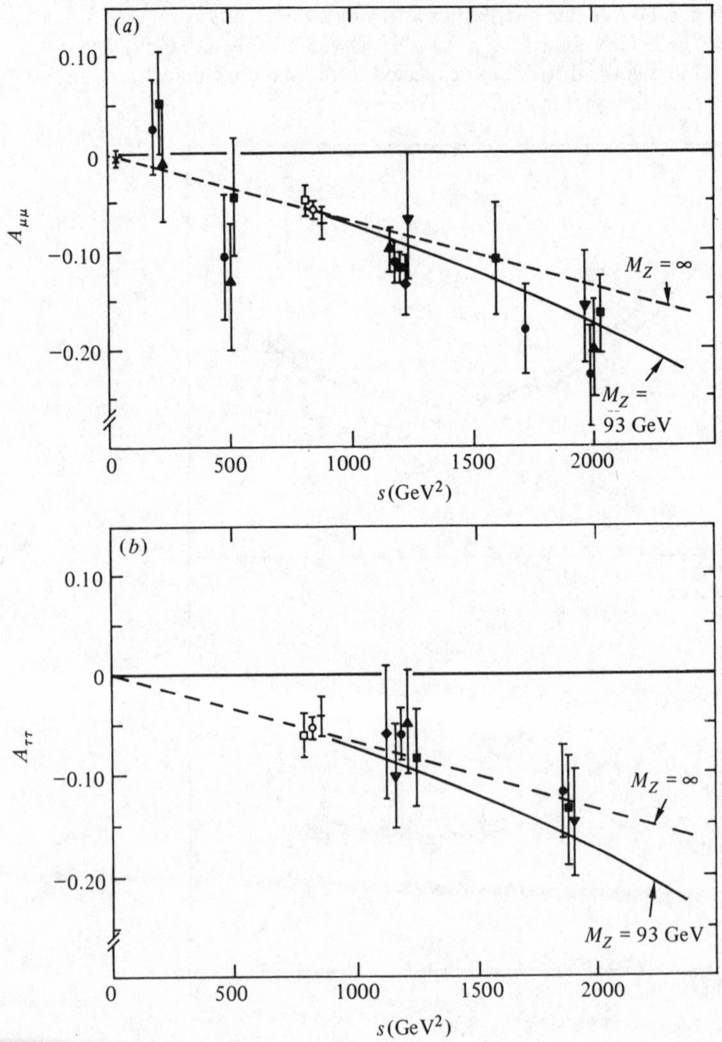

Table 6.4. *Weak neutral current couplings extracted from measurements on* $e^+e^- \to f\bar{f}$, *compared to those expected from the standard model (SM)*

f	$g_V^e g_V^f$ (from R)	$g_A^e g_A^f$ (from A)
μ	0.051 ± 0.023	0.284 ± 0.018
τ	0.025 ± 0.032	0.205 ± 0.026
SM ($\sin^2 \theta_W = 0.22$)	0.004	0.25

The sensitivity of measurements on R_{ff} or A_{ff} obviously increases as $s^{1/2}$ increases. For example, at the PEP energy $s^{1/2} = 29$ GeV, $F/2 = -0.31$, compared to the value at the highest PETRA energy $s^{1/2} \sim 45$ GeV of $F/2 = -0.88$. From a compilation of the data on $e^+e^- \to \mu^+\mu^-$ and $e^+e^- \to \tau^+\tau^-$ Marshall (1985) extracts the values for $g_V^e g_V^f$ and $g_A^e g_A^f$ given in Table 6.4. The results are in reasonable agreement with the standard model, with $\sin^2 \theta_W = 0.22$. The data show some hint of a deviation from the hypothesis of μ–τ universality, which predicts identical couplings for μ and τ leptons. However, the experiments are difficult, so that some residual systematic effects in the results cannot be excluded. (Indeed, with the inclusion of some more recent data, Amaldi *et al.* (1987) find that this difference diminishes). Assuming e–μ–τ universality, then the data can be used to extract the squares of the couplings $g_V^2 = 0.012 \pm 0.029$ and $g_A^2 = 0.268 \pm 0.024$. Thus the e^+e^- electroweak measurements yield four solutions for (g_V, g_A), namely $[\pm(0.11 \pm 0.13), \pm(0.518 \pm 0.023)]$. These values, together with the results from neutrino–electron scattering shown in Fig. 6.8, show that the solution at $g_V \sim 0$, $g_A \sim -0.5$ is favoured. The same conclusion is obtained, although with less precision from a combined study of $e^+e^- \to e^+e^-$ and $\bar{\nu}_e e^- \to \bar{\nu}_e e^-$; reactions which involve only electron-type leptons (Fernandez *et al.*, 1987).

In conclusion, the results of all the data gathered from the studies of purely leptonic interactions are consistent with the lowest order standard electroweak model. The charged current reaction data are consistent with a pure V − A structure, whereas the neutral current reaction data are all compatible with a single value $\sin^2 \theta_W \simeq 0.22$. However, the data are not, as yet, precise enough to test the higher order predictions of the model, or to rule out possible V + A charged vector bosons with masses $\gtrsim 400$ GeV.

7

Deep inelastic scattering and quantum chromodynamics

Deep inelastic scattering has played an important role in our present understanding of quarks, gluons and QCD. In this chapter these topics are briefly reviewed. We start by writing down the formalism to describe elastic and quasi-elastic lepton–nucleon scattering. This formalism is then extended to deep inelastic scattering, and the evidence for underlying quark substructure of the nucleon is discussed. Next, a brief outline of QCD is made, with particular reference to lepton–nucleon scattering.

7.1 ELASTIC CHARGED LEPTON–NUCLEON SCATTERING

The cross-section for the electromagnetic interaction of an unpolarised electron with an unpolarised point-like proton was derived in Section 4.2. The matrix element for the $e^- p \to e^- p$ process contains the Dirac current for the proton, $\bar{u}_4 \gamma^\mu u_2$ (see Fig. 4.2). In general the proton current is of the form (4.188). The general matrix element is thus

$$\begin{aligned}\mathcal{M} &= -\frac{ie^2}{q^2}\,\bar{u}_3\gamma_\mu u_1 \cdot \bar{u}_4\left[\gamma^\mu F_1(q^2) + \frac{i\sigma^{\mu\nu}}{2M}\,q_\nu F_2(q^2)\right]u_2,\\ &= -\frac{ie^2}{q^2}\,\bar{u}_3\gamma_\mu u_1 \cdot \bar{u}_4\left[(F_1+F_2)\gamma^\mu - \frac{F_2}{2M}\,(p_2+p_4)^\mu\right]u_2,\end{aligned} \tag{7.1}$$

where M is the proton mass, and the second line is obtained by using the Gordon identity (4.189). The functions F_1 and F_2 are called the proton *form factors* and are both functions of q^2, the only non-trivial invariant quantity at the pγp vertex (Fig. 4.2). For a point-like proton (Section 4.2) we have $F_1(q^2) = 1$, $F_2(q^2) = 0$. In order to calculate the cross-section we need to evaluate

$$|\mathcal{M}|^2 = \frac{e^4}{q^4}\{\bar{u}_3\gamma_\mu u_1\bar{u}_1\gamma_\nu u_3\}$$

$$\times\left\{\bar{u}_4\left[(F_1+F_2)\gamma^\mu - \frac{F_2}{2M}P^\mu\right]u_2\bar{u}_2\left[(F_1+F_2)\gamma^\nu - \frac{F_2}{2M}P^\nu\right]u_4\right\}$$

$$= \frac{e^4}{q^4}L_{\mu\nu}W^{\mu\nu}, \tag{7.2}$$

where $P = p_2 + p_4$.

The evaluation of $L_{\mu\nu}$ is the same as for the point-like case. The tensor $W^{\mu\nu}$ has three different terms

(i) $(F_1+F_2)^2\,\mathrm{tr}[(\not{p}_4+M)\gamma^\mu(\not{p}_2+M)\gamma^\nu]$
$= 4(F_1+F_2)^2[p_4^\mu p_2^\nu + p_4^\nu p_2^\mu - g^{\mu\nu}p_2\cdot p_4 + M^2 g^{\mu\nu}]$,

(ii) $\dfrac{-F_2}{2M}(F_1+F_2)\,\mathrm{tr}[(\not{p}_4+M)\gamma^\mu(\not{p}_2+M)P^\nu] = -2F_2(F_1+F_2)P^\mu P^\nu$,

(iii) $\dfrac{F_2^2}{4M^2}\,\mathrm{tr}[(\not{p}_4+M)P^\mu(\not{p}_2+M)P^\nu] = \dfrac{F_2^2}{2M^2}P^2P^\mu P^\nu$. (7.3)

The other cross-term gives the same result as (ii). The product $L_{\mu\nu}W^{\mu\nu}$ can now be formed. Writing the result in terms of the Lorentz invariant quantities $s = (p_1+p_2)^2$ and $t = q^2 = (p_1-p_3)^2$, and averaging over the initial spin states (i.e. $\frac{1}{4}$), we obtain after some algebra

$$|\mathcal{M}|^2 = \frac{4e^4}{t^2}\left(A[(s-M^2)^2+ts] + B\frac{t^2}{2}\right), \tag{7.4}$$

where

$$A = F_1^2 - \frac{t}{4M^2}F_2^2, \qquad B = (F_1+F_2)^2. \tag{7.5}$$

The differential cross-section for elastic scattering, in invariant form, is as follows

$$\frac{d\sigma}{dt} = \frac{|\mathcal{M}|^2}{16\pi(s-M^2)^2}. \tag{7.6}$$

This is obtained by starting with the cms cross-section (4.29) and using $d\Omega = 2\pi\, d\cos\theta^*$, $t = -2E_1E_3(1-\cos\theta^*)$ and $E_1 = E_3 = (s-M^2)/(2s^{1/2})$. For elastic ep scattering this gives

$$\frac{d\sigma}{dt} = \frac{4\pi\alpha^2}{(s-M^2)^2t^2}\left\{A[(s-M^2)^2+ts] + B\frac{t^2}{2}\right\}. \tag{7.7}$$

The term in parentheses becomes $(s^2+u^2)/2$ in the point-like limit $A=B=1$ (cf. (4.46)).

In the lab system, $p_1=E_1(1,0,0,1)$ and $p_3=E_3(1,0,\sin\theta,\cos\theta)$, thus, $s=M^2+2ME_1$, $t=q^2=-Q^2=-4E_1E_3\sin^2\theta/2$, where θ is the lab scattering angle. The angular distribution is found by starting with $t=-2E_1E_3(1-\cos\theta)$, so that

$$\frac{dt}{d\cos\theta}=2E_1E_3-2E_1(1-\cos\theta)\frac{dE_3}{d\cos\theta}=2E_3^2, \tag{7.8}$$

where the relation $ME_1=E_3(E_1+M-E_1\cos\theta)$ has been used. Further, using the form $t=2M(E_3-E_1)$, so that $(s-M^2)^2+ts=4M^2E_1E_3\cos^2\theta/2$, we obtain the differential cross-section in the lab system

$$\frac{d\sigma}{d\Omega}(\text{lab})=\frac{\alpha^2}{4E_1^2\sin^4\theta/2}\frac{E_3}{E_1}\left[A\cos^2\frac{\theta}{2}-\frac{t}{2M^2}B\sin^2\frac{\theta}{2}\right]. \tag{7.9}$$

For the point-like case we have $A=B=1$, and (7.9) reduces to the form (4.43) as required. The cross-section formula (7.9) (or (7.7)) is known as the '*Rosenbluth*' cross-section. The target structure, as seen by the virtual photon, is contained in the form factors $F_1(q^2)$ and $F_2(q^2)$.

In the analysis of lepton–nucleon scattering, the *Breit* (or brick-wall) *frame* is particularly useful. In this Lorentz frame (see Fig. 7.1) the target particle, $p_2=(E_2,0,0,p_2)$ has its three-momentum reversed in direction by the collision, whilst its energy remains constant. From the conservation of four-momentum at the vertex, $q+p_2=p_4$, we obtain

$$q=(0,0,0,(Q^2)^{1/2}), \qquad p_2=(E_B,0,0,-(Q^2)^{1/2}/2),$$
$$p_4=(E_B,0,0,(Q^2)^{1/2}/2), \tag{7.10}$$

where $Q^2=-q^2$ and $E_B=(Q^2+4M^2)^{1/2}/2$. Note that the virtual photon has zero energy in the Breit frame which, in the limit $Q^2\to 0$, becomes the rest frame of the proton.

Fig. 7.1 Lepton nucleon scattering in the Breit frame of the virtual photon.

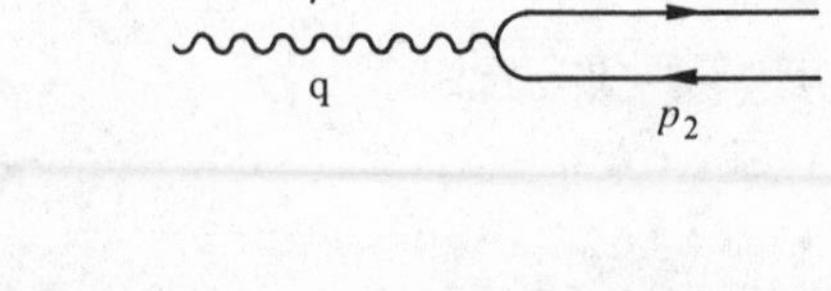

In order to understand the role of the form factors, let us consider the proton current j_p^μ in the Breit frame. The current j_p^μ is given by

$$j_p^\mu = \exp[-i(p_2 - p_4)\cdot x]\bar{u}(p_4)\left[(F_1 + F_2)\gamma^\mu - \frac{F_2}{2M}(p_2 + p_4)^\mu\right]u(p_2). \tag{7.11}$$

Ignoring, for the moment, the plane wave part $\exp(-iq\cdot x)$, then the $\mu = 0$ component, which is related to the electric charge, is

$$\begin{aligned} j_p^0 &= (F_1 + F_2)\bar{u}(p_4)\gamma^0 u(p_2) - \frac{F_2 E_B}{M}\bar{u}(p_4)u(p_2) \\ &= (F_1 + F_2)2M - \frac{F_2 E_B}{M}2E_B \\ &= 2M\left(F_1 - \frac{Q^2}{4M^2}F_2\right), \end{aligned} \tag{7.12}$$

where the spinor normalisation $\bar{u}(p_4)u(p_2) = 2E_B$ and $\bar{u}(p_4)\gamma^0 u(p_2) = 2M$ has been used (using the results of Section 3.2.3, with p_2 and p_4 given by (7.10)). The spatial components of the current ($i = 1, 2, 3$) are given by

$$j_p^i = (F_1 + F_2)\bar{u}_4(p_4)\gamma^i u(p_2) = (F_1 + F_2)u^\dagger\begin{pmatrix} 0 & \sigma_i \\ \sigma_i & 0 \end{pmatrix}u, \tag{7.13}$$

and are thus related to the magnetic properties of the proton (Section 4.9.4). From the above considerations it is natural to define the *electric* (G_E) and *magnetic* (G_M) *form factors* of the nucleon as follows

$$\begin{aligned} G_E(Q^2) &= F_1(Q^2) - \frac{Q^2}{4M^2}F_2(Q^2), \\ G_M(Q^2) &= F_1(Q^2) + F_2(Q^2). \end{aligned} \tag{7.14}$$

The charge of the proton is given by integrating j_p^0 over all space, in the $Q^2 \to 0$ limit, and thus gives $G_E^p(0) = F_1^p(0) = 1$. Similarly, for the neutron $G_E^n(0) = F_1^n(0) = 0$. In the $Q^2 \to 0$ limit, the magnetic form factor $G_M(0)$ is the total magnetic moment of the nucleon, that is

$$\begin{aligned} G_M^p(0) &= F_1^p(0) + F_2^p(0) = \mu_p = 1 + \mu_p^a \\ G_M^n(0) &= F_1^n(0) + F_2^n(0) = \mu_n = \mu_n^a. \end{aligned} \tag{7.15}$$

where $\mu_p^a = 1.793$ and $\mu_n^a = -1.913$ are the anomalous magnetic moments of the proton and neutron respectively (Section 4.9.4). The non-zero values of these quantities immediately shows that the nucleon has structure, in contrast to the charged leptons which appear to be point-like. The differential cross-section in terms of G_E and G_M is, from (7.9) and putting $\tau = Q^2/4M^2$ ($Q^2 = -t$),

$$\frac{d\sigma}{d\Omega}(\text{lab}) = \frac{\alpha^2}{4E_1^2 \sin^4 \theta/2} \frac{E_3}{E_1} \left[\left(\frac{G_E^2 + \tau G_M^2}{1+\tau} \right) \cos^2 \frac{\theta}{2} + 2\tau G_M^2 \sin^2 \frac{\theta}{2} \right]. \tag{7.16}$$

Thus the values of G_E and G_M at a particular Q^2 can be extracted from a study of the dependence of the cross-section on θ.

The measurements of the nucleon form factors can be interpreted in terms of the nucleon size, as seen by the virtual photon. The relationship between the form factor $F(q^2)$ and the rms radius ($r_0 = \langle r^2 \rangle^{1/2}$) was discussed in Section 2.2.5. The radii derived from F_1 and G_E are not the same, but are related by (from (7.14))

$$\langle r^2 \rangle_{G_E} = \langle r^2 \rangle_{F_1} + \frac{1}{4M^2} F_2(0). \tag{7.17}$$

It has been found empirically that, at small Q^2, the form factors $G_E(Q^2)$ and $G_M(Q^2)/\mu_N$ can be reproduced by

$$G_D(Q^2) = \frac{1}{(1 + Q^2/Q_0^2)^2}, \qquad Q_0^2 = 0.71 \text{ GeV}^2 = 18.2 \text{ fm}^{-2}. \tag{7.18}$$

The spatial charge distribution is given by the Fourier transform of (7.18) (using (2.94)), giving $\rho(r) = \exp(-Q_0 r)$. However, substantial deviations from this '*dipole*' form are found at large Q^2 ($\gtrsim$ few GeV^2). Proton form factor data on G_E at low Q^2 ($\lesssim 0.2$ GeV^2) have been used by Simon *et al.* (1980) to extract the proton radius, giving a value $r_0^p = (0.862 \pm 0.012)$ fm. For the neutron one obtains from G_E^n that $\langle r^2 \rangle_n = -0.12$ fm^2! One can see from (7.17) that the anomalous magnetic moment $F_2^n(0)$ contributes substantially to this.

For $Q^2 \gg Q_0^2$, the form factors of the type (7.18) fall off roughly as Q^{-4}. This quantity enters squared into the cross-section and, together with the Q^{-4} propagator term, leads to a fall-off of approximately Q^{-12} at large Q^2. A rapid fall-off at large Q^2 is, perhaps, not too surprising, since the form factor represents the probability that the nucleon remains intact during the violent collision. If the scattering takes place on the constituents of the proton (i.e. the quarks) then the constituents must all be scattered in roughly the same direction, so that the proton does not break up. Brodsky and Chertok (1976) argue that the spin-averaged form factor, $[A(Q^2)]^{1/2}$, has a power dependence $(Q^2)^{1-n}$, where n is the number of elementary constituents (i.e. $n=3$ for quarks). This prediction is based on the framework of a scale invariant quark model. In order that all three quark constituents share the violent collision, an exchange of two gluons (at least) is required (Fig. 7.2). Each quark propagator contributes a power $(Q^2)^{-1}$, but the gluon propagator is compensated by its coupling to the quark currents. Thus for the nucleon one would expect $Q^4 F_N(Q^2)$ to be constant (modulo logarithmic corrections).

If the pion, rather than the proton, could be used as the target particle, then the above ideas would predict that $Q^2 F_\pi(Q^2)$ was constant at large Q^2, since $n=2$ in this case. The form factor $F_\pi(Q^2)$, for space-like four-momenta, can be measured directly from a study of $\pi^\pm e$ scattering or indirectly from $e^- p \to e^- \pi^+ n$. The data are compatible with a pion form factor, at low Q^2,

$$F_\pi(Q^2) = \frac{1}{1 + \langle r_\pi^2 \rangle Q^2/6}. \tag{7.19}$$

Amendolia *et al.* (1984), using $\pi^- e$ data, find a value $\langle r_\pi^2 \rangle^{1/2} = 0.657 \pm 0.012$ fm for the charged pion radius. The pion form factor can also be measured for time-like q^2, for example using $e^+e^- \to \pi^+\pi^-$. Some experimental data for electron scattering off pion, nucleon and deuteron targets is shown in Fig. 7.3, and it can be seen that the values $n=2$, 3 and 6 respectively are consistent with the data at large Q^2, supporting the above ideas.

Some models of hadrons predict that the size of a hadron decreases as the mass of its constituent quarks increases. Hence, it is of interest to compare the charged kaon and pion radii. Amendolia *et al.* (1986) find a

Fig. 7.2 Leading gluon exchange diagram in elastic lepton–nucleon scattering.

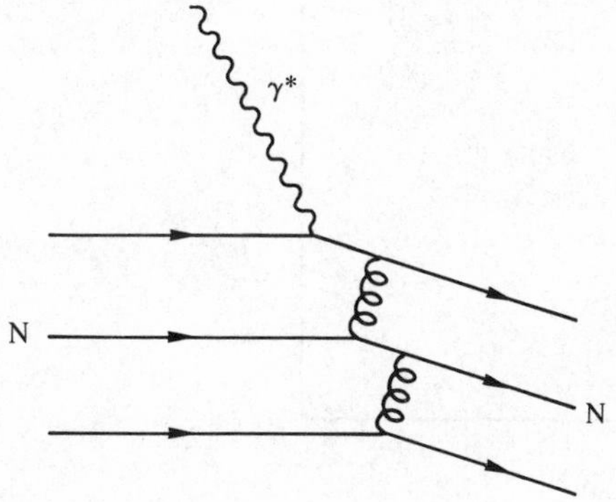

value $\langle r_K^2 \rangle^{1/2} = 0.58 \pm 0.04$ fm and a difference $\Delta = \langle r_\pi^2 \rangle - \langle r_K^2 \rangle = 0.10 \pm 0.045\ \text{fm}^2$. Combining this result with that of Dally *et al.* (1980) gives $\Delta = 0.12 \pm 0.04\ \text{fm}^2$; thus, it appears that the charged kaon is about 0.1 fm smaller than the charged pion.

Electron scattering from nuclear targets is more complex than scattering off pions or nucleons. For the pion (spin 0) there is only one form factor, the electric form factor $F_\pi(Q^2)$. For the nucleon (spin $\frac{1}{2}$) there are both electric and magnetic form factors. In the case of a nucleus with spin $J = 1$ or higher, there are electric quadrupole or higher multipole form factors. The nucleus can either be scattered back to its ground state (coherent elastic scattering) or inelastically to an excited state. The Born approximation, which is used in the analysis, is good only for rather light nuclei ($Z \lesssim 10$), because the electron wave function can be severely distorted in a heavy nucleus such as lead. A detailed discussion is given by Scheck (1983).

7.2 DEEP INELASTIC CHARGED LEPTON–NUCLEON SCATTERING

When subject to violent collision at large Q^2 the proton usually breaks up. At modest energies the excitation and subsequent decay of nucleon resonances frequently occurs. In Fig. 7.4 the distribution of the

Fig. 7.3 The elastic form factors F_n, scaled by $(Q^2)^{n-1}$, against Q^2 for different hadrons.

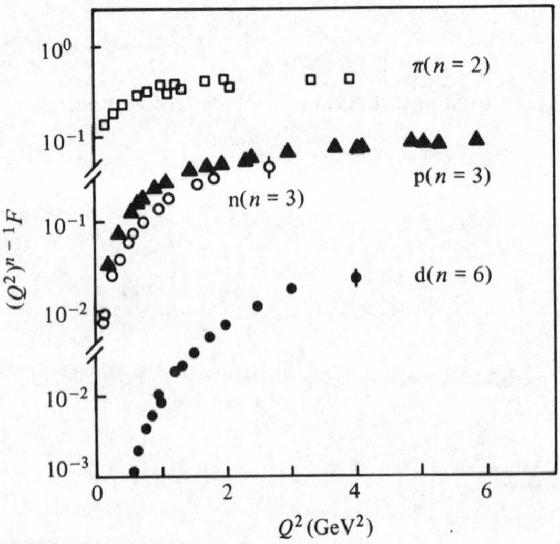

invariant mass of the recoiling hadronic system is shown from a study of $e^- + p \rightarrow e^- + X$, where X represents any kinematically possible final state. The data were collected at SLAC using a 13 GeV electron beam, and were obtained by measuring the energy of the scattered electron (E') at a fixed angle θ (single arm measurement). In addition to the elastic peak at $W = M$ (W is the invariant mass of system X), several nucleon resonances are visible in the region up to $W \sim 2$ GeV. For $W \gtrsim 2$ GeV, the so-called *continuum region*, little structure is apparent. These observations were surprising, since it was expected that each nucleon resonance would have a form factor falling rapidly with Q^2. The value of Q^2 varies rather slowly across the plot (from 0.8 GeV^2 at $W = M$ to 0.5 GeV^2 at $W = 3$ GeV), and a rapid decrease with W was anticipated.

Before discussing the formalism for inelastic scattering it is useful to rewrite the differential cross-section (7.9) for elastic scattering in the lab system in a different way. We can write q^2, evaluated in the lab frame, as follows

$$q^2 = (p_4 - p_2)^2 = 2M(M - E_4) = 2M(E_3 - E_1) = -2M\nu, \tag{7.20}$$

where $\nu = E_1 - E_3 = E_4 - M$ is the energy transfer carried by the virtual photon. In terms of $Q^2 = -q^2$ (Q^2 is a positive quantity), we have

$$Q^2 = 2M\nu. \tag{7.21}$$

Writing equation (7.7) in terms of lab variables we obtain

$$\frac{d\sigma}{dQ^2}(\text{lab}) = \frac{\pi\alpha^2}{4E_1^3 E_3 \sin^4\theta/2}\left(A\cos^2\frac{\theta}{2} + B\frac{Q^2}{2M^2}\sin^2\frac{\theta}{2}\right), \tag{7.22}$$

Fig. 7.4 Sketch of measurements of the differential cross-section for $e^-p \rightarrow e^-X$ ($E = 13$ GeV, $\theta = 4^0$) as a function of W.

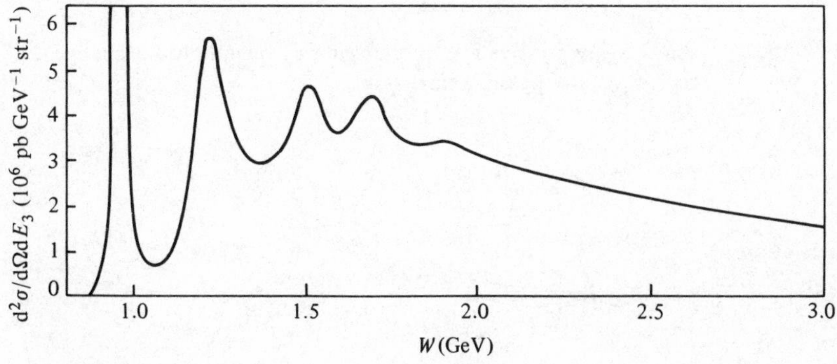

and, using the relation (7.21), this can be written as

$$\frac{d^2\sigma}{dQ^2\,d\nu}(\text{lab}) = \frac{\pi\alpha^2}{4E_1^3E_3\sin^4\theta/2}\left(A\cos^2\frac{\theta}{2} + B\frac{Q^2}{2M^2}\sin^2\frac{\theta}{2}\right)\delta\left(\nu - \frac{Q^2}{2M}\right). \tag{7.23}$$

This latter form will be of reference for future use.

The tensor $W^{\mu\nu}$ (7.2), describing the proton current in elastic ep scattering, can be written in the following form

$$W^{\mu\nu} = \frac{A}{M^2}\left[p_2^\mu - \frac{(p_2\cdot q)q^\mu}{q^2}\right]\left[p_2^\nu - \frac{(p_2\cdot q)q^\nu}{q^2}\right] - \frac{Bq^2}{4M^2}\left[-g^{\mu\nu} + \frac{q^\mu q^\nu}{q^2}\right]. \tag{7.24}$$

This is obtained by starting with (i) + 2 × (ii) + (iii) from (7.3) and using $P = p_2 + p_4$, $P^2 = 4M^2(1+\tau)$, $q^2 = -2p_2\cdot q$ and including a factor $\frac{1}{2}$ for the spin average of the initial proton. A factor $\frac{1}{4}M^2$ has also been introduced to agree with the normal convention used in the case of inelastic scattering. The tensor $W^{\mu\nu}$ in (7.24) is, in fact, the most general tensor which can be constructed from the available four momenta at the pγp vertex and which satisfies the condition $q_\mu W^{\mu\nu} = q_\nu W^{\mu\nu} = 0$. This latter condition is a consequence of conservation ($\partial_\mu j_p^\mu = 0$) of the proton current (7.11), with $q = p_4 - p_2$.

Turning now to the inelastic case (see Fig. 7.5), the leptonic tensor $L_{\mu\nu}$ has the same form as for elastic scattering (i.e. as in (7.2)). The hadronic tensor $W^{\mu\nu}$ has in general six contributing terms, namely

$$W^{\mu\nu} = -g^{\mu\nu}W_1 + \frac{p_2^\mu p_2^\nu}{M^2}W_2 - \frac{i\varepsilon^{\mu\nu\alpha\beta}p_{2\alpha}q_\beta}{2M^2}W_3 + \frac{q^\mu q^\nu}{M^2}W_4 + \frac{(p_2^\mu q^\nu + p_2^\nu q^\mu)}{2M^2}W_5 + \frac{i(p_2^\mu q^\nu - p_2^\nu q^\mu)}{2M^2}W_6. \tag{7.25}$$

Fig. 7.5 Deep inelastic e–p scattering by one-photon exchange. X represents any possible final state.

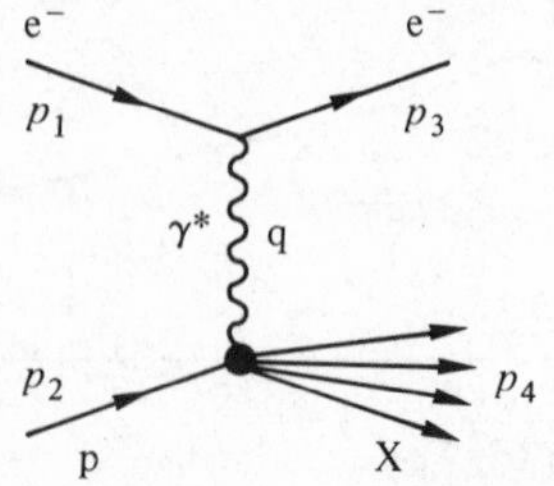

The terms in W_3 and W_6 are antisymmetric, and give zero contribution when contracted with the symmetric tensor $L_{\mu\nu}$ (they can, however, contribute to the weak interaction case where $L_{\mu\nu}$ is no longer symmetric). Conservation of the proton current gives $q_\mu W^{\mu\nu} = q_\nu W^{\mu\nu} = 0$. Applying this to (7.25), gives the identities

$$W_5 = -\frac{2(p_2\cdot q)}{q^2} W_2,$$

$$W_4 = \frac{M^2}{q^2} W_1 + \left(\frac{p_2\cdot q}{q^2}\right)^2 W_2. \tag{7.26}$$

Hence $W^{\mu\nu}$ in (7.25) can be cast in a similar form to (7.24), namely

$$W^{\mu\nu} = \frac{W_2}{M^2}\left[p_2^\mu - \frac{(p_2\cdot q)}{q^2} q^\mu\right]\left[p_2^\nu - \frac{(p_2\cdot q)}{q^2} q^\nu\right] + W_1\left[-g^{\mu\nu} + \frac{q^\mu q^\nu}{q^2}\right]. \tag{7.27}$$

For inelastic scattering the relation (7.21) no longer holds. Instead we have

$$W^2 = p_4^2 = (p_2 + q)^2 = M^2 + 2p_2\cdot q + q^2, \tag{7.28}$$

where W is the centre-of-mass energy of the outgoing hadrons. In the lab system (7.28) becomes

$$W^2 = M^2 + 2M\nu - Q^2. \tag{7.29}$$

The differential cross-section in the lab system can now be written down. Comparing (7.27) with (7.24), and using (7.23) (now Q^2 and ν are independent variables), we have, for an unpolarised electron beam,

$$\frac{d^2\sigma}{dQ^2\, d\nu}(\text{lab}) = \frac{\pi\alpha^2}{4E_1^3 E_3 \sin^4\theta/2}\left(W_2\cos^2\frac{\theta}{2} + 2W_1\sin^2\frac{\theta}{2}\right), \tag{7.30}$$

or, equivalently, noting that $Q^2 = 4E_1E_3\sin^2\theta/2$ and so $dQ^2 = E_1E_3\, d\Omega/\pi$

$$\frac{d^2\sigma}{d\Omega\, dE_3}(\text{lab}) = \frac{\alpha^2}{4E_1^2\sin^4\theta/2}\left(W_2\cos^2\frac{\theta}{2} + 2W_1\sin^2\frac{\theta}{2}\right). \tag{7.31}$$

The equivalence between W_1 and W_2 and the elastic form factors A and B is

$$W_2 \equiv A = (F_1^2 + \tau F_2^2)\,\delta\left(\nu - \frac{Q^2}{2M}\right) = \left(\frac{G_E^2 + \tau G_M^2}{1+\tau}\right)\delta\left(\nu - \frac{Q^2}{2M}\right),$$

$$W_1 \equiv \tau B = \tau(F_1 + F_2)^2\,\delta\left(\nu - \frac{Q^2}{2M}\right) = \tau G_M^2\,\delta\left(\nu - \frac{Q^2}{2M}\right), \tag{7.32}$$

where the elastic form factors are evaluated at $\nu = Q^2/2M$.

The differential cross-section can be conveniently expressed in terms of two dimensionless variables, the so-called *Bjorken scaling* variables x (or x_{BJ}) and y, defined as

$$x = \frac{-q^2}{2p_2\cdot q} = \frac{Q^2}{2M\nu}, \qquad y = \frac{p_2\cdot q}{p_2\cdot p_1} = \frac{\nu}{E_1}. \tag{7.33}$$

The Jacobian for the transformation is $dQ^2\, d\nu = 2ME_1^2 y\, dx\, dy$. Hence we can rearrange (7.30) as follows

$$\frac{d^2\sigma}{dx\, dy}(\text{lab}) = \frac{\pi\alpha^2 My}{2E_1E_3\sin^4\theta/2}\left(W_2\cos^2\frac{\theta}{2} + 2W_1\sin^2\frac{\theta}{2}\right). \tag{7.34}$$

The 'structure functions' W_1 and W_2 of the proton are, in general, functions of any two independent kinematic variables, e.g. (Q^2, ν), (Q^2, x), (x, W^2).

Bjorken, in 1969, proposed that in the *deep inelastic limit*, namely large ν and Q^2 but Q^2/ν finite, the structure functions W_1 and νW_2 would only depend on the ratio Q^2/ν (or equivalently on x) and not on Q^2 or ν separately. In order to pursue this discussion it is useful to introduce the structure functions F_1 and F_2 defined* as

$$\begin{aligned} F_1(x, Q^2) &= MW_1(Q^2, \nu), \\ F_2(x, Q^2) &= \nu W_2(Q^2, \nu). \end{aligned} \tag{7.35}$$

Bjorken's hypothesis would mean that F_1 and F_2 would *scale*. That is, they would both become functions of x only. The differential cross-section in terms of F_1 and F_2 is, from (7.34), and using $Q^2 = 4E_1E_3\sin^2\theta/2 = 2ME_1xy = (s - M^2)xy$,

$$\begin{aligned} \frac{d^2\sigma}{dx\, dy}(\text{lab}) &= \frac{8\pi\alpha^2 ME_1}{Q^4}\left[F_2\left(1 - y - \frac{Mxy}{2E_1}\right) + \frac{y^2}{2}\cdot 2xF_1\right] \\ &= \frac{4\pi\alpha^2(s - M^2)}{Q^4}\left\{2xF_1\left[\frac{1 + (1-y)^2}{2}\right]\right. \\ &\qquad\left. + (1-y)(F_2 - 2xF_1) - \frac{M^2xyF_2}{s - M^2}\right\}. \end{aligned} \tag{7.36}$$

The second line of (7.36) is written in a form which is both covariant and useful for the quark model interpretation (Section 7.3).

* The same symbols are conventionally (and confusingly) used for the elastic form factors and the deep inelastic structure functions.

As can be seen from Fig. 7.5, the lowest order (one photon exchange) diagram for deep inelastic scattering involves the interaction of a virtual photon carrying four-momentum q with a proton, leading to a final hadronic state X. The interaction of a real photon of momentum q with a proton, leading to the same final state X, is shown in Fig. 7.6. The difference between these two diagrams is that the virtual photon has a space-like four-momentum and has both longitudinal and transverse polarisation states, whereas the real photon has $q^2 = 0$ and only transverse polarisation states (see Section 3.3.2). In order to try and relate these processes we start with equation (7.31) and rewrite it in the following form

$$\mathrm{d}\sigma = \frac{1}{4p_1 \cdot p_2} \frac{e^4}{Q^4} (L_{\mu\nu} 4\pi M W^{\mu\nu}) \frac{\mathrm{d}^3 p_3}{(2\pi)^3 2E_3}. \tag{7.37}$$

This is obtained by using the result

$$L_{\mu\nu} W^{\mu\nu} = 4E_1 E_3 (W_2 \cos^2 \theta/2 + 2W_1 \sin^2 \theta/2),$$

which is an intermediate step in the calculation of the differential cross-section. Now the matrix element squared has the form $|\mathcal{M}|^2 = e^4/Q^4 L_{\mu\nu} j^\mu_{\mathrm{p}} j^\nu_{\mathrm{p}}$. Hence, comparison with the general form of the cross-section formula (Appendix C) shows that

$$W^{\mu\nu} \equiv \frac{(2\pi)^4}{4\pi M} \delta^4(p_2 + q - p_4) j^\mu_{\mathrm{p}} j^\nu_{\mathrm{p}} \prod_k \frac{\mathrm{d}^3 p_k}{(2\pi)^3 2E_k}, \tag{7.38}$$

where the product runs over the final state particles in the hadronic system X.

In configuration space, the hadronic tensor is the Fourier transform of (7.38), namely

$$\begin{aligned} W^{\mu\nu} &= \frac{1}{4\pi M} \sum_s \int \mathrm{d}^4\xi \exp(\mathrm{i}q \cdot \xi) \langle p, s | j_\mu(\xi) j_\nu(0) | p, s\rangle \\ &= \frac{1}{4\pi M} \sum_s \int \mathrm{d}^4\xi \exp(\mathrm{i}q \cdot \xi) \langle p, s | [j_\mu(\xi), j_\nu(0)] | p, s\rangle, \end{aligned} \tag{7.39}$$

Fig. 7.6 Interaction of a real photon with a proton. X represents any possible final state.

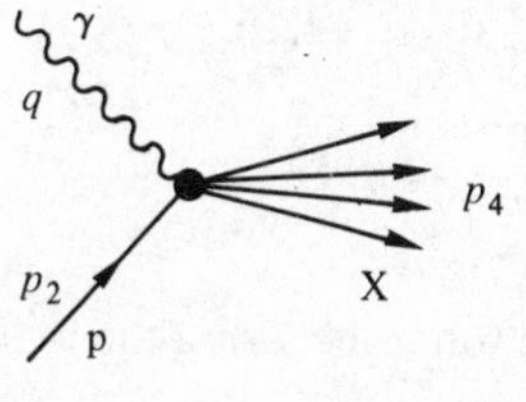

where s is the spin of the incident nucleon state, and $j_\mu(\xi)$ the electromagnetic current at space–time point ξ. The current commutator appears because there is zero contribution from $j_\nu(0)j_\mu(\xi)$, since this corresponds to $p_4 = p_2 - q$ at the lower vertex, which is kinematically forbidden.

Some further insight into Bjorken scaling can be obtained using the so-called *light-cone variables*. If the current is along the z-axis, then we can define $q_\pm = q_0 \pm q_3$ and $\xi_\pm = \xi_0 \pm \xi_3$, and hence $q \cdot \xi = (q_+\xi_- + q_-\xi_+)/2$. In the deep inelastic limit (Q^2, $\nu \to \infty$, x fixed), the light-cone variables of the photon in the lab frame, in which $q = (\nu, 0, 0, (\nu^2 + Q^2)^{1/2})$ are $q_+ \sim 2\nu$ ($\to \infty$) and $q_- \sim -Q^2/(2\nu) = -xM$ (i.e. finite). The dominant contributions from the $\exp(iq \cdot \xi)$ term in (7.39) are expected to come from values of $q \cdot \xi \sim 1$ (where oscillations are least rapid). This corresponds to $\xi_0 - \xi_3 \sim 1/\nu$ and $|\xi_0 + \xi_3| \sim 2/(xM)$ or $\xi_0^2 - \xi_3^2 \sim O(1/Q^2)$. Thus ξ^2 vanishes as $Q^2 \to \infty$, so that the scaling limit corresponds to the properties of the current commutator on the light-cone. The distance scales involved are $\xi_0 \sim |\xi_3| \sim 1/(xM)$.

Using the various rules of Appendix C, the cross-section for the diagram of Fig. 7.6 can be written down, giving

$$\sigma_{\text{tot}}(\gamma \text{p} \to \text{X}) = \frac{4\pi^2\alpha}{k} \varepsilon_\mu^\lambda \varepsilon_\nu^{\lambda *} W^{\mu\nu}, \tag{7.40}$$

where k is the lab energy of the photon, and the expression for $W^{\mu\nu}$ in (7.38) has been used. This expression for the total γp cross-section is strictly only valid for real photons. For virtual photons there are two additional considerations. Firstly, the flux factor K of virtual photons, needed to define the cross-section, cannot be measured directly, and so is somewhat arbitrary. We shall use the convention due to Hand (1963), which takes K to be the energy of the real photon needed to create the state X, i.e. $K = (W^2 - M^2)/2M$ from (7.29). An alternative convention, due to Gilman (1967), is to take $K = (Q^2 + \nu^2)^{1/2}$, i.e. the three-momentum of the virtual photon. These definitions reduce to $K = k$ in the limit $Q^2 \to 0$ as required.

The second difference to real photons is that virtual photons can also have a longitudinal polarisation. We can write the polarisation states as (see Sections 3.3 and 4.5)

$$\varepsilon_{\text{R}}^\mu = \frac{1}{2^{1/2}}(0, 1, i, 0), \qquad \varepsilon_{\text{L}}^\mu = \frac{1}{2^{1/2}}(0, 1, -i, 0),$$

$$\varepsilon_{\text{S}}^\mu = \frac{1}{(Q^2)^{1/2}}((Q^2 + \nu^2)^{1/2}, 0, 0, \nu), \tag{7.41}$$

where we take the virtual photon propagation to be along the z-axis, with

$q = (\nu, 0, 0, (Q^2 + \nu^2)^{1/2})$, that is we have $\varepsilon_\lambda \cdot q = 0$. The polarisation states ε_R and ε_L are transverse (T) polarisation states, whereas ε_S is longitudinal or *scalar* (S). Using $W^{\mu\nu}$ defined by (7.27), it is easy to show that the total γp cross-sections for the different polarisation states are

$$\sigma_T = \frac{\sigma_R + \sigma_L}{2} = \frac{4\pi^2\alpha}{K} W_1, \tag{7.42}$$

$$\sigma_S = \sigma_0 \quad = \frac{4\pi^2\alpha}{K}\left[\left(\frac{Q^2 + \nu^2}{Q^2}\right) W_2 - W_1\right]. \tag{7.43}$$

The cross-sections σ_R and σ_L are equal, as would be expected for a parity-conserving interaction. Alternatively, we can write,

$$F_1 = \frac{MK}{4\pi^2\alpha}\sigma_T = \frac{(2M\nu - Q^2)}{8\pi^2\alpha}\sigma_T, \tag{7.44}$$

and

$$F_2 = \frac{K}{4\pi^2\alpha}\frac{Q^2\nu}{(Q^2 + \nu^2)}(\sigma_S + \sigma_T) = \frac{(2M\nu - Q^2)\nu Q^2}{8\pi^2\alpha M(Q^2 + \nu^2)}(\sigma_S + \sigma_T). \tag{7.45}$$

The differential cross-section in the lab system, equation (7.31), can be written in the form

$$\frac{d^2\sigma}{d\Omega\, dE_3}(\text{lab}) = \Gamma(\sigma_T + \varepsilon\sigma_S), \tag{7.46}$$

where

$$\Gamma = \frac{\alpha K}{2\pi^2 Q^2}\frac{E_3}{E_1}\frac{1}{(1-\varepsilon)}, \tag{7.47}$$

and

$$\varepsilon = \left[1 + 2\left(\frac{\nu^2 + Q^2}{Q^2}\right)\tan^2\frac{\theta}{2}\right]^{-1} \simeq \frac{2(1-y)}{1 + (1-y)^2}. \tag{7.48}$$

The cross-section (7.30), expressed in terms of σ_T, σ_S and ε, is

$$\frac{d^2\sigma}{dQ^2\, d\nu} = \frac{\alpha}{4\pi}\frac{(1-x)}{xME_1^2}\frac{1}{(1-\varepsilon)}(\sigma_T + \varepsilon\sigma_S). \tag{7.49}$$

The quantity ε is called the *polarisation parameter*. The approximate form of ε in (7.48) is valid for $Q^2/E_1^2 = 2Mxy/E_1 \ll 1$. It can be seen that ε, and hence the σ_S contribution, is largest for $y \sim 0$. For large y values $\varepsilon \to 0$, so only the transverse polarisation states contribute.

The ratio of the cross-sections for longitudinal and transverse virtual photons is, from (7.44) and (7.45),

$$R(x, Q^2) = \frac{\sigma_S}{\sigma_T} = \frac{F_2(x, Q^2)}{2xF_1(x, Q^2)}\left(1 + \frac{4M^2x^2}{Q^2}\right) - 1$$

$$\simeq \frac{F_2(x, Q^2) - 2xF_1(x, Q^2)}{2xF_1(x, Q^2)}. \tag{7.50}$$

It is instructive to consider the value of R for elastic ep scattering. From (7.32) and (7.35) we can write, changing the argument of the δ-function to x (Appendix D),

$$F_2(x, Q^2) = \frac{2MG_E^2(Q^2) + yE_1G_M^2(Q^2)}{2M + yE_1}\delta(1-x),$$

$$F_1(x, Q^2) = \frac{G_M^2(Q^2)}{2}\delta(1-x). \tag{7.51}$$

From which R for elastic scattering ($x = 1$) can be expressed as

$$R_{el} = \frac{4M^2}{Q^2}\frac{G_E^2(Q^2)}{G_M^2(Q^2)} = \frac{4M^2}{Q^2}\frac{1}{\mu_p^2}, \tag{7.52}$$

where the relation (7.15) has been used. For a point-like Dirac particle $R = 4M^2/Q^2$, and so tends to zero for large Q^2. For elastic ep scattering we obtain $R_{el} \simeq 0.45/Q^2(\text{GeV}^2)$. Use of this estimate shows that G_E contributes little to elastic scattering for $Q^2 \gtrsim 5\ \text{GeV}^2$, so that it is essentially G_M^2 which is measured in this region.

The relation between the kinematic variables in deep inelastic scattering is shown in Fig. 7.7. The lab energy of the virtual photon is in the range $0 < \nu < E_1$. The maximum value of Q^2 is $2ME_1$, and corresponds to $x = 1$ (elastic scattering). Contours of fixed W are lines parallel to the $x = 1$ boundary. A line of constant scattering angle θ is also shown. Note that, in practice, the $1/Q^4$ propagator term means that most events occur at low Q^2 and that a sizeable θ selection is required to probe high Q^2.

7.3 THE QUARK–PARTON MODEL

The discovery of hard point-like scattering centres in the nucleon came about in a similar way to the discovery of the nucleus by Rutherford. The scattering cross-section was discovered to be much greater than expected at large scattering angles. These expectations for the nucleon were based on the Mott scattering formula, modified by the form factors of the nucleon (or excited nucleons). Figure 7.8 shows some results from

SLAC on measurements of the proton structure function F_2 at $x = 0.25$. The data span angles between 6° and 26° and have $W > 2$ GeV, so that the resonance region does not contribute. A detailed account of the various measurements can be found in Friedman and Kendall (1972). It can be seen that the structure function F_2 remains constant at large Q^2 (or scattering angles). This is indicative that the proton has some substructure.

Fig. 7.7 Kinematics of deep inelastic lepton–nucleon scattering.

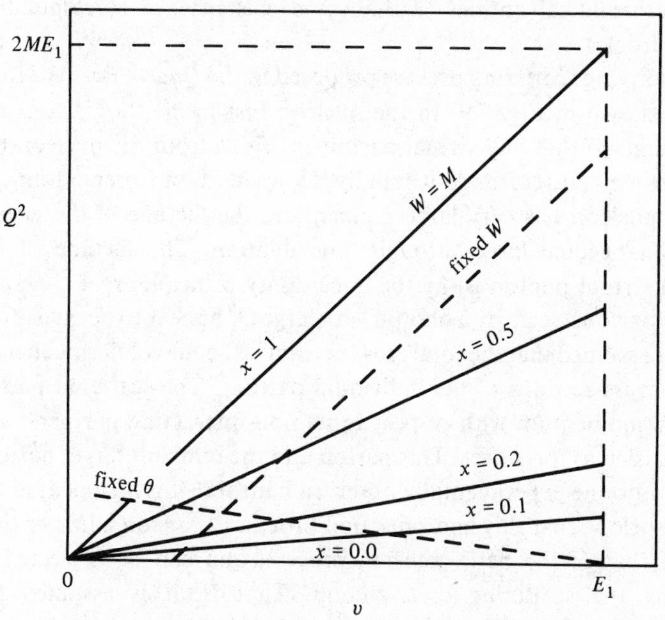

Fig. 7.8 Some results of measurements made at SLAC on the proton structure function F_2 at $x = 0.25$, as a function of Q^2.

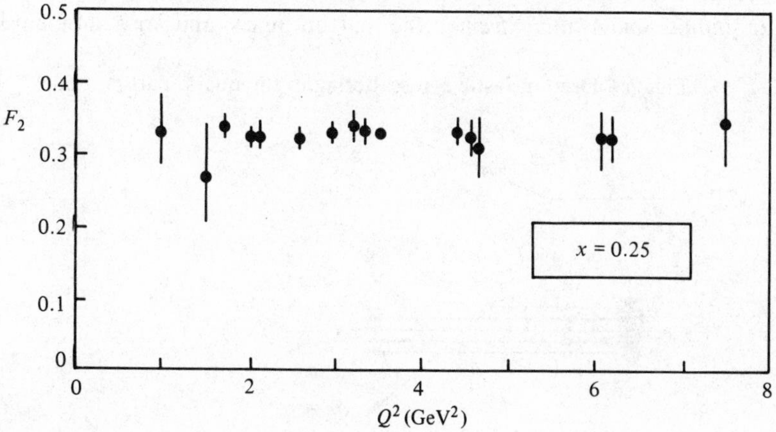

A major advance in our understanding of matter came from the interpretation of these (and other) results by Feynman, who proposed that the nucleon consisted of dynamic point-like scattering centres, called *partons*: see Feynman (1972). The parton scattering centres were later identified as quarks. It is worth emphasising that, at that time, the idea of quarks to explain hadron spectroscopy was widely accepted but, since free quarks did not appear to exist, many people thought that the quarks must be very massive and tightly bound. Others took the quarks to be merely mathematical entities, without the existence of corresponding physical particles.

The underlying scattering process proposed in the *Quark–Parton Model* (QPM) is shown in Fig. 7.9. In the nucleon rest frame the partons are tightly bound, so that any virtual parton emerging from the nucleon has a short lifetime and returns to it rapidly. However, in a Lorentz frame in which the nucleon has very large momentum, the lifetime of the virtual partons can become large through time dilation. The lifetime of the exchanged virtual photon is, by the uncertainty principle, $\tau_\gamma \sim 1/(Q^2)^{1/2}$. Hence, the partons 'seen' by a photon with large Q^2 appear to be quasi-free. It is further assumed that the total cross-section is the sum of the incoherent scattering cross-sections of the individual partons. The scattered parton has a large momentum with respect to its non-interacting partners, and can be regarded as quasi-free. This parton and the remnant target partons 'fragment' into the experimentally observed hadrons. It is assumed, in the discussion below, that this hadronisation process occurs on a longer time scale than that of the hard collision process, and can be neglected in determining the scattering cross-section. That is, it is assumed the hadronisation occurs with unit probability once the partons have been set in motion. It is, however, difficult to check this hypothesis experimentally.

In the frame in which the nucleon momentum is very large (often called the *infinite momentum frame*) the parton mass and any momentum

Fig. 7.9 Deep inelastic e–p scattering in the quark–parton model.

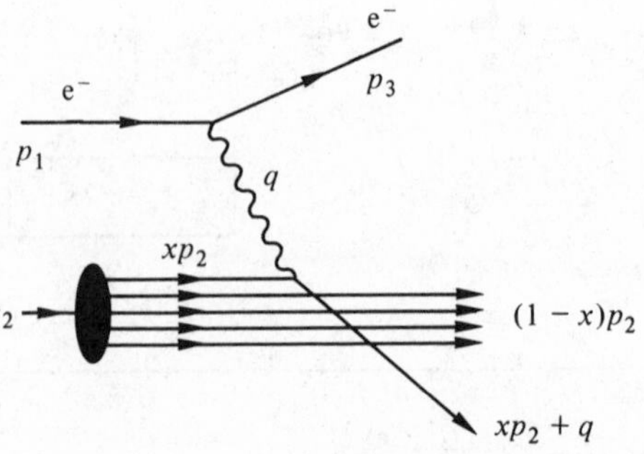

transverse to the nucleon direction can (for the moment) be neglected. Hence, the four-momentum of the parton i is just $p_i = xp_2$, where p_2 is the nucleon four-monentum and x is the fraction of this momentum carried by the parton. Note that, with these assumptions, the mass of the parton is $m_i = xM$

In the rest frame of the target $p_2 = (M, 0, 0, 0)$ and, in the Bjorken limit, $p_i = (p_i^0, 0, 0, -p_i^0)$ and $q = (\nu, 0, 0, \nu + Q^2/2\nu)$. Kinematics requires that the light-cone momenta satisfy $q_- + (p_i)_- = 0$, thus $p_i^0 = xM/2$ and $(p_i)_- = xM$. That is, $x = (p_i)_-/(p_2)_-$ is the fraction of the nucleon's light-cone momentum carried by the parton, and this description is frame-invariant.

The differential cross-section for a point-like quark i of charge e_i, mass m_i and spin $\frac{1}{2}$ is, from (7.23),

$$\frac{d^2\sigma}{dQ^2\,d\nu} = \frac{\pi\alpha^2}{4E_1^3E_3\sin^4\theta/2} \times \left[e_i^2\cos^2\frac{\theta}{2} + e_i^2\frac{Q^2}{4m_i^2}2\sin^2\frac{\theta}{2}\right]\delta\left(\nu - \frac{Q^2}{2m_i}\right). \tag{7.53}$$

The corresponding structure functions, from (7.30) in the point-like limit, are

$$W_1^i = \frac{F_1^i}{M} = e_i^2\frac{Q^2}{4M^2x^2}\delta\left(\nu - \frac{Q^2}{2Mx}\right),$$
$$W_2^i = \frac{F_2^i}{\nu} = e_i^2\,\delta\left(\nu - \frac{Q^2}{2Mx}\right). \tag{7.54}$$

Using the properties of the δ-function (Appendix (D.4)), we find

$$F_2^i = e_i^2x\,\delta\left(x - \frac{Q^2}{2M\nu}\right), \qquad F_1^i = \frac{e_i^2}{2}\delta\left(x - \frac{Q^2}{2M\nu}\right). \tag{7.55}$$

If $q_i(x)$ is the probability density function for a quark of type i, then

$$F_2(x, Q^2) = \sum_i\int_0^1 dx q_i(x)e_i^2x\,\delta\left(x - \frac{Q^2}{2M\nu}\right) = \sum_i e_i^2xq_i(x). \tag{7.56}$$

That is, the structure function $F_2(x, Q^2)$ corresponds to the total momentum fraction carried by all the quarks and antiquarks in the nucleon, weighted by the squares of the quark charges.

The structure function F_1 is related to F_2 by the *Callan–Gross* relation

$$F_2(x, Q^2) = 2xF_1(x, Q^2). \tag{7.57}$$

This can be obtained directly from (7.55), but was originally derived using current algebra (Callan and Gross, 1969). The relation stems from the spin $\frac{1}{2}$ nature of the quarks. If we would have taken spin 0 quarks then, using (4.139), we can see that the F_2 (or W_2) term has the same form as for spin $\frac{1}{2}$, but the transverse $F_1(W_1)$ term, proportional to $\sin^2 \theta/2$, is zero. Hence, from (7.50) we obtain the following predictins for R for different quark spin assignments

$$\text{spin } 0 \qquad R \to \infty,$$

$$\text{spin } \tfrac{1}{2} \qquad R = Q^2/\nu^2 = \frac{4M^2x^2}{Q^2} \xrightarrow{\nu^2 \gg Q^2} 0. \tag{7.58}$$

These results can be seen directly by considering the helicity configurations in the Breit frame (Fig. 7.10), in which the γ^* has zero energy, so the momentum of the struck quark is reversed. A spin $\frac{1}{2}$ quark can absorb a transverse but not a longitudinal photon ($R \to 0$), whereas a spin 0 quark can absorb a longitudinal but not a transverse photon ($R \to \infty$).

In order to measure R experimentally, at a given x and Q^2 point, one must vary y (see (7.46) and (7.48)). Since $Q^2 = 2ME_1xy$, this means measuring the cross-section at different incident beam energies. These measurements are far from easy. Some results on the measurement of R in e^-p collisions are shown in Fig. 7.11 (data from Mestayer *et al.*, 1983). It can be seen that $R \sim 0.2$, thus excluding the possibility that all the quarks have spin zero.

Non-zero values of $R(x, Q^2)$ are expected from deviations from the simplistic assumptions made above. For example, the parton mass will give a contribution to $R \sim 4m^2/Q^2$ (see (7.52)). There is also a contribution to R from the momentum component (k_T) of the struck parton, transverse to the direction of the virtual photon, which we will now consider. It is assumed that k_T is relatively small and bounded in value and we neglect here the parton mass. We choose the z-axis to be along the direction of the virtual photon. Thus in the lab frame $q = (\nu, 0, 0, (\nu^2 + Q^2)^{1/2})$, $p_1^{\text{lab}} = E_1(1, \sin\alpha, 0, \cos\alpha)$ and $p_3^{\text{lab}} = E_3(1, \sin\beta, 0, \cos\beta)$, where the angles

Fig. 7.10 Virtual photon–quark scattering in the Breit frame. The configuration shown is for a righthanded spin $\frac{1}{2}$ quark absorbing a transverse photon.

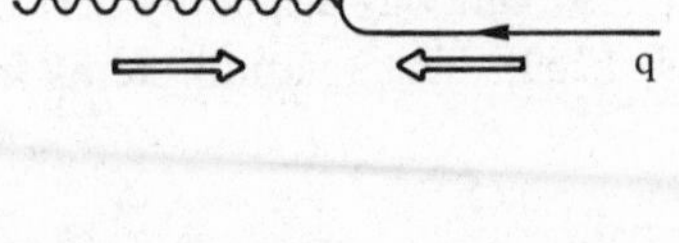

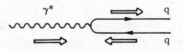

α and β are given by (using $p_3 \cdot q = -p_1 \cdot q = Q^2/2$)

$$\sin\alpha = \left(\frac{Q^2 E_3}{(Q^2+\nu^2)E_1}\right)^{1/2} \cos\frac{\theta}{2},$$

$$\sin\beta = \left(\frac{Q^2 E_1}{(Q^2+\nu^2)E_3}\right)^{1/2} \cos\frac{\theta}{2}. \tag{7.59}$$

In the Breit frame $q^{\mathrm{B}} = (0, 0, 0, Q)$ with $Q = (Q^2)^{1/2}$, so that the Lorentz transformation from the lab to the Breit frame has $\beta = \nu/(\nu^2+Q^2)^{1/2}$, $\gamma = (\nu^2+Q^2)^{1/2}/Q$. The momenta of the particles in the Breit frame are

$$p_1^{\mathrm{B}} = E_1[((\nu^2+Q^2)^{1/2} - \nu\cos\alpha)/Q, \sin\alpha, 0, ((\nu^2+Q^2)^{1/2}\cos\alpha - \nu)/Q],$$

$$p = \left[\frac{Q}{2}(1+4k_{\mathrm{T}}^2/Q^2)^{1/2}, k_{\mathrm{T}}\cos\phi, k_{\mathrm{T}}\sin\phi, -\frac{Q}{2}\right]. \tag{7.60}$$

The scattered electron four-momentum (p_3^{B}) is the same as p_1^{B} but with the z-component reversed. The four-momentum p is that of the struck parton, which is assumed to have $p_z = x(p_2^{\mathrm{B}})_z = -Q/2$ (from momentum conservation). The cross-section for elastic electron–parton scattering is

Fig. 7.11 Measurements of $R = \sigma_{\mathrm{L}}/\sigma_{\mathrm{T}}$ as a function of x from ep scattering. Q^2 (in GeV)^2): ●, 3; ○, 6; ▲, 9; ▽, 12; ■, 15; □, 18.

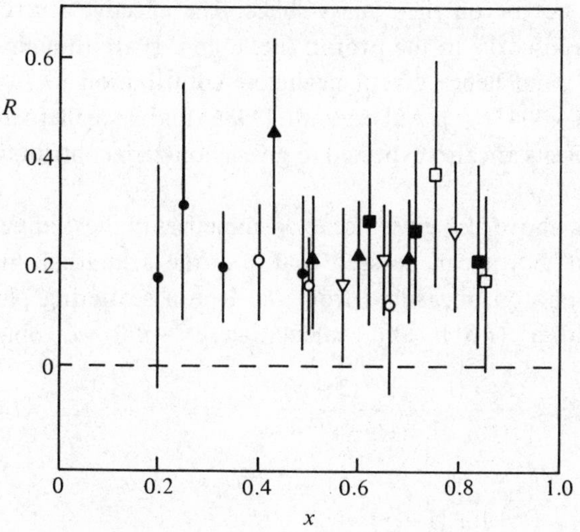

proportional to $\hat{s}^2 + \hat{u}^2$ (see (4.46)), where $\hat{s} \simeq 2p_1^{\mathrm{B}} \cdot p$ and $\hat{u} \simeq -2p_3^{\mathrm{B}} \cdot p$ (the hat ^ denotes a subprocess). From (7.59) and (7.60) we find, for $\nu^2 \gg Q^2$,

$$\hat{s} = 2ME_1 x\left[1 + (2-y)\frac{k_{\mathrm{T}}^2}{Q^2} - 2\frac{k_{\mathrm{T}}}{Q}(1-y)^{1/2}\cos\phi\right],$$

$$\hat{u} = -2ME_1 x(1-y)\left[1 + \left(\frac{2-y}{1-y}\right)\frac{k_{\mathrm{T}}^2}{Q^2} - \frac{2k_{\mathrm{T}}}{Q(1-y)^{1/2}}\cos\phi\right]. \quad (7.61)$$

The contribution to the total cross-section is found by integrating $\hat{s}^2 + \hat{u}^2$ over ϕ, from 0 to 2π. Terms odd in $\cos\phi$ give zero contribution, hence

$$\sigma \sim \hat{s}^2 + \hat{u}^2 = 8\pi M^2 E_1^2 x^2\left\{\left(1 + \frac{2k_{\mathrm{T}}^2}{Q^2}\right)[1 + (1-y)^2] + \frac{8k_{\mathrm{T}}^2}{Q^2}(1-y)\right\}. \quad (7.62)$$

Note that we recover the form $1 + (1-y)^2$ in the limit $k_{\mathrm{T}} \to 0$. Comparison with (7.36) (with $Mxy/2E_1 \to 0$) and (7.48) and (7.49) shows that the intrinsic transverse momentum k_{T} gives a longitudinal contribution (coefficient of $2(1-y)$) such that, when averaged over many events,

$$R = \frac{\sigma_{\mathrm{S}}}{\sigma_{\mathrm{T}}} = 4\frac{\langle k_{\mathrm{T}}^2\rangle}{Q^2}. \quad (7.63)$$

If this is the source of the measured value of $R \sim 0.2$ for $Q^2 \simeq 10\ \mathrm{GeV}^2$ (Fig. 7.11) then $\langle k_{\mathrm{T}}^2\rangle \sim 0.5\ \mathrm{GeV}^2$. More generally, we have $R \simeq 4(\langle k_{\mathrm{T}}^2\rangle + m^2)/Q^2$, as discussed above. Note that, by introducing the quark transverse momentum, we can assign a unique mass to the parton, and no longer require that $m_i = xM$. However, since the partons are confined in the nucleon, they need not be on their mass shells. The effective (current) masses of the valence quarks in the proton (i.e. u and d) are thought to be only a few MeV, and hence give a negligible contribution to R. At larger values of Q^2 ($\sim 30\ \mathrm{GeV}^2$), Aubert *et al.* (1986) find a smaller value of R ($\lesssim 0.1$). QCD effects are also expected to give a non-zero contribution to R (Section 7.7.4).

The k_{T} mechanism above also gives rise to asymmetries in the ϕ angular distribution. From (7.60) it can be seen that ϕ is the azimuthal angle around the virtual photon, measured from the lepton scattering plane ($\phi = 0$). Starting from (7.61), and calculating $\hat{s}^2 + \hat{u}^2$, we obtain

$$\langle\cos\phi\rangle = -2\frac{\langle k_{\mathrm{T}}\rangle}{Q}\frac{(1-y)^{1/2}(2-y)}{1 + (1-y)^2}, \quad (7.64)$$

$$\langle\cos 2\phi\rangle = 2\frac{\langle k_{\mathrm{T}}^2\rangle}{Q^2}\frac{(1-y)}{1 + (1-y)^2}. \quad (7.65)$$

These considerations imply that there should be sizeable asymmetries at the parton level. However, partons are not observed directly, and such effects can only be inferred from studying the properties of the jets of hadrons into which the partons fragment. The hadronisation process,

Fig. 7.12 Measurements of the structure function F_2 as a function of x, for different Q^2 values, for (*a*) F_2^p and (*b*) $F_2^D/2$. The data are from SLAC (Bodek *et al.*, 1979) and the EMC (Aubert *et al.*, 1986). The broken and full lines are drawn through the F_2^p and $F_2^D/2$ points at $Q^2 \sim 5$ GeV^2. Q^2 (GeV^2): ●, 2; ○, 4; ■, 6; □, 8; ▲, 15; △, 48.

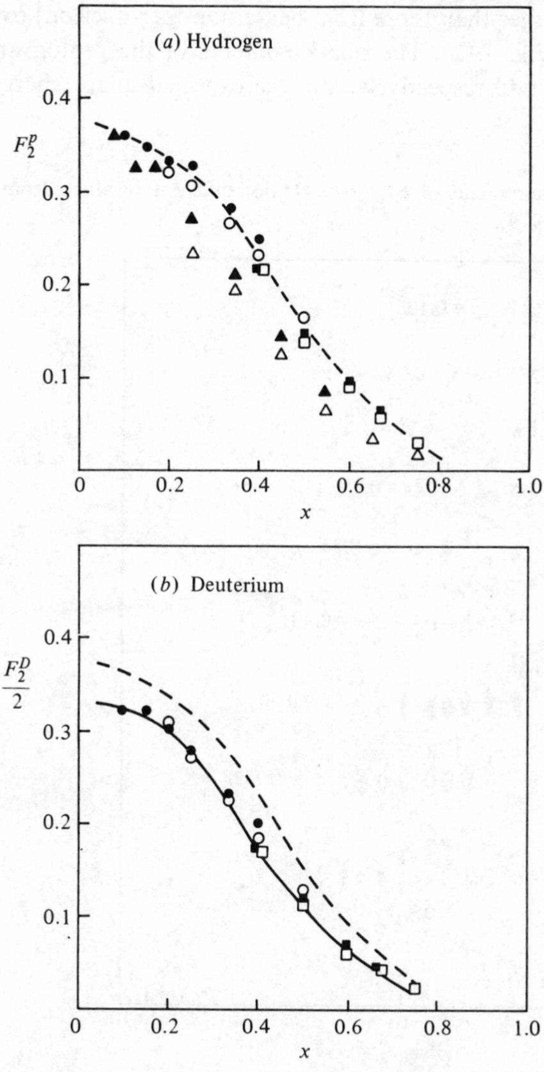

which is discussed further in Section 7.9, leads to considerable dilution of the asymmetries.

Some data on the structure function F_2, obtained from proton and deuterium targets, are shown in Figs. 7.12 and 7.13. It can be seen that F_2 falls off rapidly at large x. At low x, there is some increase with increasing Q^2 and, at large x, F_2 falls with increasing Q^2. For $x \sim 0.25$, F_2 is roughly Q^2 independent. Thus these data (together with those of other data sets) show significant *scaling violations*. However, these deviations from scaling are relatively small, and the ideas proposed to explain scaling are expected to form the basis of a more complete explanation. The values of F_2 from a proton target are larger than those from deuterium (per nucleon) over the whole x range (Fig. 7.12). The quark contents of the proton and neutron are uud and udd respectively, and the expected quark charges

Fig. 7.13 Measurements of F_2^p versus Q^2, for different x values; from Bodek *et al.* (1979).

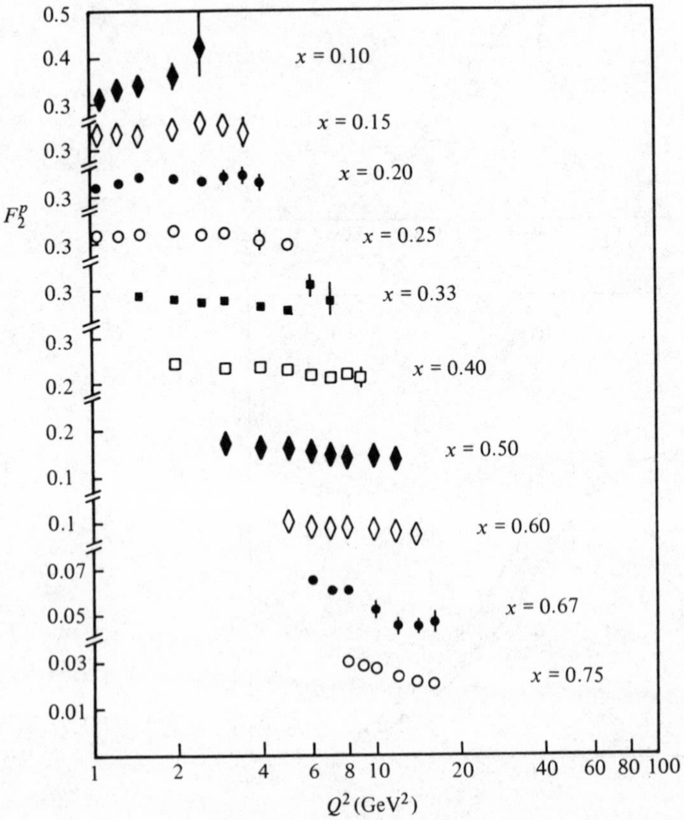

are $e_u = \frac{2}{3}$ and $e_d = -\frac{1}{3}$. Hence our first guess for F_2 from (7.56) might be

$$F_2^p = \tfrac{4}{9}x\,u(x) + \tfrac{1}{9}x\,d(x), \qquad F_2^n = \tfrac{1}{9}x\,u(x) + \tfrac{4}{9}x\,d(x),$$
$$F_2^D/2 = (F_2^p + F_2^n)/2 = \tfrac{5}{18}[x\,u(x) + x\,d(x)]. \tag{7.66}$$

Here $u(x)$ and $d(x)$ are the density distributions of u and d quarks, and it is assumed, from isospin invariance, that the distribution of u(d) quarks in a proton is equal to that of d(u) quarks in a neutron. Hence the quark distributions all refer to the proton. Now we have seen that $F_2^p > F_2^D/2$, from which we deduce that $u(x) > d(x)$. Note that, in (7.66), it is assumed that F_2^D is the average of F_2^p and F_2^n. This ignores the effects of nuclear binding in deuterium (Fermi motion), which can lead to significant deviations from this assumption, particularly at large x. Furthermore, one nucleon may 'shadow' the other from the virtual photon (this may be important at low x; see Section 7.7.4). The possibility also exists that, at least part of the time, the deuteron is in a six-quark state or 'bag'.

The integral of $F_2(x, Q^2)$, over x from 0 to 1, gives the total momentum fraction of the proton carried by its charged constituents at a particular Q^2, weighted by the squares of the parton charges. Using the broken and full lines in Fig. 7.12 as rough guides to the data at $Q^2 \sim 5\ \text{GeV}^2$, we obtain

$$\int_0^1 F_2^p(x)\,dx = 0.18, \qquad \int_0^1 \frac{[F_2^p(x) + F_2^n(x)]}{2}\,dx = 0.15. \tag{7.67}$$

If ε_u and ε_d are the momentum fractions carried by the u and d quarks in a proton, then $F_2^p = (4\varepsilon_u + \varepsilon_d)/9$ and $F_2^D/2 = (5\varepsilon_u + 5\varepsilon_d)/18$. Using the values from (7.67) gives $\varepsilon_u = 0.36$ and $\varepsilon_d = 0.18$, and hence $\varepsilon_u + \varepsilon_d = 0.54$. Although these values should be treated as a rather crude estimate they, nevertheless, illustrate the important point that the charged constituents of the proton carry only approximately 50% of its momentum. Thus, if the parton picture is correct, then almost half of the momentum is carried by electrically neutral constituents. These neutral partons are the gluons.

The above observations on the total momentum fraction, and on the pattern of scaling violations, have led to the following picture of deep inelastic scattering. The 'resolution' of the virtual photon, that is the distance down to which the scattering cross-section is sensitive, is given roughly by the uncertainty principle $\Delta x = 0.2/Q(\text{GeV})$ fm. For low values of Q^2 ($\lesssim 0.5\ \text{GeV}^2$) the distance resolved is of the order of the nucleon radius, and the scattering occurs on the whole nucleon, leading predominantly to elastic scattering. At higher values of Q^2 ($\sim$few GeV^2), the virtual photon probes the internal structure of the nucleon, and scattering from the three constituent (or valence) quarks occurs. At still higher values of

Q^2 a new phenomenon is resolved. The gluons, which are assumed responsible for the forces confining the quarks inside the nucleon, participate in the vertex transitions $q \to q + g$, $g \to q + \bar{q}$, $\bar{q} \to \bar{q} + g$ and $g \to g + g$, where $\bar{q}$ represents an antiquark. The effect of these transitions is that the nucleon's momentum appears to be shared between more and more partons as the Q^2 of the probe is increased. This is illustrated in Fig. 7.14. Thus the valence quark distribution becomes softer as Q^2 increases, and there is a corresponding growth in the 'sea' (or 'ocean') of q–$\bar{q}$ pairs, which are mainly at low x (Fig. 7.15). There is no bound on the number of very soft partons, but this does not lead to potentially infinite observable quantities. In principle the gluon can produce a quark–antiquark pair of any flavour, but the production of heavy quarks is suppressed relative to light quarks because of the additional energy required to create the pair. The gluon momentum distribution is expected to be between that of the sea and valence quark distributions. In some models the valence quark blobs of Fig. 7.14(*b*) represent the 'bare' valence quarks, surrounded by their own clusters of associated gluons and quark–antiquark pairs.

A spatial picture of the nucleon, as seen by a γ^* with $Q^2 \sim$ few GeV^2, can be built up by considering the Fourier transform of the quark momentum distribution, from which the distribution of the quark light-cone correlation length ξ can be found (Llewellyn Smith, 1985). For valence quarks with $x \gtrsim 0.1$, typical lengths are $\xi \lesssim 1$ fm (Fig. 7.15(*c*)). However, for sea quarks (or gluons) the peaking of the momentum distribution at small x leads to a long tail in ξ (Fig. 7.15(*d*)). Even for sea quarks with $x \gtrsim 0.05$, a typical correlation length is $\langle \xi \rangle \sim 2.5$ fm, with a sizeable tail up to approximately 10 fm. For lower x, these distances become longer. Hence, we can picture the nucleon as composed of a core of hard ($x \gtrsim 0.1$) valence quarks with $(R^2)^{1/2} \lesssim 0.5$ fm, surrounded by à large halo of sea quarks and gluons, the tails of which extend to infinity.

Fig. 7.14 Schematic diagram showing the change in the resolution of the virtual photon with increasing Q^2.

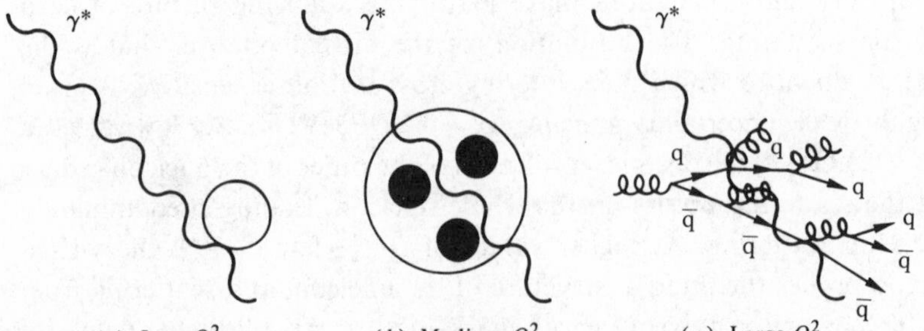

(*a*) Low Q^2 (*b*) Medium Q^2 (*c*) Large Q^2

The structure function F_2 can, in general, be written as (cf. (7.66))

$$F_2^{\mathrm{ep}}(x, Q^2) = \tfrac{4}{9}xu_{\mathrm{v}} + \tfrac{1}{9}xd_{\mathrm{v}} + \tfrac{4}{9}x(u_{\mathrm{s}} + \bar{u}_{\mathrm{s}}) + \tfrac{1}{9}x(d_{\mathrm{s}} + \bar{d}_{\mathrm{s}}) + \tfrac{1}{9}x(s + \bar{s}) + \tfrac{4}{9}x(c + \bar{c}) + \ldots, \quad (7.68)$$

where $u_{\mathrm{v}} = u_{\mathrm{v}}(x, Q^2)$, etc. The subscripts v and s for the u and d quarks denote valence and sea respectively. Thus the total u-quark density is $u = u_{\mathrm{v}} + u_{\mathrm{s}}$. Similarly,

$$F_2^{\mathrm{en}}(x, Q^2) = \tfrac{1}{9}xu_{\mathrm{v}} + \tfrac{4}{9}xd_{\mathrm{v}} + \tfrac{1}{9}x(u_{\mathrm{s}} + \bar{u}_{\mathrm{s}}) + \tfrac{4}{9}x(d_{\mathrm{s}} + \bar{d}_{\mathrm{s}}) + \tfrac{1}{9}x(s + \bar{s}) + \tfrac{4}{9}x(c + \bar{c}) + \ldots, \quad (7.69)$$

where the u and d quark distributions refer to the proton ($u_{\mathrm{p}} = d_{\mathrm{n}}$, etc.).

From the mechanism $\mathrm{g} \to \mathrm{q}\bar{\mathrm{q}}$ one would expect the sea distributions of u and d quarks to be the same, as they have a similar mass; however, this equality is not, in general, required. It can be seen from (7.68) and (7.69) that the difference between F_2^{ep} and F_2^{en} (the latter is obtained from

Fig. 7.15 Change in the valence and sea-quark momentum distributions from (*a*) medium Q^2 to (*b*) large Q^2. In (*c*) and (*d*) the estimated light-cone correlation lengths for valence and sea-quarks respectively are shown.

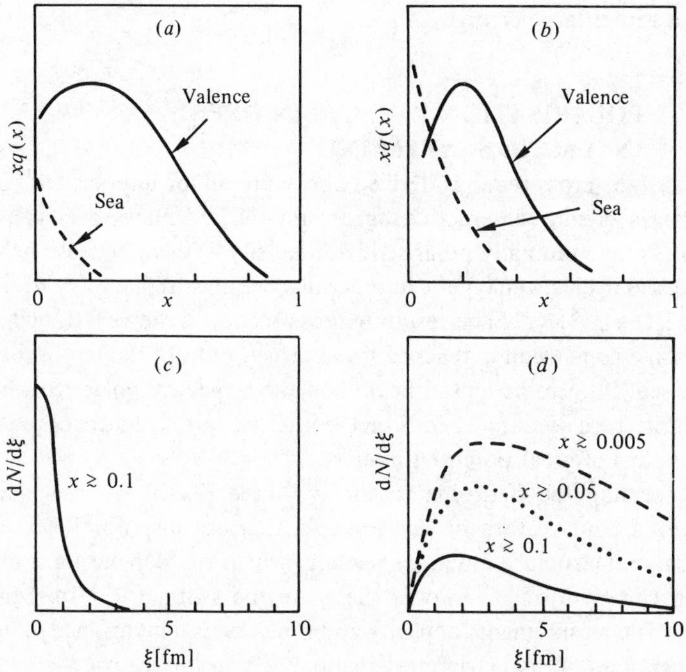

$F_2^{\text{eD}} - F_2^{\text{ep}}$, and applying corrections for Fermi motion), is

$$F_2^{\text{ep}}(x, Q^2) - F_2^{\text{en}}(x, Q^2) = \tfrac{1}{3}x(u_\text{v} - d_\text{v}). \tag{7.70}$$

That is, the sea contribution (assuming it is the same for p and n) cancels. A test of the validity of the above ideas can be made by constructing the integral

$$I_\text{G} = \int_0^1 \frac{\text{d}x}{x}[F_2^{\text{ep}}(x) - F_2^{\text{en}}(x)] = (e_\text{u}^2 - e_\text{d}^2)\int_0^1 \text{d}x[u_\text{v}(x) - d_\text{v}(x)] = \tfrac{1}{3}. \tag{7.71}$$

That is, I_G is proportional to the difference in the number of u and d valence quarks in the proton (*Gottfried sum rule*), as measured by the dynamic properties of the nucleon. Experimentally the data cover only part of the total x range. For high energy (120–280 GeV) μp and μn data (Aubert *et al.*, 1983a), it is found that, for $\langle Q^2 \rangle \sim 20\ \text{GeV}^2$,

$$I'_\text{G} = \int_{0.03}^{0.65} \frac{[F_2^{\mu\text{p}}(x) - F_2^{\mu\text{n}}(x)]}{x}\,\text{d}x = 0.18 \pm 0.01 \pm 0.07, \tag{7.72}$$

where the first error is statistical and the second systematic. Extrapolating the measured values to low x to complete the integral yields $I_\text{G} = 0.24 \pm 0.02 \pm 0.13$; in agreement with the QPM prediction of $e_\text{u}^2 - e_\text{d}^2 = \tfrac{1}{3}$, but with a rather large error.

7.4 POLARISATION EFFECTS IN DEEP INELASTIC SCATTERING

The cross-sections derived above are all for unpolarised electron (or muon) beams, interacting on an unpolarised nucleon target. If the lepton is longitudinally polarised with helicity λ, then, because helicity is conserved in electromagnetic interactions, we must replace $\bar{u}_3\gamma_\mu u_1$, in (7.1) by $\bar{u}_3\gamma_\mu(1 + \lambda\gamma^5)u_1/2$. The resulting cross-section is, however, unchanged. The same conclusion is reached if the target, but not the lepton beam, is polarised. If both the lepton beam and the target are polarised, then the resulting cross-section depends on the internal spin structure of the target, and is thus of great potential interest.

In forming the hadronic tensor $W^{\mu\nu}$ (see (7.25)) we now have an additional four-vector, the nucleon spin s, at our disposal. There are, in general, ten structure functions (each of which can depend on x and Q^2) which can be formed. Two of these are the symmetric terms given in (7.27). The requirements of parity and time-reversal invariance reduce the number of additional structure functions to two (for a review see Hughes and Kuti, 1983). The additional (antisymmetric) piece to the tensor $W^{\mu\nu}$ can be written

$$W_{\mathrm{A}}^{\mu\nu} = \mathrm{i}MG_1\varepsilon^{\alpha\mu\beta\nu}q_\alpha s_\beta + \mathrm{i}\frac{G_2}{M}\varepsilon^{\alpha\mu\beta\nu}q_\alpha[(p_2\cdot q)s_\beta - (s\cdot q)p_{2\beta}]. \tag{7.73}$$

This term changes sign if s is reversed. The lepton vertex tensor is

$$L_{\mu\nu} = \tfrac{1}{4}\,\mathrm{tr}[\not{p}_3\gamma_\mu(1+\lambda\gamma^5)\not{p}_1\gamma_\nu(1+\lambda\gamma^5)], \tag{7.74}$$

which gives the additional antisymmetric piece

$$L_{\mu\nu}^{\mathrm{A}} = 2\mathrm{i}\lambda p_3^\theta p_1^\phi \varepsilon_{\theta\mu\phi\nu}. \tag{7.75}$$

Following the usual techniques we obtain, for a longitudinally polarised lepton beam ($\lambda = 1$),

$$|\mathcal{M}|^2 = \frac{e^4}{q^4}\left\{\frac{2W_2}{M^2}[2(p_1\cdot p_2)(p_2\cdot p_3) - M^2(p_1\cdot p_3)] - 2W_1q^2 \right.$$
$$\left. - 2q^2MG_1 s\cdot(p_1+p_3) - \frac{4q^2G_2}{M}[(p_1\cdot p_2)(s\cdot p_3) - (p_2\cdot p_3)(s\cdot p_1)]\right\}. \tag{7.76}$$

In the lab frame we have $p_1 = E_1(1, 0, 0, 1)$, $p_3 = E_3(1, 0, \sin\theta, \cos\theta)$ and the polarisation vector of the proton (or neutron) is $s = (0, \hat{\mathbf{s}})$ with $s\cdot p_2 = 0$. If the nucleon is polarised longitudinally with $s = (0, 0, 0, 1)$ (i.e. along the incident beam direction), then we have two possible spin configurations, namely parallel ($\uparrow\uparrow$) or antiparallel ($\uparrow\downarrow$). The sum of the differential cross-sections (cf. (7.31)) for these two configurations is

$$\frac{d^2\sigma}{\mathrm{d}\Omega\,\mathrm{d}E_3}(\uparrow\downarrow + \uparrow\uparrow) = \frac{8\alpha^2E_3^2}{Q^4}\left[W_2\cos^2\frac{\theta}{2} + 2W_1\sin^2\frac{\theta}{2}\right], \tag{7.77}$$

which is just twice the unpolarised cross-sections. Equation (7.77) can be converted into other variables as before. The structure functions G_1 and G_2 drop out of this sum because the sign of s changes. However, if we take the difference, then W_1 and W_2 drop out, and we obtain

$$\frac{d^2\sigma}{\mathrm{d}\Omega\,\mathrm{d}E_3}(\uparrow\downarrow - \uparrow\uparrow) = \frac{4\alpha^2E_3}{Q^2E_1}[MG_1(E_1 + E_3\cos\theta) - Q^2G_2]. \tag{7.78}$$

In the case of the nucleon being transversely polarised, with $s = (0, s_x, s_y, 0)$, then we find a further combination of G_1 and G_2 from which these structure functions can be extracted, namely

$$\frac{d^2\sigma}{\mathrm{d}\Omega\,\mathrm{d}E_3}(\uparrow\rightarrow - \uparrow\leftarrow) = \frac{-4\alpha^2E_3^2s_y\sin\theta}{Q^2E_1}[MG_1 + 2E_1G_2]. \tag{7.79}$$

From the structure functions $G_1(\nu, Q^2)$, $G_2(\nu, Q^2)$ it is usual to define

$$g_1(x, Q^2) = M^2 \nu G_1(\nu, Q^2),$$
$$g_2(x, Q^2) = M \nu^2 G_2(\nu, Q^2), \tag{7.80}$$

which are expected, from the parton model, to scale in the Bjorken limit ($\nu, Q^2 \to \infty$, x finite). In order to interpret these structure functions in terms of the parton model, it is useful first to consider the possible polarisation states of the photon and nucleon. As discussed in Section 7.2, we must evaluate the contribution to σ_{tot} from

$$\sigma_{\text{tot}} \propto \varepsilon_\mu^\lambda \varepsilon_\nu^{\lambda *} W^{\mu\nu} = \varepsilon_\mu^\lambda \varepsilon_\nu^{\lambda *} (W_{\text{S}}^{\mu\nu} + W_{\text{A}}^{\mu\nu}), \tag{7.81}$$

where $W_{\text{S}}^{\mu\nu}$ and $W_{\text{A}}^{\mu\nu}$ are given by (7.27) and (7.73) respectively.

If the photon and nucleon spins are aligned along the z-axis (i.e. $\sigma_{3/2}$), then we must use ε_{R} and $s^\mu = (0, 0, 0, 1)$. If the nucleon spin is in the opposite direction, then we obtain $\sigma_{1/2}$. Starting from (7.81), and using the photon polarisation states given by equation (7.41), we obtain, in the Bjorken limit,

$$\sigma_{3/2} \propto W_1 - M\nu G_1 - q^2 G_2,$$
$$\sigma_{1/2} \propto W_1 + M\nu G_1 + q^2 G_2. \tag{7.82}$$

Using (7.35) and (7.80), these can be rewritten as

$$\sigma_{3/2} \propto F_1(x, Q^2) - g_1(x, Q^2) = \sum_i e_i^2 q_i^\downarrow(x, Q^2),$$
$$\sigma_{1/2} \propto F_1(x, Q^2) + g_1(x, Q^2) = \sum_i e_i^2 q_i^\uparrow(x, Q^2), \tag{7.83}$$

where the contribution of g_2 is neglected in the deep inelastic limit. The final expressions on the righthand side of (7.83) are the QPM expectations, using helicity conservation, for quarks (antiquarks) i of charge e_i. The functions $q_i^\uparrow$ ($q_i^\downarrow$) represent the distributions of quarks with spins parallel (antiparallel) to the nucleon spin direction. Thus, from (7.83), we can express g_1 as

$$g_1(x, Q^2) = \tfrac{1}{2} \sum_i e_i^2 [q_i^\uparrow(x, Q^2) - q_i^\downarrow(x, Q^2)]. \tag{7.84}$$

Hence G_1 is a potential measure of the parton helicity structure of the nucleon. Note that, at the parton level, it is easy to show that the cross-sections for elastic polarised electron–quark scattering are

$$\frac{d\sigma^{\uparrow\downarrow}}{dQ^2} = e_i^2 \frac{4\pi\alpha^2}{Q^4}, \qquad \frac{d\sigma^{\uparrow\uparrow}}{dQ^2} = e_i^2 \frac{4\pi\alpha^2}{Q^4}(1 - y)^2. \tag{7.85}$$

Experimentally, the quantity which has been measured is

$$A_{\parallel}^{\exp} = \frac{d\sigma(\uparrow\downarrow - \uparrow\uparrow)}{d\sigma(\uparrow\downarrow + \uparrow\uparrow)}. \tag{7.86}$$

This quantity is to be compared to the virtual photon asymmetry

$$A_{\parallel} = \frac{\sigma_{1/2} - \sigma_{3/2}}{\sigma_{1/2} + \sigma_{3/2}} = \frac{g_1(x, Q^2)}{F_1(x, Q^2)}. \tag{7.87}$$

The relationship between $A_{\parallel}$ and $A_{\parallel}^{\exp}$ involves not only a kinematic factor, but also dilution factors from the fact that neither lepton beams nor nucleon targets can be constructed with 100% polarisation. The resulting measured asymmetry is only a few percent, making the experiments very difficult. The first measurements were obtained from an ep experiment (Hughes and Kuti, 1983) and are shown in Fig. 7.16.

From the theoretical point of view an obvious starting point is that the nucleon obeys an SU(6) symmetry. This would arise from invariance under transformations between $u^{\uparrow}$, $u^{\downarrow}$, $d^{\uparrow}$, $d^{\downarrow}$, $s^{\uparrow}$ and $s^{\downarrow}$. Explicit calculations for

Fig. 7.16 Measurements of the asymmetry $A_{\parallel}$ as a function of x_{BJ}, from the SLAC–Yale and EMC collaborations. ◇, E–80; ○, E–130; ■, EMC. The broken line indicates the Carlitz and Kaur (1977) model.

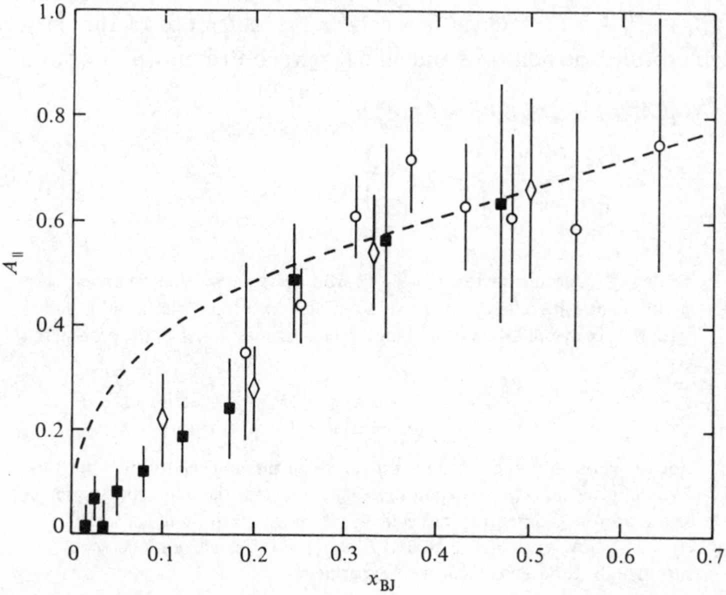

the proton (see Close, 1979) give $u^{\uparrow} = \frac{5}{9}$, $u^{\downarrow} = \frac{1}{9}$, $d^{\uparrow} = \frac{1}{9}$, $d^{\downarrow} = \frac{2}{9}$.# In the neutron $u^{\uparrow}$ is replaced by $d^{\uparrow}$ etc. Weighting these factors with the charge-squared factors $\frac{4}{9}$ and $\frac{1}{9}$ for the u and d quarks, gives the predictions

$$A_{\parallel}^{p} = \tfrac{5}{9}, \qquad A_{\parallel}^{n} = 0. \tag{7.88}$$

In a more realistic model $A_{\parallel}$ will be a function of x. It is generally expected that $A_{\parallel} \to 0$ as $x \to 0$, where the scattering is decoupled from the nucleon spin.

The QPM can be used to construct the sum rule, originally derived by Bjorken (1966, 1970) using current algebra. Integrating $g_1(x)$, from (7.84), over all x gives, in terms of the spin population numbers,

$$\int_0^1 g_1(x)\,\mathrm{d}x = \tfrac{1}{2}\sum_i e_i^2(n_i^{\uparrow} - n_i^{\downarrow}), \tag{7.89}$$

If we consider the following axial-vector quark current in the Breit frame of a proton of helicity 1, then we obtain, using similar methods to those of Section 7.1,

$$\langle p|A^{\mu}|p\rangle = 2p^{\mu}\sum_i e_i^2(n_i^{\uparrow} - n_i^{\downarrow}), \tag{7.90}$$

where

$$A^{\mu} = \sum_i e_i^2 \bar{\psi}_i \gamma^{\mu}\gamma^5 \psi_i.$$

Now, for u, d and s quarks, the quark quantum numbers satisfy $e_i^2 = (4B_i + 2I_3^i + Y_i)/6$. Hence, if we take the difference of the currents between protons and neutrons, only the I_3 part contributes. So we have

$$\begin{aligned}\langle p|A^{\mu}|p\rangle - \langle n|A^{\mu}|n\rangle &= \langle p|A_{+}^{\mu}|n\rangle \\ &= 2p^{\mu}\left|\frac{g_A}{g_V}\right|,\end{aligned} \tag{7.91}$$

The wave functions for the $J^P = \frac{1}{2}^+$ baryon octet are symmetric under simultaneous interchange of the spin and flavour of any quark pair, and must then be symmetrised by making a cyclic permutation. For the proton, this gives

$$|p^{\uparrow}\rangle = \frac{1}{18^{1/2}}(2u^{\uparrow}u^{\uparrow}d^{\downarrow} + 2u^{\uparrow}d^{\downarrow}u^{\uparrow} + 2d^{\downarrow}u^{\uparrow}u^{\uparrow} - u^{\uparrow}u^{\downarrow}d^{\uparrow} - u^{\downarrow}u^{\uparrow}d^{\uparrow} \\ - u^{\uparrow}d^{\uparrow}u^{\downarrow} - u^{\downarrow}d^{\uparrow}u^{\uparrow} - d^{\uparrow}u^{\downarrow}u^{\uparrow} - d^{\uparrow}u^{\uparrow}u^{\downarrow}).$$

For the neutron u and d are interchanged. In the non-relativistic limit this formula gives $\mu_p = \frac{4}{3}\mu_u - \frac{1}{3}\mu_d$ and $\mu_n = \frac{4}{3}\mu_d - \frac{1}{3}\mu_u$ for the magnetic moments of the proton and neutron respectively. Putting $\mu_q = (e_q/2m_q)s_q$, with $m_q \simeq 0.34$ GeV, gives $\mu_p = 2.76$ and $\mu_n = -1.84$. This naive model gives surprisingly good agreement with experiment.

where the current A_+^μ is the isospin raising current (see Section 5.2). This latter current is just the axial part of the weak current appearing in beta decay, thus giving the term g_A/g_V. Hence, putting the above relations together, we obtain the Bjorken sum rule

$$I_{BJ} = \int_0^1 [g_1^p(x) - g_1^n(x)]\,dx = \frac{1}{6}\left|\frac{g_A}{g_V}\right|. \tag{7.92}$$

This sum rule is modified by QCD effects and, to order α_S, the righthand side is multiplied by $(1-\alpha_S(Q^2)/\pi)$; see Kodaira *et al.* (1979), giving an expected value $I_{BJ} = 0.194 \pm 0.002$. The integrals depend on the quark distribution functions (rather than their momentum distributions) and so are very sensitive to the low x region. The contribution from g_1^n is (naively) expected to be small (see equation (7.88)), so that measurement of g_1^p over the entire x range is required.

Using SU(3) current algebra, and the further assumption that the strange quark sea is unpolarised, Ellis and Jaffe (1974) derived separate sum rules for the proton and neutron, namely,

$$I_{EJ}^{p(n)} = \int_0^1 g_1^{p(n)}(x)\,dx = \frac{1}{12}\left|\frac{g_A}{g_V}\right|\left[+1(-1) + \frac{5}{3}\left(\frac{3F-D}{F+D}\right)\right]. \tag{7.93}$$

Using the values of the SU(3) couplings F and D discussed in Section 8.3.1, this gives expected values of $I_{EJ}^p = 0.198 \pm 0.005$ and $I_{EJ}^n = -0.009 \pm$ ± 0.005 respectively. Using the ep SLAC data, which cover an x range 0.10 to 0.64, Hughes and Kuti (1983) quote $I_{EJ}^p = 0.095 \pm 0.008$. This saturates less than 50% of the sum rule. Recent data from the EMC (Ashman *et al.*, 1988) cover the low x region more precisely (Fig. 7.16). The data are at higher Q^2 than the SLAC data, but no significant Q^2 dependence is observed. The measured value $(0.01 < x < 0.70)$ is $I_{EJ}^p = 0.114 \pm 0.012$ (stat) ± 0.026 (syst), and is thus considerably below the expected value. If the more fundamental Bjorken sum rule holds then $I_{EJ}^n \simeq -0.09$, a value much more negative than quark model predictions. A re-examination of the sum rule by Jaffe (1987a) suggests that depolarisation as a function of Q^2 via the U(1) axial current (anomaly) in QCD (Section 10.7.1) is a possible explanation.

Another measurement which gives information on the internal spin structure of the nucleon, although in a less direct way, is the ratio of d to u valence quarks in the proton at large x. Neglecting the small contribution of sea quarks at large x, we obtain, from (7.68) and (7.69)

$$\frac{F_2^{en}(x,Q^2)}{F_2^{ep}(x,Q^2)} = \frac{u_v + 4d_v}{4u_v + d_v}. \tag{7.94}$$

Hence measurement of this ratio gives d_v/u_v at large x. Some measurements of F_2^n/F_2^p are shown in Fig. 7.17(a). Measurements using neutrino beams are also sensitive to this ratio, since the main valence quark transitions are $\nu d_v \to \mu^- u_v$ and $\bar{\nu} u_v \to \mu^+ d_v$, as discussed in Section 7.5. It can be seen from Fig. 7.17 that d_v falls much faster than u_v at large x.

7.5 DEEP INELASTIC NEUTRINO–NUCLEON SCATTERING

In this section we consider the deep inelastic scattering of neutrinos and antineutrinos from nucleon targets by charged current interactions. Note that, since the exchanged W is charged, there is no elastic channel. Discussion on some aspects of the quasi-elastic reactions, $\nu n \to \mu^- p$ and $\bar{\nu} p \to \mu^+ n$, is given in Section 8.4. The lowest order diagram for $\nu(p_1) + N(p_2) \to \mu^-(p_3) + X(p_4)$ is similar to that shown in Fig. 7.5. However, at the upper vertex, the current $\bar{u}\gamma^\mu u$ for the $e\gamma e$ vertex is replaced by $\bar{u}\gamma^\mu(1-\gamma^5)/2u$ for the $\nu W \mu^-$ vertex. The essential differences in the Feynman rules are that the coupling e at each vertex is replaced by $g/2^{1/2}$ (Appendix C), the propagator is that of a massive particle of mass M_W, and there are (in the standard model) both V and A currents possible. The upper vertex leads, in the usual way, to the lepton vertex tensor#

$$L_{\mu\nu} = \mathrm{tr}\left[\not{p}_3 \gamma_\mu \frac{(1-\gamma^5)}{2} \not{p}_1 \gamma_\nu \frac{(1-\gamma^5)}{2} \right]$$

$$= 2[p_{3\mu} p_{1\nu} + p_{3\nu} p_{1\mu} - g_{\mu\nu}(p_1 \cdot p_3) - i p_3^\theta p_1^\phi \varepsilon_{\theta\mu\phi\nu}]. \tag{7.95}$$

The calculation of the differential cross-section requires the evaluation of $L_{\mu\nu} W^{\mu\nu}$ (see (7.37)), where $W^{\mu\nu}$ is given by (7.25). The terms W_4, W_5 and W_6 are proportional to lepton masses, and are thus negligible. Note that the antisymmetric term W_3 contributes in the weak case because of the corresponding antisymmetric term $\varepsilon_{\theta\mu\phi\nu}$ in (7.95). This term comes from the V – A interference, and is parity violating. Contracting W_1, W_2 and W_3 in turn with $L_{\mu\nu}$ from (7.95) gives

$$L_{\mu\nu} W^{\mu\nu} = 4 p_1 \cdot p_3 W_1 + \frac{2}{M^2}[2 p_1 \cdot p_2 \, p_2 \cdot p_3 - M^2 p_1 \cdot p_3] W_2$$

$$- \frac{2}{M^2}[p_2 \cdot p_3 \, q \cdot p_1 - q \cdot p_3 \, p_1 \cdot p_2] W_3. \tag{7.96}$$

Measurement of the helicity $h(\mu^+) = 1.10 \pm 0.24$ of the outgoing μ^+ from $\bar{\nu}_\mu N \to \mu^+ X$ (Jonker *et al.*, 1983) is consistent with the V–A theory ($h = 1$).

Fig. 7.17 (*a*) Measurements of F_2^n/F_2^p as a function of x_{BJ}. The scale for d/u on the righthand side is applicable at large x where sea-quarks are negligible. ○, SLAC–MIT ($2 < Q^2 < 20$ GeV^2); ●, EMC ($7 < Q^2 < 170$ GeV^2). (*b*) Measurements of the valence and sea quark distributions extracted from $\nu(\bar{\nu})D_2$ interactions. The curves are fits to the data, which have $\langle Q^2 \rangle \sim 5$ GeV^2.

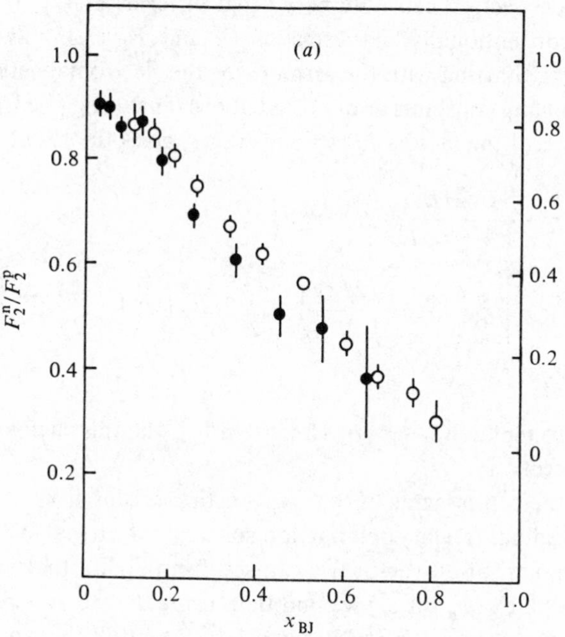

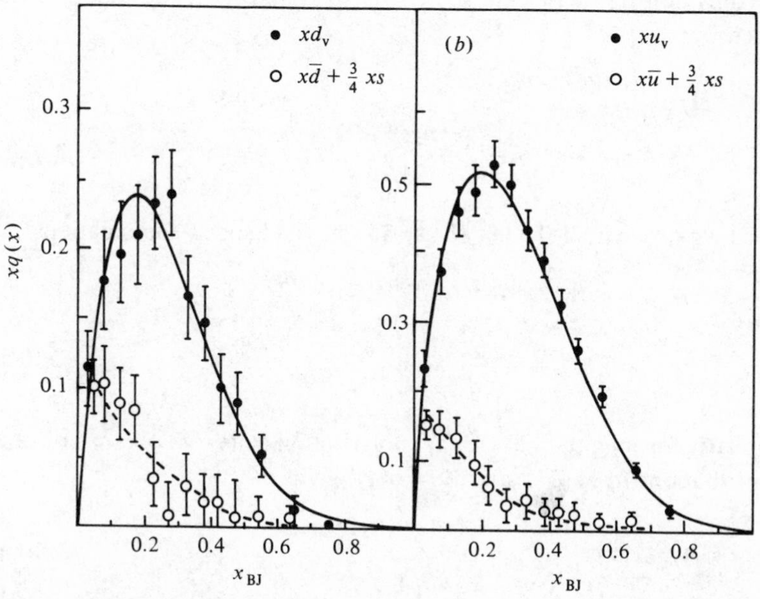

In the lab system $p_2 = (M, 0, 0, 0)$, $p_1 = (E_1, 0, 0, E_1)$, so this simplifies to

$$L_{\mu\nu}W^{\mu\nu} = 4ME_1xyW_1 + 2[2E_1(E_1 - \nu) - ME_1xy]W_2 + 2E_1xy(2E_1 - \nu)W_3. \tag{7.97}$$

The differential cross-section can now be written in terms of W_1, W_2 and W_3 or, more conventionally, in terms of F_1 and F_2 (7.35) and $F_3(x, Q^2) = \nu W_3(x, Q^2)$. Starting with the formula for the electromagnetic case, replacing the coupling constants as discussed above (and using (5.63)), and also keeping track of the factors for spin averages, gives the result

$$\frac{\mathrm{d}^2\sigma}{\mathrm{d}x\,\mathrm{d}y}\begin{pmatrix}\nu\mathrm{N}\\ \bar{\nu}\mathrm{N}\end{pmatrix} = \frac{G_\mathrm{F}^2 ME_1}{\pi(1 + Q^2/M_\mathrm{W}^2)^2} \times \left[F_2\left(1 - y - \frac{Mxy}{2E_1}\right) + \frac{y^2}{2}2xF_1 \pm y\left(1 - \frac{y}{2}\right)xF_3\right]. \tag{7.98}$$

For $\bar{\nu}$N interactions we replace $(1 - \gamma^5)$ by $(1 + \gamma^5)$ in $L_{\mu\nu}$, and this changes the sign of the xF_3 term.

We next consider the components of the cross-section arising from the transverse and longitudinal (scalar) polarisation states of the virtual $\mathrm{W}^{\pm}$. Using (7.41) for the polarisation vector, we can see, from (7.40), that we must evaluate $\varepsilon_\mu^\lambda \varepsilon_\nu^{\lambda *} W^{\mu\nu}$. For ε_R (or ε_L) we find that, using the lab system, the only non-zero terms are W^{11}, W^{22}, W^{12} and W^{21}, and that these are given by

$$W^{11} = W^{22} = W_1$$

$$W^{12} = -W^{21} = -\mathrm{i}\frac{(\nu^2 + Q^2)^{1/2}}{2M}W_3. \tag{7.99}$$

Thus we can extract the left (L) and right (R) handed combinations

$$W_{\mathrm{L,R}} = \tfrac{1}{2}[W^{11} + W^{22} \pm \mathrm{i}W^{12} \mp \mathrm{i}W^{21}]$$

$$= W_1 \pm \frac{(\nu^2 + Q^2)^{1/2}}{2M}W_3. \tag{7.100a}$$

Similarly for longitudinal (scalar) polarisation states we obtain non-zero contributions from W^{00} and W^{33} only, giving

$$W_\mathrm{S} = -W_1 + \left(1 + \frac{\nu^2}{Q^2}\right)W_2. \tag{7.100b}$$

W-absorption cross-sections can be defined in a similar way to those for photons ((7.42), (7.43)), giving

$$W_1 = \frac{K}{2^{1/2}G_F\pi}(\sigma_R + \sigma_L),$$

$$W_2 = \frac{K}{2^{1/2}G_F\pi}\frac{Q^2}{(Q^2+\nu^2)}(\sigma_R + \sigma_L + 2\sigma_S),$$

$$W_3 = \frac{K}{2^{1/2}G_F\pi}\frac{2M}{(\nu^2+Q^2)^{1/2}}(\sigma_L - \sigma_R). \tag{7.101}$$

Thus, unlike the electromagnetic case (where parity conservation requires $\sigma_L = \sigma_R$), W_3 does not vanish for W-boson exchange.

In the Bjorken scaling limit ($\nu, Q^2 \to \infty$, Q^2/ν fixed) we thus have the structure function combinations

$$MW_{L,R} = F_1 \pm F_3/2 = (2xF_1 \pm xF_3)/2x = F_{L,R}/x$$

$$2MW_S = (F_2 - 2xF_1)/x = F_S/x. \tag{7.102}$$

Using (7.102), we can rewrite (7.98) in the following form

$$\frac{d^2\sigma}{dx\,dy}(\nu N) = \frac{G_F^2 M E_1}{\pi(1+Q^2/M_W^2)^2}[F_S(1-y) + F_L + F_R(1-y)^2]. \tag{7.103}$$

That is, we have a longitudinal term, with a $(1-y)$ dependence, and two transverse terms. For $\bar{\nu}N$, F_L and F_R are interchanged.

In total there are 12 nucleon structure functions: F_1, F_2 and F_3 for each of νp, νn, $\bar{\nu}p$ and $\bar{\nu}n$. Charge symmetry implies that $F_i^{\nu p} = F_i^{\bar{\nu}n}$ and $F_i^{\nu n} = F_i^{\bar{\nu}p}$ for $i = 1, 2, 3$, reducing the number to 6. Experimentally, massive neutrino detectors are constructed of materials of medium atomic number, e.g. CHARM use marble ($A = 40$) and CDHS use Fe ($A = 56$). Calcium is isoscalar (equal numbers of protons and neutrons) and iron is approximately so. Hence structure functions are often measured per nucleon, reducing the number to be measured to three. In contrast, a large bubble chamber such as BEBC (Big European Bubble Chamber) filled with deuterium allows measurement of all structure functions, but with more modest statistics.

These general formulae can be interpreted in terms of the QPM, in a similar way to l^+N scattering. The exchanged $W^\pm$ changes the flavour of the struck quark, and the probability of a particular transition is given by the appropriate KM matrix term (Appendix (C.3.1)). We consider below the simpler case of four quark flavours, in which case there is just one Cabibbo angle. Utilising the results derived for νe^- and $\bar{\nu}e^-$ scattering

(Section 5.1.3), we can see that a νq (or $\bar{\nu}\bar{\text{q}}$) reaction will be flat in y (5.54), whereas for $\bar{\nu}$q (or $\nu\bar{\text{q}}$) there is a $(1-y)^2$ dependence (5.57). This is just a statement of helicity conservation, together with the assumption that neutrinos are purely lefthanded. Hence, starting from (5.54) and (5.57), we can write the QPM cross-section for νN scattering (neglecting the propagator term) as

$$\frac{\mathrm{d}^2\sigma}{\mathrm{d}x\,\mathrm{d}y}(\nu\mathrm{N}) = \frac{2G_\mathrm{F}^2 M E_1 x}{\pi}\left[\sum_i g_i^2 q_i(x) + \sum_j g_j^2 \bar{q}_j(x)(1-y)^2\right]. \quad (7.104)$$

In this equation we have assumed that the quark has four-momentum xp_2, so we can write $s = (p_1 + xp_2)^2 \simeq 2xp_1 \cdot p_2 \simeq 2ME_1 x$. The quantities $q_i(x)$ and $\bar{q}_j(x)$ are the probability distributions of quark i and antiquark j respectively, and g_i and g_j are Cabibbo factors ($\cos\theta_\mathrm{C}$ or $\sin\theta_\mathrm{C}$ as appropriate).

A comparison of the general form of the cross-section (7.103) with that from the QPM, shows that the naive QPM requires $F_\mathrm{S} = 0$. This implies that $F_2 = 2xF_1$, which is the Callan–Gross relation (7.57). Assuming this to be true, then comparison of (7.103) and (7.104) gives

$$F_2(x) = 2\sum_{i,j}[g_i^2 x q_i(x) + g_j^2 x \bar{q}_j(x)],$$

$$xF_3(x) = 2\sum_{i,j}[g_i^2 x q_i(x) - g_j^2 x \bar{q}_j(x)]. \quad (7.105)$$

For νN the possible quark transitions are the Cabibbo-favoured ($\cos^2\theta_\mathrm{C}$) reactions d → u, s → c, ū → d̄, and c̄ → s̄, and the unfavoured ($\sin^2\theta_\mathrm{C}$) reactions d → c, s → u, c̄ → d̄, and ū → s̄. For $\bar{\nu}$N the corresponding transitions are for $\cos^2\theta_\mathrm{C}$, u → d, c → s, d̄ → ū, and s̄ → c̄, and for $\sin^2\theta_\mathrm{C}$, u → s, c → d, s̄ → ū, and d̄ → c̄. Hence, (7.105) gives

$$\nu\mathrm{p}:\ F_2 = 2x[\mathrm{d} + \mathrm{s} + \bar{\mathrm{u}} + \bar{\mathrm{c}}], \qquad xF_3 = 2x[\mathrm{d} + \mathrm{s} - \bar{\mathrm{u}} - \bar{\mathrm{c}}],$$

$$\nu\mathrm{n}:\ F_2 = 2x[\mathrm{u} + \mathrm{s} + \bar{\mathrm{d}} + \bar{\mathrm{c}}], \qquad xF_3 = 2x[\mathrm{u} + \mathrm{s} - \bar{\mathrm{d}} - \bar{\mathrm{c}}],$$

$$\bar{\nu}\mathrm{p}:\ F_2 = 2x[\mathrm{u} + \mathrm{c} + \bar{\mathrm{d}} + \bar{\mathrm{s}}], \qquad xF_3 = 2x[\mathrm{u} + \mathrm{c} - \bar{\mathrm{d}} - \bar{\mathrm{s}}],$$

$$\bar{\nu}\mathrm{n}:\ F_2 = 2x[\mathrm{d} + \mathrm{c} + \bar{\mathrm{u}} + \bar{\mathrm{s}}], \qquad xF_3 = 2x[\mathrm{d} + \mathrm{c} - \bar{\mathrm{u}} - \bar{\mathrm{s}}], \quad (7.106)$$

where isospin symmetry (i.e. $u^\mathrm{p} = d^\mathrm{n}$, $d^\mathrm{p} = u^\mathrm{n}$, etc.) has been assumed. Averaging the proton and neutron results gives, per nucleon,

$$\nu\mathrm{N}:\quad F_2 = x[\mathrm{u} + \mathrm{d} + \bar{\mathrm{u}} + \bar{\mathrm{d}} + 2\mathrm{s} + 2\bar{\mathrm{c}}],$$

$$xF_3 = x[\mathrm{u} + \mathrm{d} - \bar{\mathrm{u}} - \bar{\mathrm{d}} + 2(\mathrm{s} - \bar{\mathrm{c}})]$$

$$= x[u_\mathrm{v} + d_\mathrm{v} + 2(\mathrm{s} - \bar{\mathrm{c}})], \quad (7.107)$$

$$\bar{\nu}\text{N}: \quad F_2 = x[\text{u} + \text{d} + \bar{\text{u}} + \bar{\text{d}} + 2\text{c} + 2\bar{\text{s}}],$$
$$xF_3 = x[\text{u} + \text{d} - \bar{\text{u}} - \bar{\text{d}} + 2(c - \bar{\text{s}})]$$
$$= x[u_\text{v} + d_\text{v} + 2(\text{c} - \bar{\text{s}})]. \tag{7.108}$$

Assuming that $\text{s} = \bar{\text{s}}$ and $\text{c} = \bar{\text{c}}$, then $F_2^{\nu\text{N}}$ and $F_2^{\bar{\nu}\text{N}}$ give the total momentum fraction carried by quarks plus antiquarks. The structure function xF_3 depends mainly on the valence quark distributions ($u_\text{v} = \text{u} - \bar{\text{u}}, d_\text{v} = \text{d} - \bar{\text{d}}$). Note that, at currently accessible neutrino energies ($E_1 \lesssim 300$ GeV), the charm-quark sea is suppressed relative to the strange sea, because the charm-quark is much more massive. Furthermore, at small values of the invariant mass of the hadronic final state, the reactions producing charm-quarks (and hence charmed hadrons) will also be suppressed. Hence equations (7.106) to (7.108) are valid only well above such thresholds. From (7.107) and (7.108) it can be seen that c($\bar{\text{c}}$) quarks will be produced in νN($\bar{\nu}$N) interactions with the cross-sections

$$\frac{\text{d}^2\sigma^{\nu\text{N}}}{\text{d}x\,\text{d}y}(\text{c}) = \frac{G_\text{F}^2 M E_1 x}{\pi}[(\text{u} + \text{d})\sin^2\theta_\text{C} + 2\text{s}\cos^2\theta_\text{C}],$$
$$\frac{\text{d}^2\sigma^{\bar{\nu}\text{N}}}{\text{d}x\,\text{d}y}(\bar{\text{c}}) = \frac{G_\text{F}^2 M E_1 x}{\pi}[(\bar{\text{u}} + \bar{\text{d}})\sin^2\theta_\text{C} + 2\bar{\text{s}}\cos^2\theta_\text{C}], \tag{7.109}$$

where the u and d quark distributions again refer to the proton. There is, therefore, a substantial (~5%) production of charm quarks expected in $\nu(\bar{\nu})$ interactions. For comparison, in hadronic interactions of similar energies, the charm production rate is only about 0.1%.

The presence of a c($\bar{\text{c}}$) quark in the final state can be detected, for example, through the semileptonic decays of the charmed hadrons ($\text{c} \to \text{s}\mu^+\nu, \bar{\text{c}} \to \bar{\text{s}}\mu^-\bar{\nu}$). Note that the experimental signature for this sequence is a dimuon event with muons of opposite charge. In bubble chambers, the decays $\text{c} \to \text{se}^+\nu, \bar{\text{c}} \to \bar{\text{s}}\text{e}^-\bar{\nu}$ can also be detected. These are indeed observed accompanied by a strange particle as predicted. Measurement of the $\mu^+\mu^-$ rate in $\bar{\nu}$N interactions gives a measure of the amount of strange sea (7.109). The CDHS experiment (Abramowicz *et al.*, 1982) finds that $2\text{s}/(\bar{\text{u}} + \bar{\text{d}}) = 0.52 \pm 0.09$. That is, the strange sea is suppressed by a factor of about two compared to the value expected from SU(3) flavour symmetry ($\bar{\text{u}} = \bar{\text{d}} = \bar{\text{s}}$). In contrast, Allasia *et al.* (1984) find $\bar{\text{d}} - \bar{\text{u}} = 0.05 \pm 0.06 \pm 0.11$, from an analysis of $\nu/\bar{\nu}$D events, compatible with equal sea contributions for u and d quarks.

By measuring the cross-sections for νp, νn, $\bar{\nu}p$ and $\bar{\nu}n$ in BEBC filled with deuterium, Allasia *et al.* (1984) extracted the quark density distributions shown in Fig. 7.17(*b*). The sea-quark distributions fall off as $x\bar{q}(x) \sim (1 - x)^4$, whereas the valence quark distributions have the shape $xq_v(x) \sim x^\alpha(1-x)^\beta$, with $\alpha \sim 0.8$ and $\beta = 3.3 \pm 0.1$ for u_v and $\beta = 4.4 \pm 0.5$ for d_v. The d_v distribution falls off more rapidly than u_v for large x. Other data and parameterisations, for $Q^2 \sim 15\,\text{GeV}^2$, agree with these results to within 5–15% (Feltesse, 1985), which is about the level of uncertainty on all these measurements of absolute cross-sections.

There are various sum rules which have been derived for the weak currents, and these have simple interpretations in the QPM. The sum rule due to Adler (1966) is

$$I_A = \int_0^1 \frac{(F_2^{\bar{\nu}p} - F_2^{\nu p})}{2x}\,dx = \int_0^1 \frac{(F_2^{\nu n} - F_2^{\nu p})}{2x}\,dx = \int_0^1 (u_v - d_v)\,dx = 1. \tag{7.110}$$

Experimentally, Allasia *et al.* (1984) find $I_A = 1.01 \pm 0.08 \pm 0.18$. This sum rule is independent of Q^2 and hence of QCD corrections. Although the interpretation is simple in the QPM, the derivation of the sum rule was originally made using the local commutation relations of the axial current (current algebra). Gross and Llewellyn Smith (1969) derived a sum rule for xF_3, namely

$$I_{GLS} = \int_0^1 \frac{(xF_3)}{x}\,dx = \int_0^1 (u_v + d_v)\,dx = 3\left(1 - \frac{\alpha_S}{\pi}\right), \tag{7.111}$$

where an order $O(\alpha_S)$ QCD correction term has been added. Sciulli (1985), reviewing the experimental situation, gives the value $I_{GLS} = 2.81 \pm 0.16$. Modulo small QCD corrections, these sum rules are a check of the number of valence quarks (N_v) in the nucleon. The results give $N_v = 3.00 \pm 0.17$ ($\alpha_S = 0.2$) and are thus compatible with the QPM expectations. Note that the quantity to be measured in each case is a structure function divided by x, and so very detailed measurements at low x are needed.

In the QPM, the structure function F_2 for νN (7.107) and for $l^\pm$N (the average of (7.68) and (7.69)), have a simple relationship. If q_i and $\bar{q}_i$ are the quark and antiquark distributions ($i = $ u, d, s, c), then

$$\frac{F_2^{l^\pm N}}{F_2^{\nu N}} = \frac{(e_u^2 + e_d^2)/2\left[\sum_i x(q_i + \bar{q}_i) - 3/5x(s + \bar{s} - c - \bar{c})\right]}{\sum_i x(q_i + \bar{q}_i)} \simeq \frac{5}{18}. \tag{7.112}$$

That is, neglecting any small differences between the strange and charm sea, this ratio measures the sums of the squares of the quark charges.

Some data on the ratio for $Q^2 \sim 15\,\text{GeV}^2$ are shown in Fig. 7.18. It can be seen that the QPM expectation is satisfied to 10–15%, which is about the size of the relative systematic errors. This test, together with that of equation (7.71), constitutes evidence that the dynamic scattering centres, which we have referred to as quarks, have charges $e_u = +2/3$ and $e_d = -1/3$. The above relations yield, in fact, e_u^2 and e_d^2. That the absolute sign of $e_u(e_d)$ is positive (negative) can be confirmed by studying the charges of the fastest hadrons (i.e. those most likely to be produced directly from the fragmentation of the struck quark). These are predominantly positive (negative) for $\nu N(\bar{\nu}N)$ events, as expected for the fragments of a u(d) quark. In μp scattering (Albanese *et al.*, 1984), the net charge of the forward jet of hadrons is found to increase from a rather small value (~ 0.1), for $x_{BJ} < 0.05$, to around 0.5 at large x_{BJ}, as shown in Fig. 7.19. This can be understood as a decrease in the relative importance of the quark–antiquark sea (which leads on average to a jet of charge zero) as

Fig. 7.18 Some measurements of the ratio $(5/18)F_2^{\nu N}/F_2^{\mu N}$ as a function of x_{BJ}. The value of unity is expected in the QPM. ●, CCFRR/EMC; ▲, CDHSW/EMC.

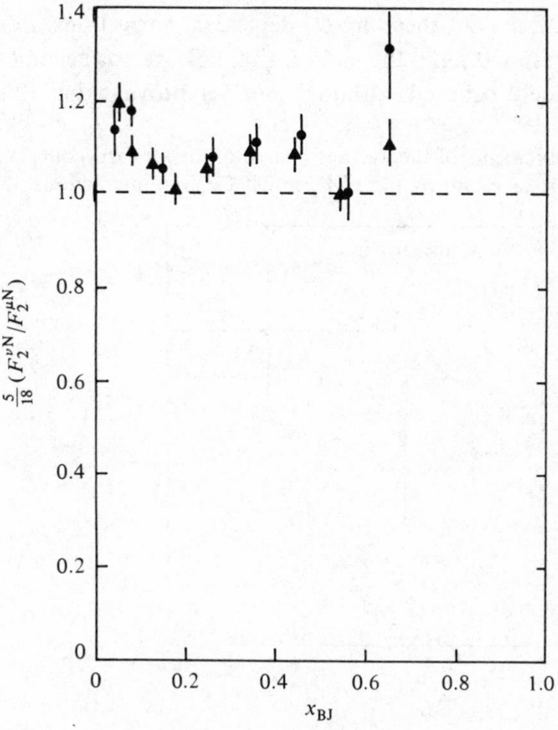

x_{BJ} increases. A combination of parameterised quark distributions and the Lund string fragmentation model reproduces the data.

7.6 QUANTUM CHROMODYNAMICS (QCD)

QCD is a theory dealing with the colour interactions of quarks and gluons. Colour plays a role in QCD somewhat analogous to that of electric charge in QED. Each quark flavour is assumed to come in three colours, $q_i(x)$, $i=1, 2, 3$. In addition to the spin $\frac{1}{2}$ quarks, there are eight massless vector gluons, which are the analogues of the photon in QED. The colour composition of the gluons $G_a^\mu(x)$, $a=1, \ldots, 8$, was given in Sections 1.7 and 2.5.5. Under the SU(3) group of colour transformations, q transforms as the fundamental (**3**) representation and G as the adjoint (**8**) representation. Invariance under local phase transformations leads to the non-Abelian theory of QCD. The $SU(3)_{\mathrm{colour}}$ symmetry is assumed to be exact, and is not to be confused with the approximate flavour symmetry $SU(3)_{\mathrm{flavour}}$ (Section 2.5.3).

QCD has the important property of being asymptotically free. If Q^2 represents some large momentum scale in the interaction process, then the effective QCD coupling constant $\alpha_S(Q^2) \to 0$, as $Q^2 \to \infty$. Thus, quarks become 'free' asymptotically, and we recover the quark parton model in this limit. However, at finite Q^2, there are Q^2-dependent corrections. For momentum scales corresponding to the hadron size, however, α_S becomes large. Indeed, it is widely believed, although not yet proven, that this

Fig. 7.19 Measurements of the average charge of the forward-going hadrons ($x_F > 0$), as found by the EMC in 280 GeV μp interactions.

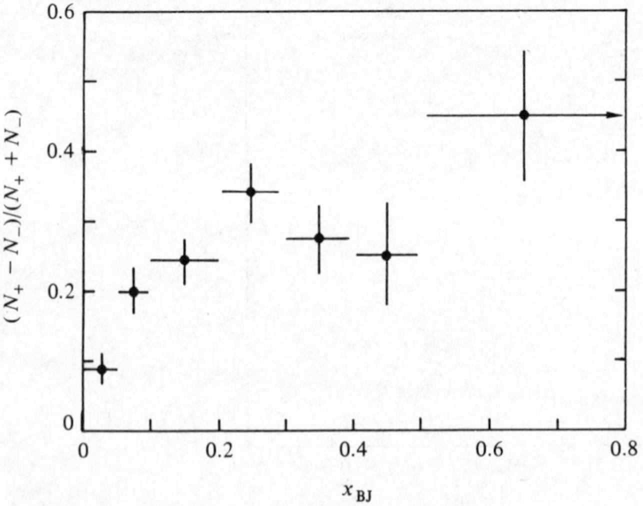

increase in the $Q^2 \to 0$ limit leads to the confinement of quarks inside hadrons (infrared slavery).

QCD is important, not only because of the hope that it will prove to be the theory of strong interactions, but also because its basic properties are assumed (directly or indirectly) in making deductions about quark properties from the observed hadron properties. A detailed discussion of QCD is beyond the scope of this book, and only some of the main results are outlined below. Where possible, analogies with SU(2) $\otimes$ U(1) are used. For a more thorough exposé the reader is referred to Politzer (1974), Buras (1980), Altarelli (1982), Pennington (1983), or to a more advanced text, e.g. Leader and Predazzi (1982), or Cheng and Li (1984).

7.6.1 The QCD Lagrangian

Under local transformations of the colour axes the quark fields $q_i(u_i, d_i, s_i, \text{etc.})$, where $i = 1, 2, 3$ are the three colour indices, transform as

$$q_i(x) \to q_i'(x) = Uq_i(x) \qquad U = \exp\left(-ig\alpha_a(x)\frac{\lambda_a}{2}\right) \tag{7.113}$$

where λ_a (a = 1, . . ., 8) are the eight SU(3) matrices (see (2.250) and (2.251)), and a summation over a is understood. This is similar in form to that for the U(1) and SU(2) transformations (5.107). The SU(3) generators do not commute (see (2.262)), hence the theory is non-Abelian. Note that observables such as the charge

$$Q = e\sum_{i=1}^{3}\int d^3x\,\frac{1}{3^{1/2}}\left(\tfrac{2}{3}u_i^\dagger u_i - \tfrac{1}{3}d_i^\dagger d_i - \tfrac{1}{3}s_i^\dagger s_i + \cdots\right), \tag{7.114}$$

are invariant under (7.113), because they are bilinear in the quark fields.

The free quark Lagrangian $\mathscr{L}_0$ is of the form (5.101). Invariance under global colour transformations leads, by Noether's theorem, to the existence of eight conserved currents

$$j_a^\mu(x) = \bar{q}(x)\frac{\lambda_a}{2}\gamma^\mu q(x) \qquad (a = 1, \ldots, 8). \tag{7.115}$$

The free Lagrangian $\mathscr{L}_0$ is not invariant under the local transformations (7.113), which give the extra term

$$\delta\mathscr{L} = g\bar{q}U^{-1}\frac{\lambda_a}{2}\gamma^\mu(\partial_\mu\alpha_a)Uq. \tag{7.116}$$

Local gauge invariance is restored if we couple eight massless vector boson fields $G_a^\mu(x)$, the gluon fields, to the conserved currents (7.115) and add to $\mathscr{L}_0$ the interaction term

$$\mathscr{L}_{\text{int}} = -g\bar{q}\frac{\lambda_a}{2}\gamma^\mu q G^\mu_a, \tag{7.117}$$

which cancels the term (7.116), provided the gauge fields are also transformed by

$$G^\mu_a(x) \to G^\mu_a(x) + \partial^\mu \alpha_a(x) - g f_{abc} \alpha_b(x) G^\mu_c(x). \tag{7.118}$$

This can be shown by substituting (7.118) into the Lagrangian, and using the expansion $U \simeq 1 - \mathrm{i}g\alpha_b\lambda_b/2$ and the commutation relation (2.262) ($F_b = \lambda_b/2$).

In analogy with QED and the Weinberg–Salam model (Section 5.4), we can introduce a gluon field tensor

$$G^{\mu\nu}_a = \partial^\mu G^\nu_a - \partial^\nu G^\mu_a + g f_{abc} G^\mu_b G^\nu_c, \tag{7.119}$$

so that the complete QCD Lagrangian# is

$$\mathscr{L}_{\text{QCD}} = \bar{q}(\mathrm{i}\gamma_\mu D^\mu - m_q)q - \tfrac{1}{4}G^{\mu\nu}_a (G_a)_{\mu\nu}, \tag{7.120}$$

where the derivation D^μ is given by

$$D^\mu = \partial^\mu + \mathrm{i}g\frac{\lambda_a}{2}\mathrm{G}^\mu_a. \tag{7.121}$$

The Lagrangian (7.120) thus contains both trilinear and quadrilinear self-couplings of the gluon fields; a consequence of the non-Abelian nature of the theory. From the gauge invariance of $\mathscr{L}_{\text{QCD}}$, it follows that there are no higher multigluon couplings.

The Feynman rules for QCD can be obtained from the Lagrangian (7.120). The rules for tree-level graphs can essentially be deduced from the form of the Lagrangian and from previous considerations; they are given in Appendix C. The gluon fields can be expressed in a variety of gauges. Graphs involving gluon loops, and in particular helicity 0 contributions, introduce some new problems. These lead to a potential violation of unitarity, which can be restored by introducing into the theory negative metric particles which are scalars, but which obey Fermi statistics (*ghosts*). This applies to a covariant gluon gauge. In a physical or axial gauge,

There is a further possible term $\mathscr{L}_\theta = \theta g^2/(32\pi^2) G^{\mu\nu}_a (\tilde{G}_a)_{\mu\nu}$, where $\tilde{G}^{\mu\nu} = \frac{1}{2}\varepsilon^{\mu\nu\lambda\rho}G_{\lambda\rho}$ and θ is a parameter. This term violates P, T and CP symmetries, and leads to a neutron electric dipole moment $d_{\text{n}} \sim 4 \times 10^{-16}\theta$ e cm. Comparison with the experimental value ($d_{\text{n}} < 6 \times 10^{-25}$ e cm) implies $\theta \lesssim 10^{-9}$. If the electroweak sector is also included, then θ is modified to $\bar{\theta} = \theta + \arg\det M$ where M is the quark mass matrix.

which has a more complicated gluon propagator but only helicity 1 contributions, these ghosts do not appear. Such problems do not arise in QED. The essential difference is that in QED the photon is coupled to a conserved current, whereas in QCD the gluon couples to other gluons, as well as to the quark current. Finally, we note that, for massless quarks, there is no scale in the QCD Lagrangian. The naive expectation is that, in a massless theory, there will be scaling in the asymptotic limit, i.e. a dependence only on the dimensionless ratios of the large momenta. However, the requirement of regularisation and renormalisation means that a finite mass scale must be introduced in order to define the renormalised coupling constant. This destroys the naive scaling laws.

7.6.2 Renormalisation and the effective coupling constant

In the renormalisation of QED (Section 4.9) we saw that the divergences could be rendered harmless by regularising and renormalising the coupling constant, the fermion mass and the associated fields. Before discussing QCD, let us first consider the simpler case of an electrically neutral scalar particle, which has self-interactions (ϕ^4 theory), giving the Lagrangian

$$\mathscr{L}(x)=\tfrac{1}{2}[\partial_\mu\phi_B(x)]^2-\tfrac{1}{2}m_B^2\phi_B^2(x)-\frac{g_B}{4!}\phi_B^4(x). \tag{7.122}$$

The first two terms are the free particle Lagrangian $\mathscr{L}_0$, and the final term the self-interaction part $\mathscr{L}_{\text{int}}$. The constants m_B and g_B are the bare mass and coupling constant respectively, and ϕ_B is the bare field. The factor 4! is introduced in order to make the Feynman rules simpler. The propagator has the usual form $\text{i}G(p)=\text{i}/(p^2-m^2)$, and the vertex term for four fields is $\Gamma_0^{(4)}=-\text{i}g$.

The graphs which cause problems in the evaluation of the propagator are those with loops. The method used to handle these problems is similar to that for QED. The propagator is expanded as shown in Fig. 7.20. The blobs on the righthand side are called *'one-particle irreducible' diagrams* and cannot be further split into two parts by cutting any single internal line. Such diagrams isolate all the problems with loops. The sum of the

Fig. 7.20 Propagator $G^{(2)}$ in ϕ^4 theory.

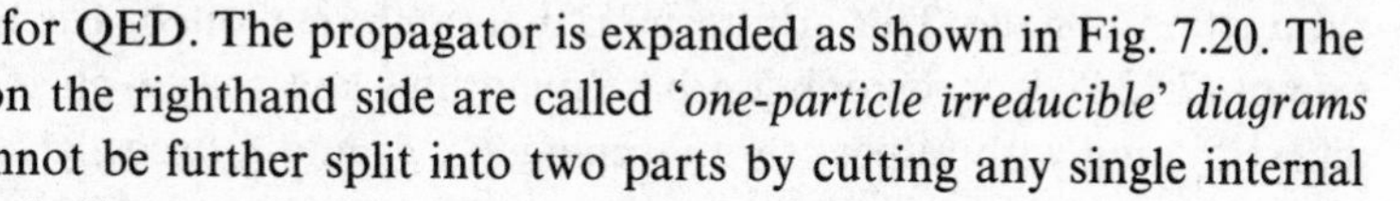

geometric series gives the total propagator

$$iG^{(2)}(p) = \frac{iG_0}{1-(iG_0)(i\Sigma)} = \frac{i}{p^2 - m_\mu^2 + \Sigma_\mu(p^2)}. \tag{7.123}$$

The superscript (2) denotes the fact that there are two external legs for the propagator. The quantity Σ is found by summing over the graphs appearing in $G^{(2)}$, but with their external propagator lines '*truncated*', the latter being taken care of by the G_0 factors (see Fig. 7.20). The mass m_μ in (7.123) is associated with the renormalisation point μ (i.e. the scale at which the infinities are subtracted out), and is defined such that the propagator has the form $iG^{(2)}(p) = i/(p^2 - m_\mu^2)$ as $p^2 \to -\mu^2$ (i.e. space-like and thus free of poles) and also such that $\Gamma^{(4)} = -ig_\mu$. The μ indices are dropped below, but they are implicit.

The Green function $G^{(2)}$, describing the propagator, is given by the Fourier transform of the vacuum expectation value of the time-ordered product $\langle 0|T[\phi(x_1)\phi(x_2)]|0\rangle$. The relationship between the bare and physical fields is (see Section 4.9.3) $\phi_B = Z_\phi^{1/2}\phi$, where Z_ϕ depends on some parameter (λ, say) such that the result is finite as $\lambda \to \infty$. If we generalise to the case of n external legs, we have $G_B^{(n)} = Z_\phi^{n/2} G^{(n)}$. The truncated vertex form $\Gamma^{(n)}$ is obtained from $G^{(n)}$ by dividing it by the product of n single particle propagators, each of which involves two fields. If p_i represents the set of the n particle momenta, we can write

$$\Gamma_B^{(n)}(p_i; m_B, g_B, \lambda) = \left[Z_\phi\left(\frac{\lambda}{\mu}, \frac{m}{\mu}, g\right)\right]^{-n/2} \Gamma^{(n)}(p_i; m, g). \tag{7.124}$$

The arguments of Z_ϕ have been written as ratios, since Z_ϕ is dimensionless. The vertex term $\Gamma^{(n)}$ corresponds to a physical quantity, which is measurable at some momentum scale p_i.

The scale μ at which the infinities are subtracted out (*regulated*) is arbitrary, thus the finite term $\Gamma^{(n)}$ will depend on the way in which the subtraction is made. This arbitrariness in μ can be exploited. The lefthand side of (7.124) depends only on bare quantities and so not on μ; hence, differentiating with respect to μ (keeping m_B, g_B and λ fixed), gives

$$0 = Z_\phi^{-n/2}\left[\frac{-n}{2Z_\phi}\Gamma^{(n)}\frac{dZ_\phi}{d\mu} + \left(\frac{\partial}{\partial\mu} + \frac{dm}{d\mu}\frac{\partial}{\partial m} + \frac{dg}{d\mu}\frac{\partial}{\partial g}\right)\Gamma^{(n)}\right],$$

i.e.

$$\left(\mu\frac{\partial}{\partial\mu} + \beta\frac{\partial}{\partial g} - n\gamma + \mu\frac{dm}{d\mu}\frac{\partial}{\partial m}\right)\Gamma^{(n)} = 0, \tag{7.125}$$

where

$$\beta = \mu \frac{\mathrm{d}g}{\mathrm{d}\mu}, \qquad \gamma = \frac{\mu}{2Z_\phi}\frac{\mathrm{d}Z_\phi}{\mathrm{d}\mu} = \frac{\mu}{2}\frac{\mathrm{d}}{\mathrm{d}\mu}(\ln Z_\phi). \tag{7.126}$$

The quantities β and γ are, in fact, finite as $\lambda \to \infty$ and, since they are dimensionless, do not depend on μ after taking this limit; so that $\beta = \beta(g)$ and $\gamma = \gamma(g)$. Equation (7.125) is known as the *renormalisation group equation* (RGE).

We will consider only a massless theory (since QCD is massless). In this case the RGE is simplified, since the final term in (7.125) is zero. In order to investigate the momentum dependence of $\Gamma^{(n)}$ we introduce a scaling factor for the momenta, σp_i, so that by varying σ the momenta can be taken to arbitrarily large values. We wish to relate the dependence of $\Gamma^{(n)}(\sigma p_i; g, \mu)$ on μ to its dependence on momenta (i.e. on σ). This can be found by considering the dimensions of $G^{(n)}$ and $\Gamma^{(n)}$. Since $G^{(n)}$ is the Fourier transform of the product of n fields ϕ (each having the dimension of mass M) with respect to $(n-1)$ variables, and $\Gamma^{(n)}$ corresponds to this n-field product divided by n single-particle propagators (each being the product of two fields ϕ), we obtain the following dimensions

$$G^{(n)} = [M]^{4-3n}, \qquad \Gamma^{(n)} = [M]^{4-n}. \tag{7.127}$$

From this dimensional form it follows that

$$\left[\sigma \frac{\partial}{\partial \sigma} + \mu \frac{\partial}{\partial \mu}\right] \Gamma^{(n)} = (4-n)\Gamma^{(n)}, \tag{7.128}$$

since, in $\Gamma^{(n)}(\sigma p_i; g, \mu)$, the coupling constant g is dimensionless. The form of (7.128) can be checked explicitly by setting $\Gamma^{(n)}$ equal to some arbitrary function of σ and μ. Eliminating $\mu\partial/\partial\mu$ from (7.125) (for the massless case) and (7.128), gives

$$\left[\sigma \frac{\partial}{\partial \sigma} - \beta(g)\frac{\partial}{\partial g} + n\gamma(g) + n - 4\right]\Gamma^{(n)}(\sigma p_i; g, \mu) = 0,$$

or

$$\frac{\partial \Gamma^{(n)}}{\partial t} - \beta(g)\frac{\partial \Gamma^{(n)}}{\partial g} = [4 - n - n\gamma(g)]\Gamma^{(n)}, \tag{7.129}$$

where $t = \ln \sigma$. This is a first-order partial differential equation, which can be written in the form

$$\frac{\partial \Gamma^{(n)}}{\partial t}\frac{\mathrm{d}t}{\mathrm{d}s} + \frac{\partial \Gamma^{(n)}}{\partial g}\frac{\mathrm{d}g}{\mathrm{d}s} = \frac{\mathrm{d}\Gamma^{(n)}}{\mathrm{d}s}. \tag{7.130}$$

Comparison of (7.129) and (7.130) yields three differential equations, which must satisfy the appropriate boundary conditions. The first term yields $\mathrm{d}t/\mathrm{d}s = 1$, and hence, choosing the integration constant to be zero, we have $s = t$. The second term gives

$$\frac{\mathrm{d}g}{\mathrm{d}t} = -\beta(g) \Rightarrow g = g(\bar{g}, t) \qquad \text{or} \qquad \bar{g} = \bar{g}(g, t), \tag{7.131}$$

where $\bar{g}$ is the value of g at $t = 0$ (or $\sigma = 1$). Integrating (7.131) gives

$$t = -\int_{\bar{g}}^{g} \frac{\mathrm{d}g'}{\beta(g')}, \tag{7.132}$$

and, by differentiating, (7.132) can be recast as

$$\frac{\mathrm{d}\bar{g}(t)}{\mathrm{d}t} = \beta(\bar{g}), \qquad \text{with } \bar{g}(0) = g. \tag{7.133}$$

The third term in the comparison of (7.129) and (7.130) gives

$$\frac{\mathrm{d}\Gamma^{(n)}}{\mathrm{d}t} = [4 - n - n\gamma(g)]\Gamma^{(n)}, \tag{7.134}$$

which has the solution

$$\Gamma^{(n)}(\sigma p_i; g, \mu) = \sigma^{4-n} F \Gamma^{(n)}(p_i, \bar{g}(g, t), \mu), \tag{7.135}$$

where

$$F = \exp\left\{ -n \int_0^t \gamma[\bar{g}(g, t')]\, \mathrm{d}t' \right\}. \tag{7.136}$$

Equation (7.135) is a powerful result; it relates the value of $\Gamma^{(n)}$, evaluated at a momentum scale $\sigma p_i (\sigma > 1)$, to that at p_i, but with the latter computed using the effective coupling constant $\bar{g}$. In addition, there is a factor approximately equal to σ^{4-n}. This factor would be obtained in a 'scaling' (e.g. parton) model. The additional term F violates the naive dimensional scaling, and gives rise to an '*anomalous*' *dimension* $n\gamma(0)$.

In QCD the situation is somewhat more complicated, because we have both quark and gluon fields, and these particles have spin. There are also the further problems of the choice of the gluon gauge and ghosts, as discussed previously. Nevertheless, features similar to those of the ϕ^4 theory emerge. A similar RGE is found, but it has two γ-functions, γ_{F} and γ_{A}, for the quarks and gluons in the fundamental (F) and adjoint (A) representations respectively. There is also a similar formulation of the effective coupling constant.

Let us consider the effective q$\bar{\text{q}}$g vertex coupling $\bar{g}$ which is, in lowest order, just the bare coupling g appearing in the Lagrangian. The diagrams (Fig. 7.21), contributing to the next order, all contain one loop over which integration must be performed. Hence, these are of order g^3, and yield a result $\sim\ln(C^2/Q^2)$, where C is an ultraviolet cut-off and $Q^2 = -q^2$, with q denoting the gluon momentum. Thus, in terms of these *leading logs*, we have

$$\bar{g} = g - Ag^3 \ln(Q^2/C^2) + O[g^5 \ln^2(Q^2/C^2)], \tag{7.137}$$

where A is calculable. The effective constant $\bar{g}$ is dependent on Q^2, but is infinite as $C \to \infty$. It can be renormalised by defining $\bar{g}$ to be $\bar{g}(\mu^2)$ at $Q^2 = \mu^2$, giving

$$\bar{g}(Q^2) = \bar{g}(\mu^2) - A\bar{g}(\mu^2)^3 \ln(Q^2/\mu^2) + O(\bar{g}^5 \ln^2(Q^2/\mu^2)), \tag{7.138}$$

which is independent of C, and so finite. Applying the RGE (7.133) to (7.138), we obtain

$$\beta(\bar{g}) = \frac{\mathrm{d}\bar{g}}{\mathrm{d}\ln Q} = -2A\bar{g}^3 + O(\bar{g}^5), \tag{7.139}$$

which has the solution

$$\bar{g}^2(Q^2) = \frac{\bar{g}^2(\mu^2)}{1 + 2A\bar{g}^2(\mu^2)\ln(Q^2/\mu^2)}. \tag{7.140}$$

From (7.140), we can make an expansion of $\bar{g}(Q^2)$ as a series in $\bar{g}^2(\mu^2) \ln Q^2/\mu^2$, and we obtain equation (7.138). Hence, equation (7.140), which dictates the Q^2 evolution of $\bar{g}$, gives the coefficients term by term in the perturbation series in terms of $\bar{g}(\mu^2)$. It represents a summation, *to all orders*, of these leading log terms.

The evolution of the coupling constant ((7.131), (7.133)) depends critically on the nature of the theory, and hence on the form of β. At the

Fig. 7.21 Order g and g^3 diagrams contributing to the effective q$\bar{\text{q}}$g coupling.

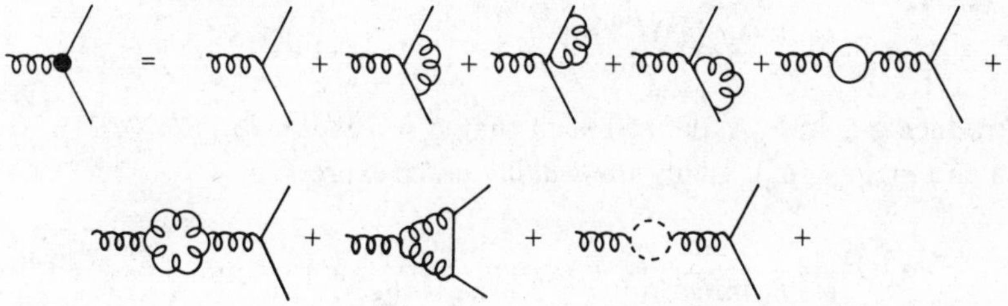

one-loop level, calculations give (e.g. Cheng and Li, 1984) the following

$$\text{QED:} \qquad \beta = +\frac{e^3}{12\pi^2} + O(e^5), \tag{7.141}$$

$$\phi^4: \qquad \beta = \frac{3g^2}{16\pi^2} + O(g^3), \tag{7.142}$$

$$\text{QCD:} \qquad \beta = -\frac{g^3}{16\pi^2}\left[\tfrac{11}{3}C_F(G) - \tfrac{4}{3}T(R)\right] + O(g^5)$$

$$= -\frac{g^3}{16\pi^2}\left[11 - \tfrac{2}{3}n_F\right] + O(g^5). \tag{7.143}$$

For QED, the calculation of β is that of the vacuum polarisation term for the photon propagator, and this was discussed in Section 4.9.2. In QCD there are additional propagator contributions from gluon (and ghost) loops. The term $C_F(G)$ depends on the colour group and $T(R)$ depends on the gluon representation and on the number of quark flavours n_f. For $SU(3)_{col}$, with the gluons in the octet representation, $C_F(G) = \sum_{c,d} \delta_{ab} f_{acd} f_{bcd} = 3$, and $T = \delta_{ab} n_f \,\mathrm{tr}(\lambda^a \lambda^b)/4 = n_f/2$ (see Section 2.5.5). Note that, at one-loop level, β_{QCD} does not depend on the gauge.

The main difference between QCD and QED (or a ϕ^4 theory) is that $\beta(\bar{g})$ is negative (provided $n_f < 17$), and so $\bar{g}$ decreases with increasing t, leading to asymptotic ($t \to \infty$) freedom. In this limit we obtain (perturbatively) the zeroth order amplitude and, at finite t, there are corrections to this 'free particle' amplitude of the form indicated by equation (7.135). The evolution of the coupling constant $\bar{g}$ with some large scale Q^2 ($=\sigma^2\mu^2$, with $g = \bar{g}$ for $\sigma = e^t = 1$) is given by (7.140). Comparison of (7.139) with (7.143) shows that

$$A = \frac{\beta_0}{32\pi^2}, \qquad \beta_0 = \frac{33 - 2n_f}{3}. \tag{7.144}$$

It is customary to quote the QCD coupling constant in terms of $\alpha_S = \bar{g}^2/4\pi$. From (7.140) and (7.144) we obtain

$$\alpha_S(Q^2) = \frac{\alpha_S(\mu^2)}{1 + \alpha_S(\mu^2)\dfrac{\beta_0}{4\pi}\ln\left(\dfrac{Q^2}{\mu^2}\right)}. \tag{7.145}$$

Introducing a scale Λ, defined such that $\Lambda = \mu \exp\{-4\pi/[\beta_0 \alpha_S(\mu^2)]\}$, we can eliminate $\alpha_S(\mu^2)$, giving the *leading order* expression

$$\alpha_S(Q^2) = \frac{4\pi}{\beta_0 \ln(Q^2/\Lambda^2)} = \frac{12\pi}{(33 - 2n_f)\ln(Q^2/\Lambda^2)}. \tag{7.146}$$

The analogous variation of α_{QED} with Q^2 is very small, since $\Lambda_{QED} = m_e \exp(3\pi/2\alpha) \sim 10^{277}$ GeV.

In next-to-leading order (i.e. two-loop level) of QCD, the renormalisation scheme must be specified. In the *momentum–space* (MOM) scheme, a vertex function (e.g. ggg) is defined, with the invariant masses of particles having the values $-\mu^2$ (i.e. similar to the ϕ^4 example). However, calculations are difficult in this scheme. In t'Hooft's *minimal subtraction* (MS) scheme, loop integrals, etc. are calculated in $n = 4 - \varepsilon$ dimensions, and finite results are obtained by subtracting the divergent terms, which are poles in $1/\varepsilon$, etc. Cumbersome terms in $(\ln 4\pi - \gamma_E)$, where γ_E = Euler constant $\simeq 0.577$, appear in this scheme. They are removed in the *modified-minimal-subtraction* ($\overline{\text{MS}}$) scheme, which is, in practice, the most commonly used. The quantity Λ is not the same in the different schemes, and roughly $\Lambda_{MOM}/\Lambda_{\overline{MS}}/\Lambda_{MS} \simeq 5.5/2.7/1.0$. The $\overline{\text{MS}}$ scheme will be used here. The expansion of α_S to next-to-leading order is (e.g. Buras, 1980)

$$\alpha_S(Q^2) = \frac{4\pi}{\beta_0 \ln(Q^2/\Lambda^2)} \left[1 - \frac{\beta_1}{\beta_0^2} \frac{\ln\ln(Q^2/\Lambda^2)}{\ln(Q^2/\Lambda^2)} \right]. \tag{7.147}$$

where $\beta_1 = (306 - 38n_f)/3$. In addition to summing the leading logs $(g^2 \ln Q^2/\mu^2)^n$, this also sums the next-to-leading logs $(g^2)^n(\ln Q^2/\mu^2)^{n-1}$.

It is of interest to compare the effective QED constant (4.196) with that of QCD. In QED β is positive, so that $\alpha(Q^2)$ increase as Q^2 increases (or distance decreases). The reason why β_{QCD} is negative is the presence of the term C_F in (7.143), and this arises from the gluons. Indeed, this would happen even if there were no quarks. The QED form for β corresponds to $C_F = 0$ (no photon loops) and $T(R) = 1$. In the latter case, the virtual e^+e^- pairs effectively 'screen' the electron charge. In QCD, however, the gluons 'anti-screen' the charge, so that the effective coupling increases as the distance between two quarks increases. In Fig. 7.22 the colour flows corresponding to some of the lowest order corrections to the quark–gluon vertex are shown. In Fig. 7.22(*a*) there is a closed loop, and this gets counted three times (once for each colour). In the vertex diagram (Fig. 7.22(*c*)), the gluon does not 'see' the colour of the intermediate quark c, so that the effective colour charge decreases as the distance probed gets smaller, due to the spreading out of the colour charge. The above ideas, although providing a simple intuitive picture, are somewhat over-simplified, since they depend, for example, on the gluon gauge.

An alternative (but more cumbersome) approach to having an effective coupling constant, would be to have a fixed coupling constant, defined (say) for on-shell particles at $q^2 = 0$. Then, for $q^2 \neq 0$, all the vertex diagrams which 'shield' the colour charge would need to be specifically calculated. If the effective coupling constant is used, however, then the vertices are simple.

The expansion of α_S in (7.146) is valid for $Q^2 \gg \Lambda^2$. At low values of Q^2, α_S can become large, thus invalidating perturbative methods. The parameter Λ is not specified in perturbative QCD, and must be determined from experiment. If the strong rise in $\alpha_S(Q^2)$ as $Q^2 \to \Lambda^2$ is related to partion confinement, then we would roughly expect that Λ corresponds to a typical hadron size ($\sim$100–300 MeV). Experimental results are, at present, still somewhat imprecise, but are not incompatible with these values.

7.7 QCD AND DEEP INELASTIC SCATTERING

7.7.1 Operator product expansion

The behaviour of the deep inelastic lepton–nucleon structure functions at large Q^2 can be derived from the RGE of QCD. The method used (Wilson, 1969) is to consider the light-cone behaviour of the products of currents of the type $J_\mu(\xi/2)J_\nu(-\xi/2)$. These arise in the expression for the hadronic tensor (7.39) (and making a translation to a symmetric form). Such products contain singularities in their light-cone behaviour ($\xi^2 \to 0$), but can be expanded as

$$J\left(\frac{\xi}{2}\right)J\left(-\frac{\xi}{2}\right) \simeq \sum_{k,N} C_k^N(\xi^2)\xi^{\mu_1}\cdots\xi^{\mu_N}O^{k\,N}_{\mu_1\ldots\mu_N}(0), \tag{7.148}$$

Fig. 7.22 Corrections to the quark–gluon interaction. The upper graphs show the Feynman diagrams, and the corresponding colour flow is shown below.

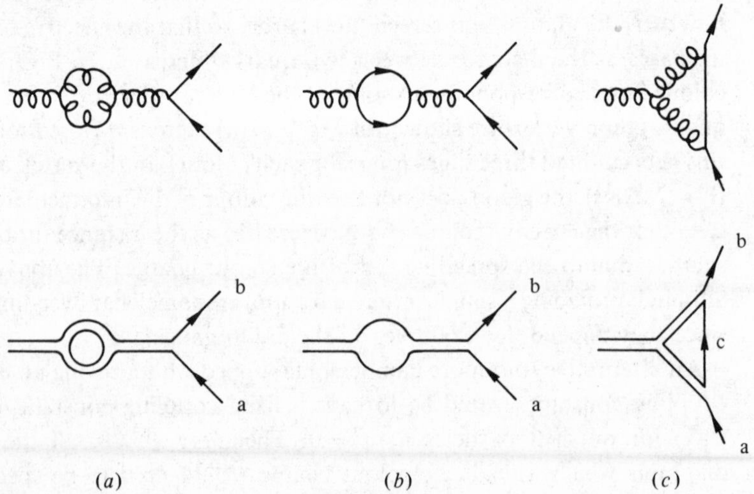

where the operators $O^{k,N}$ are non-singular and are called spin N operators. The coefficient functions C_k^N are singular as $\xi^2 \to 0$. Their behaviour can be found from dimensional counting. If d_J and $d_{k,N}$ are the mass dimensions of J and $O^{k,N}$ respectively, then

$$[C_k^N] = [M]^{2d_J + N - d_{k,N}}, \tag{7.149}$$

so that, up to log terms,

$$C_k^N(\xi^2) \xrightarrow{\xi^2 \to 0} \left(\frac{1}{\xi^2}\right)^{\frac{1}{2}(2d_J - \tau_{k,N})}, \tag{7.150}$$

where $\tau_{k,N} = d_{k,N} - N$ is called the '*twist*' of operator $O^{k,N}$, i.e. the twist is the mass dimension minus the spin N of $O^{k,N}$. Operators of low twist dominate the light-cone expansion. In the case of the structure functions, the lowest twist is 2, and higher twist terms (4, 6, etc.) become negligible for $Q^2 \to \infty$.

The Fourier transforms of $C^N(\xi^2)$ can be related to the Nth moment of the structure functions. For F_2, the Nth moment is defined as

$$M_2^N(Q^2) = \int_0^1 x^{N-2} F_2(x, Q^2)\,\mathrm{d}x, \tag{7.151}$$

and there are similar expressions for xF_1 and xF_3. The simplest predictions are for flavour non-singlet structure function combinations. These contain only the valence quark distributions, in contrast to the singlet case where there are quark and antiquark sea terms. Using the RGE of QCD, it can be shown that a non-singlet (NS) moment satisfies

$$M_{\mathrm{NS}}^N(Q^2) = \left[\frac{\bar{g}^2(t)}{g^2}\right]^{d_{\mathrm{NS}}^N} (M_0)_{\mathrm{NS}}^N(\mu^2), \tag{7.152}$$

where M_0 is the asymptotic free parton (i.e. QPM) result. If we divide $M_{\mathrm{NS}}^N(Q^2)$ by $M_{\mathrm{NS}}^N(Q_0^2)$, then we obtain, using also (7.146),

$$M_{\mathrm{NS}}^N(Q^2) = \left[\frac{\ln(Q^2/\Lambda^2)}{\ln(Q_0^2/\Lambda^2)}\right]^{-d_{\mathrm{NS}}^N} M_{\mathrm{NS}}^N(Q_0^2), \tag{7.153}$$

where Q_0^2 is a (large) value of Q^2, from which the evolution of the moments is predicted. Note that there is no prediction for the moment itself, only for its Q^2 evolution. The constants d_{NS}^N are given directly by QCD. The moments of singlet structure function combinations have three such terms, corresponding essentially to valence, sea and gluon contributions. Thus, the non-singlet structure functions provide, in principle, the most direct tests of QCD.

Experimentally, at large values of Q^2, the moments are difficult to measure. For example, in the μN measurements of the BCDMS and EMC collaborations at CERN, only values of F_2 up to $x_{BJ} \sim 0.8$ have been measured reliably. Thus moments, which require measurements up to $x_{BJ} = 1$, can be only partially constructed. Furthermore, the contributions to the observed scaling violations of the $1/Q^2$ twist-four terms are, *a priori*, unknown. However, such terms are expected to be most important for large x_{BJ}.

7.7.2 The Altarelli–Parisi equations

An alternative approach to the analysis of structure function measurements is to use the *Altarelli–Parisi equations*. These equations are built out of the leading order parton cross-sections for the processes $q \to q + g$, $g \to q + \bar{q}$ and $g \to g + g$. The first of these processes ($q \to q + g$) has, apart from the colour factor, the same form as $e^- \to e^- + \gamma$. In Section 4.4 the Compton cross-section $\gamma(k) + e^-(p) \to \gamma(k') + e^-(p')$ was evaluated. Let us now consider this latter process for the case of an incident virtual photon having $k^2 = -Q^2$. The matrix element has the form (cf. (4.89))

$$\mathcal{M}_{fi} = -ie^2 \varepsilon_\alpha \varepsilon'_\beta \bar{u}_f \left[\frac{\gamma^\beta(\not{p} + \not{k})\gamma^\alpha}{s} + \frac{\gamma^\alpha(\not{p} - \not{k}')\gamma^\beta}{u} \right] u_i$$
$$= \mathcal{M}_s + \mathcal{M}_u. \tag{7.154}$$

Using the usual trace techniques, together with summing over the photon polarisation states (i.e. a factor $-g_{\alpha\beta}$) and averaging over initial spins, we obtain

$$|\mathcal{M}_s|^2 = \frac{e^4}{4s^2} \mathrm{tr}[\not{p}'\gamma^\beta(\not{p} + \not{k})\gamma^\alpha \not{p} \gamma_\alpha (\not{p} + \not{k})\gamma_\beta]$$
$$= 2e^4 \left(\frac{-u}{s} \right), \tag{7.155}$$

where $s = (k + p)^2$, $u = (k - p')^2$ and $t = (k - k')^2$. Similarly,

$$|\mathcal{M}_u|^2 = 2e^4 \left(\frac{-s}{u} \right), \qquad 2|\mathcal{M}_u \mathcal{M}_s^*| = 4e^4 \frac{tQ^2}{su}. \tag{7.156}$$

The presence of the interference term is the only change from the real photon case ($Q^2 = 0$). The sum of these three terms is

$$|\mathcal{M}_{fi}|^2 = 32\pi^2\alpha^2 \left[\frac{(-u)}{s} + \frac{s}{(-u)} + \frac{2tQ^2}{su} \right]. \tag{7.157}$$

It is now a relatively simple matter to find the cross-section for $\gamma^*(q) + q(p) \to q(p') + g(k)$ (Fig. 7.23(*a*) and (*b*)). Note that with these (conventional) assignments for the outgoing four-momenta (i.e. $t = (q - p')^2$, $u = (q - k)^2$), the roles of t and u are reversed compared to $\gamma^* e \to \gamma e$. A further, and more important, difference is that the product of the vertex coupling constants is $C_F e_q^2 \alpha \alpha_S$ instead of α^2. The colour factor $C_F = \frac{4}{3}$ is obtained by summing over the eight octet states and averaging over the three quark colours. The additional factor $\frac{1}{2}$ comes from the normalisation of the λ-matrix which appears at the qg vertex (see Section 2.5.5 and Appendix C). Thus, the spin and colour-averaged matrix element squared is

$$|\mathscr{M}_{fi}|^2 = 32\pi^2 (C_F e_q^2 \alpha \alpha_S) \left[\frac{(-t)}{s} + \frac{s}{(-t)} + \frac{2uQ^2}{st} \right]. \tag{7.158}$$

The differential cross-section for $\gamma^* q \to qg$ is, using (4.29),

$$\frac{d\sigma}{d\Omega} = \frac{C_F e_q^2 \alpha \alpha_S}{2s} \left[\frac{(-t)}{s} + \frac{s}{(-t)} + \frac{2uQ^2}{st} \right]. \tag{7.159}$$

In deriving equation (7.159), the Hand convention, namely that the flux factor corresponds to the momentum of a real photon giving the same value of $s^{1/2}$, has been used.

Fig. 7.23 Lowest order graphs contributing to scaling violations in deep inelastic scattering.

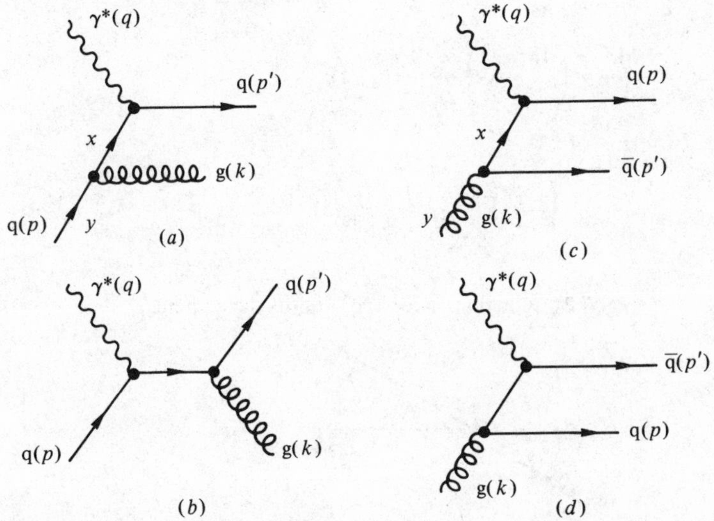

In Section 7.3, we considered the modifications to the simple quark–parton model predictions arising from the intrinsic transverse momentum (k_T). We will now investigate the transverse momentum arising from the gluon emission process described above. The kinematics of this process in the γ^*q cms are shown in Fig. 7.24. The variables t and u are related to the quark scattering angle θ by

$$t = -\frac{(s+Q^2)}{2}(1-\cos\theta), \qquad u = -\frac{(s+Q^2)}{2}(1+\cos\theta). \tag{7.160}$$

If p^* is the cms momentum of the final state quark (assumed massless) or gluon, then the transverse momentum of the quark is $p_T = p^* \sin\theta$, and so $d\Omega = (4\pi/s)\, dp_T^2$ (for $\cos\theta \simeq 1$). The differential cross-section (7.159) is largest for small values of θ (i.e. $t \to 0$) and, in this limit, is

$$\frac{d\sigma}{dp_T^2} = \frac{2\pi}{s^2} C_F e_q^2 \alpha\alpha_S \frac{1}{(-t)}\left[s + \frac{2(s+Q)Q^2}{s}\right]. \tag{7.161}$$

In deriving (7.161) the approximation $(-u) \simeq s + Q^2$ has been used, and the term in t^2/s has been neglected. This neglected term corresponds to the graph shown in Fig. 7.23(*b*), in which the gluon is emitted after absorption of the γ^*.

In analogy with the Bjorken variable x, we may define

$$z = \frac{Q^2}{2p\cdot q} = \frac{Q^2}{s+Q^2}, \tag{7.162}$$

using $s = (p+q)^2$. Now, from (7.160), we have $\sin^2\theta = 4tu/(s+Q^2)^2$, so that $p_T^2 = stu/(s+Q^2)^2 \simeq s(-t)/(s+Q^2)$, for small θ. Therefore, (7.161) may be written as

$$\frac{d\sigma}{dp_T^2} = \left(\frac{4\pi^2\alpha e_q^2}{s}\right)\frac{\alpha_S}{2\pi}\frac{1}{p_T^2} P_{qq}(z), \tag{7.163}$$

or, in terms of t,

$$\frac{d\sigma}{dt} = \left(\frac{4\pi^2\alpha e_q^2}{s}\right)\frac{\alpha_S}{2\pi}\frac{1}{(-t)} P_{qq}(z), \tag{7.164}$$

Fig. 7.24 Kinematics for $\gamma^*q \to qg$ in the γ^*q cms.

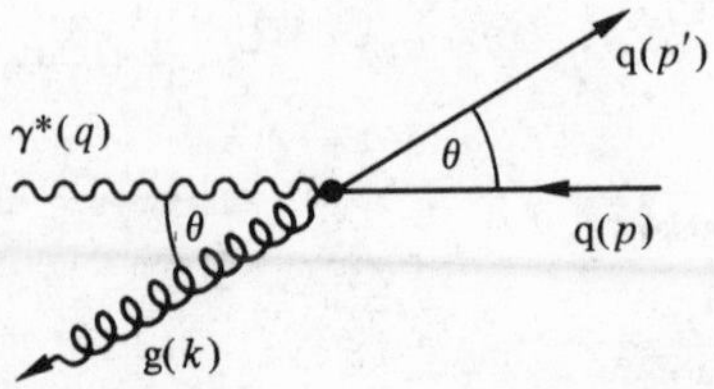

where

$$P_{qq}(z) = C_F\left(\frac{1+z^2}{1-z}\right). \tag{7.165}$$

Note that t is the virtual mass of the quark which absorbs the current, and that $-t \simeq p_T^2/(1-z)$.

We are interested in the effect that the $\gamma^*q \to qg$ interaction has on the longitudinal momentum fraction of the nucleon which is carried by the quark. Hence, we must integrate (7.163), and this gives

$$\hat{\sigma} = \hat{\sigma}_0 \frac{\alpha_S}{2\pi} P_{qq}(z) \ln(Q^2/\lambda^2); \qquad \hat{\sigma}_0 = 4\pi^2\alpha e_q^2/\hat{s}. \tag{7.166}$$

The hat (ˆ) indicates that this is really a subprocess in the γ^*–nucleon collision. The integral over $p_T^2(t)$ in (7.166) is from λ^2 (an infrared cut-off) to the maximum allowed value, namely $\hat{s}/4$, which we take to be about Q^2 for large Q^2. The term in $\ln Q^2$ is an example of a leading logarithm. The neglected term becomes relatively unimportant at high Q^2 compared to the leading term.

The quark in the above process must be considered as part of the target nucleon. Without gluon emission we have $q = (0, 0, 0, Q)$ and $p = (Q/2, 0, 0, -Q/2)$, so that, from (7.162), $z = 1$. The cross-section $\hat{\sigma}(z)$ in this case can be found using (7.45) and recalling that $\sigma_S = 0$; hence, $\hat{\sigma}(z) = \hat{\sigma}_0\, \delta(1-z)$. In general, we may interpret z as the fraction by which the relative momentum of the quark is changed by gluon emission. This is illustrated in Fig. 7.23(*a*). A quark with momentum fraction y radiates a gluon, reducing its momentum fraction to x, i.e. $z = x/y$. The nucleon cross-section $\sigma(x, Q^2)$ is related to the parton cross-section $\hat{\sigma}(z, Q^2)$ by

$$\begin{aligned} \sigma(x, Q^2) &= \int_0^1 \mathrm{d}z \int_0^1 \mathrm{d}y\, q(y)\, \delta(x - zy)\hat{\sigma}(z, Q^2) \\ &= \int_x^1 \frac{\mathrm{d}y}{y} q(y)\hat{\sigma}\left(\frac{x}{y}, Q^2\right). \end{aligned} \tag{7.167}$$

Using the cross-sections for the naive QPM process and for $\gamma^*q \to qg$, we can see that the quark density distributions (i.e. F_2/x) should obey

$$q(x, Q^2) = \int_x^1 \frac{\mathrm{d}y}{y} q(y)\left[\delta\left(1 - \frac{x}{y}\right) + \frac{\alpha_S}{2\pi} P_{qq}\left(\frac{x}{y}\right) \ln\left(\frac{Q^2}{\lambda^2}\right)\right]. \tag{7.168}$$

Thus, the effect of gluon emission is to introduce an effective Q^2 dependence into the quark density functions, arising from the second term in (7.168).

The change or evolution of the quark density distribution (7.168) with $\ln Q^2$ is

$$\frac{dq(x, Q^2)}{d\ln Q^2} = \frac{\alpha_S}{2\pi}\int_x^1 \frac{dy}{y}\, q(y, Q^2)P_{qq}(x/y). \tag{7.169}$$

Equation (7.169) is one of the Altarelli–Parisi equations (Altarelli and Parisi, 1977). The '*splitting*' function $P_{qq}(x/y)$ gives the probability of finding a quark, with momentum fraction x, inside a quark with momentum fraction y. Thus, equation (7.168) gives the sum of all such contributions with $y > x$. The $\ln Q^2$ scaling violating term comes from integrating the $t^{-1}(p_T^{-2})$ dependence of the cross-section. For large s, the largest contribution to the cross-section comes from the region $(-t) \to 0$ (see (7.158)), i.e. the t-channel exchange graph, Fig. 7.23(a), which leads to a strong forward peak in the cross-section. The p_T^{-2} dependence comes from the $t^{-2} \sim p_T^{-4}$ propagator, multiplied by a matrix element factor $t \sim p_T^2$. This latter factor arises from helicity conservation at the $q \to q + g$ vertex, since a quark cannot emit a gluon at $\theta = 0$ without flipping its helicity.

The evolution of the quark density function with Q^2 also occurs because of the process $g \to q + \bar{q}$. In order to calculate this process we can again start by considering the analogous electromagnetic process $\gamma^*(q) + \gamma(k) \to e^-(p) + e^+(p')$. The matrix element is

$$\begin{aligned}\mathcal{M}_{fi} &= -ie^2\varepsilon_\alpha(q)\varepsilon_\beta(k)\bar{u}\left[\frac{\gamma^\alpha(\not{p} - \not{q})\gamma^\beta}{t} - \frac{\gamma^\beta(\not{p}' - \not{q})\gamma^\alpha}{u}\right]v \\ &= \mathcal{M}_t + \mathcal{M}_u.\end{aligned} \tag{7.170}$$

Using the usual techniques, the spin-averaged contributions are

$$\begin{aligned}&|\mathcal{M}_t|^2 = 2e^4(-u)/(-t), \qquad |\mathcal{M}_u|^2 = 2e^4(-t)/(-u), \\ &|\mathcal{M}_{ut}| = -4e^4sQ^2/tu.\end{aligned} \tag{7.171}$$

Thus,

$$|\mathcal{M}_{fi}|^2 = 32\pi^2\alpha^2\left[\frac{(-u)}{(-t)} + \frac{(-t)}{(-u)} - \frac{2sQ^2}{tu}\right]. \tag{7.172}$$

To obtain the cross-section for $\gamma^*(q) + g(k) \to q(p) + \bar{q}(p')$, we replace α^2 by $\alpha e_q^2 C_F \alpha_S$, where C_F is the appropriate colour factor. The graphs corresponding to the matrix elements $\mathcal{M}_t$ and $\mathcal{M}_u$ are shown in Figs. 7.23(c) and (d) respectively. Following similar steps to the $\gamma^* q \to qg$ calculation, the differential cross-section is

$$\frac{d\sigma}{dt} \simeq \left(\frac{4\pi^2\alpha e_q^2}{s}\right)\frac{\alpha_S}{2\pi}\frac{1}{(-t)}P_{qg}(z), \tag{7.173}$$

where

$$P_{qg}(z) = C_F[z^2 + (1-z)^2]. \tag{7.174}$$

The colour factor is $C_F = (8\ q\bar{q}$ colour states)/(8 gluon states) $\times \frac{1}{2} = \frac{1}{2}$ (see (2.292)). The splitting function $P_{qg}(z)$ represents the probability of finding a quark, with relative momentum fraction z, inside a gluon. Equation (7.173) can also be expressed in terms of p_T^2 by using $dp_T^2/p_T^2 = dt/t$.

To complete the formulae for the Q^2 evolution we must also consider the gluon evolution. The required quantities are (e.g. Pennington, 1983)

$$P_{gq}(z) = \frac{4}{3}\left[\frac{1+(1-z)^2}{z}\right], \tag{7.175}$$

$$P_{gg}(z) = 6\left[\frac{z}{1-z} + \frac{1-z}{z} + z(1-z)\right]. \tag{7.176}$$

The probability of finding a gluon in a quark P_{gq} can be found by noting that, from momentum conservation, $P_{gq}(z) = P_{qq}(1-z)$, and using (7.165). The probability of finding a gluon in a gluon $P_{gg}(z)$ is found by evaluating the triple gluon vertex $g \to gg$. The evolution of the density functions, to leading order, is thus

$$\frac{dq_i(x, Q^2)}{d\ln Q^2} = \frac{\alpha_S(Q^2)}{2\pi}\int_x^1 \frac{dy}{y}\left[P_{qq}\left(\frac{x}{y}\right)q_i(y, Q^2) + P_{qg}\left(\frac{x}{y}\right)g(y, Q^2)\right], \tag{7.177}$$

$$\frac{dg(x, Q^2)}{d\ln Q^2} = \frac{\alpha_S(Q^2)}{2\pi}\int_x^1 \frac{dy}{y}\left[\sum_j P_{gq}\left(\frac{x}{y}\right)q_j(y, Q^2) + P_{gg}\left(\frac{x}{y}\right)g(y, Q^2)\right], \tag{7.178}$$

where the sum j is over all q and $\bar{q}$ flavours (i.e. $j = 1, \ldots, 2n_f$).

7.7.3 Singularities and higher orders

There are several points which have been skipped over in the above discussion. Firstly the $(1-z)$ terms in the denominators of the P functions are singular in the limit $z \to 1$. Furthermore, we have considered only the first-order term in α_S. From (7.166) it can be seen that this enters as $\alpha_S \ln Q^2$, which is $O(1)$, since $\alpha_S \sim 1/\ln Q^2$. Hence, we should not stop the calculation at the first order and terms of the type $\alpha_S^n(\ln Q^2)^n$, which appear as the leading logs in a calculation to order n, must be summed to all orders.

The infrared singularity $(z \to 1)$ in the $O(\alpha_S)$ calculation of $\gamma^* q \to qg$ is cancelled (in a similar way to that discussed for QED in Section 4.9) by adding the vertex correction graph and graphs with gluon loops on the external legs; these graphs contributing at $z = 1$. The form of the virtual terms can be found by imposing the requirement, arising from flavour conservation, that $\int_0^1 P_{qq}(z)\,dz = 0$. This latter formula can be shown by integrating (7.169), evaluated for the non-singlet $q_{NS}(x, Q^2)$ over all x, which yields zero, since the total number of quarks minus antiquarks is zero. In order to integrate (7.165) over all z, we need a prescription to regulate the $z \to 1$ divergence. This can be achieved by using (Altarelli, 1982)

$$\int_0^1 dz \frac{\phi(z)}{(1-z)_+} = \int_0^1 dz \frac{[\phi(z) - \phi(1)]}{(1-z)}. \tag{7.179}$$

Thus, if we define

$$P_{qq}(z) = \frac{4}{3}\left(\frac{1+z^2}{1-z}\right)_+ = \frac{4}{3}\left[\frac{1+z^2}{(1-z)_+} + \tfrac{3}{2}\,\delta(1-z)\right], \tag{7.180}$$

then the desired result is achieved. Hence, the required form of the vertex correction term is proportional to $\delta(1-z)$.

In the calculation of the graphs shown in Fig. 7.23, the *Feynman gauge* for the gluon propagator was used. In this *covariant gauge* a sum is made over the helicity 0 states of the massless gluons. The leading log term has contributions from both the t-channel exchange (Fig. 7.23(a)) and from the interference term. However, if a *physical gauge* is used, in which a sum is made only over the helicity states $\lambda = \pm 1$, then only the t-channel graph contributes. This gauge is useful in obtaining some insight into the effects of higher order graphs. Using the optical theorem, the gluon emission graph may be drawn as shown in Fig. 7.25(a). The broken line signifies that all the particles which it cuts are external on-shell particles. Thus, the evaluation of the cross-section, and its contribution to F_2, amounts to computing the imaginary part of the Compton scattering amplitude. The extension of this 'ladder' diagram to higher orders is shown (for the non-singlet case) in Fig. 7.25(b). Each rung i, with four-momentum squared t_i, involves an integral of the type

$$\int_{t_{i-1}}^{t_{i+1}} \frac{dt_i}{t_i} \sim \ln\left(\frac{t_i + 1}{t_i - 1}\right).$$

In the physical gauge we can interpret this diagram in a probabilistic way, with the dominant contribution coming from ordered momenta, so that the virtual photon sees successive layers of off-mass shell quarks, all of which contribute to the cross-section. The Altarelli–Parisi formalism corresponds to the sum to all orders of the ladder contributions, since repeated insertion of the lefthand-side in the righthand side generates the successive ladder diagrams.

The equivalence of the Altarelli–Parisi formalism with the operator product expansion and RGE method can be seen by taking moments of equation (7.169). This gives

$$\begin{aligned}\frac{\mathrm{d}M^N(Q^2)}{\mathrm{d}\ln Q^2} &= \frac{\alpha_\mathrm{S}}{2\pi}\int_0^1 \mathrm{d}x\, x^{N-1}\int_x^1 \frac{\mathrm{d}y}{y} P_{\mathrm{qq}}(x/y)q(y,Q^2)\\ &= \frac{\alpha_\mathrm{S}}{2\pi}\int_0^1 \mathrm{d}y\, y^{N-1}q(y,Q^2)\int_0^1 \mathrm{d}z\, z^{N-1}P_{\mathrm{qq}}(z)\\ &= \frac{\alpha_\mathrm{S}}{2\pi}M^N(Q^2)A^N. \qquad (7.181)\end{aligned}$$

In deriving (7.181) a change of variables has been made and the convolution property of the Mellin transform used. The term A^N is (Altarelli, 1982)

$$\begin{aligned}A^N &= \int_0^1 \mathrm{d}z\, z^{N-1}P_{\mathrm{qq}}(z)\\ &= C_\mathrm{F}\left[-\frac{1}{2}+\frac{1}{N(N+1)}-2\sum_{k=2}^{N}\frac{1}{k}\right], \qquad (7.182)\end{aligned}$$

Fig. 7.25 Non-singlet ladder diagrams in deep inelastic scattering.

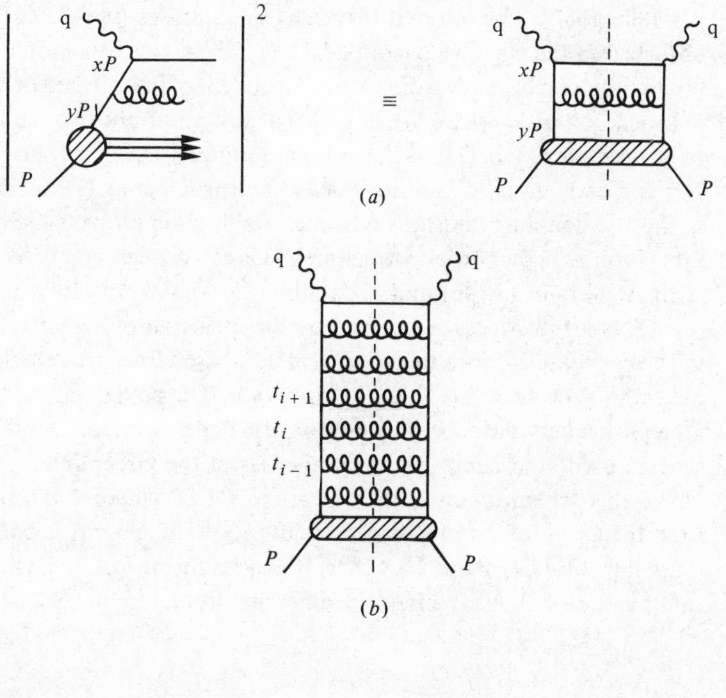

where the prescription outlined above for handling the $(1-z)$ singularity in equation (7.165) for $P_{qq}(z)$ has been used.

The solution of (7.181) has the same form as (7.153). The *anomalous dimension* d^N is given by $d^N = -2A^N/\beta_0$. Anomalous dimensions for the other splitting functions can be calculated in a similar way. Note that $M_g^2 + M_\Sigma^2 = 1$, where g and Σ represent the gluon and the sum over all q, $\bar{q}$ flavours respectively. This is a statement of momentum conservation. In the calculation of the $\gamma^*q \to qg$ process leading to equation (7.168), the coupling constant α_S was taken as constant. However, in order to obtain equivalence between (7.181) and the OPE and RGE approach we should have used the effective coupling constant (i.e. renormalisation group improved). Using $\alpha_S(p_T^2)$ in (7.163) (or $\alpha_S(t)$ in (7.164)) leads to $\hat{\sigma} \sim \ln \ln Q^2$, and (7.169) is obtained when the variation with respect to $\ln Q^2$ is found, since this gives $1/\ln Q^2$, i.e. $\sim \alpha_S(Q^2)$. The scaling violations given by the rigorous QCD approach (equation (7.153)), correspond to a summation to all orders of the leading logarithms. In the Altarelli–Parisi approach these summed terms are re-absorbed into effective Q^2 dependent parton density functions. The Q^2 dependence to *leading order* is then given either by the moment equations or by the Altarelli–Parisi equations themselves. This latter approach has the advantage of being applicable to processes where the OPE and RGE method is inapplicable. To do this it is also necessary to show that effective parton densities can be defined, which are dependent only on the target, and not on the nature of the probe (see Altarelli, 1982). The infrared (or mass) singularities (see (7.166)) are also absorbed in the effective parton densities. These can be factored out, and predictions can be made which are independent of the infrared behaviour.

In leading order (LO) QCD (i.e. that discussed above), the argument to be used for α_S (i.e. Q^2, p_T^2, t, etc.) is not uniquely specified. For example, $2Q^2$ is clearly equally as good a scale in a problem as Q^2. Such changes in the LO coupling constant introduce further log terms, which have the same form as higher order corrections. This can be seen simply by changing Λ to $\Lambda' = c\Lambda$ in (7.146) and expanding the 'old' α_S in terms of the new one. This arbitrariness is lessened by the inclusion of higher order terms in the calculation, since the coefficients of these terms are calculated with the same scale as α_S (see, e.g., Buras, 1980). The actual value of Λ which fits a particular set of experimental data also depends on the renormalisation scheme used, and hence on the properties of the Green function used to define this scheme. This is in contrast to QED, where α can be defined from the $Q^2 \to 0$ limit of Compton scattering. Therefore Λ is not, as such, a fundamental constant. However, it is a stringent test of QCD that the same value of Λ is obtained in different processes, provided the same renormalisation scheme is used and sufficient higher order terms are included. In addition to higher order QCD terms (which are in principle calculable), there are uncalculable higher twist terms (see, e.g., Fig. 7.26) which are important mainly at large x and fall-off as $1/Q^2$ ($1/Q^4$, etc.) with respect to the leading-twist (2) term. Since practical calculations involve inserting measured (or parameterised) quark and gluon density distributions into the righthand side of (7.177) and (7.178), care must be taken that higher twist scale breaking terms are not attributed to leading twist QCD.

7.7.4 Comparison with the data

The analysis of non-singlet flavour combinations (e.g. $q_{NS} = q_i - q_j$) greatly simplifies the theory, because both the gluon density (which must necessarily be found indirectly) and the gluon 'generated' sea-quark terms cancel in (7.177). Note, however, that this cancellation will not occur if part of the sea is generated by a non-charge symmetric source. The determination of the non-singlet structure function xF_3 requires a subtraction of the differential cross-sections measured using neutrinos and antineutrinos (see (7.98)). Some measurements of xF_3 off calcium nuclei are shown in Fig. 7.27. In terms of quark densities, $xF_3 = \sum x(q - \bar{q}) = u_v + d_v$ (see (7.107)), and the pattern of scaling violations indicated by Fig. 7.15 can clearly be seen in these data. The structure function combination $F_2^p - F_2^n$, measured by charged leptons, is also a non-singlet.

In order to extract the structure function $F_2(x, Q^2)$, the value of $R = \sigma_S/\sigma_T$ (or alternatively, the function $2xF_1$) must be measured (see equations (7.36) and (7.98)). Data on R at low Q^2 are shown in Fig. 7.11 and at higher Q^2 in Fig. 7.28. Although the errors are rather large, R is smaller for the higher Q^2 data sets. A decrease is expected from the intrinsic k_T of the struck quark (7.63), which fall as Q^{-2}, and also from QCD

Fig. 7.26 Example of a higher twist diagram in deep inelastic scattering.

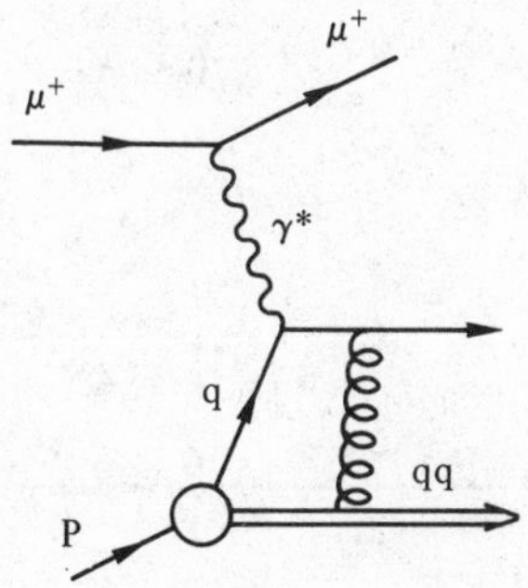

effects, which fall as approximately $(\ln Q^2)^{-1}$. The QCD calculation for R as a function of x_{BJ}, with $\Lambda = 90$ MeV, is also shown in Fig. 7.28.

Analysis of the structure function F_2 is sensitive, not only to the value of R, but also to assumptions on the sea-quark and gluon distributions (i.e. flavour singlet terms). A detailed QCD analysis by the EMC (Aubert *et al.*, 1986) shows reasonable agreement with the QCD expectations. These, and fits to other data, yield a value of Λ (in the commonly used modified minimal subtraction or $\overline{\text{MS}}$ scheme) of $50 \lesssim \Lambda_{\overline{\text{MS}}} \lesssim 300$ MeV. Note that the value of Λ varies considerably with the renormalisation scheme (see, e.g. Pennington, 1983).

A comparison of the $F_2^{\mu N}$ structure function measured off iron and deuterium by the EMC (Aubert *et al.*, 1983b), for the range $9 < Q^2 < 170$ GeV^2, showed that they had a strikingly different behaviour as a

Fig. 7.27 Measurements of the structure function xF_3, as a function of Q^2, for different x values, obtained using a calcium target by the CHARM collaboration (Bergsma *et al.*, 1984).

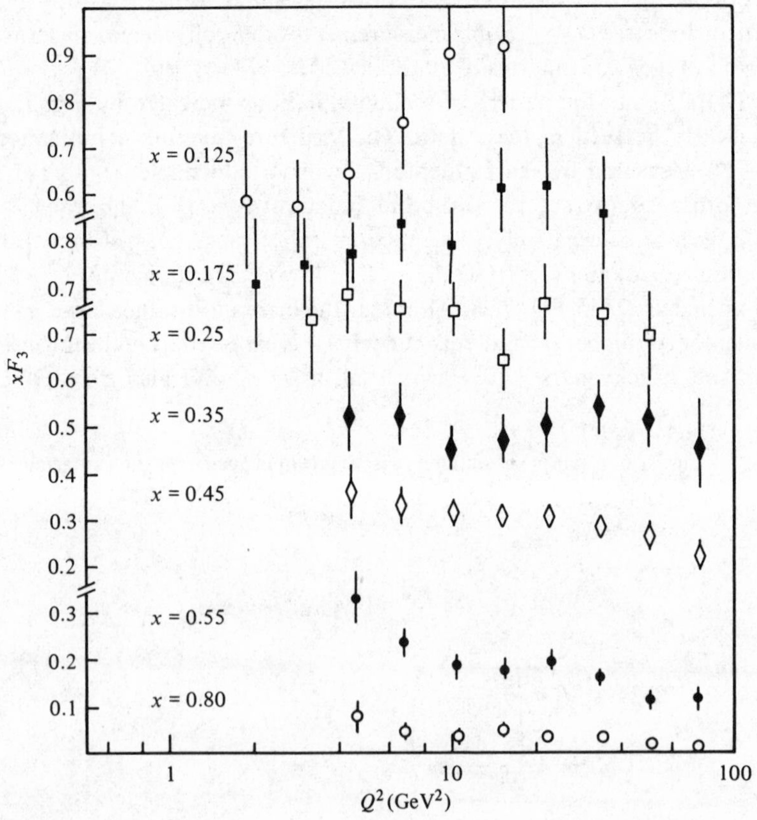

function of x. Prior to this observation it was assumed that, because of the high Q^2 nature of the γ^* probe, the interactions off quarks in a nucleon inside an iron nucleus would not be substantially different to those off quarks in the quasi-free deuterium nucleons. However, some differences were expected due to the Fermi motion of the nucleons, which can significantly change the event kinematics with respect to those of a free nucleon. This leads to an effective increase in F_2 for the nuclear target at large values of x_{BJ}. If the whole nucleus acts as the target, then one should really define x_{BJ} as $x_{BJ} = Q^2/(2M_A\nu)$, where M_A is the mass of the nucleus. Thus, in principle, if the conventional definition of x_{BJ} is used, then values up to $x_{BJ} \simeq A$ are possible. In practice, the structure functions fall rapidly at large x_{BJ}, so that the effects of Fermi motion are the only ones yet apparent.

Some data on this effect, subsequently called the 'EMC effect', are shown in Fig. 7.29. The structure function data are corrected for higher

Fig. 7.28 Measurements of R as a function of x from high energy μN scattering experiments. The errors shown are statistical, systematic errors are similar in size. ○, EMC μH_2; ▲, EMC μFe; ■, BCDMS μC.

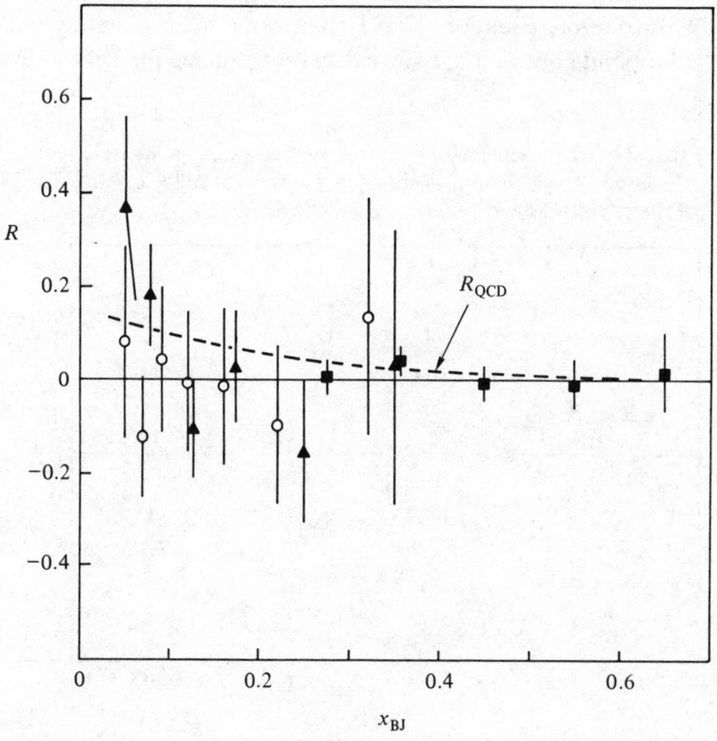

order QED effects, so that they correspond to the single-photon exchange process. These corrections are most important for the low x_{BJ} and high y regions. A clear fall of the ratio F_2^A/F_2^D below unity is seen for $x_{BJ} \gtrsim 0.3$, with a rise for $x_{BJ} \gtrsim 0.7$, this rise being attributable to Fermi motion. In the low x_{BJ} region the agreement between the different experiments is less good; however, they are reasonably compatible when the systematic errors (not shown) are taken into account. However, the ratio seems to rise above unity. Many theoretical models have been proposed to explain some, or all, of this effect (for a review see Berger and Coester, 1987). Two broad classes of models have emerged for the region $x \gtrsim 0.10$. In *conventional nuclear physics* models it is assumed that

$$F_2^A(x, Q^2) = \sum_c \int_x^1 dy\, f_c^A(y) F_2^c(x, Q^2), \tag{7.183}$$

where the sum is over all possible nuclear constituents c (which are taken as N, π, Δ, etc., as well as the more exotic possibilities of 'bags' of 6 (9, 12, etc.) quarks). The function $f_c^A(y)$ denotes the probability of finding constituent c in A, with momentum fraction y, and $F_2^c(x)$ is the free particle structure function of c. That is, it is assumed that the structure functions of the constituents are not changed by the nuclear environment. The assumed convolution equation (7.183) has been questioned by Jaffe (1987b), who points out that specific calculations (made for light nuclei)

Fig. 7.29 Measurements of the cross-section ratios $\sigma(A)/\sigma(D_2)$ as a function of x_{BJ}; the errors shown are statistical only. ▲, EMC Cu/D; ●, BCDMS Fe/D; ○, SLAC Fe/D; □, SLAC Cu/D.

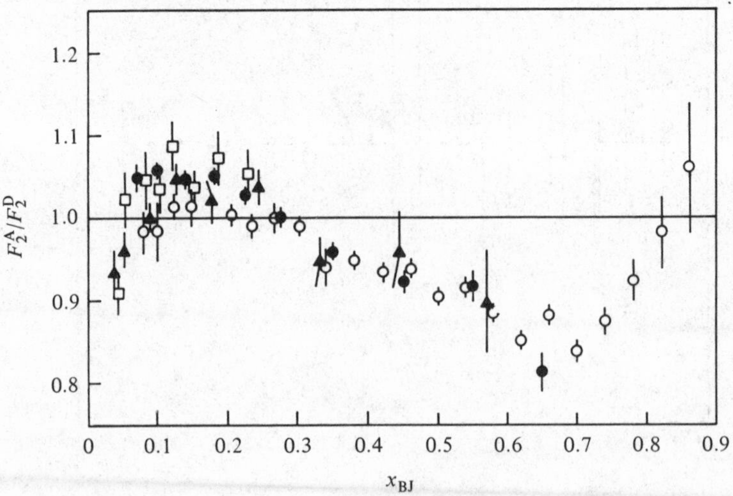

of multiquark effects (from Fermi statistics) violate the convolution form, and give a sizeable contribution to the effect.

A different interpretation of the data has been given by Jaffe *et al.* (1984). These authors noted that the EMC structure functions on iron and deuterium become compatible if the Q^2 value of iron is *rescaled* by some factor (~ 2). The QCD renormalisation scale μ^2 is taken to be larger for heavy nuclei than for deuterium. (This corresponds effectively to a swelling of the nucleons in a nucleus.) In this approach, the nuclear environment (in the sense of (7.183)) is assumed unaltered. Close *et al.* (1987) have argued that the two approaches outlined above may be equivalent, in that they are 'dual' statements of the same physics. Neither model reproduces the decrease in the ratio F_2^A/F_2^D at low x ($\lesssim 0.08$). In this region the data are reasonably consistent with the model of Nicolaev and Zakarov (1975), in which, due to the long correlation lengths corresponding to the low x region, the partons from different nucleons 'fuse' together, leading to a reduction in their density, and hence interaction probability. This reduction is referred to as *shadowing*, in that the cross-section depends on A more like $\sigma \propto A^{2/3}$ than $\sigma \propto A$, i.e. a surface effect shadowing the remaining nucleons. The Q^2 dependence of nuclear shadowing has been considered by Qiu (1987) using modified Altarelli–Parisi equations, which take into account parton recombination effects. Qiu concludes that shadowing will vanish only slowly with increasing Q^2.

Despite the theoretical uncertainties on the details of deep inelastic scattering in nuclear matter, and the possibly confusing experimental situation at low x_{BJ}, there is nevertheless good evidence for accepting the basic ideas of the Quark Parton Model and the existence of the scaling violations predicted by QCD.

7.8 TOTAL CROSS-SECTION FOR $e^+e^- \to$ hadrons

In this section we will sketch the methods by which $\sigma(e^+e^- \to$ hadrons) is calculated using renormalisation group techniques. From the optical theorem (2.123), the total cross-section for e^+e^-, $\sigma_{tot}(e^+e^-)$, is linearly related to the imaginary part of the forward scattering amplitude for the elastic process $e^+e^- \to e^+e^-$, averaged over spins. The relevant diagram is shown in Fig. 7.30. The blob in the propagator is one photon irreducible, and some of the lowest order contributing diagrams are shown. In general the photon propagator term can be written

$$\Gamma^{\mu\nu}(q^2) = \left(-g^{\mu\nu} + \frac{q^\mu q^\nu}{q^2}\right)\Pi(q^2). \tag{7.184}$$

We are interested in calculating the ratio R_h of the total cross-section for $e^+e^- \to$ hadrons to the point-like cross-section $e^+e^- \to \mu^+\mu^-$. In this ratio common factors, including the particle flux, cancel, so that, for $Q^2 = -q^2$ (and ignoring the subtleties of going from space-like to time-like q^2) we have

$$R_h = \frac{\operatorname{Im} \Pi_h(Q^2)}{\operatorname{Im} \Pi_\mu(Q^2)}. \tag{7.185}$$

In (7.185), Π_h and Π_μ are the parts of $\Pi(Q^2)$ with intermediate hadrons (i.e. all possible final state hadrons arising from the fragmentation of the final state partons) and with intermediate $\mu^+\mu^-$ states respectively. The method used to compute $\Pi(Q^2)$, is to solve the QCD renormalisation group equation (cf. (7.125)), with the γ-factor computed using the graphs shown in Fig. 7.30, corresponding to $n = 2$. The solution is (e.g. Leader and Predazzi, 1982)

$$\Pi(Q^2) \simeq Q^2 \exp\left\{-2 \int_0^t \gamma[\bar{g}(t')]\, dt'\right\}, \tag{7.186}$$

where (as above) $t = \frac{1}{2} \ln Q^2/\mu^2$, and

$$\gamma(\bar{g}) \simeq Ke^2 \left\{\sum_l e_l^2 + \sum_q e_q^2 \left[1 + 3C_2(R)\frac{\bar{g}^2}{16\pi^2}\right] \cdots\right\}. \tag{7.187}$$

In (7.187), K is a constant which will cancel in the ratio R_h, e_l and e_q are the charges of the leptons and quarks respectively and $C_2(R) = \frac{4}{3}$ is the colour factor for the gluon octet representation. Since γ is of order e^2 (i.e. small), we can expand the exponential in (7.186), giving

$$\Pi(Q^2) \simeq Q^2\left\{1 - 2Ke^2\left[\left(\sum_l e_l^2\right)t + \left(\sum_q e_q^2\right)\left(t + \int_0^t \frac{\bar{g}^2}{4\pi^2}\, dt'\right)\right] + \cdots\right\}. \tag{7.188}$$

Fig. 7.30 Elastic scattering $e^+e^- \to e^+e^-$, used to evaluate $\sigma_{tot}(e^+e^-)$.

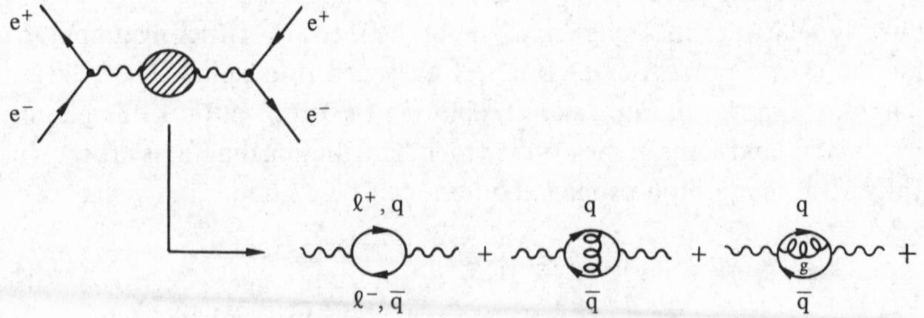

The remaining integral can be computed using equation (7.140), and is equal to $(2/\beta_0)\ln[\bar{g}^2(\mu^2)/\bar{g}^2(t)]$. Using (7.140), we can express $\bar{g}^2(\mu^2)$ in terms of $\bar{g}^2(t)$, then, expanding the log terms, get the result $\bar{g}^2(t)t/4\pi^2$ for the integral. Hence, picking out the absorptive (imaginary) parts of (7.188), we obtain the ratio

$$R_h = \frac{\sigma(e^+e^- \to \text{hadrons})}{\sigma(e^+e^- \to \mu^+\mu^-)} = \sum_q e_q^2\left[1 + \frac{\alpha_S(Q^2)}{\pi} + \cdots\right]. \tag{7.189}$$

The leading order term in (7.189) is that expected in the quark parton model, and has the same form as that found for $e^+e^- \to f\bar{f}$ in Section 4.3. The sum over quarks includes all possible flavors and colours. The colour factor of three is often written explicitly, leaving only the sum over flavours. Note that the simple parton model result is only expected asymptotically ($Q^2 \to \infty$). At finite Q^2, R_h is modified by the effective coupling constant $\alpha_S(Q^2)$. The reason that the effective coupling constant appears (rather than g^2) again stems from the RGE. Note also that the result for $\sigma(e^+e^- \to \text{hadrons})$ does not contain any mass singularities.

An alternative approach to the RGE method is to explicitly calculate the cross-section for the process $e^+ + e^- \to \gamma^*(q) \to q(p_1) + \bar{q}(p_2) + g(p_3)$ (Fig. 1.10). We have already calculated a similar process ($\gamma^* + q \to q + g$), which led to (7.158). For the e^+e^- case we obtain the analogous matrix element squared

$$|\mathcal{M}|^2 = 32\pi^2 C_F e_q^2 \alpha\alpha_S\left(\frac{t}{s} + \frac{s}{t} + \frac{2uQ^2}{st}\right), \tag{7.190}$$

where the definitions of the kinematic invariant quantities now become $s = (q - p_1)^2$, $t = (q - p_2)^2$, and $u = (q - p_3)^2$. Note that the centre-of-mass energy of the e^+e^- system is Q ($Q^2 = q^2$ is positive).

In order to calculate the differential cross-section for the q, $\bar{q}$, g final state, we must also evaluate the three-body phase space factor (see Appendix C).

$$N_3 = \int \frac{d^3p_1\, d^3p_2\, d^3p_3}{E_1E_2E_3}\, \delta^4(q - p_1 - p_2 - p_3), \tag{7.191}$$

where the final state particles are assumed massless. Integrating over d^3p_3, then using $E_3\, dE_3 = p_1p_2\, d\cos\theta_{12}$ (from $p_3^2 = p_1^2 + p_2^2 + 2p_1p_2\cos\theta_{12}$), we obtain

$$\begin{aligned} N_3 &= \int \frac{4\pi p_1^2\, dp_1\, 2\pi p_2^2\, dp_2\, d\cos\theta_{12}\, \delta(Q - E_1 - E_2 - E_3)}{E_1E_2E_3} \\ &= 8\pi^2 \int dE_1\, dE_2. \end{aligned} \tag{7.192}$$

It is useful to introduce the (Dalitz plot) variables

$$x_i = \frac{2E_i}{Q}, \qquad i = 1, 2, 3, \tag{7.193}$$

in which the parton energies are scaled by the maximum possible parton energy $Q/2$ (i.e. with $\mathrm{q\bar{q}}$ only). The direction of the parton with the maximum energy in a given event (x_1, say) is called the *thrust axis.* Thus, in the γ^* cms we have $q = Q(1, 0, 0, 0)$, $p_1 = Q/2(x_1, 0, 0, x_1)$, $p_2 = Q/2(x_2, 0, x_2 \sin\theta, -x_2\cos\theta)$, and $p_3 = Q/2(x_3, 0, -x_2\sin\theta, -x_1 + x_2\cos\theta)$, where the thrust axis is taken to be along $0z$. The kinematic invariants can therefore be written

$$s = Q^2(1 - x_1), \qquad t = Q^2(1 - x_2), \qquad u = Q^2(1 - x_3). \tag{7.194}$$

Energy conservation gives the relation $x_1 + x_2 + x_3 = 2$. In terms of these variables the matrix element squared (7.190) becomes

$$|\mathcal{M}|^2 = 32\pi^2 C_{\mathrm{F}} e_{\mathrm{q}}^2 \alpha\alpha_{\mathrm{S}} \left[\frac{x_1^2 + x_2^2}{(1 - x_1)(1 - x_2)}\right]. \tag{7.195}$$

In order to find the differential cross-section for the entire $\mathrm{e^+e^-} \to \gamma^* \to \mathrm{q\bar{q}g}$ process, we need to graft on the $\mathrm{e^+e^-} \to \gamma^*$ vertex to the calculation. This can be achieved by comparing the $\mathrm{e^+e^-} \to \mathrm{q\bar{q}g}$ cross-section to that for $\mathrm{e^+e^-} \to \mathrm{q\bar{q}}$, i.e. $\sigma_0 = 4\pi\alpha^2 e_{\mathrm{q}}^2/3Q^2$ (4.62). The result is

$$\frac{\mathrm{d}^2\sigma}{\mathrm{d}x_1\,\mathrm{d}x_2} = \sigma_0 \frac{\alpha_{\mathrm{S}}}{2\pi}\frac{4}{3}\left[\frac{x_1^2 + x_2^2}{(1 - x_1)(1 - x_2)}\right]. \tag{7.196}$$

The reason for this form can be seen by making a change of variable from x_2 to t (7.194), under which we have $t\,\partial/\partial t = -(1 - x_2)\,\partial/\partial x_2$. In the limit $x_2 \simeq 1$, equation (7.196) can thus be written

$$\frac{\mathrm{d}^2\sigma}{\mathrm{d}x_1\,\mathrm{d}t} \simeq \sigma_0 \frac{\alpha_{\mathrm{S}}}{2\pi}\frac{1}{(-t)} C_{\mathrm{F}}\left(\frac{1 + x_1^2}{1 - x_1}\right), \tag{7.197}$$

which has the form of (7.164) and consequently of an Altarelli–Parisi equation. This method of factorisation of the cross-section into the product of those of subprocesses is, in fact, rather general (*Weizäcker–Williams formula*).

The differential cross-section (7.196) is singular if either (i) $x_1 \simeq 1$ or $x_2 \simeq 1$ (in which case the gluon is colinear with the quark or antiquark), or (ii) $x_1 \simeq x_2 \simeq 1$, $x_3 \simeq 0$ (i.e. soft gluon limit). In QED these divergences can be handled by defining suitable physical quantities such as the number of electrons with energies between E and $E+\mathrm{d}E$, and angles between θ and $\theta+\mathrm{d}\theta$. At the parton level such finite quantities can be defined following the approach due to Sterman and Weinberg (see, e.g., Pennington, 1983). However, experimentally it is the hadrons, and not the partons, which are observed. The non-perturbative effects of the hadronisation of the partons must therefore be taken into account. Hence, at present, QCD predictions can only be tested in conjunction with some model for fragmentation. Ideally, these tests should be as independent as possible of the details of these models. The total cross-section, obtained by integrating (7.196) over x_1 and x_2 (over the range 0 to 1), clearly diverges. This divergence is exactly cancelled by that arising from the interference between the lowest order quark–antiquark diagram and the diagrams with a one gluon loop. This can be seen (e.g. Pennington, 1983) by giving the gluon a non-zero mass m_g and summing the contributions of all the diagrams, which gives a result independent of m_g.

7.9 HADRONISATION

As discussed above, the cross-section formulae for $e^+e^- \to$ hadrons (or for deep inelastic scattering), refer to the production of quarks and gluons, and not to the production of hadrons directly. In order that the formulae are applicable for hadrons we must further assume that the parton to hadron process is decoupled from the initial parton production. This is plausible at high energies, where the time scales for the initial hard QCD processes are expected to be significantly shorter than those of the soft fragmentation processes. Hence a parton fragments into a jet of hadrons with essentially unit probability.

Some experimental results on measurements of the total cross-section ratio R in e^+e^-, as a function of $s^{1/2}$, are shown in Fig. 1.14. Near the threshold for the production of a new $q\bar{q}$ flavour, resonances are formed (J/ψ plus excited states for $c\bar{c}$, Υ plus excited states for $b\bar{b}$). These resonances can be considered as '*hidden*' flavour production. At somewhat higher energies the production of mesons and baryons containing the new quarks occurs (*open* flavour production), these particles being slow in the cms. Only well above threshold do we expect the quark mass and bound state effects to become negligible and, hence, a comparison with QCD to be meaningful. As there is, as yet, no complete theory of hadronisation, the scale at which this should happen is not precisely known. However, from Fig. 1.14 it can be seen that, at PETRA energies ($12 \lesssim s^{1/2} \lesssim 45$ GeV), good agreement with equation (7.189), calculated for $q = u, d, s, c, b$ (each with three colours) and $\alpha_S(1000\ \mathrm{GeV}^2) \sim 0.2$, is obtained. At larger values of $s^{1/2}$, Z^0 exchange becomes important and must be taken into account.

Similar considerations apply in deep inelastic scattering. At low values of W, nucleon resonances dominate the cross-section. However, in the continuum region at higher W, it is assumed that hadronisation effects are less significant. At what values they can be considered negligible is, however, a more difficult question. There is a further phenomenon which is important at low Q^2 (and W), namely the hadronic structure of the virtual photon. We have assumed above that the photon acts as a point-like probe. However, the virtual photon can make a temporary $\gamma^* \to q\bar{q}$ transformation, and the $q\bar{q}$ can form a virtual vector meson (ρ^0, ω, ϕ, J/ψ, etc.), since these mesons have the same quantum numbers as the photon. This transition will occur (very roughly) during a fraction $\alpha(1/137)$ of the time, and the lifetime of the energy fluctuation is given by, using the uncertainty principle,

$$\Delta\tau \sim \frac{1}{\Delta E} \sim \frac{1}{\nu - (\nu^2 - Q^2 - m_v^2)^{1/2}} \sim \frac{2\nu}{Q^2 + m_v^2}, \tag{7.198}$$

where m_v is the vector meson mass. Hence, this effect (known as *vector meson dominance* of the cross-section) is expected to decrease rapidly for increasing Q^2 or m_v^2.

The vector meson dominance model can also lead to shadowing at low x of the cross-section (i.e. $\sigma \propto A^{2/3}$) in nuclei. However, this contribution is again expected to fall off rapidly with Q^2, and so is unlikely to be the main cause of the shadowing at large Q^2 discussed in Section 7.7.4.

The virtual vector meson, once formed, has a large cross-section and is expected to interact with the nucleon by mainly a diffractive-type process (Fig. 7.31) with

$$\frac{d\sigma}{dt} \simeq \exp(-b|t|), \qquad b \sim \tfrac{1}{2}(R_N^2 + R_v^2). \tag{7.199}$$

The parameter b depends on the sizes of the nucleon and vector meson, and the resulting value ($b \sim 8\ \text{GeV}^{-2}$) leads to a rapid fall-off in t, charac-

Fig. 7.31 Lepton–nucleon scattering by the vector meson dominance mechanism.

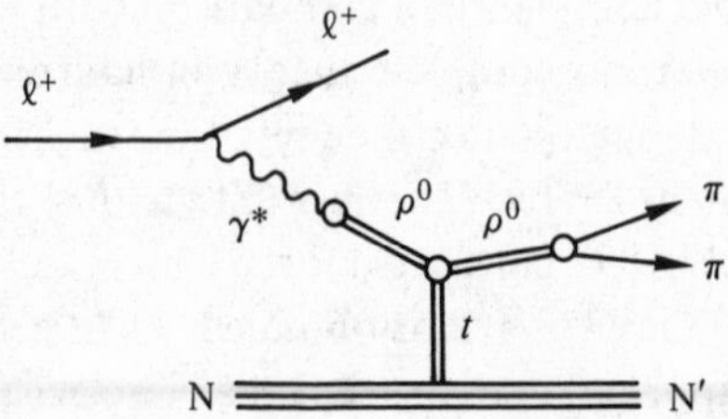

teristic of a diffractive process. At low Q^2 this picture predicts that the helicity of the scattered ρ^0 meson is the same as that of the photon (± 1), and this is confirmed experimentally. At high Q^2 ($\gtrsim 5\,\mathrm{GeV}^2$), however, the slope b increases (i.e the collision becomes harder) and the scattered ρ^0 no longer has the helicity ± 1 (Aubert *et al.*, 1985). Thus, reasonably large values of Q^2 and W are needed to ensure point-like deep inelastic scattering.

It is an empirical observation that hadron jets are observed in hard collision processes, and that these jets seem to correspond to the energetic partons in the process in question. The experiments at PETRA (JADE, MARK-J, PLUTO, TASSO and later CELLO) and also those at PEP (DELCO, HRS, MAC, MARK2, TPC) have analysed a large number of high energy hadronic final states. The dominant feature is that of two collimated back-to-back jets, corresponding to the quark and antiquark directions. The hadrons from these jets have a mean transverse momentum with respect to the jet axis of about 0.35 GeV, roughly the momentum corresponding to a typical hadronic radius.

The QCD diagrams in Fig. 1.10 can lead to a sizeable transverse momentum between the quark and antiquark. This can be seen by writing (7.197) in terms of p_{T}, the transverse momentum of the antiquark with respect to the quark axis, that is

$$\frac{\mathrm{d}^2\sigma}{\mathrm{d}x_1\,\mathrm{d}p_{\mathrm{T}}^2} \simeq \sigma_0 \frac{\alpha_{\mathrm{S}}}{2\pi} C_{\mathrm{F}} \frac{1}{p_{\mathrm{T}}^2}\left(\frac{1+x_1^2}{1-x_1}\right). \tag{7.200}$$

Keeping p_{T}^2 fixed, and integrating over x_1, we obtain

$$\begin{aligned}\frac{1}{\sigma_0}\frac{\mathrm{d}\sigma}{\mathrm{d}p_{\mathrm{T}}^2} &\simeq \frac{\alpha_{\mathrm{S}}}{2\pi} C_{\mathrm{F}} \frac{1}{p_{\mathrm{T}}^2}\int_{(x_1)_{\min}}^{(x_1)_{\max}} \frac{(1+x_1^2)}{1-x_1}\,\mathrm{d}x_1 \\ &\simeq \frac{4\alpha_{\mathrm{S}}}{3\pi}\frac{1}{p_{\mathrm{T}}^2}\ln\left(\frac{Q^2}{4p_{\mathrm{T}}^2}\right),\end{aligned} \tag{7.201}$$

where we have used $(x_1)_{\max} = 1 - 4p_{\mathrm{T}}^2/Q^2$, and retained only the leading log term (for the full expression see Pennington, 1983). Thus for a given value of p_{T}^2, the relative three-jet cross-section is predicted to increase with increasing Q^2. Some data on the p_{T}^2 distributions of final state hadrons (with respect to the jet or *sphericity axis*, defined as that minimising $\sum p_{\mathrm{T}}^2$ of the hadrons), for different values of Q ($= W$), are shown in Fig. 7.32. The predicted increase is indeed observed. The above discussion is somewhat oversimplified as the process involves two large scales (Q^2 and p_{T}^2), so that α_{S} must be defined with care (Dokshitzer *et al.*, 1980, advocate $\alpha_{\mathrm{S}}(p_{\mathrm{T}}^2)$). A similar increase in p_{T}^2 (defined with respect to the γ^* direction), with increasing centre-of-mass energy (W), is expected (and observed) in deep inelastic scattering. In this case there are three large scales namely W, Q^2 and p_T^2, in the problem.

The order α_s QCD corrections to $e^+e^- \to$ hadrons modify the QPM prediction of two jets (corresponding to the quark and antiquark) so that, in a significant fraction of the events at a cms energy $\gtrsim 30$ GeV, a third (gluon) jet is expected. Higher order diagrams should correspond to the production of four or more jets of hadrons. A detailed analysis of these events requires the use of a model for the fragmentation of the partons. The singularities due to soft and colinear gluons are handled by assuming that, below some cut-off, these configurations lead to two-jet topologies. The remaining part of the cross-section in equation (7.196) is assumed to produce three-jet topologies. A similar procedure can also be defined

Fig. 7.32 Differential p_T^2 distributions found by TASSO in e^+e^- annihilation for different values of the cms energy W: ×, 14 GeV; ●, 22 GeV; ■, 34 GeV; ▼, 41.5 GeV.

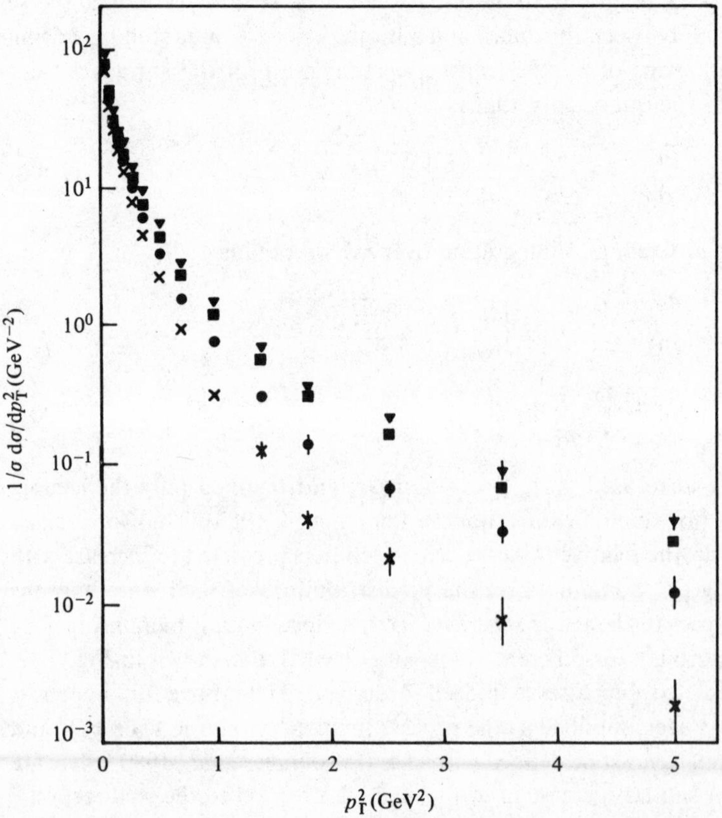

for higher numbers of jets. The 'hadronised' QCD formulae fit the data reasonably well, but uncertainties in the cut-off parameters, the dependence on the fragmentation model and the influence of higher order terms all limit the precision with which the strong coupling constant α_S can be extracted. Duke and Roberts (1985) derive a value $\alpha_S(\overline{MS}) = 0.19 \pm 0.06$ from consideration of such studies.

QCD-inspired models of hadrons have been developed to explain different aspects of the non-perturbative regimes. In the (MIT) 'bag' model, confinement is put in by hand by means of a spherical impenetrable bag and it is assumed that low order QCD calculations suffice, even though the effective coupling constant can be large. Some reasonable success in explaining the static properties of hadrons (e.g. mass spectra) is achieved. These ideas can be extended to explain some dynamic properties. If, due to a hard collision, a quark and antiquark are set in rapid relative motion, then a strong colour field will be formed between them. Because of the attractive force between the colour flux lines (due to gluons) the field is assumed to become confined into a tube-shaped bag. This narrow colour flux tube (or *string*) has constant energy density per unit length (κ), and gives rise to a long distance linear potential $\sim \kappa r$, for a separation r. The string can fission anywhere along its length (this becomes energetically more favourable with increasing separation), and this process is assumed to continue until the string pieces are about 1 fm (Fig. 7.33(*a*)), when hadrons are formed. The estimated string tension is $\kappa \sim 1$ GeV/fm (or 14 tonne weight in macroscopic units!). A detailed model based on the ideas of such massless relativistic strings has been formulated by the Lund group (Andersson *et al.*, 1983). The emission of a hard gluon is treated by

Fig. 7.33 (*a*) Break up of a 'string' colour field joining q and $\bar{q}$ into successively smaller segments. (*b*) Fragmentation scheme used in QCD branching models.

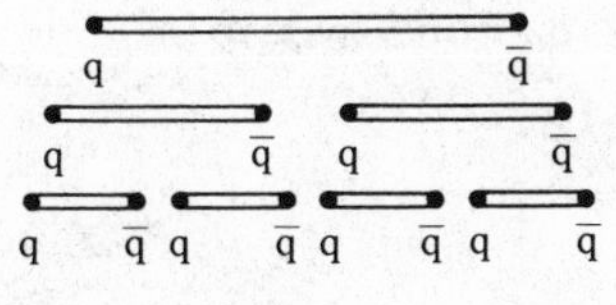

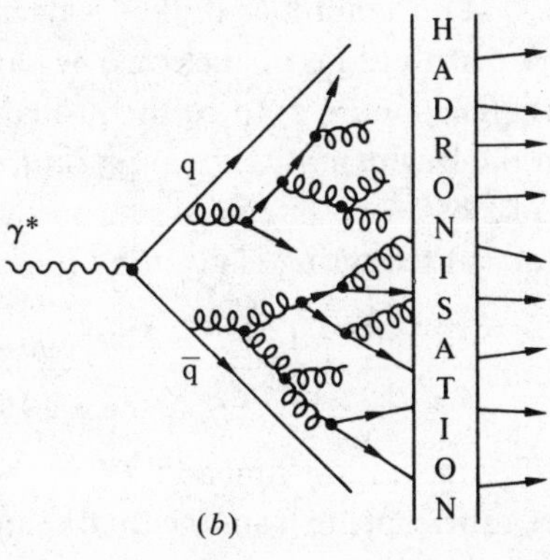

introducing a 'kink' in the string. The Lund model, which is based on semiclassical considerations, successfully reproduces a wide class of experimental data, but has a sizeable number of parameters which must be determined from the data.

An alternative approach, and one perhaps more directly related to QCD, is to simulate the QCD cascade processes. Starting with, say, a highly virtual quark, the leading log formulae for the branching processes $q \to q + g$, $g \to g + g$, $g \to q + \bar{q}$, etc., are used (Fig. 7.33(*b*)). This process is continued until colourless clusters, of mass less than some cut-off Q_0, are obtained. The value of the cut-off signifies the region below which perturbative methods are inapplicable. A value $Q_0 \sim 1$ GeV is typically used. In the Webber model (Webber, 1983), the effects of soft gluon interference are also taken into account. The colourless cluster is hadronised using phase space probabilities, calculated from the known spectrum of particles. The value of the cut-off and the method of hadronisation are, however, somewhat arbitrary. The model successfully accounts for a wide class of data in both e^+e^- annihilation and in jet production in high energy $p\bar{p}$ interactions.

Although phenomenological models, such as those outlined above, are important in that they provide a framework in which perturbative QCD can be tested, it is clearly desirable to solve the QCD Lagrangian by non-perturbative methods. Some progress in achieving this goal has been made by using lattice methods, which draw on experience obtained with statistical mechanics. Practical calculations require the use of large super-computers, or even purpose-built computers. The lattice spacing and the overall size of the lattice are important parameters in these calculations. At present, most calculations have concentrated on trying to compute static properties, such as the hadron mass spectrum.

7.10 FRAGMENTATION FUNCTIONS

Our ignorance of the fragmentation process can (like that of the quark distributions in nucleons) be parameterised. We define the *fragmentation function* $D_q^h(z)$ to be the probability to observe a hadron of type h from the fragmentation of a quark q, carrying a momentum fraction z of the quark ($z = E_h/E_q$). Thus, the production of a hadron h in deep inelastic μp (or ep) scattering is given by (for a sample of N_μ events)

$$\frac{1}{\sigma_{tot}} \frac{d\sigma^h}{dz} = \frac{1}{N_\mu} \frac{dN^h}{dz} = \frac{\sum_i e_i^2 x q_i(x) D_i^h(z)}{\sum_i e_i^2 x q_i(x)}. \tag{7.202}$$

The fragmentation function D_q^h is shown schematically in Fig. 7.34. It refers to fragments from the quark, and these are predominantly forward in the cms of the final state hadrons. A fragmentation function for the remnant diquark system, $D^{\mathrm{h}}_{\mathrm{qq}}$, can be defined in a similar way. Note that the fragmentation functions satisfy

$$\sum_{\mathrm{h}}\int_0^1 zD^{\mathrm{h}}_{\mathrm{q}}(z)\,\mathrm{d}z = 1, \qquad \sum_{\mathrm{h}}\int_0^1 D^{\mathrm{h}}_{\mathrm{q}}(z)\,\mathrm{d}z = n_{\mathrm{h}}, \tag{7.203}$$

expressing momentum conservation and the total hadronic multiplicity respectively.

The fragmentation functions $D^{\pi^+}_{\mathrm{u}}$ and $D^{\pi^-}_{\mathrm{u}}$ can be obtained from analysing the π^+ and π^- spectra in $\mu\mathrm{D}_2$ scattering. If we restrict ourselves to valence quarks only ($x_{\mathrm{BJ}} \gtrsim 0.15$ for $Q^2 \gtrsim 10\ \mathrm{GeV}^2$), then, using (7.202), we find that

$$\frac{1}{N_\mu}\left(\frac{\mathrm{d}N^{\pi^+}}{\mathrm{d}z} + \frac{\mathrm{d}N^{\pi^-}}{\mathrm{d}z}\right) = D^{\pi^+}_{\mathrm{u}} + D^{\pi^-}_{\mathrm{u}},$$

$$\frac{1}{N_\mu}\left(\frac{\mathrm{d}N^{\pi^+}}{\mathrm{d}z} - \frac{\mathrm{d}N^{\pi^-}}{\mathrm{d}z}\right) = \tfrac{3}{5}(D^{\pi^+}_{\mathrm{u}} - D^{\pi^-}_{\mathrm{u}}), \tag{7.204}$$

In deriving equation (7.204) isospin invariance is assumed, so that $u_n(x) = d_{\mathrm{p}}(x)$, $d_{\mathrm{n}}(x) = u_{\mathrm{p}}(x)$, $D^{\pi^+}_{\mathrm{d}} = D^{\pi^-}_{\mathrm{u}}$ and $D^{\pi^-}_{\mathrm{d}} = D^{\pi^+}_{\mathrm{u}}$. The fragmentation functions $D^{\pi^+}_{\mathrm{u}}(z)$ and $D^{\pi^-}_{\mathrm{u}}(z)$, determined in this way in 280 GeV μd scattering, are shown in Fig. 7.35. In deep inelastic scattering z is defined, in terms of the lab variables, as $z = E_{\mathrm{h}}/\nu$, where ν is the energy of the γ^* (and hence approximately that of the struck quark).

It can be seen that $D^{\pi^+}_{\mathrm{u}}$ is significantly harder than $D^{\pi^-}_{\mathrm{u}}$. This is expected, since the scattered u-quark can pick up a $\bar{\mathrm{d}}$ from the colour field to form a leading π^+ meson, whereas a π^- can only be formed further down the fragmentation chain. For low values of z ($\lesssim 0.2$) many of the observed $\pi^\pm$ are not produced directly in the fragmentation process, but arise from

Fig. 7.34 Schematic representation of the fragmentation function $D^{\mathrm{h}}_{\mathrm{q}}$ for a quark q to fragment into a hadron h in deep inelastic scattering. $D^{\mathrm{h}'}_{\mathrm{qq}}$ represents the fragmentation function for the remnant diquark system to produce a hadron h′.

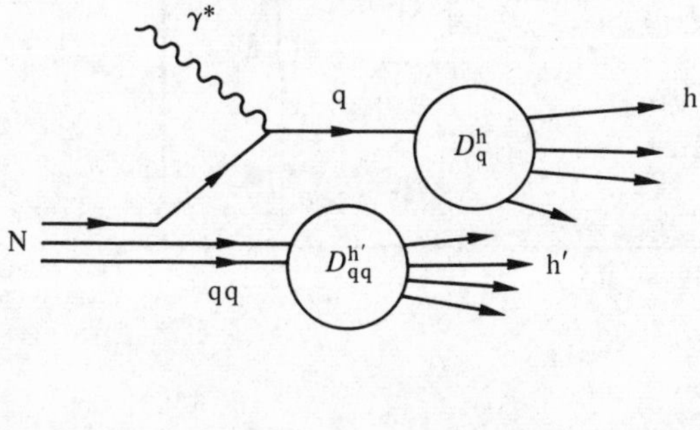

decays of resonances (e.g. ρ). It is found empirically that primary vector and pseudoscalar particles are produced in roughly equal numbers. Strange particles are suppressed by a factor of about 0.3 with respect to non-strange particles. In terms of the chain fragmentation ideas, the ratios of the probabilities of picking up an s or $\bar{s}$ quark, compared to a u or $\bar{u}$ (or d, $\bar{d}$), is $\gamma_s/\gamma_u \sim 0.3$. The values of the vector/pseudoscalar and γ_s/γ_u ratios may also depend on the cms energy. The total hadronic multiplicity is found to increase with the cms energy W, such that $\langle n \rangle \sim \ln W^2$, with the increase coming from the low z region, indicating that $D_q^h(z) \sim 1/z$ at low z. In e^+e^- annihilation a rise faster than $\ln W^2$ (for $W \gtrsim 15$ GeV) is found. Such an increase is expected from QCD considerations.

Equations (7.202) and (7.204) were calculated using the simple QPM formulae. In QCD, the scattered quark can radiate a gluon, and the gluon can subsequently fragment into the observed hadron. Since $D_g^{\pi^+} = D_g^{\pi^-}$, and the gluon fragmentation appears to be softer than that of quarks, the net result will be a softening of, for example, the $D_u^{\pi^+}$ distribution.

Fig. 7.35 Fragmentation functions for a u-quark obtained from μd scattering. The curves are $zD_u^{\pi^+}$ (full circles) $= 0.7(1-z)^{1.75}$ (full curve) and $zD_u^{\pi^-}$ (open circles) $= (1-z)/(1+z)zD_u^{\pi^+}$ (broken curve).

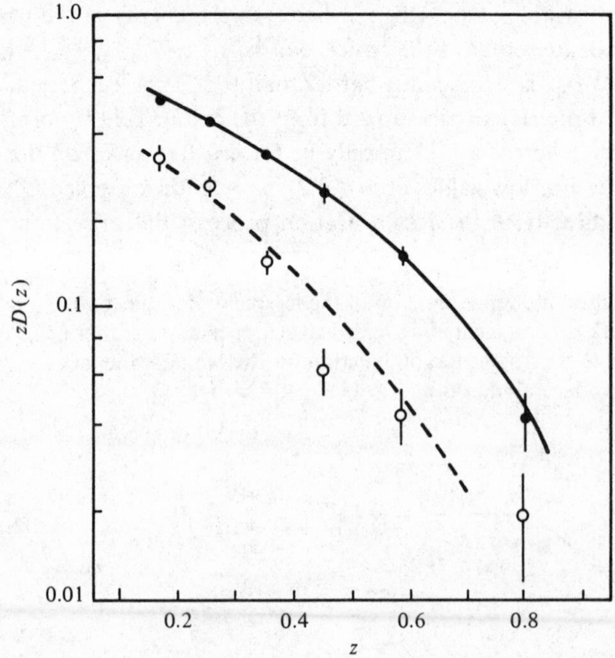

Divergences occur in a similar way to those in the quark distributions. These can be summed to all orders and reabsorbed into effective quark fragmentation functions $D_q^h(z, Q^2)$. In leading order QCD, equation (7.202) still holds, but with Q^2 dependent distributions $q_i(x, Q^2)$ and $D_i^h(z, Q^2)$. The fragmentation functions obey, to leading order, Altarelli–Parisi equations. Note that (unlike the quark scattering case) the gluon can contribute directly. In next-to-leading order QCD, the factorisation property of the cross-section (7.202) no longer holds, and the effective fragmentation function also depends on x (Altarelli *et al.*, 1979). The analysis of scaling violations for fragmentation functions has proven extremely difficult, because of the presence of (*a priori* unknown) non-perturbative hadronisation effects. However, it is established that the fragmentation functions soften at large z with increasing cms energy W.

The hadron formation mechanism envisaged in most models of fragmentation is that the energetic quark Q picks up an antiquark $\bar{q}$, from a $q\bar{q}$ pair formed from the colour field created by Q. This forms a leading meson $M = Q\bar{q}$ (or baryon if a qq system is picked up). In the case of a heavy quark (Q = c, b), it is kinematically favoured for M to retain a large fraction z $(= E_M/E_Q)$ of the quark energy, because $m_Q \gg m_q$. The leading meson is thus harder for a c-quark $(\langle z_c \rangle \sim 0.6)$ or a b-quark $(\langle z_b \rangle \sim 0.7\text{–}0.9)$ than for light quarks, as sketched in Fig. 7.36. (A commonly used parameterisation for heavy quarks is that of Peterson *et al.*

Fig. 7.36 Sketch of the fragmentation functions for a quark Q (= u, c or b) to form a leading meson.

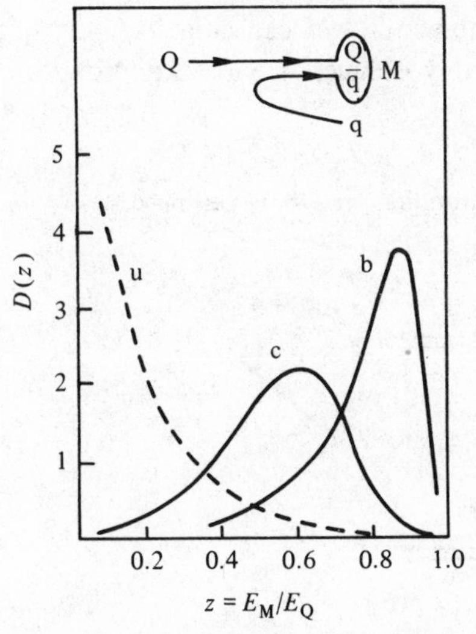

(1983)). This property of the leading hadron can be used as a convenient way to 'tag' the presence of a heavy quark. The gluon coupling to c-quarks is compatible to that to light quarks, as required by QCD. This has been shown by the TASSO collaboration (Althoff *et al.*, 1984), who find $\alpha_S^c/\alpha_S = 1.00 \pm 0.20 \pm 0.20$, where α_S^c is found for a sample enriched with charmed quarks.

7.11 HADRON–HADRON SCATTERING

The bulk of the cross-section in hadron–hadron collisions arises from soft multiple collisions. Such processes are outside the scope of perturbative QCD, which requires the presence of at least one large momentum scale. A small fraction of the interactions, however, do involve hard collisions and we will now consider such processes.

7.11.1 The Drell–Yan process

If we apply time reversal to the reaction $l^+l^- \to \gamma^* \to q\bar{q}$, then we obtain the Drell–Yan (DY) process $q\bar{q} \to \gamma^* \to l^+l^-$. We assume below that the l^+l^- pair is in the continuum region, well away from any heavy $q\bar{q}$ resonant states. The quark and antiquark 'beams' are obtained by firing high energy hadrons at each other. In pp interactions the $\bar{q}$ can only come from the sea. Hence, a study of $p\bar{p}$ or $\pi^{\pm}p$ collisions should give a higher cross-section for this process, because the antiquarks are valence quarks for $\bar{p}$ and $\pi^{\pm}$. The lowest order diagram for $h_1h_2 \to X\gamma^* \to Xl^+l^-$, where X denotes all the remaining particles produced, is shown in Fig. 7.37. In the h_1h_2 cms we have (neglecting hadron masses) the four momenta $p_1 = (p, 0, 0, p)$ and $p_2 = (p, 0, 0, -p)$, with $s = 4p^2$. Neglecting parton masses and transverse momenta, we can write $k_1 = x_1p_1$ and $k_2 = x_2p_2$, for the q and $\bar{q}$ respectively. Hence, the virtual photon (or l^+l^-

Fig. 7.37 Lowest order diagram for the Drell–Yan process $q\bar{q} \to \gamma^* \to l^+l^-$ in $h_1h_2 \to Xl^+l^-$.

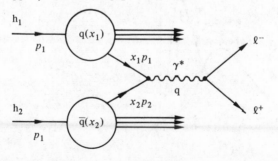

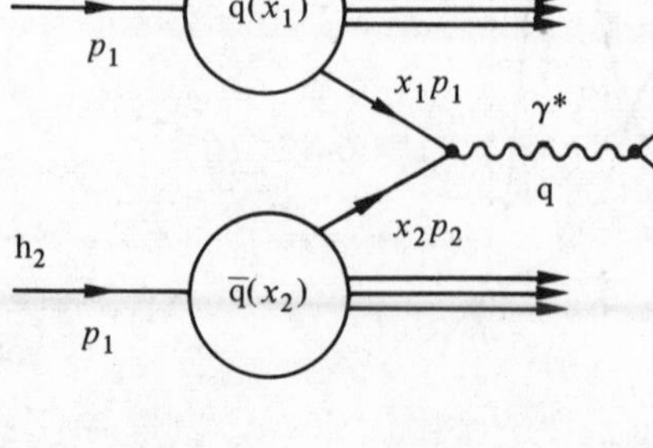

pair) has four momentum

$$q = ((x_1 + x_2)p, 0, 0, (x_1 - x_2)p), \qquad q^2 = 4x_1x_2p^2 = x_1x_2s. \tag{7.205}$$

The kinematics for the process can thus be fixed by measuring the lepton-pair mass q and the longitudinal momentum q_z of the pair, noting that the Feynman variable $x_F = 2q_z/s^{1/2} = x_1 - x_2$.

The cross-section for the parton subprocess is given by (see Section 4.3)

$$\frac{\mathrm{d}\hat{\sigma}}{\mathrm{d}q^2}(\mathrm{q}_i\bar{\mathrm{q}}_i \to l^+l^-) = \frac{4\pi\alpha^2}{3q^2}\, e_i^2\, \delta(q^2 - x_1x_2s). \tag{7.206}$$

The region where this QPM calculation is expected to be applicable is $s, q^2 \to \infty$, with $\tau = q^2/s = x_1x_2$ fixed. In order to obtain the overall cross-section for the process, we must weight (7.206) by the appropriate fluxes of quarks and antiquarks, giving

$$\begin{aligned}\frac{\mathrm{d}\sigma}{\mathrm{d}q^2}&(\mathrm{h}_1\mathrm{h}_2 \to \mathrm{X}l^+l^-)\\ &= \frac{1}{3}\sum_i \int \mathrm{d}x_1\, \mathrm{d}x_2 [q(x_1)\bar{q}(x_2) + \bar{q}(x_1)q(x_2)]\, \frac{\mathrm{d}\hat{\sigma}}{\mathrm{d}q^2}\\ &= \frac{4\pi\alpha^2}{9sq^2}\sum_i e_i^2 \int \mathrm{d}x_1\, \mathrm{d}x_2 [q(x_1)\bar{q}(x_2) + \bar{q}(x_1)q(x_2)]\, \delta(x_1x_2 - \tau),\end{aligned} \tag{7.207}$$

where the factor $\frac{1}{3}$ arises because there are three possible colourless $\mathrm{q}\bar{\mathrm{q}}$ pairings out of nine possible combinations.

The QPM formula (7.207) predicts that $q^4\, \mathrm{d}\sigma/\mathrm{d}q^2$ is a function, $F(\tau)$, of τ only, i.e. it scales. Further predictions are that, in the dilepton rest frame, the leptons have a $1 + \cos^2\theta$ angular distribution, expected for spin $\frac{1}{2}$ quarks (4.61). The ratio of the Drell–Yan cross-sections of $\pi^+(\mathrm{u}\bar{\mathrm{d}})$ and $\pi^-(\mathrm{d}\bar{\mathrm{u}})$ mesons, on an isoscalar nuclear target (e.g. $^{12}\mathrm{C}$), is a sensitive test of the model. Since $\tau \sim x_1x_2$, we expect the following ratios

$$\frac{\sigma(\pi^+\mathrm{A} \to \mu^+\mu^-\mathrm{X})}{\sigma(\pi^-\mathrm{A} \to \mu^+\mu^-\mathrm{X})} = \begin{cases} 1 & \tau \text{ small},\\ \sim (e_{\bar{\mathrm{d}}}/e_{\bar{\mathrm{u}}})^2 = \frac{1}{4} & \tau \to 1. \end{cases} \tag{7.208}$$

This arises because, for small values of τ, the interactions are mainly on sea-quarks, whereas for large τ the $\bar{\mathrm{d}}$ and $\bar{\mathrm{u}}$ valence quarks in the π^+ and π^- respectively should dominate. The cross-section for a target of atomic number A is expected to behave as $\sigma \sim A^1$, because the scattering process is hard (unlike $\sigma_{\mathrm{tot}}(\pi A)$ which varies approximately as $A^{2/3}$ because the π 'sees' only the surface nucleons). These QPM predictions are reasonably well satisfied experimentally. Some deviations from the $\sigma \propto A$ behaviour might, however, be expected because the EMC effect suggests that the quark distributions are different in nuclei. Such deviations have been reported by the NA10 experiment at CERN (Bordalo *et al.*, 1987), from a comparison of π^-W and π^-D_2 collisions. No effect is seen for the π^- (as expected), but the x distribution for W compared to D_2 shows a similar shape to that expected from deep inelastic scattering (DIS). The data are in the range $18 < Q^2 < 225\ \text{GeV}^2$ (time-like).

If we assume that the quark distributions in (7.207) are the same as those measured in deep inelastic scattering, then an absolute prediction of the cross-section for, say, $pp \to \mu^+\mu^- X$ can be made. Alternatively, the formula can be used to extract the quark distributions in mesons. For example, by measuring $\pi A \to \mu^+\mu^- X$, and taking the quark distributions in the nucleus from DIS, Badier *et al.* (1983), using $A =$ platinum, find that $v_\pi(x) \sim x^{0.5}(1-x)$ and $s_\pi(x) \sim (1-x)^8$ give good fits for the pion valence (i.e. u and $\bar{d}$ in π^+) and sea-quarks respectively. It is further found that roughly half the pion momentum is carried by gluons, as is the case in DIS off nucleons. Experimentally it is also found that the measured cross-section for the Drell–Yan process is a factor of about 2.3 (*K-factor*) larger than the QPM prediction. This can be treated as an additional parameter, by normalising the valence quark distribution in the pion to unit probability.

Theoretically, a value of $K \sim 2$ is, in fact, predicted. The leading log terms, which were re-absorbed into Q^2 dependent quark distributions in deep inelastic scattering, are process independent, and so (7.207) can be written in terms of such Q^2 dependent distributions. The K-factor arises from next-to-leading and further log terms. To $O(\alpha_S)$ it has the value $K = 1 + \alpha_S\delta \simeq 2$; that is to say that the perturbative correction is very large, and questions the validity of the convergence. Fig. 7.38 shows the lowest order diagram (*a*), together with some of the QCD corrections ((*b*) to (*d*)). Diagrams of the same type appear also in deep inelastic scattering, but there is an important difference: for DY the γ^* is time-like ($q^2 > 0$), whereas for DIS the γ^* is space-like ($q < 0$). The vertex correction graph,

Fig. 7.38 (*a*) the basic Drell–Yan process, (*b*) to (*d*) show some QCD corrections.

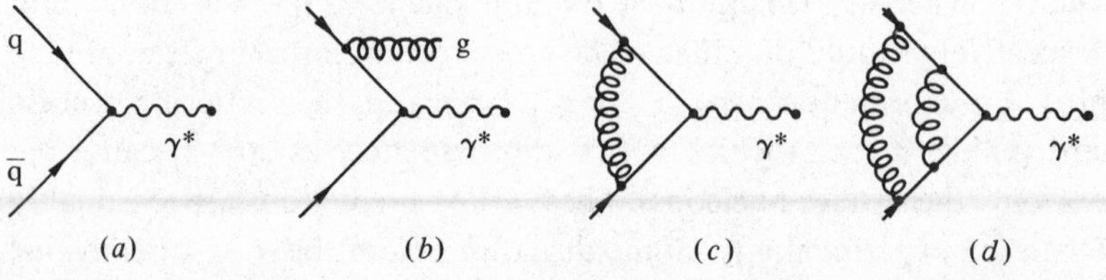

Fig. 7.38(c), contributes to $O(\alpha_S)$ through its interference with the diagram of Fig. 7.38(a). The vertex factor has the form, for space-like q^2 and with $C_F = \frac{4}{3}$,

$$A_{DIS} = -C_F \frac{\alpha_S}{2\pi} \ln^2(-q^2). \tag{7.209}$$

Hence, A_{DIS} is real, and is cancelled by the $O(\alpha_S)$ diagram for real gluon emission (Fig. 7.23(a) and (b)), leaving the $\alpha_S \ln Q^2$ terms which give rise to scaling violations. For $q^2 > 0$, we can put $-q^2 = \exp(-i\pi)q^2$, so that (7.209) can be written

$$A_{DY} = -C_F \frac{\alpha_S}{2\pi} (\ln q^2 - i\pi)^2$$
$$= -C_F \frac{\alpha_S}{2\pi} \ln^2 q^2 - \pi^2 - 2i\pi \ln q^2. \tag{7.210}$$

The $\ln^2 q^2$ term is again cancelled by diagrams with real gluon emission (Fig. 7.38(b), etc.), and the imaginary term does not contribute to the cross-section ($\sigma \propto A + A^*$). However, the term in π^2 remains, and gives a contribution $(\alpha_S/2\pi)4\pi^2/3$ to K. For $\alpha_S \sim 0.2$, this gives $K = 1 + \delta K$, with $\delta K \sim 0.42$. There are theoretical reasons to believe that higher order terms (e.g. Fig. 7.38(d)) give rise to an exponential series, i.e. $\exp(\delta K)$, which would explain the bulk of the $K \simeq 2$ factor needed. Of course, a full calculation is made in practice, and K is not a constant, but can depend on the relevant kinematic variables. These ideas are supported by a high statistics study of 194 GeV π^- on tungsten (Betev *et al.*, 1985), in which data are studied both above and below the Υ resonance. The data indicate that Q^2 dependent quark distributions are needed, and that the inclusion of next-to-leading log terms gives a reasonable representation of the data.

In deriving equation (7.207) the transverse momenta of the quark and antiquark were neglected. Non-zero values can arise from the intrinsic transverse momentum (k_T), as well as from perturbative QCD diagrams such as $gq \to \gamma^* q$ (Compton) and $\bar{q}q \to \gamma^* g$ (annihilation). The expected form (which can be obtained by dimensional analysis) for the transverse momentum of the lepton pair q_T is given by

$$\langle q_T \rangle = a + \alpha_S s^{1/2} F(\tau, \alpha_S), \tag{7.211}$$

where a is related to k_T, and F can be calculated in QCD. The Compton and annihilation graphs are computed in a similar way to the calculations described in Section 7.7.2. The resulting formulae are then plugged into an expression of the type (7.207), and an integration over the quark, antiquark and also gluon distributions is made. The Compton diagram dominates at large q_T ($\gtrsim$few GeV). The prediction that $\langle q_T \rangle$ depends linearly on $s^{1/2}$ is compared with some data in Fig. 7.39. Although this prediction is well satisfied, there are theoretical uncertainties as to the choice of the argument of α_S and again with the K-factor.

7.11.2 Parton–parton scattering

Measurements of the properties of hadrons produced in high energy pp collisions at the CERN ISR ($s^{1/2}$ up to ~60 GeV) in the 1970s showed that the transverse momentum (p_T), with respect to the incident beam direction, extended to very large values, and grew with increasing s. These results were contrary to the notion, based originally on cosmic ray measurements, that the transverse momentum was limited in hadron–hadron collisions. The ISR observations were compatible with the idea that hard parton–parton collisions ($qq \to qq$, $qg \to qg$, $gg \to gg$, etc.) were responsible for the high p_T tails. However, because of the rapidly falling p_T distributions, it proved difficult to establish clear jets of hadrons. At the much larger values of $s^{1/2}$ available at the CERN $\bar{p}p$ collider ($s^{1/2} \sim 600$ GeV), clear jet structures have been observed by the UA1 and UA2 collaborations.

Perturbative QCD can be used to calculate the parton–parton collisions, provided that at least one kinematic variable is asymptotically large. As

Fig. 7.39 Data from various experiments on the mean transverse momentum of lepton pairs $\langle q_T \rangle$, as a function of $s^{1/2}$, for pN collisions. The line is $\langle q_T \rangle = 0.45 + 0.025 s^{1/2}$ GeV.

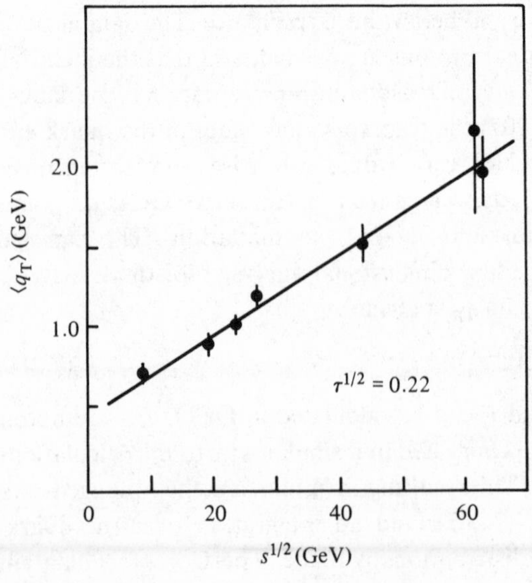

an example, let us consider the process $q_1(p_1)+q_2(p_2)\to q_1'(p_3)+q_2'(p_4)$, with $q_1\neq q_2$ (Fig. 7.40). The matrix element is (from Appendix C)

$$\mathcal{M}_{\rm fi}=(-ig)^2\frac{(\lambda^a)_{ij}}{2}\frac{(\lambda^b)_{kl}}{2}\bar{u}_1'\gamma_\mu u_1\left(\frac{-i\delta_{ab}g^{\mu\nu}}{\hat{t}}\right)\bar{u}_2'\gamma_\nu u_2$$

$$=ig^2\frac{(\lambda^a)_{ij}(\lambda^a)_{kl}}{4\hat{t}}\bar{u}_1'\gamma_\mu u_1\bar{u}_2\gamma^\mu u_2. \qquad (7.212)$$

$|\mathcal{M}_{\rm fi}|^2$ is calculated in the usual way. Each of the vertices contributes a colour factor $\mathrm{tr}(\lambda^a\lambda^a)/4=\frac{1}{2}$ to $|\mathcal{M}_{\rm fi}|^2$. Summing over the eight possible terms a, and averaging over the nine initial colour states, gives the overall colour factor $\frac{2}{9}$. Thus, summing and averaging over both spins and colours gives

$$|\mathcal{M}_{\rm fi}|^2=\tfrac{4}{9}g^4\left(\frac{\hat{s}^2+\hat{u}^2}{\hat{t}^2}\right)=\frac{64\pi^2\alpha_S^2}{9}\left(\frac{\hat{s}^2+\hat{u}^2}{\hat{t}^2}\right). \qquad (7.213)$$

The differential cross-section ($\mathrm{d}\sigma/\mathrm{d}\hat{t}=|\mathcal{M}_{\rm fi}|^2/16\pi\hat{s}^2$, from (4.29)) is

$$\frac{\mathrm{d}\sigma}{\mathrm{d}\hat{t}}=\frac{\pi\alpha_S^2}{\hat{s}^2}|A|^2, \qquad |A|^2=\frac{4}{9}\left(\frac{\hat{s}^2+\hat{u}^2}{\hat{t}^2}\right). \qquad (7.214)$$

Note that this formula is of order α_S^2, to be compared to that for $e^+e^-\to q\bar{q}g$ which is of order $\alpha^2\alpha_S$ (7.196). Other $2\to 2$ parton scattering subprocesses involve more tedious calculations, and the results for $|A|^2$ are given in Table 7.1 (Combridge *et al.*, 1977). Some computational details can be found in Leader and Predazzi (1982), including the treatment of three and four gluon vertices.

The cross-sections for these parton–parton subprocesses must be combined with the relevant quark, antiquark and gluon density distributions (in a way similar to that used for the Drell–Yan process in (7.207)), in

Fig. 7.40 Lowest order QCD diagram for $q_1+q_2\to q_1'+q_2'$. The colour indices are (i), (j), (k) and (ℓ).

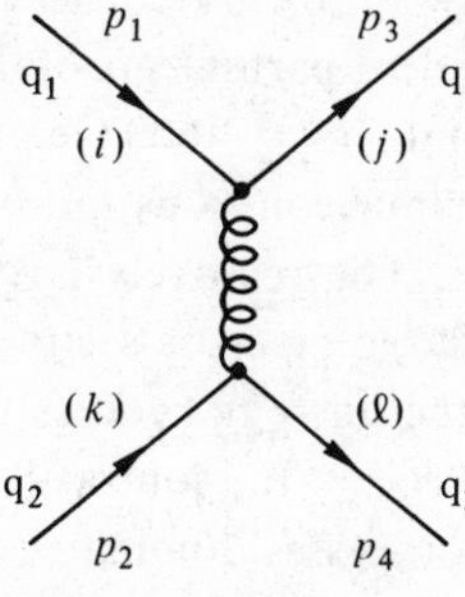

Table 7.1. *Values of* $|A|^2$ *for various* $2 \to 2$ *parton subprocesses; the differential cross-section is* $\mathrm{d}\sigma/\mathrm{d}t = (\pi\alpha_s^2/s^2)|A|^2$
The symbol $\wedge$ (denoting a subprocess) is understood for s, t and u

Process	$\|A\|^2$	$\|A\|^2$ $(\theta = \pi/2)$
$q_1q_2 \to q_1q_2$, $q_1\bar{q}_2 \to q_1\bar{q}_2$	$\frac{4}{9}\frac{s^2+u^2}{t^2}$	2.22
$q_1q_1 \to q_1q_1$	$\frac{4}{9}\left(\frac{s^2+u^2}{t^2}+\frac{s^2+t^2}{u^2}\right)-\frac{8}{27}\frac{s^2}{ut}$	3.26
$q_1\bar{q}_1 \to q_2\bar{q}_2$	$\frac{4}{9}\frac{t^2+u^2}{s^2}$	0.22
$q_1\bar{q}_1 \to q_1\bar{q}_1$	$\frac{4}{9}\left(\frac{s^2+u^2}{t^2}+\frac{t^2+u^2}{s^2}\right)-\frac{8}{27}\frac{u^2}{st}$	2.59
$q\bar{q} \to gg$	$\frac{32}{27}\frac{u^2+t^2}{ut}-\frac{8}{3}\frac{u^2+t^2}{s^2}$	1.04
$gg \to q\bar{q}$	$\frac{1}{6}\frac{u^2+t^2}{ut}-\frac{3}{8}\frac{u^2+t^2}{s^2}$	0.15
$qg \to qg$	$-\frac{4}{9}\frac{u^2+s^2}{us}+\frac{u^2+s^2}{t^2}$	6.11
$gg \to gg$	$\frac{9}{2}\left(3-\frac{ut}{s^2}-\frac{us}{t^2}-\frac{st}{u^2}\right)$	30.4

order to obtain their contribution to the total cross-section. The nucleon structure functions have been measured only up to Q^2 values of $\lesssim 100\ \mathrm{GeV}^2$. Hence, they must be evolved using QCD up to the much higher Q^2 values ($\lesssim 10^4\ \mathrm{GeV}^2$) achieved with the high energy $p\bar{p}$ colliders. A further complication is that, in leading order, there is ambiguity as to which kinematic quantity should be chosen as the 'Q^2' scale. The gluon density distribution $G(x)$ is determined only indirectly in DIS, and so is rather poorly known; roughly $G(x) \sim (1-x)^5$ at $Q^2 \sim 20\ \mathrm{GeV}^2$. The values of the momentum fractions x_1 and x_2 of the incident partons are of the order $x_T = 2p_T/s^{1/2}$. For the CERN $\bar{p}p$ collider data ($s^{1/2} \sim 600$ GeV) the p_T values are typically 30–100 GeV, so that the nucleon structure functions are sampled in the range $0.1 \lesssim x_1(x_2) \lesssim 0.3$. The relatively large gluon density in this x range, combined with the large gg cross-section due to the colour factors (Table 7.1), means that the $gg \to gg$ process provides a substantial contribution at collider energies. In contrast, for the $s^{1/2} \sim 60$ GeV range of the CERN ISR, qq processes dominate.

Jets of hadrons are generally detected experimentally using calorimetry methods (with as fine grain a resolution as can be afforded). In order to isolate a jet with any degree of confidence, it must be distinguishable from the jets coming from the beam fragments. The region around $\theta \sim 90°$ is therefore particularly clean. The p_T^2 of the parton (or jet), with respect to the beam axis, is given by $p_T^2 = \hat{u}\hat{t}/\hat{s}$; so that, from Table 7.1, we expect that

$$\frac{d\sigma}{dp_T^2} = \frac{1}{p_T^4} f(x_T, \alpha_S(p_T^2)). \tag{7.215}$$

Thus, at fixed x_T, the cross-section should fall as p_T^{-4}, with a logarithmic correction from α_S. This dependence is achieved only at the large $\bar{p}p$ collider energies, and at smaller energies the jets (or individual hadrons from the jets) fall off with p_T^2 with a power $n > 4$. The data on jets at large p_T agree reasonably well with QCD expectations, but there are several problems in making a precise comparison. Experimentally the jet energy must be measured accurately and without bias (a 5% shift on p_T can cause about a 50% change in the cross-section because it is steeply falling). Further, there is ambiguity in choosing which soft hadrons should be added to a particular jet. Uncertainties in the parton structure functions, and in the higher order QCD corrections (K-factors), limit the precision of the theoretical calculations. Despite these problems the observation of jets with energies up to 250 GeV has provided a spectacular confirmation of the QPM and QCD description of the nucleon.

Another interesting limit is that for small $\hat{t}$ (or θ). For fixed $\hat{s}$, the limit $\hat{t} \to 0$ means that $\hat{u} \to -\hat{s}$. The diagrams dominating the cross-section in this limit are those involving t-channel gluon exchange, which behave as $\hat{t}^{-2}$. From Table 7.1 it can be seen that, for small $\hat{t}$, the cross-sections for $gg \to$ anything, gq(or $\bar{q}$) $\to$ anything, and $q\bar{q} \to$ anything are in the ratio $1 : \frac{4}{9} : (\frac{4}{9})^2$. Hence, we can define the total cross-section in terms of a common subprocess cross-section and an effective structure function

$$F(x) = G(x) + \tfrac{4}{9}[Q(x) + \bar{Q}(x)]. \tag{7.216}$$

In fact, this approximation is good over a large part of the angular range. The measured values of $F(x)$, assuming $K = 2$, from the UA1 experiment (Arnison *et al.*, 1984a) are shown in Fig. 7.41 (similar results are found by the UA2 group). It can be seen that the data are well above the contribution from $Q(x) + \bar{Q}(x)$ alone in (7.216) (estimated by evolving the CDHS ν structure function results to $Q^2 \simeq 2000\ \text{GeV}^2$), especially in the region $x \lesssim 0.3$. This evidence that the gluon component of the nucleon plays an important role is supported by a study of the average charge of the hadrons in the jets, which is compatible with zero in the region where gluon jets dominate, but is significantly positive (negative) for q($\bar{q}$) jets. Note that the gg → gg process involves the triple and quartic gluon vertices, and that these are a consequence of the non-Abelian nature of QCD.

The observed angular distribution of jets for small $\hat{t}$ (or θ) agrees roughly with the $1/\hat{t}^2[1/(1-\cos\theta)^2]$ behaviour, expected from gluon exchange. This confirms the vector nature of the gluon propagator (a scalar gluon would give a $1/\hat{t}$ dependence for the gg and gq processes). A more detailed test of these ideas has been made by the UA1 collaboration. Fig. 7.42 shows the distribution of the variable $\chi = (1+\cos\theta)/(1-\cos\theta)$, for a fixed bin of $\hat{s}$. The distribution of $\mathrm{d}\sigma/\mathrm{d}\chi$ should be flat, for $\chi \gtrsim 2$, if the data have a $\hat{t}^{-2}$ distribution. Deviations from these expectations can arise because of the dependence on Q^2 of both α_s and the structure functions. As $\cos\theta \to 1$ ($\chi \to \infty$), then $Q^2 \sim p_T^2$ decreases. Hence, $\alpha_s(Q^2)$ increases and also the structure functions become harder, giving an increase in the effective parton flux; the full curve in Fig. 7.42 includes estimates of both these effects.

Fig. 7.41 Measurements of the structure function $F(x)$, as a function of x, by the UA1 collaboration. The curves are derived from the CDHS νN structure functions. A: $G(x)+\frac{4}{9}[Q(x)+\bar{Q}(x)]$, $Q^2 = 20\ \mathrm{GeV}^2$; B: $G(x)+\frac{4}{9}[Q(x)+\bar{Q}(x)]$, $Q^2 = 2000\ \mathrm{GeV}^2$; C: $\frac{4}{9}[Q(x)+\bar{Q}(x)]$, $Q^2 = 2000\ \mathrm{GeV}^2$.

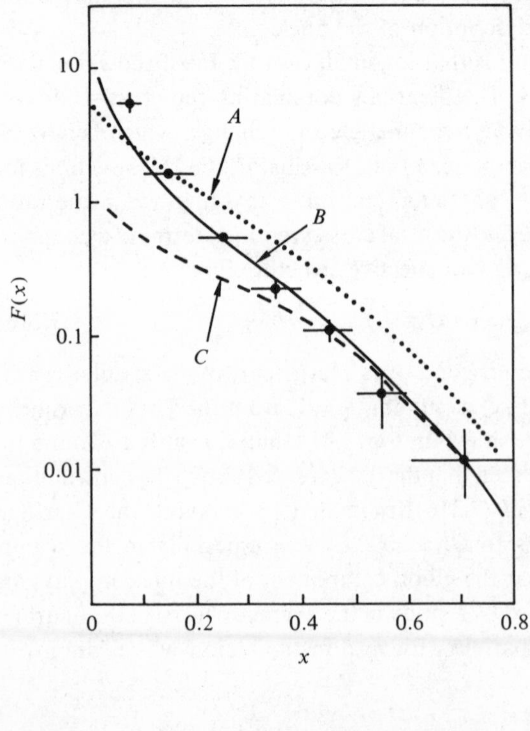

Clear examples of three-jet events have also been found, at a rate of about 10–20% of the two-jet rate. Fig. 7.43 shows the jet energies measured in a typical three-jet event (in addition, there are jets corresponding to the p and $\bar{p}$ fragments). QCD tests are, however, again complicated by several factors. The non-leading contributions to the two and three-jet cross-section give rise to K-factors, for which as yet there is no complete calculation. A further consequence of the uncalculated higher order terms is that the choice of the variable to be used as the argument for α_S is uncertain. Note also that there are two large scales in the problem.

The properties of the individual hadrons within these high energy jets also conform to the expectations of QCD. The fragmentation function for charged hadrons $D(z) = 1/N_{jet}(dN/dz)$, with $z = p_L/p_{jet}$ the scaled longitudinal momentum, is softer than that observed in e^+e^- at $s^{1/2} \sim 34$ GeV (e.g. Sass, 1985). This could arise from either scaling violations of the fragmentation functions or from the different parton content at the large Q^2 values probed by the collider. It is expected that the fragmentation of a gluon is softer than that of a quark. In order to try and distinguish between these possibilities the UA1 collaboration developed an algorithm in which a given jet was assigned a probability to be either a quark or gluon jet, based on the parton–parton scattering formulae. The fragmentation functions from jet samples rich in gluon jets, for two ranges of Q^2, are shown in Fig. 7.44 (from Sass, 1985). It can be seen that the gluon fragmentation function, defined in this way, becomes softer as the scale of the interaction increases.

Fig. 7.42 Distribution of the variable χ for two-jet events with effective mass in the interval 150 to 250 GeV, as measured by the UA1 collaboration. The full (broken) curve is the expected distribution with (without) a Q^2 dependence in α_S and the structure functions.

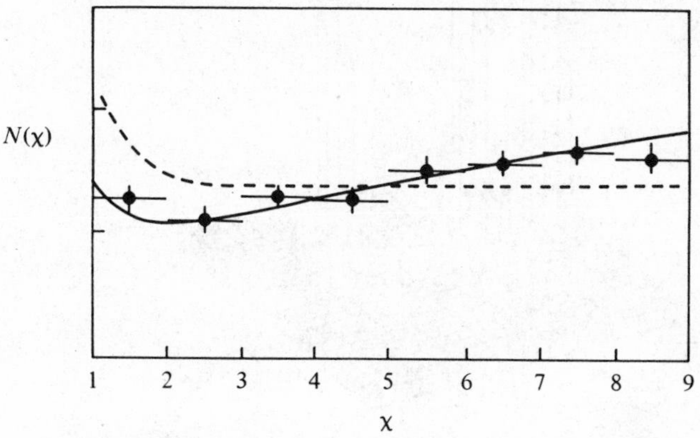

7.12 STRUCTURE FUNCTION OF THE PHOTON

In Section 7.9 the possibility that, part of the time, the photon behaves as a hadron, was discussed. The hadronic structure of the photon arises through the $\gamma \to q\bar{q}$ transition. As with other structure functions, the photon structure will depend on the Q^2 of the probe with which it is measured. Note that the photon also has a leptonic structure, which arises through the $\gamma \to l^+l^-$ transition. The photon structure functions F_1^γ and F_2^γ can be measured using the reaction $e\gamma \to eX$, as shown in Fig. 7.45. This is a subprocess in the reaction $e^+e^- \to e^+e^-\gamma\gamma$, $\gamma\gamma \to X$. Experimentally, a scattered e^+ (or e^-) is detected (tagged) at some large scattering angle θ, and this defines the Q^2 of the 'probe'. The other $e^-(e^+)$ will generally be lost down the beam pipe, and the second (target) photon is quasi-real ($P^2 \simeq 0$). If W is the cms energy of the particles produced in the $\gamma\gamma$ collision,

Fig. 7.43 Example of a three-jet event from the UA2 collaboration from the CERN S$\bar{p}$pS collider at $s^{1/2} = 540$ GeV. The calorimeter energies are displayed with respect to the polar angle ϕ and pseudo-rapidity η.

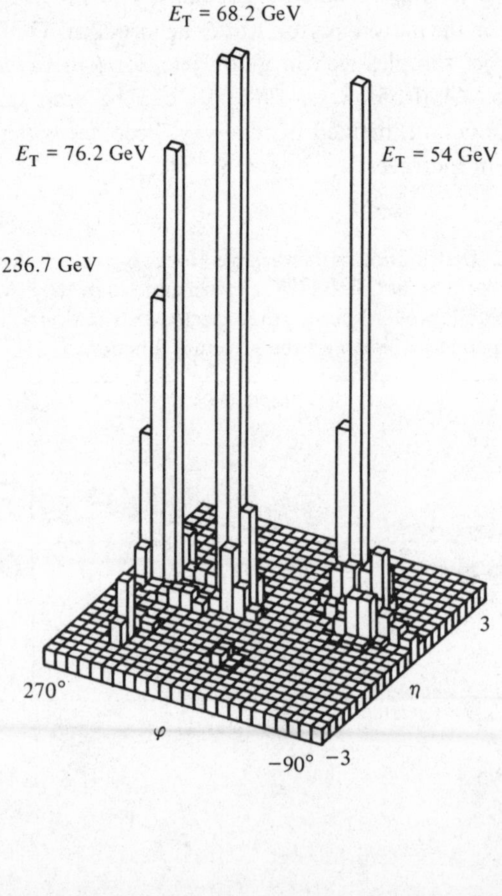

then the deep inelastic scaling variables are $x = Q^2/(Q^2 + W^2)$ and $y = 1 - (E'/E)\cos^2(\theta/2)$. Thus, the structure of the target photon can be measured as a function of x and Q^2.

The differential cross-section for the reaction $e\gamma \to eX$ has the form (see Fig. 7.45)

$$\frac{d^2\sigma}{dx\,dy} = \frac{16\pi\alpha^2 EE_\gamma}{Q^4}\left[F_2^\gamma(1-y) + \frac{y^2}{2}2xF_1^\gamma\right]. \tag{7.217}$$

Fig. 7.44 Measurements of the gluon fragmentation function by the UA1 collaboration, for two ranges of Q^2. ●, $1000 < Q^2 < 1600$ GeV^2; ○, $2600 < Q^2 < 4000$ GeV^2.

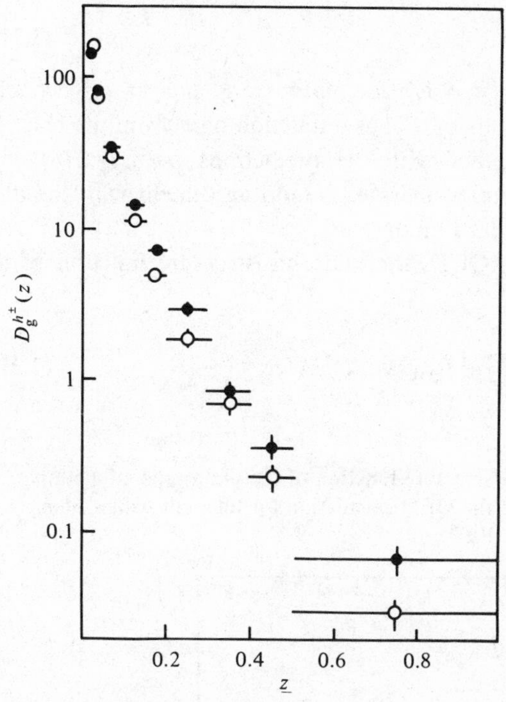

Fig. 7.45 The $\gamma\gamma$ process in e^+e^- interactions.

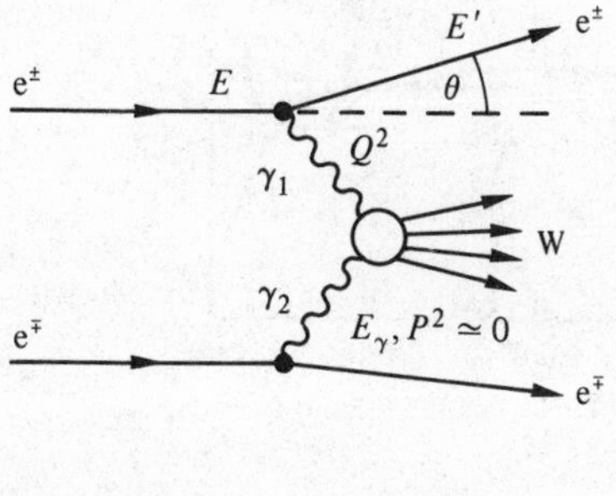

This is to be compared with (7.36) for the $eN \to eX$ case. The variable y, in practice, is small, so that only the F_2^{γ} term contributes. Measurement of the variables W, and hence x, requires the detection of all the particles produced in the $\gamma\gamma$ collision. In practice, corrections for missing hadrons must be applied, as well as radiative corrections and corrections for background processes (inelastic Compton scattering, $\tau^+\tau^-$ production, etc.).

The lowest order $\gamma\gamma$ process is the dissociation $\gamma_2 \to f\bar{f}$ ($f = l, q$), followed by the interaction of γ_1 on f (or $\bar{f}$). This process gives

$$F_2^{\gamma}(x) = n_c \frac{\alpha}{\pi} \sum_f e_f^4 \left\{ x[x^2 + (1-x)^2] \ln\left(\frac{W^2}{m_f^2}\right) + 8x^2(1-x) - x \right\}, \tag{7.218}$$

where the 'colour' factor $n_c = 1, 3$ for leptons and quarks respectively. Fig. 7.46 shows some results on $F_2^{\mu\mu}$ as a function of x (from the reaction $e^+e^- \to e^+e^-\mu^+\mu^-$), together with the predictions from (7.218) with different values of the fermion mass m_μ. Good agreement with the data is obtained with the standard muon mass.

In leading order (LO) QCD, the hadronic structure function of the photon is given by

$$F_2^{LO}(x, Q^2) = \frac{3\alpha}{\pi} \sum_q e_q^4 f_{LO}(x) \ln(Q^2/\Lambda^2), \tag{7.219}$$

Fig. 7.46 Leptonic structure function of the photon as a function of x, together with the QED calculation for different values of m_μ. The data are from PEP.

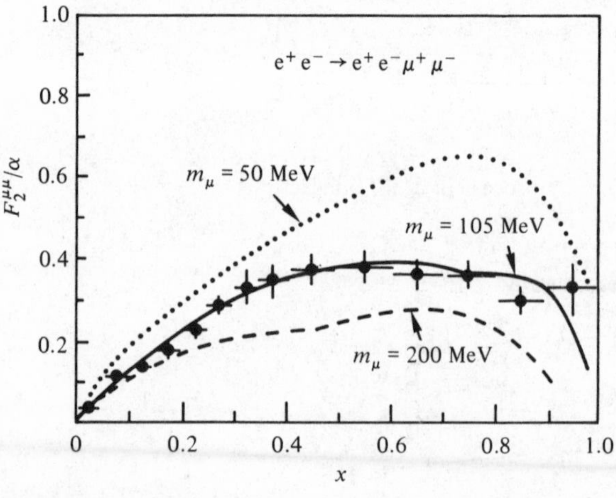

where $f_{LO}(x)$ is a known function (Witten, 1977). For low Q^2 this expression is no longer valid, and it is usual to compute F_2^γ expected in the Vector Dominance Model (VDM). A commonly used (but not unique) parameterisation is

$$F_2^{\gamma DM}(x) = (0.2 \pm 0.05)\alpha(1 - x). \tag{7.220}$$

The LO QCD form (7.219) contains no free parameters (except the scale Λ), and it thus might appear that an absolute QCD prediction of F_2^γ is made (unlike F_2^N, where only the Q^2 evolution is predicted). The photon structure function is free of confinement problems and, unlike the nucleon structure function, increases at large x as Q^2 increases. Higher order QCD calculations have also been made (Duke and Owens, 1980). However, there is a singularity in the second moment of F_2^γ, and a difficulty arises in attempts to accommodate the non-perturbative VDM piece in a consistent way, because F_2^γ becomes negative at low x. Furthermore, the higher order corrections are large for large x. Gluck and Reya (1983) argue that only the Q^2 evolution is thus reliable, and in the central x region only.

Fig. 7.47 F_2^γ/α as a function of x for different Q^2 values. The broken line is the VDM estimate, whereas the full and dotted lines are estimates of the total F_2^γ (VDM + QPM) for 5 and 100 GeV^2 respectively. ●, PLUTO $Q^2 = 5\,\text{GeV}^2$; ○, JADE $Q^2 = 100\,\text{GeV}^2$.

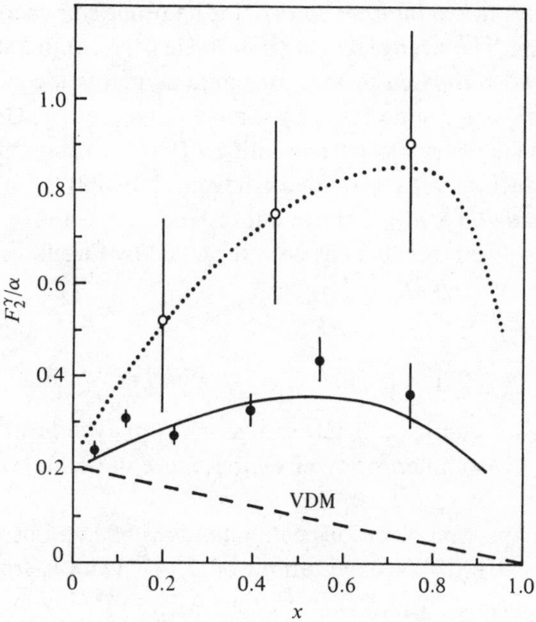

The experimental results (for a review see Alexander, 1985) are still rather imprecise, but are consistent with a logarithmic rise in Q^2, at fixed x, as expected from the form of (7.219) (see Fig. 7.47). The results are potentially a very sensitive method to determine Λ, and reasonably consistent results are obtained, leading to an average (Alexander, 1985) of $\Lambda_{\overline{MS}} = 210 \pm 45$ MeV, in fair agreement with other determinations. Studies of the hadron final states give results consistent with the above picture. For example, the transverse momentum of the hadrons increases with Q^2, consistent with the increased importance of the point-like processes.

7.13 DETERMINATION OF α_S

Much experimental effort has been spent in trying to extract reliable values of α_S from the various reactions discussed in the previous sections. As mentioned earlier, the effects of higher twist terms, of hadronisation and of the uncertainty in the values of higher order terms, all limit the precision with which α_S (or Λ) can be extracted. Systematic errors in the various experiments also play an important role.

The expected variation of α_S with Q^2 is given by (7.147). A further problem is in the interpretation of the number of flavours (n_f), which are 'active' for a particular Q^2. For an s-channel process, such as $e^+e^- \to$ hadrons, the quark composition of the final state, as a function of $Q^2 = s$, is well known. For deep inelastic scattering, interactions off heavy sea-quarks are suppressed for low values of the hadronic cms energy W, by phase space effects. The strange sea, at $Q^2 \sim 20$ GeV^2, is suppressed by a factor of about two compared to that for u and d quarks (Section 8.6.2). The charm quark sea is suppressed by a much larger factor. One common method is to increment n_f by one unit as $Q^2(W^2)$ passes the threshold for the production of a new quark flavour (possibly with a threshold behaviour factor), i.e. $n_f = 4$ above c-threshold, $n_f = 5$ above b-threshold, etc. An alternative procedure has been suggested by Engelbrecht (1986), which is to use

$$n_f(Q^2) = \sum_q \frac{1}{(1 + m_q^2/Q^2)}, \tag{7.221}$$

where the sum is over all quarks q. That is, a propagator-type form is assumed, so that the active number of flavours varies more smoothly with Q^2.

Fig. 7.48 shows a comparison of various determinations of α_S, all using the $\overline{MS}$ scheme, compared with the expectations of (7.147) (with n_f from (7.221)), for various values of $\Lambda_{\overline{MS}}$. It can be that $\Lambda_{\overline{MS}} \sim 0.2$ GeV is in reasonable accord with the data. However, the errors on the data points are still rather large, and the 'running' of α_S with Q^2, although consistent with the data, remains to be convincingly established.

Fig. 7.48 Measurements of α_S in the $\overline{\text{MS}}$ renormalisation scheme. The data for time (space)-like four-momenta are shown as closed (open) circles. The curves show α_S, as a function of Q^2, for different values of $\Lambda_{\overline{\text{MS}}}$.

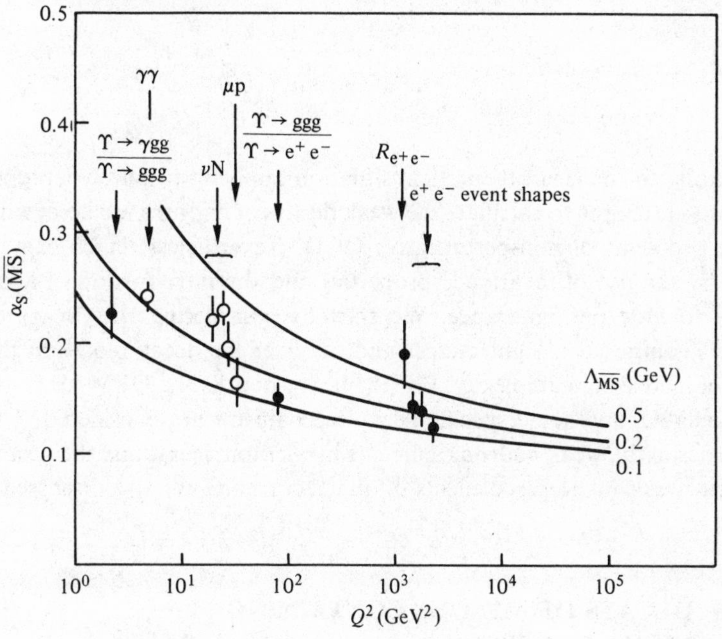

8

Weak hadronic currents, electroweak interference effects

In contrast to the clean theoretical situation applying to purely leptonic processes, attempts to calculate the weak decays of hadrons are beset with all the problems of non-perturbative QCD. Nevertheless, much can be learnt by the use of invariance properties and the introduction of form factors to hide our ignorance. We start by considering the decays of hadrons composed of light quarks and consider the decay modes in the order of increasing complexity. For the heavy quarks the QPM ideas are more reliable, and some specific decay mechanisms are considered. The properties of the weak hadronic currents in τ-lepton decay, and the results of electroweak interference effects in quark currents, are also discussed.

8.1 DECAYS OF MESONS CONTAINING LIGHT QUARKS

Historically, much work has been carried out on the study of the weak decays of π and K mesons and of the hyperons. We will not, in general, describe attempts made to calculate the probability of these decays in terms of quark interactions inside the hadrons, but merely indicate the most likely quark interactions responsible for the decay, and describe how the decays can be parameterised.

8.1.1 $\pi \to l\nu$ and $K \to l\nu$ decays

The simplest diagrams, at the quark level, for the decays of $\pi^-(\mathrm{d\bar{u}}) \to l^-\bar{\nu}_l$ and $K^-(\mathrm{s\bar{u}}) \to l^-\bar{\nu}_l$ are shown in Fig. 8.1(*a*) and (*b*) respectively. If the quarks were quasi-free, then we would have, for pion decay, the weak current $j_\mathrm{q}^\mu = \bar{u}\gamma^\mu(1-\gamma^5)d$. However, because the quarks are bound in hadrons, we cannot reliably compute the quark density functions. Instead, we write down the matrix element for $\pi^- \to l^- \bar{\nu}_l$

$$\mathcal{M}_{\mathrm{fi}} = \frac{G_{\mathrm{F}}}{2^{1/2}} [\bar{u}_1(p_1)\gamma^\mu(1-\gamma^5)v_2(p_2)]\langle 0|J^\dagger_\mu|\pi^-\rangle. \tag{8.1}$$

This form assumes the standard model (or V – A theory) for the leptonic part. The hadronic current $\langle 0|J^\dagger_\mu|\pi^-\rangle$ is constructed out of the available four-momenta. Since the π has $J^P = 0^-$ (pseudoscalar), the only four-momentum available is $q = p_1 + p_2$. Hence

$$\langle 0|J^\dagger_\mu|\pi^-\rangle = \mathrm{i}\cos\theta_{\mathrm{C}} f_\pi q_\mu = \mathrm{i}f_\pi \cos\theta_{\mathrm{C}}(p_{1\mu} + p_{2\mu}), \tag{8.2}$$

and this current is purely axial, since q is a polar vector. The relevant Cabibbo factor# is $\cos\theta_{\mathrm{C}}$, since this is a $\Delta S = 0$ current. The *pion decay constant* f_π characterises the strong interaction probability of the $\mathrm{d}\bar{\mathrm{u}}$ process. In principle, it should be calculable in QCD.

The matrix element from (8.1) and (8.2) can be written

$$\begin{aligned}\mathcal{M}_{\mathrm{fi}} &= \mathrm{i}\frac{G_{\mathrm{F}}}{2^{1/2}}\cos\theta_{\mathrm{C}} f_\pi[\bar{u}_1(p_1^\mu\gamma_\mu + p_2^\mu\gamma_\mu)(1-\gamma^5)v_2] \\ &= \mathrm{i}\frac{G_{\mathrm{F}}}{2^{1/2}}\cos\theta_{\mathrm{C}} f_\pi m_l[\bar{u}_1(1-\gamma^5)v_2],\end{aligned} \tag{8.3}$$

where the Dirac equations $\bar{u}_1 \not{p}_1 = m_l\bar{u}_1$ and $\not{p}_2(1-\gamma^5)v_2 = (1+\gamma^5)\not{p}_2 v_2 = 0$ have been used in the simplification. From (8.3) we can write down $\mathcal{M}^*_{\mathrm{fi}}$, and thus

$$\begin{aligned}|\mathcal{M}|^2 &= \frac{G_{\mathrm{F}}^2}{2}\cos^2\theta_{\mathrm{C}} f_\pi^2 m_l^2 \,\mathrm{tr}[(\not{p}_1 + m_l)(1-\gamma^5)\not{p}_2(1+\gamma^5)] \\ &= 4G_{\mathrm{F}}^2\cos^2\theta_{\mathrm{C}} f_\pi^2 m_l^2(p_1\cdot p_2).\end{aligned} \tag{8.4}$$

Fig. 8.1 Lowest order diagrams for the decay of (*a*) $\pi^- \to l^- \bar{\nu}_l$ and (*b*) $\mathrm{K}^- \to l^- \bar{\nu}_l$.

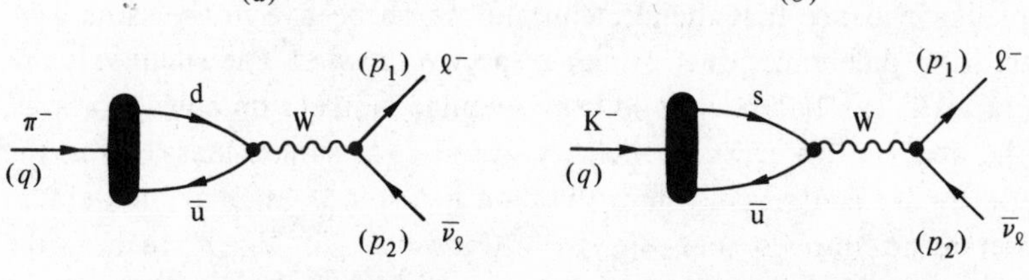

The 'Cabibbo' Lagrangian form 5.191 is assumed here. The inclusion of further quarks is discussed in Sections 8.3.1 and 8.6 and results in the Cabibbo factor being replaced by the appropriate Kobayashi–Maskawa matrix element (5.193). This, however, does not significantly affect the conclusions drawn in these sections.

The decay rate is given by (using (C.3))

$$\mathrm{d}\omega = \frac{G_F^2 \cos^2\theta_C f_\pi^2 m_l^2 (p_1\cdot p_2)\,\delta^4(q-p_1-p_2)\,\mathrm{d}^3p_1\,\mathrm{d}^3p_2}{8\pi^2 m_\pi E_1 E_2}$$

$$= \frac{G_F^2 \cos^2\theta_C f_\pi^2 m_l^2 (E_1E_2+E_2^2)E_2^2\,\mathrm{d}E_2\,\delta(m_\pi - E_1 - E_2)}{2\pi m_\pi E_1 E_2}, \qquad (8.5)$$

where the cms system of the pion, in which $p_1 = (E_1, \mathbf{p}_1)$, $p_2 = (E_2, -\mathbf{p}_1)$ with $|\mathbf{p}_1| = E_2$, has been used. Putting $y = m_\pi - E_1 - E_2$, so that $\mathrm{d}y = \mathrm{d}E_2(E_1+E_2)/E_1$, then the total decay rate is

$$\omega = \frac{G_F^2\cos^2\theta_C f_\pi^2 m_l^2}{2\pi m_\pi}\int E_2^2\,\delta(y)\,\mathrm{d}y$$

$$= \frac{G_F^2\cos^2\theta_C f_\pi^2 m_l^2 m_\pi}{8\pi}\left(1-\frac{m_l^2}{m_\pi^2}\right)^2, \qquad (8.6)$$

since $E_2 = (m_\pi^2 - m_l^2)/(2m_\pi)$. Decays to both electrons and muons are allowed kinematically. The ratio of the decay probabilities is

$$R_\pi = \frac{\omega(\pi^- \to \mathrm{e}^-\bar{\nu}_\mathrm{e})}{\omega(\pi^- \to \mu^-\bar{\nu}_\mu)} = \frac{m_\mathrm{e}^2}{m_\mu^2}\left(\frac{m_\pi^2 - m_\mathrm{e}^2}{m_\pi^2 - m_\mu^2}\right)^2 = 1.28\times 10^{-4}. \qquad (8.7)$$

Experimentally the measured value is $(1.23 \pm 0.02)\times 10^{-4}$. Radiative corrections (from $\pi \to \mathrm{e}\nu\gamma$, etc.) are important. Inclusion of these changes the V – A prediction to $R_\pi = 1.23\times 10^{-4}$ (Berman, 1958; Kinoshita, 1959), in good agreement with experiment. Care is needed in the evaluation of radiative corrections, because there are sizeable terms which are opposite in sign. They should be computed for the actual experimental set-up. As discussed in Chapter 4, a two-body decay involving a charged particle does not exist; it is always accompanied by soft photons.

The agreement between the V – A theory calculation (8.7) and experiment is further evidence that the electron and muon behave in the same way, apart from differences due to their respective masses. The small value of the ratio R_π can be understood from angular momentum considerations. In the decay $\pi^- \to \mathrm{e}^-\bar{\nu}_\mathrm{e}$, the helicity of the $\bar{\nu}_\mathrm{e}$ (assumed massless) in the V – A theory is $h(\bar{\nu}_\mathrm{e}) = 1$. Since the pion has spin 0, angular momentum conservation implies that $h(\mathrm{e}^-) = 1$. However, in V – A theory, the electron should have helicity -1 in the limit $m_\mathrm{e} \to 0$. Thus the decay $\pi^- \to \mathrm{e}^-\bar{\nu}_\mathrm{e}$ forces the e^- into the 'wrong' helicity state. This causes a suppression factor proportional to m_e^2 (see (8.3)), and so the e^- rate is suppressed, relative to the μ^- rate, by a factor $(m_\mathrm{e}/m_\mu)^2$, arising from the matrix element (squared).

If we consider the general case of all possible couplings, then there are possible contributions from both the pseudoscalar and axial currents. Before considering the pseudoscalar case it is instructive to compute the decay rate from the phase space part alone of $d\omega$. Starting with (C.3) in Appendix C, and following the steps leading to equation (8.6), we obtain

$$\omega = |\mathcal{M}_{fi}|^2 \frac{(m_\pi^2 - m_l^2)}{16\pi m_\pi^3}. \tag{8.8}$$

For the pseudoscalar case, the most general matrix element is

$$\mathcal{M}_{fi} = i\frac{G_F}{2^{1/2}} \cos\theta_C f_\pi^P \bar{u}_1(1 + a\gamma^5)v_2, \tag{8.9}$$

giving a decay rate

$$\omega = \frac{G_F^2 \cos^2\theta_C (f_\pi^P)^2}{16\pi}(1 + |a|^2)m_\pi\left(1 - \frac{m_l^2}{m_\pi^2}\right)^2. \tag{8.10}$$

Thus, for pure pseudoscalar coupling, we have

$$R_\pi = \frac{\omega(\pi^+ \to e^+\nu_e)}{\omega(\pi^+ \to \mu^+\nu_\mu)} = \left(\frac{m_\pi^2 - m_e^2}{m_\pi^2 - m_\mu^2}\right)^2 = 5.49. \tag{8.11}$$

Note that for phase space alone (i.e. $|\mathcal{M}_{fi}|^2 = \text{constant}$), equation (8.8) gives the value $R_\pi = 2.34$. Thus, the experimental value of R_π gives a stringent limit on the possible amount of pseudoscalar coupling.

The small value of R_π is not, however, evidence for the V – A theory, as opposed to other mixtures of V and A couplings. If we start with the matrix element for pure V + A coupling (or, indeed, any admixture of V and A couplings), then we obtain the same result (i.e. equation (8.7)). The l^- is again forced into the 'wrong' helicity state; in this case $h(e^-) = -1$. In order to learn more about the nature of the interaction, the helicity of the decay lepton must be measured. This has been done to very good precision: $h_{\mu^+} \leqslant -0.9959$ (90% c.l.), as discussed in Section 6.1.2, compatible with pure V – A coupling.

The two-body leptonic decays of the $K^\pm$ mesons can be treated in a way similar to π_{l2} decays. The axial current (assuming V – A theory) is written as $J_\mu^A = i \sin\theta_C f_K q_\mu$, where f_K is the *kaon decay constant*. This decay has $|\Delta S| = 1$, and hence the Cabibbo factor is $\sin\theta_C$. The ratio $R_K = \omega(K_{e2})/\omega(K_{\mu2})$ is measured to be $(2.42 \pm 0.11)10^{-5}$, compared to the lowest order theoretical value (R_K in (8.7)) of 2.57×10^{-5}. Pure pseudoscalar coupling gives $R_K = 1.098$. Note that, since the muon is significantly lighter than the kaon, the muon mode is not suppressed strongly by phase space, as is the case for pion decay. The helicity of the μ^+ in $K^+_{\mu 2}$ decay is $h_\mu = -0.970 \pm 0.047$ (Yamanaka *et al.*, 1986).

Next we consider the hadronic part of the matrix element. Using the measured value of the π^+ lifetime ($\tau^\mu_\pi = 1/\omega^\mu_\pi = 2.60 \times 10^{-8}$ s), G_F from muon decay and $\cos\theta_C = 0.975$ (see Section 8.3), we obtain from (8.6) that $f_\pi = 0.132$ GeV $= 0.95 m_\pi$. Similarly, using $\tau^\mu_K = 1/\omega^\mu_K = 1.95 \times 10^{-8}$ s, we find that $f_K = 0.160$ GeV. The values of f_π and f_K depend on the quark distribution functions. If the quark distributions were the same for $\pi^\pm$ and $K^\pm$ (i.e. SU(3) symmetry), then f_π and f_K would be equal. The ratio is computed to be

$$f_K/f_\pi = 1.22 \pm 0.01, \qquad (8.12)$$

which agrees with SU(3) to within about 20%. In the above calculation, the value of $\sin\theta_C = 0.221$, derived from the pure vector K_{e3}-decay (Section 8.1.2), is used. If SU(3) symmetry is assumed, then a value of $\sin\theta_C$ from these pure axial decays can be found, namely $\sin\theta_C = 0.257$. The result (8.12) is thus compatible with the $K^\pm$ being slightly smaller in size than the $\pi^\pm$, as was also deduced from electron scattering (Section 7.1).

8.1.2 π_{l3} and K_{l3} decays

The possible semileptonic decays of charged pions and charged and neutral kaons are

$$\pi^+(p_1) \to \pi^0(p_2) + l^+(p_3) + \nu_l(p_4), \qquad l = e, \qquad (8.13a)$$

$$K^+(p_1) \to \pi^0(p_2) + l^+(p_3) + \nu_l(p_4), \qquad l = e, \mu, \qquad (8.13b)$$

$$K^0_L(p_1) \to \pi^0(p_2) + l^+(p_3) + \nu_l(p_4), \qquad l = e, \mu. \qquad (8.13c)$$

Semileptonic decays of neutral kaons are discussed in more detail in Chapter 9. Some simple graphs for $K^+ \to \pi^0 l^+ \nu_l$ are shown in Fig. 8.2.

The hadronic matrix element is constructed out of the available four-vectors; in this case p_1 and p_2. Since the parity of the kaon and pion are the same then, amongst V and A, only a vector current can be made. The most general V, A hadronic current for $a \to b + l + \nu_l$ is

$$\langle b|V^\mu|a\rangle = V_{ab}[f_1(q^2)p_1^\mu + f_2(q^2)p_2^\mu], \qquad (8.14)$$

where V_{ab} is the Cabibbo factor, which is $\cos\theta_C$ for (8.13a) and $\sin\theta_C$ for (8.13b) and (8.13c). The quantities f_1 and f_2 are *form factors*, which in general are functions of the available invariant quantities. The only non-trivial invariant is $q^2 = (p_1 - p_2)^2$. It is usual to rewrite (8.14) in terms of two further form factors $f_+(q^2)$ and $f_-(q^2)$, as follows

$$\langle \mathrm{b}|V^\mu|\mathrm{a}\rangle = \begin{pmatrix}\cos\theta_\mathrm{C}\\ \sin\theta_\mathrm{C}\end{pmatrix}[f_+(q^2)(p_1^\mu + p_2^\mu) + f_-(q^2)(p_1^\mu - p_2^\mu)]. \quad (8.15)$$

The ratio of these form factors is $\xi(q^2) = f_-(q^2)/f_+(q^2)$.

Let us consider in more detail the simplest of these processes, namely $\pi^+ \to \pi^0 \mathrm{e}^+ \nu_\mathrm{e}$. Since the energy release is small, we may take q^2 to be zero, and neglect the f_- term in (8.15). Thus, in the V – A theory, we have

$$\mathscr{M}_{\mathrm{fi}} = \frac{G_\mathrm{F}\cos\theta_\mathrm{C}}{2^{1/2}} f_+[\bar{u}_4\gamma_\mu(1-\gamma^5)v_3][p_1^\mu + p_2^\mu]$$

$$\simeq 2^{1/2} G_\mathrm{F}\cos\theta_\mathrm{C} f_+ m_\pi[\bar{u}_4\gamma_0(1-\gamma^5)v_3], \quad (8.16)$$

where the kinetic energy of the π^0 is small and has been neglected. Using the usual methods, and neglecting the electron mass, gives

$$|\mathscr{M}^2| = 2G_\mathrm{F}^2\cos^2\theta_\mathrm{C} f_+^2 m_\pi^2 \,\mathrm{tr}[\not{p}_4\gamma_0(1-\gamma^5)\not{p}_3\gamma_0(1-\gamma^5)]$$

$$= 16G_\mathrm{F}^2\cos^2\theta_\mathrm{C} f_+^2 m_\pi^2[E_3E_4 + \mathbf{p}_3\cdot\mathbf{p}_4]. \quad (8.17)$$

The term in $\mathbf{p}_3\cdot\mathbf{p}_4$ gives zero contribution when we integrate over all angles. The total decay rate is (Appendix C)

$$\omega = \frac{16G_\mathrm{F}^2\cos^2\theta_\mathrm{C} f_+^2 m_\pi^2}{(2\pi)^5 2m_\pi}\int\frac{E_3E_4\,\mathrm{d}^3p_2\,\mathrm{d}^3p_3\,\mathrm{d}^3p_4\,\delta^4(p_1-p_2-p_3-p_4)}{8m_\pi E_3E_4}$$

$$= \frac{G_\mathrm{F}^2\cos^2\theta_\mathrm{C} f_+^2}{(2\pi)^5}\int\mathrm{d}^3p_3\,\mathrm{d}^3p_4\,\delta(\Delta - E_3 - E_4), \quad (8.18)$$

Fig. 8.2 Possible diagrams for the decay $\mathrm{K}^+ \to \pi^0 l^+ \nu_l$ (*a*) and (*b*), and $\mathrm{K}^+ \to \pi^+\pi^0$ (*c*), (*d*) and (*e*).

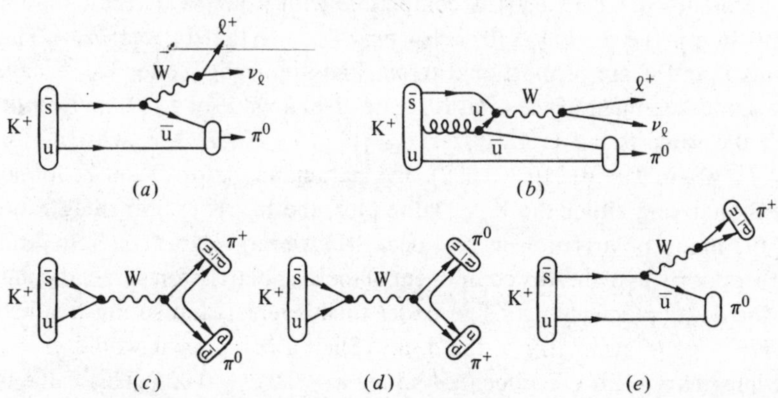

where $\Delta = m_{\pi^+} - m_{\pi^0}$ is the energy of the lepton pair. Putting $d^3p_4 = 4\pi E_4^2\, dE_4$ in (8.18), and integrating over dE_4, we obtain

$$\omega = \frac{4\pi G_F^2 \cos^2\theta_C f_+^2}{(2\pi)^5}\int_0^\Delta 4\pi E_3^2\, dE_3(\Delta - E_3)^2$$

$$= \frac{G_F^2 \cos^2\theta_C f_+^2 \Delta^5}{60\pi^3}. \tag{8.19}$$

Experimentally the decay branching ratio to π_{e3} is $(1.025 \pm 0.034) \times 10^{-8}$. Inserting the appropriate constants in (8.19), yields the value

$$|f_+(0)| = 1.37 \pm 0.02. \tag{8.20}$$

We return to the significance of this result in Section 8.2.1.

For K_{l3} decays the above approximations are not valid, and both $f_+(q^2)$ and $\xi(q_2)$ contribute. The final state properties can be studied by a variety of methods (see, e.g., Chounet *et al.*, 1972). For example, the decay Dalitz plot density is given by

$$\rho(E_\pi, E_l) \propto \sin^2\theta_C f_+^2(q^2)[A + B\xi(q^2) + C\xi^2(q^2)], \tag{8.21}$$

where A, B and C are kinematic factors. A further combination of form factors is often used, in particular in studying their q^2 variation. These are

$$f_+(q^2) = f_+(0)[1 + \lambda_+ q^2/m_\pi^2], \tag{8.22a}$$

$$f_0(q^2) = f_+(q^2) + \frac{q^2}{m_K^2 - m_\pi^2} f_-(q^2) = f_0(0)[1 + \lambda_0 q^2/m_\pi^2]. \tag{8.22b}$$

If $f_-(q^2)$ does not diverge as $q^2 \to 0$, then $f_0(0) = f_+(0)$ and there are three parameters $f_+(0)$, λ_+ and λ_0 (assuming a linear expansion is adequate). The form factors f_+ and f_0 correspond to vector and scalar meson amplitudes in the $K\pi$ system respectively. For example, the measured value of $\lambda_+ = 0.029 \pm 0.004$ is compatible with a $K^*(892)$ vector meson contribution (i.e. $f_+(q^2)/f_+(0) = m_{K^*}/(m_{K^*}^2 - q^2)$, so that $\lambda_+ \simeq m_\pi^2/m_{K^*}^2$). The terms B and C are proportional to m_l^2, and are negligible for K_{e3} decay. Thus, measurement of the total K_{e3} rate gives a value of $f_+(0)\sin\theta_C$. For K_{e3}^+ the value found is $f_+^{(+)}(0)\sin\theta_C = (0.216 \pm 0.002)/2^{1/2}$, whereas for K_L^0 $f_+^{(0)}(0)\sin\theta_C = (0.210 \pm 0.002)$. The parameter λ_0 (or ξ) can be found from analysing either the $K_{\mu 3}$ Dalitz plot, the $K_{e3}/K_{\mu 3}$ branching ratio or the muon polarisation in $K_{\mu 3}$ decay. However, λ_0 is, at present, still rather poorly known. Any component of the μ^+ polarisation perpendicular to the decay plane changes sign under time reversal, and so must vanish if T is a good symmetry of the decay. Such a component would give ξ an imaginary part; experimentally, Im $\xi = -0.017 \pm 0.025$. This value is compatible with T being conserved. The physics of this limit requires a model for interpretation. A more detailed discussion of the theoretical interest in K_{l3} decays is given by Bailin (1982).

8.2 SEMILEPTONIC DECAYS OF LIGHT BARYONS

Baryon decays introduce a further degree of complexity, since the initial and final state baryons in the decay $B_1(p_1) \to B_2(p_2) + l(p_3) + \nu_l(p_4)$ both have spin. Note that a purely leptonic decay would violate baryon number conservation. The most general hadronic V – A matrix element which can be constructed in terms of the available four-momenta is

$$\mathcal{M}^\mu = \mathcal{M}^\mu_{\mathrm{V}} - \mathcal{M}^\mu_{\mathrm{A}}, \tag{8.23}$$

where

$$\mathcal{M}^\mu_{\mathrm{V}} = \begin{pmatrix} \cos\theta_{\mathrm{C}} \\ \sin\theta_{\mathrm{C}} \end{pmatrix} \bar{u}_2 \left[f_1 \gamma^\mu + \frac{\mathrm{i} f_2}{2M_{\mathrm{B}}} \sigma^{\mu\nu} q_\nu + f_3 q^\mu \right] u_1,$$

$$\mathcal{M}^\mu_{\mathrm{A}} = \begin{pmatrix} \cos\theta_{\mathrm{C}} \\ \sin\theta_{\mathrm{C}} \end{pmatrix} \bar{u}_2 \left[g_1 \gamma^\mu \gamma^5 + \frac{\mathrm{i} g_2}{2M_{\mathrm{B}}} \sigma^{\mu\nu} \gamma^5 q_\nu + g_3 \gamma^5 q^\mu \right] u_1. \tag{8.24}$$

The form factors are all functions of q^2, where $q = p_1 - p_2$. The vector current form factors f_1, f_2 and f_3 are known as the *vector*, the *weak magnetism* and the *induced scalar* form factors respectively; whereas the axial vector counterparts g_1, g_2 and g_3 are called the *axial-vector*, the *pseudotensor* and the *induced pseudoscalar* form factors respectively. The actual decay rates of the semileptonic hyperon decays are rather small ($\lesssim 10^{-3}$). This is because the two-body weak hadronic decays are favoured by phase space.

8.2.1 Conserved vector current hypothesis

The form of the vector part of (8.24) is similar to that of the electromagnetic current (4.188), except for the additional weak term $f_3(q^2)$. If we write the matrix element for the electromagnetic current of a baryon B as

$$\begin{aligned}\mathcal{M}^\mu_{\mathrm{B}} &= \langle \mathrm{B}' | J^\mu_{\mathrm{em}} | \mathrm{B} \rangle \\ &= \exp(-\mathrm{i} q \cdot x) \bar{u}_{\mathrm{B}'} \left[F^{\mathrm{B}}_1(q^2) \gamma^\mu + \frac{\mathrm{i} F^{\mathrm{B}}_2(q^2)}{2M_{\mathrm{B}}} \sigma^{\mu\nu} q_\nu + F^{\mathrm{B}}_3(q^2) q^\mu \right] u_{\mathrm{B}},\end{aligned} \tag{8.25}$$

where the plane wave part has been included, then current conservation ($\partial_\mu J^\mu_{\mathrm{em}} = 0$) gives

$$\bar{u}_{\mathrm{B}'} \left[-F^{\mathrm{B}}_1(q^2) \gamma^\mu q_\mu - \frac{\mathrm{i} F^{\mathrm{B}}_2(q^2)}{2M_{\mathrm{B}}} \sigma^{\mu\nu} q_\nu q_\mu - F^{\mathrm{B}}_3(q^2) q^\mu q_\mu \right] u_{\mathrm{B}} = 0. \tag{8.26}$$

From the Dirac equation, $\bar{u}_{B'}\not{q}u_B = -\bar{u}_{B'}(\not{p}' - \not{p})u_B = -(M_B - M_B)\bar{u}_{B'}u_B = 0$, and the term $\sigma^{\mu\nu}q_\nu q_\mu$ also gives zero contribution. Therefore, since $q^\mu q_\mu = q^2 \neq 0$, we obtain from current conservation that

$$F_3^B(q^2) = 0. \tag{8.27}$$

Invariance under T implies that F_1^B and F_2^B are relatively real.

In Section 4.9.4, we found that $F_1^B(0)$ and $F_2^B(0)$ correspond to the charge and anomalous magnetic moment respectively of the baryon B (e.g. $F_1^p(0) = 1$, $F_2^p(0) = 1.79$, $F_1^n(0) = 0$, $F_2^n(0) = -1.91$).

For the electromagnetic current, the initial and final state nucleons are the same, whereas the weak current involves a transition between the neutron and proton. In order to try and relate these currents, we introduce a two-component nucleon spinor u, with the following isospin projections

$$\tfrac{1}{2}(1+\tau_3)u = \tfrac{1}{2}(1+\tau_3)\begin{pmatrix} u_p \\ u_n \end{pmatrix} = \begin{pmatrix} u_p \\ 0 \end{pmatrix},$$
$$\tfrac{1}{2}(1-\tau_3)u = \tfrac{1}{2}(1-\tau_3)\begin{pmatrix} u_p \\ u_n \end{pmatrix} = \begin{pmatrix} 0 \\ u_n \end{pmatrix}. \tag{8.28}$$

The electromagnetic current between two nucleons must have the form

$$\langle N'|J_{em}|N\rangle = (\bar{u}_p\, \bar{u}_n)\begin{pmatrix} a & 0 \\ 0 & b \end{pmatrix}\begin{pmatrix} u_p \\ u_n \end{pmatrix}$$
$$= \bar{u}\left[\frac{(a+b)}{2} I + \frac{(a-b)}{2}\tau_3\right]u. \tag{8.29}$$

That is, the electromagnetic current has both *isoscalar* and *isovector* parts. Identifying a and b from (8.25) (with $F_3 = 0$), these isospin parts can be written

$$\mathcal{M}_0 = \bar{u}'\left[(F_1^p + F_1^n)\gamma^\mu + \mathrm{i}\frac{(F_2^p + F_2^n)}{2M}\sigma^{\mu\nu}q_\nu\right]\frac{I}{2}u, \tag{8.30}$$

$$\mathcal{M}_1 = \bar{u}'\left[(F_1^p - F_1^n)\gamma^\mu + \mathrm{i}\frac{(F_2^p - F_2^n)}{2M}\sigma^{\mu\nu}q_\nu\right]\frac{\tau_3}{2}u. \tag{8.31}$$

The vector part of the weak n–p current can be written (the SU(3) notation is explained below)

$$\langle p|j_V^\mu|n\rangle = \cos\theta_C\langle p|j_1^\mu + \mathrm{i}j_2^\mu|n\rangle. \tag{8.32}$$

The n → p transition requires the isospin operator $\tau_+ = (\tau_1 + \mathrm{i}\tau_2)/2$. Hence, in terms of the nucleon spinor u, we find from (8.24) and (8.32) that

$$\langle p|j_1^\mu + \mathrm{i}j_2^\mu|n\rangle = \bar{u}'\left[f_1(q^2)\gamma^\mu + \mathrm{i}\frac{f_2(q^2)}{2M}\sigma^{\mu\nu}q_\nu + f_3(q^2)q^\mu\right]\tau_+ u. \tag{8.33}$$

In the limit $q^2 \to 0$, the isovector electromagnetic current (8.31) becomes

$$\langle \mathrm{N}|j^\mu_{\mathrm{IV}}|\mathrm{N}\rangle = \bar{u}'\left\{[F_1^{\mathrm{p}}(0) - F_1^{\mathrm{n}}(0)]\gamma^\mu \frac{\tau_3}{2}\right\}u = \bar{u}'\gamma^\mu \frac{\tau_3}{2}u. \tag{8.34}$$

The form of the matrix element found for nucleon beta decay (5.44) implies that $f_1(0) \simeq 1$, since q^2 is small for this decay. Hence the weak vector current in the $q^2 \to 0$ limit becomes, from (8.33),

$$\langle \mathrm{p}|j_1^\mu + \mathrm{i}j_2^\mu|\mathrm{n}\rangle \simeq \bar{u}'\gamma^\mu\tau_+ u. \tag{8.35}$$

Thus, $j_1^\mu - \mathrm{i}j_2^\mu$ (i.e. $\mathrm{p} \to \mathrm{n}$), j^μ_{IV} and $j_1^\mu + \mathrm{i}j_2^\mu$ are the members of an *isotriplet vector of current operators*.

With the further assumption (Feynman and Gell-Mann, 1958) that, since the electromagnetic current is conserved, the weak vector current is also conserved (*Conserved Vector Current*, or CVC, hypothesis), we obtain relationships between the electromagnetic and weak form factors, namely

$$\begin{aligned} f_1(q^2) &= F_1^{\mathrm{p}}(q^2) - F_1^{\mathrm{n}}(q^2) \xrightarrow{q^2 \to 0} 1, \\ f_2(q^2) &= F_2^{\mathrm{p}}(q^2) - F_2^{\mathrm{n}}(q^2) \xrightarrow{q^2 \to 0} \mu_{\mathrm{p}} - \mu_n, \\ f_3(q^2) &= 0. \end{aligned} \tag{8.36}$$

The electromagnetic charge of the proton is, to good precision, equal to that of the electron. In the proton, the charge is spread out in space (giving rise to μ_{p}), but the strong interaction effects do not 'renormalise' its total charge. The experimental observation that $f_1(0) \simeq 1$, in line with the CVC prediction, indicates that the 'weak' charge is also not renormalised by the effects of strong interactions.

At the quark level, the electromagnetic and vector weak currents for u and d quarks are

$$J^\mu_{\mathrm{em}} = \tfrac{2}{3}\bar{u}\gamma^\mu u - \tfrac{1}{3}\bar{d}\gamma^\mu d = \tfrac{1}{6}\bar{q}\gamma^\mu I q + \bar{q}\gamma^\mu \frac{\tau_3}{2} q, \tag{8.37}$$

$$J^\mu_{\mathrm{V}} = \bar{d}\gamma^\mu u \qquad = \qquad \bar{q}\gamma^\mu \tau_- q, \tag{8.38}$$

where q represents the (u, d) isospin doublet. Isospin symmetry is equivalent to the approximate degeneracy in mass of u and d quarks. The currents $\bar{q}\gamma^\mu\tau_- q$, $\bar{q}\gamma^\mu(\tau_3/2)q$ and $\bar{q}\gamma^\mu\tau_+ q$ form an isotriplet of vector currents (the current $\bar{q}\gamma^\mu I q$ is isoscalar). CVC thus implies that $\partial_\mu J^\mu_{\mathrm{em}} = \partial_\mu J^\mu_{\mathrm{V}} = 0$. For a weak transition involving a hadron of isospin I, from states I_3 to $I_3 + 1$, we can define a weak isospin raising charge

$$Q_{\mathrm{wk}} = \int u^\dagger(x)\,\mathrm{d}(x)\,\mathrm{d}\mathbf{x} = [I(I+1) - I_3(I_3+1)]^{1/2}. \tag{8.39}$$

Thus, for the decay $n \to pe^-\bar{\nu}_e$, which has $I=\frac{1}{2}$, $I_3=-\frac{1}{2}$, we have $Q_{wk}=1$. As discussed above, the experimental data support this idea that the weak charge is not renormalised. For the decay $\pi^- \to \pi^0 e^- \bar{\nu}_e$, which has $I=1$, $I_3=0$, we find $Q_{wk}=2^{1/2}$. This is just the quantity f_+ in expression (8.19), and the experimental result (8.20) is compatible with the CVC prediction.

A further test of CVC comes from measurements of the weak magnetism term (f_2) of the $A=12$ isotriplet nuclei ^{12}B, ^{12}N and $^{12}C^*$, all having $J^P=1^+$. These states decay to the $J^P=0^+$ ground state ^{12}C by the β-decays $^{12}B \to {}^{12}Ce^-\bar{\nu}_e$, $^{12}N \to {}^{12}Ce^+\nu_e$ and by the magnetic dipole transition $^{12}C^* \to {}^{12}C + \gamma$ respectively. The energy release is large (~ 15 MeV), and the electron energy (E) spectrum has the form

$$dN(e^\pm) \sim F(Z,E)pE(\Delta - E)^2(1 \mp \tfrac{8}{3}aE)\,dE. \tag{8.40}$$

The additional term with respect to the beta-decay formula (5.17) comes from f_2. This term is completely specified from the $^{12}C^*$ decay by CVC, and depends on $\mu_p - \mu_n$. Experimentally, both the expected sign and magnitude are confirmed (Lee *et al.*, 1963).

8.2.2 Partially conserved axial current

The axial-vector current has no electromagnetic analogue. It is, however, clearly not conserved. The coefficient of the axial term for beta-decay is 1.26 (equation (5.44)), and not about one as for the vector term. Furthermore, the $\pi^\pm$ would not decay if the axial current was conserved. This decay is purely axial, so that the axial current and its derivative are

$$\langle 0|A_\mu|\pi\rangle = if_\pi q_\mu \exp(-iq\cdot x), \tag{8.41}$$

$$\langle 0|\partial^\mu A_\mu|\pi\rangle = f_\pi m_\pi^2 \exp(-iq\cdot x) \qquad (q^2 = m_\pi^2). \tag{8.42}$$

The divergence is non-zero but, since the pion is light on the hadronic mass scale, can be considered small. Gell-Mann and Levy (1960) introduced the concept of a *partially conserved axial current* (PCAC), with conservation applying in the *soft pion limit* $m_\pi^2 \to 0$.

The divergence $\partial^\mu A_\mu$ appears to be related to the creation operator ϕ_π for the π-meson. It can be explicitly assumed that

$$\partial^\mu A_\mu = c\phi_\pi(x). \tag{8.43}$$

Now we have

$$\langle 0|\phi_\pi(x)|\pi(q)\rangle = \exp(-iq\cdot x), \tag{8.44}$$

so that

$$\langle 0|\partial^\mu A_\mu|\pi\rangle = c\exp(-iq\cdot x) \Rightarrow c = f_\pi m_\pi^2. \tag{8.45}$$

A further dynamical assumption can be made that the p-n matrix element is dominated by the pion pole giving, for $0 < q^2 < m_\pi^2$,

$$\langle \mathrm{p}|\partial^\mu A_\mu|\mathrm{n}\rangle \simeq f_\pi m_\pi^2 \frac{\langle \mathrm{p}\pi(q)|\mathrm{n}\rangle}{(m_\pi^2 - q^2)}. \tag{8.46}$$

At $q^2 = 0$ we have, from (8.24),

$$\langle \mathrm{p}|\partial^\mu A_\mu|\mathrm{n}\rangle = -\mathrm{i}g_1(0)\bar{u}_\mathrm{p}\not{q}\gamma^5 u_\mathrm{n} = \mathrm{i}(M_\mathrm{p} + M_\mathrm{n})g_1(0)\bar{u}_\mathrm{p}\gamma^5 u_\mathrm{n}, \tag{8.47}$$

and we can write

$$\langle \mathrm{p}\pi|\mathrm{n}\rangle = \mathrm{i}g_{\mathrm{pn}\pi}\bar{u}_\mathrm{p}\gamma^5 u_\mathrm{n}, \tag{8.48}$$

where $g_{\mathrm{pn}\pi}$ is the coupling constant obtained from pion–nucleon scattering ($g_{\mathrm{pn}\pi} = 2^{1/2}(13.6 \pm 0.3)$). From (8.46), (8.47) and (8.48) we obtain

$$g_1(0) = \frac{f_\pi g_{\mathrm{pn}\pi}}{M_\mathrm{p} + M_\mathrm{n}} \simeq 1.31 \pm 0.03. \tag{8.49}$$

This is the *Goldberger–Trieman* relation, and it can be seen that $g_1(0)$, calculated in this way, is reasonably close to the value 1.26 obtained from beta decay.

8.2.3 Second-class currents

In the hypothetical limit that we could slowly turn off strong interactions, then we would expect $\mathscr{M}_\mathrm{V}^\mu$ and $\mathscr{M}_\mathrm{A}^\mu$ in (8.24) to reduce to the currents for 'bare' particles, namely

$$\mathscr{M}_\mathrm{V}^\mu(0) = \bar{u}_2\gamma^\mu u_1, \qquad \mathscr{M}_\mathrm{A}^\mu(0) = \bar{u}_2\gamma^\mu\gamma^5 u_1. \tag{8.50}$$

Thus, it appears plausible that the terms f_2, f_3, g_2 and g_3 arise because of strong interaction (QCD) effects. If this is so, they should satisfy the known symmetry laws of strong interactions. In particular, G-parity ($G = CR_y$, where C = charge conjugation and R_y a rotation by an angle π around the isospin axis I_y) should be conserved. All six terms in (8.24) transform as isovectors under R_y, so we need only consider application of C. The transformation properties of Dirac spinors and matrices were discussed in Section 3.2.7. For the vector current, f_2 transforms with the same sign as f_1, whereas f_3 has the opposite sign. For the axial current, g_3 transforms with the same sign as g_1, whereas g_2 has the opposite sign. The currents with coëfficients f_3 and g_2 are known as *second-class currents*, whereas the others are *first-class currents* (Weinberg, 1958). The above arguments would imply the absence of second-class currents, so that f_3 and g_2 are expected to be small. Note that CVC also implies that $f_3 = 0$.

Experimentally, there is no compelling evidence for the existence of second-class currents. The possible effects are small and require careful correction for nuclear physics and radiative phenomena, thus making experiments extremely difficult.

8.2.4 $\pi^0 \to \gamma\gamma$ decay

In contrast to the weak interaction decays of the charged pion, the neutral pion decays electromagnetically. The simplest quark graph for $\pi^0 \to \gamma\gamma$ decay is similar to that shown in Fig. 10.24 (but with H^0 replaced by π^0 and with a quark (u or d) running around the internal triangle). Such graphs involve two vector and one axial-vector currents (π is a pseudoscalar) and are *anomalous* in the sense that the associated three-point functions $\Gamma^{(3)}$ do not satisfy the expected Ward identities (*Adler–Bell–Jackiw anomaly*; Adler (1969), Bell and Jackiw (1969)). This results effectively in an additional term in the divergence of the axial current.# In the soft pion limit ($m_\pi \to 0$) it can be shown, by current algebra techniques (see, e.g., Cheng and Li, 1984), that the decay rate is given entirely by the anomalous term. The result is $\Gamma(\pi^0) = 0.84 n_c^2$ eV, where n_c is the number of colours. Comparison with the experimental decay rate (7.6 ± 0.4 eV) gives $n_c = 3.01 \pm 0.08$ colours. Hence, this result supports the hypothesis that there are three colours and also the existence of the anomaly. The basic theoretical problem is in understanding the chiral symmetry properties of the Lagrangian (see also Section 10.7), which affect the renormalisability of the theory. In the standard model, cancellation of the anomaly occurs if there are equal numbers of quark and lepton doublets.

8.3 CABIBBO THEORY

We have seen that the weak hadronic current is approximately invariant under the SU(2) isospin transformations of u and d quarks. In particular, CVC implies that $\Delta I_3 = \pm 1$ for non-strangeness changing currents. The lightest meson and baryon multiplets contain u, d and s quarks, so that the question arises as to whether the weak current is invariant (or approximately invariant) under the group SU(3) of flavour transformations.

Empirically, it is observed that, for $\Delta S \neq 0$ semileptonic decays,

(i) decays with $|\Delta S| > 1$ are absent;

The divergence of the gauge invariant axial chiral current $j^\mu_A = \bar{\psi}\gamma^\mu\gamma^5\psi$ is $\partial_\mu j^\mu_A = 2im\bar{\psi}\gamma^5\psi + (e^2/8\pi^2)\tilde{F}_{\mu\nu}F^{\mu\nu}$, where $\tilde{F}^{\mu\nu} = \frac{1}{2}\varepsilon^{\mu\nu\lambda\rho}F_{\lambda\rho}$ ($F_{\mu\nu}$ is the electromagnetic field tensor). Thus j^μ_A is not conserved in the massless limit due to the presence of the second (anomalous) term.

(ii) the strangeness-changing amplitude is strongly suppressed compared to that which is strangeness non-changing;
(iii) If $\Delta Q = Q_i - Q_f$ and $\Delta S = S_i - S_f$, where i and f represent the initial and final state hadrons, then observed decays have $\Delta S = \Delta Q$. For example, the limits on possible decay modes violating this rule are

$$\frac{\Gamma(\Sigma^+ \to nl^+\nu_l)}{\Gamma(\Sigma^- \to nl^-\bar{\nu}_l)} < 0.04 \qquad (90\% \text{ c.l.})$$

$$\frac{\Gamma(K^+ \to \pi^+\pi^+e^-\bar{\nu}_e)}{\Gamma(K^+ \to \pi^+\pi^-e^+\nu_e)} < 0.0003 \qquad (90\% \text{ c.l.}). \qquad (8.51)$$

That is, the experimental results are consistent with there being only the weak transitions $u \rightleftharpoons d$ ($\Delta S = 0$) and $u \rightleftharpoons s$ ($|\Delta S| = 1$).

In the Cabibbo theory (Cabibbo, 1963) it is assumed that the vector currents V_μ^j, $j = 1$ to 8, are members of an SU(3) octet of currents. A similar octet A_μ^j exists for the axial-vector currents. The vector part is in the same octet as the electromagnetic current, so that $\partial^\mu V_\mu^j = 0$, $j = 1$ to 8. The currents $V_\mu(A_\mu)$ can be written in terms of the SU(3) matrices λ_j, $j = 1$ to 8 (Section 2.5.3) and the three-component (u, d, s) quark field q. Thus the vector part of the $d \to u$ transition, which is a $\Delta S = 0$ isospin raising current, can be written

$$\bar{\psi}_u\gamma_\mu\psi_d = (\bar{u}\ \bar{d}\ \bar{s})\begin{pmatrix} 0 & 1 & 0 \\ 0 & 0 & 0 \\ 0 & 0 & 0 \end{pmatrix}\begin{pmatrix} u \\ d \\ s \end{pmatrix} = \bar{q}\gamma_\mu \frac{(\lambda_1 + i\lambda_2)}{2} q$$

$$= V_\mu^{1+i2}. \qquad (8.52)$$

The inverse transformation $u \to d$ is given by the current V_μ^{1-i2}. Similarly the $s \to u$, $\Delta S = 1$ vector current, can be written

$$\bar{\psi}_u\gamma_\mu\psi_s = \bar{q}\gamma_\mu \frac{(\lambda_4 + i\lambda_5)}{2} q = V_\mu^{4+i5}. \qquad (8.53)$$

In the same notation the electromagnetic current is

$$J_\mu^{em} = V_\mu^3 + \frac{1}{3^{1/2}} V_\mu^8, \qquad (8.54)$$

which is equivalent to $Q = I_3 + Y/2$ ($Y = 2/3^{1/2}\, V^8$, equation (2.257)).

The total hadronic current is

$$h_\mu = \cos\theta_C(V_\mu^{1+i2} - A_\mu^{1+i2}) + \sin\theta_C(V_\mu^{4+i5} - A_\mu^{4+i5}). \qquad (8.55)$$

In terms of h_μ and the leptonic current l_μ, the effective (charged current) weak Lagrangian is

$$\mathscr{L}_{CC} = \frac{G_F}{2^{1/2}}(l_\mu + h_\mu)(l_\mu^\dagger + h_\mu^\dagger), \tag{8.56}$$

which covers all possible combinations of leptonic (e and μ when envisaged by Cabibbo) and hadronic ($\Delta S = 0, \pm 1$) currents. The $\Delta Q = \Delta S$ rule is satisfied automatically (for the lowest order graphs), from the form of (8.56).

The theory gives relationships for the various transitions between the $J^P = \frac{1}{2}^+$ baryon states. The most general matrix element between the baryon octet states B_k and B_i, of an octet O_j of currents, can be written

$$\langle B_i | O_j | B_k \rangle = \mathrm{i} f_{ijk} F + d_{ijk} D, \tag{8.57}$$

where f_{ijk} and d_{ijk} are given in Table 2.2. That is, the coupling of two octets to form a third has two reduced matrix elements and the terms F and D are thus antisymmetric and symmetric respectively. In terms of the baryon octet states, we identify

$$\mathrm{p} = \frac{1}{2^{1/2}}(\mathrm{B}_4 + \mathrm{iB}_5), \qquad \mathrm{n} = \frac{1}{2^{1/2}}(\mathrm{B}_6 + \mathrm{iB}_7), \qquad \Sigma^\pm = \frac{1}{2^{1/2}}(\mathrm{B}_1 \pm \mathrm{iB}_2),$$

$$\Xi^- = \frac{1}{2^{1/2}}(\mathrm{B}_4 - \mathrm{iB}_5), \qquad \Xi^0 = \frac{1}{2^{1/2}}(\mathrm{B}_6 - \mathrm{iB}_7), \tag{8.58}$$

$$\Sigma^0 = \mathrm{B}_3, \qquad \Lambda^0 = \mathrm{B}_8.$$

These assignments can be found by comparing (2.269) and (2.274).

The electromagnetic current between B_k and B_i has thus both F and D parts, i.e. there are corresponding form factors F_1^F, F_1^D, F_2^F and F_2^D, which are the same for all members of the octet. Inserting J^{em} from (8.54) (which has $j = 3$ and 8 contributions only) for O_j in (8.57) we obtain

$$\langle B_i | J^{em} | B_k \rangle = \mathrm{i}\left[f_{i3k} + \frac{1}{3^{1/2}} f_{i8k}\right] F + \left[d_{i3k} + \frac{1}{3^{1/2}} d_{i8k}\right] D. \tag{8.59}$$

For example, the current for the $\Sigma^0(B_3)$ is

$$\langle \Sigma | J^{em} | \Sigma \rangle = \mathrm{i}\left[f_{333} + \frac{1}{3^{1/2}} f_{383}\right] F + \left[d_{333} + \frac{1}{3^{1/2}} d_{383}\right] D = \frac{D}{3}, \tag{8.60}$$

since $f_{333} = f_{383} = d_{333} = 0$, and $d_{383} = 1/3^{1/2}$. Similarly,

$$\langle \mathrm{n} | J^{em} | \mathrm{n} \rangle = -\frac{2D}{3}, \qquad \langle \mathrm{p} | J^{em} | \mathrm{p} \rangle = F + \frac{D}{3}. \tag{8.61}$$

Hence, for the neutron, there is no contribution from F_1^F or F_2^F. For the neutron and proton we have seen that (for small q^2) we have $F_1^n = 0$, $F_2^n = \mu_n$ and $F_1^p = 1$, $F_2^p = \mu_p$ respectively. Therefore, equating the two forms, we find

$$\begin{aligned}
F_1^n &= -\tfrac{2}{3}F_1^D = 0, && \text{i.e. } F_1^D = 0, \\
F_1^p &= F_1^F + \frac{F_1^D}{3} = 1, && \text{i.e. } F_1^F = 1, \\
F_2^n &= -\tfrac{2}{3}F_2^D = \mu_n, && \text{i.e. } F_2^D = -\tfrac{3}{2}\mu_n, \\
F_2^p &= F_2^F + \frac{F_2^D}{3} = \mu_p, && \text{i.e. } F_2^F = \mu_p + \mu_n/2.
\end{aligned} \tag{8.62}$$

For large q^2 values we have, more generally, $F_2^D = -\frac{3}{2}F_2^n$, $F_2^F = F_2^p + F_2^n/2$.

The weak V – A current between B_i and B_f can be written as

$$\langle B_f|h^\mu|B_i\rangle = \bar{u}_f\left[f_1^{F,D}\gamma^\mu + i\frac{f_2^{F,D}}{2M}\sigma^{\mu\nu}q_\nu - g_1^{F,D}\gamma^\mu\gamma^5\right]u_i. \tag{8.63}$$

In addition, there is the cos θ_C or sin θ_C Cabibbo factor, as appropriate. The terms in f_3, g_2 and g_3 have been neglected in (8.63). The vector part of (8.63) is entirely specified by CVC, which gives (from (8.36) and (8.62))

$$f_1^F = 1, \qquad f_1^D = 0, \qquad f_2^F = \mu_p + \tfrac{1}{2}\mu_n, \qquad f_2^D = -\tfrac{3}{2}\mu_n. \tag{8.64}$$

The full hadronic matrix element can now be written down. For example, for the Σ^+(uus) → Λ(uds), i.e. u → d, transition we have

$$\begin{aligned}
\langle \Lambda|j_1 - ij_2|\Sigma^+\rangle &= \left\langle B_8\middle|j_1 - ij_2\middle|\frac{1}{2^{1/2}}(B_1 + iB_2)\right\rangle \\
&= \frac{1}{2^{1/2}}(d_{811} + d_{822})D = (\tfrac{2}{3})^{1/2}D,
\end{aligned} \tag{8.65}$$

since all the F terms are zero and, from Table 2.2, $d_{811} = d_{822} = 1/3^{1/2}$. The only non-zero vector term is thus f_2^D. Hence, including the Cabibbo factor, we have

$$\begin{aligned}
\langle \Lambda|j_1^\mu - ij_2^\mu|\Sigma^+\rangle &= (\tfrac{2}{3})^{1/2}\cos\theta_C\bar{u}_\Lambda\left[-i\frac{3}{2}\left(\frac{\mu_n}{2M}\right)\sigma^{\mu\nu}q_\nu - g_1^D\gamma^\mu\gamma^5\right]u_\Sigma \\
&= \cos\theta_C\bar{u}_\Lambda\left[-\left(\frac{3}{2}\right)^{1/2} i\frac{\mu_n}{2M}\sigma^{\mu\nu}q_\nu - \left(\frac{2}{3}\right)^{1/2} g_1^D\gamma^\mu\gamma^5\right]u_\Sigma.
\end{aligned} \tag{8.66}$$

Note that CVC alone predicts that the decay is purely axial.

Table 8.1. *Hadronic matrix elements for* $J^P = \frac{1}{2}^+$ *semileptonic decays*

Decay	Cabibbo factor	Vector $f_1(0)$	Weak magnetism $f_2(0)$	Axial $g_1(0)$
$n \to p$	$\cos\theta_C$	1	$\mu_p - \mu_n$	$F + D$
$\Sigma^\pm \to \Lambda$	$\cos\theta_C$	0	$-(\frac{3}{2})^{1/2}\mu_n$	$(\frac{2}{3})^{1/2}D$
$\Sigma^- \to \Sigma^0$	$\cos\theta_C$	$2^{1/2}$	$(2)^{1/2}(\mu_p + \mu_n/2)$	$2^{1/2}F$
$\Lambda \to p$	$\sin\theta_C$	$-(\frac{3}{2})^{1/2}$	$-(\frac{3}{2})^{1/2}\mu_p$	$-(\frac{3}{2})^{1/2}\left(F + \frac{D}{3}\right)$
$\Sigma^- \to n$	$\sin\theta_C$	-1	$-(\mu_p + 2\mu_n)$	$-(F - D)$
$\Xi^- \to \Lambda$	$\sin\theta_C$	$(\frac{3}{2})^{1/2}$	$(\frac{3}{2})^{1/2}(\mu_p + \mu_n)$	$(\frac{3}{2})^{1/2}\left(F - \frac{D}{3}\right)$
$\Xi^- \to \Sigma^0$	$\sin\theta_C$	$(\frac{1}{2})^{1/2}$	$(\frac{1}{2})^{1/2}(\mu_p - \mu_n)$	$(\frac{1}{2})^{1/2}(F + D)$
$\Xi^0 \to \Sigma^+$	$\sin\theta_C$	1	$\mu_p - \mu_n$	$F + D$
$\Xi^- \to \Xi^0$	$\cos\theta_C$	1	$\mu_p + 2\mu_n$	$F - D$

Table 8.1 shows some of the matrix elements for the transitions between the $J^P = \frac{1}{2}^+$ baryons. These can be derived in a way similar to that for the $\Sigma \to \Lambda$ transition described above. The vector and axial-vector terms correspond to $f_1(0)$ and $g_1(0)$ in (8.24).

8.3.1 Experimental tests of the Cabibbo theory

For the semileptonic decay $B_1 \to B_2 + l + \nu_l$, the measurements which can be made are as follows (Gaillard and Sauvage, 1984).

(i) *Total decay rate*

$$\Gamma = G^2 \begin{pmatrix} \cos^2\theta_C \\ \sin^2\theta_C \end{pmatrix} \frac{\Delta m^5}{60\pi^3} (f_1^2 + 3g_1^2)(1 - 3\delta), \tag{8.67}$$

where $\Delta m = M_1 - M_2$ and $\delta = (M_1 - M_2)/(M_1 + M_2)$.

(ii) *Lepton–neutrino correlation*

$$W(\cos\theta_{l\nu}) = \tfrac{1}{2}(1 + \alpha_{l\nu}\cos\theta_{l\nu}), \tag{8.68}$$

where

$$\alpha_{l\nu} = \frac{f_1^2 - g_1^2}{f_1^2 + 3g_1^2} - 2\delta. \tag{8.69}$$

(iii) *Measurements sensitive to the sign of* g_1/f_1. These include measuring the polarisation of B_2 with B_1 unpolarised, the lepton-decay asymmetry with B_1 polarised or the shape of the lepton spectrum with B_1 unpolarised.

The experiments can be broadly categorised into two types. In the old experiments ($\lesssim$1975), the momenta of the decaying hyperons were $\lesssim$1 GeV, with consequently small decay lengths, but with large polarisation values. The bubble chamber techniques used, however, resulted in rather low statistics. More recently, high energy and intensity hyperon beams at FNAL and CERN have been developed. For $p_B \sim 100$ GeV, the mean decay lengths in the lab frame ($\sim \gamma c\tau \sim 100c\tau$) are thus the order of metres, allowing the use of purely electronic experiments, with consequently higher statistics.

From Table 8.1 it can be seen that, in the Cabibbo theory, there are three unknown parameters θ_C, F and D. It is assumed that G_F can be taken from muon decay. The relative sign of F and D is found to be positive and, by convention, $g_1/f_1 = F + D$ is taken to be positive for neutron decay. Therefore, measurements of the types (i) and (ii) each give a relationship between F and D. Fig. 8.3 shows a compilation of the results of such measurements. The q^2 dependence of the form factors is taken into account by expanding the dipole form of the elastic nucleon form factors linearly in q^2, and assuming this form holds for the time-like decays. Thus,

Fig. 8.3 Results from hyperon and neutron decay expressed in terms of F and D. The full lines are from branching ratio measurements and the broken lines from measurements of g_1/f_1. The constraint from the neutron lifetime is also shown.

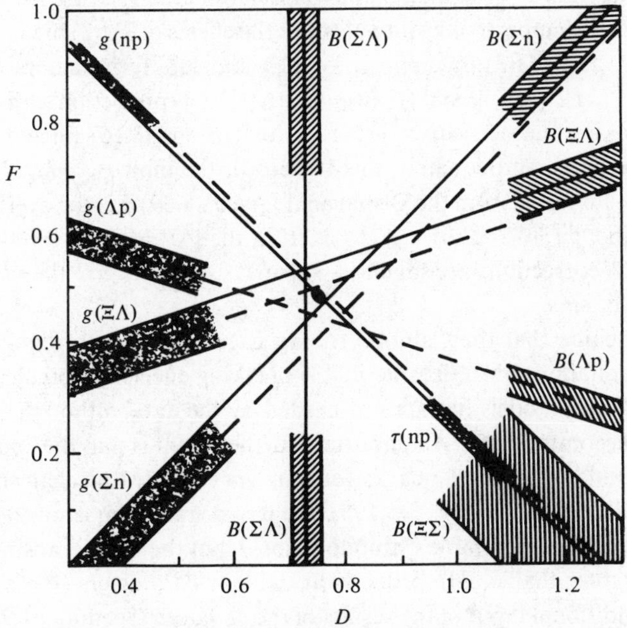

for $\Delta S = 0$,

$$f_1(q^2) = f_1(0)(1 + 2q^2/m_V^2), \qquad g_1(q^2) = g_1(0)(1 + 2q^2/m_A^2), \quad (8.70)$$

where $m_V = 0.84$ GeV and $m_A = 1.08$ GeV. More details on these assumptions, radiative corrections and a discussion of the data can be found in Gaillard and Sauvage (1984). The result of a three-parameter fit to the available data, which gives a good chi-squared probability, is

$$\sin\theta_C = 0.231 \pm 0.003, \quad F = 0.477 \pm 0.012, \quad D = 0.756 \pm 0.011. \quad (8.71)$$

Assuming the baryons are in the 56 representation of SU(6), one finds that $F + D = \frac{5}{3}$, and that $\alpha = D/(F + D) = \frac{3}{5}$, these values being in tolerable agreement with the data, and thus with this simple quark interpretation.

The Cabibbo theory might be expected to be valid in the limit of exact flavour SU(3). We know that the hadron masses in the lightest octet and decuplet of baryons violate this symmetry by about 15%. It is thus perhaps surprising that the Cabibbo theory works so well for hyperon decays. Theoretically, the SU(3) breaking effects are expected to vanish to first order in the symmetry-breaking for the vector terms (theorem of Ademollo and Gatto, 1964). However, there is no similar statement for the axial terms, and recourse must be made to hadronic (e.g. bag) models to estimate the size of the corrections. It is noteworthy that the value of $\sin\theta_C$ in (8.71) differs from that found from the total K_{e3}^+ decay rate (i.e. vector current), namely $\sin\theta_C = 0.221 \pm 0.002$. This latter analysis (Leutwyler and Roos, 1984) attempts to take into account the effects of SU(3) breaking for K_{e3}^+ decay. The SU(3) predictions for K_{e3}^+ and K_{e3}^0 form factors are $f_+^{(+)}(0) = 1/2^{1/2}$, $f_+^{(0)}(0) = 1$ and $f_-(0) = 0$; the latter prediction can be deduced by taking the derivative of the vector current (8.15) (including its plane-wave part), and equating this to zero in the limit $m_b = m_a$. The value $f_+^{(0)}(0) = 1$ is found from the V-spin analogue of (8.39) (i.e. the overlap integral $u^\dagger s$) and $f_+^{(+)}(0)/f_+^{(0)}(0) = 1/2^{1/2}$ is from the $|\Delta I| = \frac{1}{2}$ rule (Section 8.5). The SU(3) corrections are small, so K_{e3} measurements give the most reliable value of $\sin\theta_C$.

We can conclude that the Cabibbo theory is consistent with the data to a precision of about 5%. Sizeable SU(3) breaking effects are predicted in some theoretical models but are not needed by the data, although 5% effects can be accommodated. An important further point is that the above discussion includes only three quarks (or four via the GIM mechanism). In the KM matrix for six quarks (5.193), the $u \to d$ transition is given by $c_1 = \cos\theta_C$ (i.e. as for the simple Cabibbo theory), but the $u \to s$ transition term is s_1c_3 (rather than s_1). In order to fit the four parameters (F, D, θ_1 and θ_3), the additional input of the results of the ft values (Section 5.1.2.1) from $0^+ \to 0^+$ superallowed Fermi transitions (e.g. ^{14}O, ^{26}Al, ^{34}Cl, ^{38}K, etc.) is needed. This gives, after appropriate radiative corrections (Sirlin and Zucchini, 1986), $c_1 = 0.9747 \pm 0.0011$. Imposing this constraint from beta decay, Gaillard and Sauvage (1984) obtain F and D values compatible with (8.71). From the above results we have

$$s_1 = 0.224 \pm 0.005, \qquad s_1 c_3 = 0.221 \pm 0.002. \tag{8.72}$$

Thus, θ_3 is small, and at the 90% c.l. $c_3 > 0.95$ or $|s_3| < 0.32$.

The semileptonic decays of the $J^P = \frac{3}{2}^+ \Omega^-$ have also been detected. Bourquin *et al.* (1984) find

$$R_\Omega = \frac{\Gamma(\Omega^- \to \Xi^0 e\nu)}{\Gamma(\Omega^- \to \text{all})} = (0.56 \pm 0.28) \times 10^{-2}, \tag{8.73}$$

a value which is roughly an order of magnitude larger than that for other $|\Delta S| = 1$ semileptonic hyperon decays. A relatively higher rate is expected from phase space considerations. The Cabibbo theory makes no predictions for this decuplet to octet decay. However, using the Ω^- lifetime, and relating this decay to $\nu + \text{N} \to l + \Delta$ by SU(3), an estimated value of $R_\Omega \sim 10^{-2}$ is obtained.

8.4 FURTHER PROPERTIES OF THE CURRENTS OF LIGHT QUARKS

8.4.1 Quasi-elastic neutrino scattering

The range of four-momentum transfer squared which can be studied from meson and hyperon decays is small ($q^2 \lesssim (350\text{ MeV})^2$ and is time-like). The study of the quasi-elastic reactions

$$\nu_l \text{n} \to l^- \text{p}, \qquad \bar{\nu}_l p \to l^+ \text{n} \qquad (l = \text{e}, \mu), \tag{8.74}$$

extends this range to the GeV region. Denoting the four-momenta as $\nu(p_1) + \text{N}(p_2) \to \text{l}(p_3) + \text{N}'(p_4)$, then $q^2 = (p_1 - p_3)^2 = (p_2 - p_4)^2$ is the space-like four-momentum transfer squared. The hadronic part of the matrix element for the reactions in (8.74) has the same form as that in equation (8.24). That is, there are in general six form factors, all of which can be q^2 dependent. The contributions from f_3 and g_2 are proportional to m_l, and hence can be neglected.

Using the CVC relations, (8.36), together with the expressions for the electric (G_E^p, G_E^n) and magnetic (G_M^p, G_M^n) form factors from elastic electron–nucleon scattering given by (7.14), and the dipole parameterisation (7.18)

$$\frac{G_\text{E}^\text{p}(Q^2)}{G_\text{E}^\text{p}(0)} = \frac{G_\text{M}^\text{p}(Q^2)}{G_\text{M}^\text{p}(0)} = \frac{G_\text{M}^\text{n}(Q^2)}{G_\text{M}^\text{n}(0)} = \frac{1}{(1 + Q^2/m_\text{V}^2)^2}, \tag{8.75}$$

with $m_V = Q_0 = 0.84$ GeV, then we obtain

$$f_1(q^2) = \left[(G_E^p - G_E^n) + \frac{Q^2}{4M^2}(G_M^p - G_M^n) \right] / (1 + Q^2/4M^2)$$

$$= \frac{[1 + Q^2/4M^2(1 + \mu_p - \mu_n)]}{(1 + Q^2/4M^2)(1 + Q^2/m_V^2)^2}, \tag{8.76}$$

and

$$f_2(q^2) = \frac{(\mu_p - \mu_n)}{(1 + Q^2/4M^2)(1 + Q^2/m_V^2)^2}. \tag{8.77}$$

In calculating (8.76) and (8.77) equation (7.15) has been used, and G_E^n is taken to be zero.

The axial form factor $g_1(q^2)$ is parameterised in a similar way, namely

$$g_1(q^2) = 1.25/(1 + Q^2/m_A^2)^2. \tag{8.78}$$

The value at $q^2 = 0$ is that determined from beta decay and m_A is a mass specifying the q^2 dependence of the axial form factor.

The differential cross-section for $E_\nu \gg m_l$, averaging over nucleon spins, is

$$\frac{d\sigma(^{\nu}_{\bar{\nu}})}{dQ^2} = \frac{G_F^2 \cos^2\theta_C}{8\pi E_\nu^2}\left\{A(Q^2) \pm B(Q^2)\left[\frac{s-u}{M^2}\right] + C(Q^2)\left[\frac{s-u}{M^2}\right]^2\right\}, \tag{8.79}$$

where

$$A(Q^2) = \frac{Q^2}{4}\left[f_1^2\left(\frac{Q^2}{M^2} - 4\right) + f_1 f_2\left(\frac{4Q^2}{M^2}\right) + f_2^2\left(\frac{Q^2}{M^2} - \frac{Q^4}{4M^4}\right) + g_1^2\left(4 + \frac{Q^2}{M^2}\right)\right],$$

$$B(Q^2) = (f_1 + f_2) g_1 Q^2,$$

$$C(Q^2) = \frac{M^2}{4}\left(f_1^2 + f_2^2 \frac{Q^2}{4M^2} + g_1^2 \right).$$

The signs + and − refer to ν and $\bar{\nu}$ respectively. In the high energy limit $(E_\nu \to \infty)$, this simplifies to

$$\frac{d\sigma}{dQ^2} = \frac{G_F^2 \cos^2\theta_C}{2\pi}\left[f_1^2(Q^2) + \frac{Q^2}{4M^2} f_2^2(Q^2) + g_1^2(Q^2)\right]. \tag{8.80}$$

With the above assumptions on the form factors there is just one parameter m_A. Experimentally it is found that $m_A \simeq 1.0$ GeV. Studies of Δ production in, e.g., $\nu_\mu p \to \mu^- \Delta^{++}$ can also be used to study m_A, yielding a roughly similar value.

8.4.2 Current algebra and the masses of light quarks

We have assumed throughout the discussion on quarks that 'free' quark fields obey the Dirac equation and hence their creation and annihilation operators satisfy anticommutation relations of the type (3.169). In terms of the column vector $q(x)$ for the u, d, s quark fields and the SU(3) matrices λ (2.251), the vector and axial currents can be written

$$V^j_\mu(x) = \bar{q}(x)\gamma_\mu \frac{\lambda_j}{2} q(x), \qquad A^j_\mu(x) = \bar{q}(x)\gamma_\mu\gamma^5 \frac{\lambda_j}{2} q(x). \tag{8.81}$$

Starting with (3.169), it can be shown that, for equal times,

$$\begin{aligned} &\{q_\alpha(t, \mathbf{x}_1), \bar{q}_\beta(t, \mathbf{x}_2)\} = \gamma^0_{\alpha\beta}\, \delta(\mathbf{x}_1 - \mathbf{x}_2), \\ &\{q_\alpha(t, \mathbf{x}_1), q_\beta(t, \mathbf{x}_2)\} = \{\bar{q}_\alpha(t, \mathbf{x}_1), \bar{q}_\beta(t, \mathbf{x}_2)\} = 0. \end{aligned} \tag{8.82}$$

Corresponding to these currents, 16-vector and axial-vector charges can be defined as follows

$$Q^j(t) = \int \mathrm{d}^3x\, V^j_0(t, \mathbf{x}), \qquad Q^j_5(t) = \int \mathrm{d}^3x\, A^j_0(t, \mathbf{x}). \tag{8.83}$$

These charges are independent of time if the corresponding current is conserved. Using the properties (8.82), this leads to the *equal time commutation relations*

$$\begin{aligned} &[Q^j(t), Q^k(t)] = \mathrm{i}f^{jkm}Q^m(t), \\ &[Q^j(t), Q^k_5(t)] = \mathrm{i}f^{jkm}Q^m_5(t), \\ &[Q^j_5(t), Q^k_5(t)] = \mathrm{i}f^{jkm}Q^m(t). \end{aligned} \tag{8.84}$$

Further, defining $Q^j_\pm = (Q^j \pm Q^j_5)/2$, then we obtain the commutation rules

$$\begin{aligned} &[Q^j_+, Q^k_+] = \mathrm{i}f^{jkm}Q^m_+, \\ &[Q^j_-, Q^k_-] = \mathrm{i}f^{jkm}Q^m_-, \\ &[Q^j_+, Q^k_-] = 0. \end{aligned} \tag{8.85}$$

The charges $Q_\pm$ correspond to right and lefthanded charges, and the (current) algebra generated is chiral SU(3) $\otimes$ SU(3). A mass term in the Lagrangian mixes left and righthanded helicity states, therefore chiral symmetry would be expected to be exact for massless quarks.

The axial current is not conserved, but its divergence is proportional to m_π^2 (8.42), and thus small on the hadronic scale. Comparing the hadronic decays involving u$\leftrightarrow$d and u$\leftrightarrow$s transitions we have (crudely)

$$\begin{aligned}
\langle\pi^0|\partial^\mu V_\mu(S=0)|\pi^-\rangle &\simeq 2^{1/2}(m_{\pi^-}^2 - m_{\pi^0}^2) \simeq 2^{1/2}(0.001\ \mathrm{GeV}^2),\\
\langle 0|\partial^\mu A_\mu(S=0)|\pi\rangle &\simeq f_\pi m_\pi^2 \qquad\qquad \simeq f_\pi(0.02\ \mathrm{GeV}^2),\\
\langle\pi|\partial^\mu V_\mu(S=1)|\mathrm{K}\rangle &\simeq m_\mathrm{K}^2 - m_\pi^2 \qquad\quad \simeq 0.22\ \mathrm{GeV}^2,\\
\langle 0|\partial^\mu A_\mu(S=1)|\mathrm{K}\rangle &\simeq f_\mathrm{K} m_\mathrm{K}^2 \qquad\qquad \simeq f_\mathrm{K}(0.25\ \mathrm{GeV}^2).
\end{aligned} \tag{8.86}$$

Now, if the chiral symmetry of the Lagrangian is broken by the quark mass terms (represented by a 3 × 3 diagonal matrix), then

$$\begin{aligned}
\partial^\mu V_\mu(S=0) &= (m_\mathrm{d} - m_\mathrm{u})\bar{u}d,\\
\partial^\mu A_\mu(S=0) &= (m_\mathrm{d} + m_\mathrm{u})\bar{u}\gamma^5 d,\\
\partial^\mu V_\mu(S=1) &= (m_\mathrm{s} - m_\mathrm{u})\bar{u}s,\\
\partial^\mu A_\mu(S=1) &= (m_\mathrm{s} + m_\mathrm{u})\bar{u}\gamma^5 s.
\end{aligned} \tag{8.87}$$

Taking $f_\pi \simeq f_\mathrm{K}$, and assuming $\langle 0|\bar{u}\gamma^5 d|\pi^-\rangle = \langle 0|\bar{u}\gamma^5 s|\mathrm{K}^-\rangle$ i.e. SU(3) invariance, we obtain from (8.86) and (8.87)

$$\frac{m_\mathrm{u} + m_\mathrm{d}}{m_\mathrm{u} + m_\mathrm{s}} \simeq \frac{m_\pi^2}{m_\mathrm{K}^2} \simeq \frac{0.02}{0.25}. \tag{8.88}$$

For $m_\mathrm{u} \simeq m_\mathrm{d}$ (isospin invariance), this gives $m_\mathrm{s} \simeq 25 m_\mathrm{d}$. Detailed consideration of the masses of the lightest pseudoscalar mesons gives $m_\mathrm{u}/m_\mathrm{d} \simeq 0.55$. This ratio, together with a QCD estimate (Kremer *et al.*, 1983) that $m_\mathrm{u} + m_\mathrm{d} \simeq 20$ MeV at $Q^2 = 1\ \mathrm{GeV}^2$, gives $m_\mathrm{u} \simeq 7$ MeV, $m_\mathrm{d} \simeq 13$ MeV, and $m_\mathrm{s} \simeq 240$ MeV. Alternatively, using (8.88) together with $m_\mathrm{d} - m_\mathrm{u} \simeq 3$ MeV (from the n − p and other mass differences in isospin multiplets), we find $m_\mathrm{u} \simeq 4$, $m_\mathrm{d} \simeq 7$ and $m_\mathrm{s} \simeq 125$ MeV. Note that these (*current*) quark masses are somewhat different to those obtained from the non-relativistic quark model, namely $m_\mathrm{u} \simeq m_\mathrm{d} \simeq M/3 \simeq 300$ MeV and $m_\mathrm{s} = M_\Lambda - 2m_\mathrm{u} \simeq 500$ MeV. Such a *constituent* quark mass may represent some effective mass, which includes the effects of a gluon (and q$\bar{\mathrm{q}}$) cloud surrounding the quark. For a detailed discussion see Gasser and Leutwyler (1982).

The hadronic tensor $W^{\mu\nu}$ for deep inelastic scattering can be expressed in terms of current commutators (see (7.39)). The non-linear relationships (8.85) define the current normalisation, and can be used to relate the cross-sections (quadratic in the matrix element) to the current matrix elements. An example of a relationship derived in this way is the *Adler sum rule* (Adler, 1966), namely

$$\lim_{E_\nu\to\infty}\left(\frac{\mathrm{d}\sigma^{\bar{\nu}T}}{\mathrm{d}Q^2} - \frac{\mathrm{d}\sigma^{\nu T}}{\mathrm{d}Q^2}\right) = \frac{2G_\mathrm{F}^2}{\pi}\langle I_3\rangle_T, \tag{8.89}$$

where T is any target. That is, the difference in the cross-section for ν and $\bar{\nu}$ is independent of Q^2, and depends only on the target isospin. This rule was derived using the QPM in Section 7.5.

8.5 HADRONIC WEAK DECAYS

The simplest purely hadronic weak decays are of the form $M_1 \to M_2 + \pi$ or $B_1 \to B_2 + \pi$, where the mesons M_1 and M_2 and the baryons B_1 and B_2 have different flavour quantum numbers. The main weak hadronic decay modes for hadrons composed of light quarks (with respective branching ratios) are

$K^+ \to \pi^+\pi^0(21.2\%)$, $K^+ \to (3\pi)^+(7.3\%)$, $K^0_s \to (2\pi)(\sim 100\%)$, $K^0_L \to 3\pi\,(34\%)$,

$\Lambda \to p\pi^-(64.2\%)$, $\Lambda \to n\pi^0(35.8\%)$,

$\Sigma^+ \to p\pi^0(51.6\%)$, $\Sigma^+ \to n\pi^+(48.4\%)$, $\Sigma^- \to n\pi^-(\sim 100\%)$,

$\Xi^0 \to \Lambda\pi^0(\sim 100\%)$, $\Xi^- \to \Lambda\pi^-(\sim 100\%)$,

$\Omega^- \to \Xi\pi(32\%)$, $\Omega^- \to \Lambda K^-(68\%)$.

Thus the dominant decays of strange baryons are to two-body hadronic final states.

Possibly decay schemes at the quark level for $\Sigma^+ \to p\pi^0$ are shown in Fig. 8.4. Note that only quark colour configurations which lead to colourless hadrons are allowed. In terms of the Lagrangian (8.56), the currents involved are $h_\mu h_\mu^\dagger$ and $h_\mu^\dagger h_\mu$. At the quark level the possible transitions are $(\bar{u}s)(\bar{d}u)$ and $(\bar{s}u)(\bar{u}d)$. The applicability of this Lagrangian, or of the underlying quark picture in this and other decays, implies that $|\Delta S| \geqslant 2$ decays are forbidden. Experimentally this expectation is compatible

Fig. 8.4 Possible diagrams for the decay $\Sigma^+ \to p\pi^0$.

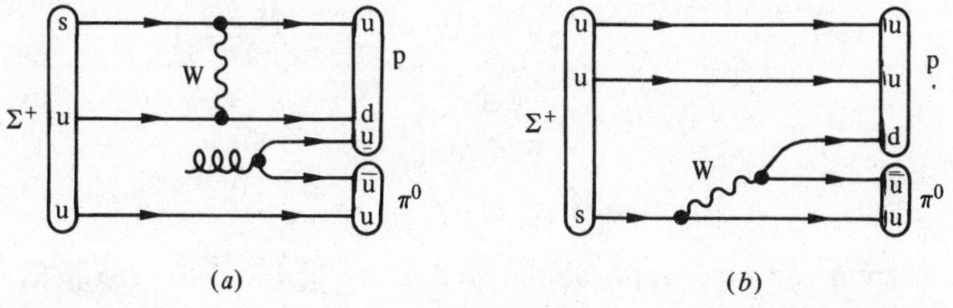

with the data. For example, the Particle Data Group (1986) quote (at 90% c.l.)

$$\Gamma(\Xi^- \to n\pi^-)/\Gamma(\Xi^- \to \Lambda\pi^-) < 1.9 \times 10^{-5},$$
$$\Gamma(\Xi^0 \to p\pi^-)/\Gamma(\Xi^0 \to \Lambda^0\pi^0) < 3.6 \times 10^{-5}, \tag{8.90}$$
$$\Gamma(\Omega^- \to \Lambda\pi^-)/\Gamma(\Omega^- \to \text{all}) < 1.9 \times 10^{-4}.$$

A further limit comes from the $K_L^0 - K_s^0$ mass difference (Chapter 9). Experimentally it is found that $\Delta m \simeq 0.5\Gamma_s$, where Γ_s is the K_s^0 decay width. The mass difference corresponds to a $|\Delta S| = 2$ transition, and the observed value is $\sim 10^{-5}$ of that expected if this was a first-order transition, i.e. the value $\Gamma \sim G_F^2$ is a second-order effect.

The main difficulty in the theoretical analysis of purely hadronic decays is that the matrix element (at the hadron level) cannot, in general, be factorised into the product of two currents which match what happens at the quark level. Hence, we must construct the decay matrix element from the available quantities. The most general decay amplitude for the hyperon decay $B_1 \to B_2 + \pi$, where B_1 and B_2 both have spin $\frac{1}{2}$, is

$$\mathcal{M}_{fi} = \bar{u}_2(A - B\gamma^5)u_1, \tag{8.91}$$

where A and B are constants, predictable, in principle, from the standard model plus QCD. Other potential terms such as $\bar{u}_2 \not{p}_\pi u_1$ and $\bar{u}_2 \not{p}_\pi \gamma^5 u_1$ can be reduced to the form of (8.91) by application of the Dirac equation. Since the pion is pseudoscalar, the term A is also pseudoscalar, whereas B (which has an extra γ^5 factor) is scalar. The terms A and B thus represent $l = 0$ (s-wave) and $l = 1$ (p-wave) angular momentum states respectively.

In the rest frame of B_1 we can write the baryon spinors as

$$B_1 = \begin{pmatrix} \chi_1 \\ 0 \end{pmatrix}, \qquad B_2 = (E_2 + m_2)^{1/2} \begin{pmatrix} \chi_2 \\ \dfrac{\boldsymbol{\sigma}\cdot\mathbf{p}_2}{E_2 + m_2}\chi_2 \end{pmatrix}, \tag{8.92}$$

where χ_1 and χ_2 are two-component spinors. Putting $\bar{u}_2 = u_2^\dagger \gamma^0$, (8.91) can be written

$$\begin{aligned} \mathcal{M}_{fi} &\propto \left(\chi_2^\dagger \quad -\chi_2^\dagger \frac{\boldsymbol{\sigma}\cdot\mathbf{p}_2}{E_2 + m_2}\right)\begin{pmatrix} A & -B \\ -B & A \end{pmatrix}\begin{pmatrix} \chi_1 \\ 0 \end{pmatrix} \\ &\propto A\chi_2^\dagger\chi_1 + B\chi_2^\dagger \frac{\boldsymbol{\sigma}\cdot\mathbf{p}_2}{E_2 + m_2}\chi_1 \\ &\propto \chi_2^\dagger(s + p\boldsymbol{\sigma}\cdot\hat{\mathbf{n}})\chi_1, \end{aligned} \tag{8.93}$$

where $\hat{\mathbf{n}}$ is a unit vector along particle 2, $s = A$ and $p = B|\mathbf{p}_2|/(E_2 + m_2)$.

Table 8.2. *Values of decay parameters for the hadronic decays of hyperons*

Decay	α	ϕ (deg)	γ
$\Lambda \to p\pi^-$	0.642 ± 0.013	-6.5 ± 3.5	0.76
$\Lambda \to n\pi^0$	0.646 ± 0.044		
$\Sigma^+ \to p\pi^0$	-0.980 ± 0.015	36 ± 34	0.16
$\Sigma^+ \to n\pi^+$	0.068 ± 0.013	167 ± 20	-0.97
$\Sigma^- \to n\pi^-$	-0.068 ± 0.008	10 ± 15	0.98
$\Xi^0 \to \Lambda\pi^0$	-0.413 ± 0.022	21 ± 12	0.85
$\Xi^- \to \Lambda\pi^-$	-0.455 ± 0.015	4 ± 5	0.89

If $\hat{\mathbf{s}}_1$ and $\hat{\mathbf{s}}_2$ are unit vectors along the baryon spins, then the transition probability (i.e. $|\mathscr{M}_{\text{fi}}|^2$) can be reduced in the usual way to

$$R = 1 + \alpha(\hat{\mathbf{s}}_1 + \hat{\mathbf{s}}_2)\cdot\hat{\mathbf{n}} + \beta(\hat{\mathbf{s}}_2 \times \hat{\mathbf{s}}_1)\cdot\hat{\mathbf{n}} + \gamma\hat{\mathbf{s}}_2\cdot\hat{\mathbf{s}}_1 + (1-\gamma)(\hat{\mathbf{s}}_2\cdot\hat{\mathbf{n}})(\hat{\mathbf{s}}_1\cdot\hat{\mathbf{n}}), \quad (8.94)$$

where $\alpha = 2\,\text{Re}(sp^*)/N$, $\beta = 2\,\text{Im}(sp^*)/N$ and $\gamma = (|s|^2 - |p|^2)/N$ with $N = |s|^2 + |p|^2$ so that $\alpha^2 + \beta^2 + \gamma^2 = 1$. Hence two parameters suffice, and these are generally chosen to be α and the angle ϕ, where $\beta = (1-\alpha^2)^{1/2}\sin\phi$ and $\gamma = (1-\alpha^2)^{1/2}\cos\phi$. Invariance under time reversal implies $\beta = 0$, i.e. $\phi = 0$ or π. Simplifying cases of (8.94) are (i) with B_1 unpolarised, so that

$$R \simeq 1 + \alpha\hat{\mathbf{s}}_2\cdot\hat{\mathbf{n}}, \quad (8.95)$$

or (ii) with B_1 polarised and summing over polarisation states of B_2, giving

$$R \simeq 1 + \alpha\hat{\mathbf{s}}_1\cdot\hat{\mathbf{n}}. \quad (8.96)$$

The experimental values of the parameters α, ϕ and γ are given in Table 8.2. Note that the observation of values of α different from zero means that both s-wave (s) and p-wave (p) amplitudes are present, so that parity is not conserved in these decays. For Λ decay, $|p|/|s| \sim 0.4$. The values of α may give some insight into which quark mechanisms are dominant. A useful aspect of the hyperon polarisation is that it can be used as a tool to study the production mechanisms in the strong, electromagnetic or weak interactions in which the hyperon is produced. For example, in the reaction $\pi^- p \to \Lambda K^0$, the Λ particles are strongly polarised perpendicular to the reaction plane. At higher energies the measurement of the Λ polarisation is one of the few handles available in studying the role of quark spin effects in the hadronisation process.

At the quark level, the Lagrangian contains the currents $(\bar{s}u)(\bar{u}d)$. Thus, in terms of isospin, the interaction has $|\Delta I_3| = \frac{1}{2}$ and $|\Delta I| = \frac{1}{2}$ or $\frac{3}{2}$. It is empirically observed that the $|\Delta I| = \frac{1}{2}$ transitions dominate ($|\Delta I| = \frac{1}{2}$ *rule*).

The predictions of the $|\Delta I| = \frac{1}{2}$ rule can be conveniently calculated by adding a fictitious $I = \frac{1}{2}$, $I_3 = -\frac{1}{2}$ particle so as to make the decay isospin invariant. Then the relevant decay amplitudes are just the ratios of Clebsch–Gordan coefficients. For example,

$$\frac{\mathscr{M}(\Lambda \to \mathrm{p}\pi^-)}{\mathscr{M}(\Lambda \to \mathrm{n}\pi^0)} = \frac{\langle \frac{1}{2}, \frac{1}{2}; 1, -1 | \frac{1}{2}, -\frac{1}{2} \rangle}{\langle \frac{1}{2}, -\frac{1}{2}; 1, 0 | \frac{1}{2}, -\frac{1}{2} \rangle}, \tag{8.97}$$

predicting a decay branching ratio

$$R = \Gamma(\Lambda \to \mathrm{p}\pi^-)/\Gamma(\Lambda \to \mathrm{n}\pi^0) = 2. \tag{8.98}$$

Experimentally, this ratio is $R = 1.8$, showing that the rule is not exact. Since perturbative QCD is not strictly valid at the momentum scales involved in hyperon decay, the understanding of the suppression of $|\Delta I| = \frac{3}{2}$ transitions is clearly difficult. However, QCD colour considerations suggest that $|\Delta I| = \frac{3}{2}$ transitions are suppressed (Shifman *et al.*, 1977).

The $|\Delta I| = \frac{1}{2}$ rule also holds to a good approximation for the hadronic decays of K mesons. Since the K has spin zero, the relative angular momentum of the two pions in the decay $\mathrm{K} \to \pi\pi$ will be $l = 0$. Pions are bosons so, in order to have overall symmetrisation, the possible isospin states are $I = 0$ or 2. The $|\Delta I| = \frac{1}{2}$ rule would then imply that only $I = 0$ is possible. In this case, we would expect the decay branching ratio

$$R = \Gamma(\mathrm{K}^0_\mathrm{s} \to \pi^+\pi^-)/\Gamma(\mathrm{K}^0_\mathrm{s} \to \pi^0\pi^0) = 2. \tag{8.99}$$

Experimentally, this branching ratio is 2.19.

The decay $\mathrm{K}^+ \to \pi^+\pi^0$ (which accounts for 21% of all K^+ decays) has $I_3 = 1$, and hence $I = 2$, and so violates the $|\Delta I| = \frac{1}{2}$ rule. However, the decay $\mathrm{K}^+ \to \pi^+\pi^0$ is slow compared to $\mathrm{K}^0_\mathrm{s} \to \pi^+\pi^-$. Defining decay amplitudes corresponding to $I = 0$ and $I = 2$ as

$$A_0 = \langle 2\pi_{I=0} | H_{\mathrm{wk}} | \mathrm{K}^0 \rangle, \qquad A_2 = \langle 2\pi_{I=2} | H_{\mathrm{wk}} | \mathrm{K}^0 \rangle, \tag{8.100}$$

then, from the measured value of

$$\Gamma(\mathrm{K}^+ \to \pi^+\pi^0)/[\Gamma(\mathrm{K}^0_\mathrm{s} \to \pi^+\pi^-) + \Gamma(\mathrm{K}^0_\mathrm{s} \to \pi^0\pi^0)],$$

we find

$$|A_2/A_0| \simeq 0.04. \tag{8.101}$$

The simplest quark graphs for $\mathrm{K}^+ \to \pi^+\pi^0$ are shown in Fig. 8.2. The *annihilation* graphs ((*c*) and (*d*)) have π^0 mesons produced with amplitudes of opposite sign ($\mathrm{d}\bar{\mathrm{d}}$ in (*c*), and $\mathrm{u}\bar{\mathrm{u}}$ in (*d*)). The annihilation graphs are thus expected to play essentially no role, leaving the *spectator* graph (*e*) as the simplest quark picture.

8.6 DECAYS OF HADRONS CONTAINING HEAVY QUARKS

The study of the short-lived ($\tau \sim 10^{-13}$ s) decays of charm and bottom/beauty flavoured hadrons has been one of great experimental ingenuity. From the weak interaction point of view, the main parameters to be extracted are the KM terms V_{cd}, V_{cs}, V_{cb}, V_{ub} and V_{tb}. The couplings to a possible top quark must, at present, be found by indirect means.

8.6.1 Decays of charmed particles

The particles of interest in the study of the weak decays of charmed hadrons are the $C' = 1$, $J^P = 0^-$ mesons $D^+(c\bar{d})$, $D^0(c\bar{u})$, and D_s^+ [or F^+] $(c\bar{s})$ and their $C' = -1$ antiparticles $D^-(\bar{c}d)$, $\bar{D}^0(\bar{c}u)$ and D_s^- [or F^-] $(\bar{c}s)$. Some properties of these mesons, and of the $J^P = \frac{1}{2}^+$ baryon $\Lambda_c(udc)$, are given in Table 8.3. These particles have a large number of possible decay modes, with branching ratios to individual channels of typically only a few per cent. Because of their short lifetimes, detectors with good vertex resolution (e.g. emulsion, bubble chambers, silicon strip detectors) are needed for direct study. Most experiments have used hadron beams ($\sigma_{charm}/\sigma_{tot} \sim 0.1\%$) or photon beams ($\sigma_{charm}/\sigma_{tot} \sim 1\%$). Neutrino beams offer a high charm content ($\sigma_{charm}/\sigma_{tot} \sim 5\%$), but have a large spatial spread, making practical experiments more difficult.

An intermediate technique in determining lifetimes is to use the so-called *impact parameter method* (see Fig. 8.5). This method does not necessarily require observation of the decay point, but relies on the fact that, on average, the absolute value of the impact parameter δ is non-zero for finite decay lengths. The method is not strongly dependent on the momentum of the decaying particle, but requires very good spatial resolution (e.g. 10–30 μm). This method has been used extensively in e^+e^- annihilation, in which there is a high cross-section ($\sim 40\%$) for charm production, and from which many of the measurements of decay branching ratios have come. A further technique which has proven useful in the selection of c-quark jets is to identify the $D^* \to D\pi$ decay sequence, which has a large branching ratio. By forming all appropriate invariant mass combinations, the constraint that the $D^* - D$ mass difference is unique (146 MeV) can be used to efficiently 'tag' c-quarks.

In the standard model the basic diagram leading to c-quark decay is simple, and is shown in Fig. 8.6. In terms of the Cabibbo theory we have

$$c \to q \begin{cases} \cos\theta_C & (q = s), \\ -\sin\theta_C & (q = d), \end{cases} \qquad W^+ \to ab \begin{cases} 1 & (e^+\nu_e, \mu^+\nu_\mu, \tau^+\nu_\tau), \\ \cos\theta_C & u\bar{d}, \\ \sin\theta_C & u\bar{s}. \end{cases} \tag{8.102}$$

Table 8.3. *Some properties of the lightest charmed particles: X represents any other particles produced and cc means charge conjugate*

Particle	Decay mode		%
D^+ (D^-)	e^+X	D^+ or	18.2 ± 1.7
	K^-X	$D^- \to$ cc	16 ± 4
$m = 1869.3 \pm 0.6$ MeV	K^+X		6.0 ± 3.3
$\tau = (9.01 \pm 0.75) \times 10^{-13}$ *s*	$\bar{K}^0X$ and K^0X		48 ± 15
	ηX		<13
	$e^+\nu$		<2.5
	$\mu^+\nu$		<0.084
	$\pi^+\pi^0$		<0.5
	$\pi^+\pi^+\pi^0$		0.5 ± 0.2
D^0 ($\bar{D}^0$)	e^+X	D^0 or	7.0 ± 1.1
	K^-X	$\bar{D}^0 \to$ cc	44 ± 10
$m = 1864.6 \pm 0.6$ MeV	K^+X		8 ± 3
	K^0X and $\bar{K}^0X$		33 ± 10
$\tau = (4.35 \pm 0.32) \times 10^{-13}$ s	ηX		<13
	$\pi^+\pi^-$		0.18 ± 0.05
	$\pi^+\pi^-\pi^0$		1.1 ± 0.4
	$\pi^+\pi^+\pi^-\pi^-$		1.5 ± 0.6
D_s^+ (D_s^-) (formerly $F^\pm$)	$\phi\pi$	D_s^+ or	seen
	$\phi\pi^+\pi^+\pi^-$	$D_s^- \to$ cc	seen
$m = 1970.5 \pm 2.5$ MeV			
$\tau = (2.8 \pm ^{1.6}_{0.7}) \times 10^{-13}$ s			
Λ_c^+	e^+X		4.5 ± 1.7
	ΛX		33 ± 29
$m = 2281.2 \pm 3.0$ MeV	$pK^-\pi^+$		2.2 ± 1.0
	$p\bar{K}^0$		1.1 ± 0.7
$\tau = (2.3 \pm ^{0.8}_{0.5}) \times 10^{-13}$ s			

Fig. 8.5 Impact parameter δ used to estimate the lifetime of short-lived particles.

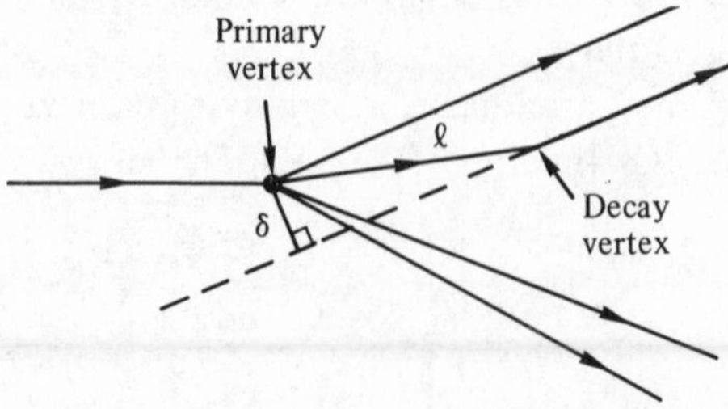

Possible leptonic and non-leptonic decay modes are given in Table 8.4. At the quark level, the (leading order) decay spectra and total rate have the same form as those of muon decay (Section 6.1). Neglecting the Cabibbo unfavoured modes (i.e. in limit $\sin^2\theta_C \sim 0$, $\cos^2\theta_C \sim 1$) and, in a cavalier fashion, any QCD corrections to Fig. 8.6, then we find that

$$\omega(c \to su\bar{d})/\omega(c \to se^+\nu_e)/\omega(c \to s\mu^+\nu_\mu)/\omega(c \to s\tau^+\nu_\tau) = 3/1/1/1, \quad (8.103)$$

where the factor of 3 comes from colour. The decay modes involving τ^+ will, however, be suppressed by phase space (see (6.25)). For D^+, the e^+X branching ratio (Table 8.3) is about the magnitude expected from (8.103), but $D^0 \to e^+X$ is considerably smaller. In the discussion below, we concentrate on D^0 and D^+ decays because of the paucity of data for other charmed particles.

The simplest D decay modes can be calculated in a similar way to those for π and K decay. The decay rate for $D^+ \to l^+\nu_l$ is (see (8.6))

$$\omega(D^+ \to l^+\nu_l) = \frac{G_F^2 \sin\theta_C f_D^2 m_l^2 m_D}{8\pi}\left(1 - \frac{m_l^2}{m_D^2}\right)^2. \quad (8.104)$$

Table 8.4. *Classification of c-quark decays*

Decay	Cabibbo factor	Examples
Leptonic		
$c \to sl^+\nu_l$	$\cos^2\theta_C$	$D^+ \to \bar{K}^0 l^+\nu_l$, $D^0 \to K^- l^+\nu_l$,
$(\Delta C = \Delta S)$		$D_s^+ \to l^+\nu_l$
$c \to dl^+\nu_l$	$\sin^2\theta_C$	$D^+ \to l^+\nu_l$, $D_s^+ \to K^0 l^+\nu_l$
$(\Delta S = 0)$		
Non-leptonic		
$c \to su\bar{d}$	$\cos^4\theta_C$	$D^+ \to \bar{K}^0(n\pi)^+$, $D^0 \to K^-(n\pi)^+$,
$(\Delta C = \Delta S)$		$\bar{K}^0(n\pi)^0$, $D_s^+ \to \eta\pi^+$
$c \to du\bar{d}$, $su\bar{s}$	$\sin^2\theta_C \cos^2\theta_C$	$D \to n\pi$, $D_s^+ \to K\pi$
$(\Delta S = 0)$		
$c \to du\bar{s}$	$\sin^4\theta_C$	$D^+ \to K^+(n\pi)^0$, $K^0(n\pi)^+$, $D^0 \to K^+(n\pi)^-$,
$(\Delta C = \Delta S)$		$K^0(n\pi)^0$, $D_s^+ \to K^+K^0$

Fig. 8.6 Lowest order diagram for c-quark decay.

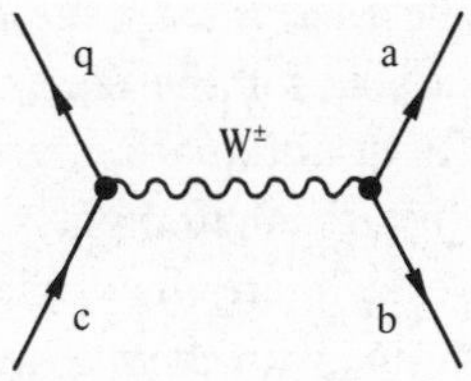

From (8.104), it follows that the leptonic decay rates are in the ratio $e/\mu/\tau = 10^{-5}/0.44/1$. Using the 90% c.l. upper limit of 8.4×10^{-4} for the $D^+ \to \mu^+ \nu_\mu$ decay mode, and comparing (8.104) with (8.6), we find (putting in values for the appropriate constants) that $f_D < 340$ MeV. This is not yet, however, a very stringent limit, since most models have $f_D \sim f_\pi$.

The decay $D_s^+(F^+) \to l^+ \nu_l$ is Cabibbo-favoured, with a resulting decay rate

$$\omega(D_s^+ \to l^+ \nu_l) = \frac{G_F^2 \cos^2 \theta_C f_{D_s}^2 m_l^2 m_{D_s}}{8\pi}\left(1 - \frac{m_l^2}{m_{D_s}^2}\right)^2. \qquad (8.105)$$

At the quark level, the simplest diagram is the annihilation of c and $\bar{s}$ (i.e. similar to Fig. 8.1). From (8.105), the relative branching ratios are $e/\mu/\tau = 2.5 \times 10^{-6}/0.11/1$. Hence, the decay $D_s^+ \to \tau\nu_\tau$ is strongly favoured and this mode offers a possible source by which direct observation of ν_τ might be made. Using the measured D_s^+ lifetime and (8.105), we find that the expected decay branching ratio is

$$\omega(D_s^+ \to \tau^+ \nu_\tau)/\omega(D_s^+ \to \text{all}) \simeq 0.45 f_{D_s}^2, \qquad (8.106)$$

where f_{D_s} is in GeV. The decay constant f_{D_s} is related to the overlap integral of c and $\bar{s}$ and, hence, to the size of the hadron. If the D_s is similar in size to the π, then $f_{D_s} \simeq f_\pi$, and the branching ratio is rather small ($\sim$0.8%). A value $f_{D_s} \sim 0.5$ GeV, however, would give a branching ratio of 11%.

A crude estimate of the magnitude of the lifetimes of charmed particles can be made using equation (6.24), yielding

$$\Gamma_{tot} \simeq (5\text{–}6)\frac{G_F^2 m_c^5}{192\pi^3} \Rightarrow \tau = \frac{1}{\Gamma} \simeq 7 \times 10^{-13}\ \text{s}. \qquad (8.107)$$

Here the quark mass is taken to be $m_c = 1.5$ GeV. The factor 5–6 is the number of possible decay modes at the quark level (8.103), the τ-mode being suppressed due to phase space (see 6.25). This crude calculation gives roughly the correct lifetimes, as can be seen from Table 8.3. Note that the $D^\pm$ and D^0 lifetimes seem, however, to be substantially different, with $\tau(D^\pm)/\tau(D^0) \simeq 2.1$. However, the leptonic decay rates, namely $\Gamma(D^+ \to e^+ \nu_e X) = (2.0 \pm 0.2) \times 10^{11}\ s^{-1}$ and $\Gamma(D^0 \to e^+ \nu_e X) = (1.6 \pm 0.3) \times 10^{11}\ s^{-1}$, are almost equal. Hence the difference must arise from the hadronic decay modes. Some possible diagrams for the decay of a $Q\bar{q}$ meson, where Q is a heavy quark and $\bar{q}$ a light antiquark, are shown in Fig. 8.7. In the *spectator* diagrams ((*a*) and (*b*)), the light $\bar{q}$ plays no direct role. In (*b*), the colours of the W decay products are the same as those of the initial Q, hence the spectator can combine with them. Naively, this diagram is suppressed by a factor 3 (i.e. 9 in the rate). The *W exchange* diagram (Fig. 8.7(c)) is possible for a neutral heavy meson (e.g. D^0), but forbidden for a charged heavy meson (e.g. D^+). The *annihilation* diagram is possible if the initial heavy meson can fluctuate directly to a virtual W, which subsequently decays (e.g. $D_s^+ = c\bar{s}$, $D^+ = c\bar{d}$ (Cabibbo-suppressed)).

The spectator diagrams correspond to the 'decay' of the heavy quark (i.e. analogous to muon decay), and this occurs at a rate proportional to m_Q^5. The hadronisation effects might be expected to become negligible if m_Q is much larger than a typical hadron mass. The non-spectator diagrams involve a factor $\psi(0)$ (which is related to f_Q), corresponding to the amplitude for finding a Q and $\bar{q}$ at the same point. If the hadronic radius is not strongly dependent on the quark mass then $|\Psi(0)|^2 \ll m_Q^3$ so that the spectator diagrams should dominate for heavy quarks.

If the only diagram operative was the spectator diagram, and there was no colour mixing, then equality of the hadronic decay widths of D^+ and D^0 would be expected. However, the hadronic width of the D^0 exceeds that of the D^+ by a factor of about 2.7. Furthermore, the D^0 decays $\bar{K}^0 + \pi^0(4.0 \pm 1.8)\%$, $\bar{K}^0 + \eta, \rho^0, \omega(6.9 \pm 2.1)\%$ and $\bar{K}^{*0} + \pi^0(2.3 \pm 2.1)\%$ cannot occur by the simple quark picture of Fig. 8.7(a) (although they clearly can occur by a more complex rearrangement of quark and gluon lines). The decay mode $D^0 \to \bar{K}^0\phi$ $(1.5 \pm 0.4)\%$ is most easily explained by the existence of an exchange diagram of the type shown in Fig. 8.7(c). Indeed, any mode without a $\bar{u}$ in the final state (e.g. $\bar{K}^0K^0$) would be best explained by this diagram. Hence, this gives some evidence for the existence of the W exchange graph.

Fig. 8.7 Possible diagrams for the decay of a heavy $Q\bar{q}$ meson: (a) spectator, (b) spectator with colour mixing, (c) W exchange and (d) annihilation.

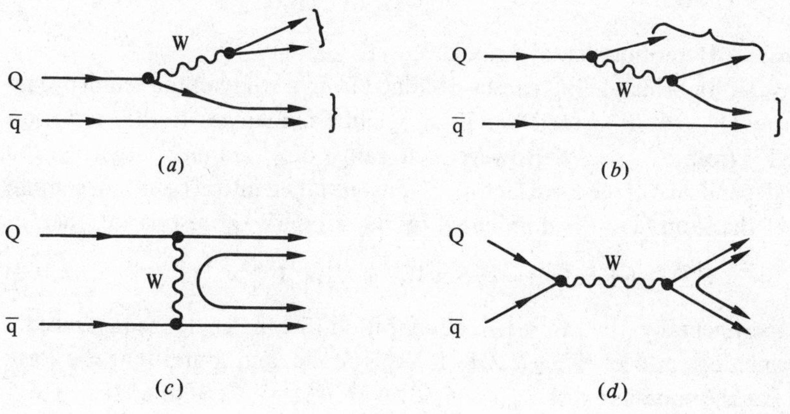

Two mechanisms have been proposed to explain the hadronic D width inequality. The first explanation relies on the observation that, for D^0 decay, Figs. 8.7(*a*) and (*b*) lead to different quark final states (and hence incoherence), whereas they lead to the same quark states for D^+ decay. If there is destructive interference, then the D^+ mode could be suppressed. Thorndike (1985) considers that this mechanism is probably not significant, although the apparently large branching ratio for $D^+ \to \phi\pi^+ (1.0 \pm 0.3\%)$, for which the simplest mechanism is graph (*b*), but is also Cabibbo-suppressed, indicates perhaps that colour suppression is not operative. In the second explanation, the difference in widths is attributed to the contribution of the exchange diagrams for D^0 decay. However, this diagram is expected to be suppressed from helicity considerations. This supression can be overcome by invoking gluons, either in the meson wave function or radiated by one of the quarks, so that the $Q\bar{q}$ system is no longer in a 0^- state. Hence, no clear explanation emerges for the observed width inequality. Perhaps the safest conclusion is that, for a quark mass ~ 1.5 GeV, these simple quark diagrams are too naive and that the multitude of higher order QCD graphs cannot be ignored.

The branching ratios for some of the Cabibbo-suppressed modes (e.g. $D^0 \to K^+K^- (0.64 \pm 0.11)\%$, $D^0 \to \pi^+\pi^- (0.18 \pm 0.05)\%$, $D^0 \to \pi^+\pi^-\pi^0 (1.1 \pm 0.4)\%$, and $D^0 \to \pi^+\pi^+\pi^-\pi^- (1.5 \pm 0.6)\%$) are again rather large in terms of the naive expectations. Note that $D^0 \to \pi^+\pi^-$ is smaller than $D^0 \to K^+K^-$, although phase space ($\propto p^*$) favours the pion mode by a factor of 1.17. However, this does not necessarily violate SU(3) symmetry at the quark level, because the extra phase space available in the pion mode could enhance the production of extra $q\bar{q}$ pairs, giving additional final state particles. Indeed, the 3π and 4π modes are significant. More details on the discussion of charm decays can be found in the review of Thorndike (1985).

8.6.2 Determination of V_{cs} and V_{cd}

In principle, V_{cs} can be obtained from a study of the semileptonic decay $D^+ \to \bar{K}^0 e^+ \nu_e$ (or $D^0 \to K^- e^+ \nu_e$), in a similar way to the extraction of V_{us} from K_{e3} decay. However, the range of q^2 is much larger, so the q^2 dependence of the form factor f_+^D must be taken into account. Assuming that the form factor is dominated by the 2.1 GeV D_s^* resonance, then

$$\Gamma(D^+ \to \bar{K}^0 e^+ \nu_e) = (1.5 \times 10^{11}\ \mathrm{s}^{-1}) \times |V_{cs}|^2 \times |f_+^D(0)|^2. \quad (8.108)$$

Experimentally, the Mark III collaboration (Schindler, 1985) measured a branching ratio of $(9.3 \pm 2.2)\%$ for this mode, and found that the form factor is compatible with $f_+(q^2)/f_+(0) = M^2/(M^2 - q^2)$, with $M \sim 2.1$ GeV.

Using this, together with the D^+ lifetime, gives

$$f_+^D(0)|V_{cs}| = 0.83 \pm 0.12. \quad (8.109)$$

Thus, if SU(4) were exact, then $|V_{cs}| = 0.83 \pm 0.12$. However, a calculation taking SU(4) breaking into account (Aliev *et al.*, 1984) gives $f_+^D(0) = 0.6 \pm 0.1$, leading to a value of $|V_{cs}|$ larger than, but consistent with, unity.

Both of the elements $|V_{cs}|$ and $|V_{cd}|$ can be obtained from an analysis of dimuon events in $\nu/\bar{\nu}$ deep inelastic scattering off nucleons. The general formulae for the production of charm in $\nu/\bar{\nu}$N scattering are those given by (7.109), with $\sin^2\theta_C$ and $\cos^2\theta_C$ replaced by $|V_{cd}|^2$ and $|V_{cs}|^2$ respectively. The relevant decay channels for dimuon studies are $c \to d(s)\mu^+\nu_\mu$ for νN and $\bar{c} \to \bar{d}(\bar{s})\mu^-\bar{\nu}_\mu$ for $\bar{\nu}$N. Thus, the final state in each case is a pair of oppositely charged muons. The observed yield thus depends on the weighted average of the branching ratios (B) of the produced charmed hadrons (known to be mainly D mesons, with $D^+ \simeq 32 \pm 11\%$, $D^0 \simeq 68 \pm 11\%$, together with some Λ_c (from the results of emulsion and bubble chamber experiments)), and on the muon detection efficiency for the decay.

Integrating over x and y, we obtain the following equations for the production of dimuons (σ^{ν}_{+-} and $\sigma^{\bar{\nu}}_{+-}$) and for all charged currents events (σ^{ν} and $\sigma^{\bar{\nu}}$)

$$\sigma^{\nu}_{+-} = KB[|V_{cd}|^2(U + D) + |V_{cs}|^2 2S], \quad (8.110a)$$

$$\sigma^{\bar{\nu}}_{+-} = KB[|V_{cd}|^2(\bar{U} + \bar{D}) + |V_{cs}|^2 2\bar{S}], \quad (8.110b)$$

$$\sigma^{\nu} = K[(U + D + 2S) + \tfrac{1}{3}(\bar{U} + \bar{D})], \quad (8.110c)$$

$$\sigma^{\bar{\nu}} = K[\tfrac{1}{3}(U + D) + (\bar{U} + \bar{D} + 2\bar{S})], \quad (8.110d)$$

where equations (7.107), (7.108) and (7.109) have been used, and where $K = G^2ME/\pi$. The quantities U, D, S $(= \bar{S})$ represent the integrals (over all x) of $xu(x)$, $xd(x)$ and $xs(x)$ respectively. Defining $r_{CC} = \sigma^{\bar{\nu}}/\sigma^{\nu}$, we can solve equations (8.110) for $|V_{cd}|^2$, giving

$$B|V_{cd}|^2 = \frac{2}{3}\left(\frac{\sigma^{\nu}_{+-} - \sigma^{\bar{\nu}}_{+-}}{\sigma^{\nu} - \sigma^{\bar{\nu}}}\right) = \frac{2}{3}\left[\frac{(\sigma^{\nu}_{+-}/\sigma^{\nu}) - r_{CC}(\sigma^{\bar{\nu}}_{+-}/\sigma^{\bar{\nu}})}{1 - r_{CC}}\right]. \quad (8.111)$$

The CDHS experiment at CERN (Abramowicz *et al.*, 1982) has measured the dimuon rates and, using the measured value of $r_{CC} = 0.48 \pm 0.02$, they extract $B|V_{cd}|^2 = (0.41 \pm 0.07) \times 10^{-2}$. The D^+/D^0 admixture mentioned above leads to an effective value of $B = (10.6 \pm 2.1)\%$, and hence

$$|V_{cd}| = 0.20 \pm 0.03. \quad (8.112)$$

A similar experiment at FNAL, by the CCFRR collaboration (see Nash, 1983), finds $|V_{cd}| = 0.25 \pm 0.07$.

The value of $|V_{cs}|$ can be extracted from measurements of the x-distribution of ν and $\bar{\nu}$ dimuon events. Since $|V_{cd}|$ is small, the x-distribution of $\mu^+\mu^-$ events in $\bar{\nu}$N gives $xs(x)$. The x-distribution $x[u(x)+d(x)]$ is obtained from a study of single muon charged current events. Thus, the x-distribution of $\mu^+\mu^-$ events in νN interactions can be fitted as a linear sum of these components (see (8.110a)). The result of the fit, shown in Fig. 8.8, is

$$\frac{|V_{cs}|^2 2S}{|V_{cd}|^2(\mathrm{U}+D)} = 1.19 \pm 0.09. \tag{8.113}$$

From this value, and the measured ratio of valence to sea quarks, they find

$$\frac{|V_{cs}|^2 2S}{|V_{cd}|^2(\bar{\mathrm{U}}+\bar{D})} = 9.3 \pm 1.6. \tag{8.114}$$

Since the s-quark is heavier than u or d, it is must likely that $2S/(\bar{U}+\bar{D})$ is less than unity. Taking the SU(3) limit $\bar{U}=\bar{D}=\bar{S}$ as the maximum possible strange sea, then (8.112) and (8.114) give the lower limit

$$|V_{cs}| > 0.48 \qquad (90\% \text{ c.l.}). \tag{8.115}$$

Turning the argument around, if the KM angles θ_2 and θ_3 are small (so that $V_{cd}/V_{cs} = \tan\theta_C$), then using $\sin\theta_C = 0.221 \pm 0.002$, (8.114) gives

$$2S/(\bar{U}+\bar{D}) = 0.48 \pm 0.08. \tag{8.116}$$

Fig. 8.8 Distributions of x_{BJ} for the reactions (*a*) $\bar{\nu}\mathrm{N}\rightarrow\mu^+\mu^-\mathrm{X}$ (the full curve is the expectation for $\bar{\nu}\mathrm{N}\rightarrow\mu^+\mathrm{X}$) and (*b*) $\nu\mathrm{N}\rightarrow\mu^-\mu^+\mathrm{X}$; the broken curve is the sum of $x(u+d)$ and xs.

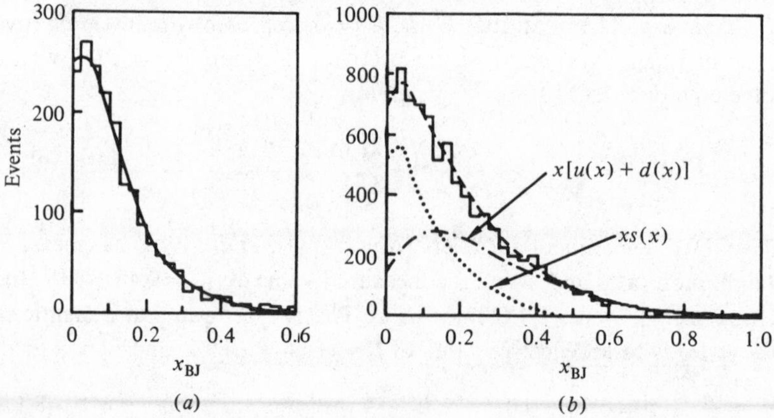

Table 8.5. *Some properties of hadrons containing b-quarks*

Particle	Decay mode	%
$B^+(B^-)$	$\bar{D}^0\pi^+$	1.1 ± 0.6
$m = 5271.2 \pm 3.0$ MeV	$D^*(2010)\pi^+\pi^-$	2.7 ± 1.7
$B^0(\bar{B}^0)$	$\bar{D}^0\pi^+\pi^-$	7 ± 5
$m = 5275.2 \pm 2.8$ MeV	$D^*(2010)^-\pi^+$	1.7 ± 0.7
	$D^*(2010)^-\rho^+$	$8 \pm ^{7}_{4}$
$B^\pm$, B^0, $\bar{B}^0$ (unseparated)	$e^\pm\nu$ hadrons	12.3 ± 0.8
	$\mu^\pm\nu$ hadrons	11.0 ± 0.9
$\tau = (14.2 \pm 2.7) \times 10^{-13}$ s	D^0X	80 ± 28
	$D^*(2010)^\pm X$	23 ± 9
	$(J/\psi)X$	1.2 ± 0.3

That is, the strange sea is suppressed to roughly half its SU(3) value at $Q^2 \simeq 10\ \text{GeV}^2$.

8.6.3 Decays of b-quark hadrons and the measurement of V_{ub} and V_{cb}

The established weakly decaying particles containing b-quarks are the $J^P = 0^-$ mesons $B^-(b\bar{u})$ and $\bar{B}^0(b\bar{d})$ and their antiparticles $B^+(\bar{b}u)$ and $B^0(\bar{b}d)$. Some properties of these decays are given in Table 8.5. The mass difference of B^0 and B^+ is 4 ± 4 MeV, giving a further constraint on the u and d quark mass difference. Much of the data comes from the e^+e^- storage rings CESR (Cornell) and DORIS (DESY), taken at the $\Upsilon(4S)$ resonance, at which the branching fractions to B^+B^- and $B^0\bar{B}^0$ are $(60 \pm 2)\%$ and $(40 \pm 2)\%$ respectively (Behrends *et al.*, 1983). The mean charged multiplicity for B decays is $\langle n_{ch}\rangle = 5.8 \pm 0.2$. For semileptonic decays $\langle n_{ch}\rangle = 4.1 \pm 0.4$, and for non-leptonic decays $\langle n_{ch}\rangle = 6.3 \pm 0.3$ (Jaros, 1983). There is a substantial production of charmed particles in B decays. The CLEO collaboration (see Jarlskog, 1983) find 0.8 ± 0.3 D^0 per B decay. The inclusive baryon branching ratios, namely $BR(B \to p + X)$ and $BR(B \to \Lambda + X)$, are about 6% and 4% respectively, and are consistent with the expectations of the model of Bigi (1981) and of arising from charmed baryon decay. A branching ratio $B \to \Lambda_c \sim 7 \pm 3\%$ is estimated.

The experimental study of B mesons at e^+e^- machines operating in the continuum is more difficult than for charm, because of the smaller production rate. However, since the decay multiplicity is not large, a useful experimental signature is the large p_T of the decay products with respect to the quark direction ($p_T \lesssim 2.3$ GeV). A crude calculation of the magnitude of the lifetime, using (8.107) (with nine possible decay channels and $m_b = 4.5$ GeV), gives $\tau \sim 2 \times 10^{-15}$ s. This value would mean that a 10 GeV B meson had an average decay length of only about 1 μm. Experimentally a typical measured impact parameter for B-decay is $\langle|\delta|\rangle \sim 100\ \mu$m. In fact, the measured lifetime (Table 8.5) is almost three orders of magnitude longer than this simple expectation. Although these B lifetime studies are still in their infancy, one can conclude that the decay is heavily suppressed by some mechanism. The KM matrix terms involved in the decay are V_{ub} and V_{cb}, and so, in this picture, both must be small. By unitarity ($|V_{ub}|^2 + |V_{cb}|^2 + |V_{tb}|^2 = 1$), the value of V_{tb} must be nearly unity. This suggests that the b quark is strongly coupled to the t-quark but, since that is not a possible decay option, it is quasi-stable. In terms of the KM angles (assuming θ_1 is small), the long b-lifetime implies that both θ_2 and θ_3 are small.

As we have seen in the study of both s and c quark decays, the analysis of semileptonic interactions is important, both in extracting the weak couplings and in the understanding of the decay mechanisms. The possible quark transitions (in the spectator model) are $b \to ul^-\bar{\nu}_l$ and $b \to cl^-\bar{\nu}_l$. In the latter case, the charmed particle produced will also decay semileptonically in about 12% of the cases. Hadronic decays, such as $b \to c\bar{u}d$, followed by $c \to sl\nu$, will also lead to a semileptonic final state configuration. As we shall see below, the ratio $|V_{ub}|/|V_{cb}|$ is small, so that the bulk of the decays are $b \to cW^-$. At the quark level, the leptonic branching fraction is that expected from $[W^- \to e^-\nu_e]/[W^- \to (\bar{u}d, \bar{c}s, e^-\nu_e, \mu^-\nu_\mu, \tau^-\nu_\tau)]$, i.e. about 11%. This estimate ignores phase space effects. However, these are important, since, in addition to the c-quark spectator, there can be a second c-quark or a τ-lepton produced, and these modes will be suppressed. Stone (1983) calculates that this effect changes the semileptonic branching ratio to about 17%.

Let us consider the decay $b \to c\bar{u}d$ in more detail (the same arguments apply also to $b \to u$ decays). At the quark level, the effective Lagrangian is, using the notation $\bar{c}b = \bar{c}\gamma^\mu(1-\gamma^5)b$, etc.,

$$\mathscr{L}_{eff} = \frac{G_F}{2^{1/2}} V_{cb}V_{ud}(\bar{c}b)(\bar{d}u). \tag{8.117}$$

From Fig. 8.7(*a*) it can be seen that the b and c quarks have the same colour, and also that the u and d quarks have the same colour (since the W is colourless). In QCD there are gluonic corrections to this basic graph. For example, one of the quarks can radiate a gluon. This is rather like the QED radiative correction to muon decay, and is expected to be fairly small ($\lesssim 15\%$). Gluons can also be exchanged between the various quarks, changing the colour flow in the diagram. These corrections can be summed, and give a contribution which, in the leading log approximation, is of the form $[\alpha_S \ln(M_W/m_b)]^n$. The net effect is to give an additional term to the Lagrangian, so that

$$\mathscr{L}'_{\text{eff}} = \frac{G_F}{2^{1/2}} V_{cb} V_{ud} \left[\frac{(f_+ + f_-)}{2} (\bar{c}b)(\bar{d}u) + \frac{(f_+ - f_-)}{2} (\bar{c}u)(\bar{d}b) \right], \quad (8.118)$$

where the QCD coefficients f_+ and f_- are given by

$$f_- = \frac{1}{f_+^2} = \left[\frac{\alpha_S(m_b)}{\alpha_S(M_W)} \right]^\gamma; \qquad \gamma = \frac{12}{33 - 2n_f}. \quad (8.119)$$

The first term in (8.118) is the original current, modified by the coefficient $(f_+ + f_-)/2$. In the second term, the b and u quarks have been interchanged. This reflects the colour flow since, for this graph, the b and d quarks, and also the u and c quarks, have the same colour. The amplitude for this colour configuration is proportional to $(f_+ - f_-)/2$. In the limit $\alpha_S \to 0$, we have $f_+ = f_- = 1$, and we regain (8.117). For $m_b = 4.5$ GeV, (8.119) gives (using $\Lambda = 0.5$ GeV)

$$f_+ \simeq 0.80, \qquad f_- \simeq 1.56.$$

That is, the first and second terms in (8.118) have coefficients 1.18 and 0.38 respectively.

Note that for charm decays these corrections are substantially larger ($f_+ \simeq 0.7$ and $f_- \simeq 2.1$ for $m_c = 1.5$ GeV). These calculations, of course, refer to quarks and not hadrons. That is to say, that neither the effects of hadronisation, nor of the hadron phase space, are taken into account. One cannot, therefore, expect the calculation to be exact, but rather a guide as to the possible magnitude of the QCD corrections.

QCD calculations of the above type, for all the spectator diagrams for b quark decays, have been made by Pham (1982), who finds, in units of $(G_F^2 m_b^5/192\pi^3)$

$$\Gamma(B \to e\nu X) = 0.38|V_{cb}|^2 + 0.77|V_{ub}|^2, \quad (8.120a)$$

$$\Gamma(B \to \text{hadrons}) = 1.9|V_{cb}|^2 + 5.1|V_{ub}|^2, \quad (8.120b)$$

$$\Gamma(B \to \text{all}) = 2.72|V_{cb}|^2 + 6.92|V_{ub}|^2. \quad (8.120c)$$

Thus the semileptonic branching ratio is

$$B_e = \frac{\Gamma(B \to e\nu X)}{\Gamma(B \to \text{all})} = \frac{0.38 + 0.77r}{2.72 + 6.92r}, \qquad r = \frac{|V_{ub}|}{|V_{cb}|}. \quad (8.121)$$

Note that, irrespective of the value of r, B is bounded to be between 0.11 and 0.14. Thus, within the context of this QCD model, a value outside this range would violate the validity of the spectator model, independently of the weak interaction parameters. The combined electron and muon branching ratio for B-mesons is $B_e = 0.117 \pm 0.006$, which can be used to set an upper limit of 30% on non-spectator diagrams. For a value $r \sim 0$ (see below), a non-spectator contribution of $(20 \pm 6)\%$ is needed. It should be stressed, however, that these deductions are valid only in the context of this QCD model, and that it is difficult to assess its reliability.

The extraction of reliable values of V_{ub} and V_{cb} is important, because they indicate the amount of quark generation mixing and, indirectly, give the coupling of t to b quarks. Further, a non-zero value of V_{ub} is important for the explanation of *CP* violation using the KM matrix (Section 9.1.7). Using the expression (8.120c) for $\Gamma(\mathrm{B} \to \text{all})$, calculated using $m_b = 4.95 \pm 0.20$ GeV (see below), together with the measured B-decay widths, gives the limits (from positivity)

$$|V_{ub}| \leqslant 0.04, \qquad |V_{cb}| \leqslant 0.06, \qquad (90\% \text{ c.l.}). \tag{8.122}$$

If we include also the measured semileptonic branching ratio for $\mathrm{B} \to l\nu\mathrm{X}$, then (8.120a) and (8.120c) can be solved to yield values for $|V_{ub}|$ and $|V_{cb}|$. These values are compatible with zero within errors, and give the following upper limits

$$|V_{ub}| \leqslant 0.04, \qquad |V_{cb}| \leqslant 0.05, \qquad (90\% \text{ c.l.}). \tag{8.123}$$

Clearly there is some model dependence in these limits, but both terms are, however, small (i.e. $\lesssim 0.1$).

Measurement of the lepton energy spectrum in $\mathrm{B} \to l\nu\mathrm{X}$ has yielded the most precise limit. The decays $\mathrm{b} \to \mathrm{u}l\nu$ and $\mathrm{b} \to \mathrm{c}l\nu$ differ because the c quark has a much larger mass than the u quark. In particular, the maximum lepton momentum is significantly larger for the $\mathrm{b} \to \mathrm{u}$ transition. A measurement of the electron spectra from the CLEO collaboration (Thorndike, 1985) is shown in Fig. 8.9. The curves come from a model by Altarelli *et al.* (1982), which includes the effects of first-order soft gluon corrections, and ascribes a Gaussian Fermi motion to the initial quarks in the B-meson. The ratio $(\mathrm{b} \to \mathrm{u})/(\mathrm{b} \to \mathrm{c})$ can be extracted by fitting the lepton energy shape to the components, or by measuring the fraction of decays which are kinematically forbidden for a c-quark. Both methods are, to some extent, model-dependent, but the latter requires the model only for $\mathrm{b} \to \mathrm{u}$. Thorndike (1985), using the most recent CLEO data and, using a rather conservative approach, concludes that $(\mathrm{b} \to \mathrm{u})/(\mathrm{b} \to \mathrm{c})$ is less than 0.08 (90% c.l.). However, with additional plausible assumptions, a more stringent limit $(b \to u)/(b \to c) < 0.04$ (90% c.l.) is obtained. The more conservative limit gives

$$|V_{ub}|/|V_{cb}| < 0.20, \qquad (90\% \text{ c.l.}). \qquad (8.124)$$

One of the most sensitive parameters in the extraction of the weak couplings from heavy quark decays is the heavy quark mass (which enters as the fifth power). There is no theoretical consensus as to the 'correct' value, and a conservative approach is to allow a reasonable range of values. A possible picture is as follows: The B-meson is considered to be a dressed b-quark, loosely bound to a light quark by gluon exchange. In turn, the dressed b-quark is composed of a bare point-like b-quark, surrounded by gluons. In the decay process, the relevant wavelengths are large compared to the size of the B-meson, so that the B-meson mass is the appropiate mass. At shorter wavelengths the dressed b-quark will be the relevant structure, and at progressively shorter wavelengths, the gluon cloud becomes resolved and the appropriate mass becomes smaller. Fig. 8.10(*a*) shows the b-quark 'running' mass (Gasser and Leutwyler, 1982), together with the corresponding $l\nu$ invariant mass distributions for $b \to cl\nu$ and $b \to ul\nu$ in Fig. 8.10(*b*) (from Thorndike, 1985). The 'measured'

Fig. 8.9 Electron momentum spectrum from the decay $B \to Xe\nu$ (CLEO experiment). The curves are spectator model calculations for (*a*) $b \to ce\nu$, (*b*) $b \to ue\nu$ (normalised by $(b \to u)/(b \to c) = 0.2$) and (*c*) $b \to c \to se\nu$. The full line is the result of a fit with $V_{ub} = 0$.

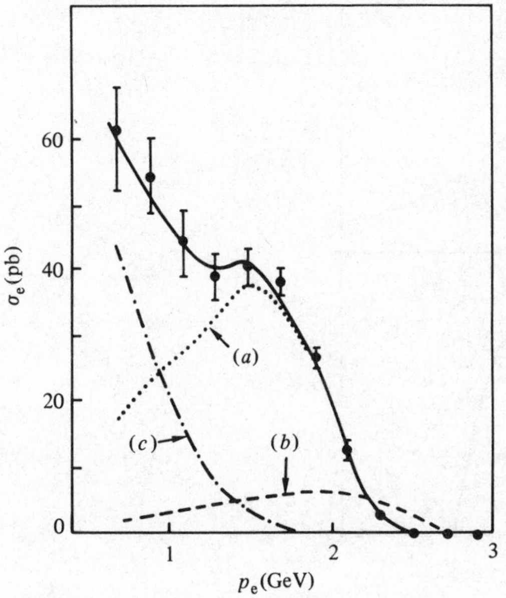

b-quark mass of $m_b = 4.99 \pm 0.20$ GeV comes from a fit by the CLEO collaboration, using the Altarelli *et al.* model with $m_c = 1.7$ GeV. The corresponding Gaussian width of the Fermi model is $P_F = 0.15 \pm 0.10$ GeV. It can be seen that, using the $l\nu$ invariant mass as the energy scale for the b-quark mass, gives a plausible picture. One may take a value $m_B = 4.95 \pm 0.20$ GeV as a reasonable estimate.

A value of $|V_{cb}|$ can now be evaluated. Using (8.120c), together with the limit (8.124), and the measured B lifetime, gives

$$|V_{cb}| = 0.050 \pm 0.007. \tag{8.125}$$

This value is rather insensitive to $|V_{ub}|$, within the limit (8.124), but is sensitive to m_B. For example, if $m_B = 4.5$ GeV were used, the value of $|V_{cb}|$ changes to 0.064. Using the value of $|V_{cb}|$ from (8.125), the limit (8.124) can be expressed as an upper limit on $|V_{ub}|$, namely

$$|V_{ub}| < 0.011, \qquad (90\% \text{ c.l.}). \tag{8.126}$$

To conclude this section on b-quark decays, we discuss one specific mode, namely $B \rightarrow (J/\psi) + X$, in the context of colour suppression. This decay proceeds, in the spectator picture, by a colour mixed diagram (Fig.

Fig. 8.10 (*a*) Running b-quark mass as a function of the scale μ. The 'measured' mass is described in the text. (*b*) The $l\nu$ effective mass distribution from $b \rightarrow Xl\nu$ decays.

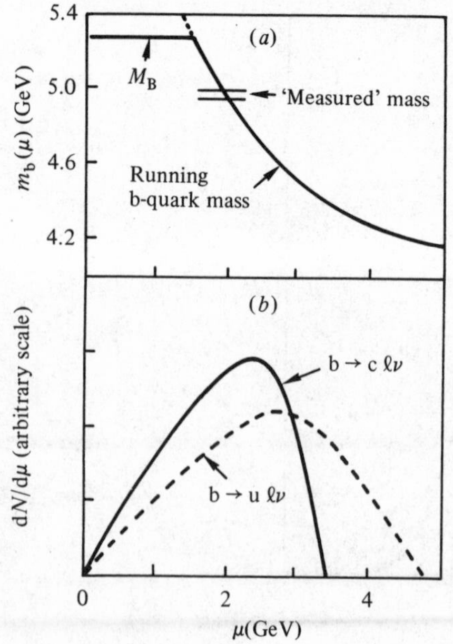

8.11). The observed branching ratio is $(1.1 \pm 0.3)\%$, and the ψ momentum spectrum (Alam *et al.*, 1986) is shown in Fig. 8.12, together with spectator model curves. The calculated spectrum is in reasonable agreement with the data. Differences could arise from gluon emission (which is not in the model) and from the relative importance of $B \rightarrow (\psi' \text{ or } \chi) + X \rightarrow \psi$, as well as from the validity of the spectator model. QCD calculations contain the colour suppression factor of $\frac{1}{9}$, and the predicted branching ratio is compatible with the data. This gives support for the existence of this mechanism; however, there remains some model dependence in the interpretation.

Fig. 8.11 Possible decay mechanism for $\bar{B} \rightarrow \Psi X$.

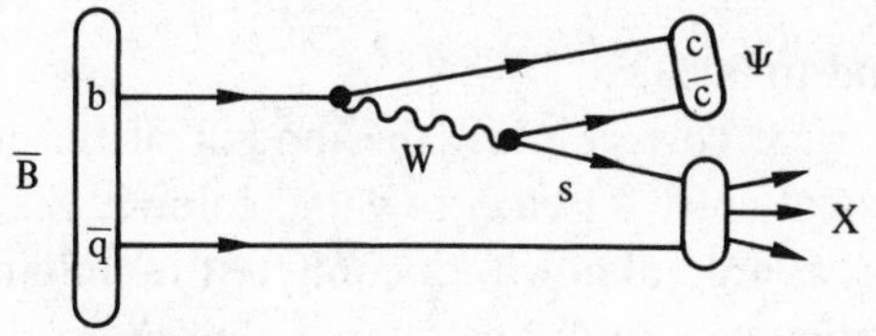

Fig. 8.12 Measured Ψ momentum distribution in the decay $B \rightarrow \Psi X$. The full line shows the result of a spectator model calculation, the individual contribution of $B \rightarrow (\Psi' \text{ or } \chi)X \rightarrow \Psi$ is shown as a broken line.

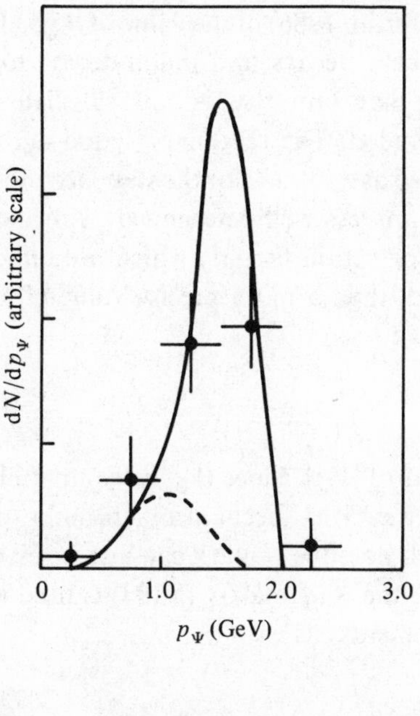

8.6.4 The Kobayashi–Maskawa (KM) matrix elements

The various KM matrix elements, the determination of which has been described in this and preceding chapters, are as follows (see (5.193))

$$\begin{aligned}
&|V_{ud}| = 0.9747 \pm 0.0011 && \text{nuclear beta decay,} \\
&|V_{us}| = 0.221 \pm 0.002 && \mathrm{K} \to \pi e \nu, \\
&|V_{ub}| < 0.011 && \mathrm{B} \to \mathrm{X} e \nu, \\
&|V_{cd}| = 0.20 \pm 0.03 && \nu(\bar{\nu})\mathrm{N} \to \mu^+\mu^-\mathrm{X}, \\
&|V_{cs}| > 0.67 && \mathrm{D} \to \mathrm{K} e \nu, \\
&|V_{cb}| = 0.050 \pm 0.007 && B \to X e \nu + \tau(B),
\end{aligned} \tag{8.127}$$

where the limits correspond to 90% c.l.

If the six-quark model were correct, then the moduli of the matrix elements squared would equal unity for each row and column. Since the t-quark couplings (if it exists) are unknown, the only test of unitarity so far possible is for the coupling of u to d-type quarks, that is

$$|V_{ud}|^2 + |V_{us}|^2 + |V_{ub}|^2 = 0.9988 \pm 0.0023. \tag{8.128}$$

This value is close to unity, although it still leaves some room for a further heavy d-type quark. The coupling to the u-quark of such a quark (d_h) can be found using unitarity from (8.128), and satisfies $|V_{ud_h}| < 0.07$ (90% c.l.). A careful analysis (Marciano and Sirlin, 1986) of the value of $|V_{ud}|$, derived from a comparison of $0^+ \to 0^+$ beta decays and muon-decay, together with the value of $|V_{us}|$, shows that unitarity is not satisfied unless $SU(2) \otimes U(1)$ corrections are included. The reasonably good agreement of (8.128) with unity can be regarded as a success for the standard model.

The couplings for the c-quark are less well determined, with the main term (V_{cs}) dependent on SU(4) correction factors, which are difficult to determine. If unitarity is assumed, then a more precise value of V_{cs} can be computed, namely

$$|V_{cs}| = 0.979 \pm 0.006. \tag{8.129}$$

This value is compatible with that of $|V_{ud}|$. Since $|V_{cd}|$ is compatible with $|V_{us}|$, the deductions made using the GIM mechanism still hold to good approximation. The couplings to a possible t-quark can also be estimated by unitarity. Using this property, the KM matrix (5.193) can be written (in terms of the moduli of the elements) as

$$V = \begin{matrix} & \text{d} & \text{s} & \text{b} \\ \begin{matrix}\text{u}\\ \text{c}\\ \text{t}\end{matrix} & \multicolumn{3}{c}{\left(\begin{matrix} 0.9747 \pm 0.0011 & 0.221 \pm 0.002 & <0.011 \\ 0.20 \pm 0.03 & 0.979 \pm 0.006 & 0.050 \pm 0.007 \\ <0.17 & <0.13 & 0.9987 \pm 0.0004 \end{matrix}\right)} \end{matrix}. \tag{8.130}$$

The t-quark couplings V_{td} and V_{ts} are roughly compatible with zero, so are expressed as upper limits. Note that only five of the nine elements are measured directly, the remaining four are obtained from unitarity. Hence, the values of the deduced couplings are very sensitive to those which are measured. In addition, the measurement of all the quark couplings depend, to a greater or lesser extent, on the quark hadronisation process. It is difficult to ascribe a realistic error to this uncertainty. Although no clear pattern amongst the various KM matrix elements has yet emerged, it appears that transitions involving a change of two generations are suppressed compared to those involving a single generation change.

8.7 HADRONIC DECAYS OF τ-LEPTONS

As we have seen in Section 6.1.3, the leptonic decays of the τ-lepton are well described by the standard model, which for the q^2 values involved is synonymous with the V − A theory. However, in contrast to muon decay, hadronic decays are possible for the τ-lepton. The principle decay modes are listed in Table 6.2. The decay mechanism is illustrated in Fig. 8.13. There are five possible basic decay modes open for the W^-, namely $e^-\bar{\nu}_e$, $\mu^-\bar{\nu}_\mu$ and $d'\bar{u}$ (in each of three colours), where $d' = d\cos\theta_C + s\sin\theta_C$. Thus, a value of the leptonic branching ratios (B_e, B_μ) in the region of 20% is expected and observed. A study of the hadronic modes is of interest because it yields information on the fragmentation of the d$\bar{\text{u}}$ and s$\bar{\text{u}}$ systems, without the complication of other quarks (which make the interpretation of heavy quark decays more difficult).

Fig. 8.13 Lowest order diagram for the decay $\tau^- \to \nu_\tau W^-$ ($W^- \to a\bar{b}$).

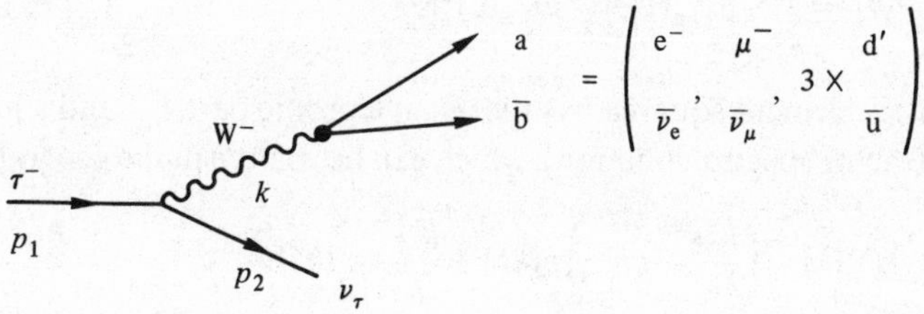

Many of the exclusive decay mode branching ratios can be calculated using the ideas previously formulated. The simplest are

(i) $\tau^- \to \nu_\tau \pi^-$. The method here is very similar to that used for $\pi \to l\nu$ in Section 8.1.1, with the hadronic current taking the form of (8.2). The usual methods yield (assuming $m_{\nu_\tau} = 0$)

$$\omega(\tau^- \to \nu_\tau \pi^-) = \frac{G_F^2 \cos^2\theta_C f_\pi^2 m_\tau^3}{16\pi}\left(1 - \frac{m_\pi^2}{m_\tau^2}\right)^2. \tag{8.131}$$

Relative to the $\tau^- \to \nu_\tau e^- \bar{\nu}_e$ rate (ω_e), which can be calculated exactly using (6.24), this predicts a branching ratio (using the values of the constants discussed previously) of $\omega(\tau^- \to \nu_\tau \pi^-)/\omega_e = 0.61$. Experimentally this ratio is 0.58 ± 0.07.

(ii) $\tau^- \to \nu_\tau K^-$. The hadronic current here contains $f_K \sin\theta_C$, giving

$$\omega(\tau^- \to \nu_\tau K^-) = \frac{G_F^2 \sin^2\theta_C f_K^2 m_\tau^3}{16\pi}\left(1 - \frac{m_K^2}{m_\tau^2}\right)^2. \tag{8.132}$$

The expected value of $\omega(\tau^- \to \nu_\tau K^-)/\omega_e = 0.0397$, compared to the experimental value of 0.039 ± 0.010. The ratios of the decay rates of $\tau^- \to \nu_\tau K^-$ to $\tau^- \to \nu_\tau \pi^-$, and of $K^- \to \mu^- \bar{\nu}_\mu$ to $\pi^- \to \mu^- \bar{\nu}_\mu$, give (ignoring the effects of radiative corrections)

$$(f_K/f_\pi)\tan\theta_C = 0.26 \pm 0.04 \qquad \tau\text{-decay},$$
$$(f_K/f_\pi)\tan\theta_C = 0.2754 \pm 0.0004 \qquad \pi\text{-, K-decay}. \tag{8.133}$$

These values are clearly consistent and, using $\sin\theta_C = 0.221 \pm 0.002$, give $f_K/f_\pi \sim 1.22$. This provides further evidence for the consistency of the Cabibbo theory.

(iii) $\tau^- \to \nu_\tau \pi^- \pi^0$. Before calculating this decay mode we consider the more general decay $\tau^- \to \nu_\tau h^-$, where h is any hadronic system with total four-momentum k. The matrix element can be written (see Fig. 8.13)

$$\mathscr{M}_{fi} = \frac{G_F}{2^{1/2}} \bar{u}_2 \gamma_\mu (1 - \gamma^5) u_1 \langle h | j^\mu | 0 \rangle. \tag{8.134}$$

The matrix element squared has the usual leptonic part $L_{\mu\nu}$ and a hadronic tensor $H^{\mu\nu}$ (or spectral function), which can be written in the general form

$$H^{\mu\nu}(k) = \left(\frac{k^\mu k^\nu}{k^2} - g^{\mu\nu}\right)\rho(k^2) + \frac{k^\mu k^\nu}{k^2}\rho'(k^2). \tag{8.135}$$

Contracting this tensor with $L_{\mu\nu}$ we obtain

$$|\mathscr{M}_{\mathrm{fi}}|^2 = 2G_{\mathrm{F}}^2\left[\left(\frac{2p_1\cdot kp_2\cdot k}{k^2}+p_1\cdot p_2\right)\rho(k^2)+\left(\frac{2p_1\cdot kp_2\cdot k}{k^2}-p_1\cdot p_2\right)\rho'(k^2)\right]$$
$$= G_{\mathrm{F}}^2 m_\tau^2[(1+2x)\rho(x)+\rho'(x)](1-x)/x, \qquad (8.136)$$

where $x = k^2/m_\tau^2$, and the evaluation of the vector products is most easily carried out in the τ rest frame.

Using the decay rate formula (C.3), we obtain the following expression for the decay width

$$\omega(\tau^- \to \nu_\tau h^-) = \frac{G_{\mathrm{F}}^2 m_\tau}{16\pi}\int_0^1 \mathrm{d}x\,\frac{(1-x)^2}{x}[(1+2x)\rho(x)+\rho'(x)]. \qquad (8.137)$$

For a stable particle $\rho(x)$ has a simple form. For example, in the decay $\tau^- \to \nu_\tau \pi^-$ we have $\mathrm{h} = \pi^-$, and inspection of the previous methods shows that $\rho(x) = 0$ and that $\rho'(x) = f_\pi^2 \cos^2\theta_{\mathrm{C}} x m_\tau^2\,\delta(x - m_\pi^2/m_\tau^2)$. Inserting these values in (8.137) gives (8.131). Next we consider the decay $\tau^- \to \nu_\tau \rho^-$ (this dominates the $\tau^- \to \nu_\tau \pi^- \pi^0$ channel) and treat, first of all, the ρ as stable. The hadronic current for the ρ is of the form $J^\mu = f_\rho \varepsilon^\mu$, where ε is the ρ^- polarisation vector. Thus, considering the polarisation sums for a massive vector particle, we can see that the hadronic tensor for the ρ^- has $\rho'(x) = 0$ and $\rho(x) = f_\rho^2 \cos^2\theta_{\mathrm{C}}\,\delta(x - m_\rho^2/m_\tau^2)$. From (8.137), this gives the decay rate

$$\omega(\tau^- \to \nu_\tau \rho^-) = \frac{G_{\mathrm{F}}^2\cos^2\theta_{\mathrm{C}} f_\rho^2 m_\tau^3}{16\pi m_\rho^2}\left(1-\frac{m_\rho^2}{m_\tau^2}\right)^2\left(1+\frac{2m_\rho^2}{m_\tau^2}\right). \qquad (8.138)$$

The magnitude of the ρ coupling to the weak vector current is given by CVC, which can be used to relate the $\mathrm{W} \to \rho$ vector current to that in $\mathrm{e}^+\mathrm{e}^- \to \gamma^* \to \rho$.

A more sophisticated treatment includes the finite ρ width. In fact, a more general approach is to use the measured $\mathrm{e}^+\mathrm{e}^- \to \pi^+\pi^-$ cross-section, $\sigma(k^2)$, over the appropriate range of k^2. This gives

$$\omega(\tau^- \to \nu_\tau \pi^- \pi^0) = \frac{G_{\mathrm{F}}^2\cos^2\theta_{\mathrm{C}} m_\tau^7}{128\pi^4\alpha^2}\int_{\sim 0}^1 x(1-x)^2(1+2x)\sigma(k^2)\,\mathrm{d}x. \qquad (8.139)$$

Putting in the measured $\mathrm{e}^+\mathrm{e}^- \to \pi^+\pi^-$ cross-section, and integrating over k^2, gives (Gilman and Rhie, 1985) a branching ratio $\omega(\tau^- \to \nu_\tau \pi^- \pi^0)/\omega_{\mathrm{e}} = 1.23$, in good agreement with the experimental value of 1.27 ± 0.12. The τ^- decay mode to $\pi^-\pi^0$ is dominated by $\tau^- \to \nu_\tau \rho^-$ $(21.8 \pm 2.0\%)$, with an additional non-resonant contribution of $0.3 \pm 0.3\%$.

(iv) $\tau^- \to \nu_\tau(\mathrm{K}\pi)^-$. Since the $\pi\pi$ mode is dominated by the ρ, this Cabibbo-suppressed mode is expected, and indeed observed, to be dominated by the K*(892). The expected rate is obtained by multiplying the $\pi\pi$ rate by $\tan^2\theta_C$, and by the phase space factor

$$P = \frac{m_\rho^2(1 - m_{\mathrm{K}^*}^2/m_\tau^2)^2(1 + 2m_{\mathrm{K}^*}^2/m_\tau^2)}{m_{\mathrm{K}^*}^2(1 - m_\rho^2/m_\tau^2)^2(1 + 2m_\rho^2/m_\tau^2)}. \tag{8.140}$$

With these corrections, the predicted ratio is $\omega(\tau^- \to \nu_\tau \mathrm{K}^{*-})/\omega_e = 0.047$. Incorporating SU(3) breaking effects, by setting the relative strengths of the vector currents as $g_{\mathrm{K}^*}^2/g_\rho^2 = m_{\mathrm{K}^*}^2/m_\rho^2$, then $\omega(\tau^- \to \nu_\tau \mathrm{K}^{*-})/\omega_e = 0.064$ (Gilman and Rhie, 1985). The experimental value of 0.10 ± 0.04 is consistent with both these values.

(v) $\tau^- \to \nu_\tau(4\pi)^-$. The hadronic current is vector, and (as for (iii)) can be related to $e^+e^- \to 4\pi$ using CVC. The possible 4π states for e^+e^- are $2\pi^+2\pi^-$ and $\pi^+\pi^-2\pi^0$ and, for τ^--decay, are $2\pi^-\pi^+\pi^0$ and $\pi^-3\pi^0$. The decay rate for $\tau^- \to \nu_\tau\pi^-3\pi^0$ is given by (8.139), where $\sigma(k^2)$ is replaced by $\frac{1}{2}\sigma(e^+e^- \to 2\pi^+2\pi^-)$. Similarly, for the decay rate $\tau^- \to \nu_\tau 2\pi^-\pi^+\pi^0$, $\sigma(k^2)$ is replaced by $[\sigma(e^+e^- \to \pi^+\pi^-2\pi^0) + \frac{1}{2}\sigma(e^+e^- \to 2\pi^+2\pi^-)]$. The $e^+e^- \to 4\pi$ cross-section is dominated by the broad ρ' resonance ($M \simeq 1.55$ GeV, $\Gamma \simeq 0.3$ GeV). Using the measured e^+e^- cross-section data Gilman and Rhie (1985) give the predictions

$$\omega(\tau^- \to \nu_\tau\pi^-3\pi^0)/\omega_e = 0.055, \quad \omega(\tau^- \to \nu_\tau 2\pi^-\pi^+\pi^0)/\omega_e = 0.275. \tag{8.141}$$

The $2\pi^-\pi^+\pi^0$ prediction has an uncertainty of about ± 0.06, due to imprecision in the $e^+e^- \to 4\pi$ data. The measured τ^- branching ratios are $\omega(\tau^- \to \nu_\tau\pi^-3\pi^0)/\omega_e = 0.17 \pm 0.16$ and $\omega(\tau^- \to \nu_\tau 2\pi^-\pi^+\pi^0)/\omega_e = 0.30 \pm 0.05$, and are consistent with these predictions.

(vi) $\tau^- \to \nu_\tau(3\pi)^-$. The hadronic current is axial, and can have either $J^P = 0^-$ or 1^+. There is no accurate theoretical prediction, but estimates using the properties of the weak axial current and the A_1 meson are consistent with the data.

One can conclude that the specific hadronic decays of the τ-lepton discussed above are reasonably well described theoretically. This gives further support to the use of these ideas in the description of heavy quark decays. However, the experimental errors are still rather large, and the breakdown of the one and three prong decays into specific decay channels is not yet complete.

The existence of the decay $\tau^- \to \omega\pi^-\nu_\tau$, with the $\omega\pi$ having $J^P = 1^+$ (either the resonance $B(1235)$ or non-resonant), would be evidence for a second-class current. The measured decay rate is $(1.5 \pm 0.4)\%$ (Albrecht *et al.*, 1987a), but a spin-parity analysis of the $\omega\pi$ system shows no evidence for a second-class current contribution. The decay $\tau^- \to \eta\pi^-\nu_\tau$ would also proceed by a second-class current (the $\eta\pi$ has $J^P = 0^+$ or 1^- but odd G-parity). Derrick *et al.* (1987) measured a branching ratio of $(5.1 \pm 1.0 \pm 1.2)\%$. However Gan *et al.* (1987) and Behrend *et al.* (1988) find no evidence for this mode, and give upper limits of 1% (95% c.l.) and 1.4% (90% c.l.) respectively, compatible with the absence of second-class currents. In the standard model a branching ratio approximately equal to α^2 is expected from isospin violating contributions.

8.8 NEUTRAL CURRENTS IN DEEP INELASTIC SCATTERING

The discovery and subsequent study of neutral current (NC) events in deep inelastic $\nu(\bar{\nu})$–nucleon scattering have played an important role in the development of the standard electroweak model. The observation of the interactions (Hasert *et al.*, 1973b)

$$\nu/\bar{\nu}(p_1) + \mathrm{N}(p_2) \to \nu/\bar{\nu}(p_3) + \mathrm{X}(p_4), \tag{8.142}$$

constituted some of the first evidence, albeit indirect, for the existence of the Z^0 boson. These events are envisaged to occur through the exchange of a Z^0 (see Fig. 1.7(*b*)) between the ν (in practice a ν_μ in high energy experiments) and the struck quark in the nucleon.

The theoretical description of these NC events is a straightforward development of that for CC events, discussed in Section 7.5, and that for νe NC events, discussed in Section 6.2. The matrix element for the NC scattering, $\nu(p_1) + \mathrm{q}(p_2) \to \nu(p_3) + \mathrm{q}(p_4)$, in the limit $q^2 \ll M_\mathrm{Z}^2$, and assuming the incident neutrino is lefthanded, is, using the rules of Appendix C plus (5.99),

$$\begin{aligned}\mathcal{M}_\mathrm{fi} &= \frac{ig^2}{8M_\mathrm{Z}^2\cos^2\theta_\mathrm{W}}\,\bar{u}_3\gamma^\mu(1-\gamma^5)u_1\cdot\bar{u}_4\gamma_\mu[C_\mathrm{L}(1-\gamma^5)+C_\mathrm{R}(1+\gamma^5)]u_2 \\ &= \frac{i\rho G_\mathrm{F}}{2^{1/2}}\,\bar{u}_3\gamma^\mu(1-\gamma^5)u_1\cdot\bar{u}_4\gamma_\mu[C_\mathrm{L}(1-\gamma^5)+C_\mathrm{R}(1+\gamma^5)]u_2,\end{aligned} \tag{8.143}$$

where (5.63) has been used and where

$$\rho = M_\mathrm{W}^2/(M_\mathrm{Z}^2\cos^2\theta_\mathrm{W}). \tag{8.144}$$

The quantity $\rho = 1$ in the standard model for the simplest choice of Higgs representation, as can be seen from (5.168). A deviation from $\rho = 1$ would also result if the ν_L coupling to the Z^0 is different to the assignment $C_L = \frac{1}{2}$ assumed here.

In order to simplify the discussion, we consider the case of an isoscalar target containing only u and d quarks and antiquarks (the expressions for additional quarks and for p and n targets separately can be written down in a similar way). The CC and NC differential cross-sections are

$$\frac{d^2\sigma^{\nu N}(CC)}{dx\,dy} = \frac{G_F^2 MEx}{\pi}[q(x) + \bar{q}(x)(1-y)^2], \tag{8.145}$$

$$\frac{d^2\sigma^{\nu N}(NC)}{dx\,dy} = \frac{G_F^2 \rho^2 MEx}{\pi}\{[|C_L^u|^2 + |C_L^d|^2][q(x) + \bar{q}(x)(1-y)^2] + [|C_R^u|^2 + |C_R^d|^2][\bar{q}(x) + q(x)(1-y)^2]\}. \tag{8.146}$$

The expressions (8.145) and (8.146) follow in a straightforward way from the results derived in Section 7.5.

Experimentally, the quantity which can be measured most easily is the ratio of neutral to charged current cross-sections. From (8.145) and (8.146), this is

$$R_\nu = \frac{\sigma^{\nu N}(NC)}{\sigma^{\nu N}(CC)} = \rho^2\{[|C_L^u|^2 + |C_L^d|^2] + r_{CC}[|C_R^u|^2 + |C_R^d|^2]\}, \tag{8.147}$$

and similarly for antineutrinos

$$\bar{R}_\nu = \frac{\sigma^{\bar{\nu} N}(NC)}{\sigma^{\bar{\nu} N}(CC)} = \rho^2\left\{[|C_L^u|^2 + |C_L^d|^2] + \frac{1}{r_{CC}}[|C_R^u|^2 + |C_R^d|^2]\right\}, \tag{8.148}$$

where r_{CC} is the ratio of $\bar{\nu}$ to ν CC cross-sections, that is

$$r_{CC}(x) = [x\bar{q}(x) + axq(x)]/[xq(x) + ax\bar{q}(x)]. \tag{8.149}$$

The term a is the integral of $(1-y)^2$ over y, and $a = \frac{1}{3}$ if the whole range of y is detected. In practice, experiments detect NC and CC events with a hadronic energy (E_H) above a certain value (e.g. 5 GeV). This is because the only experimental signature of an NC event is the hadronic shower. The lower limit arises because of problems with resolution and backgrounds from (particularly for bubble chamber experiments) n and K_L^0 interactions or from cosmic-ray interactions. Since the neutrino energy has a spread in values, this cut does not correspond to a unique value of y, so that in practice the $\nu/\bar{\nu}$ energy spectrum, plus the detector response and acceptance, all enter the calculation.

In the standard model $\rho = 1$, and the Z^0–quark couplings are given in Appendix (C.3.2). Inserting these in (8.147) and (8.148) gives

$$R_\nu = \tfrac{1}{2} - \sin^2\theta_W + \tfrac{5}{9}(1 + r_{CC})\sin^4\theta_W,$$
$$\bar{R}_\nu = \tfrac{1}{2} - \sin^2\theta_W + \tfrac{5}{9}(1 + r_{CC}^{-1})\sin^4\theta_W. \quad (8.150)$$

Since r_{CC} is less than unity (typically $r_{CC} \simeq 0.5$), it can be seen by comparing (8.147) and (8.148), or from (8.150), that $\bar{R}_\nu$ is expected to be larger than R_ν. The dependence of the ratios on the quark distributions can be removed by considering the combinations (Paschos and Wolfenstein, 1973)

$$R_\nu^\pm = \frac{\sigma^{\nu N}(NC) \pm \sigma^{\bar{\nu}N}(NC)}{\sigma^{\nu N}(CC) \pm \sigma^{\bar{\nu}N}(CC)} = \rho^2\{|C_L^u|^2 + |C_L^d|^2 \pm [|C_R^u|^2 + |C_R^d|^2]\}$$
$$= \begin{cases} \tfrac{1}{2} - \sin^2\theta_W + \tfrac{10}{9}\sin^4\theta_W, \\ \tfrac{1}{2} - \sin^2\theta_W, \end{cases} \quad (8.151)$$

where the latter expression is for the standard model. Note that the effects of other sea-quarks will cancel in the combination R_ν^-.

Experimentally, the most precise results on measurements of the NC to CC ratio have been made on isoscalar (or approximately isoscalar) targets), and using '*narrow-band*' neutrino beams. This type of beam is produced from a parent beam of pions and kaons having a small momentum spread. For a meson of mass m_M and energy E_M, decaying via $M \to \mu + \nu_\mu$ to a neutrino, of energy E_ν, subtending an angle θ with respect to the incident meson direction, then the two-body kinematics give a unique relationship between E_ν and θ, namely

$$E_\nu \simeq E_M(1 - m_\mu^2/m_M^2)(1 + \gamma^2\theta^2)^{-1}, \quad (8.152)$$

where $\gamma = E_M/m_M$. In practice (e.g. CERN neutrino beam), the detector is at a distance $L \sim 10^3$ m from the decay, and has a radius of about 1.5 m. The energy of the parent mesons ($\pi^\pm$ and $K^\pm$) has a spread of about 5%, and the central value is typically 160 GeV. This gives good separation of neutrinos from π and K decay as a function of the radial position in the detector. The neutrino flux and energy spectrum from such a beam can be determined more accurately than for the alternative *wide* (*or broad*) *band* beam, but at the expense of reduced statistics. Using the 160 GeV parent meson beam at CERN, the CDHS (Abramowicz *et al.*, 1986) and CHARM (Allaby *et al.*, 1986) collaborations obtained the following ratios

$$R_\nu = 0.3072 \pm 0.0025 \pm 0.0020 \qquad (E_H > 10\ \text{GeV})\ \text{CDHS},$$
$$R_\nu = 0.3093 \pm 0.0031 \qquad (E_H > 4\ \text{GeV})\ \text{CHARM}. \quad (8.153)$$

In addition, the CDHS collaboration (Abramowicz *et al.*, 1985), using the 200 GeV narrow band beam, have measured $\bar{R}_\nu = 0.363 \pm 0.015$, for $E_H > 10$ GeV. Fig. 8.14 shows these results, together with the expected values of R_ν and $\bar{R}_\nu$ as a function of $\sin^2\theta_W$. For $\sin^2\theta_W \simeq 0.22$ the value of R_ν is more sensitive to $\sin^2\theta_W$ than that of $\bar{R}_\nu$. In the derivation of equations (8.150), only leading order electroweak terms were included. Further, the effects of strange and charm quarks (these are important since they are produced in CC events) were neglected. Including these effects, the values of $\sin^2\theta_W$ deduced are

$$\sin^2\theta_W = 0.225 \pm 0.005\ (\text{expt}) \pm 0.003\ (\text{theor.}) + 0.013\ (m_c - 1.5)\ \text{CDHS},$$
$$\sin^2\theta_W = 0.236 \pm 0.005\ (\text{expt}) \pm 0.003\ (\text{theor.}) + 0.012\ (m_c - 1.5)\ \text{CHARM}, \tag{8.154}$$

where m_c is the charm quark mass in GeV. The theoretical 'error' excludes the uncertainty as to the value of the c-quark mass which should be used. The effects of uncertainties in the QPM/QCD model used, together with uncertainties in the radiative corrections are, however, contained in this error. The radiative corrections were calculated using the prescription of

Fig. 8.14 The NC/CC ratios as a function of $\sin^2\theta_W$ for νN and $\bar{\nu}$N. The horizontal bands correspond to the experimental results.

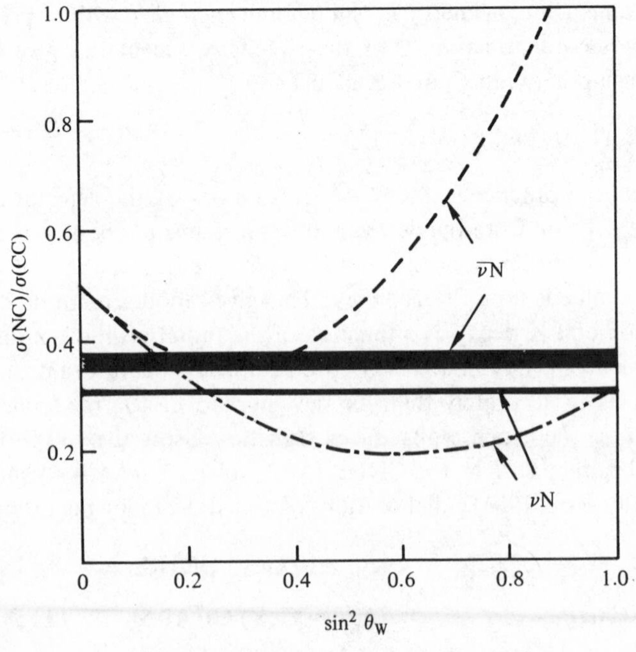

Wheater and Llewellyn Smith (1982), and $\sin^2\theta_W$ is expressed in the on-shell renormalisation scheme (Sirlin, 1980). The effect of radiative corrections is to reduce the value of $\sin^2\theta_W$ extracted by about 0.010. This shift is significant compared to the experimental errors and the corrected value is in good agreement with other determinations; giving some support to the correction procedure.

The impressive precision achieved by the above experiments raises the question as to whether the uncertainty in our understanding of structure function data (i.e. the role of higher twist, the EMC effect, nuclear shadowing, etc.), limit the precision to which $\sin^2\theta_W$ can be extracted in DIS. Llewellyn Smith (1983) has pointed out that such effects do not significantly change the extracted value of $\sin^2\theta_W$, at the level of current (and planned) experimental accuracy. This stems, essentially, from the mainly isovector nature of the weak neutral current.

Although measurements of the inclusive NC/CC ratio on isoscalar targets provide the most accurate determinations of $\sin^2\theta_W$ (using neutrino beams), they do not allow the extraction of the individual couplings C_L^u, C_L^d, C_R^u and C_R^d in (8.147) and (8.148). Extraction of these couplings requires, in addition, measurements on a target with a different ratio of u and d quarks. The BEBC deuterium collaboration (Allasia *et al.*, 1983) have measured R_ν and $\bar{R}_\nu$ (with $E_H > 5$ GeV) on both proton and neutrino targets and extract

$$|C_L^u|^2 = 0.133 \pm 0.026 \pm 0.015, \qquad |C_R^u|^2 = 0.020 \pm 0.019 \pm 0.004,$$
$$|C_L^d|^2 = 0.192 \pm 0.026 \pm 0.015, \qquad |C_R^d|^2 = 0.002 \pm 0.019 \pm 0.004. \tag{8.155}$$

In terms of a two-parameter fit (ρ and $\sin^2\theta_W$) these values give $\rho = 1.01 \pm 0.06$ and $\sin^2\theta_W = 0.202 \pm 0.054$. A similar analysis, using ν_μ and $\bar{\nu}_\mu$ beams on a hydrogen target, gives compatible values for the chiral couplings, and also $\rho = 0.989 \pm 0.029$ and $\sin^2\theta_W = 0.231 \pm 0.028$ (Jones *et al.*, 1986). Combining the available data gives a value of $\rho = 1.01 \pm 0.01$, which is compatible with the simplest Higgs assignment.

A more accurate determination of the righthanded couplings has been made using the semi-inclusive scattering data on $\nu_\mu(\bar{\nu}_\mu)\mathrm{N} \to \nu_\mu(\bar{\nu}_\mu)\pi^\pm \mathrm{X}$. In the current quark fragmentation region, the π^+ to π^- ratio for ν_μ interactions is (see Section 7.10)

$$\frac{n(\pi^+)}{n(\pi^-)} = \frac{(|C_L^u|^2 + a|C_R^u|^2)D_u^{\pi^+} + (|C_L^d|^2 + a|C_R^d|^2)D_d^{\pi^+}}{(|C_L^u|^2 + a|C_R^u|^2)D_u^{\pi^-} + (|C_L^d|^2 + a|C_R^d|^2)D_d^{\pi^-}} \tag{8.156}$$

where $a = \frac{1}{3}$ for the full y range, and the fragmentation functions satisfy

$D_d^{\pi^+} = D_u^{\pi^-}$ and $D_d^{\pi^-} = D_u^{\pi^+}$ from isospin invariance. The fragmentation functions can be measured in the corresponding ν_μ and $\bar{\nu}_\mu$ CC events. Including the effects of sea-quarks, Sehgal (1977), using the Gargamelle heavy liquid bubble chamber data of Kluttig *et al.* (1977), finds

$$|C_R^u| = 0.17 \pm 0.04, \qquad |C_R^d| = 0.00 \pm 0.12, \tag{8.157}$$

indicating that C_R^u is non-zero. The constraints from the above results are sketched in Fig. 8.15.

In order to ascertain the signs of the chiral coupling constants C_L^u, C_R^u, C_L^d and C_R^d, it is convenient to rewrite the $\bar{u}_4 \cdots u_2$ part of the matrix element as (Hung and Sakurai, 1981)

$$J_{NC}^\mu = \alpha \bar{q} \gamma^\mu \frac{\tau_3}{2} q + \beta \bar{q} \gamma^\mu \gamma^5 \frac{\tau_3}{2} q + \gamma \bar{q} \gamma^\mu \frac{I}{2} q + \delta \bar{q} \gamma^\mu \gamma^5 \frac{I}{2} q, \tag{8.158}$$

where q is the u, d quark doublet, τ_3 the Pauli spin matrix, and I the 2×2 unit matrix. The terms in τ_3 and I correspond to *isovector* (IV) and *isoscalar* (IS) currents, as discussed in Section 8.2.1. Thus, the coefficients correspond to the terms α(IV – V), β(IV – A), γ(IS – V) and δ(IS – A). The correspondence with the chiral couplings is

$$\alpha = C_L^u - C_L^d + C_R^u - C_R^d, \qquad \beta = -(C_L^u - C_L^d) + C_R^u - C_R^d,$$

$$\gamma = C_L^u + C_L^d + C_R^u + C_R^d, \qquad \delta = -(C_L^u + C_L^d) + C_R^u + C_R^d. \tag{8.159}$$

Hence, in the standard model $\alpha = 1 - 2z$, $\beta = -1$, $\gamma = -\frac{2}{3}z$ and $\delta = 0$, where $z = \sin^2 \theta_W$.

Information on the signs of the couplings comes from the analysis of single pion production in the reactions $\nu p(n) \to \nu N\pi$ and $\bar{\nu} p(n) \to \bar{\nu} N\pi$. The $N\pi$ system is dominated by Δ production, showing the mainly isovector nature of the weak neutral current. An analysis by Fogli (1982) gives

$$\alpha = 0.68 \pm^{0.24}_{0.45}, \qquad \beta = -0.99 \pm^{0.37}_{0.45},$$

$$\gamma = -0.20 \pm^{0.08}_{0.12}, \qquad \delta = 0.01 \pm^{0.10}_{0.10}. \tag{8.160}$$

Comparison with (8.155) and (8.159) shows that this requires C_L^u and C_L^d to be positive and negative respectively. Analysis of elastic $\nu(\bar{\nu})N \to \nu(\bar{\nu})N$ scattering (see, e.g., Barbiellini and Santoni, 1986) shows that C_R^u is negative. Inspection of Fig. 8.15 shows that the u and d quark chiral couplings are in agreement with the standard model.

Vector and axial-vector couplings give a y-dependence which is a mixture of flat and $(1 - y)^2$ components. The presence of S or P couplings would give an additional y^2 dependence, with an equal contribution to ν

Fig. 8.15 Allowed regions of (*a*) C_L^u, C_L^d and (*b*) C_R^u, C_R^d. The circles are limits from measurements of R_ν and $\bar{R}_\nu$ on an isoscalar target. The boxes are from R_ν and $\bar{R}_\nu$ on p and n targets in (*a*) and from semi-inclusive hadron studies in (*b*).

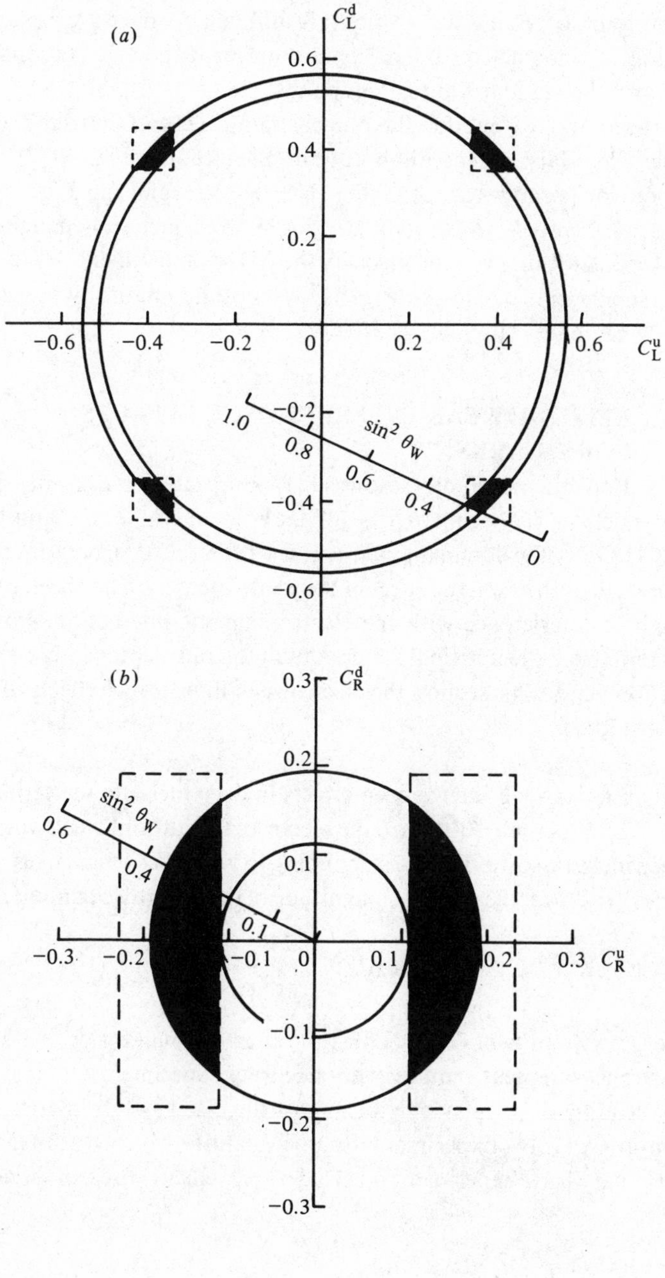

and $\bar{\nu}$. The CHARM collaboration (Jonker *et al.*, 1981) find an upper limit $g_{SP}^2/g_{VA}^2 < 0.03$ (95% c.l.) for such couplings. Thus, the weak neutral current is mainly V − A in nature, with a small V + A component. In terms of isospin components, it is mainly isovector in nature. The x_{BJ} distributions in NC events (which are difficult to measure because the outgoing ν is not detected) are compatible with the expectations of the QPM and the standard model couplings.

In the standard model flavour-changing neutral currents do not appear. The Gargamelle collaboration (Blietschau *et al.*, 1977) find no evidence for strangeness-changing NC events, and give at 90% c.l. $\sigma(\nu \to \nu\Lambda(\Sigma)X)/\sigma(\nu \to \nu X) < 5.4 \times 10^{-3}$. A 90% c.l. σ(charm-changing NC)/ σ(NC) < 0.026 limit has been given by the CDHS experiment (Holder *et al.*, 1978). Such events would show up as 'wrong-sign' muons by the sequence $\nu_\mu N \to \nu_\mu cX$, $c \to \mu^+\nu_\mu \cdots$.

8.9 ELECTROWEAK INTERFERENCE EFFECTS FOR QUARKS

For the currently accessible kinematic range in charged lepton–nucleon scattering ($Q^2 \lesssim 200$ GeV^2), and in e^+e^- annihilation ($s^{1/2} \lesssim 45$ GeV), the dominant interaction is the electromagnetic interaction. However, in both cases, the effect of the weak interaction has been observed through its interference with the electromagnetic interaction. For e^+e^-, this interference effect in final states containing only leptons, was discussed in Section 6.3. In this section the electroweak interference effects of quarks are considered.

8.9.1 Electroweak interference effects in deep inelastic scattering

If $\mathcal{M}(\gamma)$ and $\mathcal{M}(Z^0)$ are the electromagnetic and weak amplitudes for deep inelastic scattering, corresponding to γ and Z^0 exchange respectively then, for $Q^2 \ll M_Z^2$, the order of magnitude of the electroweak interference is

$$A \simeq \mathcal{M}(\gamma)\cdot\mathcal{M}(Z^0)/|\mathcal{M}(\gamma)|^2 \simeq \mathcal{M}(Z^0)/\mathcal{M}(\gamma) \simeq \frac{GQ^2}{e^2} \simeq 10^{-4}Q^2, \tag{8.161}$$

where Q^2 is in units of GeV^2. Thus, the effect is small for $Q^2 \lesssim 100$ GeV^2, and the measurement requires very accurate experiments.

The leading diagrams contributing to the electroweak interference are shown in Fig. 8.16. Experimentally, the quantity which was measured in the original SLAC experiment on electroweak effects (Prescott *et al.*, 1979) was

$$A = (\sigma_R - \sigma_L)/(\sigma_R + \sigma_L), \tag{8.162}$$

where σ_R and σ_L are the cross-sections for right and lefthanded incident polarised electrons respectively. The experiment took advantage of the change in precession of the electron spin about the beam transport system due to its anomalous magnetic moment. The amplitude for the case of a *lefthanded* e^- is (using Appendix C)

$$\mathcal{M} = \mathcal{M}(\gamma) + \mathcal{M}(Z^0)$$

$$= \frac{ie^2 Q_e Q_q}{2q^2} \bar{u}_3 \gamma^\mu (1-\gamma^5) u_1 \cdot \bar{u}_4 \gamma_\mu u_2$$

$$+ \frac{ig^2}{4\cos^2\theta_W (q^2 - M_Z^2)}$$

$$\times \bar{u}_3 \gamma^\mu C_L^e (1-\gamma^5) u_1 \cdot \bar{u}_4 \gamma_\mu [C_L^q (1-\gamma^5) + C_R^q (1+\gamma^5)] u_2. \quad (8.163)$$

Using the usual trace techniques, we obtain

$$|\mathcal{M}(\gamma)|^2 = \frac{16e^4 Q_e^2 Q_q^2}{q^4} [(p_1 \cdot p_2)(p_3 \cdot p_4) + (p_1 \cdot p_4)(p_2 \cdot p_3)],$$

$$|\mathcal{M}(Z^0)|^2 = \frac{16g^4 (C_L^e)^2}{\cos^4\theta_W (q^2 - M_Z^2)^2}$$

$$\times [(C_L^q)^2 (p_1 \cdot p_2)(p_3 \cdot p_4) + (C_R^q)^2 (p_1 \cdot p_4)(p_2 \cdot p_3)], \quad (8.164)$$

$$|\mathcal{M}_{int}| = \frac{32e^2 Q_e Q_q g^2 C_L^e}{\cos^2\theta_W q^2 (q^2 - M_Z^2)} [C_L^q (p_1 \cdot p_2)(p_3 \cdot p_4) + C_R^q (p_1 \cdot p_4)(p_2 \cdot p_3)].$$

For a *righthanded* e^- beam, we replace C_L^e by C_R^e and C_L^q by C_R^q (and *vice versa*) in (8.163) and (8.164).

The above expressions are for γq and Z^0q scattering. To obtain the cross-section for a nucleon, these must be weighted with the quark momentum distribution $xq(x, Q^2)$. For an isoscalar target consisting only

Fig. 8.16 Lowest order diagrams giving rise to electroweak interference in deep inelastic scattering.

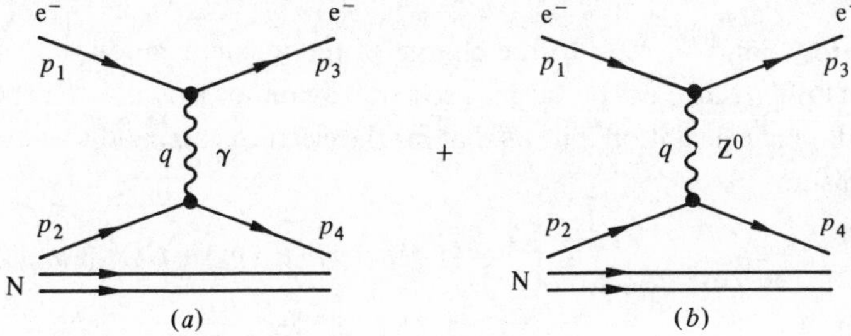

of u and d quarks (this is a reasonable approximation for the $E_e = 20$ GeV, $Q^2 \lesssim 5$ GeV^2 SLAC experiment), the denominator of (8.162) is proportional to $5/18x(u+d)$. The numerator is also proportional to $x(u+d)$. After some manipulation, (8.162) and (8.164) give, for $Q^2 \ll M_Z^2$,

$$\frac{A}{Q^2} = a_1 + a_2 g(y), \tag{8.165}$$

where

$$g(y) = [1-(1-y)^2]/[1+(1-y)^2], \tag{8.166}$$

and

$$a_1 = -B(C_R^e - C_L^e)[(C_R^u + C_L^u) - \tfrac{1}{2}(C_R^d + C_L^d)] = B g_A^e [g_V^u - \tfrac{1}{2} g_V^d],$$
$$a_2 = -B(C_R^e + C_L^e)[(C_R^u - C_L^u) - \tfrac{1}{2}(C_R^d - C_L^d)] = B g_V^e [g_A^u - \tfrac{1}{2} g_A^d]. \tag{8.167}$$

The correspondence between the chiral couplings and the V and A couplings is $g_V = C_L + C_R$ and $g_A = C_L - C_R$. The constant factor is $B = 6G_F/(5(2^{1/2})\pi\alpha) = 4.3 \times 10^{-4}$ GeV^{-2}. In the standard model (see Appendix C) we obtain

$$a_1 = -B(\tfrac{3}{8} - \tfrac{5}{6}\sin^2\theta_W), \qquad a_2 = -B(\tfrac{3}{8} - \tfrac{3}{2}\sin^2\theta_W). \tag{8.168}$$

The results of the SLAC experiment using a deuterium target are shown in Fig. 8.17. The result of a two-parameter fit gives $a_1 = (-9.7 \pm 2.6) \times 10^{-5}$ GeV^{-2} and $a_2 = (4.9 \pm 8.1) \times 10^{-5}$ GeV^{-2}. Note that, if the right-handed electron formed a doublet with a possible neutral heavy lepton E^0, symmetric with the usual lefthanded doublet, then a value $a_1 = 0$ would result. Hence, this is excluded in a model independent way. Using (8.167), the result for a_1 gives $g_A^e(g_V^u - g_V^d/2) = -0.23 \pm 0.06$ and hence $g_V^u - g_V^d/2 = 0.45 \pm 0.12$ (using the standard model value $g_A^e = -\tfrac{1}{2}$). Hence this ingenious experiment lends good support for the standard model, and gives a value $\sin^2\theta_W = 0.224 \pm 0.020$.

Electroweak interference effects have also been measured at the much higher Q^2 values obtainable at the CERN muon beam, by the BCDMS collaboration (Argento *et al.*, 1983). In this case the quantity measured was

$$\Delta = (\sigma_L^+ - \sigma_R^-)/(\sigma_L^+ + \sigma_R^-), \tag{8.169}$$

where + and − refer to the charge of the incident muon, and L and R to its polarisation. The target used was again an isoscalar (carbon). The method of calculation follows that for the electron case, and the interference terms are

$$|\mathcal{M}_{\text{int}}(^{+}_{L})| = \frac{32e^2 Q_\mu Q_q g^2 C_R^\mu}{\cos^2\theta_W q^2(q^2 - M_Z^2)} [C_L^q(p_1\cdot p_2)(p_3\cdot p_4) + C_R^q(p_1\cdot p_4)(p_2\cdot p_3)], \tag{8.170}$$

and

$$|\mathcal{M}_{\text{int}}(\bar{\text{R}})| = \frac{32e^2 Q_\mu Q_q g^2 C_R^\mu}{\cos^2\theta_W q^2(q^2 - M_Z^2)}[C_R^q(p_1\cdot p_2)(p_3\cdot p_4) + C_L^q(p_1\cdot p_4)(p_2\cdot p_3)]. \tag{8.171}$$

Using these expressions, and incorporating the u and d quark densities (for large x the restriction to u and d is still reasonable), we obtain for (8.169)

$$\frac{\Delta}{Q^2} = BC_R^\mu[2(C_R^u - C_L^u) - (C_R^d - C_L^d)]g(y)$$
$$= -\text{B}(g_V^\mu - g_A^\mu)[g_A^u - \tfrac{1}{2}g_A^d]g(y). \tag{8.172}$$

Note that, in practice, the beam does not have helicity $h = 1$ as assumed above, but $|h| = 0.66$ and 0.81 for $p_\mu = 120$ and 200 GeV respectively. In general, we must replace $(g_V^\mu - g_A^\mu)$ by $(|h|g_V^\mu - g_A^\mu)$. Since g_V^μ is small, the main contribution to Δ is (in contrast to the SLAC method) from the parity conserving axial terms. In the standard model (for $h = 1$) we obtain

$$\frac{\Delta}{Q^2} = -\tfrac{3}{2}B\sin^2\theta_W g(y). \tag{8.173}$$

Fig. 8.17 Measurements of the asymmetry parameter A, as a function of Q^2, for ed scattering (SLAC, 16.2–22.2 GeV).

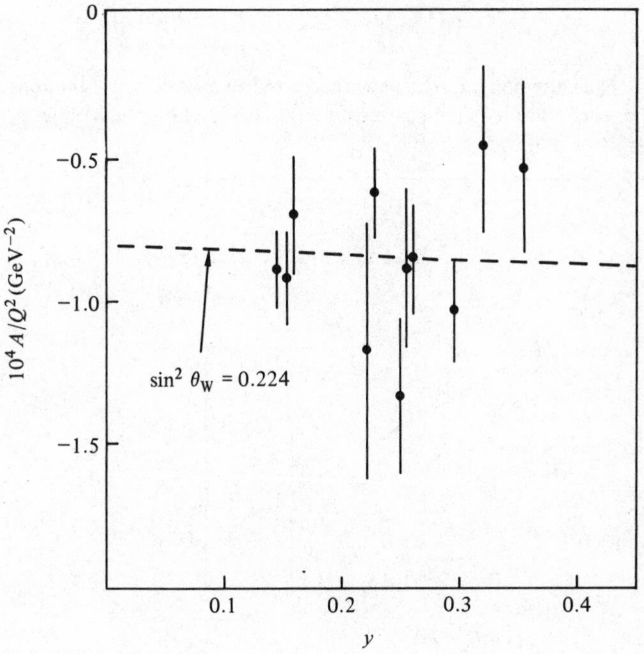

The experimental result using the 200 GeV beam is shown in Fig. 8.18. It can be seen that the result is compatible with (8.172). The measured slope is $(-1.47 \pm 0.37 \pm 0.24)10^{-4}\,\text{GeV}^{-2}$, compared to -1.51×10^{-4} for the standard model with $\sin^2\theta_W = 0.23$. The results, including the effects of radiative corrections (from the interference of one photon and two photon diagrams), give a value $\sin^2\theta_W = 0.23 \pm 0.07 \pm 0.04$. Alternatively, if we assume $C_R^{\mu} = \sin^2\theta_W$ is known from other sources, then (from (8.172)) the results give a powerful constraint on $[2(C_R^u - C_L^u) - (C_R^d - C_L^d)]$, which is consistent with the standard model assignments.

At very large values of Q^2 ($\gtrsim M_Z^2$) the $|\mathscr{M}_Z|^2$ term should become important. The cross-section can be deduced in the same way as for the electromagnetic interaction alone, using $|\mathscr{M}|^2$ from (8.163) for left-handed electrons or the equivalent expression for righthanded electrons. For unpolarised e^- beams, the weighted average of these two terms is required. The resulting differential cross-section for e^-p can be written

$$\frac{d^2\sigma(\text{NC})}{dx\,dy} = \frac{2\pi\alpha^2 s}{Q^4}[T(\gamma) + T(Z) + T(\text{int})], \tag{8.174}$$

where the contributions from the exchange graphs involving one photon, one Z^0 and the γ–Z^0 interference term are respectively

$$T(\gamma) = Y_+ F_2,$$

Fig. 8.18 Measurements of the asymmetry parameter Δ, as a function of $g(y)Q^2$, for muon–carbon scattering (200 GeV). The straight line is the result of a linear fit to the data.

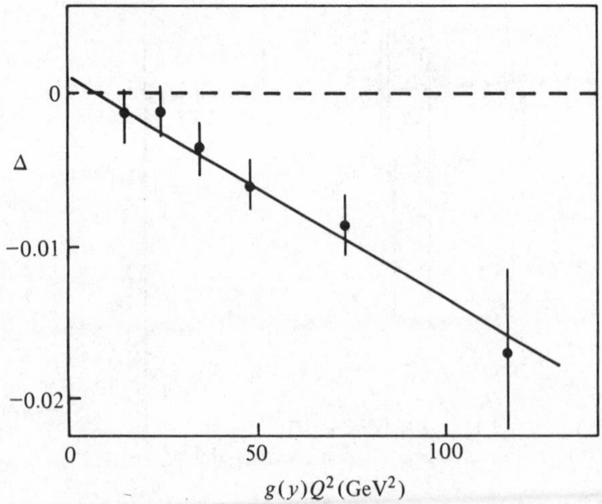

$$T(Z) = \frac{Q^4}{16 \sin^4 \theta_W \cos^4 \theta_W (Q^2 + M_Z^2)^2} \{Y_+[g_V^{e^2} + g_A^{e^2} - 2g_V^e g_A^e h_e]Z_2$$

$$+ Y_-[2g_V^e g_A^e - (g_V^{e^2} + g_A^{e^2})h_e]xZ_3\},$$

$$T(\text{int}) = \frac{Q^2}{4 \sin^2 \theta_W \cos^2 \theta_W (Q^2 + M_Z^2)} \{Y_+[-g_V^e + h_e g_A^e]I_2$$

$$+ Y_-[-g_A^e + h_e g_V^e]xI_3\}, \tag{8.175}$$

with

$$F_2 = x\Sigma(q + \bar{q})Q_q^2, \qquad Z_2 = x\Sigma(q + \bar{q})(g_V^{q^2} + g_A^{q^2}), \qquad I_2 = 2x\Sigma(q + \bar{q})g_V^q Q_q,$$

$$xZ_3 = 2x\Sigma(q - \bar{q})g_V^q g_A^q, \qquad xI_3 = 2x\Sigma(q - \bar{q})g_A^q Q_q. \tag{8.176}$$

Further, $Y_\pm = 1 \pm (1 - y)^2$ and h_e denotes the helicity of the incident electron and has the value $h_e = 1$ for e_R^- (righthanded) and $h_e = -1$ for e_L^-. The proton beam is assumed unpolarised. The structure functions in (8.176) are, in general, functions of both x and Q^2. The expressions given in (8.176) are those for the QPM. The contribution from the longitudinal polarisation states of the γ and Z^0 states have been neglected, but in general these must be considered, as indeed must the contribution of higher order electroweak graphs. However, $R(x, Q^2) = \sigma_L/\sigma_T$ is expected, from QCD, to be small. The effect of a non-zero value of R can be taken into account by redefining Y_+ to be $Y_+ = 1 + (1 - y^2) - y^2R/(1 + R)$.

The expressions given in (8.175) and (8.176) can be derived by starting with the matrix elements given by (8.164) for e_L^-–q, together with analogous terms for e_R^-–q, e_L^-–$\bar{q}$ and e_R^-–$\bar{q}$, weighting these with the appropriate quark or antiquark density distributions and using the formulae of Sections 7.2 and 7.3 for deep inelastic scattering. Note that $Q^2 \simeq sxy$ for the kinematic region of high energy ep colliders, and at these very large values of Q^2 the contribution of sea-quarks becomes increasingly important. Note also the distribution in y is no longer the $[1 + (1 - y)^2]$ distribution obtained for purely electromagnetic interactions. This is because the left and righthanded couplings of quarks are no longer equal. Fig. 8.19 shows an estimate of the cross-section at the HERA (30 GeV on 820 GeV) ep collider for $\sigma(\gamma + Z^0)$ divided by that for γ alone, $\sigma(\gamma)$, as a function of y for $x \simeq 0.25$. It can be seen that, for a lefthanded e^- beam, the ratio at large y is much larger than for a lefthanded e^+ beam.

In order to test the standard model it is necessary to extract the left- and righthanded couplings of all quarks and leptons. It is clearly desirable to use polarised electron (and also proton) beams in trying to get a handle on these couplings. From the form of (8.175) it can be seen that measurement of the asymmetry $A(h_e)=[\sigma(h_e)-\sigma(-h_e)]/[\sigma(h_e)+\sigma(-h_e)]$ is sensitive to the couplings and has the advantage that some systematic and theoretical uncertainties cancel.

At the very large values of Q^2 expected at the HERA Collider, the contribution of the $W^\pm$ exchange (i.e. charged current) reactions $e^- q_1 \to \nu_e q_2$ and $e^- \bar{q}_2 \to \nu_e \bar{q}_1$ (q_1 and q_2 are quarks of charges $\frac{2}{3}$ and $-\frac{1}{3}$ respectively) become important. The cross-section for these reactions can be derived in a similar way to that for the NC case and leads to

$$\frac{d^2\sigma(CC)}{dx\,dy}=\frac{2\pi\alpha^2 s}{Q^4}\left[\frac{Q^2}{4\sin^2\theta_W(Q^2+M_W^2)}\right]^2\left(\frac{1-h_e}{2}\right)(Y_+W_2^\pm+Y_-xW_3^\pm). \tag{8.177}$$

The $W^\pm$ structure functions $W_2^\pm$ and $xW_3^\pm$ are again functions of both x and Q^2. In the QPM, neglecting any threshold effects for the production of heavy quarks, we have

$$W_2^\pm = 2x\Sigma(q_1+\bar{q}_2), \qquad xW_3^\pm = 2x\Sigma(q_1-\bar{q}_2). \tag{8.178}$$

The ratio of the NC and CC cross-sections is a potentially further useful quantity for tests of the standard model. Further, comparison of experimental results for both NC and CC events with the above formulae, should provide a sensitive test for the existence of any further neutral or charged

Fig. 8.19 The ratio of cross-sections $\sigma(\gamma+Z^0)/\sigma(\gamma)$, as a function of y, for $x=0.25$ for various lepton beam polarisations.

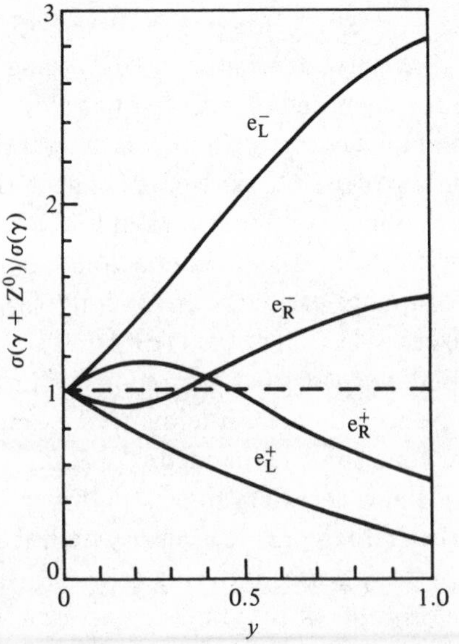

vector bosons (e.g. through modification of the Q^2 dependence of the cross-section compared to that expected from the propagator).

8.9.2 Electroweak interference effects in $e^+e^- \rightarrow q\bar{q}$

The general formulae for the electroweak asymmetry for $e^+e^- \rightarrow f\bar{f}$ were derived in Section 6.3. For hadronic final states, decisive tests are difficult because of the problems associated with identifying the quark flavour for a particular event in a reliable way. A further problem is in the reconstruction of the quark axes from the final state hadrons. The effects of hadronisation, hard QCD effects, the classification of slow particles to a particular jet and track losses, all introduce potential systematic effects in the reconstruction of the quark axes.

The quantity R_h (6.91) has been measured for all hadrons (i.e. for u, d, s, c and b quarks) up to $s^{1/2} \sim 45$ GeV. Specific quark asymmetry measurements have been made (e.g. Marshall, 1985) for c-quarks (e.g. HRS have used D* mesons, i.e. $c \rightarrow D^* \rightarrow D\pi$, as a tag) giving $A_{cc} = (-15 \pm 6)\%$ and b-quarks (e.g. JADE have used the large p_T property of the decay leptons) giving $A_{bb} = (-22.8 \pm 6.0 \pm 2.5)\%$. Note that the predicted asymmetry is larger for quarks than leptons, because of the presence of the Q_q term in (6.92) in the denominator. However, phase space suppression at PETRA energies reduces this increase from 3 to about 2.7. A fit to the asymmetry data by Marshall (1985) gives

$$g_A^c = 0.56 \pm 0.18, \qquad g_A^b = -0.50 \pm 0.10. \tag{8.179}$$

These values are consistent (within the rather large errors) with the standard model values of $g_A^c = 0.5$ and $g_A^b = -0.5$. Including the measurements on R_h gives the vector couplings

$$g_V^{uc} = 0.38 \pm 0.32, \qquad g_V^{dsb} = -0.58 \pm 0.22, \tag{8.180}$$

to be compared to the standard model values of ($\sin^2\theta_W = 0.225$), $g_V^{uc} = 0.20$ and $g_V^{dsb} = -0.35$. The contribution to R_h expected from QCD and electroweak effects is shown in Fig. 8.20.

8.9.3 Electroweak effects in atomic physics

The very small values of the momentum transfers involved in atomic transitions make the size of potential electroweak effects tiny (see (8.161)). However, electroweak effects have been studied by measuring parity violating effects in heavy atoms. The $g_V^e g_A^q$ (q = u, d) term is small compared to that for $g_A^e g_V^q$, since the nucleon spins tend to cancel in the nucleus. Thus, the parity violating terms lead to an effective weak charge of the form, for Z protons and N neutrons,

$$Q_{\mathrm{wk}} \propto g_{\mathrm{A}}^{\mathrm{e}}[Z(2g_{\mathrm{V}}^{\mathrm{u}} + g_{\mathrm{V}}^{\mathrm{d}}) + \mathrm{N}(g_{\mathrm{V}}^{\mathrm{u}} + 2g_{\mathrm{V}}^{\mathrm{d}})]. \tag{8.181}$$

In the standard model we have

$$2g_{\mathrm{V}}^{\mathrm{u}} + g_{\mathrm{V}}^{\mathrm{d}} = \tfrac{1}{2}(1 - 4\sin^2\theta_{\mathrm{W}}), \qquad g_{\mathrm{V}}^{\mathrm{u}} + 2g_{\mathrm{V}}^{\mathrm{d}} = -\tfrac{1}{2}, \tag{8.182}$$

so that only the neutron term has appreciable size.

The presence of such parity violating interactions means that the atomic levels are no longer pure parity eigenstates. The magnitude of the matrix element is $\mathcal{M} \sim 10^{-18} Z^3 e^2/a_0$, for large $Z(\sim 80)$. Crudely, one power of Z comes from the weak charge (8.181), a second from the electron momentum and the third from the wave function at the origin. The use of heavy atoms (e.g. bismuth or caesium) thus enhances the electroweak effects. Other methods of enhancing these effects are the study of (electromagnetically) forbidden transitions, or the use of an external magnetic field to adjust the hyperfine splitting, so levels of opposite parity overlap. For example, the rotation of the plane of polarisation of laser light has been measured for bismuth and lead. A measurement of the asymmetry in the absorption cross-sections for R and L circularly polarised laser light, for forbidden magnetic transitions in caesium, gives a value (Bouchiat *et al.*, 1984) of $\sin^2\theta_{\mathrm{W}} = 0.205 \pm 0.035 \pm 0.025$, with a further 'atomic physics

Fig. 8.20 Some measurements of R_{h}, as a function of $s^{1/2}$, together with the expectations of the QPM, QCD ($\alpha_{\mathrm{s}} \simeq 0.18$) and electroweak effects ($\sin^2\theta_{\mathrm{W}} = 0.25$). ●, PETRA; ○, PEP.

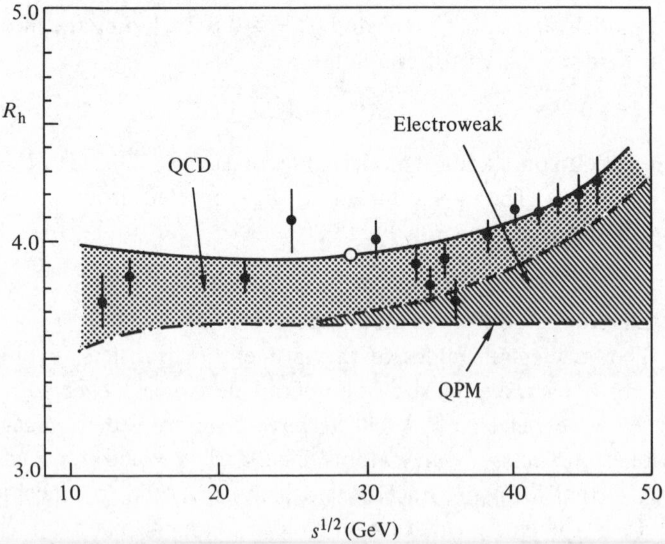

error' of ± 0.045. Results consistent with the standard model are obtained from these atomic physics experiments, but the precision has yet to rival that from other methods. A detailed discussion on the above topics, and on parity violating effects in nuclear forces, can be found in Commins and Bucksbaum (1983).

8.10 SUMMARY OF NC COUPLINGS OF LEPTONS AND QUARKS

In this section the various results on the couplings of leptons and quarks to the Z^0 described above and in Chapter 6 are summarised. Results from the decays of Z^0 bosons produced in $\bar{p}p$ collisions are described in Chapter 10. The total cross-section data from $\nu_\mu e^- \to \nu_\mu e^-$ and $\bar{\nu}_\mu e^- \to \bar{\nu}_\mu e^-$ yield two possible solutions, namely

$$g_A^e = -0.49 \pm 0.03, \qquad g_V^e = -0.05 \pm 0.05, \tag{8.183}$$

or

$$g_A^e = -0.05 \pm 0.05, \qquad g_V^e = -0.49 \pm 0.03. \tag{8.184}$$

In obtaining these numbers, the values of the cross-sections quoted in Section 6.2 for these processes, together with the formulae given in Table 6.3, have been used. The other two possible solutions are excluded by the $\nu_e e^- \to \nu_e e^-$ and $\bar{\nu}_e e^- \to \bar{\nu}_e e^-$ results (see Fig. 6.8).

The ambiguity between these solutions can be resolved by using the $e^+e^- \to l^+l^-$ data (Section 6.3). These show that it is the axial coupling g_A which is large and that $g_V \sim 0$; that is, solution (8.183). Assuming τ–μ–e universality, and taking the weighted mean of the asymmetry results for $\mu^+\mu^-$ and $\tau^+\tau^-$, gives another estimate for the axial coupling, namely

$$g_A^e = -0.508 \pm 0.015. \tag{8.185}$$

The results on R for $e^+e^- \to l^+l^-$ give the value $|g_V^e| = 0.20 \pm 0.05$ for the vector coupling, with the negative sign being preferred from the $\nu_\mu(\bar{\nu}_\mu)e$ scattering data.

For the quark sector, the moduli of the chiral couplings are given by (8.155). The results on the π^+/π^- ratio in $\nu_\mu(\bar{\nu}_\mu)$NC interactions give more accurate values for the righthanded couplings. Using additional information from the analysis of single pion production events and from elastic $\nu(\bar{\nu})$N scattering, it can be deduced that C_L^u is positive and that C_L^d and C_R^u are negative. Hence, the results of (8.155) and (8.157) give (using (6.57))

$$g_V^u = 0.20 \pm 0.06, \qquad g_A^u = 0.54 \pm 0.06, \tag{8.186}$$

and

$$g_V^d = -0.44 \pm 0.12, \qquad g_A^d = -0.44 \pm 0.12. \tag{8.187}$$

Table 8.6. *Vector and axial-vector couplings for leptons and quarks to the Z^0*
The standard model (SM) values are for $z = \sin^2\theta_W = 0.225$

	g_V		g_A	
Particle	Expt	SM	Expt	SM
ν_l		$\frac{1}{2}$		$\frac{1}{2}$
l^-	-0.05 ± 0.05	$-\frac{1}{2} + 2z$ $= -0.05$	-0.504 ± 0.012	$-\frac{1}{2}$
$q(+\frac{2}{3})$	(u) 0.20 ± 0.06	$\frac{1}{2} - \frac{4}{3}z$ $= 0.20$	(u) 0.54 ± 0.06 (c) 0.56 ± 0.18	$\frac{1}{2}$
$q(-\frac{1}{3})$	(d) -0.44 ± 0.12	$-\frac{1}{2} + \frac{2}{3}z$ $= -0.35$	(d) -0.44 ± 0.12 (b) -0.50 ± 0.10	$-\frac{1}{2}$

The chiral coupling C_R^d is poorly determined, and gives larger errors for the d-quark couplings.

The results of the electroweak interference experiments in deep inelastic scattering are compatible with the above values, but with larger errors. Using the measured value of a_1 and equation (8.167), the SLAC experiment gives $(g_V^u - g_V^d/2) = 0.45 \pm 0.12$. The axial couplings are very poorly determined. However, the BCDMS experiment is sensitive to the axial couplings, and the measured slope of the plot of Δ against $g(y)Q^2$ gives, using (8.172), $(g_A^u - g_A^d/2) = 0.76 \pm 0.22$. The axial couplings for the heavy c and b quarks are found from e^+e^- asymmetry measurements to be $g_A^c = 0.56 \pm 0.18$ and $g_A^b = -0.50 \pm 0.10$.

Table 8.6 summarises the measurements of these couplings. It can be seen that they are all compatible with the standard model predictions, with $\sin^2\theta_W = 0.225$. The most accurate determination of $\sin^2\theta_W$ is, at present, from the inclusive NC/CC ratio on isoscalar targets, giving for $m_c = 1.5 \pm 0.3$ GeV, $\sin^2\theta_W = 0.231 \pm 0.004$ (expt) ± 0.005 (theory). The combined ρ value is $\rho = 1.01 \pm 0.01$, compatible with the simplest doublet Higgs assignment.

9

Quark and lepton oscillations

9.1 THE NEUTRAL KAON SYSTEM

The valence quark composition of the K^+ and K^- mesons are $u\bar{s}$ and $\bar{u}s$ respectively. It is thus natural to assign the valence quark compositions $d\bar{s}$ and $\bar{d}s$ to their corresponding neutral counterparts, K^0 and $\bar{K}^0$. That is, the neutral kaons created in a strong, electromagnetic or weak process have, at the moment of creation (proper time $t = 0$), definite strangeness. The K^0 and $\bar{K}^0$ have $S = 1$ and -1 respectively. Since strangeness is conserved for the strong and electromagnetic interactions, and since the kaon is the lightest strange particle, the K^0 and $\bar{K}^0$ can only decay weakly. The lowest order decay processes are shown in Fig. 9.1, and both lead to a $u\bar{u}d\bar{d}$ final state.

We have seen from muon decay, and other processes, that the W only couples to lefthanded fermions or righthanded antifermions. The operation of charge conjugation C turns a lefthanded quark into a lefthanded antiquark, which has no coupling to the W. Subsequent operation of the parity transformation P gives a righthanded antiquark (P changes $\mathbf{r} \to -\mathbf{r}$, $\mathbf{p} \to -\mathbf{p}$, but $\boldsymbol{\sigma} = \mathbf{r} \times \mathbf{p} \to \mathbf{r} \times \mathbf{p}$), which can couple to the W. Thus, weak decays are expected to be eigenstates of CP. Now, the K^0 and $\bar{K}^0$ particles are not eigenstates of CP, but are transformed into one another by this operation. Taking the convention that

$$CP|K^0\rangle = |\bar{K}^0\rangle, \qquad CP|\bar{K}^0\rangle = |K^0\rangle, \tag{9.1}$$

we can construct the following linear combinations of K^0 and $\bar{K}^0$, which are CP eigenstates

$$\begin{aligned} |K_1^0\rangle &= \frac{1}{2^{1/2}}[|K^0\rangle + |\bar{K}^0\rangle] = \frac{1}{2^{1/2}}[|d\bar{s}\rangle + |s\bar{d}\rangle] \qquad CP = +1, \\ |K_2^0\rangle &= \frac{1}{2^{1/2}}[|K^0\rangle - |\bar{K}^0\rangle] = \frac{1}{2^{1/2}}[|d\bar{s}\rangle - |s\bar{d}\rangle] \qquad CP = -1. \end{aligned} \tag{9.2a}$$

Alternatively, we can express the strangeness eigenstates in terms of the states of definite CP

$$|K^0\rangle = |d\bar{s}\rangle = \frac{1}{2^{1/2}}[|K_1^0\rangle + |K_2^0\rangle],$$
$$|\bar{K}^0\rangle = |s\bar{d}\rangle = \frac{1}{2^{1/2}}[|K_1^0\rangle - |K_2^0\rangle]. \tag{9.2b}$$

9.1.1 Hadronic decay modes

The possible hadronic final states for neutral kaon decay are $2\pi(\pi^+\pi^-, \pi^0\pi^0)$ and $3\pi(\pi^+\pi^-\pi^0, \pi^0\pi^0\pi^0)$. These states have CP eigenvalues $+1$ and -1 respectively. This can be seen as follows. In the pion the quark and antiquark have zero orbital angular momentum l, and also zero net spin. The 'intrinsic' parity of the pion is $P_\pi = (-1)^{l+1} = -1$ (see Section 3.2.5). The 2π system is invariant under C and has zero orbital angular momentum, hence $CP(2\pi) = (P_\pi)^2 = 1$. In the 3π system, the pions are in states of zero relative orbital angular momentum. Hence, $CP(3\pi) = (P_\pi)^3 = -1$. Thus, the final states are eigenstates of CP, but the initial states K^0 and $\bar{K}^0$ are not. For CP-conserving decays, it is the K_1^0 and K_2^0 components which will decay and with which we can associate specific lifetimes.

The phase space factors for the 2π and 3π decay modes of the kaon are substantially different. It can be seen from the formulae given in Section 4.2, that the $K \to 3\pi$ phase space integral is much less than that for $K \to 2\pi$. Hence, it is expected that the K_2^0 component has a substantially longer lifetime than that of K_1^0. Experimentally, *short* (K_S^0) and *long* (K_L^0) lived components are indeed observed, with lifetimes $\tau_s = 0.892 \times 10^{-10}$ s and $\tau_L = 5.18 \times 10^{-8}$ s respectively.

Ignoring, for the moment, the possibility of CP violation, let us consider the proper time evolution of an initially pure beam of K^0 particles (produced for example by the reaction $\pi^- p \to K^0\Lambda^0$). At times 0 and t we have therefore

$$|\psi(0)\rangle = \frac{1}{2^{1/2}}[|K_S^0\rangle + |K_L^0\rangle], \tag{9.3a}$$

Fig. 9.1 Lowest order decay processes for (*a*) K^0 and (*b*) $\bar{K}^0$.

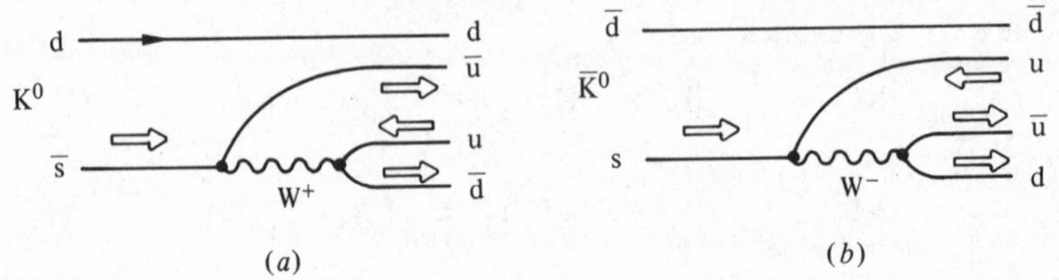

$$|\psi(t)\rangle = \frac{1}{2^{1/2}}[|K_s^0\rangle \exp(-i\lambda_s t) + |K_L^0\rangle \exp(-i\lambda_L t)], \tag{9.3b}$$

where we have identified K_1^0 and K_2^0 with the short- and long-lived components respectively. The time constants are given by (see Section 2.2.4)

$$\lambda_s = m_s - i\Gamma_s/2, \qquad \lambda_L = m_L - i\Gamma_L/2. \tag{9.4}$$

The fraction of K^0 particles in the beam, after a time t, is

$$F(K^0) = |\langle K^0 | \psi(t)\rangle|^2$$
$$= \tfrac{1}{4}[\exp(-\Gamma_s t) + \exp(-\Gamma_L t) + 2\cos(\Delta m t)\exp(-\bar{\Gamma} t)], \tag{9.5}$$

where $\bar{\Gamma} = (\Gamma_s + \Gamma_L)/2$ and $\Delta m = m_L - m_s$. The corresponding fraction of $\bar{K}^0$ is

$$F(\bar{K}^0) = |\langle \bar{K}^0 | \psi(t)\rangle|^2$$
$$= \tfrac{1}{4}[\exp(-\Gamma_s t) + \exp(-\Gamma_L t) - 2\cos(\Delta m t)\exp(-\bar{\Gamma} t)]. \tag{9.6}$$

These fractions of K^0 and $\bar{K}^0$, as a function of time, are shown in Fig. 9.2. The experimentally measured values of the relevant parameters are $\Gamma_s = 1.12 \times 10^{10}\ s^{-1}$, $\Gamma_L = 1.93 \times 10^7\ s^{-1}$ and $\Delta m = 0.535 \times 10^{10}\ s^{-1}$ (see later). The decay lengths for K_s^0 and K_L^0 are 5.4 cm/(GeV/c) and 31 m/(GeV/c)

Fig. 9.2 Variation with time of the $K^0/\bar{K}^0$ components of an initially pure K^0 beam.

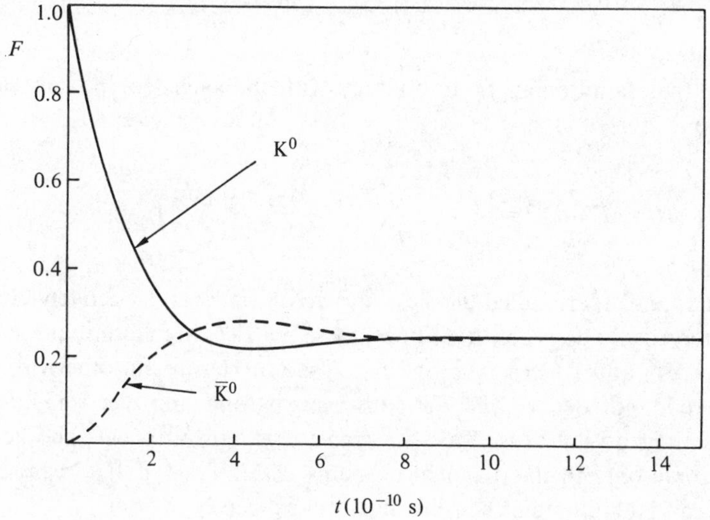

respectively. It can be seen that there is an oscillatory term, with a frequency of $\Delta m/2\pi$, which is multiplied by an exponential factor from K_s^0 decay. After several K_s^0 lifetimes, the K^0 and $\bar{K}^0$ components become equal and the beam is approximately a pure K_L^0 beam.

9.1.2 *CP* violation – general formalism

If *CP* was conserved, then, sufficiently far from the K^0 production point, neutral decays to two pions would no longer occur. However, in 1964 it was observed (Christenson *et al.*, 1964) that $K_L^0 \to \pi^+\pi^-$ indeed occurred, and with a rate

$$\Gamma(K_L^0 \to \pi^+\pi^-)/\Gamma(K_L^0 \to \text{all}) \simeq 2 \times 10^{-3}. \tag{9.7}$$

Thus, Cronin, Fitch and co-workers had discovered that *CP* was violated in this decay. This important observation has been confirmed in this and other decay modes of the K^0 system. However, no evidence for *CP* violation in other hadron decays has so far been found.

Let us reconsider the time evolution of a $K^0\bar{K}^0$ system, allowing for the possibility of *CP* violation. At time t we have, in general,

$$|\psi(t)\rangle = a(t)|K^0\rangle + \bar{a}(t)|\bar{K}^0\rangle. \tag{9.8}$$

The time evolution is given, for this coupled two-channel system, by

$$i\frac{d}{dt}|\psi(t)\rangle = H|\psi(t)\rangle = \begin{pmatrix} H_{11} & H_{12} \\ H_{21} & H_{22} \end{pmatrix}|\psi(t)\rangle, \tag{9.9}$$

where the Hamiltonian H, in analogy with the single-component case, can be expressed as

$$H = M - i\Gamma/2 = \begin{pmatrix} M_{11} - i\Gamma_{11}/2 & M_{12} - i\Gamma_{12}/2 \\ M_{21} - i\Gamma_{21}/2 & M_{22} - i\Gamma_{22}/2 \end{pmatrix}, \tag{9.10}$$

where M and Γ are called the *mass* and *decay matrices* respectively. Since they represent observable quantities, these matrices are Hermitian; hence $M_{21} = M_{12}^*$ and $\Gamma_{21} = \Gamma_{12}^*$. However, H is not Hermitian, otherwise the K^0 would not decay. The *CP*-conserving strong and electromagnetic interactions do not cause $K^0 \to \bar{K}^0$ transitions, nor K^0 decay, and hence contribute only to the diagonal elements of M. Thus, if H_{wk} represents the weak Hamiltonian, then the mass matrix can be written

$$M_{11} = m_{K^0} + \langle K^0|H_{wk}|K^0\rangle + \sum_n \frac{|\langle K^0|H_{wk}|n\rangle|^2}{m_{K^0} - E_n},$$

$$M_{22} = m_{\bar{K}^0} + \langle \bar{K}^0|H_{wk}|\bar{K}^0\rangle + \sum_n \frac{|\langle \bar{K}^0|H_{wk}|n\rangle|^2}{m_{\bar{K}^0} - E_n}, \tag{9.11}$$

$$M_{12} = M_{21}^* = \sum_n \frac{\langle K^0|H_{wk}|n\rangle\langle n|H_{wk}|\bar{K}^0\rangle}{m_{K^0} - E_n},$$

where the sum extends to the n possible zero-strangeness intermediate states of energy E_n. CPT invariance implies that $m_{K^0} = m_{\bar{K}^0}$, and that $H_{11} = H_{22}$ (i.e. $M_{11} = M_{22}$, $\Gamma_{11} = \Gamma_{22}$). In the discussion below it is assumed that the strong and electromagnetic interactions conserve CPT. The term $\langle K^0|H_{wk}|\bar{K}^0\rangle$ has been omitted from M_{12}, since direct $|\Delta S| = 2$ transitions are not allowed in the standard model.

The decay matrix Γ can be expressed as a sum of the contributions corresponding to distinct $S = 0$ physical states. Conservation of probability implies that (see (2.71))

$$\frac{d}{dt}\langle\psi|\psi\rangle = -2\pi \sum_F \rho_F |\langle F|H_{wk}|\psi\rangle|^2, \tag{9.12}$$

where F is a specific final state from K^0 and $\bar{K}^0$ decay, and ρ_F is the density of final states. From (9.9) and (9.10) we can write

$$\frac{d}{dt}\langle\psi|\psi\rangle = \langle\dot{\psi}|\psi\rangle + \langle\psi|\dot{\psi}\rangle$$
$$= -\langle\psi|\Gamma/2 - iM|\psi\rangle - \langle\psi|\Gamma/2 + iM|\psi\rangle = -\langle\psi|\Gamma|\psi\rangle. \tag{9.13}$$

Hence, from (9.12) and (9.13), we have

$$\langle\psi|\Gamma|\psi\rangle = 2\pi \sum_F \rho_F |\langle F|H_{wk}|\psi\rangle|^2,$$

$$\Gamma_{11} = 2\pi \sum_F \rho_F |\langle F|H_{wk}|K^0\rangle|^2,$$

$$\Gamma_{22} = 2\pi \sum_F \rho_F |\langle F|H_{wk}|\bar{K}^0\rangle|^2, \tag{9.14}$$

$$\Gamma_{12} = \Gamma_{21}^* = 2\pi \sum_F \rho_F \langle F|H_{wk}|K^0\rangle^* \langle F|H_{wk}|\bar{K}_0\rangle.$$

If H_{wk} is invariant under CPT (this is widely expected on theoretical grounds), then

$$\langle F|H_{wk}|K^0\rangle = \langle F|(CPT)^{-1}H_{wk}(CPT)|K^0\rangle,$$
$$= \langle \bar{K}^0|H_{wk}|\bar{F}'\rangle = \langle \bar{F}'|H_{wk}|\bar{K}^0\rangle^*, \tag{9.15}$$

where $\bar{F}'$ is the charge conjugate of final state F, but with the spins reversed. CPT invariance of H_{wk} thus implies that $\Gamma_{11} = \Gamma_{22}$, since the sum in (9.14) is over all possible final states F. If T invariance holds for H_{wk}, then the off-diagonal elements are real and equal, and so $M_{12} = M_{21} = M^*_{12}$ and $\Gamma_{12} = \Gamma_{21} = \Gamma^*_{12}$. If CPT and CP (or T) invariance both hold, then $M_{11} = M_{22}$, $\Gamma_{11} = \Gamma_{22}$, $M_{12} = M_{21}$ and $\Gamma_{12} = \Gamma_{21}$.

An alternative way to derive these relationships for the mass and decay matrices is to write

$$M - i\Gamma/2 = a_0 I_2 + a_1\sigma_1 + a_2\sigma_2 + a_3\sigma_3, \tag{9.16}$$

where σ_1, σ_2 and σ_3 are the Pauli spin matrices. For the phase convention (9.1), the operator $CP \equiv \sigma_1$. The operator $T \equiv K$, the complex conjugation operator. Thus, if $R = CPT$, then CPT invariance implies that $RH_{wk}R^{-1} = H^\dagger_w$, where the form of the righthand side follows from the anti-unitary nature of K. Applying this to (9.16), and also considering the equivalent operations for T and CP, gives (using the properties of the σ matrices) $a_3 = 0(CPT)$, $a_2 = 0(T)$ and $a_2 = a_3 = 0(CP)$.

In general we can express K_s and K_L as follows (cf. (9.2))

$$|K_s\rangle = \frac{p|K^0\rangle + q|\bar{K}^0\rangle}{(|p|^2 + |q|^2)^{1/2}}, \qquad |K_L\rangle = \frac{r|K^0\rangle + s|\bar{K}^0\rangle}{(|r|^2 + |s|^2)^{1/2}}. \tag{9.17}$$

The mass and widths of the physical (decaying) states correspond to the eigenvalues of the matrix H, given by (9.10) or (9.16), namely

$$\begin{pmatrix} a_0 + a_3 & a_1 - ia_2 \\ a_1 + ia_2 & a_0 - a_3 \end{pmatrix}\begin{pmatrix} p \\ q \end{pmatrix} = \lambda_s\begin{pmatrix} p \\ q \end{pmatrix}, \tag{9.18}$$

with a similar equation for λ_L (with r and s replacing p and q). The eigenvalues of (9.18) are $\lambda = a_0 \pm a$, so that $\lambda_s = a_0 + a$ and $\lambda_L = a_0 - a$, where $a = (a_1^2 + a_2^2 + a_3^2)^{1/2}$. Using these results, together with (9.10), we obtain

$$\begin{aligned} a_0 &= (M_{11} + M_{22})/2 - i(\Gamma_{11} + \Gamma_{22})/4 = (M_s + M_L)/2 - i(\Gamma_s + \Gamma_L)/4, \\ a_1 &= \text{Re } M_{12} - i \text{ Re } \Gamma_{12}/2, \\ a_2 &= -\text{Im } M_{12} + i \text{ Im } \Gamma_{12}/2, \\ a_3 &= (M_{11} - M_{22})/2 - i(\Gamma_{11} - \Gamma_{22})/4, \\ a &= (M_s - M_L)/2 - i(\Gamma_s - \Gamma_L)/4 \simeq a_1, \end{aligned} \tag{9.19}$$

where the latter approximation follows since a_2 and a_3 are small.

The eigenvectors p, q, r and s in (9.17) can now be found from (9.18) and its equivalent for K_L. This gives

$$p/q = \frac{(a_1 - ia_2)}{a - a_3} \simeq 1 - i\frac{a_2}{a_1} + \frac{a_3}{a_1}, \tag{9.20}$$

$$r/s = -\frac{(a_1 - ia_2)}{a + a_3} \simeq -\left(1 - i\frac{a_2}{a_1} - \frac{a_3}{a_1}\right). \tag{9.21}$$

Defining $\varepsilon = -ia_2/(2a_1)$ and $\Delta = a_3/(2a_1)$ (i.e. both small), we obtain, to a good approximation,

$$\begin{aligned} |K_s\rangle &= [(1+\varepsilon+\Delta)|K^0\rangle + (1-\varepsilon-\Delta)|\bar{K}^0\rangle]/2^{1/2}, \\ |K_L\rangle &= [(1+\varepsilon-\Delta)|K^0\rangle - (1-\varepsilon+\Delta)|\bar{K}^0\rangle]/2^{1/2}, \end{aligned} \tag{9.22}$$

where the parameters ε and Δ are, from (9.19),

$$\begin{aligned} \varepsilon &= \frac{[-\mathrm{Im}\, M_{12} + i\,\mathrm{Im}\,\Gamma_{12}/2]}{(\Gamma_s - \Gamma_L)/2 + i(M_s - M_L)}, \\ \Delta &= \frac{1}{2}\frac{[i(M_{11} - M_{22}) + (\Gamma_{11} - \Gamma_{22})/2]}{(\Gamma_s - \Gamma_L)/2 + i(M_s - M_L)}. \end{aligned} \tag{9.23}$$

Note that if CPT is conserved, then $a_3 = 0$, and hence $\Delta = 0$; so that (9.22) is then specified by the single parameter ε. Thus, a non-zero value of Δ (i.e. non-diagonal mass matrix) would indicate that CPT was violated. T invariance gives $a_2 = 0$, and hence $\varepsilon = 0$. If CP is conserved, then (9.22) takes the form of (9.2). The overlap of the eigenstates for K_L and K_s is given by (using (9.22))

$$\langle K_L | K_s \rangle = 2(\mathrm{Re}\,\varepsilon + i\,\mathrm{Im}\,\Delta). \tag{9.24}$$

9.1.3 *CP* violation in 2π and 3π decays

Because of the definite CP eigenvalues of the final state pions, the decay modes $K^0_L \to \pi^+\pi^-$ and $K^0_L \to \pi^0\pi^0$ are CP violating. The quantities which can be extracted experimentally are

$$\eta_{+-} = \frac{\langle \pi^+\pi^- | H_{wk} | K^0_L \rangle}{\langle \pi^+\pi^- | H_{wk} | K^0_s \rangle} = |\eta_{+-}| \exp(i\phi_{+-}), \tag{9.25}$$

$$\eta_{00} = \frac{\langle \pi^0\pi^0 | H_{wk} | K^0_L \rangle}{\langle \pi^+\pi^- | H_{wk} | K^0_s \rangle} = |\eta_{00}| \exp(i\phi_{00}). \tag{9.26}$$

From Bose symmetry, the $\pi\pi$ state must have $I = 0$ or 2. The $\pi^+\pi^-$ and $\pi^0\pi^0$ states are related to the states of definite isospin by the usual

Clebsch–Gordan coefficients

$$\langle\pi^+\pi^-| = (\tfrac{1}{3})^{1/2}\langle(\pi\pi)_2| + (\tfrac{2}{3})^{1/2}\langle(\pi\pi)_0|, \tag{9.27}$$

$$\langle\pi^0\pi^0| = (\tfrac{2}{3})^{1/2}\langle(\pi\pi)_2| - (\tfrac{1}{3})^{1/2}\langle(\pi\pi)_0|. \tag{9.28}$$

The decay system is specified by four weak decay amplitudes, namely

$$\langle(\pi\pi)_I|H_{\rm wk}|{\rm K}^0\rangle = A_I \exp({\rm i}\delta_I), \qquad \langle(\pi\pi)_I|H_{\rm wk}|\bar{\rm K}^0\rangle = \bar{A}_I \exp({\rm i}\delta_I), \tag{9.29}$$

where $I = 0, 2$. The phase shifts δ_0 and δ_2 arise from final state interactions amongst the pions. If *CPT* invariance holds, then $\bar{A}_I = A_I^*$ and, since the overall phase is unobservable, we can choose A_0 to be real. If *CPT* is violated then $\bar{A}_0 \neq A_0$, but one can still choose A_0 and $\bar{A}_0$ to have the same phase with $\bar{A}_0/A_0$ real. A measure of *CPT* violation for $I = 0$ is given by

$$\lambda_0 = (\bar{A}_0 - A_0)/(\bar{A}_0 + A_0). \tag{9.30}$$

The corresponding parameter λ_2 is complex. Note that the amplitude A_2 violates the $|\Delta I| = \frac{1}{2}$ rule, and is thus only about 4% of A_0 (see Section 8.5). If we let $z = (A_2 + \bar{A}_2)/(A_0 + \bar{A}_0)$, then using (9.25), together with (9.22), (9.27) and (9.29), we obtain

$$\eta_{+-} = \frac{(A_2 - \bar{A}_2)\exp({\rm i}\delta_2) + (\varepsilon - \Delta)(A_2 + \bar{A}_2)\exp({\rm i}\delta_2) + 2^{1/2}[(A_0 - \bar{A}_0)\exp({\rm i}\delta_0) + (\varepsilon - \Delta)(A_0 + \bar{A}_0)\exp({\rm i}\delta_0)]}{(A_2 + \bar{A}_2)\exp({\rm i}\delta_2) + (\varepsilon + \Delta)(A_2 - \bar{A}_2)\exp({\rm i}\delta_2) + 2^{1/2}[(A_0 + \bar{A}_0)\exp({\rm i}\delta_0) + (\varepsilon + \Delta)(A_0 - \bar{A}_0)\exp({\rm i}\delta_0)]}$$

$$\text{i.e. } \eta_{+-} \simeq \varepsilon_0 + \varepsilon' \tag{9.31}$$

where

$$\varepsilon_0 = \varepsilon - \Delta - \lambda_0,$$

$$\varepsilon' \simeq \frac{1}{2^{1/2}}\frac{[(A_2 - \bar{A}_2) - z(A_0 - \bar{A}_0)]}{A_0 + \bar{A}_0}\exp[{\rm i}(\delta_2 - \delta_0)]. \tag{9.32}$$

Similarly, for the $\pi^0\pi^0$ decay mode

$$\eta_{00} = \varepsilon_0 - 2\varepsilon'. \tag{9.33}$$

For the case of *CPT* invariance, $A_0 = \bar{A}_0$ and $\Delta = \lambda_0 = 0$, so that

$$\varepsilon_0 = \varepsilon, \qquad \varepsilon' \simeq \frac{\rm i}{2^{1/2}A_0}\,{\rm Im}\,A_2\exp[{\rm i}(\delta_2 - \delta_0)]. \tag{9.34}$$

Note that η_{+-} and η_{00} can be non-zero, even if there is no K_s^0–K_L^0 mixing ($\varepsilon = 0$). Such *direct CP* violation would give a non-zero value of ε'. The quantity ε, however, arises from *CP* violation in the mass matrix.

In Section 9.1.1, the time evolution of an initially pure K^0 beam was considered, assuming *CP* (and *CPT*) invariance. If we allow for *CP*(*CPT*) violation, then we must express $|K^0\rangle$ in terms of $|K_s^0\rangle$ and $|K_L^0\rangle$, using (9.22). The time dependence is thus given by (cf. (9.3))

$$|\psi(t)\rangle = \frac{1}{2^{1/2}}[(1-\varepsilon+\Delta)|K_s\rangle \exp(-i\lambda_s t) + (1-\varepsilon-\Delta)|K_L\rangle \exp(-i\lambda_L t)]. \tag{9.35}$$

Similarly, if we start with an initially pure $\bar{K}^0$ beam, we obtain

$$|\bar{\psi}(t)\rangle = \frac{1}{2^{1/2}}[(1+\varepsilon-\Delta)|K_s\rangle \exp(-i\lambda_s t) - (1+\varepsilon+\Delta)|K_L\rangle \exp(-i\lambda_L t)]. \tag{(9.36)}$$

In order to calculate the $\pi\pi$ decay probability for an initially pure K^0 beam, we require the matrix element

$$\langle\pi\pi|H_{wk}|\psi\rangle = \frac{1}{2^{1/2}}\langle\pi\pi|H_{wk}|K_s\rangle \times [(1-\varepsilon+\Delta)\exp(-i\lambda_s t) + \eta_{\pi\pi}(1-\varepsilon-\Delta)\exp(-i\lambda_L t)]. \tag{9.37}$$

Squaring this gives the relative decay probabilities, which contain a factor $N = |\langle\pi\pi|H_{wk}|K_s\rangle|^2/2$. The relative decay probabilities for initially pure K^0 and $\bar{K}^0$ beams are thus given by

$$R(t) = N[A\exp(-\Gamma_s t) + B\exp(-\bar{\Gamma}t)\cos(\Delta m t) + C\exp(-\bar{\Gamma}t)\sin(\Delta m t) + D\exp(-\Gamma_L t)], \tag{9.38}$$

where (using (9.37), etc.)

$$A = 1 \mp 2(\mathrm{Re}\,\varepsilon - \mathrm{Re}\,\Delta),$$

$$B = \pm[2(1 \mp 2\,\mathrm{Re}\,\varepsilon)\,\mathrm{Re}\,\eta_{\pi\pi} \pm 4\,\mathrm{Im}\,\Delta\,\mathrm{Im}\eta_{\pi\pi}], \tag{9.39}$$

$$C = \pm[2(1 \mp 2\,\mathrm{Re}\,\varepsilon)\,\mathrm{Im}\,\eta_{\pi\pi} \mp 4\,\mathrm{Im}\,\Delta\,\mathrm{Re}\,\eta_{\pi\pi}],$$

$$D = |\eta_{\pi\pi}|^2[1 \mp 2(\mathrm{Re}\,\varepsilon + \mathrm{Re}\,\Delta)].$$

In (9.39), the upper and lower signs refer to K^0 and $\bar{K}^0$ beams respectively, $\bar{\Gamma} = (\Gamma_s + \Gamma_L)/2$ and $\Delta m = m_L - m_s$. Thus *CP* non-invariance leads to a difference in the $\pi\pi$ decay distribution of initially pure K^0 and $\bar{K}^0$ beams, due to the presence of constructive or destructive K_s–K_L interference (see Fig. 9.3).

If we assume CPT invariance holds (i.e. $\Delta = 0$), then (from (9.38) and (9.39)) the 2π mode has a decay probability, as a function of proper time t (assuming pure K^0 at $t = 0$), as follows

$$I(t) = I(0)[\exp(-\Gamma_s t) + |\eta_{\pi\pi}|^2 \exp(-\Gamma_L t) + 2|\eta_{\pi\pi}| \exp(-\bar{\Gamma} t) \cos(\Delta m t + \phi_{\pi\pi})]. \quad (9.40)$$

Hence, observation of the $\pi\pi$ decay probability as a function of time gives the magnitude and phase of the CP violating amplitude $\eta_{\pi\pi}$, provided Δm is known. Fig. 9.4 shows an example of the interference term obtained by observing the $K^0 \to \pi^+\pi^-$ decay mode from an initial K^0 beam (Geweniger *et al.*, 1974), subtracting off the $\exp(-\Gamma_s t)$ and $|\eta_{\pi\pi}|^2 \exp(-\Gamma_L t)$ terms and dividing by $2|\eta_{\pi\pi}| \exp(-\bar{\Gamma} t)$.

Measurement of the K_L–K_s mass difference can be made using a phenomenon known as *regeneration*. For simplicity, in order to illustrate the method, we neglect (for the moment) CP violation. Let us consider the set-up shown in Fig. 9.5, where an essentially pure beam of K_L^0 is incident on the first of two plates in which there will be strong interactions. Now the relevant eigenstates for the strong interactions of neutral kaons

Fig. 9.3 Effect of CP non-invariance on the $\pi\pi$ decay rates of initially pure K^0 and $\bar{K}^0$ beams.

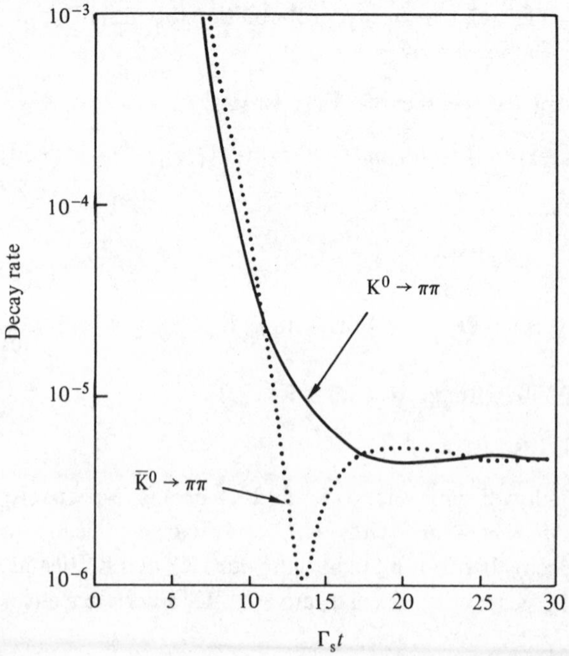

are the $K^0(S=1)$ and $\bar{K}^0(S=-1)$ states. The $\bar{K}^0$ component has more interaction channels open than the K^0 component, since one can have $\bar{K}^0 p \to \Lambda^0 \pi^+$ etc. (i.e. the production of $S=-1$ hyperons); hence $\sigma(\bar{K}N) > \sigma(KN)$. Let us consider the (unrealistic) limit, in which all the $\bar{K}^0$ component is absorbed in plate 1, leaving only K^0 mesons on exit from the plate ($t=0$). Just before the second plate, the wave function will be

$$|\psi(t)\rangle = \frac{\exp(-\mathrm{i}m_s t)}{2}\{|K^0\rangle[\exp(-\Gamma_s t/2) + \exp(-\mathrm{i}\,\Delta m t)] + |\bar{K}^0\rangle[\exp(-\Gamma_s t/2) - \exp(-\mathrm{i}\,\Delta m t)]\}, \tag{9.41}$$

which is obtained by starting with (9.3) and substituting for K_s and K_L

Fig. 9.4 Interference term in $K^0 \to \pi\pi$ decay.

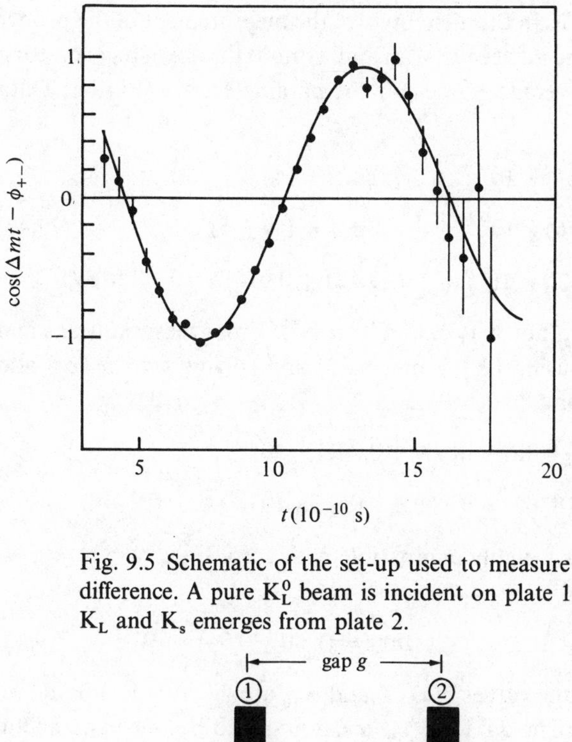

Fig. 9.5 Schematic of the set-up used to measure the $K_L - K_s$ mass difference. A pure K_L^0 beam is incident on plate 1, and a mixture of K_L and K_s emerges from plate 2.

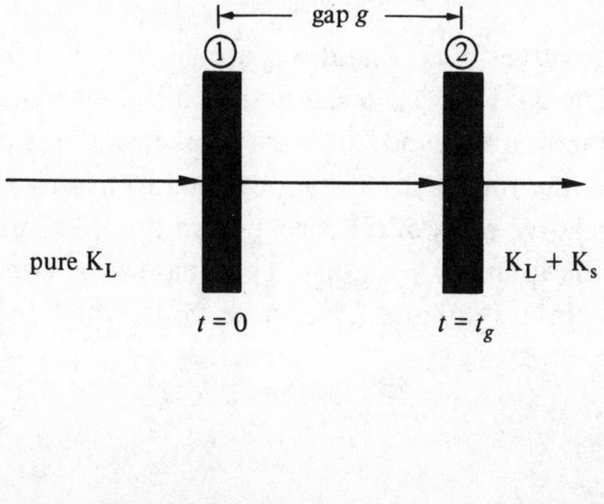

in terms of K^0 and $\bar{K}^0$ (using (9.22), with $\varepsilon = \Delta = 0$). It is further assumed that $\exp(-\Gamma_L t/2) \sim 1$.

In the second plate, the $\bar{K}^0$ component is again removed. Re-expressing (9.41) in terms of K_s and K_L we have, on emergence from the second plate (time t_g), an intensity of K_s given by

$$I(t_g) = |\langle K_s|\psi(t_g)\rangle|^2$$

$$\simeq \frac{I(0)}{4}[1 + \exp(-\Gamma_s t_g) + 2\exp(-\Gamma_s t_g/2)\cos(\Delta m t_g)]. \tag{9.42}$$

Hence, by changing the gap distance g, and hence t_g, Δm can be measured. In practice the analysis is more complicated, since the $\bar{K}^0$ component is not completely absorbed and there is an additional phase factor arising from *CP* violation (for a more complete discussion see Commins and Bucksbaum, 1983).

Experimental studies in this field involve the measurement of the proper time distribution of the $\pi\pi$ decays, with and without the use of regenerators. The current 'world average' values for the parameters are (Particle Data Group, 1986)

$$\begin{aligned} |\eta_{+-}| &= (2.275 \pm 0.021) \times 10^{-3}, \qquad & \phi_{+-} &= (44.6 \pm 1.2)^\circ, \\ |\eta_{00}| &= (2.299 \pm 0.036) \times 10^{-3}, \qquad & \phi_{00} &= (54 \pm 5)^\circ, \end{aligned} \tag{9.43}$$
$$\Delta m = (0.5349 \pm 0.0022) \times 10^{10}\ \mathrm{s}^{-1} = (3.521 \pm 0.014) \times 10^{-12}\ \mathrm{MeV}.$$

Note that $\Delta m \sim 0.5\Gamma_s$, but that $\Delta m/m_K \sim 10^{-14}$. From these values, ε and ε' can be computed using (9.31) and (9.33) and solving for the real and imaginary parts, giving

$$\begin{aligned} \varepsilon' &= [(0.09 \pm 0.06) - i(0.09 \pm 0.04)] \times 10^{-3}, \\ \varepsilon_0 &= [(1.53 \pm 0.06) + i(1.69 \pm 0.05)] \times 10^{-3}. \end{aligned} \tag{9.44}$$

That is, $\varepsilon' \ll \varepsilon_0$ and is roughly compatible with zero. If we take $\varepsilon' = 0$, we obtain

$$\mathrm{Re}\,\varepsilon = (1.62 \pm 0.04) \times 10^{-3}, \qquad \mathrm{Im}\,\varepsilon = (1.60 \pm 0.04) \times 10^{-3}. \tag{9.45}$$

Note, however, that the values of ϕ_{+-} and ϕ_{00} are different by about two standard deviations. The data on ϕ_{00} are dominated by one experiment, so that more accurate data are needed to clarify this important point.

An accurate limit on the ratio ε'/ε can be obtained from the results of measurements of the relative rates of *CP* violation in the $\pi^+\pi^-$ and $\pi^0\pi^0$ decay modes, that is the quantity $|\eta_{+-}/\eta_0|$. The phase of ε' (see (9.34)) depends on the values of the $I = 0$ and 2 $\pi\pi$ phase shifts, which have been determined to be $\delta_0 = (46 \pm 5)^\circ$ and $\delta_2 = (-7.2 \pm 1.3)^\circ$. The phases of η_{+-} and η_{00} are both about 45°, therefore the phases of ε and ε' are the same to within 10°. Taking the phases to be the same gives, from (9.33) and (9.34),

$$\varepsilon'/\varepsilon = [|\eta_{+-}/\eta_{00}|^2 - 1]/6 = (-3 \pm 4) \times 10^{-3}. \tag{9.46}$$

The numerical value in (9.46) is computed using the world average value of $|\eta_{00}/\eta_{+-}| = 1.010 \pm 0.013$. The preliminary results from the latest generation of experiments are $\varepsilon'/\varepsilon = (3.5 \pm 3 \pm 2) \times 10^{-3}$ for the Chicago–Fermilab–Princeton–Saclay experiment at Fermilab and $\varepsilon'/\varepsilon = (3.5 \pm 1.4) \times 10^{-3}$ for the NA31 experiment at CERN (Mannelli, 1987). Including these data gives a world average value of $\varepsilon'/\varepsilon = (2.9 \pm 1.2) \times 10^{-3}$, and hence may indicate that ε' is not zero.

The analysis of the 3π decay modes of the neutral kaon system is a further potential handle on *CP* violation. However, at present, there is no evidence for *CP* violation in these modes, and the upper limits are as follows:

$$\begin{aligned} |\eta_{+-0}|^2 &= \frac{\Gamma(K_s^0 \to \pi^+\pi^-\pi^0)^{CP\,\mathrm{viol}}}{\Gamma(K_L^0 \to \pi^+\pi^-\pi^0)} < 0.12 \qquad (90\%\ \mathrm{c.l.}), \\ |\eta_{000}|^2 &= \frac{\Gamma(K_s^0 \to \pi^0\pi^0\pi^0)^{CP\,\mathrm{viol}}}{\Gamma(K_L^0 \to \pi^0\pi^0\pi^0)} < 0.1 \qquad (90\%\ \mathrm{c.l.}). \end{aligned} \tag{9.47}$$

9.1.4 *CP* violation in K_{l3} decays

The branching ratios for the semileptonic K_L^0 decays are

$$\begin{aligned} &\Gamma(K_L^0 \to \pi^\pm\mu^\mp\nu_\mu)/\Gamma(K_L^0 \to \mathrm{all}) = 0.271 \pm 0.004, \\ &\Gamma(K_L^0 \to \pi^\pm e^\mp\nu_e)/\Gamma(K_L^0 \to \mathrm{all}) = 0.387 \pm 0.006. \end{aligned} \tag{9.48}$$

The muon rate is smaller because of the more limited phase space available. The possible transition matrix elements for the semileptonic decays are

$$\begin{aligned} f &= \langle \pi^- l^+ \nu_l | H_{\mathrm{wk}} | K^0 \rangle, \qquad & f^* &= \langle \pi^+ l^- \bar{\nu}_l | H_{\mathrm{wk}} | \bar{K}^0 \rangle, \\ g &= \langle \pi^- l^+ \nu_l | H_{\mathrm{wk}} | \bar{K}^0 \rangle, \qquad & g^* &= \langle \pi^+ l^- \bar{\nu}_l | H_{\mathrm{wk}} | K^0 \rangle, \end{aligned} \tag{9.49}$$

where *CPT* conservation has been assumed (the general case without assuming *CPT* invariance has been given by Tanner and Dalitz (1986)). The amplitude $g(g^*)$ violates the $\Delta Q = \Delta S$ rule. This violation is usually measured in terms of the parameter $x = g/f$, which is expected to be small ($\lesssim 10^{-14}$ in the standard model).

We next consider the charged asymmetry of the decay lepton in semileptonic decays, which is defined as

$$\delta(t) = \frac{N^+(t) - N^-(t)}{N^+(t) + N^-(t)}, \tag{9.50}$$

where $N^+(N^-)$ are the number of $l^+(l^-)$ decay leptons observed at time t. For an initially pure K^0 beam, the wave function at time t is given by (9.35). Assuming CPT conservation ($\Delta = 0$), then the matrix elements for l^+ and l^- decays are (neglecting ε^2 terms)

$$\langle l^+|H_{\text{wk}}|\psi(t)\rangle = \frac{f}{2}[(1 + x - 2\varepsilon x)\exp(-\text{i}\lambda_s t) + (1 - x + 2\varepsilon x)\exp(-\text{i}\lambda_L t)], \tag{9.51}$$

and

$$\langle l^-|H_{\text{wk}}|\psi(t)\rangle = \frac{f^*}{2}[(x^* + 1 - 2\varepsilon)\exp(-\text{i}\lambda_s t) + (x^* - 1 + 2\varepsilon)\exp(-\text{i}\lambda_L t)]. \tag{9.52}$$

Squaring these gives the relative decay rates N^+ and N^- respectively, and hence the decay asymmetry is approximately

$$\delta(t) = \frac{2(1 - |x|^2)\{\text{Re}\,\varepsilon[\exp(-\Gamma_s t) + \exp(-\Gamma_L t)] + \exp(-\bar{\Gamma} t)\cos(\Delta m t)\}}{|1 + x|^2 \exp(-\Gamma_s t) + |1 - x|^2 \exp(-\Gamma_L t)}. \tag{9.53}$$

If $t \gg 1/\Gamma_s$ (i.e. no K_s^0), then (9.53) simplifies to

$$\delta(t) \simeq \frac{2(1 - |x|^2)}{|1 - x|^2}[\exp(-\bar{\Gamma} t)\cos(\Delta m t) + \text{Re}\,\varepsilon]. \tag{9.54}$$

The results of a measurement of $\delta(t)$ for K_{e3}^0 decays by Gjesdal *et al.* (1974) are shown in Fig. 9.6, and a fit to the data gives a value $\Delta m = (0.533 \pm 0.004) \times 10^{10}\ \text{s}^{-1}$. The quantity x, which measures any violation of the $\Delta Q = \Delta S$ rule, is compatible with zero. The world average value is

$$\text{Re}\,x = 0.009 \pm 0.020, \qquad \text{Im}\,x = -0.004 \pm 0.026. \tag{9.55}$$

For large times, only the second term in (9.54) is important, and the measurements of $\delta(t)$ give a value of $(3.30 \pm 0.12) \times 10^{-3}$, which for $x = 0$ and using $\text{Re}\,\varepsilon = \delta/2$, gives

$$\text{Re}\,\varepsilon = (1.65 \pm 0.06) \times 10^{-3}. \tag{9.56}$$

This value is compatible with the value derived from $K_{\pi 2}$ results (9.45). Note that if *CPT* is not assumed, then, for $x = 0$, $\delta = 2\,\mathrm{Re}(\varepsilon - \Delta)$ at very large times.

9.1.5 Further limits on the *CP* and *CPT* parameters

The overlap $\langle K_L | K_s \rangle$ depends on the values of both ε and Δ, as can be seen from equation (9.24). In order to evaluate this, we start with (9.13) and insert into this equation an arbitrary state

$$|\psi\rangle = a_s \exp(-i\lambda_s t)|K_s\rangle + a_L \exp(-i\lambda_L t)|K_L\rangle. \tag{9.57}$$

Equating coefficients we find that

$$-i(\lambda_L^* - \lambda_s)\langle K_L|K_s\rangle = \langle K_L|\Gamma|K_s\rangle, \tag{9.58}$$

which, using Schwartz's inequality, gives

$$|\lambda_L^* - \lambda_s||\langle K_L|K_s\rangle| \leqslant [\langle K_L|\Gamma|K_L\rangle\langle K_s|\Gamma|K_s\rangle]^{1/2} = (\Gamma_s\Gamma_L)^{1/2}. \tag{9.59}$$

Substituting for the experimental values gives the limit

$$|\langle K_L|K_s\rangle| \lesssim 0.06. \tag{9.60}$$

By considering the individual decay channels separately in (9.58), a more stringent limit of 0.006 can be deduced (Bell and Steinberger, 1965). This,

Fig. 9.6 Decay asymmetry, as a function of time, measured using K^0_{e3} decays. The curve is a fit to the data (Gjesdal *et al.*, 1974).

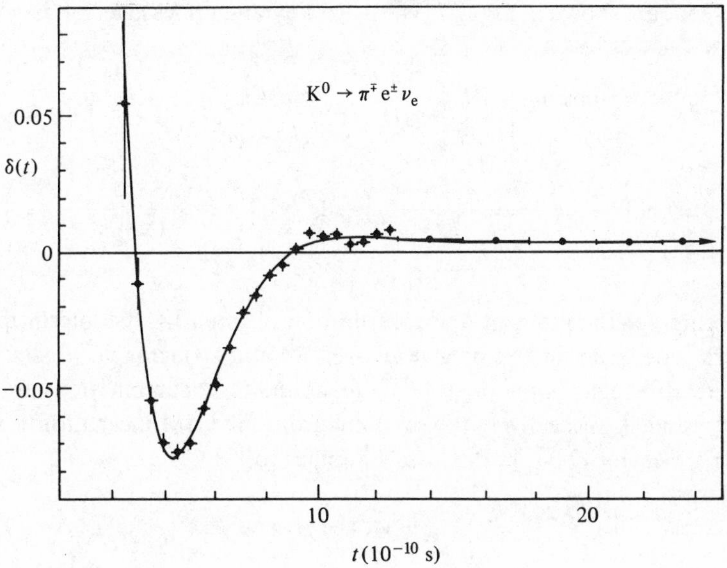

together with (9.24) and (9.54), gives

$$|\text{Im}\,\Delta| \lesssim 2.5 \times 10^{-3}. \tag{9.61}$$

A study of the $K_s^0 \to \pi\pi$ decays in the reactions $\bar{p}p \to K^+\bar{K}^0 + \text{pions}$ and $\bar{p}p \to K^-K^0 + \text{pions}$ also gives information on ε and Δ, since the strangeness of the initial K^0 (or $\bar{K}^0$) is known. Combining the existing data, Tanner and Dalitz (1986) estimate that

$$\text{Re}\,\Delta - \text{Re}\,\varepsilon = (-2.3 \pm 6) \times 10^{-3}, \tag{9.62}$$

and hence, using (9.56), the value of Re Δ is

$$\text{Re}\,\Delta = (-0.7 \pm 6) \times 10^{-3}. \tag{9.63}$$

This value, together with $\text{Im}\,\Delta = (-0.11 \pm 0.10) \times 10^{-3}$ (see Tanner and Dalitz, 1986) allows a limit on the value of $|a_3| = |2a_1\Delta|$ to be made. Since $2a_1 \approx 2a = \lambda_s - \lambda_L$ then, inserting all these values, gives

$$|a_3| \lesssim 0.007\Gamma_s \lesssim 10^{-16} m_K. \tag{9.64}$$

A first sight this appears to be a rather stringent limit on *CPT* conservation; however, the scale with which any violation should be judged is maybe that of a second-order weak effect such as Γ_s. In this case, the limit is only about 1%, i.e comparable with *CP* violation.

9.1.6 The $K_L - K_s$ mass difference

The lowest order diagrams contributing to the $K_L - K_s$ mass difference are shown in Fig. 9.7. The usual Feynman rules give, for diagram (*a*),

$$\mathcal{M}_a = \frac{ig^4 \sin^2\theta_C \cos^2\theta_C}{64} \int \frac{d^4k}{(2\pi)^4} \Bigg[\bar{s}\gamma_\mu(1-\gamma^5)\left(\frac{\not{k}+m_u}{k^2-m_u^2} - \frac{\not{k}+m_c}{k^2-m_c^2}\right)\gamma_\nu(1-\gamma^5)d$$
$$\times\, \bar{s}\gamma_\theta(1-\gamma^5)\left(\frac{\not{k}+m_u}{k^2-m_u^2} - \frac{\not{k}+m_c}{k^2-m_c^2}\right)\gamma_\phi(1-\gamma^5)d\left(\frac{-ig^{\mu\phi}}{k^2-M_W^2}\right)\left(\frac{-ig^{\theta\nu}}{k^2-M_W^2}\right)\Bigg]. \tag{9.65}$$

Ignoring, for the moment, the contribution of t-quarks, the intermediate quark states are u and c, and give an overall Cabibbo factor $\sin^2\theta_C \sin^2\theta_C$. The relative signs of the (u, u) (u, c), (c, u) and (c, c) combinations are +, −, − and + respectively (these follow from the GIM mechanism). The matrix element (9.65) in this case simplifies to

$$\mathcal{M}_a = \frac{-ig^4 \sin^2\theta_C \cos^2\theta_C}{16} [\bar{s}\gamma_\mu\gamma^\alpha\gamma_\nu(1-\gamma^5)d \cdot \bar{s}\gamma^\nu\gamma^\beta\gamma^\mu(1-\gamma^5)d] I_{\alpha\beta}, \tag{9.66}$$

where

$$I_{\alpha\beta} = \int \frac{d^4k}{(2\pi)^4} \frac{k_\alpha k_\beta (m_c^2 - m_u^2)^2}{(k^2 - M_W^2)^2 (k^2 - m_u^2)^2 (k^2 - m_c^2)^2}$$

$$\simeq \frac{g_{\alpha\beta} m_c^2}{64\pi^2 i M_W^4}. \tag{9.67}$$

The limit $M_W \gg m_c \gg m_u$ has been assumed in the evaluation of the integral $I_{\alpha\beta}$ (if m_u is retained, then the expression contains $m_c^2 - m_u^2$). Substituting $g^4 = 32 G_F^2 M_W^4$, and adding an identical contribution for diagram (*b*), we obtain (for details, see Gaillard and Lee, 1974)

$$\mathcal{M} = -\frac{G_F^2 \sin^2\theta_C \cos^2\theta_C m_c^2}{16\pi^2} [\bar{s}\gamma_\mu \gamma_\alpha \gamma_\nu (1-\gamma^5) d \cdot \bar{s}\gamma^\nu \gamma^\alpha \gamma^\mu (1-\gamma^5) d]$$

$$= -\frac{G_F^2 \sin^2\theta_C \cos^2\theta_C m_c^2}{4\pi^2} [\bar{s}\gamma_\mu (1-\gamma^5) d \cdot \bar{s}\gamma^\mu (1-\gamma^5) d]. \tag{9.68}$$

If we neglect the effects of any *CP* or *CPT* non-invariance, then from (9.19) the mass difference is $\Delta m = m_L - m_s = -2M_{12}$, where M_{12} is given by (9.11). After some simplification of (9.68), we obtain from (9.11) and (9.68)

$$\Delta m = \frac{G_F^2}{2\pi^2} m_c^2 \sin^2\theta_C \cos^2\theta_C [\langle \bar{K}^0 | \bar{s}\gamma_\mu (1-\gamma^5) d | 0 \rangle \langle 0 | \bar{s}\gamma^\mu (1-\gamma^5) d | K^0 \rangle]$$

$$\simeq \frac{G_F^2}{4\pi^2} m_c^2 \sin^2\theta_C \cos^2\theta_C f_K^2 m_K, \tag{9.69}$$

where the term in square brackets has been replaced in the second line, by $f_K^2 m_K^2 / 2m_K$, since this term can be identified with the $|K\rangle \to |0\rangle$ current in K_{l2} decay. The calculation outlined above is rather crude, but nonetheless gives a reasonable value for Δm, providing $m_c \sim 1.5$ GeV. If

Fig. 9.7 Diagrams contributing to the $K_L - K_s$ mass difference. The intermediate quark states i, j are the charge $\frac{2}{3}$ quarks u, c, t.

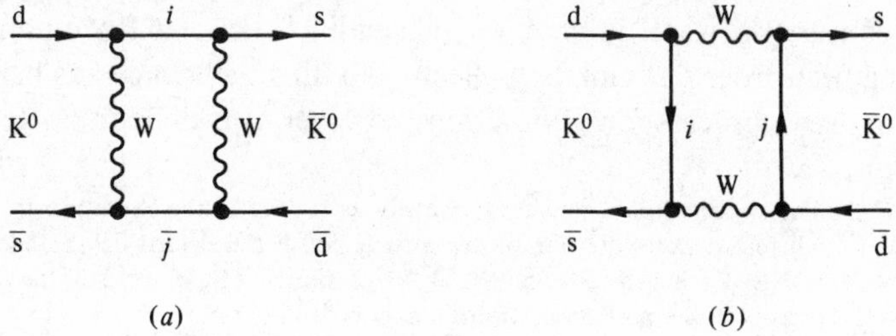

the diagrams including the c-quark had not been included, then there would be no cancellation of the type resulting from the GIM couplings, and the expression for Δm contains M_W^2, instead of m_c^2. Gaillard and Lee (1974) used just this approach (before the discovery of charm) to predict (correctly) the c-quark mass.

9.1.7 Models of *CP* violation

CP violation has so far only been observed in the K^0 system. Hence, given the rather limited data (despite the improved accuracy), potential models have proven difficult to rule out entirely. The relative strength of the *CP* violating interaction can usually be adjusted to give the correct magnitude of *CP* violation in K^0 decays. Some of the various suggested origins of *CP* violation are as follows (for a detailed review see Wolfenstein (1986)):

(i) *CP violation in electromagnetic interactions.* If the origin of *CP* violation is in the electromagnetic interaction, then the required violation is large (~ 0.1). However, the experimental limits on *C*, *P* and *T* violations in electromagnetic interactions are better than 10^{-12}. A non-zero neutron electric dipole moment (d_n),# can only arise if both *P* and *T* are violated. If the electromagnetic interaction is *T*-violating and *P*-conserving, then the normal *T*-conserving and *P*-violating weak interaction is needed to generate d_n. Very roughly, one might expect $d_n \sim (G_F m_N^2/4\pi)(e/m_N) \sim 10^{-20}$ e cm, or, with a more sophisticated treatment, $d_n \sim 10^{-23}$ e cm. The current experimental limit is $d_n < 6 \times 10^{-25}$ e cm.

(ii) *Millistrong interaction.* A small ($\sim 10^{-3}$) violation of *CP* in the strong interaction is needed, with the $K_L^0 \to 2\pi$ occurring via a second order transition in this *CP*-violating interaction. Since *P* is conserved to better than 10^{-5} in strong interactions, there must be *C* (and *T*) violation in strong interactions of about 10^{-3}. An electric dipole moment for the neutron of $d_n \sim 10^{-21}$ e cm is expected.

(iii) *Milliweak interactions.* In this model there is, in addition to the *CP*-conserving weak interaction, a small ($\sim 10^{-3}$) *CP*-violating piece, giving first-order *CP*-violating effects. No direct evidence has been found in any weak process for *T* violation; however, it is difficult, in general, to

The electric dipole moment d contributes a term to the Hamiltonian $H = d\mathbf{J}\cdot\mathbf{E}$, where $\mathbf{J}$ denotes the spin of the particle and $\mathbf{E}$ the electric field. Under P $\mathbf{J} \to \mathbf{J}$ and $\mathbf{E} \to -\mathbf{E}$, whereas under T $\mathbf{J} \to -\mathbf{J}$ and $\mathbf{E} \to \mathbf{E}$. In terms of the cms charge density of the particle ρ_{JJ} (third component $m = J$ along 0z), $d = \int \rho_{JJ} z \, d\tau$.

turn these limits into rigorous limits on a possible milliweak interaction. A value of $d_n \sim 10^{-23}$ e cm is expected in this model.

(iv) *Superweak interaction.* In this model (Wolfenstein, 1964) a $\Delta S = 2$ interaction is postulated, which causes $K^0 \leftrightarrow \bar{K}^0$, and hence $K_L^0 \leftrightarrow K_s^0$, transitions. These transitions are first order in this interaction, which has some coupling strength F. Taking the scale of both F and G_F to be m_N, then the relative value of a_2 (first order in F) to a_1 (second order in G) is

$$|a_2/a_1| \sim Fm_N^2/(Gm_N^2)^2. \tag{9.70}$$

Now this ratio must be approximately 10^{-3} in order to explain CP violation, hence $F \sim 10^{-8}G_F$. The CP violation arises entirely from the mass matrix, and not from the ordinary weak interactions which give rise to the $K \to 2\pi$ transitions, hence $\bar{A}_2 = A_2$ so that $\varepsilon' = 0$ and, further, a_2 is real. Hence the model predicts that

$$\eta_{+-} = \eta_{00} = \varepsilon, \qquad \phi_{+-} = \phi_{00} = \tan^{-1}\left(\frac{2\Delta m}{\Gamma_s}\right). \tag{9.71}$$

The prediction for the phase follows from noting that $a_2 = 2ia_1\varepsilon \simeq i\varepsilon(\lambda_s - \lambda_L)$ is real, and thus the imaginary part of $i\varepsilon(\lambda_s - \lambda_L)$ can be equated to zero. Putting in the measured values for Δm and Γ_s gives $\phi = (43.7 \pm 0.1)°$. These predictions are in tolerable agreement with the data. However the non-zero value of ε', suggested by the most recent data, would imply that the superweak model is not the only source of CP violation. Note that, the superweak theory implies that any effects outside the K^0 system are very small. The electric dipole moment of the neutron, for example, is predicted to be less than or equal to 10^{-29} e cm.

(v) *Standard model.* In the six-quark scheme depicted by the KM matrix (5.193), the phase δ represents CP violation. From (9.23) we see that ε arises from non-zero values of Im M_{12} or Im Γ_{12}. From the experimental results ($\phi \sim 44°$), one can deduce that Im $\Gamma_{12} \ll$ Im M_{12}. Thus, the Re M_{12} gives rise to the mass difference Δm, whereas Im M_{12} gives rise to CP violation. If we reconsider the diagrams of Fig. 9.7, including the t-quark contributions, then we obtain (cf. (9.65)) for either diagram (*a*) or (*b*)

$$\mathcal{M}_a = -i\frac{G_F^2}{2}\int \frac{d^4k}{(2\pi)^4}\sum_{i,j} a_i^{KM} a_j^{KM}\left[\bar{s}\gamma_\mu(1-\gamma^5)\frac{\not{k}+m_i}{k^2-m_i^2}\gamma_\nu(1-\gamma^5)d \right.$$
$$\left. \times\, \bar{s}\gamma^\nu(1-\gamma^5)\frac{\not{k}+m_j}{k^2-m_j^2}\gamma^\mu(1-\gamma^5)d\right], \tag{9.72}$$

where $i, j = $ u, c, t and a_i^{KM} is the product of the couplings of the ith quark to d and s quarks. That is, from (5.193)

$$\begin{aligned} a_u^{KM} &= c_1 s_1 c_3, \\ a_c^{KM} &= -s_1 c_2 [c_1 c_2 c_3 - s_2 s_3 \exp(i\delta)], \\ a_t^{KM} &= -s_1 s_2 [c_1 s_2 c_3 + c_2 s_3 \exp(i\delta)]. \end{aligned} \tag{9.73}$$

To obtain ε, it is necessary to separate the real and imaginary parts of M, and evaluate them by contour integration (neglecting m_u). This gives (Ellis *et al.*, 1976b)

$$\operatorname{Im} M_{12}/\operatorname{Re} M_{12} = 2 s_2 c_2 s_3 \sin\delta \cdot F(\eta), \tag{9.74}$$

where $\eta = m_c^2/m_t^2$ and, defining $\eta' = \eta/(1-\eta)$,

$$F(\eta) = \frac{s_2^2(1 + \eta' \ln \eta) - c_2^2(\eta + \eta' \ln \eta)}{c_1 c_3 (c_2^4 \eta + s_2^4 - 2 s_2^2 c_2^2 \eta' \ln \eta)}. \tag{9.75}$$

The parameters in (9.75) are rather poorly known, but the predicted value of ε ($|\varepsilon| \simeq \operatorname{Im} M_{12}/\Delta m$) is about the correct magnitude.

The above calculations are for second-order processes. There are also possible first-order diagrams, for example the so-called *penguin* diagram of Fig. 9.8, which contribute to ε'. Fig. 9.9 shows the predicted value of ε'/ε, as a function of the t-quark mass, from a standard model calculation by Gilman and Hagelin (1983). The parameter B is a measure of the deviation of the hadronic part of the matrix element from the vacuum insertion form of equation (9.69). Calculations give values of B in the range 0.3 to 1, with a value of $B \sim 0.7$ to 0.8 perhaps favoured. The prediction is calculated assuming $\Gamma(b \to ue\nu)/\Gamma(b \to ce\nu)$ is less than 0.05 (see Section 8.6.3; this gives a constraint on θ_2 and θ_3 in the KM matrix). The contributions of these and the other diagrams, in particular the computation of QCD corrections and of bound state effects, is a matter

Fig. 9.8 Penguin diagram contribution to CP violation, $i = $ u, c, t. The $K^0 \to \pi\pi(I = 0)$ decay amplitude can be complex through this $|\Delta S| = 1$ transition.

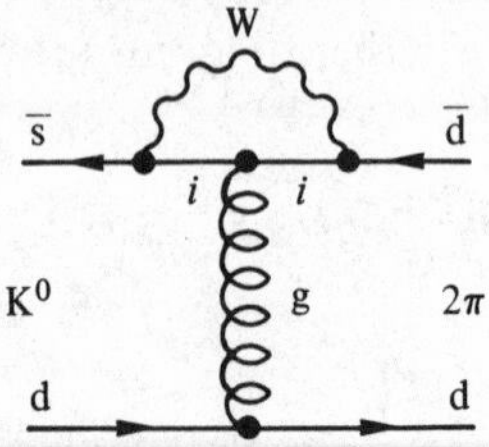

of much theoretical debate. For a review, see Buras (1985), who concludes that if $|\varepsilon'/\varepsilon| < 10^{-3}$, $m_t < 40$ GeV and $\Gamma(b \to ue\nu)/\Gamma(b \to ce\nu) < 0.01$, then one must also look outside the minimal standard model for an explanation of *CP* violation (e.g. extra Higgs doublets, more generations, left–right symmetric models, supersymmetry, etc.). The experimental goal is therefore to measure ε'/ε to better than 10^{-3}.

The neutron dipole moment can also be predicted in the standard model. Estimates give (Ellis *et al.*, 1976b) $d_n \lesssim 10^{-27}$ e cm; however, these are sensitive to the input parameters (e.g. m_t), and many estimates give $d_n < 10^{-30}$ e cm.

9.2 MIXING AND *CP* VIOLATION IN HEAVY QUARK SYSTEMS

The $K^0\bar{K}^0$ system can undergo $|\Delta S| = 2$, $d\bar{s} \leftrightarrow \bar{d}s$ transitions. In general, one would expect other neutral combinations of quarks of different flavours to exhibit similar phenomena. Thus, mixing and *CP* violation might be expected for the $D^0\bar{D}^0(c\bar{u} \leftrightarrow \bar{c}u)$, $B^0\bar{B}^0(b\bar{d} \leftrightarrow \bar{b}d$, $b\bar{s} \leftrightarrow \bar{b}s)$ and $T^0\bar{T}^0(t\bar{u} \leftrightarrow \bar{t}u$, $t\bar{c} \leftrightarrow \bar{t}c)$ meson systems. The formalism is the same as for the K^0 system; hence, assuming *CPT* invariance, we have two states of

Fig. 9.9 Predictions for ε'/ε, as a function of the t-quark mass, for different values of the bag parameter B. The hatched region indicates the uncertainty in the calculation resulting from the b-lifetime. The experimental value is plotted at $m_t = 50$ GeV.

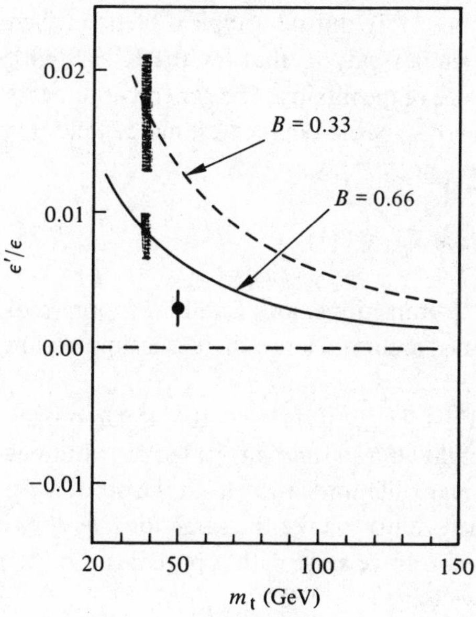

definite mass (M_s and M_L), which are linear combinations of the states of definite flavour ($M^0, \bar{M}^0$). Assuming that ε is small, so that in the normalisation $1 + |\varepsilon_M|^2 \simeq 1$, we have following (9.22)

$$|M_s\rangle = [(1 + \varepsilon_M)|M^0\rangle + (1 - \varepsilon_M)|\bar{M}^0\rangle]/2^{1/2},$$
$$|M_L\rangle = [(1 + \varepsilon_M)|M^0\rangle - (1 - \varepsilon_M)|\bar{M}^0\rangle]/2^{1/2}. \tag{9.76}$$

For the K^0 system, the $2\pi(CP = 1)$ and $3\pi(CP = -1)$ decay modes have very different decay probabilities because the 3π mode is suppressed by phase space. For the heavy quark meson systems there are many decay channels open, and these are often not CP eigenstates. For example, in the decays of the D^0 meson, the CP states must have $S = 0$, but these are Cabibbo-suppressed (e.g. $D^0 \to \pi^+\pi^-, K^+K^-$ have branching ratios much less than 10^{-2}).

The $B^0\bar{B}^0$ system seems to be the most favourable system for the observation of mixing phenomena. For a heavy quark doublet (H, L), i.e. (c, s) or (t, b), we expect $\Gamma_L \ll \Gamma_H$ because the phase space is larger for H than L, and also because the decays of L involve a generation change and are suppressed. The mass differences can be roughly estimated using box diagrams, in a similar way to that for the K^0 meson, giving

$$\Delta m_L \propto m_H^2, \qquad \Delta m_H \propto m_L^2. \tag{9.77}$$

Thus, we expect that $\Delta m_L > \Delta m_H$. If Δm is large enough that the phase between M_s and M_L changes appreciably during a typical lifetime, then sizeable mixing can occur (in a similar way to that for the K^0 system). That is, $\Delta m/\Gamma$ is a measure of the size of the mixing. The general arguments above suggest that $(\Delta m/\Gamma)_L > (\Delta m/\Gamma)_H$. More detailed estimates yield (for B mesons this is discussed below)

$$(\Delta m/\Gamma)_D \approx 10^{-3}, \qquad (\Delta m/\Gamma)_B \lesssim 0(1). \tag{9.78}$$

Note that D^0–$\bar{D}^0$ (like K^0–$\bar{K}^0$) transitions are Cabibbo-suppressed, however D decays (unlike K decays) are not. Hence there is a suppression factor $(\Delta m/\Gamma)_D \propto \sin^2\theta_C/\cos^2\theta_C$.

Because Δm is small, and Γ is large (typically $10^{13}\ s^{-1}$), mixing phenomena are probably best sought after in time-integrated distributions (i.e. those observed in a detector many lifetimes away from the production point). The formulae for the time evolution take the same form as those derived for the neutral kaon system. If we start with a pure beam of M^0 mesons at $t = 0$, then at time t we have (cf. (9.35))

$$|\psi(t)\rangle = \frac{1}{2^{1/2}}[(1-\varepsilon_M)|M_s^0\rangle \exp(-i\lambda_s t) + (1-\varepsilon_M)|M_L^0\rangle \exp(-i\lambda_L t)]$$

$$= \tfrac{1}{2}[(1-\varepsilon_M^2)(\exp(-i\lambda_s t) + \exp(-i\lambda_L t))|M_0\rangle + (1-\varepsilon_M)^2(\exp(-i\lambda_s t) - \exp(-i\lambda_L t))|\bar{M}_0\rangle]. \tag{9.79}$$

Thus the ratio of the M^0 to the $\bar{M}^0$ content, for large times, is

$$r = \frac{(M^0 \to \bar{M}^0)}{(M^0 \to M^0)} = \frac{\int_0^\infty |\langle \bar{M}^0|\psi(t)\rangle|^2\, dt}{\int_0^\infty |\langle M^0|\psi(t)\rangle|^2\, dt} = \left|\frac{1-\varepsilon_M}{1+\varepsilon_M}\right|^2 \delta, \tag{9.80}$$

The integration term δ is given by, defining $\Delta m = m_L - m_s$, $\bar{\Gamma} = (\Gamma_s + \Gamma_L)/2$ and $\delta\Gamma = \Gamma_s - \Gamma_L$,

$$\delta = \frac{(\Delta m/\bar{\Gamma})^2 + (\delta\Gamma/2\bar{\Gamma})^2}{2 + (\Delta m/\bar{\Gamma})^2 - (\delta\Gamma/2\bar{\Gamma})^2}. \tag{9.81}$$

Similarly, for an initial $\bar{M}^0$ beam

$$\bar{r} = \frac{(\bar{M}^0 \to M^0)}{(\bar{M}^0 \to \bar{M}^0)} = \left|\frac{1+\varepsilon_M}{1-\varepsilon_M}\right|^2 \delta. \tag{9.82}$$

There are two conditions which result in maximal mixing (i.e. $\delta = 1$). Firstly, if $|\delta\Gamma/2| \approx \bar{\Gamma}$ (i.e. $\Gamma_s \gg \Gamma_L$), then the M_s^0 component will quickly decay, leaving M_L^0 (i.e. as for the K^0 system). Secondly, if $\Delta m \gg \bar{\Gamma}$, the system oscillates quickly between M^0 and $\bar{M}^0$ before decaying. Mixing causes a change in the asymmetry in the semileptonic decay modes. For example, the decay of the D^0 meson is via $c \to s(d)l^+\nu_l$, whereas for $\bar{D}^0$ it is $\bar{c} \to \bar{s}(\bar{d})l^-\bar{\nu}_l$. The asymmetry can be measured, for example, in the reaction

$$e^+e^- \to \psi'' \to D^0\bar{D}^0$$

$$\quad\quad\quad\quad\quad\quad \llcorner\!\!\longrightarrow l^- \ldots$$

$$\quad\quad\quad\quad\quad \llcorner\!\!\longrightarrow l^+ \ldots \tag{9.83}$$

If $N^{++}(N^{--})$ is the number of $l^+l^+(l^-l^-)$ events detected, then a measurement of the relative fraction f, of like-sign to opposite-sign dilepton events, can be expressed as

$$f = \frac{r + \bar{r}}{1 + r\bar{r}} \simeq \frac{2\delta}{1 + \delta^2}. \tag{9.84}$$

This method can be applied to any reaction in which $D^0\bar{D}^0$ pairs are produced. For example, from a study of dimuon events in hadron–iron collisions, Bodek *et al.* (1982) find that the $D^0\bar{D}^0$ mixing is less than 4%. A similar study of dimuons produced in 225 GeV πN collisions (Louis *et* *al.*, 1986) gives a more precise limit of 5.6×10^{-3} (90% c.l.). An alternative method is to study the $D^0 \to K^-\pi^+$ and $\bar{D}^0 \to K^+\pi^-$ decays in $e^+e^- \to D^0\bar{D}^0$. Mixing, in this case, shows up as an apparent violation of the $\Delta C = \Delta S$ rule. Goldhaber *et al.* (1977) find that any such violation is less than 18% (90% c.l.). From a study of the well constrained decay sequence $D^{*+} \to D^0\pi^+$ etc. in e^+e^- annihilations, Albrecht *et al.* (1987c) extract a limit $\Gamma(D^0 \to \bar{D}^0)/\Gamma(D^0 \to \text{all}) < 1.4\%$ (90% c.l.). Neglecting any *CP* violation, the above results can then be used to set the limits $\Delta m/\bar{\Gamma} < 0.11$, $\delta\Gamma/\bar{\Gamma} < 0.22$ and also that $\Delta m < 1.6 \times 10^{-4}$ eV.

The $B^0\bar{B}^0$ system is expected to be the most fruitful experimental possibility for mixing. An estimate of the mass difference Δm_q for B^0_q–$\bar{B}^0_q$ (q = d, s) can be made using the box diagram (in a similar way to that described in Section 9.1.6). The t-quark contribution dominates,# giving

$$\Delta m_q \simeq \frac{G_F^2}{6\pi^2}\eta^B f^2_{B_q} B m_B m_t^2 |V^*_{tb}V_{tq}|^2, \tag{9.85}$$

where $\eta^B(\sim 0.8)$ is a QCD correction, B the bag constant and f_{B_q} is the B-meson decay constant. Estimates give roughly $Bf^2_{B_d} \sim (0.11\ \text{GeV})^2$ and $Bf^2_{B_s} \sim (0.15\ \text{GeV})^2$. The KM matrix elements in the Wolfenstein parameterisation are approximately $V_{ts} \sim \lambda^2$ and $V_{td} \lesssim \lambda^3$ ($\lambda = \sin^2\theta_C = 0.23$), so that $x_q = (\Delta m/\bar{\Gamma})_q$ is expected to be larger for B^0_s than for B^0_d. Two measurements yielding non-zero values of f (in (9.84)) have been reported (see, e.g., Bigi (1987) for a discussion). The Argus experiment (Albrecht *et al.*, 1987b), working at the Υ(4S) resonance (i.e. $B^0_d\bar{B}^0_d$), find a value of $f = 0.21 \pm 0.08$. At the Υ(4S), the B–$\bar{B}$ are in a state $l = 1$, and $f \simeq r_d$ (instead of (9.84)), due to the effects of Bose–Einstein statistics. The UA1 group (Albajar *et al.*, 1987b) report a value of $f = 0.42 \pm 0.07 \pm 0.03$ from the analysis of a sample of $\bar{p}p \to \mu^\pm\mu^\pm X$, containing an (unknown) mixture of B^0_d and B^0_s (an estimate is 15% B^0_s, 40% B^0_d). Part (but probably not all) of this signal is due to processes other than $B^0\bar{B}^0$ mixing (e.g. $\bar{p}p \to b\bar{b}X$, $b \to c \to \mu^+$, $\bar{b} \to \mu^+$), leaving f(mixing) $\sim 0.14 \pm 0.06$, compatible with all values of r_s for $r_d \sim 0.2$. If these results become firmly established, then (using equations (9.80) to (9.85)) a limit $m_t \gtrsim 34$ GeV, and a value $x_d \simeq (\Delta m/\bar{\Gamma})_d = 0.73 \pm 0.18$ can be extracted. Thus, these experiments are consistent with mixing; however, the corresponding *CP* violation is expected to be small in the standard model. A non-zero value of $a = (l^{++} - l^{--})/(l^{++} + l^{--}) = (r - \bar{r})/(r + \bar{r})$, would be evidence of *CP*

This calculation makes use of the bag model in which the hadronic part of the matrix element (i.e. the analogue of the $\bar{s}\cdots d$ term in (9.68)) is replaced by $\frac{4}{3}Bf^2_B m^2_B/(2m_B)$.

violation. It may be that the K^0 system alone will be the only one which will provide data on *CP* violation for some time. For a detailed review of the potential sources and size of mixing and *CP* violation effects, see, for example, Chau (1983) and Bigi and Sanda (1987).

Our understanding of the intriguing phenomenon of *CP* violation is, after more than 20 years' work, still far from satisfactory. The phenomenon itself, however, allows a unique definition of what we refer to as matter and antimatter. The baryon to antibaryon excess in the universe may also have some connection with *CP* violation.

9.3 NEUTRINO MASS AND OSCILLATIONS

The properties of the neutrinos (ν_l, $l = \mathrm{e}, \mu, \tau$), as so far portrayed, are that they are zero-mass particles which have only lefthanded couplings (or only righthanded for $\bar{\nu}_l$), and that for each generation there is a conserved additive lepton quantum number. In general, however, the neutrino mass eigenstates and lepton-number eigenstates are not necessarily the same. Thus, there can be mixing between the generations in a way analogous to the KM scheme for quarks. Such mixing gives rise to neutrino oscillations beween eigenstates of definite lepton number (Maki *et al.*, 1962; Pontecorvo, 1968a).

Transitions between neutrino states imply that the associated lepton number is violated. In attempts to unify the electroweak model with QCD, such as the SU(5) model of Georgi and Glashow (1974), quarks and leptons are grouped together in multiplets, and there can be transitions between them. These are mediated by very massive ($\sim 10^{15}$ GeV) X and Y bosons, with charges $\pm\frac{4}{3}$ and $\pm\frac{1}{3}$ respectively. Baryon number conservation is also violated, and the proton can decay by $\mathrm{p} \to \mathrm{e}^+\pi^0$, etc., but with a long lifetime ($\gtrsim 10^{28}$ yr), due to the large mass of the X boson. The lepton number violation can occur via $|\Delta L| = 0$ or 2 transitions, where $L = L_\mathrm{e} + L_\mu + L_\tau$. For both cases the oscillations between neutrino types require that the neutrino from at least one generation has finite mass, and that the masses of at least two generations are not equal.

9.3.1 $\Delta L = 0$ oscillations

For simplicity we consider only the electron and muon generations. The relevant part of the standard model Lagrangian is that given in (5.181). This can be written in the form

$$\mathscr{L}_{\mathrm{LH}} = \bar{l}' M_l' l' + \bar{\nu}' M_\nu' \nu'; \qquad l' = \begin{pmatrix} e \\ \mu \end{pmatrix}, \; \nu' = \begin{pmatrix} \nu_\mathrm{e} \\ \nu_\mu \end{pmatrix}, \tag{9.86}$$

where (with $\kappa = v/2^{1/2}$)

$$M'_\nu = \kappa \begin{pmatrix} h_{ee} & h_{e\mu} \\ h_{e\mu} & h_{\mu\mu} \end{pmatrix}.$$

That is, the mass matrix for the neutrino has, in general, off-diagonal elements.

The mass eigenvalues correspond to a basis in which the mass matrix $(M_\nu)_d$ is diagonal, and this is achieved by a rotation from ν' to ν. That is,

$$\begin{pmatrix} \nu_e \\ \nu_\mu \end{pmatrix} = \begin{pmatrix} \cos\theta & \sin\theta \\ -\sin\theta & \cos\theta \end{pmatrix} \begin{pmatrix} \nu_1 \\ \nu_2 \end{pmatrix} \qquad \text{i.e. } \nu' = U^\dagger \nu. \tag{9.87}$$

Now, for physical equivalence between the two bases, we require that $\bar{\nu}' M'_\nu \nu' = \bar{\nu}(M_\nu)_d \nu$, so that

$$M'_\nu = U^\dagger (M_\nu)_d U, \qquad (M_\nu)_d = U M'_\nu U^\dagger, \tag{9.88}$$

where, if m_1 and m_2 are the mass eigenvalues, we have

$$M_\nu = \begin{pmatrix} m_1 & 0 \\ 0 & m_2 \end{pmatrix}. \tag{9.89}$$

Defining $m_{\nu_e} = \kappa h_{ee}$, $m_{\nu_\mu} = \kappa h_{\mu\mu}$ and $m_{\nu_e \nu_\mu} = \kappa h_{e\mu}$, we obtain from (9.86) to (9.89)

$$\begin{aligned} m_{\nu_e} &= m_1 \cos^2\theta + m_2 \sin^2\theta, \\ m_{\nu_\mu} &= m_1 \sin^2\theta + m_2 \cos^2\theta, \\ m_{\nu_e\nu_\mu} &= (m_2 - m_1)\sin\theta\cos\theta. \end{aligned} \tag{9.90}$$

If we start with a pure ν_e beam at $t = 0$, then oscillations between the lepton number eigenstates can occur in a similar way to those between strangeness eigenstates in the K^0 system. At time t the state becomes

$$\begin{aligned} |\psi(t)\rangle &= \cos\theta \exp(-iE_1 t)|\nu_1\rangle + \sin\theta \exp(-iE_2 t)|\nu_2\rangle \\ &= [\cos^2\theta \exp(-iE_1 t) + \sin^2\theta \exp(-iE_2 t)]|\nu_e\rangle \\ &\quad + \sin\theta\cos\theta[\exp(-iE_2 t) - \exp(-iE_1 t)]|\nu_\mu\rangle. \end{aligned} \tag{9.91}$$

Table 9.1. *Summary of the most abundant neutrino sources*
For each, the mean energy and typical length L from the source is given, together with the mass values to which the method is sensitive

Source	Flavour	$\langle E_\nu \rangle$	L(m)	Δm (eV)
Accelerator	ν_μ	2–100 GeV	50–1000	~ 10
Reactor	$\bar{\nu}_e$	~ 3 MeV	5–100	~ 1
Sun	ν_e	1–10 MeV	10^{11}	$\sim 10^{-5}$

Hence, the probability that the state becomes a ν_μ is

$$\begin{aligned} P(\nu_e \to \nu_\mu) &= |\langle \nu_\mu | \psi(t) \rangle|^2 \\ &= \frac{\sin^2 2\theta}{2}[1 - \cos(E_2 - E_1)t] \\ &\simeq \frac{\sin^2 2\theta}{2}\left\{1 - \cos\left[\left(\frac{m_2^2 - m_1^2}{2E}\right)L\right]\right\} \\ &= \sin^2 2\theta \sin^2\left(\frac{1.27\Delta m^2 L}{E}\right), \end{aligned} \tag{9.92}$$

where, in the last line of (9.92), the approximation that the beam momentum $p \gg m_i$ has been used and the quantity L is the distance travelled by the beam. In the final expression the units are L in metres, E in MeV and $\Delta m^2 = m_2^2 - m_1^2$ in eV^2. From (9.92), one can define an oscillation length

$$\Lambda = \frac{4\pi E}{|m_2^2 - m_1^2|} \simeq \frac{2.5E\ (\text{MeV})}{\Delta m^2\ (\text{eV}^2)} \text{ metres.} \tag{9.93}$$

Because the potential range of possible values for any non-zero neutrino mass is large, any particular experiment is only sensitive to a small window around $\Delta m \sim (2.5E/L)^{1/2}$. A summary of possible methods, and their ranges of sensitivity, is given in Table 9.1.

The above formulation can easily be extended to the case of three generations. The matrix U (cf. (9.87)) becomes a 3×3 unitary matrix and, as for the KM matrix, there are three possible mixing angles and one CP-violating phase. Given the large number of parameters, and the limited number of 'windows' available in practice, the unravelling of any neutrino oscillations is clearly difficult. In the following only a very brief discussion is given; for more extensive reviews see, for example, Vergados (1986), Winter (1986).

In the reactor experiments the $\bar{\nu}_e$s produced have energies below the threshold for the charged current interactions of $\bar{\nu}_\mu$ and $\bar{\nu}_\tau$. Hence, effects can only be observed by *disappearance* experiments, and these are not sensitive to small values of $\sin^2 2\theta$. The reaction $\bar{\nu}_e p \to e^+ n$ is measured by detecting the recoil neutron and measuring the energy of the positron. The event rates and the spectra are then compared at two distances. However, this method requires good knowledge of the reactor flux. This dependency could be greatly reduced by simultaneously measuring the $\bar{\nu}_e$ flux at two distances.

Neutrino beams at high energy accelerators are mainly ν_μ (or $\bar{\nu}_\mu$; if negative parent mesons are used). The beam energies are large enough that both *appearance* and disappearance of flavours can be studied. Furthermore, beams rich in ν_e and $\bar{\nu}_e$ can be produced from the so-called *beam dump* experiments. In this arrangement, the conventional flux arising from $\pi^\pm$ and $K^\pm$ decays is largely eliminated by using a target consisting of many interaction lengths of material. The ν_e, $\bar{\nu}_e$ arise from the decays of short-lived particles (e.g. D^0, $\bar{D}^0$, Λ_c). The production of ν_τ and $\bar{\nu}_\tau$ is also possible from the semileptonic decays $D_s \to \tau\nu_\tau$, etc. If the source of the ν and $\bar{\nu}$ giving rise to the observed reactions in the various detectors is from the semileptonic decays of charmed particles, then essentially equal rates for ν_e and ν_μ (or $\bar{\nu}_e$ and $\bar{\nu}_\mu$) are expected. Some indication that the flux $(\nu_e) \simeq 0.6 \times \text{flux}(\nu_\mu)$ was obtained from the CERN beam dump experiments (see, e.g., Hulth, 1984); however, this anomaly has largely disappeared with a repeat of the experiments.

A more direct method of studying possible oscillations is to use two ν_μ beams of widely different energies. Such an experiment has been carried out at CERN, using the usual SPS ν_μ beam ($\langle E_\nu \rangle \sim 50$ GeV), and a specially constructed beam from the low energy PS accelerator ($\langle E_\nu \rangle \sim 1.5$ GeV). The accelerator neutrino experiments are sensitive to rather small values of $\sin^2 2\theta$. Fig. 9.10 shows some limits on $\Delta m^2(\nu_e \leftrightarrow \nu_\mu)$ from both reactor and accelerator experiments. The reactor experiments are incompatible. The Bugey experiment (Cavaignac *et al.*, 1984) claims a signal for oscillation, but the Gösgen experiment (Zacek *et al.*, 1985) is compatible with a null result, over a wider range of Δm^2 and $\sin^2 2\theta$. The limits on possible transitions from ν_e or ν_μ to ν_τ are less precise.

In the third possible method listed in Table 9.1, that using electron neutrinos produced by the thermonuclear reactions in the sun, the sensitivity is to very small mass differences. Since the experiments are of the disappearance type, the ν_e flux produced by the complex nuclear chain reactions must be well known. The incident flux on earth is approximately $10^{11} \nu_e/(\text{cm}^2\ \text{s})$. Experimentally, a series of measurements on solar neutrinos, detected by the reaction $\nu_e + {}^{37}\text{Cl} \to e^- + {}^{37}\text{Ar}$, has been carried out since 1970 by Davis and coworkers (Bahcall *et al.*, 1983). The resulting mean value from these data for the solar neutrino flux is

$$\text{experiment} \quad R_{\text{exp}}({}^{37}\text{Cl}) = 2.1 \pm 0.3 \text{ SNU},$$
$$\text{theory} \quad R_{\text{th}}({}^{37}\text{Cl}) = 5.8 \pm 2.2 \text{ SNU}, \quad (3\sigma \text{ limit}) \qquad (9.94)$$

where the 'theory' is computed using the standard solar model. One solar neutrino unit (SNU) is equal to $10^{-36}\nu_e$ captures per target nucleus per second. The model has some sensitivity to the input parameters, and lower values can be obtained (Filippone, 1981). The difference between experiment and theory is generally referred to as the *solar neutrino problem.*

Before any conclusion on ν_e oscillations can be drawn from the results, the correctness of both the experiment and the solar model must be firmly established. It should be noted that the threshold for the ^{37}Cl interaction

Fig. 9.10 Some results on the 90% confidence limits for $(\nu_e \leftrightarrow \nu_\mu)$ transitions from experiments at accelerators (BEBC, CHARM, BNL) and at reactors (GÖSGEN, BUGEY). All the experiments exclude values above the contour except the BUGEY experiment, which claims a positive result.

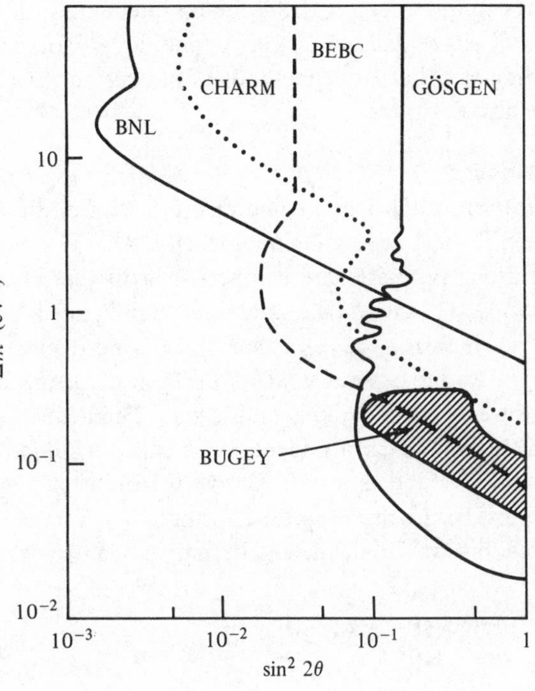

is considerable (814 KeV), and that only about 10^{-4} of the solar neutrinos (mainly from ^8B decay) have sufficient energy for this interaction. Even then, other possibilities exist. If the neutrino has a magnetic moment then, on passing through the solar magnetic fields, some decrease in the number of lefthanded neutrinos could occur. Such neutrinos would not be detected in the Davis experiment. A further consideration to be taken into account is that discussed by Wolfenstein (1978), and developed by Mikheyev and Smirnov (1986); namely that any neutrino oscillations will be influenced, if the neutrinos traverse a sufficient path length in dense matter ($L \gtrsim 10^3$ km, thickness $\gtrsim 10^9$ g cm^{-2}). This arises because the cross-section for low energy ν_e, which can have both neutral and charged current interactions, is not equal to that for ν_μ, for which only the NC channel is open. Furthermore, oscillations can occur, even if all the neutrinos are massless, if the coherent forward scattering is partially off-diagonal, i.e. neutral current interactions in which the ingoing and outgoing neutrinos are of different types. The solar matter through which the emerging ν_e particles pass could induce transitions to ν_μ and thus may explain, at least in part, the solar neutrino problem (Bethe, 1986).

Measurement of the solar neutrino flux is clearly important, not only in the understanding of neutrino oscillations but also as a test-bed for stellar models. New methods of detection are actively being pursued, and these should be sensitive to low energy ν_e from the pp interactions in the solar chain. Detectors of either gallium (^{71}Ga) or indium (^{115}In) have emerged as good candidates. Alternative methods of measuring the high energy component are also desirable.

9.3.2 $|\Delta L| = 2$ oscillations

From consideration of the Dirac equation we have seen that, if the $\nu(\bar{\nu})$ is massless, then it must have a definite helicity. Analyses of the charged currents in various processes are compatible with there being only lefthanded neutrinos, i.e. $(1-\gamma^5)u_\nu$, and only righthanded anti-neutrinos, i.e. $\bar{v}_\nu\ (1+\gamma^5)$. If this is the case, then there is no distinction between ν and $\bar{\nu}$ which will lead to experimentally observable consequences. That is, we could have the equivalence $\bar{\nu}_L \equiv \nu_L$ and $\bar{\nu}_R \equiv \nu_R$. This description of the neutrino, in which the ν and $\bar{\nu}$ are identical, is called a *Majorana neutrino* ν_M (Majorana, 1937) and is an alternative to the usual Dirac description (see Appendix B). In the Majorana formulation, the lepton number is not conserved, but we can define an alternative quantity which is conserved, namely

$$L' = \begin{cases} 1 & \text{for } \tfrac{1}{2}(1-\gamma^5)\nu_M,\ l^-, \\ -1 & \text{for } \tfrac{1}{2}(1+\gamma^5)\nu_M,\ l^+. \end{cases} \tag{9.95}$$

If the neutrino has mass, then the Dirac and Majorana descriptions are no longer identical. However, the expected differences are very small. For example, if the $\bar{\nu}_e$ produced in neutron decay ($n \to pe^-\bar{\nu}_e$) then initiates the interaction $\nu_e + {}^{37}Cl \to e^- + {}^{37}Ar$ (which is possible for Majorana neutrinos), the cross-section will be small because of helicity considerations.. The neutrino produced in neutron decay is righthanded, whereas (in the $m_\nu = 0$ limit) a lefthanded neutrino is needed for the subsequent interaction. Thus the amplitude is proportional to m_ν and is consequently small.

A potentially sensitive way of confirming whether or not neutrinos are massive Majorana particles is through the study of *double-beta decay*. The pairing energy of nuclei with even numbers of protons and neutrons is stronger than that in neighbouring odd–odd nuclei. Hence, in some cases, the even–even nucleus cannot decay to the neighbouring odd–odd nucleus but can decay, by a second-order weak interaction, to the next even–even nucleus, i.e. $(A, Z) \to (A, Z+2)2e^-2\bar{\nu}_e$ (see Elliot *et al.* (1987)). Of particular interest is the possibility of *neutrinoless double-beta decay*, i.e. $(A, Z) \to (A, Z+2)2e^-$, which can occur by the mechanism shown in Fig. 9.11. This requires that the virtual $\bar{\nu}$ from the first e^- emission is reabsorbed as a ν in order that the second e^- is emitted. This is possible for a Majorana neutrino, but it must have either a non-zero mass or the weak charged current process must have some righthanded admixture. In the latter case, the lepton current has the form

$$j^\mu \propto \bar{e}\gamma^\mu[(1-\gamma^5) + \eta(1+\gamma^5)]\nu_M. \tag{9.96}$$

Experimentally, two methods have been used. In the first, which is *passive*, the fraction of daughter nuclei trapped in ores rich in the parent nuclei is determined by geochemical means. The two main problems

Fig. 9.11 Mechanism for neutrinoless double-beta decay by the exchange of a virtual Majorana neutrino.

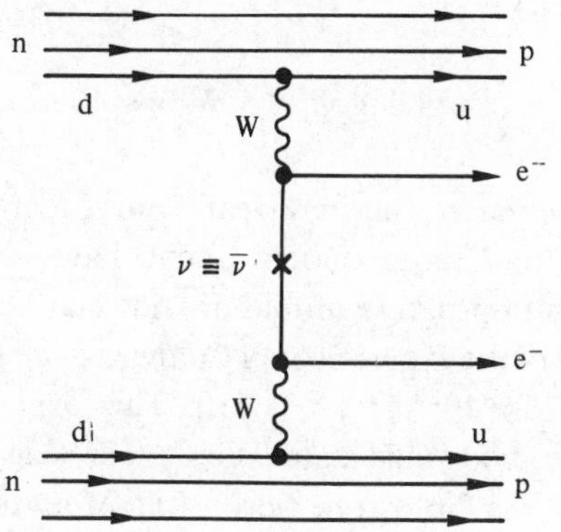

with this method are that there may have been some leakage over the long time-scales ($\sim 10^9$ yr) involved and, further, that there is no distinction between the 0ν and 2ν modes. A measurement gives the sum of the two modes, i.e. $\lambda_{tot} = \lambda_{0\nu} + \lambda_{2\nu}$. Hence, an estimate for $\lambda_{2\nu}$ is necessary. However, the phase space available for the two-body 0ν final state is much larger (by a factor of approximately 10^6) than for the four-body 2ν state. For example, for the nucleus ^{130}Te, it is estimated that the respective half-lives are

$$\tau_{1/2}^{2\nu} \sim 10^{22\pm 2}\text{ yr}, \qquad \tau_{1/2}^{0\nu} \sim 10^{15\pm 2}\eta^{-2}\text{ yr}. \tag{9.97}$$

Some of the uncertainty in the theoretical predictions can be removed if, as suggested by Pontecorvo (1968b), the ratios of the decays of two isotopes are compared. The relative decay energy, or Q value, of ^{128}Te (869 KeV) is much lower than that of ^{130}Te (2533 KeV). The ratio of the 2ν decay modes is estimated to be $\rho_{2\nu} = {}^{128}\lambda_{2\nu}/{}^{130}\lambda_{2\nu} \sim 2\times 10^{-4}$. Experimentally, Kirsten *et al.* (1983) find a value for the ratio of the total decay rates which is compatible with this estimate, namely

$$\rho_{tot} = {}^{128}\lambda_{tot}/{}^{130}\lambda_{tot} = (1.03 \pm 1.13)\times 10^{-4}. \tag{9.98}$$

If the 0ν mode occurs, then a ratio much greater than this is expected. A limit $\hat{m}_\nu \leqslant 5.6$ eV and $\eta \leqslant 2.4\times 10^{-5}$ (95% c.l.) is extracted. (The notation $\hat{m}_\nu$ denotes a Majorana neutrino mass.)

In the second method, which is *active*, the two decay electrons are detected. The requirements of low backgrounds and noise levels are essential, and germanium detectors are used. The sum of the energies of the electrons should be a constant for the 0ν mode; with lines at 2040.9 and 1481.8 keV for the $0^+ \to 0^+$ and $0^+ \to 2^+$ transitions respectively. The $0^+ \to 2^+$ transition is only possible for righthanded currents, whereas $0^+ \to 0^+$ is possible both for righthanded currents and $m_\nu \neq 0$. One of the most sensitive experiments is the Mont Blanc experiment of Bellotti *et al.* (1984), which reports (at 68% c.l.)

$$\left.\begin{aligned} \tau_{1/2}(0^+ \to 0^+) &> 7.2\times 10^{22}\text{ yr} \\ \tau_{1/2}(0^+ \to 2^+) &> 1.2\times 10^{22}\text{ yr} \end{aligned}\right\} \text{giving } \hat{m}_\nu \leqslant 5\text{ eV},\ \eta \leqslant 10^{-5}. \tag{9.99}$$

The experimental situation is, however, not entirely clear. Some experiments with potentially less accuracy than those quoted above have claimed positive signals. The experiments in progress should provide clarification. For example, preliminary results of the experiment of Caldwell *et al.* (1987) give, for ^{76}Ge, $\tau_{1/2}(0^+ \to 0^+) > 5\times 10^{23}$ yr (90% c.l.). This experiment also finds no evidence for the three-body final state $^{76}\text{Ge} \to {}^{76}\text{Se} + 2e^- + \text{B}$, where B denotes a possible massless Goldstone boson (*the Majoron*, see, e.g., Chikashige *et al.*, 1981), in contrast to the potentially less sensitive experiment of Avignone *et al.* (see Caldwell *et al.*, 1987, for a discussion). This process is of interest because a Majoron would result from spontaneous breaking of baryon minus lepton number symmetry, a process which would also give rise to light Majorana neutrinos. However, a null (or very small) result for 0ν need not mean that massive Majorana neutrinos do not exist, since there could be destructive interference between different *CP* states (Haxton and Stephenson, 1985).

9.3.3 Direct measurements of the neutrino mass

A measurement of the neutrino mass, or an upper limit on the mass, can be made by performing precise kinematic measurements, using well constrained decay processes. The currently best upper limit on the mass of the ν_τ comes from a study of the decays $\tau \to 3\pi\nu_\tau$ and $\tau \to 6\pi\nu_\tau$ by the Argus collaboration (Albrecht *et al.*, 1985; Koltick, 1986). These give a limit $m_{\nu_\tau} < 70$ MeV (95% c.l.). The result for ν_μ is somewhat more precise. By studying the muon momentum in the two-body decay $\pi \to \mu\nu$, from pions decaying at rest, the SIN group (Jeckelmann *et al.*, 1986) obtain $m_{\nu_\mu} < 270$ keV (90% c.l.). This limit is, however, only marginally better than the mass of the electron.

The most accurate neutrino mass measurements are for the $\bar{\nu}_e$ produced in the beta decay of tritium, ${}^3H \to {}^3He + e^- + \bar{\nu}_e$, which has a half-life of 12.3 yr. The decay involves an energy transition of 18.6 KeV, and measurement of the electron energy near the end-point E_0 is sensitive to the $\bar{\nu}_e$ mass. The method consists, in essence, in looking for deviations from a straight line on a Kurie plot, and is sketched in Fig. 9.12. It can be seen that the effects of the finite resolution of the experiments, and a non-zero value of m_ν, work in opposite directions. Tritium can decay to a number of atomic states of ^{3}He and hence atomic and molecular effects, at about the same order as the sensitivity to $m_\nu(\sim 20$ eV), are present. In the lab system of the parent nucleus (mass M_i), energy conservation gives $M_i = M_f + T_f + E_e + E_\nu$, where M_f and T_f are the mass and recoil energy of the daughter nucleus respectively. Thus the factor $p_e E_e p_\nu E_\nu$ in (5.16) becomes $p_e E_e[(E' - E_e)^2 - m_\nu^2]^{1/2}(E' - E_e)$, where $E' = M_i - M_f - T_f = E_0 + m_\nu$. The counting rate near the end point is very small ($\lesssim 10^{-7}$ of the total) and the sensitivity goes roughly as m_ν^3.

All experiments, with the exception of the ITEP experiment of Lubimov and coworkers (Boris *et al.*, 1985), are compatible with $m_\nu = 0$. The ITEP experiment finds a finite value and quote

$$20 < m_{\bar{\nu}_e} < 45\,\text{eV}, \qquad \text{or} \qquad m_{\bar{\nu}_e} > 9\,\text{eV}, \qquad 90\%\ \text{c.l.} \tag{9.100}$$

The upper limits from the various other experiments in operation are gradually improving. The SIN experiment of Fritschi *et al.* (1986) gives an upper limit of $m_{\bar{\nu}_e} < 18$ eV, 95% c.l. Furthermore, there have been various criticisms raised of the ITEP experiment (for a review, see Bergkvist (1985) and Fiorini (1986)), which have been answered, at least partly, by the experimenters. However, confirmation of the ITEP result is needed before this very important result can be accepted.

A claim for a rather large neutrino mass ($m_\nu = 17.1 \pm 0.2$ keV) has been made by Simpson (1985), based on a measured distortion of the Kurie plot for tritium in the low energy part of the spectrum. The corresponding mixing parameter to such a neutrino ν_i is $|U_{ei}|^2 = 0.03$. However, three other experiments have given evidence against a 17 keV neutrino with such a mixing angle. Furthermore, some doubt remains on the effects of Coulomb screening and chemical binding energy in the calculation of the low energy part of the tritium spectrum. In a review, Winter (1986) dismisses the 17 keV neutrino claim. Searches for the possible decays of massive neutrinos have also been made, particularly using beam dump data. Decays such as $\nu_i \to l^+ l^- \nu_e$ and $\nu_i \to l +$ hadrons ($l = \mathrm{e}, \mu$) are searched for. No significant signal is found in the range of sensitivity of the experiments (roughly 10–500 MeV).

The terrestrial detection on 23 February 1987 at 7 h 35 min universal time of low energy $\nu/\bar{\nu}$, which were produced by the collapse of a supernova

Fig. 9.12 Kurie plot for tritium near the end-point E_0. The curves are for different assumptions about m_ν and the electron energy resolution $\sigma(E)$. The full line is for $m_\nu = \sigma(E) = 0$, the curve (a) is for $m_\nu = 0$, $\sigma(E) \neq 0$ and curve (b) is for $\sigma(E) = 0$, $m_\nu \neq 0$.

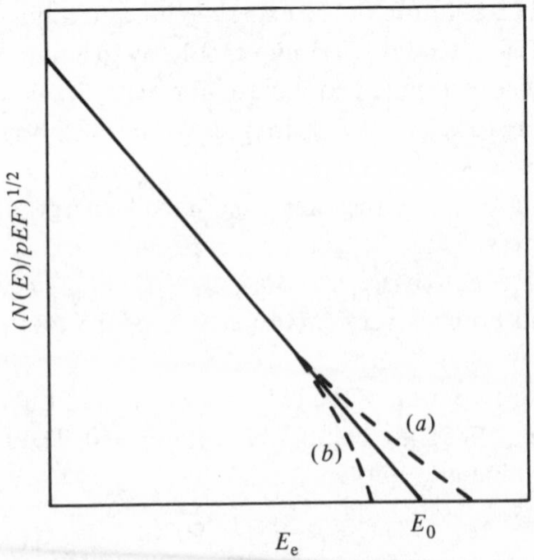

(SN 1987a in the Large Magellanic Cloud), is important in that it provides strong constraints, not only on models of supernova collapse, but also on the properties of the neutrinos produced. It is estimated that about 10^{58} neutrinos were produced, with $\langle E_\nu \rangle \sim 10$ MeV. Events were detected in the large water Cerenkov (nucleon decay) detectors of Kamiokande II (Hirata *et al.*, 1987) and IMB (Bionta *et al.*, 1987). A signal which may be associated with the same phenomenon was also detected (some 4 h 43 min earlier) in the (smaller) Mont Blanc detector (Aglietta *et al.*, 1987). Events are detected by measuring the $e^\pm$ energy in the reactions $\bar{\nu}_e p \to e^+ n$ (e^+ isotropic) and $\nu e^- \to \nu e^-$ (e^- along ν direction). In the Kamiokande detector (the most sensitive to low energy $e^\pm$), events with energies 20, 14, 8, 9, 13, 6, 35, 21 and 20 MeV (error $\sim \pm 20\%$) were detected in the first two seconds, with three further events (9, 13 and 9 MeV) between 9 and 13 s after the first event. The first two events 'point' in the expected direction ($\nu e^- \to \nu e^-$?), and the remaining ten are compatible with isotropy ($\bar{\nu}_e p \to e^+ n$?).

Bahcall and Glashow (1987) argue that the relatively narrow spread in arrival times (i.e. 9 in 2 s) implies that $m_{\nu_e} < 11$ eV, provided that the pulse has not sharpened by more than a factor two in traversing the 50 kpc flight path ($\equiv 5 \times 10^{12}$ s). If the ν_e mass were finite, it is expected that the higher energy neutrinos would arrive first; however, no strong ordering is observed. However, a 'model independent' analysis by Kolb *et al.* (1987) gives a somewhat higher limit, $m_{\nu_e} \lesssim 20$ eV. Whilst this historic event will undoubtedly give rise to much speculation, it is arguably a safer strategy to settle the issue of a finite neutrino mass (at least for $\bar{\nu}_e$) using laboratory experiments and use this as input in learning about the mechanism of supernova collapse.

In summary, although the possibility of a non-zero neutrino mass (Dirac or Majorana) exists in the data, this has not yet been proven beyond reasonable doubt. Only painstaking experimental work will resolve this problem, and this could take many years.

9.3.4 Electromagnetic properties of the neutrino

The neutrino can undergo virtual transitions to charged particles (e.g. $\nu \to e^- W^+ \to \nu$) and thus, in general, the mean square charge radius $\langle r^2 \rangle$ is expected to be non-zero. The electromagnetic current for a neutrino can be written as (see Section 7.1)

$$\begin{aligned} &\bar{u}'(p')\left[\gamma^\mu F_1(q^2) + \frac{i}{2m_e}\sigma^{\mu\nu}q_\nu F_2(q^2)\right]u(p) \\ &= \bar{u}'(p')\left[\gamma^\mu F_1(q^2) - (p+p')^\mu \frac{F_2(q^2)}{2m_e}\right]u(p), \end{aligned} \tag{9.101}$$

where $F_1(0)=0$ (assuming the neutrino is neutral) and so $F_1(q^2)\simeq -\langle r^2\rangle q^2/6$. The term $F_2(q^2)$ is related to the neutrino magnetic moment, with $F_2 \simeq \mu$ (measured in units of electron Bohr magnetons $e/2m_e$).

In $\nu_\mu e^-$ elastic scattering, for example, this will lead (in addition to the Z^0 exchange term) to a one-photon term and also a γ–Z^0 interference term. The differential cross-section can be computed using the usual methods, giving (for $E_\nu \gg m_e$, $|q^2| \ll M_Z^2$)

$$\frac{d\sigma}{dy}=\frac{d\sigma}{dy}(Z^0)+\frac{2\alpha G_F m_e E_\nu\langle r^2\rangle}{3(2)^{1/2}}[(g_V^e+g_A^e)+(g_V^e-g_A^e)(1-y)^2]$$

$$+\frac{\alpha^2 m_e E_\nu \pi\langle r^2\rangle^2}{9}[1+(1-y)^2]+\frac{\alpha^2\pi\mu^2}{m_e^2}\left(\frac{1-y}{y}\right), \qquad (9.102)$$

where $d\sigma(Z^0)/dy$ is given by (6.61).

In the standard model (or any model incorporating the two component neutrino theory), or for Majorana neutrinos, the neutrino magnetic moment $\mu=0$. However, in extensions to the standard model with massive Dirac neutrinos (i.e. righthanded states), non-zero values of μ are expected (typically $\mu \simeq 10^{-19} m_\nu$ (eV)). For $\langle r^2\rangle$, typical expectations are $\langle r^2\rangle \sim 10^{-32}$ cm^2. By comparing the measurements of the cross-sections for elastic $\nu_\mu e^-$ and $\bar{\nu}_\mu e^-$ interactions to the expectations of the standard model, Abe *et al.* (1987) obtain the limits $|\mu| \lesssim 10^{-9}$ and $-7.3\times 10^{-32} \leqslant \langle r^2\rangle \leqslant 0.8\times 10^{-32}$ cm^2. Thus the neutrino exhibits no internal structure down to about 10^{-16} cm. However the experimental limits are approaching the theoretical expectations, so that improved precision is desirable. One interesting possible consequence of a non-zero magnetic moment is that, on traversing a magnetic field, the neutrino spin could flip, with the resulting neutrino having substantially reduced cross-sections.

10

The standard model and beyond

In this chapter the production and properties of the massive gauge bosons $W^{\pm}$ and Z^0 are discussed. The properties of the other gauge boson, the massless photon, were described in Chapters 4, 5 and 7. The properties of time-like virtual $W^{\pm}$ particles were discussed in Chapters 6 and 8, in the context of the decays of leptons and hadrons. Both the photon and the $W^{\pm}$ can be used as probes of the elementary constituents of hadrons. Here, for space-like four momenta such that $Q^2 \gtrsim 5\,\text{GeV}^2$, the gauge bosons behave essentially as point-like probes. For much lower values of Q^2, the $W^{\pm}$, Z^0 and the photon (in particular, the real photon) have hadronic components, arising from their fluctuations to $q\bar{q}$ pairs. Because of the difficulty in measuring the variables x_{BJ} and Q^2 in deep inelastic scattering in the case of neutral current interactions, these reactions have mainly been used to study the couplings of the Z^0 to quarks (Section 8.8), rather than to study QCD effects. Interference phenomena between γ and Z^0 exchange have been observed for space-like momenta in charged lepton deep inelastic scattering (Section 8.9.1), and for time-like momenta in e^+e^- annihilation (Section 6.3 and 8.9.2).

Experimentally, the most important missing ingredient of the standard model is the Higgs scalar, which is needed to give masses to the particles. The complete list of quarks and leptons must also be established. The expected properties of the Higgs particle, and of the top quark (which is widely, but not universally, needed in models and for which the current experimental position is still open), are also discussed below. Finally, the problems which arise in the standard model are outlined, and attempts at going beyond it are briefly described.

10.1 DECAY OF $W^{\pm}$

The matrix element for the decay of the spin one $W^{\pm}$ gauge boson can be written down in a straightforward manner using the Feynman rules given in Appendix C. Let us consider the decay of a W^-, with four-momentum p and polarisation vector ε^{λ} (see Section 4.5), to a final state consisting of a fermion f_a (with four-momentum p_a, spin four-vector s_a) and an antifermion $\bar{f}_b(p_b, s_b)$; that is

$$W^-(p, \varepsilon^{\lambda}) \to f_a(p_a, s_a) + \bar{f}_b(p_b, s_b). \tag{10.1}$$

To lowest order, the matrix element for this decay mode (Fig. 1.8(*a*)) is

$$\mathcal{M} = \frac{-ig}{2^{1/2}} C\varepsilon^{\lambda}_{\mu}\bar{u}_a\gamma^{\mu}\frac{(1-\gamma^5)}{2}v_b, \tag{10.2}$$

where $C = 1$ for a leptonic final state $(f_a, \bar{f}_b) = (l^-, \bar{\nu}_l)$, and for a quark–antiquark final state we have the corresponding KM matrix element, i.e. $C = V_{q_a\bar{q}_b}$ for $(f_a, \bar{f}_b) = (q_a, \bar{q}_b)$.

Let us assume that we require to sum over the spin states of f_a and $\bar{f}_b$ and sum and average over the three spin states of the W. From (10.2) we obtain, taking ε^{λ} to be real and putting $K_c = |C|^2$

$$|\mathcal{M}|^2 = \frac{g^2K_c}{3\times 4}\,\mathrm{tr}[\not{p}_a\gamma^{\mu}\not{p}_b\gamma^{\nu}(1-\gamma^5)]\sum_{\lambda}\varepsilon^{\lambda}_{\nu}\varepsilon^{\lambda}_{\mu}. \tag{10.3}$$

Now, the sum over polarisation states is given by (see (4.97))

$$\sum_{\lambda=1,2,3}\varepsilon^{\lambda}_{\mu}\varepsilon^{\lambda}_{\nu} = -g_{\mu\nu} + p_{\mu}p_{\nu}/M_W^2. \tag{10.4}$$

Inserting (10.4) into (10.3), and noting that the γ^5 term gives an antisymmetric tensor which gives zero contribution when summed with the symmetric term (10.4), we obtain

$$\begin{aligned}|\mathcal{M}|^2 &= \frac{g^2K_c}{3}[p_a^{\mu}p_b^{\nu} + p_a^{\nu}p_b^{\mu} - g^{\mu\nu}(p_a\cdot p_b)]\left(-g_{\mu\nu} + \frac{p_{\mu}p_{\nu}}{M_W^2}\right)\\ &= \frac{g^2K_c}{3}\left[p_a\cdot p_b + 2\frac{(p\cdot p_a)(p\cdot p_b)}{M_W^2}\right]\\ &= \frac{g^2K_c}{3}M_W^2\left(1-\frac{m_a^2}{M_W^2}\right)\left(1+\frac{m_a^2}{2M_W^2}\right).\end{aligned} \tag{10.5}$$

In the last line of (10.5) it has been assumed, for simplicity, that $m_b = 0$ (i.e. b is a neutrino or light quark).

The differential decay rate is (Appendix C)

$$d\Gamma = \frac{|\mathcal{M}|^2\, d^3p_a\, d^3p_b\, \delta^4(p_i - p_f)}{(2\pi)^2 8 M_W E_a E_b}$$

$$= \frac{|\mathcal{M}|^2 p_a^2\, dp_a\, \delta(M_W - E_a - E_b)}{8\pi M_W E_a E_b}$$

$$= \frac{|\mathcal{M}|^2 p_a\, dE_a\, \delta(M_W - E_a - E_b)}{8\pi M_W E_b},$$

i.e.

$$\Gamma = \frac{|\mathcal{M}|^2 p_a}{8\pi M_W^2}, \tag{10.6}$$

where the δ-function is integrated in a similar way to that leading to (8.6).

Hence the decay rate for the channel $(f_a, \bar{f}_b)$ is

$$\Gamma_{ab} = \frac{g^2 K_c M_W}{48\pi}\left(1 - \frac{m_a^2}{M_W^2}\right)^2\left(1 + \frac{m_a^2}{2M_W^2}\right)$$

$$\simeq \frac{g^2 K_c M_W}{48\pi} = \frac{G_F K_c M_W^3}{6(2^{1/2})\pi}. \tag{10.7}$$

In the approximate form it is assumed that $m_a \sim 0$, and also equation (5.63) is used. For $M_W = 82$ GeV, this gives $\Gamma_{ab} = 0.241$ GeV. For a heavy quark or lepton, this rate is somewhat suppressed. For example, a top quark mass $m_t = 40$ GeV leads to $\Gamma_{tb} \simeq 0.65 \times 0.241 \simeq 0.157$ GeV.

The total decay rate is obtained by summing the partial widths of all kinematically allowed decay modes. For each of the three generations of quarks, we must sum the moduli of the terms in the KM matrix, giving $K_c = 1$ in each case. Each quark can appear in each of $N_c = 3$ colour states and, in addition, there are $N_l = 3$ possible lepton states. Hence, there are 12 possible contributions in total, giving

$$\Gamma_W \simeq N_l \Gamma_{l\nu} + N_c(2 + 0.65)\Gamma_{l\nu}$$

$$\simeq 2.64 \text{ GeV}. \tag{10.8}$$

The above estimates lead to the partial decay branching ratios given in Table 10.1. The effects of higher order corrections on the partial widths have been considered, for example, by Jegerlehner (1986).

The corresponding lifetime for the W is very short; $\tau_W \sim 2.5 \times 10^{-25}$ s. Note that, in deriving (10.7), it has been assumed that the hadronisation process does not significantly modify the decay probability. That is, it is

Table 10.1. *Expected decay branching ratios for W^- and Z^0 to lowest order in the standard model, taking $M_W = 82$ GeV, $M_Z = 93$ GeV, $m_t = 40$ GeV and $\sin^2\theta_W = 0.225$. For the W decays the KM angles are taken to be $\theta_1 = \theta_2 = \theta_3 = 0$*

	Mode	Γ (GeV)		Branching ratio (%)	
W^-					
	$e^-\bar{\nu}_e$	0.241		9.1	
	$\mu^-\bar{\nu}_\mu$	0.241	0.723	9.1	27.4
	$\tau^-\bar{\nu}_\tau$	0.241		9.1	
	$d\bar{u}$	0.723		27.4	
	$s\bar{c}$	0.723		27.4	
	$b\bar{t}$	0.470		17.8	
	total	2.64			
Z^0					
	$\nu_e\bar{\nu}_e$	0.176		6.6	
	$\nu_\mu\bar{\nu}_\mu$	0.176	0.528	6.6	19.9
	$\nu_\tau\bar{\nu}_\tau$	0.176		6.6	
	e^+e^-	0.089		3.4	
	$\mu^+\mu^-$	0.089	0.266	3.4	10.1
	$\tau^+\tau^-$	0.089		3.4	
	$u\bar{u}$	0.306		11.5	
	$c\bar{c}$	0.306	0.678	11.5	25.5
	$t\bar{t}$	0.066		2.5	
	$d\bar{d}$	0.303		14.8	
	$s\bar{s}$	0.303	1.179	14.8	44.5
	$b\bar{b}$	0.303		14.8	
	total	2.65			

assumed that the hadronisation process of a quark happens with unit probability, in a similar way to that assumed in deep inelastic scattering or in $e^+e^- \to \bar{q}q$.

The treatment of the polarisation states of the W follows that developed for the virtual W in Sections 7.2 and 7.5. We can define the polarisation states as follows

$$\varepsilon^\mu\begin{pmatrix}R\\L\end{pmatrix} = \frac{1}{2^{1/2}}(0, 1, \pm i, 0), \qquad \varepsilon^\mu(S) = \frac{1}{M_W}(|\mathbf{p}|, 0, 0, E). \qquad (10.9)$$

In the W rest system, the longitudinal or scalar (helicity zero) term reduces to $\varepsilon^\mu(S) = (0, 0, 0, 1)$. Note that the polarisation vectors satisfy $\varepsilon^\lambda_\mu \cdot p^\mu = 0$, as required. If the W^- is in a specific polarisation state, then the expression (10.4) cannot be used; instead, a specific summation using the appropriate terms in (10.9) must be made. Using the W^- rest system, and taking

$m_a = m_b = 0$, we find that for a polarisation state λ,

$$|\mathcal{M}|^2 = g^2 K_c[p_a^\mu p_b^\nu + p_a^\nu p_b^\mu - g^{\mu\nu} p_a \cdot p_b - i(p_a)_\alpha (p_b)_\beta \varepsilon^{\alpha\mu\beta\nu}]\varepsilon_\mu^\lambda \varepsilon_\nu^{\lambda *}. \quad (10.10)$$

For $\lambda = R$ or L, the totally antisymmetric tensor ε gives a finite contribution (equal but opposite for R and L). For $\lambda = S$, the ε term gives zero. Defining $p_a^\mu = (1, \sin\theta \cos\phi, \sin\theta \sin\phi, \cos\theta) M_W/2$, we obtain

$$\begin{aligned} |\mathcal{M}_R|^2 &= g^2 K_c M_W^2 (1 - \cos\theta)^2/4, \\ |\mathcal{M}_L|^2 &= g^2 K_c M_W^2 (1 + \cos\theta)^2/4, \\ |\mathcal{M}_S|^2 &= g^2 K_c M_W^2 \sin^2\theta/2. \end{aligned} \quad (10.11)$$

Summing the terms in (10.11), and dividing by three, gives the spin averaged term (10.5). The outgoing fermion and antifermion are, in all cases, left and righthanded respectively. This can be seen directly from the form of (10.2).

10.2 DECAY OF Z^0

In the standard model, the lowest order decay mode of the Z^0 is to $f_a \bar{f}_a$ (Fig. 1.8b), where f_a is any kinematically allowed fermion (lepton or quark). The decay matrix element for $Z^0(p, \varepsilon^\lambda) \to f(p_1) + \bar{f}(p_2)$ is (see Appendix C)

$$\mathcal{M} = \frac{-ig}{\cos\theta_W} \bar{u}_1 \gamma^\mu \left[C_L \frac{(1-\gamma^5)}{2} + C_R \frac{(1+\gamma^5)}{2} \right] v_2 \varepsilon_\mu^\lambda. \quad (10.12)$$

The calculation of $|\mathcal{M}|^2$ is similar to that for W^- decay and gives, averaging over the Z^0 spin states,

$$|\mathcal{M}|^2 = \frac{2g^2 M_Z^2}{3\cos^2\theta_W} [(1-\eta)(|C_L|^2 + |C_R|^2) + 3\eta(C_L C_R^* + C_R C_L^*)], \quad (10.13)$$

where $\eta = m_a^2/M_Z^2$, with m_a the mass of f_a. The resulting decay rate for this channel is

$$\begin{aligned} \Gamma_{ff} &= \frac{G_F M_Z^3}{3(2^{1/2})\pi} (1-4\eta)^{1/2} [(1-\eta)(|C_L|^2 + |C_R|^2) + 3\eta(C_L C_R^* + C_R C_L^*)] \\ &\simeq \frac{g^2 M_Z}{24\pi \cos^2\theta_W} (|C_L|^2 + |C_R|^2) = \frac{G_F M_Z^3}{3(2^{1/2})\pi} (|C_L|^2 + |C_R|^2), \end{aligned} \quad (10.14)$$

where the final approximation is for the case $\eta \sim 0$.

An alternative way to parameterise the matrix element (10.12) is in terms of the vector and axial couplings directly, that is

$$\mathscr{M} = \frac{-ig}{2\cos\theta_W} \bar{u}_1\gamma^\mu(g_V - g_A\gamma^5)v_2\varepsilon_\mu^\lambda. \tag{10.15}$$

The correspondence between the two forms is $g_V = C_L + C_R$, $g_A = C_L - C_R$. Hence, the partial rate to $f\bar{f}$ can be written, taking g_V and g_A to be real, as

$$\Gamma_{ff} = \frac{G_F M_Z^3}{6(2^{1/2})\pi}(1-4\eta)^{1/2}[(1-\eta)(g_V^2 + g_A^2) + 3\eta(g_V^2 - g_A^2)]. \tag{10.16}$$

The matrix element squared for the decays of polarised Z^0 bosons can be derived in a similar way to those for $W^\pm$. We obtain, for massless fermions,

$$|\mathscr{M}_R|^2 = \frac{g^2M_Z^2}{2\cos^2\theta_W}[(C_R^2 + C_L^2)(1+\cos^2\theta) + 2(C_R^2 - C_L^2)\cos\theta],$$

$$|\mathscr{M}_L|^2 = \frac{g^2M_Z^2}{2\cos^2\theta_W}[(C_R^2 + C_L^2)(1+\cos^2\theta) - 2(C_R^2 - C_L^2)\cos\theta], \tag{10.17}$$

$$|\mathscr{M}_S|^2 = \frac{g^2M_Z^2}{2\cos^2\theta_W}[(C_R^2 + C_L^2)\,2\sin^2\theta].$$

The sum of these terms, divided by three, gives (10.13).

The expected decay rates and branching ratios are summarised in Table 10.1. If $m_t \sim 40$ GeV, then the $t\bar{t}$ mode is strongly suppressed by phase space. With this assumption, the total decay width is $\Gamma_Z = 2.65$ GeV, similar to that of the W. Note that for both the W and Z bosons the hadronic modes constitute about 70% of the decays, the remaining 30% being leptonic. To first order in QCD, the decay rates of the $q\bar{q}$ modes for the W and Z are modified by a multiplicative factor $(1 + \alpha_S(M^2)/\pi)$, where M is the heavy boson mass, see Albert *et al.* (1980). The effects of higher order corrections on the partial widths have been considered, for example, by Jegerlehner (1986), and Altarelli *et al.* (1986). Any new leptons or quark may also be included in the decay modes of the W and Z, provided they are kinematically allowed. Hence, a study of both the decay products, and the total decay widths of the vector bosons, is important in the search for such particles.

10.3 PRODUCTION OF $W^\pm$ AND Z^0

The method employed to create the heavy vector bosons is to use the reaction $f_1 + \bar{f}_2 \to W, Z$, where f_1 and f_2 are appropriate fermions, and is thus the inverse of that studied in previous chapters. Experimentally, it is necessary to produce high energy beams of fermions and antifermions, travelling in opposite directions, and to make these interact. The fermions used can be either quarks, that is using the reactions ($u\bar{d}$, $u\bar{s} \to W^+$), ($d\bar{u}$, $s\bar{u} \to W^-$), ($u\bar{u}$, $d\bar{d} \to Z^0$), etc. in hadron colliders such as the CERN and FNAL $\bar{p}p$ colliders or electrons, using $e^+e^- \to Z^0$ at the electron–positron colliders SLC at SLAC and LEP at CERN. Note that single $W^{\pm}$ bosons can only be produced directly using hadron colliders. At higher energies vector bosons can be produced in pairs (Section 10.3.4). The experimental problems at hadron colliders are severe. The small number of W and Z bosons produced have to be found amongst an enormous potential background of hadronic interactions. However, as discussed below, the UA1 and UA2 experiments at CERN developed triggers and analysis methods for these decays which largely overcome these problems. There still remains, however, the problem that the momenta of the incident quarks are, *a priori*, unknown.

We first consider the general case of the production of a heavy boson B (of spin J), produced by the collision of $f_1\bar{f}_2$ and subsequently decaying to $f_3\bar{f}_4$, where f_1 to f_4 are spin $\frac{1}{2}$ fermions; that is, the process (Fig. 10.1)

$$f_1(p_1) + \bar{f}_2(p_2) \to B(p) \to f_3(p_3) + \bar{f}_4(p_4). \tag{10.18}$$

This process can be considered to consist of three stages, namely the production, the propagation and finally the decay of the heavy boson B. The propagator for the virtual state B has a denominator of the form $(p^2 - M^2)^{-1}$, and thus has a pole at $p^2 = M^2$. This would correspond to a zero-width state B. Now we assume that B is unstable, and decays with a lifetime τ. Hence, from the uncertainty principle, it has a finite width $\Gamma = \tau^{-1}$. In tems of field theory, there are corrections to the propagator involving higher order diagrams (e.g. $f\bar{f}$ loops). The result is that we can replace M by $M - i\Gamma/2$ in the propagator.

We are interested in finding the total cross-section for the process (10.18) near the resonant pole. To this effect, we can factorise the matrix element

Fig. 10.1 Lowest order diagram for the production and subsequent decay of a heavy boson B.

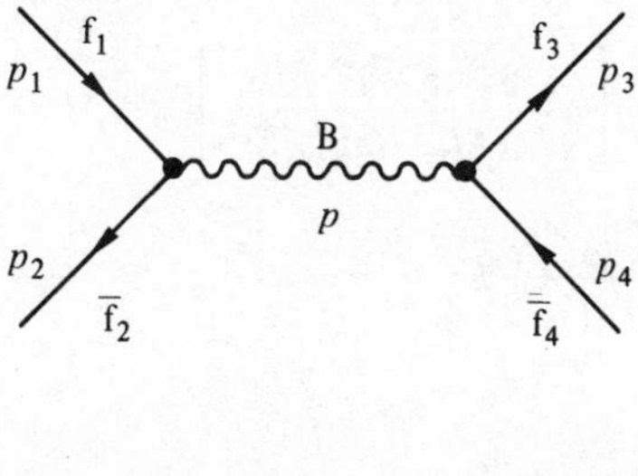

into the form

$$\mathcal{M}_{\mathrm{fi}} \simeq \mathcal{M}_{12}\left(\frac{1}{p^2-(M-\mathrm{i}\Gamma/2)^2}\right)\mathcal{M}_{34}. \tag{10.19}$$

Writing $s=p^2$, this gives

$$|\mathcal{M}_{\mathrm{fi}}|^2 \simeq |\mathcal{M}_{12}|^2\left(\frac{1}{(s-M^2)^2+M^2\Gamma^2}\right)|\mathcal{M}_{34}|^2. \tag{10.20}$$

The total cross-section is thus (Appendix C)

$$\sigma_{\mathrm{tot}}(s) = \int \frac{|\mathcal{M}_{12}|^2}{4(2s)[(s-M^2)^2+M^2\Gamma^2]} \times\left\{(2\pi)^4\,\delta^4(p_{\mathrm{f}}-p_{\mathrm{i}})|\mathcal{M}_{34}|^2\frac{\mathrm{d}^3p_3\,\mathrm{d}^3p_4}{(2\pi)^6 2E_3 2E_4}\right\}, \tag{10.21}$$

where we have neglected all fermion masses, and the factor 4 is from the spin average over the incident (assumed) unpolarised beams. Now the expression in curly brackets can be identified as $2M\Gamma_{34}$, where Γ_{34} is the decay rate of B to the channel $3\bar{4}$. A simple calculation for the decay of B to $1\bar{2}$ shows that

$$\Gamma_{12} = \frac{|\mathcal{M}_{12}|^2}{(2J+1)16\pi M}. \tag{10.22}$$

Hence, (10.21) can be written in the Breit–Wigner resonance form

$$\sigma_{\mathrm{tot}}(s) = \frac{4\pi(2J+1)\Gamma_{12}\Gamma_{34}}{(s^2-M^2)^2+M^2\Gamma^2}. \tag{10.23}$$

If the production cross-section of B to all possible decay modes is required, then Γ_{34} must be replaced by the total width Γ (which is just the sum of the partial widths to all possible channels).

10.3.1 W and Z production at hadron colliders

A W^+ can be produced by the collisions of $\mathrm{u}\bar{\mathrm{d}}$, $\mathrm{u}\bar{\mathrm{s}}$, etc. The matrix element for the process $\mathrm{q}_1+\bar{\mathrm{q}}_2 \to \mathrm{W} \to \mathrm{q}_3+\bar{\mathrm{q}}_4$ is

$$\begin{aligned}\mathcal{M}_{\mathrm{fi}} &= -\mathrm{i}\frac{g^2}{8}V_{12}V_{34}[\bar{v}_2\gamma_\mu(1-\gamma^5)u_1]\left[\frac{-g^{\mu\nu}+p^\mu p^\nu/M_{\mathrm{W}}^2}{s-(M_{\mathrm{W}}-\mathrm{i}\Gamma_{\mathrm{W}}/2)^2}\right][\bar{u}_3\gamma_\nu(1-\gamma^5)v_4]\\ &\simeq \mathrm{i}\frac{g^2}{8}V_{12}V_{34}\frac{[\bar{v}_2\gamma_\mu(1-\gamma^5)u_1][\bar{u}_3\gamma^\mu(1-\gamma^5)v_4]}{[s-(M_{\mathrm{W}}-\mathrm{i}\Gamma_{\mathrm{W}}/2)^2]},\end{aligned} \tag{10.24}$$

where V_{12} and V_{34} are the appropriate KM matrix elements, and the lepton masses have been neglected. With the usual methods, this gives

$$|\mathscr{M}_{\mathrm{fi}}|^2 = \frac{g^4}{4}\frac{|V_{12}|^2|V_{34}|^2 s^2(1+\cos\theta)^2}{[(s-M_{\mathrm{W}}^2)^2+M_{\mathrm{W}}^2\Gamma_{\mathrm{W}}^2]}, \tag{10.25}$$

where θ is the angle made by 3 with respect to the direction of 1 in the cms of 1 and 2. Using expression (4.29) for the cms cross-section, and integrating over $\mathrm{d}\Omega = 2\pi\,\mathrm{d}\cos\theta$, gives

$$\sigma_{\mathrm{tot}}(s) = \frac{g^4|V_{12}|^2|V_{34}|^2 s}{192\pi[(s-M_{\mathrm{W}}^2)^2+M_{\mathrm{W}}^2\Gamma_{\mathrm{W}}^2]}. \tag{10.26}$$

Now, from (10.7) we have $\Gamma_{\mathrm{ab}} = g^2|V_{\mathrm{ab}}|^2 M_{\mathrm{W}}/48\pi$. Hence, (10.26) becomes

$$\sigma_{\mathrm{tot}}(s) = \frac{12\pi\Gamma_{12}\Gamma_{34}}{[(s-M_{\mathrm{W}}^2)^2+M_{\mathrm{W}}^2\Gamma_{\mathrm{W}}^2]}. \tag{10.27}$$

This has the same form as (10.23), with $J = 1$.

The above formulae are valid for quarks and antiquarks of well defined momenta and of the same colour, and describe the production and subsequent decay of the $\mathrm{W}^{\pm}$. In order to obtain the cross-section $\sigma(\mathrm{p\bar{p}} \to \mathrm{W}^{\pm} + \mathrm{X})$ we must insert the approprate quark density functions and average over the colourless $\mathrm{q\bar{q}}$ initial states (see Section 7.11.1). The required cross-section can be written as

$$\sigma(\mathrm{p\bar{p}} \to \mathrm{W}^{+} + \mathrm{X}) = \int_0^1 \mathrm{d}x_1 \int_0^1 \mathrm{d}x_2\, \hat{\sigma}_{\mathrm{W}}(x_1, x_2)\mathrm{W}^{+}(x_1, x_2), \tag{10.28}$$

where x_1 and x_2 are the momentum fractions of the proton and antiproton carried by q_1 and $\bar{\mathrm{q}}_2$ respectively. The elementary cross-section $\hat{\sigma}(\mathrm{q}_1 + \bar{\mathrm{q}}_2 \to \mathrm{W}^{+})$ for the production of a W^{+} by q_1 and $\bar{\mathrm{q}}_2$ (i.e. without considering its decay) is given by

$$\hat{\sigma}_{\mathrm{W}}(\hat{s}) = \int \frac{(2\pi)^4|\mathscr{M}_{12}|^2}{8\hat{s}}\,\delta^4(p_1+p_2-p)\frac{\mathrm{d}^4p}{(2\pi)^3}\,\delta(p^2-M_{\mathrm{W}}^2)$$

$$= \frac{\pi|\mathscr{M}_{12}|^2}{4M_{\mathrm{W}}^2}\,\delta(\hat{s}-M_{\mathrm{W}}^2), \tag{10.29}$$

where the invariant form for the phase space density has been used, and where $\hat{s} = p^2 = x_1x_2s$. The matrix element $\mathscr{M}_{12} = -ig/(2(2^{1/2}))\bar{v}_2\gamma^{\mu}(1-\gamma^5)u_1\varepsilon_{\mu}^{\lambda}$, and hence $|\mathrm{M}_{12}|^2 = g^2M_{\mathrm{W}}^2$. Thus, (10.29) becomes

$$\hat{\sigma}_{\mathrm{W}}(x_1, x_2) = \frac{\pi g^2}{4}\,\delta(x_1x_2s-M_{\mathrm{W}}^2) = 2^{1/2}\pi G_{\mathrm{F}}M_{\mathrm{W}}^2\,\delta(x_1x_2s-M_{\mathrm{W}}^2). \tag{10.30}$$

The final term $W^+(x_1, x_2)$ in (10.28) contains both the quark and antiquark densities (including the colour average) and the KM matrix elements. Thus we have (neglecting the production of heavy quarks)

$$W^+(x_1, x_2) \simeq \tfrac{1}{3}\{[u(x_1)\bar{d}(x_2) + \bar{d}(x_1)u(x_2)]\cos^2\theta_C + [u(x_1)\bar{s}(x_2) + \bar{s}(x_1)u(x_2)]\sin^2\theta_C + \cdots\}. \quad (10.31)$$

For W^- production the roles of q and $\bar{q}$ are interchanged.

The equation (10.28), for the production cross-section, may be simplified by using the properties of the δ-function (Appendix D) in (10.30), and then integrating over x_2. This gives

$$\sigma(p\bar{p} \to W^+ + X) = 2^{1/2}\pi G_F \tau \int_\tau^1 \frac{dx}{x} W^+(x, \tau/x), \quad (10.32)$$

where $\tau = M_W^2/s$. In order to make numerical estimates of the cross-section, specific quark density distributions must be inserted in W^+ in (10.32). Since $x_1 x_2 = M_W^2/s \sim (0.15)^2$ at $s^{1/2} \sim 600$ GeV, valence quarks will dominate. The quark distributions are written above as a function of x only. However, they are also functions of Q^2, and they must be evolved from $Q^2 \sim 20$ GeV^2 (where they are measured) to, for example, $Q^2 \sim \hat{s} \sim 2000$ GeV^2, appropriate for the CERN $\bar{p}p$ collider with $s^{1/2} \sim 600$ GeV. An estimate of $\sigma(\bar{p}p \to W^+(W^-) + X)$ is shown in Fig. 10.2. For the conditions of the CERN collider ($s^{1/2} \sim 600$ GeV), the cross-section $(\bar{p}p \to W^+ + X) \sim 2 \times 10^{-33}$ cm^2. This is roughly 10^{-7} of the total hadronic cross-section. The $W^\pm$ cross-section increases significantly as $s^{1/2}$ is increased. The cross-sections at the Fermilab collider ($s^{1/2} \sim 1.6$ TeV) are predicted to be a factor of 4 higher. Furthermore as $s^{1/2}$ is increased, the sea quarks play an increasing role in the $W^\pm$ production, and the advantage of a $\bar{p}p$ collider over a pp collider disappears.

The cross-section $\sigma(\bar{p}p \to Z^0 + X)$ has a similar form to (10.28) (and (10.32)), but with

$$\hat{\sigma}_Z(x_1, x_2) = 2^{1/2}\pi G_F M_Z^2\, \delta(x_1 x_2 s - M_Z^2), \quad (10.33)$$

and with the function $Z(x_1, x_2)$, which represents the colour-averaged quark density distributions, weighted by their respective couplings, given by

$$Z(x_1, x_2) = \tfrac{1}{3}\{[u(x_1)\bar{u}(x_2) + \bar{u}(x_1)u(x_2)][\tfrac{1}{2} - \tfrac{4}{3}\sin^2\theta_W + \tfrac{16}{9}\sin^4\theta_W] + [d(x_1)\bar{d}(x_2) + \bar{d}(x_1)d(x_2) + s(x_1)\bar{s}(x_2) + \bar{s}(x_1)s(x_2)] \times [\tfrac{1}{2} - \tfrac{2}{3}\sin^2\theta_W + \tfrac{4}{9}\sin^4\theta_W]\}. \quad (10.34)$$

The estimated cross-section $\sigma(\bar{p}p \to Z^0 + X)$ is shown in Fig. 10.2, and it can be seen that it is less than for $W^\pm$. At $s^{1/2} \sim 600$ GeV, $\sigma(\bar{p}p \to Z^0 + X) \sim 10^{-33}$ cm^2.

The above discussion only describes the momentum spectra in the longitudinal direction (i.e. in the directions of the incident beams). Gluon radiation also produces significant momenta in the transverse direction ($\langle p_T^W \rangle \sim 5$–$10$ GeV at $s^{1/2} \sim 600$ GeV), which are much greater than the intrinsic transverse momenta of the quarks. A further complication in the estimate of the W and Z cross-sections is the QCD K factor ($K \sim 1.5$); see the discussion in Section 7.11.1.

Experimentally, detection of the $W^\pm$ and Z^0 is easiest by the leptonic decay modes $W^\pm \to l^\pm \nu_l$ and $Z^0 \to l^+ l^-$. Despite the low branching ratios to the leptonic modes, the backgrounds are smaller and the momenta of the charged leptons can be measured with reasonable accuracy. Detection through the observation of hadronic jets (e.g. $W^+ \to u\bar{d}$) is much more difficult, as the quark momenta are shared amongst many particles. These hadrons must then be reliably assigned to the correct jet and the jet

Fig. 10.2. Estimated cross-section for W^+ (or W^-) and Z^0 production in $p\bar{p}$ collisions as a function of $s^{1/2}/M_W$.

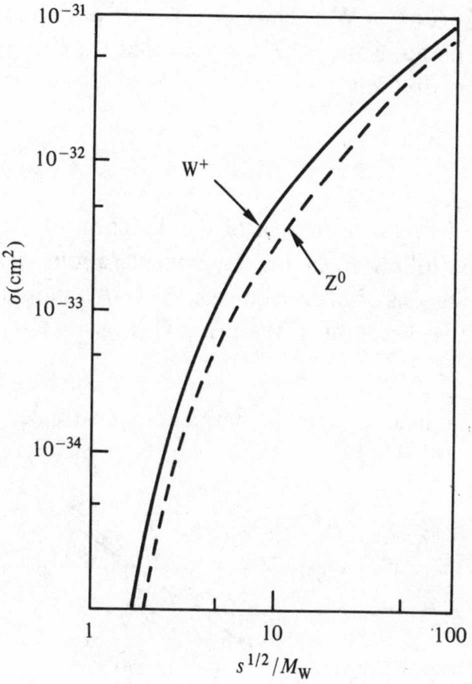

momentum accurately determined, in the presence of a large background from QCD processes.

The decay $Z^0 \to l^+ l^-$ can be identified by detecting charged leptons of opposite charge and plotting the spectrum of their invariant mass. For the decay $W \to l\nu_l$, only the charged lepton can be detected, so a different technique must be used. In order to illustrate the method used, we consider the simplified case of the decay of an unpolarised $W^+(\to l^+\nu_l)$, produced at rest in the $\bar{p}p$ cms. If θ is the angle made by l^+ with respect to the incident proton direction, then $p_T^W = (M_W/2)\sin\theta$. The decay angular distribution is isotropic for unpolarised decays, i.e. $dN/\cos\theta \sim \frac{1}{2}$. Therefore, defining $x_T = 2p_T/M_W$,

$$\frac{dN}{dx_T} = \frac{x_T}{2(1 - x_T^2)^{1/2}}, \tag{10.35}$$

which gives rise to a characteristic *Jacobian peak* towards the maximum allowed value of $p_T^W \simeq 40$ GeV.

Several effects modify the above result. The W has a finite width, and hence the value of M_W is not unique. Also QCD effects give the $W^\pm$ a sizeable transverse momentum. Furthermore, because of the $V - A$ structure of the W couplings, the $W^\pm$ are polarised. If the W is produced from valence quarks in the p and $\bar{p}$, then the main production mechanisms are $d(p) + \bar{u}(\bar{p}) \to W^-$ and $u(p) + \bar{d}(\bar{p}) \to W^+$. Thus, the $W^\pm$ are polarised in the direction of the incident $\bar{p}$ and, from (10.25), we see that the decay leptons will have an angular distribution

$$\frac{dN^\pm}{d\cos\theta} = \tfrac{3}{8}(1 \mp \cos\theta)^2, \tag{10.36}$$

where θ is the angle made by the $l^\pm$ with respect to the direction of the incident p. These distributions follow from helicity considerations, as shown in Fig. 10.3. Thus, for collisions of valence quarks, the $l^-(l^+)$ should be produced preferentially in the direction of the $p(\bar{p})$. This asymmetry

Fig. 10.3 Helicity configurations in (*a*) initial state and (*b*) final state in the reaction $d\bar{u} \to W^- \to l^-\bar{\nu}_l$.

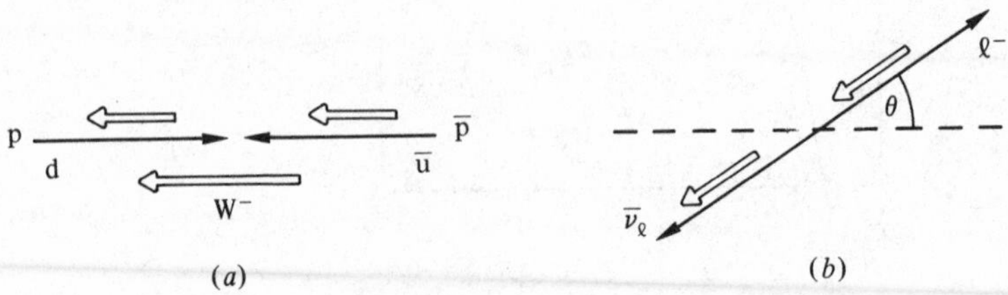

will not appear for interactions involving sea quarks (i.e. at low x) or around $\theta \sim 90°$ (i.e. at the Jacobian peak). With the above effects included, the expected p_T spectrum for $s^{1/2} = 540$ GeV is shown in Fig. 10.4.

10.3.2 Discovery and properties of $W^\pm$ and Z^0

The $W^\pm$ was discovered by the UA1 (Arnison *et al.*, 1983) and UA2 (Banner *et al.*, 1983) collaborations, who isolated samples of events containing a high p_T $e^\pm$, together with a large missing transverse energy opposite the $e^\pm$ (interpreted as neutrino emission). The UA1 experiment was equipped with both electron and muon identification, whereas the UA2 experiment had only electron detection. Both experiments were equipped with calorimeters, with which the presence of non-interacting particles could be inferred and, from transverse momentum imbalance, their transverse momenta could be estimated. This method has also been exploited by the UA1 experiment to detect the decay mode $W \to \tau\nu_\tau$, with the τ being detected by $\tau \to \nu_\tau +$ hadrons, i.e. the signature is a single jet of hadrons of low multiplicity.

Since the initial discovery, the sample of $W^\pm$ decays has been substantially increased (for a summary see Locci, 1986). The numbers of $W \to e\nu_e$ events

Fig. 10.4 Transverse momentum spectrum for l^+ in the process $p\bar{p} \to X + W^+ \to l^+\nu_l$ at $s^{1/2} = 540$ GeV.

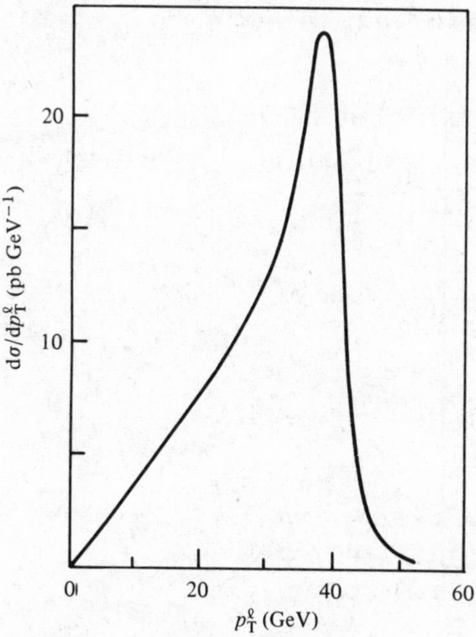

(corrected for backgrounds, which amount to about 10%) are 53 ± 8 (187 ± 15) for UA1 and 37 ± 6 (192 ± 15) for UA2, where the first (second) number refers to $s^{1/2} = 546$ (630) GeV. The UA1 collaboration also has 10 ± 3 (54 ± 8) $W \to \mu\nu_\mu$ events and (30 ± 6) $W \to \tau\nu_\tau$ events. The measured cross-sections and p_T distributions agree with the theoretical predictions (e.g. Altarelli *et al.*, 1984), to within the 30% error assigned to the theory. The UA1 experiment has measured the relative rates for e, μ and τ leptons from W decay. Expressing $\Gamma(W \to l\nu) = Ag_l^2$, where A is a common kinematic factor, they find that $g_\mu/g_e = 1.05 \pm 0.07 \pm 0.08$, and $g_\tau/g_e = 1.01 \pm 0.09 \pm 0.05$ (Albajar *et al.*, 1987a). These results are consistent with the hypothesis of e–μ–τ universality, testing it at a scale $Q^2 \simeq M_W^2$.

The angular distribution of the $e^\pm$ from the decay $W \to e\nu_e$ in the W cms, as measured by the UA1 collaboration (Locci, 1986), is shown in Fig. 10.5. The transverse momentum of the neutrino is found by momentum conservation, $\mathbf{p}_T^\nu = -\mathbf{p}_T^e - \mathbf{p}_T^{had}$. Knowing p_T^ν, the longitudinal component p_L^ν is known to within a quadratic ambiguity (corresponding to the neutrino going forwards or backwards in the W cms), and this can be resolved in about 50% of the cases (one solution unphysical or both solutions close). Hence, for a sample of events, the momentum of the W can be determined. The results are in good agreement with the V – A prediction (equation (10.36)). The measured value of the forward–backward

Fig. 10.5 Electron angular distribution in the W cms.

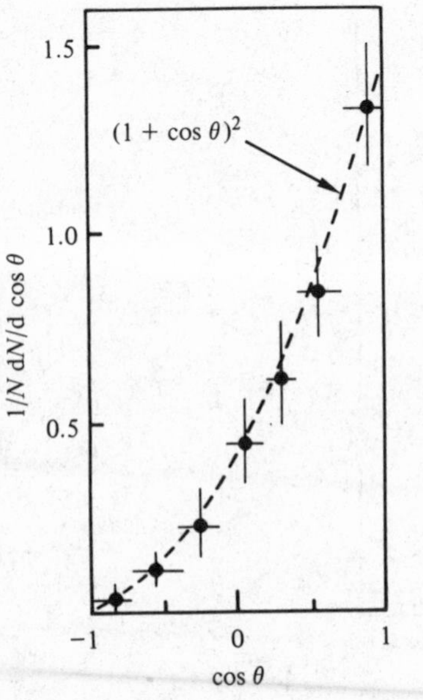

asymmetry $A_W = (N_F - N_B)/(N_F + N_B) = 0.77 \pm 0.04$, to be compared with the V − A prediction of 0.75. Further, the mean value of $\cos\theta$ gives a measure of the spin of the parent particle (Jacob, 1958), since $\langle\cos\theta\rangle = \langle\lambda\rangle\langle\mu\rangle/J(J+1)$, where $\langle\mu\rangle$ and $\langle\lambda\rangle$ are the helicities of the initial system $(u\bar{d})$ and decay system $(e\bar{\nu})$ respectively. For $\langle\mu\rangle = \langle\lambda\rangle = -1$, $\langle\cos\theta\rangle = 0.5$ for $J = 1$, compared to the UA1-measured value of 0.43 ± 0.07. Thus, the production and decay of the W is consistent with the expectations of the standard model. Note, however, that since two weak vertices enter (i.e. for the W production and decay), it is not possible to distinguish V − A from V + A coupling. This ambiguity could be resolved by measuring the τ polarisation in $W \to \tau\nu_\tau$ decays.

The Z^0 was initially discovered through its e^+e^- decay mode (Arnison *et al.*, 1983, and Banner *et al.*, 1983), but the $\mu^+\mu^-$ mode has also been detected by the UA1 group. The current statistics (Locci, 1986) for the decay mode $Z^0 \to e^+e^-$ are 32 and 37 events for UA1 and UA2 respectively, and 19 events in $Z^0 \to \mu^+\mu^-$ (UA1); these events are mostly at $s^{1/2} = 630$ GeV. One of the main backgrounds to the Z^0 is the production of e^+e^- pairs by the Drell–Yan process. In order to estimate roughly the size of this background, we can compare $\sigma(u\bar{u} \to Z^0 \to e^+e^-)$ to the Drell–Yan process $\sigma(u\bar{u} \to \gamma^* \to e^+e^-)$. The relevant matrix elements are

$$\mathcal{M}_Z = \frac{ig^2}{4\cos^2\theta_W}\frac{[\bar{v}_2\gamma^\mu(g_V^u - g_A^u\gamma^5)u_1][\bar{u}_3\gamma_\mu(g_V^e - g_A^e\gamma^5)v_4]}{[s - (M_Z - i\Gamma_Z/2)^2]}, \quad (10.37a)$$

$$\mathcal{M}_\gamma = ie^2e_u\bar{v}_2\gamma^\mu u_1\bar{u}_3\gamma_\mu v_4/s. \quad (10.37b)$$

Computing $|\mathcal{M}_Z|^2$ and $|\mathcal{M}_\gamma|^2$, integrating over decay angles of the outgoing particles and including a colour factor $\frac{1}{3}$ for the Z^0 case, gives

$$\begin{aligned}\frac{\sigma(u\bar{u} \to Z \to e^+e^-)}{\sigma(u\bar{u} \to \gamma^* \to e^+e^-)} &= \frac{[(g_V^u)^2 + (g_A^u)^2][(g_V^e)^2 + (g_A^e)^2]}{16e_u^2\sin^4\theta_W\cos^4\theta_W}\frac{s^2}{[(s - M_Z^2)^2 + M_Z^2\Gamma_Z^2]} \\ &\xrightarrow{s \sim M_Z^2} \frac{9}{4}\frac{[(g_V^u)^2 + (g_A^u)^2][(g_V^e)^2 + (g_A^e)^2]}{16\sin^4\theta_W\cos^4\theta_W}\frac{M_Z^2}{\Gamma_Z^2} \\ &\sim 100. \end{aligned} \quad (10.38)$$

Another potential background is from heavy quark Q (=c, b) production from $\bar{p}p \to \bar{Q}Q + X$, followed by the semileptonic decays of $\bar{Q}$ and Q. The various processes contributing to the l^+l^- mass spectrum are shown in Fig. 10.6 (Pakvasa *et al.*, 1979). It can be seen that the expected background under the Z^0 peak is small. The observed event rates for Z^0 are found to be compatible with the standard model calculations.

The angular distribution of the final state e^- in $q + \bar{q} \to Z^0 \to e^- + e^+$ can be found by starting with the matrix element (10.37a). This leads to

$$|\mathcal{M}_Z|^2 = \frac{g^4 s^2 \{[(g_V^q)^2 + (g_A^q)^2][(g_V^e)^2 + (g_A^e)^2](1 + \cos^2\theta) + 8 g_V^q g_A^q g_V^e g_A^e \cos\theta\}}{4\cos^4\theta_W[(s - M_Z^2)^2 + M_Z^2\Gamma_Z^2]}. \tag{10.39}$$

The forward–backward asymmetry measured by UA1 is $A_Z = 0.30 \pm 0.15$, which gives a value of $\sin^2\theta_W = 0.18 \pm 0.04$.

The mass and width of the $W^{\pm}$ are determined by a fit to the transverse mass $m_T(m_T^2 = 2p_T^e p_T^\nu(1 - \cos\phi)$, where ϕ is the azimuthal separation between p_T^e and $p_T^\nu)$. The fit takes into account both physics and detector effects. The Z^0 mass and width are found by fitting a Breit–Wigner shape to the 1^+1^- effective mass spectrum. The resulting average values are

$$M_Z = 92.1 \pm 1.7\,\text{GeV}, \qquad \Gamma_Z < 4.6\,\text{GeV (90\% c.l.)},$$
$$M_W = 81.3 \pm 1.4\,\text{GeV}, \qquad \Gamma_W < 6.5\,\text{GeV (90\% c.l.)}. \tag{10.40}$$

Accurate determinations of the masses of the W and Z are important in testing the standard model, as discussed below. Since the W and Z are massive, they can decay to any kinematically allowed lepton or quark decay mode, hence the total decay widths are also important quantities.

Fig. 10.6 Contributions to the mass spectrum of l^+l^- pairs produced in $p\bar{p}$ collisions at $s^{1/2} = 540$ GeV at $y = 0$.

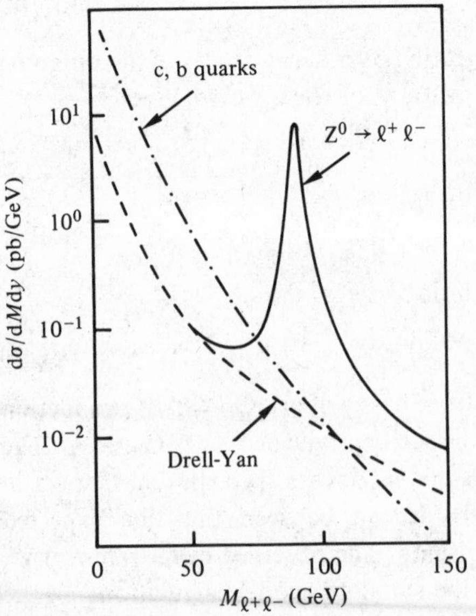

A direct search for a possible fourth generation sequential lepton has been made by the UA1 collaboration (Albajar *et al.*, 1987a), by looking for the decay sequence $W \to L\nu_L$ and $L \to q\bar{q} + \nu_L$. The signature used to search for such events is $E_T(\text{missing}) > 15$ GeV (and greater than 4σ significance), together with at least one jet with $E_T > 12$ GeV. The candidates remaining after these, plus further, analysis cuts can be explained by conventional sources. The number of additional events expected from a lepton L, with universal coupling, depends on its mass. From the observed distributions, a limit $m_L > 41$ GeV (90% c.l.) is computed.

Precise measurement of Γ_Z is difficult because of the rather small statistical samples and the need to understand fully the resolution of the detector (which is comparable to the expected width). From Table 10.1 it can be seen that the measured limits in (10.40) are significantly greater than the standard model predictions. Each additional generation of massless (or light) neutrinos contributes 0.176 GeV to Γ_Z, hence the data allow $\delta N_\nu \leqslant 11$ such additional generations. A more stringent limit can be obtained by using some model dependent assumptions. Experimentally, the ratio $R_{\text{exp}} = \sigma_W^e/\sigma_Z^e$ can be accurately measured (e.g. UA1 have $R_{\text{exp}} = 8.9^{+1.6}_{-1.3}$). This quantity can be expressed as a product of the W and Z production cross-sections and their respective ratios to electron modes. That is,

$$R_{\text{exp}} = \frac{\sigma_W B_W^e}{\sigma_Z B_Z^e} = \frac{\sigma_W \Gamma_W^{e\nu} \Gamma_Z}{\sigma_Z \Gamma_W \Gamma_Z^{ee}}. \tag{10.41}$$

In order to calculate σ_W/σ_Z, use is made of the standard model, together with the measured values of the structure functions (in particular the ratio d_v/u_v which is not very precisely known), evolved using QCD formulae. Using these ingredients, together with the standard model values for B_W^e and Γ_Z^{ee}, a value of Γ_Z can be computed. This is clearly dependent on the top quark mass. UA1 find, for $m_t \sim 20$ GeV, that $N_\nu \leqslant 7$ (90% c.l.), whereas, for a heavy t-quark mass, the limit becomes $N_\nu \leqslant 4$. The UA2 collaboration (Appel *et al.*, 1986) also finds $N_\nu \leqslant 4$ (for $m_t = 40$ GeV). An independent estimate has been made by the UA1 collaboration (Albajar *et al.*, 1987a), from a search for events with a hadron (gluon) jet produced opposite to a decay $Z^0 \to \nu\bar{\nu}$ (giving missing E_T). A limit $N_\nu \leqslant 7$ (90% c.l.) is established. Note that cosmological arguments, based on the helium abundance in the universe, give a limit $N_\nu \leqslant 4$.

10.3.3 Z^0 production at e^+e^- colliders

The matrix element for the process $e^+e^- \to f\bar{f}$ is given in equation (6.85). At, or near, the Z^0 pole ($s \sim M_Z^2$) the Z^0 term (Fig. 6.10) dominates.

For unpolarised incident beams, the differential cross-section, retaining terms in m_f, is

$$\frac{d\sigma}{d\cos\theta}(\text{cms}) = \frac{\pi\alpha^2}{32\sin^4\theta_W\cos^4\theta_W}\frac{s(1-4m_f^2/s)^{1/2}}{[(s-M_Z^2)^2+M_Z^2\Gamma_Z^2]}$$
$$\times [a_1(1+\beta^2\cos^2\theta)+a_2\beta\cos\theta+a_3m_f^2/s], \tag{10.42}$$

where β is the velocity of f, θ is the angle of f with respect to the incident e^-, and

$$a_1 = [(g_V^e)^2+(g_A^e)^2][(g_V^f)^2+(g_A^f)^2],$$
$$a_2 = 8g_V^e g_A^e g_V^f g_A^f, \tag{10.43}$$
$$a_3 = 4[(g_V^e)^2+(g_A^e)^2][(g_V^f)^2-(g_A^f)^2].$$

The total cross-section to produce $f\bar{f}$ is

$$\sigma_Z^{ff}(s) = \frac{\pi\alpha^2}{16\sin^4\theta_W\cos^4\theta_W}\frac{s(1-4m_f^2/s)^{1/2}}{[(s-M_Z^2)^2+M_Z^2\Gamma_Z^2]}\left[a_1\left(1+\frac{\beta^2}{3}\right)+a_3\frac{m_f^2}{s}\right]. \tag{10.44}$$

At, or near, the Z^0 resonance, the e^+e^- cross-section becomes large. For $m_f \sim 0$ ($\beta \sim 1$) we have

$$\sigma_Z^{ff}(s) \simeq \frac{4\pi\alpha^2 M_Z^2}{3s\Gamma_Z^2}\frac{[(g_V^e)^2+(g_A^e)^2][(g_V^f)^2+(g_A^f)^2]}{16\sin^4\theta_W\cos^4\theta_W}$$
$$\simeq \left(\frac{4\pi\alpha^2}{3s}\right)\frac{9\Gamma_{ee}\Gamma_{ff}}{\alpha^2\Gamma_Z^2}, \tag{10.45}$$

where use has been made of (10.14). To get the s-channel production cross-section to all channels, we replace Γ_{ff} by the total width Γ_Z. Expressing the result in units of the point-like s-channel electromagnetic cross-section (i.e. $4\pi\alpha^2/3s$), we obtain

$$R_Z(s) \simeq \frac{9\Gamma_{ee}}{\alpha^2\Gamma_Z} \simeq 1.7\times 10^5\frac{\Gamma_{ee}}{\Gamma_Z}, \tag{10.46}$$

which, using the values in Table 10.1, is roughly $R_Z \sim 5600$. A plot of the expected cross-section for $e^+e^- \to$ hadrons, as a function of $s^{1/2}$, is shown in Fig. 10.7(a), together with the point-like cross-section. The effects of radiative corrections are very important, and substantially alter the Breit–Wigner shape, shifting the peak to a higher value of $s^{1/2}$ and giving a significant production of Z^0 well above the pole (see Fig. 10.7(b)). If one of the incident beam particles radiates a photon, then the e^+e^- effective mass at the annihilation vertex can be on, or near, the Z^0 pole, even though the initial $s^{1/2}$ is well above the pole. The cross-sections given above refer only to the *s*-channel process, with a Z^0 (or γ) propagator. However, for the reaction $e^+e^- \to e^+e^-$, there are also graphs of the type shown in Fig. 6.7(*a*), with a *t*-channel exchange of either γ or Z^0 (see Böhm *et al.*, 1984). These latter graphs, particularly photon exchange, are important for small angle scatters, i.e. $\theta \sim 0$ (see equation (4.58)). This Bhabha scattering process at small angles is essentially given by QED, and can be used to measure and monitor the luminosity at e^+e^- machines.

Fig. 10.7 (*a*) Expected cross-section in GSW model for $e^+e^- \to$ hadrons, as a function of $s^{1/2}$, after correcting for first-order radiative effects. The number of events/day expected for $\mathscr{L} = 10^{31}\ \mathrm{cm}^{-2}\ \mathrm{s}^{-1}$ is also indicated. (*b*) Cross-section for $e^+e^- \to \mu^+\mu^-(\gamma)$, as a function of $s^{1/2}$, for the Born term (broken line) and including $0(\alpha)$ corrections (full line).

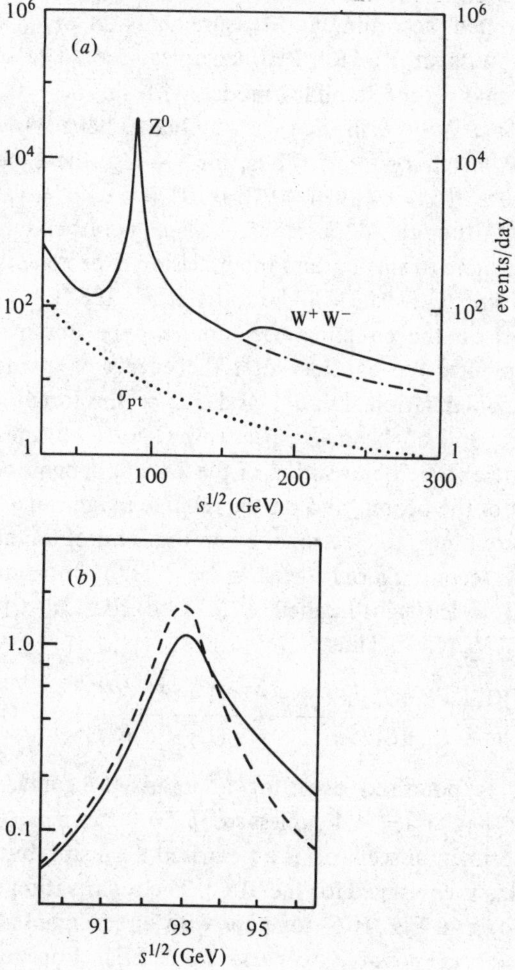

The experiments which are planned to be carried out at the e^+e^- colliders at the SLC (MARK 2, followed by SLD) and at LEP (ALEPH, DELPHI, L3 and OPAL), will attempt to measure, as accurately as possible, the various quark and lepton couplings at $s \sim M_Z^2$. From (10.45), it can be seen that measurement of the total rate for $f\bar{f}$ production yields the term a_1; that is, the product of the sums of the squares of the couplings. From (10.42) (taking $m_f = 0$), it can be seen that the *forward* (F i.e. $\cos\theta > 0$)-*backward* (B) asymmetry is

$$A_{FB} = \frac{N_F - N_B}{N_F + N_B} = \frac{3a_2}{8a_1} = \frac{3g_V^e g_A^e g_V^f g_A^f}{[(g_V^e)^2 + (g_A^e)^2][(g_V^f)^2 + (g_A^f)^2]} = \tfrac{3}{4}A^e A^f, \quad (10.47)$$

where $A^f = 2g_V^f g_A^f/[(g_V^f)^2 + (g_A^f)^2]$, etc.

In practice, the full range of $\cos\theta$ is not covered experimentally, and (10.47) must be modified accordingly. Measurements of A_{FB}, when combined with the results at PETRA/PEP energies (see (6.92)), will determine $g_V^e g_V^f$. Note that, in the standard model (with $\sin^2\theta_W = 0.225$), charged leptons have $A^l \simeq 2(1 - 4\sin^2\theta_W) = 0.20$. Quarks have $A^q = 0.69$ (charge $\frac{2}{3}$) and $A^q = 0.94$ (charge $-\frac{1}{3}$). Thus, for $s \sim M_Z^2$, the expected values for $e^+e^- \to f\bar{f}$ are $A_{FB} = 0.029$ ($f = l^-$), 0.103 ($f = q$, $e_q = \frac{2}{3}$), and 0.140 ($f = q$, $e_q = -\frac{1}{3}$). Although A^{FB} for $e^+e^- \to \mu^+\mu^-$ is rather small, it is nevertheless very sensitive to $\sin^2\theta_W$ and the precision expected at LEP from this method is $\delta \sin^2\theta_W \simeq 0.0015$ (Alexander *et al.*, 1987).

Further information on the coupling constants can be obtained by polarisation measurements. We consider first the effects of giving the beams a longitudinal polarisation. Let P^- and P^+ be the longitudinal polarisations of the e^- and e^+ beams respectively ($P = 1(-1)$ means a right(left)handed particle). The polarisation of the Z^0 will depend on the degree of polarisation of the beams, and on the relative magnitudes of its left and righthanded couplings to electrons. From the form of the matrix element (6.85) it can be seen that a right(left)handed $e^-(e^+)$ contributes to the term in C_R^e and a left(right)handed $e^-(e^+)$ to that in C_L^e, i.e. $\sigma \propto N_R^- N_L^+ (C_R^e)^2 + N_L^- N_R^+ (C_L^e)^2$. Thus,

$$P_Z = \frac{(1 + P^-)(1 - P^+)(C_R^e)^2 - (1 - P^-)(1 + P^+)(C_L^e)^2}{(1 + P^-)(1 - P^+)(C_R^e)^2 + (1 - P^-)(1 + P^+)(C_L^e)^2}. \quad (10.48)$$

Note that the Z^0 is polarised even for unpolarised beams, $P_Z = [(C_R^e)^2 - (C_L^e)^2]/[(C_R^e)^2 + (C_L^e)^2] = -A^e$, unless $\sin^2\theta_W = 0.25$ when $(C_R^e)^2 = (C_L^e)^2$. A possible experimental scenario is a polarised e^- beam, but with the e^+ unpolarised (this is envisaged for the SLC). The polarisation of the Z^0 for this case is shown in Fig. 10.8, together with another possibility, namely $P^- = -P^+ = P$ (a possible scenario for LEP). For circular machines, it is somewhat easier technically, in fact, to produce transverse rather than longitudinal polarisation of the colliding beams and magnetic rotators are needed to achieve the latter. A more complete discussion, including the effects of radiative corrections, can be found in Böhm and Hollick (1982).

The total cross-sections for e^+e^- in different polarisation states are thus not, in general, equal. From measurements with $P^- = 1(\sigma_R)$ and $P^- = -1(\sigma_L)$, and either $P^+ = 0$ or $P^+ = -P^-$, we can express the right and lefthanded total cross-sections in terms of a *longitudinal asymmetry*#

$$A_{LR} = \frac{\sigma_L - \sigma_R}{\sigma_L + \sigma_R} = \frac{(C_L^e)^2 - (C_R^e)^2}{(C_L^e)^2 + (C_R^e)^2} = \frac{2g_V^e g_A^e}{g_V^{e2} + g_A^{e2}} = A^e. \tag{10.49}$$

That is, such a measurement gives a value of $g_V^e g_A^e$ or g_V^e/g_A^e. Note that A_{LR} is considerably more sensitive to the values of the coupling constants than A_{FB}. For example, for $\sin^2\theta_W \simeq 0.23$ and a longitudinal polarisation P, $\delta A_{LR}/\delta \sin^2\theta_W \simeq 8P$, whereas $\delta A_{FB}/\delta \sin^2\theta_W \simeq -2.0$. For $P = 0.5$, an

Fig. 10.8 Polarisation of Z^0 as a function of the e^- polarisation P^-, for $P^+ = 0$ (full line) and $P^+ = -P^-$ (broken line).

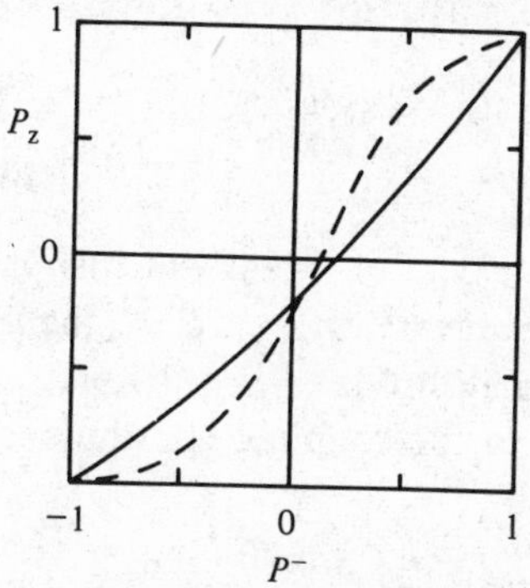

The *longitudinal asymmetry* $A_L = (\sigma_R - \sigma_L)/(\sigma_R + \sigma_L) = -A_{LR}$ is also commonly used in the literature. Note that, in practice, values of polarisation significantly below 100% are anticipated. Even for a perfect circular machine, the polarisation is only 92.4%; transverse polarisation arises from a spin-flip term in synchrotron-radiation which aligns the magnetic moment along the magnetic field (Sokolov–Ternov effect). Practical expectations for LEP, for example, are in the region of 30 to 70%. This significantly reduces the sensitivity to the coupling constants. A potentially interesting scheme for LEP (Blondel *et al.*, 1987) is to polarise the e^- bunches 1 and 4 (with 2 and 3 unpolarised) and the e^+ bunches 2 and 4 (with 1 and 3 unpolarised). The four bunch–bunch cross-sections are then $\sigma_1 = \sigma_u(1 - P^- A_{LR})$, $\sigma_2 = \sigma_u(1 + P^+ A_{LR})$, $\sigma_3 = \sigma_u$ and $\sigma_4 = \sigma_u[1 - P^+P^- + (P^+ - P^-)A_{LR}]$, where σ_u is the cross-section for the unpolarised case. From these equations, a cross-check of external measurements of P^+ and P^-, together with the extraction of A_{LR}, can be made.

accuracy of $\delta \sin^2 \theta_W \simeq 0.0003$ could be achieved with the LEP accelerator for an integrated luminosity of 40 pb^{-1} (Alexander *et al.*, 1987). The 'dilution factor' f_P, compared to $P = 1$, is $f_P = (P^- - P^+)(1 - P^-P^+)$. Hence, for the linear collider scenario ($P^- = P$, $P^+ = 0$) $f_P = P$, whereas for a circular collider ($P^- = -P^+ = P$) $f_P = 2P/(1 + P^2)$. Hence, large values of P are clearly desirable.

The quantity A_{LR} gives a determination of A^e, since the contribution from the final state fermions cancels. All final states can be used in the determination (provided radiative corrections are appropriately included), so that good statistics can be obtained. A measurement of A^f can be achieved from the quantity $A_{LR}^{FB} = [(\sigma_L^F - \sigma_L^B) - (\sigma_R^F - \sigma_R^B)]/(\sigma_R + \sigma_L) = \frac{3}{4}A^f$, where σ_L^F is the cross-section for finding f in the forward direction for a lefthanded incident beam, etc.

The dependence of the cross-section (10.42) on the initial beam polarisation can be seen more clearly by expressing this equation in terms of $C_R = (g_V - g_A)/2$ and $C_L = (g_V + g_A)/2$ as follows ($m_f = 0$ is assumed)

$$\begin{aligned}\frac{d\sigma}{d\cos\theta} = &\frac{\pi\alpha^2}{8\sin^4\theta_W\cos^4\theta_W\Gamma_Z^2}\\ &\times\{(C_R^e)^2[((C_R^f)^2 + (C_L^f)^2)(1+\cos^2\theta) + 2((C_R^f)^2 - (C_L^f)^2)\cos\theta]\\ &+ (C_L^e)^2[((C_R^f)^2 + (C_L^f)^2)(1+\cos^2\theta) - 2((C_R^f)^2 - (C_L^f)^2)\cos\theta]\}.\end{aligned} \tag{10.50}$$

The term in $C_R^e(C_L^e)$ corresponds to an incident e^- beam which is purely right(left)handed. With the standard parameters $|C_L^e| > |C_R^e|$, so the value of R_Z is greater for left than righthanded electrons.

A further quantity which is sensitive to the coupling constants is the *charge asymmetry*#, defined as

$$\begin{aligned}A_{ch}(\cos\theta) &= [N_f(\theta) - N_{\bar f}(\theta)]/[N_f(\theta) + N_{\bar f}(\theta)]\\ &= \frac{[(C_R^e)^2 - (C_L^e)^2][(C_R^f)^2 - (C_L^f)^2]2\cos\theta}{[(C_R^e)^2 + (C_L^e)^2][(C_R^f)^2 + (C_L^f)^2](1+\cos^2\theta)} = A^eA^f\frac{2\cos\theta}{1+\cos^2\theta}\\ &= -P_Z\frac{(2g_V^f g_A^f)2\cos\theta}{(g_V^{f2} + g_A^{f2})(1+\cos^2\theta)} = -P_Z A^f\frac{2\cos\theta}{(1+\cos^2\theta)}.\end{aligned} \tag{10.51}$$

where the evaluation of A_{ch} follows from (10.50), with the Z^0 polarization taken from (10.48) (for the unpolarised case). It is easy to show that the form (10.51) also holds for polarised beams. For the standard model

The quantity A_{ch} is related (assuming CP invariance holds) to A_{FB} in (10.47)

couplings, with $\sin^2\theta_W = 0.225$, then for $e^+e^- \to \mu^+\mu^-$ at $\theta = 0$, $A_{ch} = 0.039$ for $P^- = P^+ = 0$, but increases to $A_{ch} = \mp 0.20$ for $P^- = \pm 1 (P_Z = \pm 1)$. The form of (10.51) arises because, to leading order, the result can be expressed as the product of the probabilities to create the Z^0 in a certain polarisation state multiplied by that for its decay to $f\bar{f}$. The longitudinal polarisation of the Z^0 is given by (10.48), whereas the Z^0 decay probability depends on the quantities $|\mathcal{M}_R|^2$ and $|\mathcal{M}_L|^2$ in (10.17). Thus, for example, for a Z^0 polarisation P_Z, the forward–backward asymmetry (10.47) is $A_{FB} = -\frac{3}{4} A^f P_Z$. Equation (10.50) can also be generalised in a similar way.

The above discussion includes only the s-channel Z^0 diagram. However, for $|s^{1/2} - M_Z| \gtrsim \Gamma_Z$, the γ and γ–Z interference terms should again be considered, in particular for asymmetry measurements. The cross-section for $|s^{1/2} - M_Z| \gtrsim \Gamma_Z$, the γ and γ–Z interference terms should again be considered, in particular for asymmetry measurements. The cross-section for unpolarised beams was given in (6.87). Near the resonance pole, the become $(Q_e = -1)$

$$C_1 = Q_f^2 + \frac{|F|^2}{16}[(g_V^f)^2 + (g_A^f)^2][((g_V^e)^2 + (g_A^e)^2) - 2Pg_V^e g_A^e]$$

$$- \frac{(\mathrm{Re}\,F)}{2} Q_f g_V^f (g_V^e - P g_A^e),$$

$$C_2 = \frac{|F|^2}{2}(g_V^f g_A^f)\left\{(g_V^e g_A^e) - \frac{P}{2}[(g_V^e)^2 + (g_A^e)^2]\right\} - (\mathrm{Re}\,F) Q_f g_A^f (g_A^e - P g_V^e), \tag{10.52}$$

with

$$F = \frac{1}{\sin^2\theta_W \cos^2\theta_W} \frac{s}{[(s - M_Z^2) + iM_Z\Gamma_Z]}. \tag{10.53}$$

In this case, the charge asymmetry is given by $A_{ch}(\cos\theta) = (C_2/C_1)\cos\theta/(1+\cos^2\theta)$, which reduces to (10.51) at the Z^0 resonance for $P = 0$, and noting that $P_Z = (-2g_V^e g_A^e)/[(g_V^e)^2 + (g_A^e)^2]$.

The diversity of possible measurements available with polarised beams is apparent. Note that the γ–Z^0 interference terms (i.e. those in Re F) change sign on passing through the Z^0 pole (Re F is negative below, zero on, and positive above the Z^0). Some estimates of $A_{ch} = 3C_2/8C_1$, for three different e^- polarisations and as a function of $s^{1/2}$, are shown in Fig. 10.9 (taking $\sin^2\theta_W = 0.225$ and $M_Z = 93$ GeV). The quantity plotted is averaged over $\cos\theta$, for a 4π detector, and so is equivalent to A_{FB} in (10.47). Note that since the angular dependence of A_{ch} is $2\cos\theta/(1+\cos^2\theta)$, the bulk of the sensitivity comes from the forward cone ($\theta \lesssim 40°$). However, radiative effects are important in the forward direction for A_{ch}. Fig. 10.10 shows the effect of radiative corrections (Bohm and Hollik, 1982) on A_{ch} and A_L $(=-A_{LR})$, the latter being rather insensitive to such corrections. The calculation includes one-loop electromagnetic corrections as well as bremsstrahlung soft photons with $\Delta E_\gamma / E_{beam} < 0.01$.

Measurement of the polarisation of the outgoing fermion (e.g. τ^-), using either polarised or unpolarised beams, gives information on the weak

Fig. 10.9 Expected distributions of $A_{ch}(=A_{FB})$, as a function of $s^{1/2}$, for (*a*) $e^-e^+ \to \mu^-\mu^+$, (*b*) $e^-e^+ \to u\bar{u}$ and (*c*) $e^-e^+ \to d\bar{d}$. The values shown correspond to three different electron polarisations and are computed for $\sin^2\theta_W = 0.225$, $M_Z = 93$ GeV. The distribution of $R_{\mu\mu}$ is also given in (*a*).

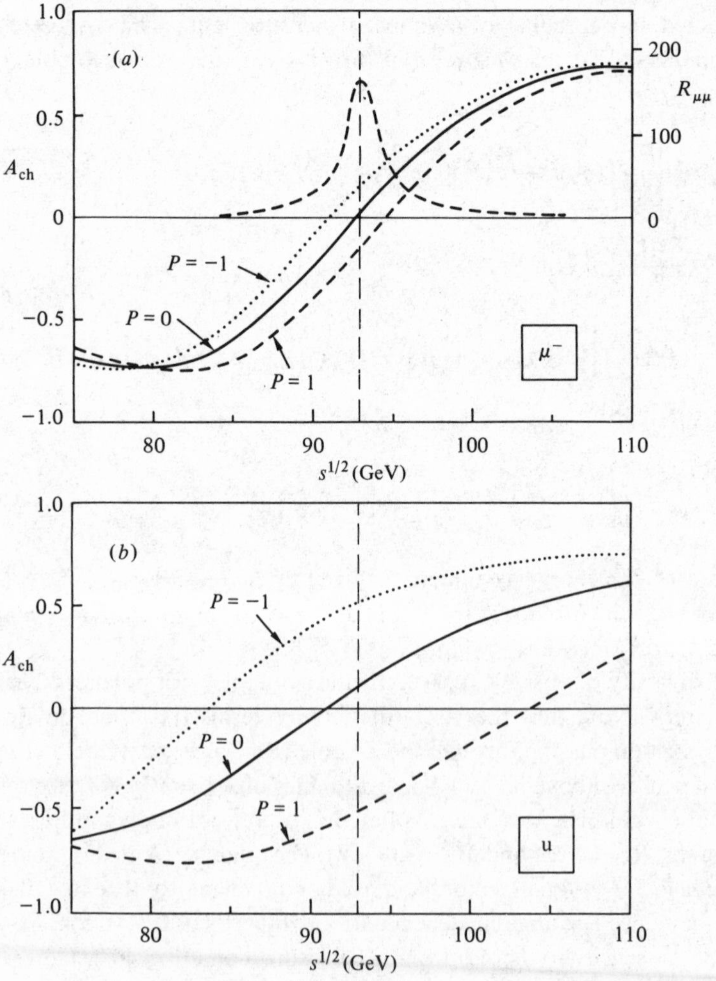

couplings of f. For example, the total number of right and lefthanded fermions produced, from *unpolarised beams*, at $s^{1/2} \sim M_Z$ is, from (10.50),

$$N_R = K_Z C_R^{f^2}(C_R^{e^2} + C_L^{e^2})\tfrac{8}{3}, \qquad N_L = K_Z C_L^{f^2}(C_R^{e^2} + C_L^{e^2})\tfrac{8}{3}, \qquad (10.54)$$

where K_Z is a constant, and the coefficient $\frac{8}{3}$ arise from integration over all

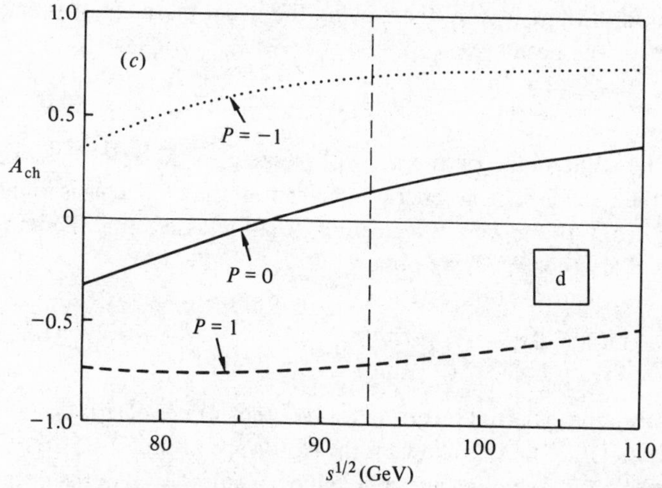

Fig. 10.10 Expected distributions of (*a*) A_{ch} and (*b*) $A_L = -A_{LR}$ as a function of $\cos\theta$ for $e^+e^- \to f\bar{f}$. For (*a*) the lowest order is shown as a broken line and the corrected results as a full line, for (*b*) the results are essentially identical.

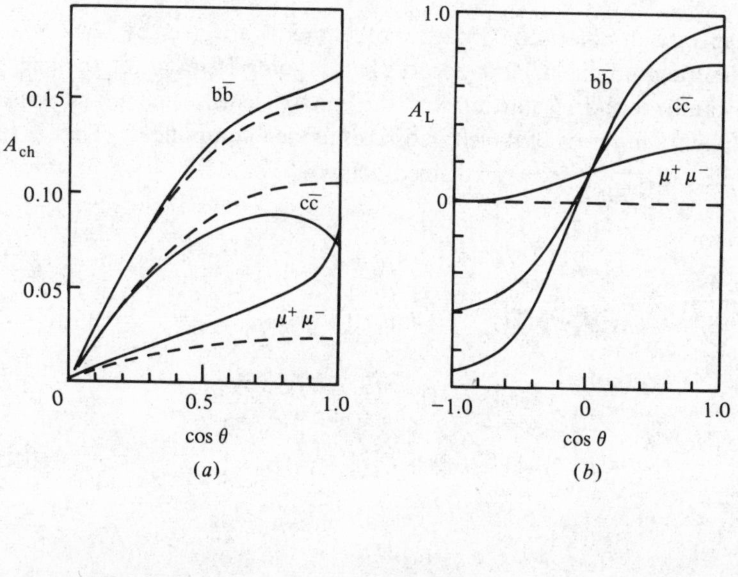

angles. Thus, the net *fermion polarisation* is

$$P_{\mathrm{f}} = \frac{(C^{\mathrm{f}}_{\mathrm{R}})^2 - (C^{\mathrm{f}}_{\mathrm{L}})^2}{(C^{\mathrm{f}}_{\mathrm{R}})^2 + (C^{\mathrm{f}}_{\mathrm{L}})^2} = \frac{-2g^{\mathrm{f}}_{\mathrm{V}}g^{\mathrm{f}}_{\mathrm{A}}}{[(g^{\mathrm{f}}_{\mathrm{V}})^2 + (g^{\mathrm{f}}_{\mathrm{A}})^2]} = -A^{\mathrm{f}}. \tag{10.55}$$

The electron couplings can be obtained by combining P_{f} with A_{FB} (from (10.47)), yielding $A_{\mathrm{FB}}/P_{\mathrm{f}} = -(\frac{3}{2})g^{\mathrm{e}}_{\mathrm{V}}g^{\mathrm{e}}_{\mathrm{A}}/[(g^{\mathrm{e}}_{\mathrm{V}})^2 + (g^{\mathrm{e}}_{\mathrm{A}})^2] = -\frac{3}{4}A^{\mathrm{e}}$.

An analysis of the decay modes of the τ-lepton appears to be a possible method of measuring the polarisation of the produced fermion. If P_τ is the polarisation of the τ-lepton, then, using the usual methods, it is easy to show that for the decay $\tau^- \to \nu_\tau \pi^-$

$$|\mathscr{M}|^2 \propto 1 + P_\tau \cos\theta_\pi, \tag{10.56}$$

where θ_π is the angle made by the π^- with respect to the spin direction of the τ^-, in the τ^- cms. The π^- energy spectrum in the $\mathrm{e^+e^-}$ cms is found by boosting from the τ cms. Neglecting small masses, and defining $x_\pi = 2E_\pi/s^{1/2} \simeq (1 + \cos\theta_\pi)/2$, we obtain

$$\frac{\mathrm{d}N}{\mathrm{d}x_\pi} = [1 + P_\tau(2x_\pi - 1)]. \tag{10.57}$$

Fig. 10.11 shows the expected spectrum for different values of $\sin^2\theta_{\mathrm{W}}$ (and hence of P_τ from (10.55)). Also shown is the lepton spectrum ($x_l = 2E_l/s^{1/2}$) for the decay $\tau \to l\nu\bar{\nu}$, again for different values of $\sin^2\theta_{\mathrm{W}}$. Here the decay lepton spectrum is given by

$$\frac{\mathrm{d}N}{\mathrm{d}x_l} = \tfrac{1}{3}[5 - 9x_l^2 + 4x_l^3 + P_\tau(1 - 9x_l^2 + 8x_l^3)]. \tag{10.58}$$

The spectra for polarised τ decays to other modes are given by Tsai (1971).

The longitudinal polarisation of the outgoing fermion for the general case can be found by starting with the matrix element (6.85), and writing it explicitly in terms of its left and righthanded components. That is (the meaning of $f^{\mathrm{e}}_{\mathrm{R}}$, $f^{\mathrm{f}}_{\mathrm{R}}$ etc. is explained below)

$$\begin{aligned}\mathscr{M} = &-\frac{\mathrm{i}e^2Q_{\mathrm{f}}}{4s}\bar{v}_2\gamma_\mu[f^{\mathrm{e}}_{\mathrm{R}}(1+\gamma^5) + f^{\mathrm{e}}_{\mathrm{L}}(1-\gamma^5)]u_1\\ &\times \bar{u}_3\gamma^\mu[f^{\mathrm{f}}_{\mathrm{R}}(1+\gamma^5) + f^{\mathrm{f}}_{\mathrm{L}}(1-\gamma^5)]v_4\\ &+\frac{\mathrm{i}e^2F}{4s}\bar{v}_2\gamma_\mu[C^{\mathrm{e}}_{\mathrm{R}}(1+\gamma^5) + C^{\mathrm{e}}_{\mathrm{L}}(1-\gamma^5)]u_1\\ &\times \bar{u}_3\gamma^\mu[C^{\mathrm{f}}_{\mathrm{R}}(1+\gamma^5) + C^{\mathrm{f}}_{\mathrm{L}}(1-\gamma^5)]v_4.\end{aligned} \tag{10.59}$$

With the usual methods, this gives

$$|\mathscr{M}|^2 = 4e^4[C_1(1+\cos^2\theta) + C_2\cos\theta], \tag{10.60}$$

where

$$\begin{aligned}C_1 = {} & Q_f^2[(f_R^e)^2+(f_L^e)^2][(f_R^f)^2+(f_L^f)^2] + |F|^2[(C_R^e)^2+(C_L^e)^2][(C_R^f)^2+(C_L^f)^2] \\ & - 2Q_f \,\mathrm{Re}\, F(f_R^e C_R^e + f_L^e C_L^e)(f_R^f C_R^f + f_L^f C_L^f) \\ C_2 = {} & 2Q_f^2[(f_R^e)^2-(f_L^e)^2][(f_R^f)^2-(f_L^f)^2] + 2|F|^2[(C_R^e)^2-(C_L^e)^2][(C_R^f)^2-(C_L^f)^2] \\ & - 4Q_f \,\mathrm{Re}\, F(f_R^e C_R^e - f_L^e C_L^e)(f_R^f C_R^f - f_L^f C_L^f).\end{aligned} \tag{10.61}$$

From these expressions we can compute $|\mathscr{M}_R|^2$, corresponding to a righthanded electron beam (i.e. this involves picking out the righthanded

Fig. 10.11 Energy spectrum of (*a*) *l* from $\tau \to l\nu\bar{\nu}$ and (*b*) π from $\tau \to \nu\pi$ for $e^+e^- \to \tau^+\tau^-$ at $s^{1/2} \sim M_Z$ for unpolarised beams.

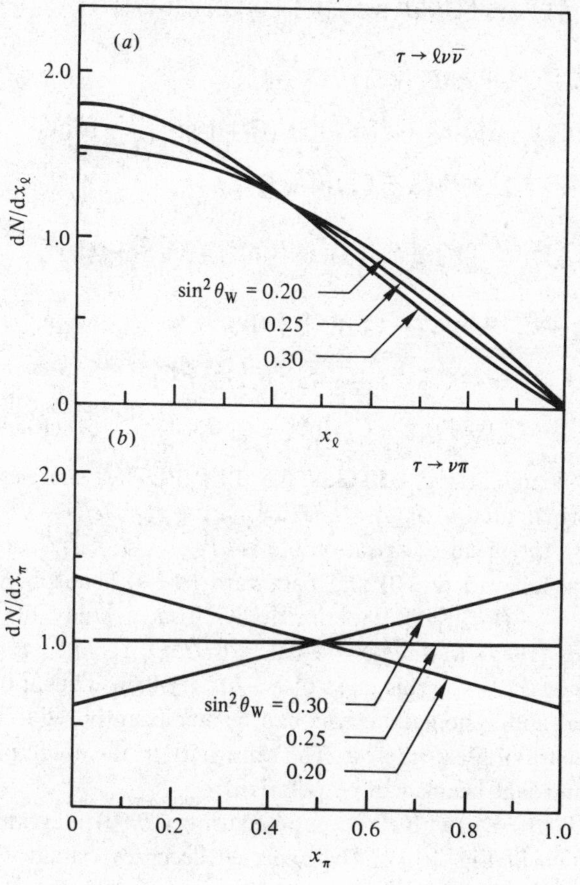

electron terms in (10.59) and amounts to setting $f_R^e = 1, f_L^e = C_L^e = 0$), and $|\mathcal{M}_L|^2$ ($f_L^e = 1, f_R^e = C_R^e = 0$). Hence, for an electron beam of polarisation P, we have

$$|\mathcal{M}|^2 = \frac{(1+P)}{2}|\mathcal{M}_R|^2 + \frac{(1-P)}{2}|\mathcal{M}_L|^2. \tag{10.62}$$

In order to find the polarisation of f, we require the relative probabilities of producing a righthanded f (i.e. (10.62) evaluated with $f_R^f = 1, f_L^f = C_L^f = 0$, giving $|\mathcal{M}(+)|^2$) and a lefthanded f ($f_L^f = 1, f_R^f = C_R^f = 0$, giving $|\mathcal{M}(-)|^2$). The longitudinal polarisation of f is then

$$P_f(\cos\theta) = \frac{|\mathcal{M}(+)|^2 - |\mathcal{M}(-)|^2}{|\mathcal{M}(+)|^2 + |\mathcal{M}(-)|^2} = \frac{a_1(1+\cos^2\theta) + a_2\cos\theta}{a_3(1+\cos^2\theta) + a_4\cos\theta}, \tag{10.63}$$

where

$$\begin{aligned}
a_1 &= \frac{|F|^2}{2}\{[(C_R^e)^2 + (C_L^e)^2] + P[(C_R^e)^2 - (C_L^e)^2]\}[(C_R^f)^2 - (C_L^f)^2] \\
&\quad - Q_f \operatorname{Re} F[(C_R^e + C_L^e) + P(C_R^e - C_L^e)](C_R^f - C_L^f), \\
a_2 &= 4PQ_f^2 + |F|^2\{[(C_R^e)^2 - (C_L^e)^2] + P[(C_R^e)^2 + (C_L^e)^2]\}[(C_R^f)^2 + (C_L^f)^2] \\
&\quad - 2Q_f \operatorname{Re} F[(C_R^e - C_L^e) + P(C_R^e + C_L^e)](C_R^f + C_L^f), \\
a_3 &= 2Q_f^2 + \frac{|F|^2}{2}\{[(C_R^e)^2 + (C_L^e)^2] + P[(C_R^e)^2 - (C_L^e)^2]\}[(C_R^f)^2 + (C_L^f)^2] \\
&\quad - Q_f \operatorname{Re} F[(C_R^e + C_L^e) + P(C_R^e - C_L^e)](C_R^f + C_L^f), \\
a_4 &= |F|^2\{[(C_R^e)^2 - (C_L^e)^2] + P[(C_R^e)^2 + (C_L^e)^2]\}[(C_R^f)^2 - (C_L^f)^2] \\
&\quad - 2Q_f \operatorname{Re} F[(C_R^e - C_L^e) + P(C_R^e + C_L^e)](C_R^f - C_L^f).
\end{aligned} \tag{10.64}$$

For unpolarised beams at $s^{1/2} \sim 30$ GeV (i.e. PEP/PETRA energies), (10.64) can be approximated to $P_f(c) = -(F/2)[g_V^e g_A^e + g_A^e g_V^f 2c/(1+c^2)]$, with $c = \cos\theta$. Thus, the mean polarisation of f is $\langle P_f \rangle = -F g_V^e g_A^f/2$, and the difference in the forward ($c > 0$) and backward ($c < 0$) hemispheres is $A_f^P = (P_f^F - P_f^B)/2 = -3F g_A^e g_V^f/8$. Experimentally, from a study of τ-decays, Ford *et al.* (1987) find $\langle P_\tau \rangle = -0.02 \pm 0.07 \pm 0.11$ and $A_\tau^P = 0.06 \pm 0.07$. Assuming $g_A^e = -\frac{1}{2}$, this gives $g_V^\tau = -0.52 \pm 0.62$. Thus, providing the statistical and systematic errors can be significantly reduced, such measurements are of great interest. The sensitivity to the couplings is enhanced if the incident beams can be polarised.

The variation of P_f (f $= \mu, \tau$) with $s^{1/2}$, computed using (10.60) integrated over angles, is shown in Fig. 10.12. The expected accuracy on $\sin^2\theta_W$

from τ-polarisation measurements at LEP is $\delta \sin^2 \theta_W \simeq 0.0019$ (Alexander *et al.*, 1987). Extra information can be obtained from a study of the angular distribution of P_f (10.63). Further information which can be extracted by using polarised beams is discussed by Prescott (1983). If both the e^- and e^+ beams are polarised, then (10.62) must be modified, as discussed above (see (10.48)). It should be mentioned again, however, that values of the beam polarisation P significantly less than 100% are expected in practice, and that the values $P = \pm 100\%$ are used above for the purposes of illustration.

In summary, therefore, the use of polarised beams, and the measurement of the polarisation of the final state fermions, is a means of significantly improving our knowledge of the coupling constants.

10.3.4 $W^{\pm}$ production at e^+e^- colliders

The two most important diagrams for the process $e^+e^- \to W^- + X$ are shown in Figs. 10.13(*a*) and (*b*). An estimate of the cross-sections for these processes can be obtained using the Weizäcker–Williams approximation of quasi-real photons, which relates $\sigma(e^+e^- \to e^{\pm} \nu W^{\mp})$ to $\sigma(\gamma e^{\pm} \to \nu W^{\pm})$. Numerical estimates (Gaillard, 1979) give $\sigma(e^+e^- \to We\nu) \sim 8 \times 10^{-38}$ cm^2, for $\sin^2 \theta_W = 0.20$, $M_W = 84$ GeV and $s^{1/2} = 150$ GeV. Another possible mechanism is that of Fig. 10.13(*c*). However, the estimated branching ratio is $\Gamma(Z \to W + X)/\Gamma_Z \sim 10^{-7}$ (Alles *et al.*, 1977), with the corresponding cross-section (at its peak) $\sim 10^{-38}$ cm^2. Thus, the cross-section for the production of a single W is estimated to be small.

Fig. 10.12 Average polarisation P_f ($f = \mu^-, \tau^-$), as a function of $s^{1/2}$, computed for $\sin^2 \theta_W = 0.225$, $M_Z = 93$ GeV.

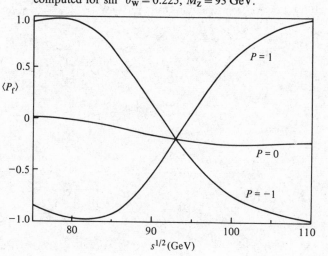

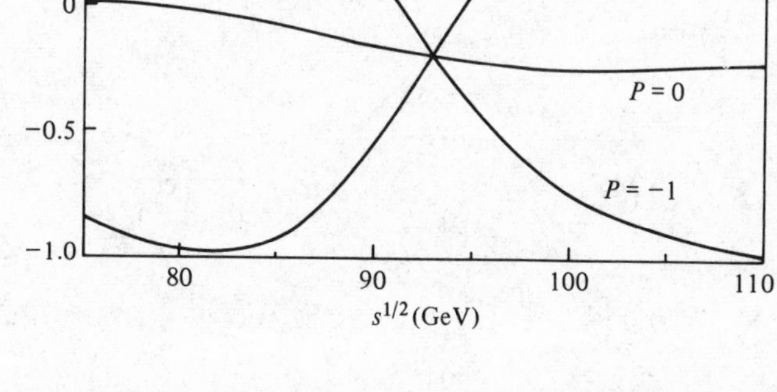

Diagrams leading to the production of a W^+W^- pair are shown in Fig. 10.14. These diagrams involve the trilinear couplings of the gauge bosons (see Appendix C), and constitute a potentially fundamental test of the standard model. The cross-section is (Alles *et al.*, 1977), with $x = \sin^2 \theta_W$

Fig. 10.13 Diagrams contributing to $e^+e^- \to W^-X$.

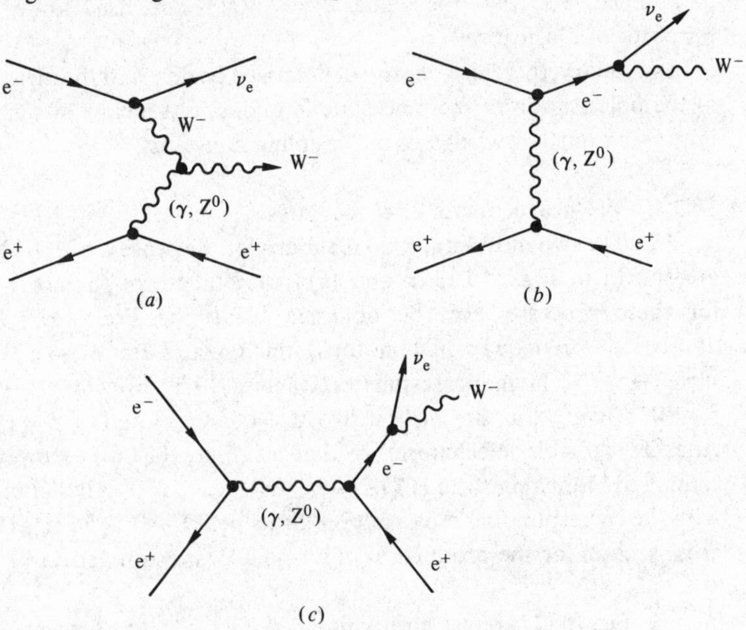

Fig. 10.14 Lowest order diagrams for $e^+e^- \to W^+W^-$.

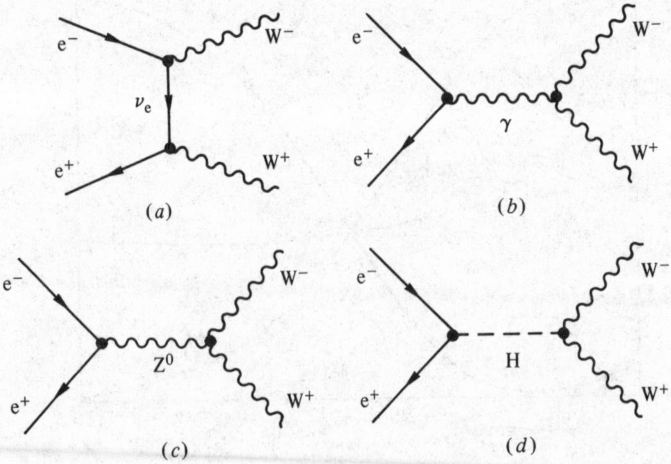

and $\beta = (1 - 4M_W^2/s)^{1/2}$

$$\frac{d\sigma}{d\Omega} = \frac{\alpha^2 \beta}{32x^2 s} \sum_{i,j} M_{ij}, \tag{10.65}$$

where

$$M_{\nu\nu} = F_1, \qquad M_{\gamma\gamma} = x^2 F_2,$$

$$M_{ZZ} = \left(x^2 - \frac{x}{2} + \frac{1}{8}\right)\frac{s^2}{(s - M_Z^2)^2} F_2, \qquad M_{Z\gamma} = 2x(\tfrac{1}{4} - x)\frac{s}{s - M_Z^2} F_2,$$

$$M_{\nu Z} = (x - \tfrac{1}{2})\frac{s}{s - M_Z^2} F_3, \qquad M_{\gamma\nu} = -xF_3, \tag{10.66}$$

with

$$F_1(\theta, s) = \frac{2s}{M_W^2} + \frac{\sin^2\theta\beta^2}{2}\left[\left(\frac{s}{t}\right)^2 + \frac{s^2}{4M_W^4}\right],$$

$$F_2(\theta, s) = \beta^2\left[\frac{16s}{M_W^2} + \left(\frac{s^2}{M_W^4} - \frac{4s}{M_W^2} + 12\right)\sin^2\theta\right], \tag{10.67}$$

$$F_3(\theta, s) = 16\left(1 + \frac{M_W^2}{t}\right) + \frac{8\beta^2 s}{M_W^2} + \beta^2\frac{\sin^2\theta}{2}\left(\frac{s^2}{M_W^4} - \frac{2s}{M_W^2} - \frac{4s}{t}\right).$$

In these expressions $t = M_W^2 - s/2(1 - \beta\cos\theta)$ is the four-momentum transfer squared between the e^- and W^-, and θ is the cms scattering angle of the W^- with respect to the incident e^-. The diagram with a Higgs propagator (Fig. 10.14(*d*)) does not contribute in the limit $m_e = 0$; however, it is needed to avoid unitarity problems at very high energies in a full calculation.

The resulting total cross-section, obtained by integrating (10.65), is

$$\sigma_{tot} = \frac{\pi\alpha^2\beta}{8x^2 s}\sum_{i,j}\bar{\sigma}_{ij}, \tag{10.68}$$

where

$$\bar{\sigma}_{\nu\nu} = \sigma_1, \qquad \bar{\sigma}_{\gamma\gamma} = x^2\sigma_2,$$

$$\bar{\sigma}_{ZZ} = \left(x^2 - \frac{x}{2} + \frac{1}{8}\right)\frac{s^2}{(s - M_Z^2)^2}\sigma_2, \qquad \bar{\sigma}_{Z\gamma} = \left(\frac{x}{2} - 2x^2\right)\frac{s}{s - M_Z^2}\sigma_2,$$

$$\bar{\sigma}_{\nu Z} = (x - \tfrac{1}{2})\frac{s}{s - M_Z^2}\sigma_3, \qquad \bar{\sigma}_{\gamma\nu} = -x\sigma_3, \tag{10.69}$$

with

$$\sigma_1 = \frac{2s}{M_W^2} + \frac{\beta^2}{12}\frac{s^2}{M_W^4} + 4\left[\left(1 - \frac{2M_W^2}{s}\right)\frac{L}{\beta} - 1\right],$$

$$\sigma_2 = 16\beta^2 \frac{s}{M_W^2} + \tfrac{2}{3}\beta^2\left[\left(\frac{s}{M_W^2}\right)^2 - 4\left(\frac{s}{M_W^2}\right) + 12\right], \tag{10.70}$$

$$\sigma_3 = 16 - 32\left(\frac{M_W^2}{s}\right)\frac{L}{\beta} + 8\beta^2\left(\frac{s}{M_W^2}\right) + \frac{\beta^2}{3}\left(\frac{s}{M_W^2}\right)^2\left(1 - \frac{2M_W^2}{s}\right)$$
$$+ 4\left(1 - \frac{2M_W^2}{s}\right) - 16\left(\frac{M_W^2}{s}\right)^2\frac{L}{\beta},$$

in which $L = \ln[(1+\beta)/(1-\beta)]$.

The total $e^+e^- \to W^+W^-$ cross-section, computed using the standard GSW model (i.e. (10.68)), is shown in Fig. 10.15, as a function of $s^{1/2} = 2E_{\text{beam}}$. The cross-section peaks at a value of about 20 pb, some 40 GeV above threshold. This corresponds, very roughly, to the production of about $10^4 W^+W^-$ pairs/yr/intersection region, for a luminosity of

Fig. 10.15 Cross-section for $e^+e^- \to W^+W^-$, as a function of $s^{1/2}$, as expected in the standard GWS model. The individual contributions which make up the total cross-section are also shown.

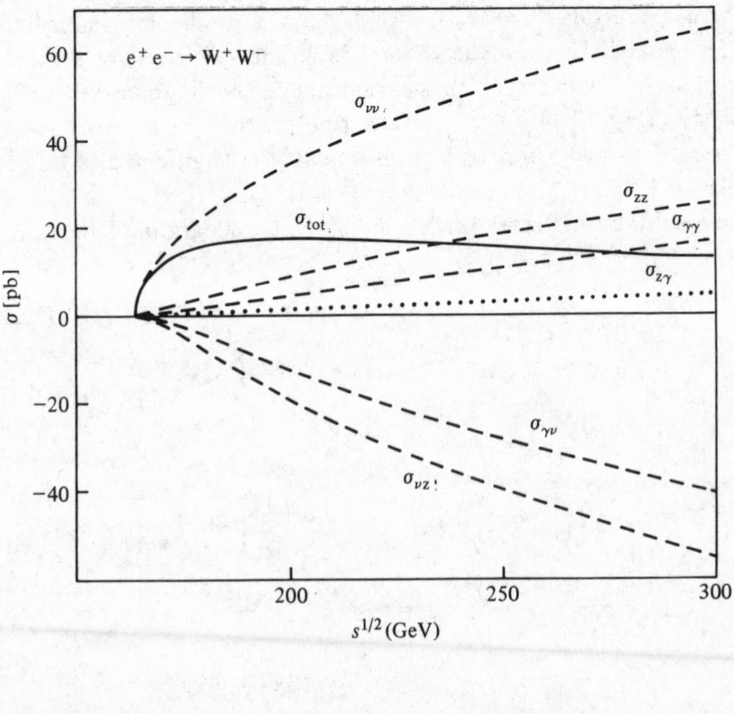

5×10^{31} cm^{-2} s^{-1}, giving an integrated luminosity of 500 pb^{-1}. The rapid rise around threshold is a sensitive method with which to determine the W mass, but the finite W width must be taken into account. Also shown in Fig. 10.15 is the energy dependence of $\sigma_{\nu\nu}$, which arises from the ν exchange graph (Fig. 10.14(*a*)). For large s ($\gg M_W^2$), the ν exchange cross-section takes the form (from (10.69))

$$\sigma_{\nu\nu} \overset{s \gg M_W^2}{\simeq} \frac{\pi\alpha^2 s}{96x^2 M_W^4}. \tag{10.71}$$

The cross-section from the ν exchange diagram increases linearly with s, and will clearly eventually exceed the unitarity bound. In the GSW model, this bad high energy behaviour is cancelled when all the graphs are considered (the individual components of the cross-section are shown in Fig. 10.15), the coefficients being dictated by gauge invariance, giving

$$\sigma(e^+e^- \to W^+W^-) \overset{s \gg M_W^2}{\simeq} \frac{\pi\alpha^2}{2x^2 s} \ln\left(\frac{s}{M_W^2}\right). \tag{10.72}$$

Precision tests of the GSW theory require accurate measurement of the cross-section. Radiative corrections are important, and must be included in such a comparison. Experimentally, the signature for W^+W^- is either through four jets in the final state (QCD four-jet production is expected to give a background of approximately 5%), or through a leptonic decay mode. Detection of W^+W^- in $\bar{p}p$ colliders is also possible, but the backgrounds are more severe.

In addition to measurements of the total cross-section, measurements of the W helicity and angular distribution permit further sensitive tests of the model. If the W^+W^- process occurs through V and A couplings only, then the cross-section is zero if the helicities of the e^- and e^+ are the same, i.e. $\sigma_{LL} = \sigma_{RR} = 0$ (this can be seen by inserting $u\binom{R}{L} = \frac{1}{2}(1 \pm \gamma^5)u$ in a general V, A current, and then using the properties of γ^5). For the neutrino exchange diagram (Fig. 10.14(*a*)), the incident e^- must be lefthanded, i.e. $\sigma_{RL}^{\nu\nu} \simeq 0$. In fact, $\sigma_{RL} \sim 10^{-2}\sigma_{LR}$ in total since, for the γ, Z^0 (or, equivalently, W_3, B) exchange, the W_3 boson contributes only to σ_{LR} and B does not couple to W^+W^-. The dominant helicity configuration is shown in Fig. 10.16. For $\theta \sim 0$, this means that we have either $W^+(\lambda = +1)$, $W^-(\lambda = 0)$, or $W^+(\lambda = 0)$, $W^-(\lambda = -1)$. The angular distribution peaks at $\theta \sim 0$, so the dominant contribution contains a significant production of helicity zero (i.e. longitudinal) W bosons. This is important, because it is the longitudinal modes which give rise to the W mass through the Higgs mechanism. Experimentally, the W polarisation can be found by measuring the angular distribution of the decay fermions in the W cms (equation (10.11)). The decays $W \to l\nu$ are probably the most suitable for this. The W cms can easily be found, provided the W^+W^- axis has been determined.

The W^- angular distribution, expected in the GSW model at $s^{1/2} =$ 200 GeV, is given in Fig. 10.17. The distribution peaks at small values of θ (this can be seen from (10.67)), and, from the breakdown of the cross-section in terms of the WW spin states, it can be seen that the potentially very interesting LL term is rather small. A study of $d\sigma/d\cos\theta$ for W^+W^- events should provide a sensitive test on the form of the trilinear gauge boson couplings, i.e. γWW and ZWW. For the standard model, these are given in Appendix (C.3.3). Gaemers and Gounaris (1979) have considered the most general form of the three-vector boson coupling.

Fig. 10.16 Helicity configuration for the dominant process in $e^+e^- \to W^+W^-$.

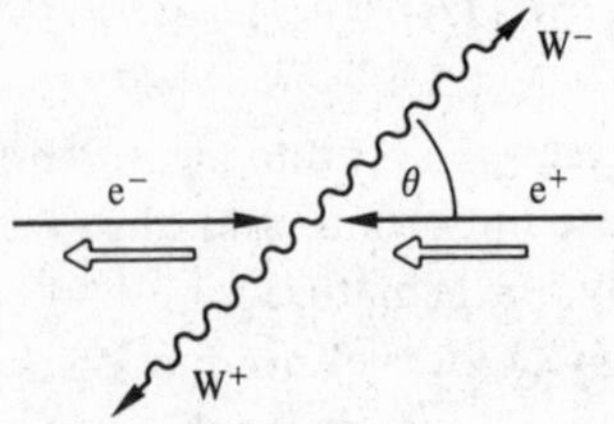

Fig. 10.17 Angular distribution of W^- in $e^+e^- \to W^+W^-$, at $s^{1/2} =$ 200 GeV, expected in the GSW model. The breakdown of the cross-section in terms of W helicity states is also given.

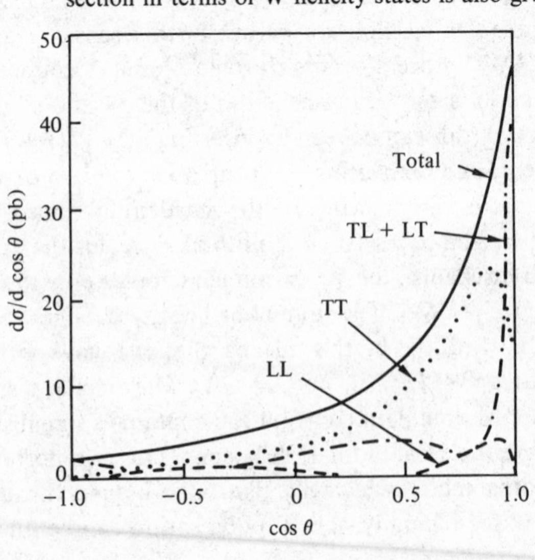

The experimental prospects for the sensitivity to the parameters of this general form (which can be formulated in terms of the W magnetic dipole and electric quadruple moments) are discussed in some detail by Barbiellini *et al.* (1986). The possibility of studying the four-boson coupling, using the processes $e^+e^- \to W^+W^-\gamma$ and $e^+e^- \to e^+e^-W^+W^-$, also exists for LEP II (.e. the second stage of LEP with beams of approximately 100 + 100 GeV).

The process $e^+e^- \to Z^0Z^0$ can occur in lowest order through the *t*-channel exchange of an electron. Near threshold, the cross-section is (Barbiellini *et al.*, 1986)

$$\sigma(e^+e^- \to Z^0Z^0) \simeq \frac{\pi\alpha^2[1 + 6(1-4x)^2 + (1-4x)^4]\beta_Z}{16x^2(1-x)^2 s}, \qquad (10.73)$$

where $x = \sin^2\theta_W$. This gives roughly $\sigma \sim (5\ \text{pb})\beta_Z$, which is about an order of magnitude less than the W^+W^- cross-section (because $x \sim 0.25$), and is of less inherent interest because there are no new vertices involved.

Accurate measurements of the Z^0 and W masses should give a stringent test of the GSW theory, beyond leading order (see Section 10.4). Measurement of the Z^0 mass (and width) can be made by carefully scanning over the Z^0 peak, and measuring the total cross-section at each energy. The limiting factors expected for LEP (Altarelli *et al.*, 1986), assuming a luminosity of $\mathscr{L} \sim 10^{31}\ \text{cm}^{-2}\,\text{s}^{-1}$ and for about 30 days running, are:

(i) statistical errors; $\delta M_Z \simeq \pm 15$ MeV, $\delta\Gamma_Z \simeq \pm 15$ MeV;
(ii) uncertainty in machine energy spread ($\Delta s^{1/2} \simeq 70$ MeV); this gives a small effect provided it is well known;
(iii) uncertainty in luminosity ($\Delta\mathscr{L} \simeq 2\%$); $\delta M_Z \simeq \pm 10$ MeV, $\delta\Gamma_Z \simeq 10$ MeV;
(iv) uncertainty in the absolute energy scale; $\delta M_Z \simeq \delta\Gamma_Z \simeq \pm 30$ MeV;
(v) uncertainty in radiative corrections; $\delta M_Z \simeq \delta\Gamma_Z \lesssim 10$ MeV.

Altarelli *et al.* (1986) estimate that the resulting overall uncertainty should be $\delta M_Z \simeq \delta\Gamma_Z \simeq \pm 50$ MeV, and that a reduction to 20 MeV is possible by improvements on the measurements of the beam parameters. (This requires a measurable beam polarisation, however a transverse polarisation will suffice.) Roughly similar accuracy is expected at the SLC (Gilman, 1986). Note that Γ_Z will accurately specify N_ν (Section 10.3.2).

Measurement of the W mass is more difficult than that of the Z^0. At least four methods seem possible (for details see Barbiellini *et al.*, 1986). These are measurement of

(i) the threshold dependence of $\sigma(e^+e^- \to W^+W^-)$;
(ii) the end point of the lepton spectrum for $W \to l\nu_l$;

(iii) the jet–jet invariant mass for $W \rightarrow q\bar{q}$;
(iv) the $l\nu_l$ invariant mass for $W \rightarrow l\nu_l$.

In each of these methods the effect of the finite W width, and also radiative corrections, must be taken into account. The methods suffer from different systematic effects, but an error of $\delta M_W \lesssim 100$ MeV should be feasible at LEP II.

10.4 TESTING THE STANDARD MODEL

The calculations in the previous sections have been confined to the leading order (Born) terms. However, higher order terms are generally significant, and these must be considered in precision tests of the theory. As discussed in Section 10.3.3, the line shape of the resonance is considerably distorted by QED radiative effects, which give shifts to $\delta\Gamma_Z/\Gamma_Z \sim \delta M_Z/\Gamma_Z \sim 0.1$ (or $\delta\Gamma_Z \sim \delta M_Z \sim 200$ MeV). The main effect is from the large logarithms associated with the real and virtual radiation from the incident particles. These give rise to terms in $t' = \ln(M_Z^2/m_l^2) \simeq 24.3$, of the order $\alpha t'/\pi \sim 0.06$. In a fully inclusive measurement there are, however, no similar logs associated with the final states, since a sum is made over all of these. However, this is not the case for a specific channel such as $e^+e^- \rightarrow \mu^+\mu^-$, where, if ΔE_γ is the maximum energy of an undetected γ, terms in $\ln(M_Z/\Delta E_\gamma)$ will arise.

The calculation of QED corrections on the incident legs requires the evaluation of the sum of leading and next-to-leading terms $(\alpha t')^n$ and $\alpha(\alpha t')^n$ for all n, in order to match the expected accuracy of the experiments. The formalism of the Altarelli–Parisi equations (Section 7.7.2) can be used (Altarelli and Martinelli, 1986). In the calculation, the probability $P_{ee}(z)$, to find an electron inside an electron carrying a fraction z of its momentum, appears (in a similar way to the splitting function P_{qq} (7.180), of finding a quark within a quark). Likewise, there are the analogous splitting functions $P_{e\gamma}(z)$ and $P_{\gamma e}(z)$ (P_{ee}, $P_{e\gamma}$ and $P_{\gamma e}$ have the same form as their QCD analogues, but without the colour factor $\frac{4}{3}$).

Compared to the Born diagram result, the $O(\alpha)$ and $O(\alpha^2)$ corrections raise the maximum of the cross-section for $e^+e^- \rightarrow \mu^+\mu^-$ by 184 and 96 MeV respectively (Berends *et al.*, 1987). The exponentiated versions of the $O(\alpha)$ and $O(\alpha^2)$ corrections differ from the full calculation by about 20 MeV, and this indicates the remaining theoretical uncertainty. One of the main uncertainties arises from loop contributions containing quarks. The measured values of $\sigma(e^+e^- \rightarrow \text{hadrons})$ for $1 \lesssim s^{1/2} \lesssim 10$ GeV are needed as input (see Section 7.8), and these are not known to the desired precision.

As discussed in Section 5.5.3, there are a considerable number of parameters in the standard GSW model. To leading order, θ_W, e, g and g' satisfying $\sin\theta_W = e/g$ and $\tan\theta_W = g'/g$ (5.99) and also the masses of the W and Z are related by $M_W = M_Z \cos\theta_W$ (5.168). Further, the vector boson masses are given in terms of the vacuum expectation value of the Higgs field $v = (2^{1/2}G_F)^{1/2}$ by $M_W = gv/2$ and $M_Z = (g^2 + g'^2)^{1/2}v/2$ (5.167), and the Fermi coupling constant G_F, which can be accurately determined from muon decay, satisfies $8G_F M_W^2 = 2^{1/2}g^2$. Thus, knowledge of $e = (4\pi\alpha)^{1/2}$, G_F and one of θ_W, M_W or M_Z is adequate to calculate the other parameters at tree level for the minimal standard model (i.e. one neutral Higgs doublet). In addition to the above parameters, there are the lepton masses and the quark masses and mixing angles (denoted generically as $\{m_f\}$).

The value of M_W^2 can be written, using (5.63) and (5.99), in terms of α, G_F and $\sin^2\theta_W$,

$$M_W^2 = \frac{\pi\alpha}{2^{1/2}G_F \sin^2\theta_W} = \frac{A_0^2}{\sin^2\theta_W} = \frac{(37.2810(3)\,\text{GeV})^2}{\sin^2\theta_W}. \tag{10.74}$$

In this expression the coupling constants $\alpha = \alpha(m_e)$, $G_F = G_F(m_\mu)$, (given by (6.46)), refer to the $Q^2 \to 0$ limit. At some scale $Q^2 \sim M^2(M_W^2, M_Z^2)$, however, the effective (running) coupling constants for this scale must be used.

The relationships $\sin\theta_W = e/g$ and $M_W = M_Z\cos\theta_W$, etc. are only valid at tree level. In higher orders, the coupling constants and masses receive potentially infinite corrections, and must be renormalised. The absolute value of $\sin^2\theta_W^X(\mu)$, will depend on which scheme X is used, and on the scale μ at which it is expressed. Values of $\sin^2\theta_W$ in the various schemes differ by $O(\alpha)$. The loop corrections also alter the tree level relationships discussed above by finite terms of $O(\alpha)$. The corrections to the measured quantities in the particular physical process under study must also be calculated in the desired scheme. A common method, the '*on-shell*' scheme (Sirlin, 1980; Marciano and Sirlin, 1984), is to use the masses M_W, M_Z (as well as m_H, $\{m_f\}$ and e) as parameters of the theory and to define

$$\sin^2\theta_W = 1 - M_W^2/M_Z^2, \tag{10.75}$$

where M_W and M_Z are the physical vector boson masses. That is, this relationship is defined to be exact, unmodified by radiative corrections. However, higher order corrections modify (10.74), giving

$$M_W^2 = \frac{\pi\alpha(0)}{2^{1/2}G_F(1-\Delta r)\sin^2\theta_W} = \frac{A^2}{\sin^2\theta_W} = M_Z^2\left[\frac{1 + (1 - 4A^2/M_Z^2)^{1/2}}{2}\right]. \tag{10.76}$$

where Δr is the electroweak radiative correction factor.

An alternative renormalisation scheme is the $\overline{\text{MS}}$ scheme (Wheater and Llewellyn Smith, 1982), in which the renormalisation constants are fixed by subtracting the singular parts of two- and three-point functions at a scale $\mu = M_W$. The scheme is useful for the analysis of Grand Unified Theories (Section 10.7) and the corrections are less sensitive to the Higgs mass m_H. The relationship between the two schemes (Marciano and Parsa, 1986) is that, for $\sin^2\theta_W \simeq 0.22$ and $m_H \simeq M_Z$, $\sin^2\theta_W = 1.006\sin^2\hat{\theta}_W(M_W)$, where $\hat{\theta}_W$ corresponds to the $\overline{\text{MS}}$ scheme and where the relating factor depends (weakly) on m_t and m_H. Once the basic parameters of the Lagrangian have been chosen for a particular renormalisation scheme, the other quantities can be calculated (perturbatively) in terms of these.

One of the main contributions to the correction Δr in (10.76) arises from the 'running' of α to $Q^2 = M^2$. This gives (the calculation involves summing all possible vacuum polarisation contributions in the derivation of (4.196))

$$\begin{aligned}\alpha^{-1}(M) &= \alpha^{-1}(0) - \frac{1}{3\pi}\sum_f Q_f^2 \ln(M^2/m_f^2) \qquad (m_f < M)\\ &= (127.2 \pm 0.3) + \tfrac{8}{9}\pi \ln(m_t/35),\end{aligned} \tag{10.77}$$

where m_t is the mass of the top quark in GeV and for $m_H = 100$ GeV and $M_Z = 93$ GeV (Jegerlehner, 1986). Thus, $\alpha(M)/\alpha(m_e) = 1.073 \pm 0.003$. As discussed in Sections 4.9.2 and 4.9.5, this arises from the vacuum polarisation diagrams and the corresponding large logs associated with the QED β function. A crude estimate of the shift in α can be made by inserting in (10.77) the appropriate lepton and quark masses. Using $m_u = 4$ MeV, $m_d = 7$ MeV, $m_s = 125$ MeV, $m_c = 1.5$ GeV, and $m_b = 4.95$ GeV (i.e the current quark masses discussed in Section 8.4.2) gives a shift $\Delta\alpha^{-1} = 10.3$. This result is very sensitive to the light quark masses and increasing m_u and m_d to 15 MeV gives better agreement with the numerical result given in (10.77) (which is from Jegerlehner, 1986). In this latter calculation, dispersion relations are used for the low energy contribution of $e^+e^- \to$ hadrons. This avoids the use of the QCD in the low energy region and the use of the ill-defined light quark masses. A disadvantage, however, is that these low-energy cross-sections are not known with adequate precision. This variation in α can, in principle, be studied directly by measuring the energy dependence of the total cross-section from threshold to the Z^0 resonance. The analogous correction $G_F(M)/G_F(m_\mu)$ contains no large logarithms, and is small. At the leading log level G_F does not 'run'.

The electroweak correction term Δr (in (10.76)) has been computed by Jegerlehner (1986). The value of Δr depends on the mass of the top quark, the existence of any further quark doublets or heavy leptons and also the mass of the Higgs scalar particle(s). A value $\Delta r = 0.0713 \pm 0.0013$ is found for $M_Z = 93$ GeV, $m_t = 35$ GeV and $m_H = 100$ GeV. Note that the correction, other than that given by (10.77), is only a few parts per mil. The corrected values of M_W and M_Z can be computed from (10.75) and (10.76). In the standard model there is a unique relationship between M_W and M_Z, as can be seen from (10.76). The effect of electroweak radiative corrections is to change the value of A from 37.281 GeV (with no corrections) to 38.65 (with corrections to order α; this value depends to some extent on m_t and m_H). Thus, for example, for $M_Z = 93$ GeV, we find $M_W = 82.0$ with corrections and $M_W = 83.1$ GeV without corrections. Hence, precision measurements of M_Z and M_W should give a sensitive test of the standard model, and in particular of the electroweak radiative corrections.

A precision measurement of $M_Z - M_W$ is sensitive to the mass of the top quark m_t, and somewhat less so to the mass of the Higgs. Marciano and Sirlin (1984) give, for the variation of Δr with m_t ($m_t \gg M_W$)

$$\delta\Delta r \simeq \frac{-3\alpha}{16\pi}\frac{\cos^2\theta_W}{\sin^4\theta_W}\frac{m_t^2}{M_W^2}, \tag{10.78}$$

and for the variation with m_H ($m_H \gg M_Z$)

$$\delta\Delta r \simeq \frac{11\alpha}{48\pi\sin^2\theta_W}\ln\left(\frac{m_H^2}{M_Z^2}\right) = 0.0024\ln\left(\frac{m_H^2}{M_Z^2}\right). \tag{10.79}$$

Thus, a 1 TeV (rather than 100 GeV) Higgs would change Δr by about 0.01. Fig. 10.18 shows the dependence of Δr on m_t and m_H, as well as on

Fig. 10.18 Variation of radiative correction Δr as a function of the Higgs (H), t-quark (t) and heavy lepton (L) masses.

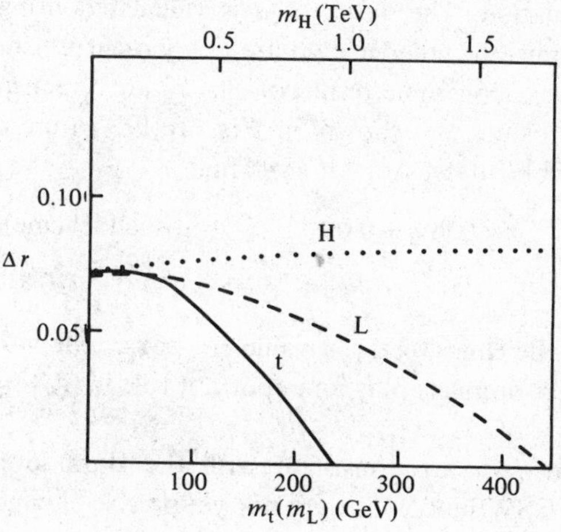

Table 10.2. *Comparison between the experimentally observed W and Z masses and the values expected from low energy neutral current measurements [from Section 8.9 and labelled Theory]*

Quantity	Theory	UA1	UA2
M_W [GeV]	80.4 ± 1.6	$83.5 \pm 1.1 \pm 2.7$	$80.2 \pm 0.8 \pm 1.3$
M_Z [GeV]	91.7 ± 1.3	$93.0 \pm 1.4 \pm 3$	$91.5 \pm 1.2 \pm 1.7$
$\sin^2\theta_W$ (a)	0.231 ± 0.009	$0.214 \pm 0.006 \pm 0.014$	$0.232 \pm 0.004 \pm 0.007$
$\sin^2\theta_W$ (b)	0.231 ± 0.009	0.194 ± 0.032	0.232 ± 0.025
ρ	1	$1.026 \pm 0.037 \pm 0.019$	$1.000 \pm 0.033 \pm 0.009$

the mass of a possible heavy lepton L. A new family of heavy quarks, with a large mass difference $m_{t'} - m_{b'}$, would cause Δr to decrease, leading to $\Delta r \sim 0$ for $m_{t'} - m_{b'} \sim 250$ GeV. Note that, in general, precision measurements of quantities involving loops can provide a sensitive probe to new particles.

Using all the available neutral current data, and applying electroweak radiative corrections to $O(\alpha)$, Amaldi *et al.* (1987) extract a value

$$\sin^2\theta_W = 0.230 \pm 0.005 \qquad \text{(on-shell scheme)}, \tag{10.80}$$

assuming $\rho(= M_W^2/M_Z^2\cos^2\theta_W) = 1$. The error also includes the various theoretical uncertainties, and is computed assuming $n_f = 3$, $m_t \leqslant 100$ GeV and $m_H \leqslant 1$ TeV. Using this value, the masses of the W and Z can be predicted using (10.75) and (10.76).

In Table 10.2 a comparison of the standard model predictions is made, with the measured M_W and M_Z values of the UA1 and UA2 collaborations. The value $\sin^2\theta_W(a)$ is derived from (10.76), whereas $\sin^2\theta_W(b)$ is obtained from the ratio M_W/M_Z (10.75) and is thus free of common systematic errors in the mass determination. The value of ρ is calculated using $\rho = \cos^2\theta_W(b)/\cos^2\theta_W(a)$. Thus, the collider data are in good agreement with the standard model. Indeed, consistent results on $\sin^2\theta_W$ are obtained over a large range of Q^2 values, as shown in Fig. 10.19. From a two-parameter fit to all available data, Amaldi *et al.* find

$$\sin^2\theta_W = 0.229 \pm 0.006, \qquad \rho = 0.998 \pm 0.009 \qquad \text{(on-shell scheme)}. \tag{10.81}$$

The value of ρ is sensitive to the Higgs sector. A value of ρ consistent with unity suggests that only Higgs doublets play an important role in $SU(2)_L$ breaking.

The present experimental data are consistent with the Born level predictions of the standard GSW model, but are not yet precise enough to test accurately the higher order (in particular the weak rather than QED) terms in the theory. Such tests must be made before the model can be fully accepted as the gauge theory of the electroweak interaction. Experimental tests are planned in several areas. An improved measurement in $\sin^2\theta_W$ from $\bar{\nu}_\mu e^-$ interactions is being made by the CHARM-2 collaboration, which should achieve an error of ± 0.005 on $\sin^2\theta_W$. Improvements in the precision of the W and Z masses from the CERN and FNAL $\bar{p}p$ collider experiments will also be made. As discussed in Section 8.8.1 studies of the processes $e^-q \to e^-q$ (γ, Z^0 exchange) and $e^-q \to \nu_e q'$ (W exchange) at the HERA e^-p collider will test the model at enormous values of Q^2. The most precise tests of the model are expected to come from the e^+e^- collider experiments at SLC and LEP. These should achieve an accuracy which is sensitive to the structure of the higher order corrections.

10.5 THE MISSING INGREDIENTS

In terms of the particles which are necessary in the GSW model, the discovery of the t-quark and that of the Higgs particle, are the most important experimental tasks.

10.5.1 t-quark and toponium

Careful studies at the highest energies available at the PETRA machine show no evidence for a t-quark state, giving a lower limit $m_t \gtrsim 23$ GeV. Some evidence for the existence of a t-quark has been

Fig. 10.19 Results for $\sin^2\theta_W$ as a function of the Q^2 value of the vector boson.

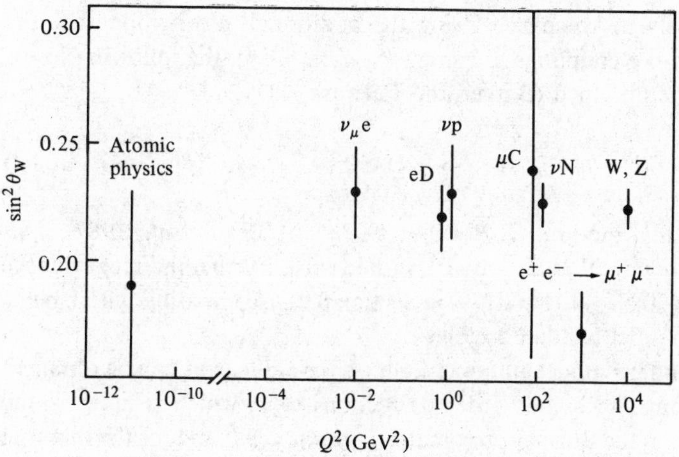

published by the UA1 collaboration (Arnison *et al.*, 1984b), using data taken in 1983. The aim of the study was to search for the decay $W \to t\bar{b}$. A sample of events was selected, which contained a high p_T lepton (e or μ), together with two hard jets (j_1, j_2). Six such events (three electron and three muon) were found. The missing p_T in these events was found to be small ($\lesssim 10$ GeV), and was identified with the presence of a neutrino. The invariant mass $m_4(l\nu j_1 j_2)$, calculated assuming $p_L^\nu = 0$, was consistent with M_W for all these events, and the mass $m_3(l\nu j_2)$ was concentrated in the region around 40 GeV. These events were interpreted as the sequence $W \to t\bar{b}$, followed by $t \to bl\nu$. A further six electron candidates were found in the 1984 running. However, from the W production rates, it is expected that the number of events with $W \to t\bar{b}(t \to be\nu)$ is less than or equal to 3, whereas nine are observed. If the observed signal is due to t-quarks, then it appears that some additional mechanism, such as direct $t\bar{t}$ production (which cannot be estimated very reliably) is needed.

The accumulation of more data, together with a more detailed study of the backgrounds, has led to a re-evaluation of the original observations. There is now no evidence for a significant signal and a lower limit $m_t > 44$ GeV (95% c.l.) is given (Albajar *et al.*, 1987c). Higher limits can be extracted with more model dependent assumptions.

A t-quark is expected in order to complete the third generation of quarks. If there is no t-quark, then we must put the weak eigenstate b′ in an SU(2) singlet. In this case, the only charged current interactions are the transformations $u \to d'$ and $c \to s'$; however, d′ and s′ would contain some small admixture of the mass eigenstate b. Hence, the neutral current is no longer automatically diagonal in flavour, and strangeness-changing neutral currents can appear. These can be removed by suitable choice of the mixing angles; however, b flavour-changing neutral currents cannot simultaneously be removed. Using the additional information that the $b \to u$ and $b \to c$ couplings are small (Section 8.6.4), the following lower bound can be obtained (Barger and Pakvasa, 1979)

$$R = \Gamma(b \to l^+ l^- X)/\Gamma(b \to l\nu X) \gtrsim 0.12. \tag{10.82}$$

Experimentally, the limit is $R < 0.01$ (90% c.l.), Bean *et al.* (1987). This is indirect evidence that the t-quark should exist. Further indirect evidence is that $g_A^b = -0.50 \pm 0.10$ ((8.179) in Section 8.9.2), compatible with b being a member of a lefthanded doublet.

In the standard model an upper limit on the value of m_t can be obtained by considering the higher order corrections to ρ, which is unity in the GSW model in the Born approximation. For $m_t \gg m_b$, and for the minimal

Higgs model (Veltman, 1977),

$$\rho \simeq 1 + \frac{3G_F m_t^2}{8(2^{1/2})\pi^2} \simeq 1 + 0.0125\left(\frac{m_t\,\text{GeV}}{200}\right)^2. \tag{10.83}$$

Inserting the upper limit of ρ from (10.81), gives $m_t \lesssim 200$ GeV.

Our theoretical understanding of heavy quark systems is expected to improve as the quark mass increases, since $\alpha_S(m_Q^2)$ becomes smaller, making perturbation theory more reliable. The short distance part of the $Q\bar{Q}$ potential for bound states is generated by gluon exchange, and has a Coulomb-like (i.e. $1/r$) behaviour (equation (2.277)). The long-distance (confining) potential is less well constrained. It is useful, therefore, to review the general properties of heavy $Q\bar{Q}$ systems.

A bound $Q\bar{Q}$ system, with orbital angular momentum L and spin S, has $P = (-1)^{L+1}$ and $C = (-1)^{L+S}$ (this is general for any fermion–antifermion pair). The states are labelled as $n^{2S+1}L_J$, where n is the radial quantum number and J the total angular momentum. For $c\bar{c}$ we have, for example, the states $\psi(3.097\text{ GeV})$ with $^{2S+1}L_J(J^{PC}) = {}^3S_1(1^{--})$, $\eta_c(2.98) = {}^1S_0(0^{-+})$, $\chi_0(3.41) = {}^3P_0(0^{++})$, etc., as well as excited vector states such as $\psi'(3.70)$. Similarly, for $b\bar{b}$ states, the sequence $\Upsilon(9.46\text{ GeV} = 1S)$ up to $\Upsilon(11.02\text{ GeV} = 6S)$, all with $J^{PC} = 1^{--}$, together with other states, have been identified.

Possibly decay mechanisms for a quarkonium vector resonance $V(Q\bar{Q})$ are shown in Fig. 10.20. The mechanism shown in Fig. 10.20(*a*) is the *annihilation* of the $Q\bar{Q}$ pair. Since the $Q\bar{Q}$ is near threshold, the problem can be treated non-relativistically. The matrix element for $V \to l^+l^-$ is

Fig. 10.20 Decay of quarkonium state $V(Q\bar{Q})$ to (*a*) l^+l^- or $q\bar{q}$, (*b*) ggg or ggγ and (*c*) via charged current decay of Q.

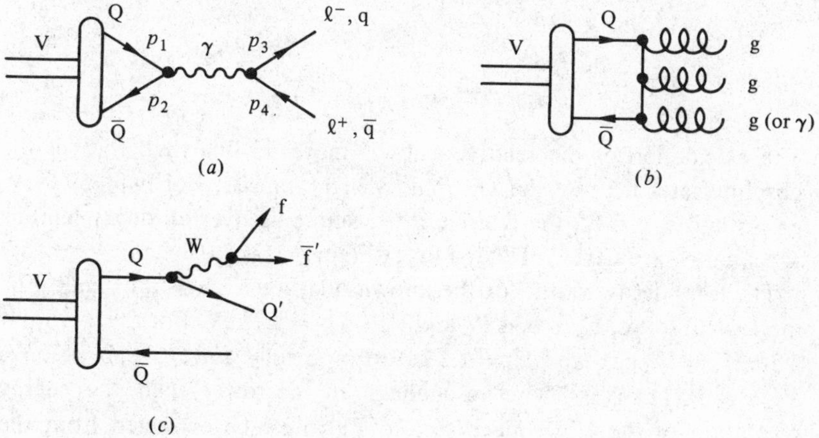

constructed from that for the process $Q+\bar{Q}\to\gamma^*\to l^+l^-$, evaluated for $\mathbf{p}_Q \simeq \mathbf{p}_{\bar{Q}} \simeq \mathbf{0}$. This amounts to inserting $\psi(0)$, the wave function evaluated at the origin, into the matrix element (Van Royen and Weisskopf, 1967, and see also Novikov *et al.*, 1978; Poggio and Schnitzer, 1979). Inserting this, and the appropriate volume normalisation, gives the decay rate for $V\to l^+l^-$, namely

$$\Gamma(l^+l^-) = e_Q^2 16\pi\alpha^2|\psi(0)|^2/m_V^2 = e_Q^2\Gamma_0. \tag{10.84}$$

The dependence on m_V in (10.84) follows from the dimensions of $|\psi(0)|^2$ (mass^3) and of Γ (mass).

In the case where the decaying vector meson is not a pure $Q\bar{Q}$ state, e_Q^2 is replaced by $|\sum a_i Q_i|^2$, where a_i are the appropriate Clebsch–Gordan coefficients. Applying this to the e^+e^- decays of $\rho^0=(u\bar{u}-d\bar{d})/2^{1/2}$, $\omega^0=(u\bar{u}+d\bar{d})/2^{1/2}$, $\phi=s\bar{s}$, $\psi=c\bar{c}$ and $\Upsilon=b\bar{b}$ gives the expected ratios $\rho^0/\omega^0/\phi/\psi/\Upsilon = 9/1/2/8/2$. This is in excellent agreement with the experimental ratios for Γ_{ee} namely

$$\rho^0/\omega^0/\phi/\psi/\Upsilon = (10.5\pm1.1)/1/(1.98\pm0.13)/(7.1\pm0.6)/(1.85\pm0.14).$$

If these ideas are correct, this suggests that $|\psi(0)|^2 \propto m_V^2$. The same ratios should hold for the vector dominance transitions $\gamma^*\to\rho^0$, ω^0, etc. However, in this case, the prediction should be modified by the propagator term (7.198).

For the $q\bar{q}$ decay modes there is an extra colour factor of three. Thus, the result for $V\to q\bar{q}$, together with the rates calculated for the processes $V\to ggg$ and $V\to gg\gamma$ (Fig. 10.20(*b*)) (e.g. Sehgal, 1985), are as follows

$$\begin{aligned}
\Gamma(q\bar{q}) &= 3e_q^2 e_Q^2 \Gamma_0, \\
\Gamma(ggg) &= \frac{10(\pi^2-9)}{81\pi}\frac{\alpha_S^3}{\alpha^2}\Gamma_0, \\
\Gamma(gg\gamma) &= \frac{36e_Q^2}{5}\frac{\alpha}{\alpha_S}\Gamma(ggg).
\end{aligned} \tag{10.85}$$

The calculation of the relative rates is more reliable than that of the absolute rates because the uncertainty in Γ_0 cancels. For $m_Q = 50$ GeV, $e_Q=\frac{2}{3}$ and $\alpha_S\simeq 0.12$, the relative rates, summing over all decay leptons and quarks, are $\Gamma(l^+l^-)/\Gamma(q\bar{q})/\Gamma(ggg)/\Gamma(gg\gamma) = 1/1.2/0.8/0.16$.

The total decay widths for the known quarkonia states are small. For the $J/\psi(c\bar{c})$ states, $\Gamma_{tot}=63$ keV and $\Gamma_{ee}=\Gamma_{\mu\mu}=4.7$ keV. For the $\Upsilon(1S)$ $b\bar{b}$ state, $\Gamma_{tot}=43$ keV and $\Gamma_{ee}=1.2$ keV. Empirically, for $Q=c, b$ we have $\Gamma_{ee}\simeq e_Q^2\times(11\text{ keV})$. These resonances are narrower than the energy resolution of the e^+e^- machines, so Γ_{tot} must be extracted from the integrated cross-section over the resonance, together with the measured branching ratio to e^+e^- (equation (10.46) illustrates this).

An estimate of the value of α_S can be made from the values of Γ_{tot} and Γ_{ee} for the $\Upsilon(1S)$ state. The total width is given by

$$\Gamma_{tot} = \sum_l \Gamma(l^+l^-) + \sum_q \Gamma(q\bar{q}) + \Gamma(ggg) + \Gamma(gg\gamma). \tag{10.86}$$

The sum is over $l = e, \mu, \tau$ and $q = u, d, s, c$. Dividing throughout by $\Gamma(e^+e^-)$ and, noting that $\Gamma(gg\gamma)/\Gamma(e^+e^-) \simeq 0.8$ for $\alpha_S \simeq 0.15$, we obtain $\Gamma(ggg)/\Gamma(e^+e^-)$. In the $\overline{\text{MS}}$ renormalisation scheme

$$\Gamma(ggg)/\Gamma(e^+e^-) = \frac{10(\pi^2-9)}{81\pi e_Q^2}\frac{\alpha_S^3}{\alpha^2}\left(1 + 9.1\frac{\alpha_S}{\pi} + \cdots\right), \tag{10.87}$$

where $\alpha_S = \alpha_S(m_\Upsilon^2)$. Inserting the experimental numbers yields

$$\alpha_S(m_\Upsilon^2) = 0.151 \pm 0.007 \quad \text{or} \quad \Lambda_{\overline{\text{MS}}} = 165 \pm 35 \text{ MeV}. \tag{10.88}$$

The error does not include the effects of hadronisation, which will modify equations (10.85). Note that, for the $\Upsilon(1S)$ state, the ratios of the decay rates are (from (10.84) and (10.85)) $\Gamma(l^+l^-)/\Gamma(q\bar{q})/\Gamma(ggg)/\Gamma(gg\gamma) = 1/1.1/6.5/0.25$. The three-gluon mode is relatively more important for the b-quark case, as the photon terms are reduced by a factor of four compared to $e_Q = \frac{2}{3}$. A further difference is that α_S decreases as Q^2 increases, so that $\Gamma(ggg)$, which is proportional to α_S^3, becomes relatively smaller. Thus, the Υ decays are dominated by three-gluon final states. Unfortunately, the energy of the gluons is rather small, limiting their usefulness in the study of the gluon fragmentation properties. A detailed study of the hadronic decays of toponium should thus be useful in this respect. The differential energy distribution of the decay gluons ($x_i = 2E_i/s^{1/2}$) is

$$\frac{1}{\Gamma_{3g}}\frac{d\Gamma}{dx_1\,dx_2} = \frac{1}{(\pi^2-9)}\left[\frac{(1-x_1)^2}{x_2^2x_3^2} + \frac{(1-x_2)^2}{x_3^2x_1^2} + \frac{(1-x_3)^2}{x_1^2x_2^2}\right]. \tag{10.89}$$

A massive $t\bar{t}$ system V (often denoted θ) can also have a significant fraction of decays through the weak interaction. In addition to the photon propagator diagram of Fig. 10.20(*a*), there is also a Z^0 term. The process $Q\bar{Q} \to \gamma^* + Z^0 \to f\bar{f}$ is analogous to $e^+e^- \to f\bar{f}$, which was considered in Section 10.3.3, except that the initial $Q\bar{Q}$ are bound in a $J^P = 1^-$ vector state. The resulting decay width becomes

$$\Gamma(f\bar{f}) = K\Gamma_0\left[e_Q^2e_f^2 + \frac{|F|^2}{16}g_V^{Q^2}(g_V^{f^2} + g_A^{f^2}) + \frac{\text{Re}\,F}{2}e_Qe_fg_V^Qg_V^f\right], \tag{10.90}$$

where $K = 1(3)$ for $f\bar{f} = l^- l^+ (q\bar{q})$, and F is given by (10.53). The decay $V(Q\bar{Q}) \to \nu_l \bar{\nu}_l$ is also described by (10.90), with a width

$$\Gamma(\nu\bar{\nu}) = \Gamma_0 |F|^2 g_V^{Q^2}(g_V^{\nu 2} + g_A^{\nu 2}).$$

The branching ratio reaches $B_{\nu\bar{\nu}} \sim 10\%$, for $m_V \sim 80$ GeV.

In addition to these neutral current decays, the charged current decay of a Q (or $\bar{Q}$) is also possible (Fig. 10.20(*c*)). Since the decay rate is proportional to m_Q^5, this mode should be significant for toponium. In this case the decay sequence is $t \to b + W^+$ ($W^+ \to l\nu$, ud' or cs'), and similarly for $\bar{t}$. Using the spectator model, this gives a rate for this *single quark decay* (using (6.24))

$$\Gamma(\mathrm{SQD}) = 18|V_{tb}|^2 \frac{G_F^2 m_t^5}{192\pi^3}, \tag{10.91}$$

assuming $m_t \gg m_b$. The expected values of these partial decay widths, as a function of the toponium mass, are sketched in Fig. 10.21 (Sehgal, 1985).

An interesting possibility is that the toponium mass is essentially degenerate with the Z_0 mass ($|M_Z - m_V| < 2\Gamma_Z$). In this case toponium–Z^0

Fig. 10.21 Estimates of partial decay widths of toponium, as a function of the $t\bar{t}$ mass.

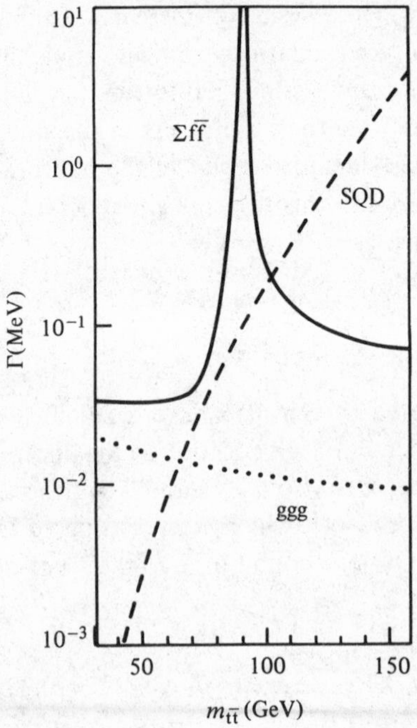

interference effects have to be taken into account, and this leads to a dip in R rather than an enhancement (Renard, 1979). This can be understood with reference to Fig. 10.22 for the process $e^+e^- \to \mu^+\mu^-$. For $m_V \ll M_Z$, the matrix element at $s^{1/2} = m_V$ has the form (from the propagators $\gamma + \gamma V \gamma$)

$$\mathcal{M}(m_V) \sim \frac{1}{m_V^2} + \frac{1}{m_V^2}\frac{g_{V\gamma}^2}{im_V\Gamma_V}\frac{1}{m_V^2}$$
$$\sim \mathcal{M}_\gamma - i\mathcal{M}_V. \tag{10.92}$$

Hence, the cross-section $|\mathcal{M}|^2 = \mathcal{M}_\gamma^2 + \mathcal{M}_V^2$ is enhanced. However, for $m_V = M_Z$, we have

$$\mathcal{M}(m_V) \sim \frac{1}{iM_Z\Gamma_Z} + \frac{1}{iM_Z\Gamma_Z}\frac{g_{VZ}^2}{im_V\Gamma_V}\frac{1}{iM_Z\Gamma_Z}$$
$$\sim \mathcal{M}_Z - \mathcal{M}'_V. \tag{10.93}$$

Hence, there is destructive interference, and a zero in the cross-section results if

$$\Gamma_V = g_{VZ}^2/(M_Z^2\Gamma_Z). \tag{10.94}$$

The total number of $t\bar{t}$ states expected is large: for $m_t \sim 40$ GeV, more than 200 states with $\Gamma < 100$ keV are predicted. The expected distribution of R for $e^+e^- \to \mu^+\mu^-$, for $m_V \sim M_Z$, is shown in Fig. 10.23. The upper figure (*a*) is the pattern expected if the beam energy resolution is approximately zero, whereas the lower distribution (*b*) shows how this pattern is largely washed out with a beam energy resolution of about 40 MeV (i.e. that of LEP). The rich (but difficult) physics which can be performed is described in detail by Buchmüller *et al.* (1986).

A spectrum of 'open' t-quark mesons and baryons is, of course, expected. The most probable decay of the t-quark should be to a b-quark (followed by $b \to c$, $c \to s$) plus $f\bar{f}'$ (from W decay), with a rate given by (10.91). The expected mechanisms for the decays are similar to those outlined for b- and c-quarks (Section 8.6). However, the much larger mass value of the

Fig. 10.22 Diagrams giving possible $\gamma - V(t\bar{t})$ and $Z - V(t\bar{t})$ interference.

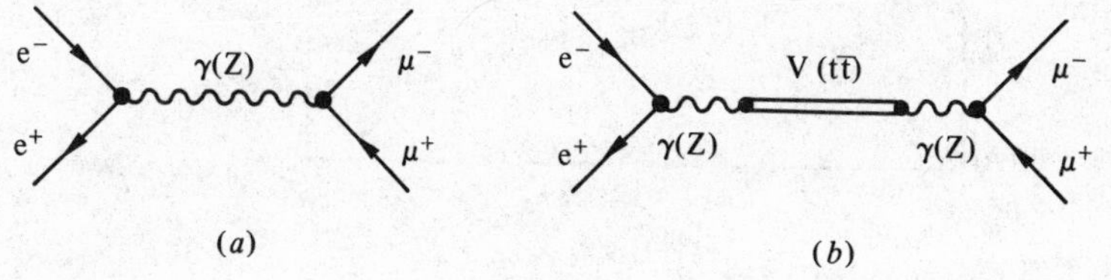

decaying quark should permit a much clearer separation of competing mechanisms. Note that, just above threshold for production of $t\bar{t}$, the final state particles will be isotropically distributed, giving a distinctive experimental signature.

Additional generations of quarks (and/or leptons) are, of course, possible. If the observed pattern in the KM matrix extends to such generations, then a further charge $\frac{2}{3}$ quark could have a relatively long lifetime, e.g. $O(10^{-12}$–10^{-13} s). Studies at PETRA have shown that any new d-type quark has a mass $\gtrsim 20$ GeV. For the minimal Higgs model, the constraint given by equation (10.83) can be applied. A more restrictive limit comes from the possible destabilisation of the vacuum by heavy quark loops, which leads to (Duncan *et al.*, 1985) $\sum m_Q^4 < (80 + 0.54 m_H)^4$ (GeV^4).

Fig. 10.23 Expected distribution of $R(e^+e^- \to \mu^+\mu^-)$ for $m_V \sim M_Z$, as observed with an energy resolution of (*a*) zero (*b*) 40 MeV.

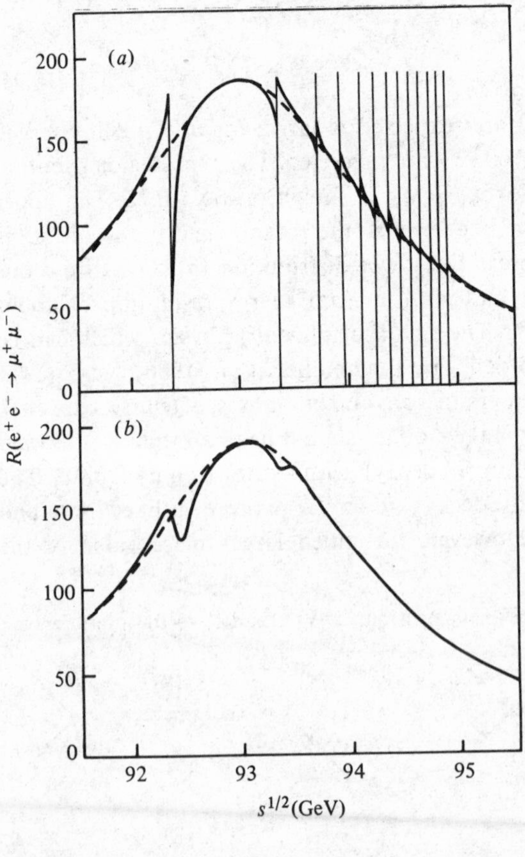

10.5.2 Higgs scalars

The experimental values of M_W and M_Z (10.40), and all the low energy neutral current phenomena, are in good agreement with the minimal standard model; that is, a model with a single Higgs doublet (see Section 5.5.2), which leads to $\rho = 1$. This complex scalar doublet of fields is needed to generate the masses of the $W^\pm$ and Z^0. One physical neutral Higgs boson H^0 survives this mechanism, the other degrees of freedom disappear into the longitudinal states of the massive bosons.

The mass of H^0 is not specified. However it is clear, from Section 5.5.2, that the Higgs will be most copiously produced in association with heavy particles, and that it will decay preferentially to the heaviest particles kinematically available. That is, depending on the value of the Higgs mass m_H, the leading order decays are $H^0 \to (l^- l^+, q\bar{q})$ or $H^0 \to (W^+W^-, Z^0Z^0)$ if $m_H > 2M_W$. An upper bound on m_H is obtained from the requirement that the self-interaction terms of the Higgs field in the Lagrangian are small enough that perturbative calculations are not jeopardised (i.e. $\lambda^2/4\pi \lesssim 1$). This gives roughly $m_H \lesssim O$ (1 TeV). A more detailed argument, using the fact that WW scattering would violate unitarity unless $m_H^2 \lesssim 16\pi v^2/3$, gives a bound $m_H \lesssim 1000$ GeV. A precise measurement of ρ is also sensitive to m_H, as discussed in Section 10.4. For small values of m_H, radiative corrections to the Higgs potential are important (Coleman and Weinberg, 1973), leading to a local minimum with $v \neq 0$ for $\lambda \sim O(\alpha^2)$. The requirement of an absolute minimum gives a lower limit

$$m_H^2 \simeq \frac{3\alpha^2 v^2}{8}\left[\frac{2+\sec^4\theta_W}{\sin^4\theta_W} - O\left(\frac{m_f}{M_W}\right)^4\right]. \tag{10.95}$$

Hence, for $m_f \ll M_W$, we have $m_H \gtrsim 7$ GeV. However, for $m_t \gtrsim M_W$ there is no lower bound. Experimentally, no Higgs signal has been found. However, establishing a reliable lower limit is somewhat problematic (see Langacker, 1985). The possibility of a very light Higgs is not yet completely ruled out.

The main potential *decay modes* of the *Higgs* are as follows.

(i) $H^0 \to f\bar{f}$. Using the Feynman rules of Appendix C, the matrix element for $H^0(p) \to f(p_1) + \bar{f}(p_2)$ is

$$\mathcal{M} = \frac{-iem_f}{2\sin\theta_W M_W}\bar{u}_1 v_2 = -im_f(2^{1/2}G_F)^{1/2}\bar{u}_1 v_2. \tag{10.96}$$

Computing $|\mathcal{M}|^2$, and noting that $\Gamma = |\mathcal{M}|^2 p_1^*/(8\pi m_H^2)$, we obtain

$$\Gamma(H^0 \to f\bar{f}) = \frac{N_c G_F m_f^2 m_H}{4(2^{1/2})\pi}\left(1 - \frac{4m_f^2}{m_H^2}\right)^{3/2}, \tag{10.97}$$

where the colour factor $N_c = 1$ for $l\bar{l}$, and 3 for $q\bar{q}$. The rate is small unless m_f (and m_H) are large. However, near threshold the rate is suppressed by phase space.

(ii) $H^0 \to W^+W^-$. The matrix element for $H(p) \to W_1^+(p_1) + W_2^-(p_2)$ is

$$\mathscr{M} = \frac{ie}{\sin\theta_W} M_W \varepsilon_1^{\mu}(\lambda)\varepsilon_{2\mu}(\lambda). \tag{10.98}$$

To find the decay rate, we must find $|\mathscr{M}|^2$, and sum over the three polarisation states of the two Ws (i.e. RR + LL + SS). The polarisation vectors are given by (10.9). Note that if particle 1 is along $+z$, then 2 is along $-z$, and this amounts to inverting the sign of the y and z components of ε_2 (eg $\varepsilon_1(S)\cdot\varepsilon_2(S) = (p_1^2 + E_1^2)/M_W^2$). Defining $x_W = 4M_W^2/m_H^2$, this leads to the decay rate

$$\Gamma(H^0 \to WW) = \frac{G_F M_W^2 m_H}{8(2^{1/2})\pi} \frac{(1 - x_W)^{1/2}}{x_W}(3x_W^2 - 4x_W + 4). \tag{10.99}$$

(iii) $H^0 \to Z^0Z^0$. This is similar to (ii) and gives, for $x_Z = 4M_Z^2/m_H^2$

$$\Gamma(H^0 \to ZZ) = \frac{G_F^2 M_Z^2 m_H}{16(2^{1/2})\pi} \frac{(1 - x_Z)^{1/2}}{x_Z}(3x_Z^2 - 4x_Z + 4). \tag{10.100}$$

The extra factor of two for the ZZ mode arises because there are two identical particles. Decays through a virtual Z (or W) $\to f\bar{f}$ are also possible (Baer *et al.*, 1986).

As an example of the expected decay widths for $m_H = 200$ GeV, we obtain $\Gamma(f\bar{f}) \simeq 0.01$ (0.5) GeV for $m_f = 5(40)$ GeV, whereas the vector boson widths are about 0.5 GeV. In addition, there are possible higher order decay modes.

(iv) $H^0 \to \gamma\gamma$. There is no direct $H^0\gamma\gamma$ coupling, but a possible mechanism is shown in Fig. 10.24. The decay rate is (Ellis *et al.*, 1976a)

$$\Gamma(H^0 \to \gamma\gamma) = \frac{\alpha^2}{8(2^{1/2})\pi^3} G_F m_H^3 |I|^2, \tag{10.101}$$

Fig. 10.24 Diagram contributing to $H^0 \to \gamma\gamma (x = f, W, \ldots)$ and $H^0 \to gg(x = q, \ldots)$.

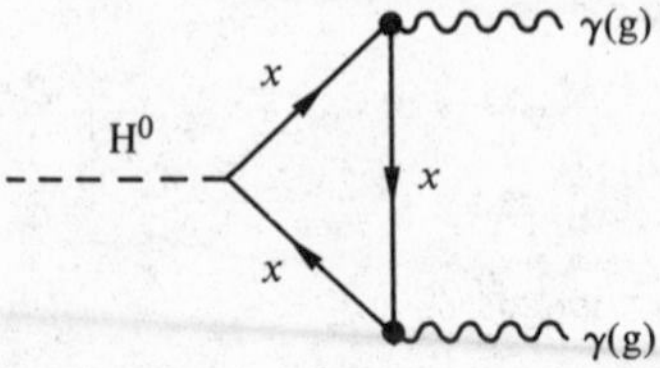

where I has contributions from the various charged particles which can run around the triangle (i.e. l, q, W, ...), and results in $I = O(1)$. Hence, the $\gamma\gamma$ decay rate is very small.

(v) $H^0 \to gg$. The mechanism here is again that of Fig. 10.24, but with the γ replaced by g and with an internal quark loop. The rate is

$$\Gamma(H^0 \to gg) = \frac{2^{1/2} G_F \alpha_S^2 m_H^3}{72\pi^3} |N|^2. \tag{10.102}$$

Only heavy quarks with $m_q \gtrsim O(m_H)$ contribute significantly to N (Baer *et al.*, 1986). The term $2^{1/2} G_F m_H^3/(72\pi^3)$ is about 20 MeV for $m_H \sim 150$ GeV, so a value of the ratio $\Gamma(H^0 \to gg)/\Gamma(H^0 \to q\bar{q})$ in the region 1–10% might be anticipated.

The experimental prospects for detecting the H^0 scalar particle depend strongly on its mass. Possible *production mechanisms* are listed below (see, e.g., Baer *et al.*, 1986):

(i) $Z^0 \to H\gamma$. The simplest mechanisms are those shown in Fig. 10.25(*a*). The relative branching ratio, compared to $\mu^+\mu^-$, is roughly

$$R_H = \frac{\Gamma(Z^0 \to H\gamma)}{\Gamma(Z^0 \to \mu^+\mu^-)} \simeq 6 \times 10^{-5} \left[1 - \left(\frac{m_H}{M_Z}\right)^2\right]^3 \left[1 + 0.1\left(\frac{m_H}{M_Z}\right)^2\right]. \tag{10.103}$$

This gives a value $R_H \sim 6 \times 10^{-5}$ for $m_H = 10$ GeV, dropping rapidly to $R_H \sim 10^{-5}$ for $m_H = 60$ GeV.

Fig. 10.25 Possible production mechanisms for neutral Higgs H^0.

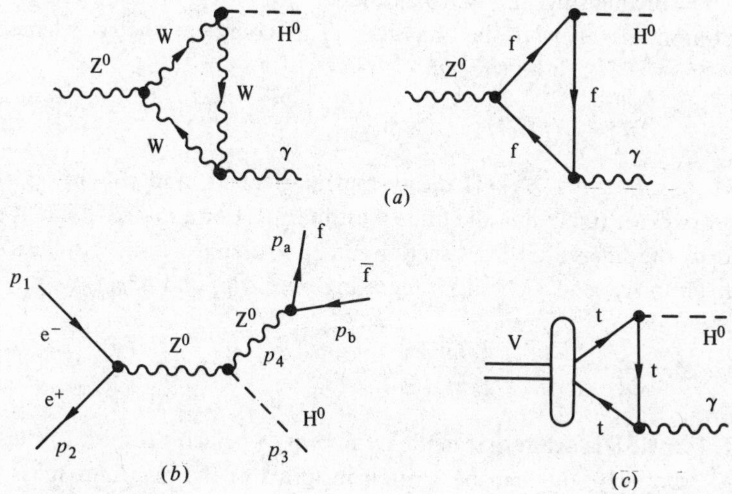

(ii) $Z^0 \to H^0 + Z^0 \to H^0 f\bar{f}$. The reaction $e^+e^- \to Z^0 + H^0 \to H^0 f\bar{f}$ (Fig. 10.25(*b*)) is one of the most promising in the search for the Higgs. We can consider this reaction as a two-stage process. The first stage is $e^-(p_1) + e^+(p_2) \to H^0(p_3) + Z^0(p_4)$, which is a two-body final state with matrix element

$$\mathcal{M}_2 = \frac{-\mathrm{i}e^2 M_Z \bar{v}_2 \gamma_\mu [C_R^e(1+\gamma^5) + C_L^e(1-\gamma^5)] u_1 \varepsilon_\lambda^\mu}{2\sin^2\theta_W \cos^2\theta_W [(s - M_Z^2) + \mathrm{i}M_Z\Gamma_Z]}, \tag{10.104}$$

where ε_λ^μ is the polarisation vector of the outgoing Z^0. In the cms we define momenta $p_1 = (s^{1/2}/2)(1, 0, 0, 1)$, $p_2 = (s^{1/2}/2)(1, 0, 0, -1)$ and $p_4 = (E_4, 0, p_4 \sin\theta, p_4 \cos\theta)$. A straightforward calculation, using equation (10.4) for the sum over Z^0 polarisation states, gives the spin-averaged matrix element squared (assuming unpolarised beams)

$$|\mathcal{M}_2|^2 = \frac{e^4 M_Z^2 (C_R^{e2} + C_L^{e2})|F|^2}{2s}\left(1 + \frac{p_4^2 \sin^2\theta}{2m_4^2}\right). \tag{10.105}$$

The differential cross-section is thus (using (4.29))

$$\frac{d\sigma_2}{d\cos\theta} = \frac{e^4 (C_R^{e2} + C_L^{e2}) M_Z^2 |F|^2 p_4}{32\pi s^2 s^{1/2}}\left(1 + \frac{p_4^2 \sin^2\theta}{2m_4^2}\right). \tag{10.106}$$

Putting $x_3 = 2E_3/s^{1/2}$ (so that $m_4^2 = s(1 - x_3 + m_H^2/s)$), and integrating over $\cos\theta$, gives the total cross-section

$$\sigma_2 = \frac{e^4 (C_R^{e2} + C_L^{e2}) M_Z^2 |F|^2 p_4}{16\pi s (s^{1/2}) m_4^2}\left(1 - x_3 + \frac{x_3^2}{12} + \frac{2}{3}\frac{m_H^2}{M_Z^2}\right). \tag{10.107}$$

The second stage in Fig. 10.25(*b*) is $Z^0(p_4) \to f(p_a) + \bar{f}(p_b)$. The outgoing fermion is taken to be massless. The complete matrix element for $e^+e^- \to H^0 f\bar{f}$ can be written

$$\mathcal{M}_3 = \mathcal{M}_2 \mathrm{i}G_V(p_4^2) \cdot \mathcal{M}_f, \tag{10.108}$$

where $\mathcal{M}_f$ is the $Z^0 \to f\bar{f}$ decay matrix element, and $\mathrm{i}G_V(p_4^2)$ is the Z^0 propagator (including the finite width term). For a matrix element of this form, the differential cross-section can be written (starting with the formula given in Appendix C, and integrating over d^3p_a and d^3p_b)

$$\frac{d\sigma_3}{d\cos\theta\, dE_3} = \frac{|\mathcal{M}_3|^2 p_3}{(2\pi)^3 16s}. \tag{10.109}$$

The angle θ is again that made by $\mathbf{p}_4 (= \mathbf{p}_a + \mathbf{p}_b)$ with respect to the beam. Alternatively, this can be written in terms of the invariant mass of the lepton pair (m_4), against which the Higgs recoils,

$$\frac{d\sigma_3}{d\cos\theta\, dm_4^2} = \frac{|\mathcal{M}_3|^2 p_3}{(2\pi)^3 32 s(s^{1/2})}. \qquad (10.110)$$

The equivalent form for the parent process $e^-e^+ \to H^0Z^0$ is

$$\frac{d\sigma_2}{d\cos\theta} = \frac{|\mathcal{M}_2|^2 p_3}{(2\pi)8s(s^{1/2})}. \qquad (10.111)$$

Thus, from (10.108), (10.110) and (10.111), we obtain

$$\frac{d\sigma_3}{d\cos\theta\, dm_4^2} = \frac{|\mathcal{M}_f|^2 |G_V(p_4^2)|^2}{16\pi^2}\left(\frac{d\sigma_2}{d\cos\theta}\right)$$

$$= \frac{\Gamma_{f\bar{f}} m_4}{\pi[(m_4^2 - M_Z^2)^2 + M_Z^2\Gamma_Z^2]}\left(\frac{d\sigma_2}{d\cos\theta}\right), \qquad (10.112)$$

where Γ_{ff} is the $Z^0 \to f\bar{f}$ decay width (see equation (10.6)).

The expected cross-section for $e^+e^- \to H^0 l^+ l^-$ (Fig. 10.26) depends strongly on both the mass of H^0 and on the value of $s^{1/2}(e^+e^-)$. Three regimes can be distinguished. For $s^{1/2} \sim M_Z$, the pole factor $|F|^2$ is important and gives a peak in the cross-section, which reaches approximately 2 pb for $m_H = 10$ GeV, but only about 6×10^{-2} pb for $m_H = 50$ GeV. This rate is equivalent to a relative Z^0 branching ratio $\Gamma(H^0 l^+ l^-)/\Gamma(\mu^+\mu^-)$ of about 2×10^{-3} (5×10^{-5}) for $m_H = 10(50)$ GeV, corresponding to H^0 event samples of roughly 60(1.5) events for 10^6 Z^0 decays. The decay mode $H^0 \to b\bar{b}$ would be the most probable for this mass range of m_H.

Fig 10.26 Cross-section for $e^+e^- \to H^0 l^+ l^-$, as a function of $s^{1/2}$.

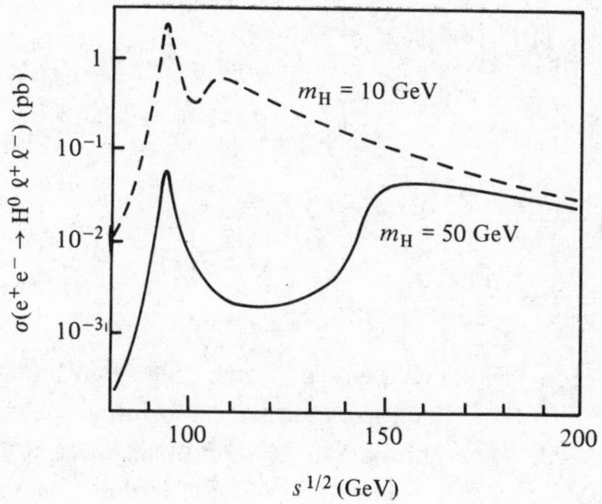

Above the Z^0 pole, both Z^0 propagators become off-shell and the cross-section decreases. However, for larger $s^{1/2}(\sim M_Z + 2^{1/2}m_H)$, the rate gets enhanced as $p_4^2 \simeq M_Z^2$; that is, the second Z^0 becomes on-shell. Once again, radiative corrections are important (see, e.g., Berends and Kleiss 1985).

(iii) Toponium $V \to H^0\gamma$ (Fig. 10.25(*c*)). This mode is potentially important because of the large coupling of H^0 to a heavy constituent quark. The decay branching ratio for vector toponium to $H^0 + \gamma$, relative to $V \to \gamma^* \to \mu^+\mu^-$, is

$$\frac{\Gamma(V \to H^0\gamma)}{\Gamma(V \to \mu^+\mu^-)} = \frac{G_F m_V^2}{4(2^{1/2})\pi\alpha}\left(1 - \frac{m_H^2}{m_V^2}\right). \tag{10.113}$$

For $m_V = 80$ GeV, this branching ratio is about 0.5(0.3) for $m_H = 10(50)$ GeV. Hence this mode is potentially sensitive for a Higgs search, provided $m_H < m_V$.

(iv) Production at hadron colliders. Possible mechanisms are $q_i\bar{q}_i \to H^0 + X$ and $gg \to H^0 + X$ (the inverse of Fig. 10.24). These mechanisms could be particularly important if the Higgs particle turns out to be very heavy, as discussed by Eichten *et al.* (1984).

In the standard Weinberg–Salam model, a single Higgs doublet ($y = 1$) is assumed, which gives rise to one physical particle, the neutral H^0. However, the Higgs sector could be more complicated than this minimal model, and indeed is in many models (e.g. supersymmetry). There are, however, constraints on the Higgs multiplets which can be introduced. In general, if one has several Higgs particles, each with weak isospin (I^i, I_3^i) and vacuum expectation value v_i, then (Ellis *et al.*, 1976a)

$$M_W^2 = \tfrac{1}{2}g^2 \sum_i [I^i(I^i+1) - (I_3^i)^2]v_i^2,$$

$$M_Z^2 \cos^2\theta_W = g^2 \sum_i (I_3^i)^2 v_i^2,$$

i.e.

$$\rho = \frac{\sum_i [I^i(I^i+1) - (I_3^i)^2]v_i^2}{2\sum_i (I_3^i)^2 v_i^2}. \tag{10.114}$$

The experimental value of ρ (10.81) is near to unity. The righthand side of (10.114) is unity for $I = \frac{1}{2}$. From the experimental value given in equation (10.81), the 90% c.l. upper limits for $I = 1$ contributions are $v(I = 1, I_3 = 0)/v(\frac{1}{2}, \frac{1}{2}) < 0.047$ and $v(1, \pm 1)/v(\frac{1}{2}, \frac{1}{2}) < 0.081$. Unless the vs happen to have certain specific values, the next lowest multiplet giving unity is $I=3$, $I_3=\pm 2$ (Gaillard, 1979). Since the member with $v_i \neq 0$ must be neutral, this would mean one would have Higgs bosons with charges up to $|Q|=5$. The less exotic possibility is thus the introduction of a further Higgs doublet, giving in total five physical Higgs particles, two charged ($H^{\pm}$) and three neutral. The relationships between M_W, the vacuum expectation value v and the HWW coupling change from the minimal Higgs model values (i.e. $M_W^2 = g^2v^2/4$ and $g_{WWH} = g^2v/2 = gM_W$ from (5.167) and (5.166) respectively) to

$$M_W^2 = \frac{g^2}{4}\sum_i v_i^2, \qquad |g_{WWH_i}| = \frac{g^2}{2}|v_i| < gM_W. \tag{10.115}$$

Thus, the coupling of each Higgs particle H_i to the W is smaller than that of the minimal Higgs model. The fermion mass is given by (cf. (5.169))

$$m_f = \sum_i g_{ffH_i} v_i. \tag{10.116}$$

The signs of v_i and g_{ffH_i} are not specified, and so there is no bound on the size of the Yukawa couplings. In general, the Yukawa couplings give rise to a mass matrix which must be diagonalised to define the strong interaction eigenstates. This can lead to $\Delta S \neq 0$ neutral currents, so the experimental absence of these gives a constraint (Gaillard, 1979).

Possible production mechanisms for $H^{\pm}$ are as follows.

(i) $e^+e^- \to \gamma(Z^0) \to H^+H^-$. The cross-section for a pair of scalar particles is given by (4.142), and exhibits a slow rise ($\propto \beta^3$) near threshold. If $m_H < M_Z/2$, the rate should be reasonable. Note that in the minimal Higgs model, $e^+e^- \to H^0H^0$ is not allowed by Bose symmetry.

(ii) $e^+e^- \to Z^0 \to W^{\pm}H^{\mp}$. The $W^{\pm}H^{\mp}Z^0$ coupling is zero in models with only weak isospin doublets and small for other multiplets, hence the expected rate is small.

(iii) $Q \to H^{\pm}q$, $V \to H^+H^-b\bar{b}$. If the heavy quark or toponium decays are not suppressed by mixing angles, these processes could be important because of the large coupling to a heavy quark.

(iv) $Z^0 \to H^+H^-H^0$. This does not occur at the tree level, but can occur at the one-loop level.

Thus, the experimental rates foreseen for both neutral and charged Higgs are rather small, and the detection of these particles (if they exist) represents a formidable experimental challenge. This could be particularly problematic if $100 \lesssim m_{H_0} \lesssim 2M_W$, a range where none of the planned accelerators are very sensitive.

10.6 LEFT–RIGHT SYMMETRIC MODELS

Although the standard electroweak model is very successful, it unifies only two of the four interactions. There has been much theoretical effort on attempts to include further interactions into the unification. In this, and the subsequent sections, the shortcomings of the standard model and the various models which have been developed with a view to further unification, are briefly reviewed.

In addition to the uncomfortably large number of parameters in the GW model (see Section 5.5.3), the model also begs the question as to why doublets are lefthanded and singlets righthanded. Left–right symmetric models have been proposed, in which the Lagrangian is both P and C invariant before spontaneous symmetry breaking, and the observed violations are attributed to non-invariance of the vacuum.

In the left–right symmetric model (for a review, see Mohapatra, 1985), the group $SU(2)_L \otimes U(1)$ is replaced by $SU(2)_L \otimes SU(2)_R \otimes U(1)$. The model has two gauge couplings prior to symmetry breaking, which are $g_2(=g_{2L}=g_{2R})$ and g'. One can define $\sin\theta_W = e/g_{2L}$, and parameterise neutral current phenomena in terms of θ_W. The righthanded leptons and quarks, which are singlets in the standard model, become doublets under $SU(2)_R$, leading to the following assignments in terms of I_L, I_R, y

$$\begin{aligned} l_L &= (\tfrac{1}{2}, 0, -1), \qquad & l_R &= (0, \tfrac{1}{2}, -1), \\ q_L &= (\tfrac{1}{2}, 0, \tfrac{1}{3}), \qquad & q_R &= (0, \tfrac{1}{2}, \tfrac{1}{3}). \end{aligned} \tag{10.117}$$

The U(1) generator (which lacks direct physical meaning in the standard model), has the attractive property in the left–right model, that it is proportional to $B-L$, where B and L are the baryon and lepton numbers respectively. The electric charge is given by

$$Q = I_{3L} + I_{3R} + (B-L)/2. \tag{10.118}$$

The symmetry can be assumed to be broken in two stages. In the first stage the parity symmetry is broken, leaving the weak gauge symmetry unbroken and W_L and W_R massless (but with different gauge couplings $g_{2L} \neq g_{2R}$. In the second stage of symmetry breaking the masses are acquired. The minimal set of Higgs multiplets to break the symmetry down to $U(1)_{em}$ is

$$\Delta_L(1, 0, 2) + \Delta_R(0, 1, 2) \qquad \text{and} \qquad \phi(\tfrac{1}{2}, \tfrac{1}{2}, 0), \tag{10.119}$$

where, under left–right symmetry $\Delta_L \leftrightarrow \Delta_R$ and $\phi \leftrightarrow \phi^\dagger$. For certain values of the relevant parameters, one obtains parity-violating minima for a left–right symmetric potential. The corresponding vacuum expectation values are

$$\langle \Delta_{L,R} \rangle = \begin{pmatrix} 0 & 0 \\ v_{L,R} & 0 \end{pmatrix}, \qquad \langle \phi \rangle = \begin{pmatrix} \kappa & 0 \\ 0 & \kappa' \end{pmatrix}, \tag{10.120}$$

in which $v_L = \gamma \kappa^2 / v_R$, where γ depends on the Higgs self-couplings. Note that $v_L \to 0$ as $v_R \to \infty$. The relation between the charged W mass eigenstates W_1 and W_2, and the states W_L and W_R, is given by (6.41), with

$$\tan \chi = \frac{-\kappa\kappa'}{\kappa^2 + \kappa'^2 + 8v_R^2}, \qquad M^2_{W_{1,2}} \simeq \frac{g^2}{2}(\kappa^2 + \kappa'^2 + 2v^2_{L,R}). \tag{10.121}$$

Existing neutral current data gives the constraints that $M_{W_1} \simeq M_W$, $M_{W_2} \gg M_{W_1}$, $\chi \ll 1$ and $v_R \gg (\kappa^2 + \kappa'^2)^{1/2} \gg v_L$.

The mass of the lighter of the two Z^0 bosons is lighter than the standard model Z^0 mass (M_Z), with

$$M^2_{Z_L} \simeq M^2_Z \left[1 - \frac{\cos 2\theta_W}{2} (1 - \tan^4 \theta_W) \left(\frac{M_{W_L}}{M_{W_R}} \right)^2 \right]. \tag{10.122}$$

Hence, for $\sin^2 \theta_W = 0.225$ and (for example) $M_{W_R} = 4M_{W_L}$, then $M_{Z_L} = 0.992 M_Z$. This is a small difference, and might prove difficult to untangle from radiative corrections. A lower limit on the mass of the heavy Z^0 can be obtained from the requirement of consistency with the neutral current data (Barger *et al.*, 1983), giving

$$M_{Z_2} \gtrsim 4M_{Z_1} \Rightarrow M_{W_R} \gtrsim 220 \text{ GeV}. \tag{10.123}$$

This limit is independent of the nature (Dirac or Majorana) of the neutrino. A somewhat greater limit is obtained from muon decay (Section 6.1.2), from which it is found that $M_{W_2} \geqslant 380$ GeV; however, this limit is valid only for light Dirac neutrinos.

A more stringent limit has been obtained from a theoretical analysis of the $K_L - K_S$ mass difference (Beall *et al.*, 1982), which receives a large contribution from the $W_L - W_R$ box diagram. Assuming that the box diagram approximation is reasonable, and that the KM matrices for q_L and q_R are the same (or one is the complex conjugate of the other), a limit of $M_{W_2} \geqslant 1.6$ TeV is found. As discussed by Mohapatra (1985), the existence of $m_{\nu_e} \sim$ few eV, $\Delta B = 2$ transitions or generation changing processes such as $\mu \to 3e$, would constitute indirect evidence for these left–right models.

10.7 GRAND UNIFIED THEORIES

In addition to the problem of why P and C are violated, there are further problems in the standard electroweak model. Each fermion is accompanied by a separate parameter specifying its Higgs coupling, and hence its mass. Thus, there is no explanation in the model as to why the ratios of the masses of quarks and leptons have their particular values (both of which span about three orders of magnitude). Although in the minimal standard model it is required that $m_{\nu_L} = 0$, in general the neutrino masses are not specified. Furthermore, there is no explanation as to the number of generations of leptons and quarks, nor to the relationship between them. The observed quantisation of charge, in that the lepton and quark charges have simple integer ratios, is also without explanation. The standard model contains two gauge group constants g and g', and the value of the angle θ_W is not specified. The presence of the Abelian U(1) group means that the charges are not specified; one can only fix $Q_u - Q_d = 1$, $Q_\nu - Q_e = 1$, etc. Whereas the conservation of charge is related to a gauge symmetry, the conservation of baryon number, and of the three types of lepton number, have no associated gauge symmetries, and thus, perhaps, no satisfactory explanation.

The gauge group $SU(3)_c$ of QCD gives a satisfactory account of strong interactions, so it is natural to include $SU(3)_c \otimes SU(2)_L \otimes U(1)$ in any attempt to unify the strong and electroweak interactions. However, in addition to the questions as to whether colour symmetry is exact, and confinement can be proven, there are further problems in QCD.

10.7.1 Problems in the QCD sector

In the limit of massless quarks, the QCD Lagrangian exhibits *chiral symmetry*. For simplicity we consider only u and d quarks, in which case $\mathscr{L}_{QCD}$ is invariant under the global symmetry $SU(2)_L \otimes SU(2)_R$. This symmetry, however, does not appear to exist for the hadron states, there being no approximate parity doubling of the lowest lying states. If the chiral symmetry is spontaneously broken, then there is no parity doubling and the pion may be identified as the approximately massless Goldstone boson associated with the symmetry breaking. The QCD Lagrangian is also invariant under the following transformations

$$\begin{aligned} q &\to \exp(i\alpha)q, & j_B^\mu &= \bar{u}\gamma^\mu u + \bar{d}\gamma^\mu d \;:\; U(1)_V \\ q &\to \exp(i\beta\gamma^5)q, & j_5^\mu &= \bar{u}\gamma^\mu\gamma^5 u + \bar{d}\gamma^\mu\gamma^5 d \;:\; U(1)_A \end{aligned} \tag{10.124}$$

The current j_B^μ, corresponding to the first transformation, can be identified with baryon number (for two flavours). However, there does not appear to be any symmetry in the hadron spectrum corresponding to the axial current. If this symmetry was realised in the Goldstone mode, another pseudoscalar meson with about the same mass as the pion should result (since it has the same quark content). No such state exists, the η' being too heavy. This is known as the $U(1)_A$ problem (see, e.g., Olive *et al.*, 1979). A solution is the existence of QCD instantons# (see, e.g., Cheng and Li, 1984).

A further problem arises from the complicated structure of the QCD vacuum state. The presence of instantons introduces an effective term in the QCD Lagrangian with some coefficient θ (see Section 7.6.1). This term violates P and T and conserves C, and hence violates CP (*strong CP violation*). *A priori*, a value of $\theta \sim O(1)$ would be expected; however, the observed limit on the neutron–electron dipole moment implies $\theta \lesssim 10^{-9}$. If the standard model Lagrangian has a further global U(1) symmetry (Peccei and Quinn, 1977), which is spontaneously broken, then the minimum of the potential gives $\theta = 0$. This requires at least two Higgs doublets ϕ_1 and ϕ_2, and the appearance of a light pseudo-Goldstone boson, the *axion*, with mass $m_a \simeq 25 n_f (x_\phi + 1/x_\phi)$ keV, where n_f is the number of families and $x_\phi = \langle\phi_2\rangle/\langle\phi_1\rangle$ is the ratio of the vacuum expectation values (vevs) and with a coupling proportional to v_{PQ}^{-1}, with $v_{PQ} \sim G_F^{-1/2}$.

A possible candidate for the axion has been reported (Schweppe *et al.*, 1983) in a study of e^+e^- correlations in heavy ion collisions (e.g. U–Cm). The mass value $m_a \simeq 1.7$ MeV needed to 'explain' the observed correlations is relatively large, and implies that either x_ϕ or x_ϕ^{-1} is large. The axion–quark couplings a–q($\frac{2}{3}$) and a–q($-\frac{1}{3}$) are proportional to x_ϕ and x_ϕ^{-1} respectively. Hence, the expected branching ratios for $\psi \to a\gamma$ or $\Upsilon \to a\gamma$ should be large, in conflict with experiment.

Variant axion models in which, for example, the couplings a–e and a–u are proportional to x_ϕ, with all others proportional to x_ϕ^{-1}, have been proposed to avoid these quarkonia limits. They also predict that $a \to e^+e^-$ is enhanced and that a is short-lived, offering an explanation as to why conventional beam dump experiments have failed to see the axion. As discussed by Peccei (1986), a recent and more sensitive beam dump experiment rules out all values of the a–e^+–e^- coupling not in conflict with $g-2$. Further, the measured branching ratio for $\pi^+ \to e^+e^-e^+\nu_e$, i.e. $(3.4 \pm 0.5)10^{-9}$, gives $\pi^+ \to a e^+ \nu_e < 2 \times 10^{-10}$. Subsequent e^+e^- correlation measurements in heavy ion collisions also do not favour the axion interpretation, and Peccei concludes that the variant axion can be

Instantons are solutions to Yang–Mills theories with non-trivial topological structure in Euclidean four-dimensional space-time, i.e. with finite space-time extension.

excluded.# If a chiral U(1) group is the solution to the strong *CP* problem, then it must be broken at very high energy scales, leading to the *invisible axion* (i.e. very large v_{PQ} and so small coupling).

10.7.2 Properties of Grand Unified Theories (GUTS)

In $SU(3) \otimes SU(2) \otimes U(1)$ the fundamental Lagrangian must contain the following fields:

(i) *Gauge fields*. These are the eight massless vector gluons and the four electroweak bosons, of which the $W^{\pm}$ and Z^0 acquire a mass after spontaneous symmetry breaking.

(ii) *Fermion fields*. There are three families of fermions, each of which contains 15 members (e.g. ν_e, e_L, u_L^i, d_L^i, e_R, u_R^i, d_R^i, where $i = 1, 2, 3$ represents the colour). This neglects the possibility of ν_{eR}.

(iii) *Higgs scalar fields*. These include at least one Higgs doublet.

In order to simplify the discussion below the possibility of further families of fermions is not considered. In the minimal $SU(3) \otimes SU(2) \otimes U(1)$ model the only unbroken global symmetries are those corresponding to the conservation of baryon number (B) and lepton number (L). Evidence for the violation of B and/or L would constitute evidence for a new interaction beyond $SU(3) \otimes SU(2) \otimes U(1)$.

Leaving aside gravity for the moment, the aim of *grand unified theories* (GUT) is to find a gauge group G, with a single gauge coupling constant, which describes all interactions and thus contains $SU(3) \otimes SU(2) \otimes U(1)$ as a subgroup. This implies that a scale must exist where the SU(3), SU(2) and U(1) couplings become equal. Since the couplings are very different at a scale of approximately 10^2 GeV, and since they vary only logarithmically with energy, this unification scale is very large; $M_{GUT} \sim 10^{15}$ GeV (see Fig. 10.27). Given the large uncertainty in this scale, the question arises as to whether gravity (whose effects should come into play at the Planck mass $\sim 10^{19}$ GeV) can indeed be neglected.

The GUT group G has certain required properties. It should contain the group $SU(3) \otimes SU(2) \otimes U(1)$ as a subgroup. The requirement that G contains electromagnetism means that the photon must be one of the gauge bosons of G. Hence, the electric charge operator Q_{ch} must be one

In addition to the interpretation in terms of one or more new elementary particles, other suggestions are that the peaks are from nuclear transitions or, more exotically, that they are manifestations of a new perturbative phase of QED. Less exotically, Scharf and Twerenbold (1987) claim that a calculation involving the dipole field of the moving ions can explain the peaks.

of the generators of G and, in fact (since G is required to be semi-simple#), all its generators are represented by traceless matrices. Thus, $\mathrm{tr}(Q_{\mathrm{ch}})=0$ in any irreducible representation of G, i.e. the sum of the electric charges of all particles in a given irreducible representation vanishes. Note that this requirement is satisfied by the members in each fermion family and thus gives a relationship between the charges of quarks and leptons. This is not the case if we consider leptons and quarks separately, so G must contain bosons which can transform leptons to quarks and vice versa. Hence, non-conservation of both B and L is possible. The group G must also admit complex representations, since the observed left and righthanded particles do not transform in the same way. For a general review of grand unified theories, see Langacker (1981).

The smallest group with the required properties is SU(5) which has rank 4 and was first considered by Georgi and Glashow (1974). In SU(5), the eight Gell-Mann matrices of SU(3) are replaced by 24 matrices λ^a, which are Hermitian and traceless. The SU(5) symmetry is assumed to be local, and each of the generators corresponds to a gauge boson. Because all the gauge couplings (being vector) conserve helicity, the left and righthanded fermions cannot be in the same representation. The way around this problem is to replace righthanded fermions by charge-conjugated lefthanded fermions (e.g. replace e_R^- by e_L^+, see Appendix (B.8)). The label L can then be dropped. A fermion family can be accommodated

Fig. 10.27 Dependence of the SU(n) coupling constants α_1, α_2 and α_3 on the energy scale.

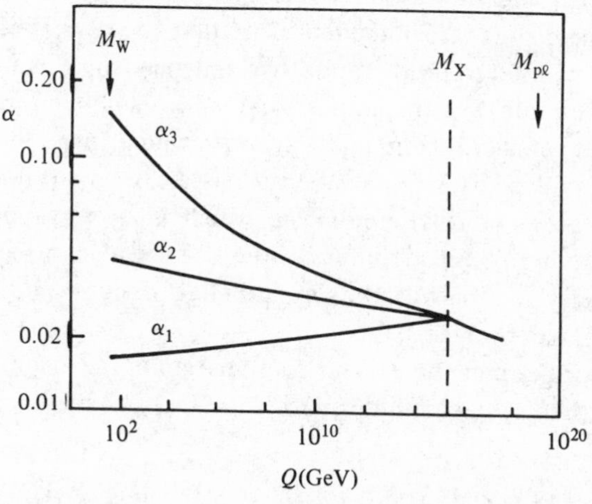

A group is *simple* (e.g. SU(n)) if it does not contain any non-trivial invariant subgroup. Direct products of simple groups, without any Abelian factors, are called *semi-simple*.

in the $\bar{5}$ and 10 representations of SU(5). Their SU(3) ⊗ SU(2) content is as follows (for each family)

$$\begin{aligned} \bar{5} &= (\bar{3}, 1) + (1, 2) = \bar{\mathrm{d}}^i + (\nu, \mathrm{e}^-), \\ 10 &= (\bar{3}, 1) + (3, 2) + (1, 1) = \bar{\mathrm{u}}^i + (\mathrm{u}^i, \mathrm{d}^i) + \mathrm{e}^+. \end{aligned} \tag{10.125}$$

Note that the requirement $\mathrm{tr}(Q_{\mathrm{ch}}) = 0$ for the $\bar{5}$, implies that $3Q_{\bar{\mathrm{d}}} + Q_\nu + Q_{\mathrm{e}^-} = 0$, so that $Q_{\bar{\mathrm{d}}} = -\frac{1}{3}Q_{\mathrm{e}^-}$. That is, the factor $\frac{1}{3}$ arises from there being three colours.

The gauge bosons belong to the 24 adjoint representation,

$$24 = \underbrace{(3, 2) + (\bar{3}, 2)}_{\begin{pmatrix} \mathrm{X} \\ \mathrm{Y} \end{pmatrix} \begin{matrix} Q_{\mathrm{X}} = \frac{4}{3} \\ Q_{\mathrm{Y}} = \frac{1}{3} \end{matrix}} + \underbrace{(8, 1)}_{\text{gluons}} + \underbrace{(1, 3) + (1, 1)}_{\gamma,\, \mathrm{Z}^0,\, \mathrm{W}^\pm}. \tag{10.126}$$

The combination of SU(5) representations above is free of axial anomalies, as required for renormalisation. There are twelve new gauge bosons, X and Y. These have charges $Q_{\mathrm{X}} = \pm\frac{4}{3}$ and $Q_{\mathrm{Y}} = \pm\frac{1}{3}$, and belong to a doublet of SU(2) and a triplet and antitriplet of SU(3).

The first stage of symmetry breaking, from SU(5) down to SU(3) ⊗ SU(2) ⊗ U(1), can be most simply accomplished by a 24-plet of Higgs bosons. Twelve get eaten up, leaving the X and Y superheavy. The remaining twelve physical Higgs are also superheavy. In the second stage it is required to break SU(3) ⊗ SU(2) ⊗ U(1) down to SU(3) ⊗ $\mathrm{U}(1)_{\mathrm{em}}$, these two latter symmetries remain unbroken, i.e. they are exact. A further set of Higgs is required, and these should also give the fermions their masses. This implies these Higgs are in the representation $\bar{5} \otimes 10 = 5 + 45$. If only the quintuplet is used, then there are two independent Yukawa couplings for each generation, and this implies that $m_{\mathrm{d}} = m_{\mathrm{e}}$, $m_{\mathrm{s}} = m_\mu$ and $m_{\mathrm{b}} = m_\tau$. This predicted equality is valid in the symmetry limit (mass scale $\mu \sim 10^{15}$ GeV). When evolved back to the laboratory scale, then one finds that $m_{\mathrm{d}}/m_{\mathrm{s}} = m_{\mathrm{e}}/m_\mu \simeq 1/200$ (a result for $m_{\mathrm{d}}/m_{\mathrm{s}}$ which is in very poor agreement with the result 1/20 (Section 8.4.2) from current algebra) and $m_{\mathrm{b}}/m_\tau \sim 3$ (i.e. reasonably good). Thus, the simplest Higgs assignment (i.e. minimal SU(5) model) has problems.

The charge generator Q_{ch} must be a linear combination of the diagonal generators in SU(5). Since Q_{ch} commutes with the $\mathrm{SU}(3)_{\mathrm{c}}$ elements, we have

$$Q_{\mathrm{ch}} = T_3 + \frac{y}{2} = T_3 + CT_0, \tag{10.127}$$

where T_3 and T_0 are the diagonal generators of SU(5) belonging to the

SU(2) and U(1) subgroups. The coefficient C relates the operators y and T_0, and has the value $C = -(\frac{5}{3})^{1/2}$. This can be shown by comparing the explicit form of the matrix λ^0 with the hypercharge assignments for the representations of (10.125) (see Cheng and Li, 1984).

At the unification scale (Q), all the coupling constants are equal. That is,

$$g_G(Q) = g_3(Q) = g_2(Q) = g_1(Q). \tag{10.128}$$

The coupling g_3 is that of $SU(3)_c$, i.e. $g_S = g_3$ with $\alpha_S = g_S^2/4\pi$. Similarly, $g_2 \equiv g$, the weak coupling constant. The coupling constant of the Abelian U(1) subgroup is given by

$$g'(Q) = g_1(Q)/C. \tag{10.129}$$

Having specified g and g' in terms of the SU(5) coupling constant, we find that

$$\sin^2\theta_W = g'^2/(g^2 + g'^2) = 1/(1 + C^2) = \tfrac{3}{8}. \tag{10.130}$$

This prediction is valid at the SU(5) scale, $Q \sim M_X$.

In order to relate (10.130) to experiment, and to estimate M_X, we need to consider the Q^2 evolution of the coupling constants; that is, to relate $g_n(M_X)$ to the values $g_n(Q)$, measured at $Q \sim M_W$. The subscript n means that the coupling is for the group SU(n). The evolution of g_n is given by (see (7.139))

$$\frac{dg_n}{d\ln Q} = -b_n g_n^3, \tag{10.131}$$

where, for $n_g(=n_f/2)$ generations,

$$b_1 = -n_g/12\pi^2, \qquad b_n = (11n - 4n_g)/48\pi^2 \qquad n \geqslant 2. \tag{10.132}$$

The expression for the QED coefficient b_1 follows from (7.141). The coefficient b_n is for SU(n), $n \geqslant 2$. For the case $n = 3$, noting that the number of quark flavours $n_f = 2n_g$, then expression (7.143) is obtained. Note that the sign of b_1 is opposite to that of b_2 and b_3.

The solution to (10.131) is

$$1/g_n^2(Q) = 1/g_n^2(Q_0) + 2b_n \ln(Q/Q_0). \tag{10.133}$$

Putting $\alpha_n(Q) = g_n^2(Q)/4\pi$, and taking $Q_0 = M_X$, so that from (10.128) $g_n(M_X) = g_G(M_X)$, (10.133) becomes

$$1/\alpha_n(Q) = 1/\alpha_G + 8\pi b_n \ln(Q/M_X), \tag{10.134}$$

where $\alpha_G = g_G^2(M_X)/4\pi$. Now, from (10.129), we have $g_1 = Cg' = Cg\tan\theta_W = Ce/\cos\theta_W$ and, hence $\alpha_1 = C^2\alpha/\cos^2\theta_W$ ($\alpha = \alpha_{QED}$). Also, since $g_2 = g =$

$e/\sin\theta_W$, we have $\alpha_2 = \alpha/\sin^2\theta_W$ and, further, $\alpha_3 = \alpha_S$. Hence, (10.134) can be written

$$\frac{\cos^2\theta_W}{C^2\alpha(Q)} = \frac{1}{\alpha_G} + 8\pi b_1 \ln(Q/M_X), \tag{10.135a}$$

$$\frac{\sin^2\theta_W}{\alpha(Q)} = \frac{1}{\alpha_G} + 8\pi b_2 \ln(Q/M_X), \tag{10.135b}$$

$$\frac{1}{\alpha_S(Q)} = \frac{1}{\alpha_G} + 8\pi b_3 \ln(Q/M_X), \tag{10.135c}$$

From equations (10.135), we can express $\sin^2\theta_W$ in terms of α and α_S by adding (10.135a) to two times (10.135c) and subtracting three times (10.135b), which gives

$$\frac{2}{\alpha_S} - \frac{3\sin^2\theta_W}{\alpha} + \frac{\cos^2\theta_W}{C^2\alpha} = 8\pi[2b_3 - 3b_2 + b_1]\ln\left(\frac{Q}{M_X}\right). \tag{10.136}$$

Now, from (10.132), we have $b_n - b_1 = 11n/48\pi^2$, so that the righthand side of (10.136) is zero. Thus, noting that $C^2 = \frac{5}{3}$,

$$\sin^2\theta_W = \frac{1}{1+3C^2}\left[1 + 2C^2\frac{\alpha(Q)}{\alpha_S(Q)}\right] = \frac{1}{6} + \frac{5}{9}\frac{\alpha(Q)}{\alpha_S(Q)}. \tag{10.137}$$

An expression for M_X is obtained by taking ($\frac{8}{3}$ times equation (10.135c) and subtracting equation (10.135b) and $\frac{5}{3}$ times equation (10.135a)) giving

$$\ln\left(\frac{M_X}{Q}\right) = \frac{\pi}{11}\left[\frac{1}{\alpha(Q)} - \frac{8}{3\alpha_S(Q)}\right]. \tag{10.138}$$

The evolution of $\sin^2\theta_W$, from the value $\sin^2\theta_W = \frac{3}{8}$ at $Q = M_X$ (10.130), can be found by eliminating α_S from (10.137) and (10.138), giving

$$\sin^2\theta_W = \tfrac{3}{8} - (55/24\pi)\alpha(Q)\ln(M_X/Q). \tag{10.139}$$

Taking the values of α (10.77) and α_S ($\Lambda \sim 0.15$ GeV, $n_f = 6$), at the scale $Q = M_W$, the following (very rough) numerical estimates are obtained

$$\begin{aligned} &\sin^2\theta_W \simeq 0.20,\\ &M_X \simeq 10^{15}\ \text{GeV},\\ &\alpha_G(M_X) \simeq 0.024. \end{aligned} \tag{10.140}$$

The energy dependence of the coupling constants α_1, α_2 and α_3 is shown in Fig. 10.27. A more sophisticated treatment (Marciano and Senjanović, 1982) gives

$$\begin{aligned}\sin^2 \hat{\theta}_{\mathrm{W}}(M_{\mathrm{W}}) &= 0.216 + 0.006 \ln(0.10\ \mathrm{GeV}/\Lambda_{\overline{\mathrm{MS}}}) \\ &= 0.214 \pm 0.004, \qquad \text{for } \Lambda_{\overline{\mathrm{MS}}} = 0.15^{+0.15}_{-0.08}\ \mathrm{GeV},\end{aligned} \tag{10.141}$$

in tolerable agreement with the experimental value $\sin^2 \hat{\theta}_{\mathrm{W}}(M_{\mathrm{W}}) = 0.229 \pm 0.005$.

Grand unified theories predict transitions between quarks and leptons, and thus can imply that the proton (and bound neutron) will decay, and (if $B-L$ is not a symmetry of the model) that nucleon–antinucleon oscillations should occur. These transitions are mediated by the gauge bosons $\mathrm{X}^i_{4/3}$, $\bar{\mathrm{X}}^i_{-4/3}$, $\mathrm{Y}^i_{1/3}$, $\bar{\mathrm{Y}}^i_{-1/3}$, where the subscript is the charge and i is the colour index, giving 12 bosons in total. The basic vertices coupling the X and Y bosons to quarks and leptons are

$$\begin{aligned}&\mathrm{X} \leftrightarrow \mathrm{e}^+\bar{\mathrm{d}},\ \mathrm{u}+\mathrm{u}, \\ &\mathrm{Y} \leftrightarrow \mathrm{e}^+\bar{\mathrm{u}},\ \bar{\nu}\bar{\mathrm{d}},\ \mathrm{u}+\mathrm{d}.\end{aligned} \tag{10.142}$$

Thus, X and Y have both lepto-quark and diquark couplings. Some diagrams contributing to proton decay are shown in Fig. 10.28. Note that, whereas B and L are separately violated, $B-L$ is conserved. In addition to these s-channel diagrams, there are also t-channel processes. The vertex coupling is proportional to $g_G \mathrm{X}^\mu_i(\bar{\mathrm{d}}^i\gamma_\mu \mathrm{e}^+)$, etc., so that we obtain an effective four-fermion contact interaction (i.e. as for muon decay), with a matrix element $\mathscr{M} \sim g_G^2/M_{\mathrm{X}}^2$. A crude estimate of the proton lifetime is thus

$$\tau_{\mathrm{p}} \sim M_{\mathrm{X}}^4/(\alpha_G^2(M_{\mathrm{X}})m_{\mathrm{p}}^5) \sim 10^{32}\ \mathrm{yr}. \tag{10.143}$$

The decay modes indicated by Fig. 10.28 are $\mathrm{p} \to \mathrm{e}^+\pi^0(\eta^0, \rho^0, \omega^0)$ and $\mathrm{p} \to \bar{\nu}_{\mathrm{e}}\pi^+(\rho^+)$. The heavier states will be suppressed by phase space.

The X and Y bosons can also couple to second generation fermions (e.g. $\mathrm{X} \to \mu^+\bar{\mathrm{s}}$, $\mathrm{Y} \to \bar{\nu}_\mu\bar{\mathrm{s}}$). Note that $\mathrm{Y} \to \mu^+\bar{\mathrm{c}}$ is kinematically forbidden. From this, and the smaller phase space available, it is expected that $\mathrm{p} \to \mu^+\mathrm{K}^0$ and $\mathrm{p} \to \bar{\nu}_\mu\mathrm{K}^+$ are much less probable than the first generation modes. Similar decay schemes for the neutron can also be written down. Note that the decay $\mathrm{n} \to \mathrm{e}^-\pi^+$ is not allowed. This decay would violate the conservation of $B-L$ (this can be seen from (10.142)). The estimate (10.143) for the lifetime is extremely crude; however, it illustrates the sensitivity to the input parameters, in particular α_{S}. A detailed theoretical review is given by Langacker (1981), and the experimental situation is reviewed by Totsuka (1985). The status can be summarised by

$$\begin{aligned}&\text{Minimal SU(5)} \quad \tau_{\mathrm{e}^+\pi^0} = 6.6 \times 10^{28\pm0.9}(\Lambda_{\overline{\mathrm{MS}}}/0.1\ \mathrm{GeV})^4\ \mathrm{yr}. \\ &\text{Experiment} \quad \tau_{\mathrm{e}^+\pi^0} > 3.3 \times 10^{32}\ \mathrm{yr}\ (90\%\ \text{c.l.}).\end{aligned} \tag{10.144}$$

Thus, barring a very large value of $\Lambda_{\overline{MS}}$, the minimal SU(5) model can be essentially ruled out by the observed limit. A value for M_X significantly larger than the SU(5) prediction is required. Experiments sensitive to more decay modes are currently underway.

If SU(5) is excluded, then a possible rank 5 group for Grand Unification is SO(10). This is the group of orthogonal rotations in a 10-dimensional space, with the determinant of the transformation matrices equal to unity. This group contains SU(5) as a subgroup, and has a 16-dimensional representation. The additional member is most naturally taken to be ν_R. Non-zero neutrino masses can be accommodated in SO(10). Indeed, if the large neutrino mass found in tritium decay were confirmed, then this could not be accommodated in minimal SU(5).

There are many other interesting phenomena associated with GUTs, in particular in their possible role in the early stages of the universe. These include the apparent baryon–antibaryon asymmetry ($(n_B - \bar{n}_B)/n_\gamma \sim 10^{-9}$) in the universe and the question of the existence of heavy magnetic monopoles, which appear in GUTs[#] (see, for example, Langacker, 1981). However, there are also theoretical difficulties. The symmetry breaking occurs in (at least) two stages. In the first stage, in which the GUT group G is broken down to $G_1(\equiv SU(3)\otimes SU(2)\otimes SU(1))$, the Higgs have vacuum expectation values of $v_1 \sim 10^{14}$ GeV, and in the second (which is required to reproduce the observed $G_0 \equiv SU(3)_c \otimes U(1)_{em}$ symmetry), the scale is $v_2 \sim 250$ GeV. These two scales are vastly different ($v_2^2/v_1^2 \sim 10^{-24}$) and, theoretically, this causes problems. Any particle 'feeling' the mechanism triggering the $G \to G_1$ breaking will get a superheavy mass and, hence, this must be avoided for the known low energy particles. For example, whereas the self-mass of a fermion diverges logarithmically ($\delta m_f = O(\alpha/\pi)m_f \ln \Lambda$ – see (4.161)), the divergence for a scalar particle is quadratic. The Higgs particles ϕ, which give the X and Y bosons their masses, contribute to the radiative corrections of the Higgs particles H (i.e. those needed to

Fig. 10.28 Diagrams with X and Y bosons leading to proton decay.

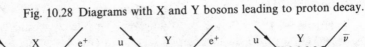

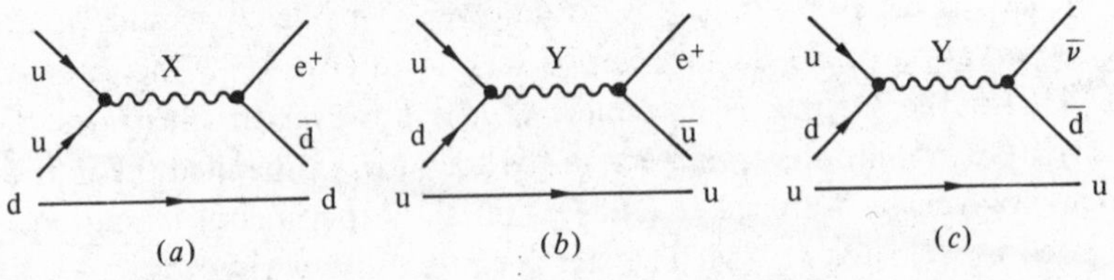

[#] Monopoles appear in GUTs whenever a simple gauge group G (e.g. SU(5)) is spontaneously broken to a group with an Abelian U(1) factor (e.g. $SU(3)_c \otimes U(1)_{em}$), and have masses $\sim M_X/\alpha \sim 10^{16}$ GeV.

give the electroweak vector bosons their masses). These corrections have a 'natural' scale of order v_1, giving typical mass corrections (see Fig. 10.29) $\delta m^2 \sim g^2\Lambda^2 = g^2 v_1^2$. Unlike the case of the fermion mass, where the unbroken SU(2) chiral symmetry ensures the corrections remain small in the massless limit, there is no such additional symmetry for scalar particles. Thus, if loop corrections change the tree-level mass m_0 to m, such that $m^2 = m_0^2 + \delta m^2$, then a small value of m^2/Λ^2 can be maintained only if there is an extremely precise cancellation between m_0^2/Λ^2 and $\delta m^2/\Lambda^2$. Hence, divergences can only be avoided by *fine tuning* the parameters (to 24 places of decimals) so that cancellations occur, and this must be repeated in each order of perturbation. The presence of two such widely different scales in the theory is known as the *hierarchy problem.* As discussed below, this problem can be avoided either by devising a theory in which there are no point-like couplings to fundamental scalars (composite models such as technicolour) or by cancelling the divergence directly with additional particles (as is the case for supersymmetry). In terms of experimental predictions, it should be noted that in the simplest GUTs there is a 'desert' in the particle spectrum between the mass scales $O(M_W)$ and $O(M_X)$.

10.8 COMPOSITENESS

Historically, many of the particles which were initially thought to be fundamental, have revealed substructure when probed at larger energy scales, and this has been central to our understanding of matter. It is natural, therefore, to explore the possibility that some, or all, of the 'elementary' particles in the standard model may be composite.

Since it is the presence in the theory of a fundamental scalar particle, and the associated quadratically divergent self-mass terms, which lead to the hierarchy problem, it is worth reviewing the reasons why the Higgs particles are necessary. These are to provide a spontaneous symmetry-breaking mechanism to give masses to the W and Z bosons and the fermions, while keeping the theory renormalisable. Further the Higgs sector must be compatible with neutral current results (i.e. $\rho = 1$).

Fig. 10.29 Loop corrections to the self-mass of the Higgs particle.

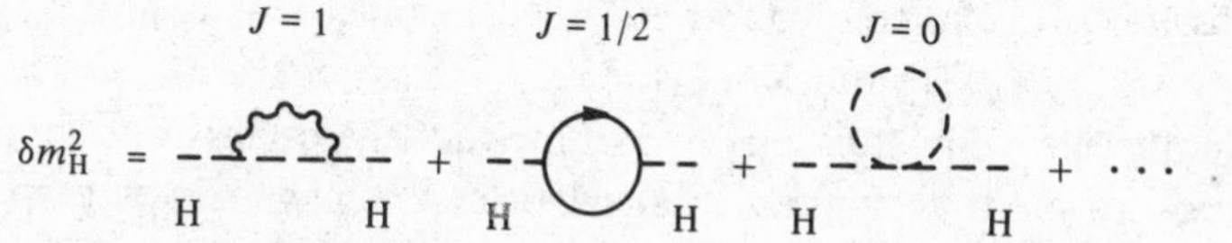

$$\simeq g^2 \int^{\Lambda} \frac{d^4k}{k^2} \sim O\left(\frac{\alpha}{\pi}\right)\Lambda^2$$

A possible remedy to the hierarchy problem is to postulate a theory without a fundamental scalar. The QCD Lagrangian for massless quarks, as discussed above, has chiral symmetry. It is possible to imagine a theory in which this symmetry is spontaneously broken, with three associated massless Goldstone bosons ($\pi^{\pm}$, π^0). In this case, the electroweak $SU(2) \otimes U(1)$ symmetry is spontaneously broken, with the π being eaten to give the three gauge bosons $W^{\pm}$, Z^0. In such a model the remaining isospin symmetry of the u and d quarks ($SU(2)_{L+R}$) ensures that the relation $M_W = M_Z \cos\theta_W$ is satisfied. However, the magnitude of the gauge boson masses is related to the vacuum expectation value of the $SU(2) \otimes SU(2)$ breaking, and in this model is related to the pion decay constant f_π. Hence, the predicted $W^{\pm}$ and Z^0 masses are much too small and, furthermore, we know that the pion exists.

In *technicolour* (or *hypercolour*) models, a further QCD-like theory is postulated. This theory has a gauge symmetry $SU(N)$ (e.g. $N = 3$), but the coupling constant is much larger than in QCD. There are N identical families of techniquarks, $(U, D)_L$, U_R, D_R, with the same electroweak properties as ordinary quarks. For example, a spectrum of technihadrons is therefore expected. The technipions (i.e. $\bar{U}\gamma^5 D$, etc.) are the approximate Goldstone bosons of chiral symmetry. However, in order to give the correct W mass, a technipion decay constant $F_\pi \sim 250$ GeV is needed, corresponding to $\Lambda_{TC} \sim 1$ TeV (compared to $\Lambda_{QCD} \sim 0.2$ GeV). The relation $M_W = M_Z \cos\theta_W$ ($\rho = 1$) is automatically reproduced.

The problems with this model is in giving masses to the fermions. The $\bar{f}f H$ couplings in the standard model are replaced by $\bar{f}f\bar{F}F$ couplings, since H is a bound state of techniquarks F. However, a four-fermion interaction leads to renormalisation difficulties. A possible solution to this problem is to extend the gauge group to give very heavy vector bosons (*extended technicolour*). However, this leads to an unpleasant proliferation of mass scales and unobserved pseudo-Goldstone bosons, some of which should be reasonably light. Furthermore, the theory leads to (unobserved) flavour-changing neutral currents at a rate greater than that observed.

Possible experimental signatures for technicolour would be the production of technipions in $W \to p^+ p^0$ or $t \to bp^+$, with the technipions (p) producing heavy leptons and quarks ($p^- \to \nu_\tau \tau^-$, $\bar{c}b$, $\bar{c}s$, etc.). A detailed discussion of technicolour and extended technicolour is given by Farhi and Susskind (1981) and Ross (1985).

An alternative postulate is to assume that the quarks and leptons are composites of more fundamental particles (*preons*). Since, at presently available energies, the fermions appear to be point-like, the scale of compositeness Λ_c must be large. Estimates, based on possible corrections to the anomalous magnetic moment $g-2$ for the muon, give $\Lambda_c \gtrsim 1$ TeV. If the muon and electron have common constituents (and solving the generation problem is clearly one of the goals of such models), then the absence of $\mu \to e\gamma$, etc., gives a more stringent limit of $\Lambda_c \gtrsim 100$ TeV.

A further idea is that the gauge bosons W and Z are composite, in a way analogous to the ρ and ω vector meson states being made up of quarks. This eliminates the need for a fundamental weak isospin gauge symmetry, and hence the need to break it. However, difficulties arise in obtaining the correct value of $\sin^2\theta_W$ and the observed $W^\pm$ and Z^0 masses, as well as in the spectrum of excited composite states.

10.9 SUSY

In Grand Unified Theories there is no relationship between particles of different spin. In *supersymmetry* (SUSY), bosons can be transformed into fermions and *vice versa*,

$$Q|F\rangle = |B\rangle, \qquad Q|B\rangle = |F\rangle. \tag{10.145}$$

The generators Q induce spin $\frac{1}{2}$ changes and, in the simplest ($N=1$) supersymmetry, Q is a spinorial charge. This charge is conserved, and Q satisfies the anticommutation relation

$$\{Q_\alpha, \bar{Q}_\beta\} = -2\gamma^\mu_{\alpha\beta}P_\mu, \tag{10.146}$$

where P_μ is the energy-momentum four-vector, and this generates space-time translations. It can be shown that SUSY is the only way of combining internal symmetry with Lorentz invariance. Consideration of invariance under space-time transformations, and of making the SUSY transformations local, takes us into the realm of gravity. Local SUSY can be accommodated in an extended version of gravity, called *supergravity*. The supersymmetry generators can be extended to N mutually anticommuting spinorial charges Q^i_α, $i=1$ to N. Each spin operation of the type (10.145) corresponds to a change in helicity of $\frac{1}{2}$. Hence, the inclusion of helicity states $|h| \leqslant 2$ (graviton has spin 2) means that $N \leqslant 8$. Such theories are called *N-extended supergravity*, and sow the seeds for the possible unification of all interactions.

In the following, only the simplest ($N=1$) supersymmetry is considered. Each particle belongs to a super-multiplet containing an equal number of fermionic and bosonic degrees of freedom. In addition to a multiplet containing the graviton, which also includes its spin $\frac{3}{2}$ supersymmetric partner, the *gravitino*, there are *gauge* (containing spin 1 and $\frac{1}{2}$) and *chiral* (spin $\frac{1}{2}$ and spin 0) *multiplets*. A list of the ordinary particles, and their supersymmetric partners (denoted by tilde), is given in Table 10.3.

Table 10.3. *Particles and their superpartners in the standard model*

Particle		Spin	Super-particle		Spin
quark	q	$\frac{1}{2}$	scalar-quark	$\tilde{q}_L, \tilde{q}_R$	0
lepton	l	$\frac{1}{2}$	scalar-lepton	$\tilde{l}_L, \tilde{l}_R$	0
neutrino	ν_l	$\frac{1}{2}$	scalar-neutrino	$\tilde{\nu}_l$	0
gluon	g	1	gluino	$\tilde{g}$	$\frac{1}{2}$
photon	γ	1	photino	$\tilde{\gamma}$	$\frac{1}{2}$
Z	Z^0	1	Z-ino	$\tilde{Z}^0$	$\frac{1}{2}$
W	$W^{\pm}$	1	W-ino	$\tilde{W}^{\pm}$	$\frac{1}{2}$
Higgs	H	0	Higgsino	$\tilde{H}$	$\frac{1}{2}$

The particles and their superpartners have the same quantum numbers (e.g. charge, colour), except for spin. In particular, all the couplings of these supersymmetry (or s)-particles are known. If the supersymmetry is unbroken, they are degenerate in mass with the ordinary particles. This follows from the property $[Q_\alpha, p^\mu] = 0$, which means that E, $\mathbf{p}$ and, hence, m are unchanged by the transformation. Note that both the left and righthanded leptons and quarks have spin 0 superpartners ($\tilde{q}_L$, $\tilde{q}_R$, etc.); the helicity labels are those of the spin $\frac{1}{2}$ particles. There are no transformations between left and righthanded fermions, so the weak V − A structure is preserved. The mass eigenstates can be mixtures of ($\tilde{q}_L$, $\tilde{q}_R$), ($\tilde{l}_L, \tilde{l}_R$), ($\tilde{\gamma}$, $\tilde{Z}^0$, $\tilde{H}^0$) or ($\tilde{W}^{\pm}, \tilde{H}^{\pm}$). Because of the restrictions of supersymmetry on the couplings, two Higgs doublets are required in order to give masses to all the charged fermions. Spontaneously broken symmetry should also give rise to *Goldstino* particles ($\tilde{G}$), which can be absorbed to give the gravitino a mass. The *gaugino* ($\tilde{W}^{\pm}$, $\tilde{Z}$) properties resemble those of heavy leptons, but with weak isospin $I = 1$. The *gluino* ($\tilde{g}$) properties are quark-like, but with colour **8** rather than **3**. One of the general hopes of supersymmetric theories is that the somewhat artificial separation into matter fields (fermions) and force fields (gauge bosons) is obviated. However, for $N = 1$, fermions and bosons cannot be assigned to the same supermultiplet. None of the supersymmetric particles can be identified with the known particles (e.g. there is no known elementary scalar particle), so a large proliferation of fundamental states ensues.

This large number of unobserved states does, however, lead to one very desirable property of supersymmetry theories; that is, a possible solution to the hierarchy problem. In the evaluation of the self-mass of the Higgs particle, diagrams containing loops (which lead to potentially quadratically divergent corrections) will have contributions in SUSY from both the particle and its superpartner. From Fermi statistics, the fermion term is opposite in sign to that for the boson, and would cancel it in the limit of degenerate masses. This implies that the supersymmetric particles cannot be too heavy, or else this desired cancellation will not occur. Thus, it is required, in order to avoid the hierarchy problem, that for some relevant coupling constant α_i,

$$\delta m_{\mathrm{H}}^2 \propto |m_{\mathrm{B}}^2 - m_{\mathrm{F}}^2| \lesssim 0\left(\frac{1\ \mathrm{TeV}}{\alpha_i}\right)^2. \tag{10.147}$$

The non-observance of supersymmetric particles gives some lower limits on their masses. For example, typical limits quoted for scalar leptons are greater than or equal to 20 GeV and for scalar quarks greater than or equal to a few GeV; however, such limits depend on further assumptions about potential decay modes, etc.

The forms of the couplings usually assumed, mean that s-particles are produced in pairs. Particles can be assigned a multiplicative quantum number (R-parity) $R = 1$, with their s-partners having $R = -1$. A consequence of this, is that the lightest s-particle should be stable; assuming it to be neutral (cosmological evidence supports this). Candidates are the $\tilde{\nu}$, $\tilde{\gamma}$, $\tilde{\mathrm{H}}$ and $\tilde{\mathrm{G}}$, with the $\tilde{\gamma}$ common in many models. Such a stable particle is expected to have a very small cross-section in matter, and its experimental signature is thus that of missing energy (i.e. neutrino-like). For example, in Fig. 10.30 some possible diagrams for the interaction of a photino with (*a*) an electron and (*b*) a quark are shown. In each case there is a massive internal scalar electron or quark propagator, and the resulting cross-section is roughly that of the usual weak interaction.

Supersymmetric particles can be produced, for example, by the reactions $\mathrm{e}^+\mathrm{e}^- \rightarrow \gamma/\mathrm{Z} \rightarrow \tilde{l}^+\tilde{l}^-$, $\tilde{\mathrm{q}}\mathrm{q}$, etc. The angular distribution and threshold behaviour for the production of s-fermions are those appropriate for scalar particles. The potential decay modes depend on the relative masses of particles. Possible decay schemes for the s-lepton are $\tilde{l} \rightarrow l + \tilde{\gamma}$ and

Fig. 10.30 Possible interactions of supersymmetric particles (*a*) $\tilde{\gamma} + \mathrm{e} \rightarrow \tilde{\gamma} + \mathrm{e}$ and (*b*) $\tilde{\gamma} + \mathrm{q} \rightarrow \tilde{\mathrm{g}} + \mathrm{q}$.

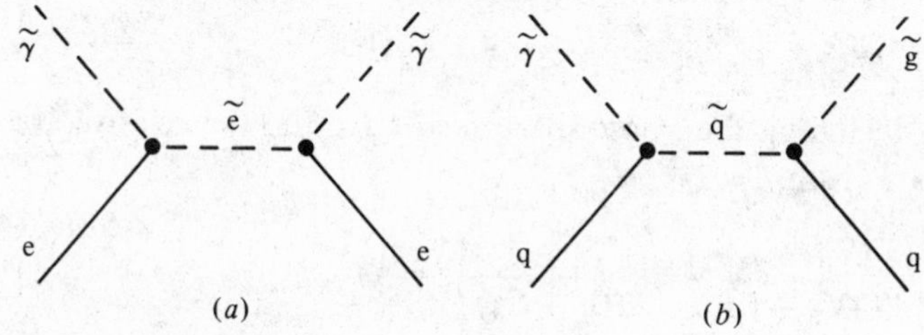

$\tilde{l} \to l + \tilde{G}$, for the s-quark $\tilde{q} \to q + \tilde{g}(g \to q\bar{q}\tilde{\gamma})$ and for the s-neutrino $\tilde{\nu} \to \nu + \tilde{\gamma}$. The expected lifetimes are short; maybe too short for a detection in vertex detectors.

As a very simple example, we consider the decay of a scalar lepton $\tilde{l}_L^-(p) \to l^-(p_1) + \tilde{\gamma}(p_2)$. The Feynman rule for this vertex (Haber and Kane, 1985) gives a factor, $ie(1+\gamma^5)/2^{1/2}$ in the matrix element. Otherwise we proceed as before, giving, after a short calculation,

$$\Gamma(\tilde{l}^\pm \to l^\pm \tilde{\gamma}) = \alpha(\tilde{M}_{\tilde{l}}^2 - m_l^2 - \tilde{M}_\gamma^2) p_{\tilde{l}}^* / \tilde{M}_{\tilde{l}}^2, \tag{10.148}$$

where $\tilde{M}_l$, m_l and $\tilde{M}_\gamma$ are the masses of the s-lepton, lepton and photino respectively. It is assumed in this calculation that the photino is a mass eigenstate. A comprehensive summary of the Feynman rules and calculation techniques, for a wide range of SUSY processes, can be found in Haber and Kane (1985).

If one (or more) of the s-fermions is light enough, then the decays of the $W^\pm$ and Z^0 are sources for their detection. For example, possible decay modes are $W \to \tilde{e}\tilde{\nu}$ with $\tilde{e} \to e\tilde{\gamma}$, $Z^0 \to \tilde{e}^+\tilde{e}^-$, $\tilde{\gamma}\tilde{\gamma}$, $\tilde{\nu}\bar{\tilde{\nu}}$, etc. Decays to gauginos (e.g. $W^\pm \to \tilde{W}^\pm\tilde{\gamma}$) may also be possible. Some of these channels would give spectacular experimental signatures. For example, the sequence $Z^0 \to \tilde{\nu}_e\bar{\tilde{\nu}}_e$ with $\tilde{\nu}_e \to \nu\tilde{\gamma}$, $\bar{\tilde{\nu}}_e \to \nu_\mu e^- \mu^+ \tilde{\gamma}$ would produce a final state with a μ^+ and e^- in one hemisphere and nothing detected in the other.

The branching ratios for $V_B(p) \to S_1(p_1) + S_2(p_2)$, compared to $V_B(p) \to f_1(p_1) + \bar{f}_2(p_2)$, where V_B is a heavy vector boson ($W^\pm$ or Z), S_1 and S_2 are s-fermions and f_1, f_2 are ordinary fermions, can easily be calculated by using the vertex factors

$$\begin{aligned} V_B S_1 S_2&: \quad -iC_S(p_1 - p_2)_\mu, \\ V_B f_1 \bar{f}_2&: \quad \gamma_\mu\left[C_L\frac{(1-\gamma^5)}{2} + C_R\frac{(1+\gamma^5)}{2}\right], \end{aligned} \tag{10.149}$$

where C_S represents the appropriate SUSY coupling constant. ($V_B S_1 S_2$ represents the three particles (fields) interacting at the vertex and the term on the righthand side is the vertex factor.) This gives, summing over the spin states of V_B (using (10.4) and of f_1 $\bar{f}_2$ (assuming f_1, f_2 to be massless)

$$\begin{aligned} |\mathcal{M}(V_B \to S_1 S_2)|^2 &= 4C_S^2 p_S^{*2}, \\ |\mathcal{M}(V_B \to f_1 \bar{f}_2)|^2 &= 2M_{V_B}^2(C_L^2 + C_R^2). \end{aligned} \tag{10.150}$$

Hence, including the phase-space factor ($\propto p^*$), the relative branching ratios are (with $\beta = 2p^*/M_{V_B}$)

$$\frac{\Gamma(V_B \to S_1 S_2)}{\Gamma(V_B \to f_1 \bar{f}_2)} = \frac{C_S^2}{2(C_L^2 + C_R^2)}\beta^3 = \tfrac{1}{2}\beta^3. \tag{10.151}$$

The latter simplification follows from the supersymmetry principle, that particles have identical couplings to their s-partners.

Equation (10.151) indicates that sizeable SUSY branching ratios can occur, provided the s-particles are resonably light. The method described in Section 10.3.2, to determine the possible number of light neutrino species, is extended in SUSY to include all light non-interacting particles. The most probable candidate for a light s-particle is the $\tilde{\gamma}$, and cosmological estimates suggest its mass is less than or equal to 100 eV. The $\tilde{\gamma}$ is one of the candidates to explain the *dark matter* in the universe; that is, matter which is indicated from gravitational properties but which is not 'visible' otherwise.

Nucleon decay is another potential testing ground for SUSY. In the standard model Lagrangian, terms of dimension $d \leqslant 4$ which satisfy $SU(3) \otimes SU(2) \otimes U(1)$, conserve both B and L separately. Terms with $d \geqslant 6$ are needed for violation of baryon number, and dimensional analysis shows these contribute as $(1/M_X^2)^{d-4}$, where M_X is the scale at which these terms become relevant (i.e. the mass of the exchanged boson or the unification scale). In SUSY, terms of dimension 4 (e.g. $\varepsilon_{ijk}\tilde{u}_R^i d_R^j s_R^k$, with i, j, k being colour indices) can lead to baryon number violation. These give a decay rate which is much too fast, but can be 'forbidden' by the requirement that $\mathscr{L}$ is invariant under some discrete transformation such as R-parity. If R-parity conservation is valid, supersymmetric particles must be produced in pairs. Note that the K-meson decays $K_L \to \mu e$, ee and $\mu\mu$ are possible in some models in which R-parity conservation is broken. Due to the presence of additional particles, which contribute to loops in the evolution of the coupling constants, the unification scale at which the couplings become equal is pushed to higher values with SUSY (e.g. $M_X \sim 10^{16}$–10^{17} GeV). The value of $\sin^2\theta_W$ is somewhat increased, in better agreement with the data.

Operators of dimension 6 thus give a lifetime much greater than the experimental limit. However, there is also a possible contribution from operators of dimension 5 (e.g. $\varepsilon_{ijk}\varepsilon_{\alpha\beta}\varepsilon_{\gamma\delta}\tilde{q}_\alpha^i \tilde{q}_\beta^j q_\gamma^k L_\delta$, where ijk and $\alpha, \beta, \gamma, \delta$ are SU(3) and SU(2) indices respectively, i.e. $q_\alpha = (u, d)$, $L_\delta = (\nu, l)$). Such terms have a heavy fermion rather than a heavy boson propagator, and contribute as M_X^{-2}. Hence they give potentially too fast a decay rate for the proton. These terms, however, appear in diagrams with loops and involving Yukawa couplings (e.g. Fig. 10.31), and the rate is, in fact, suppressed to roughly that of the standard model. The decay modes expected, however, are significantly different. The antisymmetric properties of the couplings lead to vanishing contributions if all the particles participating are in the same generation. Hence, generation-changing decays ($K\bar{\nu}_\mu$, $K\bar{\nu}_\tau$, $K\mu^+$, etc.) are favoured in this model. Specific SUSY models are not in conflict with the data, but are rather flexible. A detailed account of SUSY, and its associated phenomenology, can be found in Nilles (1984), Ross (1985), Haber and Kane (1985) and Brennan (1985). It should be recalled, however, that in spite of the many attractive theoretical properties, SUSY is not yet supported by any reliable direct evidence.

10.10 SUPERSTRINGS

A complete theory of nature must include all the forces, including gravity.* This implies that such a unified theory must work at distance scales which are smaller than $\Delta x \simeq G_N^{1/2} \simeq 10^{-33}$ cm (G_N is the Newtonian gravitational constant), which is where quantum gravity fluctuations become large. A further desirable property of such a universal theory is that it should be free of the infinities which plague the usual field theories (in which particles are treated as point-like entities). The supersymmetric quantum theory of strings, or *superstring theory*, has recently emerged as a candidate for solving all of these problems. (A detailed account of the theory is given by Green, Schwarz and Witten (1987).)

The ideas of massless relativistic strings in particle theory is not new. In the 1960s, studies of the particle spectra and elastic scattering amplitude properties (in particular their poles), lead to the idea of Regge trajectories, relating the spin J and mass M of particles by $J = \alpha_0 + \alpha' M^2$. The connection with string theory is that a classical relativistic string, with zero mass density, gives the relation $J = \alpha' M^2$, for a string of tension $T = (2\pi\alpha')^{-1}$. Further, the amplitudes developed by Veneziano (see, e.g., Collins and Martin, 1984) to describe meson–meson scattering, and which are 'dual' with s and t-channel Regge exchange, could also be interpreted as the scattering of string-like objects having a one-dimensional extension.

Fig. 10.31 Possible dominant mechanism for proton decay in SUSY. The Higgsino is super-heavy.

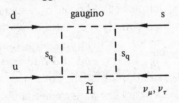

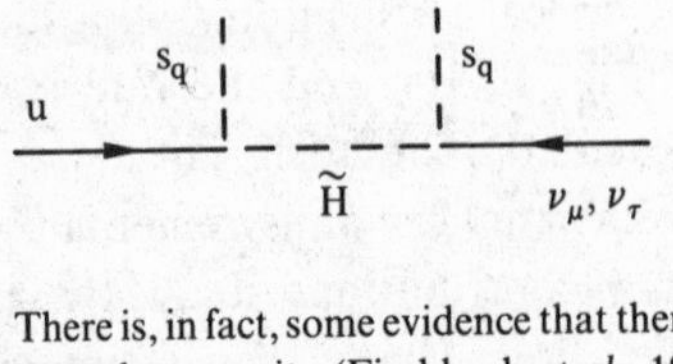

* There is, in fact, some evidence that there may be a fifth force in nature, weaker even than gravity (Fischbach *et al.*, 1986; see Peccei, 1986 for a review).

A meson, in this context, can be thought of as a quark and antiquark on the ends of a connecting string. The vibrations of a classical string can be Fourier-analysed as a sum of harmonic oscillations, and hence quantised. This gives an infinite set of states on a Regge trajectory, with $\Delta M^2 = 2\pi T$. Relativistic strings (or flux tubes) also give a reasonable description of certain long distance ($\gtrsim 1$ fm) QCD phenomena.

The various gauge theories so far discussed here have all been formulated to describe phenomena in Minkowski four-space, which consists of three-space and one-time dimensions. Einstein's classical theory of general relativity is based on the postulate that the laws of physics are the same to all observers (and not just those in uniform relative motion). This leads to curved space-time with an element $\mathrm{d}s^2 = g_{\mu\nu}(x)\,\mathrm{d}x^\mu\,\mathrm{d}x^\nu$; that is, the metric (which is a field) depends on the space-time coordinate x.

Around 1920, Kaluza and (later) Klein investigated an extension of these ideas to five dimensions, leading to a metric $\hat{g}_{\alpha\beta}(z)$, with $\alpha, \beta = 0, 1, 2, 3, 4$, and $z = (x^\mu, y)$, where y is the fifth dimension. The unobserved fifth coordinate is taken to have a different topology to the other four, and satisfies $0 \leqslant my \leqslant 2\pi$ (i.e. periodic), and is thus '*compactified*' into a circle of radius $1/m$. An analogy of this idea in ordinary space is a long thin tube, which at each point has a circular topology but, when viewed from a distance, appears to be a one-dimensional line. The dependence of the metric $\hat{g}_{\alpha\beta}$ on the coordinates can be separated by the expansion

$$\hat{g}_{\alpha\beta}(x, y) = \sum_n g^n_{\alpha\beta}(x)\exp(\mathrm{i}nmy). \tag{10.152}$$

The aim of the work of Kaluza and Klein was to unify gravity with electromagnetism. For $n = 0$, they showed that $\hat{g}_{\alpha\beta}(z)$ could be identified with the spin-2 graviton $g_{\mu\nu}(x)$, the spin-1 photon $\hat{g}_{\mu 4}(x) = A_\mu(x)$ and a spin-0 field $\hat{g}_{44}(x) = \phi(x)$. These extra fields satisfy the Maxwell and Klein–Gordan equations respectively. Thus, through this extension of general relativity, gravity and electromagnetism are related. Furthermore, invariance under general coordinate transformations (i.e. energy–momentum conservation) leads to gauge invariance of A_μ, and hence charge conservation. There are an infinite number of $n \neq 0$ modes, and these can be given a four-dimensional interpretation as particles which are massive ($m_\mathrm{n} = nm$) and charged ($e_\mathrm{n} = n\kappa m$), where $\kappa = G_\mathrm{N}^{1/2}$. Thus, the charges are quantised in units of a fundamental charge $e = \kappa m$ and, furthermore, e is related to $G_\mathrm{N}^{1/2}$. Numerically this implies that $m \sim 10^{19}$ GeV, so that the fifth dimension is curled up with a size $m^{-1} \sim 10^{-33}$ cm. These conclusions, coming from this five-dimensional Kaluza–Klein theory, survive in somewhat more realistic models, where compactification from higher dimensions down to four occurs spontaneously from the form of the underlying field equations. The internal symmetries (charge, colour, etc.) in our apparently four-dimensional world, are equated with coordinate invariance in the extra dimensions.

In superstring theories, all particles (quarks, leptons, gauge bosons, gravitons, plus their supersymmetric partners) are built out of the same fundamental entities, namely elementary strings. A relativistic string is an idealised one-dimensional extended object, that lies along some curve in space and sweeps out a two-dimensional space-time surface (world-sheet), as opposed to the world line swept out by a point particle.# The theory deals with the dynamics of such a string. The vibrational modes can be expressed as harmonic oscillations (by Fourier transformation), and then quantised. The theory naturally contains the gauge invariances of Yang–Mills theory and gravity. Investigations of the quantum theory of bosonic strings have been shown to be consistent only for a 'critical' number $d = 26$ of space-time dimensions. The superstring theory requires that space-time be fundamentally supersymmetric. The requirement that both bosonic and fermionic modes can be accommodated results in a critical dimension $d = 10$ (9 space and one time). The superstrings have spatial dimensions of the order of the Planck scale. The corresponding string tension is given by $T^{1/2} \sim m_{pl} \sim 10^{19}$ GeV. The spinors representing fermions in $d = 10$ can be both 'Weyl' (i.e. single-handed) and 'Majorana' (particle equivalent to antiparticle, see Appendix B). For energy scales $\ll m_{pl}$, the higher mass states in the theory decouple, leaving an effective low energy theory, with fields of spin 2 (graviton), $\frac{3}{2}$ (gravitino), 1 (gauge boson), $\frac{1}{2}$ (matter fields) and 0 (Higgs).

A major hurdle in the evolution of superstring theories was the problem of anomalies; that is, a breakdown of a symmetry of the Lagrangian through quantum (loop) corrections. This problem was overcome by Green and Schwarz (1984), whose work started a theoretical 'revolution' on a par with the experimental one ten years earlier. They showed that, for a $d = 10$ superstring theory with Yang–Mills interactions, the anomalies cancelled, provided the gauge group was SO(32), the 32-dimensional rotational group, or $E_8 \times E_8$, the product of the largest (rank 8) exceptional Lie group. These possibilities arise *geometrically*, as solutions of the space-time structures allowed by the theory. Further, the theory appears to be *finite*; this has so far been shown explicitly to one-loop level.

The parameters of the theory are the gravitational constant κ, the Yang–Mills coupling g and the string tension T. These constants are

A further possibility is that the fundamental entity is a 'membrane'. This would be the maximum dimension compatible with SUSY.

related. There are three consistent superstring theories known (or, more precisely, six if variants of these types are counted). In *type I* theories, there are *open strings* (as required for the Yang–Mills sector) and these can join at their ends and, hence, form a *closed string* (needed for the gravity sector). The constants are related by $\kappa \propto g^2 T$. In *type II* theories the strings are closed. In the theoretically preferred *heterotic* (i.e. 'crossbreed') string theories, which describe closed strings which carry internal symmetry, the relation is $\kappa \propto gT^{1/2}$. Type II and heterotic strings have an intrinsic orientation sense, whereas type I strings are unoriented. Each of these three theories consistently incorporates quantum gravity and is free of adjustable parameters or other arbitrariness. It is hoped that one amongst this choice of three theories can be shown to be unique. For each of the possible theories the solutions must, of course, be found in order to determine the spectrum of low-lying states and compare these with experiment. Many such solutions may be possible.

String interactions can be represented by Feynman diagrams and type II and heterotic string theories each have a single fundamental interaction of the type shown in Fig. 10.32. The diagram represents the world-sheets (as opposed to the world-lines in conventional point particle field theory) of two strings joining together (or, alternatively, splitting). A plane of constant time cutting the two-dimensional surface reveals two strings before the interaction and one afterwards. Unlike the interaction of point particles (where different observers agree on a common interaction point), the space-time point of the interaction is not unique. For a given observer the interaction occurs at the point where the time slice is tangential to the surface and, as can be seen from Fig. 10.32, this is different for different

Fig. 10.32 Feynman diagram for the interaction of strings. The surface topologically resembles a pair of pants.

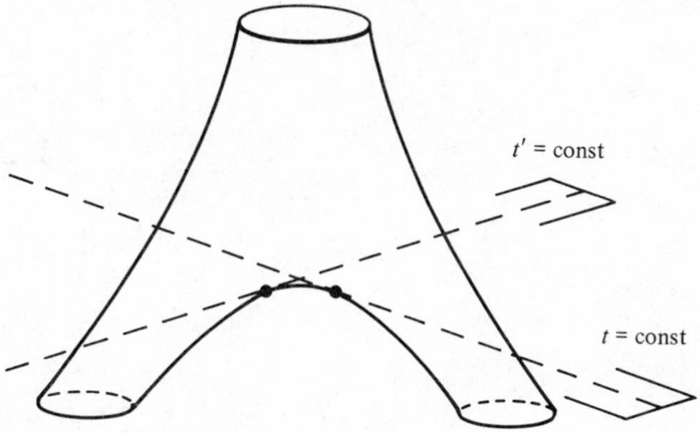

observers. Whereas the type of interaction associated with the interaction point in ordinary quantum field theory is somewhat arbitrary, that in string theory is much more constrained. The world-sheet of the string is a smooth manifold with no preferred points, and that it describes an interaction is a consequence of the topology of the surface.

If $d = 10$ is indeed the starting point for the ultimate theory, then six dimensions must be compactified to a size which is probably that of an elementary string (i.e. $\sim 10^{-33}$ cm). Furthermore, contact must be made between the gauge groups SO(32) or $E_8 \times E_8$ (this latter possibility is theoretically favoured) and the observed low energy symmetry group $SU(3) \otimes SU(2) \otimes U(1)$. No unique route has so far emerged and, indeed, many possibilities may exist. $E_8 \times E_8$ superstring models can be compactified using Calabi–Yau and related schemes. This compactification may involve rather complex manifolds. A possible scenario is that one of the E_8 groups can be broken to a subgroup of E_6 (e.g. SO(10), SU(5)); thus, possibly accommodating a Grand Unified Theory. The other E_8 group (shadow matter) interacts with the E_6 sector only through gravity. The number of fermion families, the Yukawa couplings, the values of the strong and electroweak couplings and the number of SUSY partners are all expected to emerge when this compactification is understood. All of this should be given in terms of the constants discussed above.

Until the phenomenological predictions of superstring theories become more precise, it is difficult to judge the possibility of significant experimental checks. Since the relevant gauge groups are large, additional gauge bosons, quarks, Higgs and SUSY particles are generally expected. Never before, however, have such ambitious theories been contemplated, and the potential impact of such theories may be far-reaching. The ultimate tests are experimental and, with the planned generation of future accelerators, many of the exciting questions on the nature of matter may be answered.

Appendix A

Units

The convention $\hbar = c = 1$ is used unless these symbols explicitly appear in the formula. Some useful constants are

$$\hbar = h/2\pi = 6.5822 \times 10^{-22}\ \text{MeV} \qquad \text{(Planck's constant)}$$
$$= 1.9733 \times 10^{-11}\ \text{MeV cm} = 197.33\ \text{MeV fermi},$$
$$\hbar^2 = 0.38939\ \text{GeV}^2\ \text{mb},$$
$$\alpha = e^2/(4\pi) = 1/137.036.$$

Example of usage. The total cross-section for the process $e^+e^- \to \mu^+\mu^-$ is given to lowest order in QED by

$$\sigma(e^+e^- \to \mu^+\mu^-) = \frac{4\pi\alpha^2}{3s}.$$

If the centre-of-mass energy squared, s, is given in GeV^2 then σ is in units of GeV^{-2}. To express σ in cm^2 we first use the equivalence

$$1\ \text{GeV}^{-1} = 1.9733 \times 10^{-14}\ \text{cm},$$

so

$$(1\ \text{GeV})^{-2} = 3.8939 \times 10^{-28}\ \text{cm}^2.$$

Hence

$$\sigma = \frac{4\pi \times 3.8939 \times 10^{-28}}{3 \times (137.036)^2 s(\text{GeV}^2)}\ \text{cm}^2$$
$$= \frac{8.686 \times 10^{-32}}{s(\text{GeV}^2)}\ \text{cm}^2 = \frac{86.86}{s(\text{GeV}^2)}\ \text{nb}.$$

The units of cross-section commonly used are

$$1 \text{ mb (milibarn)} = 10^{-27} \text{ cm}^2,$$

$$1 \ \mu\text{b (microbarn)} = 10^{-30} \text{ cm}^2,$$

$$1 \text{ nb (nanobarn)} = 10^{-33} \text{ cm}^2,$$

$$1 \text{ pb (picobarn)} = 10^{-36} \text{ cm}^2.$$

The useful flux of particles in colliding beam machines is measured in terms of the *luminosity* $\mathscr{L}$, defined such that the number of events n obtained in time t for a reaction of cross-section σ is

$$n = \mathscr{L} t \sigma.$$

$\mathscr{L}$ is measured in units of $\text{cm}^{-2}\,\text{s}^{-1}$. For a circular collider the beams, after initial injection, gradually die away (e.g. over a period of a few hours for LEP). The *integrated luminosity* $\mathscr{L}t$ (cm^{-2}) is a useful measure of an experimental run. For example, for $\mathscr{L} = 10^{31}\ \text{cm}^{-2}\,\text{s}^{-1}$ and $t = 10^7$ s (e.g. approximately one year's running) $\mathscr{L}t = 10^{38}\ \text{cm}^{-2} = 100\ \text{pb}^{-1}$. For a process with a cross-section $\sigma = 1.25$ nb (e.g. $e^+e^- \to \mu^+\mu^-$) the number of events $n = 1.25 \times 10^5$ for 100 inverse pb.

Appendix B

Properties of γ-matrices

B.1 The trace of an odd number of γ-matrices is zero.

B.2 $\mathrm{Tr}(\gamma^\mu\gamma^\nu) = 4g^{\mu\nu}$, $\quad \mathrm{tr}(\not{a}\not{b}) = 4a\cdot b$, where a and b are four vectors.

B.3 $\mathrm{Tr}(\gamma^\alpha\gamma^\mu\gamma^\beta\gamma^\nu) = 4[g^{\alpha\mu}g^{\beta\nu} + g^{\alpha\nu}g^{\mu\beta} - g^{\alpha\beta}g^{\mu\nu}]$,

$$\mathrm{Tr}(\not{a}_1\not{a}_2\not{a}_3\not{a}_4) = 4[(a_1\cdot a_2)(a_3\cdot a_4) + (a_1\cdot a_4)(a_2\cdot a_3) - (a_1\cdot a_3)(a_2\cdot a_4)],$$

$$\mathrm{Tr}(\not{a}_1\cdots\not{a}_n) = (a_1\cdot a_2)\,\mathrm{tr}(\not{a}_3\cdots\not{a}_n) - (a_1\cdot a_3)\,\mathrm{tr}(\not{a}_2\not{a}_4\cdots\not{a}_n) + \cdots$$
$$+ (a_1\cdot a_n)\,\mathrm{tr}(\not{a}_2\cdots\not{a}_{n-1}).$$

B.4 $\mathrm{Tr}(\gamma^5) = 0$, $\quad \mathrm{tr}(\gamma^5\not{a}\not{b}) = 0$,

$$\mathrm{Tr}(\gamma^\alpha\gamma^\mu\gamma^\beta\gamma^\nu\gamma^5) = 4\mathrm{i}\varepsilon^{\alpha\mu\beta\nu},$$

$$\mathrm{Tr}(\not{a}\not{b}\not{c}\not{d}\gamma^5) = 4\mathrm{i}\varepsilon^{\alpha\mu\beta\nu}a_\alpha b_\mu c_\beta d_\nu,$$

where $\varepsilon_{\alpha\mu\beta\nu}$ is the totally antisymmetric tensor which has the value $+1(-1)$ if the indices are an even (odd) number of permutations of 0, 1, 2, 3 and is zero if two or more indices are equal. Note that if we define $\varepsilon_{0123} = +1$ then $\varepsilon^{0123} = -1$ since $g^{00} = -g^{ii} = 1$ $(i = 1, 2, 3)$.

B.5 The following properties of γ-matrices are useful in the reduction of matrix elements

$$\gamma_\mu\gamma^\mu = 4, \qquad \gamma_\mu\not{a}\gamma^\mu = -2\not{a}, \qquad \gamma_\mu\not{a}\not{b}\gamma^\mu = 4a\cdot b,$$

$$\gamma_\mu\not{a}\not{b}\not{c}\gamma^\mu = -2\not{c}\not{b}\not{a}, \qquad \not{a}\not{b} + \not{b}\not{a} = 2a\cdot b, \qquad \not{a}\not{a} = a^2.$$

B.6 In calculating the matrix element squared the following trace combinations frequently appear and these can be evaluated with the general result

$$\begin{aligned}T &= \mathrm{tr}[\gamma^\alpha\gamma^\mu\gamma^\beta\gamma^\nu(C_1 - C_2\gamma^5)]\,\mathrm{tr}[\gamma_\theta\gamma_\mu\gamma_\phi\gamma_\nu(C_3 - C_4\gamma^5)] \\ &= 16[C_1(g^{\alpha\mu}g^{\beta\nu} + g^{\alpha\nu}g^{\mu\beta} - g^{\alpha\beta}g^{\mu\nu}) - \mathrm{i}C_2\varepsilon^{\alpha\mu\beta\nu}] \\ &\quad\times [C_3(g_{\theta\mu}g_{\phi\nu} + g_{\theta\nu}g_{\mu\phi} - g_{\theta\phi}g_{\mu\nu}) - \mathrm{i}C_4\varepsilon_{\theta\mu\phi\nu}] \\ &= 32[C_1C_3(\delta^\alpha_\theta\delta^\beta_\phi + \delta^\alpha_\phi\delta^\beta_\theta) + C_2C_4(\delta^\alpha_\theta\delta^\beta_\phi - \delta^\alpha_\phi\delta^\beta_\theta)].\end{aligned}$$

In proving this, the permutation properties of the totally antisymmetric tensor ε and the metric tensor g have been used. For a pure V − A interaction, $C_1 = C_2 = C_3 = C_4 = 1$ and the terms in $\delta^\alpha_\phi\delta^\beta_\theta$ cancel, giving

$$T = 64\delta^\alpha_\theta\delta^\beta_\phi.$$

B.7 *Representations of γ-matrices.* In addition to the Pauli–Dirac representation (Section 3.2.1), a further useful representation is that in which γ^5 is diagonal, namely the *chiral representation*

$$\gamma^0 = \begin{pmatrix} 0 & -I \\ -I & 0 \end{pmatrix}, \qquad \gamma^i = \begin{pmatrix} 0 & \sigma^i \\ -\sigma^i & 0 \end{pmatrix}, \qquad \gamma^5 = \begin{pmatrix} I & 0 \\ 0 & -I \end{pmatrix}.$$

Note that the relations (3.82) and (3.87) are satisfied in this representation. If $u = \begin{pmatrix} \phi_+ \\ \phi_- \end{pmatrix}$ is a solution defined in terms of two component Weyl spinors ϕ_+ and ϕ_-, then (cf. (3.102))

$$p_0\phi_+ + m\phi_- = (\boldsymbol{\sigma}\cdot\mathbf{p})\phi_+, \qquad p_0\phi_- + m\phi_+ = -(\boldsymbol{\sigma}\cdot\mathbf{p})\phi_-.$$

That is, in the massless limit, ϕ_+ and ϕ_- are eigenstates (u_R and u_L) of the helicity operator $h = (\boldsymbol{\sigma}\cdot\mathbf{p})/|\mathbf{p}|$, with eigenvalues +1 and −1 respectively. They are also eigenstates of chirality (γ^5), again with eigenvalues +1 and −1. However, the helicity eigenvalues are opposite to those of chirality for negative energy solutions. For $m \neq 0$, ϕ_+ and ϕ_- are still eigenstates of chirality, e.g. $\gamma^5\begin{pmatrix} \phi_+ \\ 0 \end{pmatrix} = \begin{pmatrix} \phi_+ \\ 0 \end{pmatrix}$, but not of helicity.

Under the *parity* transformation ($P = \gamma^0$), the spinor u transforms to

$$u^P = Pu = P\begin{pmatrix} \phi_+ \\ \phi_- \end{pmatrix} = \begin{pmatrix} -\phi_- \\ -\phi_+ \end{pmatrix},$$

that is, P exchanges left and righthanded states. (Thus, a parity invariant theory must treat with both chiralities symmetrically.)

Under charge conjugation ($\psi_C = i\gamma^2\psi^*$), the spinor u transforms to

$$u^C = \begin{pmatrix} 0 & i\sigma_2 \\ -i\sigma_2 & 0 \end{pmatrix}\begin{pmatrix} \phi_+^* \\ \phi_-^* \end{pmatrix} = \begin{pmatrix} i\sigma_2\phi_-^* \\ -i\sigma_2\phi_+^* \end{pmatrix}.$$

Hence, $(u_R)^C = (u^C)_L$, etc. That is, a righthanded spinor may be written as the lefthanded projection of a charge conjugate spinor, and *vice versa*.

Under the combined operation CP, u transforms to

$$u^{CP} = \begin{pmatrix} i\sigma_2\phi_+^* \\ -i\sigma_2\phi_-^* \end{pmatrix}.$$

Hence, $(u_L)^{CP} = (u^{CP})_L$, so that a theory with one lefthanded spinor can be CP invariant, but not separately C or P invariant.

B.8 *Majorana spinor*. This is a four-component spinor such that $\psi_M = \psi_M^C$. Hence, (using the representation B.7), $\phi_+ = i\sigma_2\phi_-^*$ and $\phi_- = -i\sigma_2\phi_+^*$, so that the Majorana spinor is built out of only one Weyl spinor, whereas the Dirac spinor for a massive fermion requires two (a Dirac spinor can be expressed as the sum of two Majorana spinors). An important difference between Majorana and Dirac spinors arises in the consideration of the mass term in the Lagrangian, which has the form (4.106)

$$m\bar{\psi}\psi = m(\bar{\psi}_L\psi_R + \bar{\psi}_R\psi_L).$$

Now, from (3.167), $\psi^C = i\gamma^2\gamma^0\bar{\psi}^t = C\bar{\psi}^t$ (t = transpose), so that, if $\psi_R = (\psi_L)^C$, then

$$\bar{\psi}_R\psi_L = (\psi_R^C)^t C\psi_L = \psi_L^t C\psi_L.$$

That is, a mass term can be constructed for a single chirality fermion.

In general, for spinors $\psi_L^{(i)}$ $(i = 1, \ldots, n)$, the most general mass term takes the form

$$M^{ij}\psi_L^{(i)t}C\psi_L^{(j)} + \text{h.c.}$$

That is, the Dirac and Majorana masses are particular cases of the general form of the mass matrix M^{ij}.

Appendix C

Cross-sections, decay rates and Feynman rules

C.1 CROSS-SECTIONS AND DECAY RATES

The results derived in the text are collected here for convenient reference. The cross-section for the interaction of two particles 1 and 2, to a final state containing n_f particles (fermions or bosons), is

$$d\sigma = \frac{(2\pi)^4\,\delta^4(p_i - p_f)|\mathcal{M}_{fi}|^2}{4[(p_1\cdot p_2)^2 - m_1^2 m_2^2]^{1/2}} \prod_{j=1}^{n_f} \frac{d^3 p_j}{(2\pi)^3 2E_j}, \tag{C.1}$$

where p_i and p_f are the total initial and final four-momenta respectively.

For the case of the scattering of a particle by a static field this equation becomes

$$d\sigma = \frac{(2\pi)\,\delta(E_i - E_f)|\mathcal{M}_{fi}|^2}{\beta_i 2E_i} \frac{d^3 p_f}{(2\pi)^3 2E_f}. \tag{C.2}$$

The differential decay rate for a particle of mass M to decay to n_f final state particles is, in the rest frame of the decaying particle,

$$d\omega = \frac{(2\pi)^4\,\delta^4(p_i - p_f)|\mathcal{M}_{fi}|^2}{2M} \prod_{j=1}^{n_f} \frac{d^3 p_j}{(2\pi)^3 2E_f}, \tag{C.3}$$

In each case $|\mathcal{M}_{fi}|^2$, the invariant matrix element squared, must be summed over final spin states and averaged over initial states, as dictated by the actual calculation to be performed.

For the elastic scattering process $1 + 2 \to 3 + 4$, neglecting the particle masses, (C.1) has the simple form

$$\frac{d\sigma}{d\Omega} = \frac{|\mathcal{M}_{fi}|^2}{64\pi^2 s}, \tag{C.4}$$

where $d\Omega$ is an element of solid angle in the cms for particle 3. In terms of $t=(p_1-p_3)^2$, this equation has the (invariant) form

$$\frac{d\sigma}{dt}=\frac{|\mathcal{M}_{fi}|^2}{16\pi^2 s}. \tag{C.5}$$

Alternatively, defining $y=p_2\cdot q/p_2\cdot p_1$, where $q=p_1-p_3$, this becomes

$$\frac{d\sigma}{dy}=\frac{|\mathcal{M}_{fi}|^2}{16\pi s}. \tag{C.6}$$

C.2 FEYNMAN RULES FOR QED

The rules below are for tree graphs only. The problems with loops and renormalisation are discussed in the text.

C.2.1 External particle factor

For each spin $\frac{1}{2}$ fermion (f), antifermion ($\bar{f}$) or photon (γ), include the following factors

	f	$\bar{f}$	γ
in	$u(p, s)$	$\bar{v}(p, s)$	$\varepsilon_\mu(k, \lambda)$
out	$\bar{u}(p', s')$	$v(p', s')$	$\varepsilon^*_\mu(k', \lambda')$

The spinors u, v, $\bar{u}$, $\bar{v}$ are given in terms of the four-momentum p and spin four-vector s. For the photon, the polarisation vector ε is given in terms of the four-momentum k and polarisation state λ.

There is a relative factor -1 between graphs which differ only by the interchange of two external identical fermion lines, including interchange of an initial fermion with a final antifermion.

C.2.2 Propagators†

spin 0 $\quad \overset{p}{- - - \rightarrow} \equiv \dfrac{i}{p^2-m^2}=iG_S(p),$

spin $\frac{1}{2}$ $\quad \overset{p}{\longrightarrow} \equiv \dfrac{i}{\not{p}-m}=\dfrac{i(\not{p}+m)}{p^2-m^2}=iG_D(p),$

photon $\quad \overset{q}{\sim\sim\sim} \equiv -\dfrac{ig^{\mu\nu}}{q^2}=iG_P^{\mu\nu}(q^2),$

massive vector $\quad \overset{q}{\sim\sim\sim} \equiv \dfrac{i(-g^{\mu\nu}+q^\mu q^\nu/M^2)}{q^2-M^2}=iG_V^{\mu\nu}(q^2).$

† The symbol G_j is used for the propagator with $j=S$, D, P for scalar particles, Dirac particles and photons respectively. The notation Δ_F, S_F and D_F are often used in the literature for these.

The photon propagator is given in the *Lorentz* or *Feynman* gauge. This is a particular choice ($\xi = 1$) of the general ξ gauge

$$iG_{\mathrm{P}}^{\mu\nu}(q^2) = \frac{i}{q^2}\left(-g^{\mu\nu} + (1-\xi)\frac{q^\mu q^\nu}{q^2}\right).$$

This form of the propagator gives physical results, since the $q^\mu q^\nu$ term gives zero contribution if the propagator is coupled to a conserved current.

More generally, the denominator for each of the propgators can be written with m^2 replaced by $m^2 - i\varepsilon$. Thus the photon propagator becomes $-ig^{\mu\nu}/(q^2 + i\varepsilon)$, and the limit $\varepsilon \to 0$ is taken at the end of the calculation.

C.2.3 Vertices ($e > 0$)

Spin 0

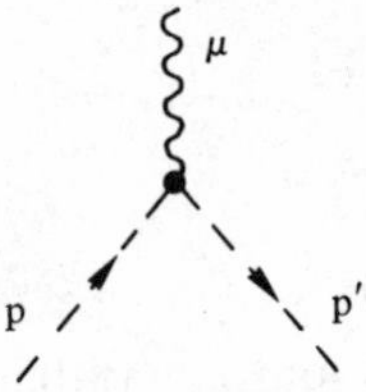

$-ie(p+p')_\mu$

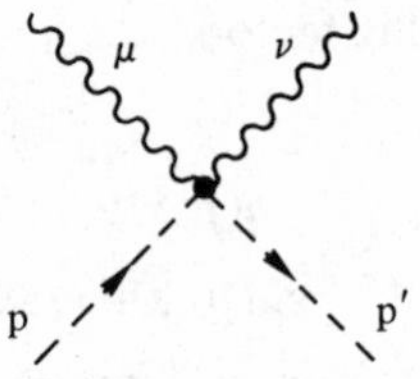

$2ie^2 g_{\mu\nu}$

Spin $\frac{1}{2}$

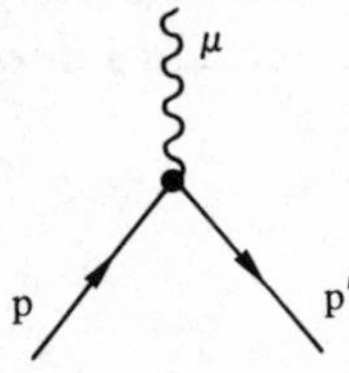

$-ie\gamma_\mu$

The four-momentum is conserved at each vertex. If the internal momentum q is not fixed by momentum conservation, then the integration $\int d^4q/(2\pi)^4$ must be carried out; one such integration arises for each closed loop.

C.3 FEYNMAN RULES FOR THE STANDARD ELECTROWEAK MODEL

The rules are again for tree graphs only, and are given in the *unitary* gauge; a gauge in which there are no unphysical particles (ghosts). The factors for external fermions and antifermions are the same as those in QED. The fermions are the lepton doublets (e, ν_e), (μ, ν_μ), (τ, ν_τ) the quark doublets (u, d′), (c, s′) and (t, b′), where the prime indicates that the weak eigenstates are mixtures of different quark flavours. The electroweak interaction does not distinguish colour, so there are three identical contributions for a given transition. The external polarisation vectors for the massive gauge bosons are similar to those of the photon in QED. The summation over polarisation vectors is given by equation (10.4), and specific forms for these vectors by (10.9). The propagators for the Higgs scalar, then spin $\frac{1}{2}$ fermions, the massless photon and the massive vector particles are as given in (C.2).

The parameters in the minimal standard model are related by equations (5.63), (5.99), (5.167) and (5.168), namely

$$\frac{G_F}{2^{1/2}} = \frac{g^2}{8M_W^2},$$

$$g \sin\theta_W = g' \cos\theta_W = e,$$

$$M_W = \frac{gv}{2}, \qquad M_Z = (g^2 + g'^2)^{1/2}\frac{v}{2},$$

$$M_W = M_Z \cos\theta_W.$$

C.3.1 Charged current vertices

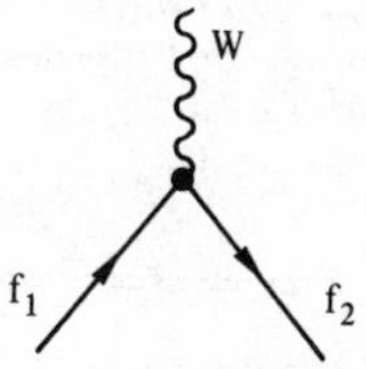

f_1	f_2	Factor
ν_l	l^-	$-\frac{ig}{2^{1/2}}\gamma_\mu\frac{(1-\gamma_5)}{2}$
q_1	q_2	$-\frac{ig}{2^{1/2}}V_{q_1q_2}\gamma_\mu\frac{(1-\gamma_5)}{2}$

where the coefficient $V_{q_1q_2}$ for the transition between q_1 (charge $\frac{2}{3}$) and q_2 (charge $-\frac{1}{3}$) is given by the KM matrix (5.193). In the case of two generations of quarks, (5.191) is used.

C.3.2 Neutral current vertices

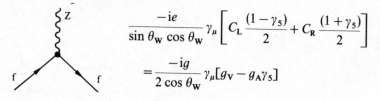

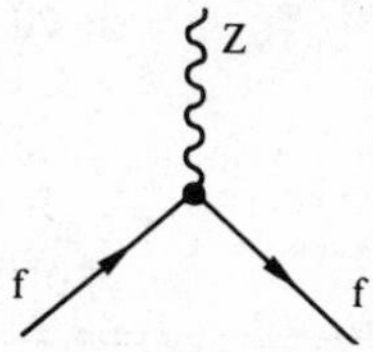

$$\frac{-ie}{\sin\theta_W \cos\theta_W}\gamma_\mu\left[C_L\frac{(1-\gamma_5)}{2} + C_R\frac{(1+\gamma_5)}{2}\right]$$

$$= \frac{-ig}{2\cos\theta_W}\gamma_\mu[g_V - g_A\gamma_5]$$

f	C_L	C_R	$g_V = C_L + C_R$	$g_A = C_L - C_R$
ν	$\frac{1}{2}$	0	$\frac{1}{2}$	$\frac{1}{2}$
l^-	$-\frac{1}{2} + \sin^2\theta_W$	$\sin^2\theta_W$	$-\frac{1}{2} + 2\sin^2\theta_W$	$-\frac{1}{2}$
$q_1(+\frac{2}{3})$	$\frac{1}{2} - \frac{2}{3}\sin^2\theta_W$	$-\frac{2}{3}\sin^2\theta_W$	$\frac{1}{2} - \frac{4}{3}\sin^2\theta_W$	$\frac{1}{2}$
$q_2(-\frac{1}{3})$	$-\frac{1}{2} + \frac{1}{3}\sin^2\theta_W$	$\frac{1}{3}\sin^2\theta_W$	$-\frac{1}{2} + \frac{2}{3}\sin^2\theta_W$	$-\frac{1}{2}$

The notation $v_f = 2g_V^f$ and $a_f = 2g_A^f$ is also frequently used in the literature.

C.3.3 Trilinear vector boson couplings $V_1V_2V_3$

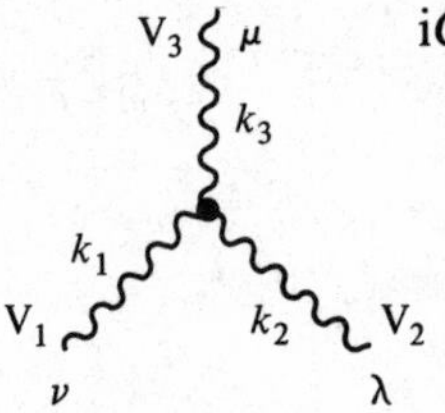

$$iC[g_{\nu\lambda}(k_1 - k_2)_\mu + g_{\lambda\mu}(k_2 - k_3)_\nu + g_{\mu\nu}(k_3 - k_1)_\lambda]$$

V_1	V_2	V_3	C
W_+^ν	W_-^λ	γ^μ	e
W_+^ν	W_-^λ	Z^μ	$e\cot\theta_W$

C.3.4 Quadrilinear vector boson couplings $V_1V_2V_3V_4$

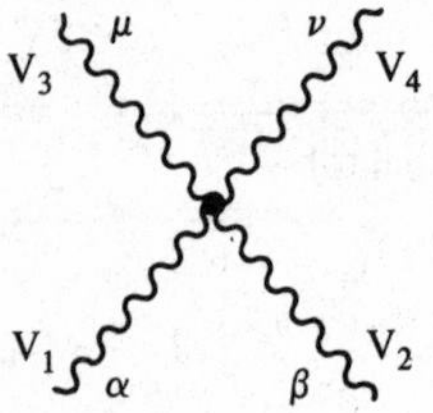

$V_1V_2V_3V_4$	Factor
$\gamma^\alpha\gamma^\beta W_+^\mu W_-^\nu$	$-ie^2(2g_{\alpha\beta}g_{\mu\nu} - g_{\alpha\mu}g_{\beta\nu} - g_{\alpha\nu}g_{\beta\mu})$
$\gamma^\alpha Z^\beta W_+^\mu W_-^\nu$	$-ie^2\cot\theta_W(2g_{\alpha\beta}g_{\mu\nu} - g_{\alpha\mu}g_{\beta\nu} - g_{\alpha\nu}g_{\beta\mu})$
$Z^\alpha Z^\beta W_+^\mu W_-^\nu$	$-ie^2\cot\theta_W(2g_{\alpha\beta}g_{\mu\nu} - g_{\alpha\mu}g_{\beta\nu} - g_{\alpha\nu}g_{\beta\mu})$
$W_+^\alpha W_-^\beta W_+^\mu W_-^\nu$	$\frac{ie^2}{\sin^2\theta_W}(2g_{\mu\alpha}g_{\nu\beta} - g_{\mu\beta}g_{\alpha\nu} - g_{\mu\nu}g_{\alpha\beta})$

C.3.5 Trilinear Higgs couplings

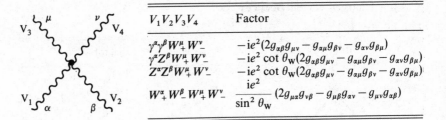

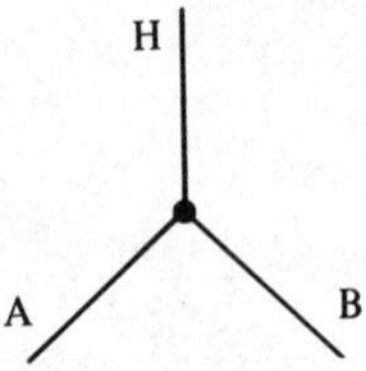

A	B	Factor
W_-^ν	W_+^μ	$\frac{ie}{\sin\theta_W}M_W g_{\mu\nu}$
Z^ν	Z^μ	$\frac{2ie}{\sin 2\theta_W}M_Z g_{\mu\nu}$
f	f	$-\frac{iem_f}{2\sin\theta_W M_W}$
H	H	$\frac{3i\mu^2 e}{M_W\sin\theta_W}$

C.3.6 Quadrilinear Higgs couplings

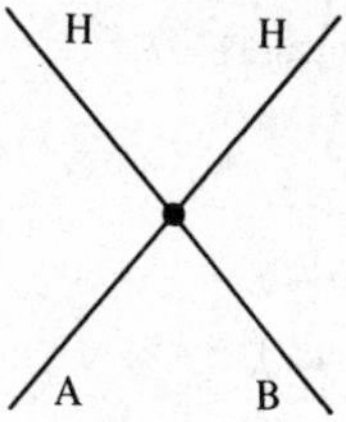

A	B	Factor
W_-^ν	W_+^μ	$\dfrac{ie^2}{2\sin^2\theta_W} g_{\mu\nu}$
Z^ν	Z^μ	$\dfrac{2ie^2}{\sin^2 2\theta_W} g_{\mu\nu}$
H	H	$\dfrac{3i\mu^2 e^2}{2M_W^2 \sin^2\theta_W}$

C.4 FEYNMAN RULES FOR QCD

For external quarks the usual Dirac spinors u and v (in) or $\bar{u}$ and $\bar{v}$ (out) are used, each however has three possible colour states. For external gluons the contributions are

$$\text{ingoing: } \varepsilon_\mu^{(\lambda)}(k)C^a; \qquad \text{outgoing: } \varepsilon_\nu^{(\lambda)*}(k)(C^a)^*$$

where C^a $(a = 1, \ldots, 8)$ represents the 'colour polarisation'.

The quark propagator has the same form as for QED. For the gluon the propagator in a covariant gauge is

a 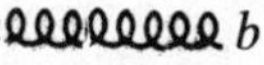 b

$$= \frac{i\delta_{ab}}{q^2}\left[-g^{\mu\nu} + (1-\xi)\frac{q^\mu q^\nu}{q^2}\right] \qquad (\xi = 1 \text{ in Feynman gauge}).$$

In the axial (or physical) gauge a vector n_μ satisfies $n_\mu G^\mu = 0$ and the propagator is

$$iG_G(q^2) = \frac{i\delta_{ab}}{q^2}\left[-g^{\mu\nu} + \frac{n^\mu q^\nu + n^\nu q^\mu}{n \cdot q} - \frac{n^2 q^\mu q^\nu}{(n \cdot q)^2}\right]$$

The vertices are as follows

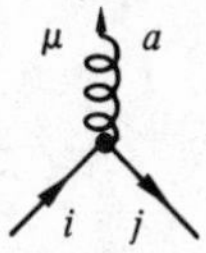

$$-ig\gamma_\mu \lambda_{ij}^a/2$$

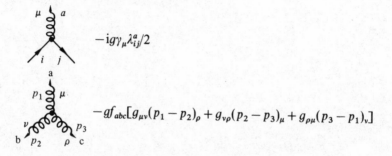

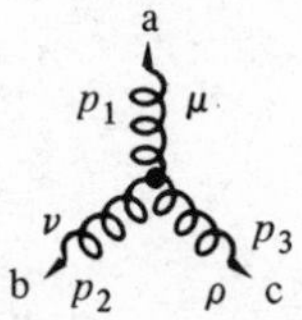

$$-gf_{abc}[g_{\mu\nu}(p_1 - p_2)_\rho + g_{\nu\rho}(p_2 - p_3)_\mu + g_{\rho\mu}(p_3 - p_1)_\nu]$$

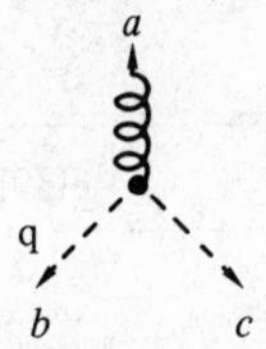

$-gf_{abc}q^{\mu}$ for scalar 'ghost' particle (absent in axial gauge)

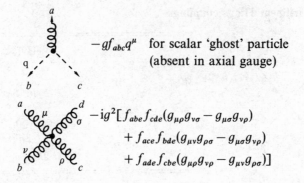

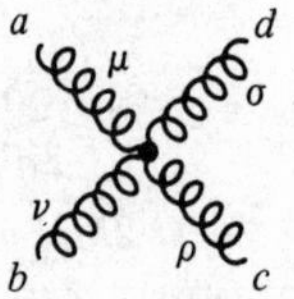

$$-ig^2[f_{abe}f_{cde}(g_{\mu\rho}g_{\nu\sigma}-g_{\mu\sigma}g_{\nu\rho}) + f_{ace}f_{bde}(g_{\mu\nu}g_{\rho\sigma}-g_{\mu\sigma}g_{\nu\rho}) + f_{ade}f_{cbe}(g_{\mu\rho}g_{\nu\rho}-g_{\mu\nu}g_{\rho\sigma})]$$

Note that these rules are sufficient only for tree level graphs.

Appendix D

Dirac δ-function

Definition

$$\delta(x-a)=0 \qquad \text{for } x \neq a,$$

$$\int_{x_1}^{x_2} \mathrm{d}x\, \delta(x-a) = \begin{cases} 1 & \text{if } x_1 < a < x_2, \\ 0 & \text{otherwise.} \end{cases} \tag{D.1}$$

The δ-function can be conveniently expressed as follows

$$\frac{1}{2\pi}\int_{-\infty}^{\infty} \mathrm{d}x \exp[\mathrm{i}(p_x - p_x')x] = \delta(p_x - p_x'), \tag{D.2}$$

or in four dimensions

$$\frac{1}{(2\pi)^4}\int_{-\infty}^{\infty} \mathrm{d}^4x \exp[\mathrm{i}(p-p')\cdot x] = \delta^4(p-p'). \tag{D.3}$$

Calculations with δ-functions can be simplified with the following properties

$$x\delta(x)=0, \qquad \delta(ax)=\frac{1}{|a|}\delta(x), \tag{D.4}$$

$$\delta(x^2-a^2)=\frac{1}{2|a|}[\delta(x-a)+\delta(x+a)], \tag{D.5}$$

$$\int \mathrm{d}x\, \delta[g(x)] = |\mathrm{d}g(x)/\mathrm{d}(x)|^{-1}_{x=x_0} \qquad \text{where } g(x_0)=0. \tag{D.6}$$

For a more detailed discussion see, for example, Gasiorowicz (1974).

Appendix E

Fierz transformation

For a contact interaction of four fermions the matrix element can be expressed in two equivalent forms

$$\mathcal{M} = \sum_i \bar{u}_3 O_i u_1 \bar{u}_4 O^i (C_i + C_i' \gamma^5) v_2 \equiv -\sum_j \bar{u}_4 O_j u_1 \bar{u}_3 O^j (g_j + g_j' \gamma^5) v_2. \tag{E.1}$$

The factor -1 arises from the anticommutation properties of the Dirac field operators, i.e. Fermi statistics. The reordering does not affect the parity (i.e. γ^5 part) of the matrix element, so we have

$$\sum_i C_i \bar{u}_3 O_i u_1 \bar{u}_4 O^i v_2 \equiv -\sum_j g_j \bar{u}_4 O_j u_1 \bar{u}_3 O^j v_2, \tag{E.2}$$

where

$$O^i = 1,\ \gamma^\mu,\ \sigma^{\mu\nu},\ \mathrm{i}\gamma^5\gamma^\mu,\ \gamma^5. \tag{E.3}$$

If we consider the explicit wave function components $u_{1\alpha}$, $v_{2\beta}$, $u_{3\gamma}$, $u_{4\delta}$ then we have

$$\sum_i C_i (O_i)_{\gamma\alpha} (O^i)_{\delta\beta} = -\sum_j g_j (O_j)_{\delta\alpha} (O^j)_{\gamma\beta}. \tag{E.4}$$

Multiplying by $(O_k)_{\beta\delta}$ gives

$$\sum_i C_i (O_i)_{\gamma\alpha} (O^i)_{\delta\beta} (O_k)_{\beta\delta} = -\sum_j g_j (O^j)_{\gamma\beta} (O_k)_{\beta\delta} (O_j)_{\delta\alpha}. \tag{E.5}$$

Using the property

$$\sum_{\beta,\delta} (O^i)_{\delta\beta} (O_k)_{\beta\delta} = \mathrm{tr}(O^i O_k) = 4\delta_{ik}. \tag{E.6}$$

This gives, returning to matrix form,

$$C_i O_i = -\frac{1}{4}\sum_j g_j (O^j O_i O_j). \tag{E.7}$$

The relationship between the coefficients can be written in the form

$$C_i = \sum_j F_{ij} g_j, \tag{E.8}$$

where the coefficients of the matrix F can be evaluated explicitly from (E.7) giving

$$F = -\frac{1}{4}\begin{pmatrix} 1 & 4 & 6 & 4 & 1 \\ 1 & -2 & 0 & 2 & -1 \\ 1 & 0 & -2 & 0 & 1 \\ 1 & 2 & 0 & -2 & -1 \\ 1 & -4 & 6 & -4 & 1 \end{pmatrix}. \tag{E.9}$$

A similar result holds for the C'_i and g'_i. Note that there is no connection between the (V, A) and T interactions. However, if there are V and A interactions in one form then there are, in general, S, V, A and P interactions in the other form. Hence, even if the interactions are mediated by vector bosons there can be effective S and P terms in the other form.

If the interaction is pure V − A ($g_V = 1, g'_V = -1, g_A = -1, g'_A = 1$, others zero) then the interaction is also pure V − A in terms of the Cs. A pure V + A interaction is also the same in both forms.

Appendix F

Creation and annihilation operators for a Dirac field

For a Dirac field the following operators can be defined, in a similar way to those for the scalar field (Section 3.1.2),

for e^-	creation $a_r^\dagger$,	annihilation a_r,
for e^+	creation $b_r^\dagger$,	annihilation b_r.

These obey the following anticommutation relations, with $a_r \equiv a_r(\mathbf{p})$, $a_s \equiv a_s(\mathbf{p}')$ etc.

$$\{a_r(\mathbf{p}), a_s^\dagger(\mathbf{p}')\} = \{b_r(\mathbf{p}), b_s(\mathbf{p}')\} = \delta_{rs}\,\delta_{pp'},$$

$$\{a_r, a_s\} = \{a_r^\dagger, a_s^\dagger\} = \{b_r, b_s\} = \{b_r^\dagger, b_s^\dagger\} = 0,$$

$$\{a_r, b_s\} = \{a_r, b_s^\dagger\} = \{a_r^\dagger, b_s\} = \{a_r^\dagger, b_s^\dagger\} = 0.$$

The field functions are constructed in terms of these operators as follows

$$\psi(x) = \psi^{(+)}(x) + \psi^{(-)}(x)$$

$$= \sum_{r,p} \frac{1}{(2E_pV)^{1/2}} [a_r(\mathbf{p})u_r(\mathbf{p}) \exp(-ip\cdot x) + b_r^\dagger(\mathbf{p})v_r(\mathbf{p}) \exp(ip\cdot x)],$$

and

$$\bar{\psi}(x) = \bar{\psi}^{(+)}(x) + \bar{\psi}^{(-)}(x)$$

$$= \sum_{r,p} \frac{1}{(2E_pV)^{1/2}} [b_r(\mathbf{p})\bar{v}_r(\mathbf{p}) \exp(-ip\cdot x) + a_r^\dagger(\mathbf{p})\bar{u}_r(\mathbf{p}) \exp(ip\cdot x)].$$

References

UA1, ARGOS, etc. refer to particular experimental collaborations

Abe K. *et al.* (1987), *Phys. Rev. Lett.* **58**, 637.
Abramowicz H. *et al.* [CDHS] (1982), *Z. Phys.* **C15**, 19.
Abramowicz H. *et al.* [CDHS] (1985), *Z. Phys.* **C28**, 51.
Abramowicz H. *et al.* [CDHS] (1986), *Phys. Rev. Lett.* **57**, 298.
Ademollo M. & Gatto R. (1964), *Phys. Rev. Lett.* **13**, 264.
Adler S. L. (1966), *Phys. Rev.* **143**, 1144.
Adler S. L. (1969), *Phys. Rev.* **177**, 2426.
Aglietta M. *et al.* (1987), *Europhysics Lett.* **3**, 1315; **3**, 1321.
Aitchison I. J. R. (1982), *An informal introduction to gauge field theories*, Cambridge University Press, Cambridge, UK.
Alam M. S. *et al.* (1986), *Phys. Rev.* **D34**, 3279.
Albajar C. *et al.* [UA1] (1987a), *Phys. Lett.* **185B**, 233; **185B**, 241.
Albajar C. *et al.* [UA1] (1987b), *Phys. Lett.* **186B**, 247.
Albajar C. *et al.* [UA1] (1987c), *CERN Report No* EP/87–190.
Albanese J. P. *et al.* [EMC] (1984), *Phys. Lett.* **144B**, 302.
Albert D. *et al.* (1980), *Nucl. Phys.* **B166**, 460.
Albrecht H. *et al.* [ARGUS] (1985), *Phys. Lett.* **163B**, 404.
Albrecht H. *et al.* [ARGUS] (1987a), *Phys. Lett.* **185B**, 223.
Albrecht H. *et al.* [ARGUS] (1987b), *Phys. Lett.* **192B**, 245.
Albrecht H. *et al.* [ARGUS] (1987c), *Phys. Lett.* **199B**, 447.
Alexander G. (1985), Proc. XVI Int. Symp. on Multiparticle Dynamics, ed. J. Grunhaus, World Scientific, Singapore.
Alexander, G. *et al.* (1987), *CERN Report No* LEPC/87–6.
Aliev T. M. *et al.* (1984), *Sov. J. Nucl. Phys.* **40**, 527.
Allaby J. V. *et al.* [CHARM] (1986), *Phys. Lett.* **177B**, 446.
Allasia D. *et al.* (1983), *Phys. Lett.* **133B**, 129.
Allasia D. *et al.* (1984), *Phys. Lett.* **135B**, 231.
Allen J. S. *et al.* (1959), *Phys. Rev.* **116**, 134.
Allen R. C. *et al.* (1985), *Phys. Rev. Lett.* **55**, 2401.
Alles W. *et al.* (1977), *Nucl. Phys.* **B119**, 125.
Altarelli G. & Parisi G. (1977), *Nucl. Phys.* **B126**, 298.
Altarelli G. *et al.* (1979), *Nucl. Phys.* **B160**, 301.
Altarelli G. (1982), *Phys. Reports* **81**, 1.
Altarelli G. *et al.* (1982), *Nucl. Phys.* **B208**, 365.

Altarelli G. *et al.* (1984), *Nucl. Phys.* **B246**, 12; *Z. Phys.* **C27**, 617.
Altarelli G. *et al.* (1986), *Physics at LEP, CERN Report No* 86–02.
Altarelli G. & Martinelli G. (1986), *Physics at LEP, CERN Report No* 86–02.
Althoff M. *et al.* [TASSO] (1984), *Phys. Lett.* **138B**, 317.
Amaldi U *et al.* (1987), *Phys. Rev.* **D36**, 1385.
Amendolia S. R. *et al.* (1984), *Phys. Lett.* **146B**, 116.
Amendiola S. R. *et al.* (1986), *Phys. Lett.* **178B**, 435.
Anderson C. D. (1932), *Science* **76**, 238.
Andersson B. *et al.* (1983), *Phys. Reports* **97**, 31.
Appel J. A. *et al.* [UA2] (1986), *Z. Phys.* **C30**, 1.
Argento A. *et al.* [BCDMS] (1983), *Phys. Lett.* **120B**, 245.
Arnison G. *et al.* [UA1] (1983), *Phys. Lett.* **122B**, 103; **126B**, 398.
Arnison G. *et al.* [UA1] (1984a), *Phys. Lett.* **136B**, 294.
Arnison G. *et al.* [UA1] (1984b), *Phys. Lett.* **147B**, 493.
Ash W. *et al.* [MAC] (1985), *SLAC Publication No* 3741.
Ashman J. *et al.* [EMC] (1988), *Phys. Lett.* **206B**, 364.
Aubert J. J. *et al.* (1974), *Phys. Rev. Lett.* **33**, 1404.
Aubert J. J. *et al.* [EMC] (1983a), *Phys. Lett.* **123B**, 123.
Aubert J. J. *et al.* [EMC] (1983b), *Phys. Lett.* **123B**, 275.
Aubert J. J. *et al.* [EMC] (1985), *Phys. Lett.* **161B**, 203.
Aubert J. J. *et al.* [EMC] (1986), *Nucl. Phys.* **B272**, 158.
Augustin J. E. *et al.* (1974), *Phys. Rev. Lett.* **33**, 1406.
Bacino W. *et al.* (1978), *Phys. Rev. Lett.* **41**, 13.
Bacino W. *et al.* (1979), *Phys. Rev. Lett.* **42**, 749.
Badier J. *et al.* (1983), *Z. Phys.* **C18**, 281.
Baer H. *et al.* (1986), *Physics at LEP, CERN Report No* 86–02.
Bahcall J. N. *et al.* (1983), *Astrophys. J.* **292**, L79.
Bahcall, J. N. & Glashow S. L. (1987), *Nature* **326**, 462.
Bailin D. (1982), *Weak Interactions*, Adam Hilger, Bristol.
Banner G. *et al.* [UA2] (1983), *Phys. Lett.* **122B**, 476; **129B**, 130.
Barbiellini G. & Santoni C. (1986), *Riv. Nuovo Cimento* **9**, No 2.
Barbiellini G. *et al.* (1986), *Physics at LEP, CERN Report No* 86–02.
Bardon M. *et al.* (1965), *Phys. Rev. Lett.* **14**, 449.
Barger V. & Pakvasa S. (1979), *Phys. Lett.* **81B**, 195.
Barger V. *et al.* (1983), *Phys. Rev.* **D28**, 1618.
Bartel W. *et al.* [JADE] (1985a), *Phys. Lett.* **146**, 437; *Z. Phys.* **C26**, 507.
Bartel W. *et al.* [JADE] (1985b), *Phys. Lett.* **161B**, 188.
Beall G. *et al.* (1982), *Phys. Rev. Lett.* **48**, 848.
Bean A. *et al.* (1987) *Phys. Rev.* **D35**, 3533.
Beg M. *et al.* (1977), *Phys. Rev. Lett.* **38**, 1252.
Behrend H. J. *et al.* [CELLO] (1988), *Phys. Lett.* **200B**, 226.
Behrends, S. *et al.* [CLEO] (1983), *Phys. Rev. Lett.* **50**, 881.
Bell J. S. & Steinberger J. (1965), *Proc. Oxford Int. Conf. on Elementary Particles*, ed. R. G. Moorhouse *et al.*, RHEL, Didcot, UK.
Bell J. S. & Jackiw R. (1969), *Nuovo Cimento* **51A**, 47.
Bellotti E. *et al.* (1984), *Proc. 11th Int. Conf. on Neutrino Physics and Astrophysics, Nordkirken*, ed. K. Kleinknecht & E. A. Paschos, World Scientific, Singapore.

Beltrami I. *et al.* (1987), *Phys. Lett.* **194B**, 326.
Berends F. A. & Kleiss R. (1985), *Nucl. Phys.* **B260**, 32.
Berends F. A. *et al.* (1987), *Phys. Lett.* **185B**, 395.
Berger E. L. & Coester F. (1987), *Ann. Rev. Nucl. Sci.*, **37**, 463
Bergkvist K. E. (1985), *Proc. Int. Symp. on Lepton and Photon Interactions at High Energies*, ed. M. Konuma & K. Takahashi, University of Kyoto, Japan.
Bergsma F. *et al.* [CHARM] (1983), *Phys. Lett.* **122B**, 465.
Bergsma F. *et al.* [CHARM] (1984), *Phys. Lett.* **141B**, 129.
Berman S. M. (1958), *Phys. Rev. Lett.* **1**, 468.
Betev B. *et al.* (1985), *Z. Phys.* **C28**, 15.
Bethe H. A. (1986), *Phys. Rev. Lett.* **56**, 1305.
Bigi I. I. (1981), *Phys. Lett.* **106B**, 510.
Bigi I. I. (1987), SL AC Publication Nos 4299 & 4300.
Bigi I. I. & Sanda A. I. (1987), *Nucl. Phys.* **B281**, 41.
Bionta R. M. *et al.* [IMB] (1987), *Phys. Rev. Lett.* **58**, 1494.
Bjorken J. D. (1966), *Phys. Rev.* **148**, 1467.
Bjorken J. D. (1969), *Phys. Rev.* **179**, 1547.
Bjorken J. D. (1970), *Phys. Rev.* **D1**, 1376.
Bjorken J. D. & Drell S. (1964), *Relativistic Quantum Mechanics*, McGraw-Hill, New York.
Bjorken J. D. & Drell S. (1965), *Relativistic Quantum Fields*, McGraw-Hill, New York.
Blackett P. M. S. & Occhialini G. P. S. (1933), *Proc. Roy. Soc.* **A139**, 699.
Blietschau J. *et al.* [Gargamelle] (1977), *Phys. Lett.* **71B**, 231.
Blondel A. *et al.* (1987) *CERN Report No* EP/87–214.
Bodek A. *et al.* (1979), *Phys. Rev.* **D20**, 1471.
Bodek A. *et al.* (1982), *Phys. Lett.* **133B**, 82.
Böhm M. & Hollik W. (1982), *Nucl. Phys.* **B204**, 45.
Böhm M. *et al.* (1984), *Phys. Lett.* **144B**, 414.
Bopp P. *et al.* (1986), *Phys. Rev. Lett.* **56**, 919.
Bordalo P. *et al.* (1987), *Phys. Lett.* **193B**, 368.
Boris S. *et al.* (1985), *Phys. Lett.* **159B**, 217.
Bouchiat C. & Michel L. (1957), *Phys. Rev.* **106**, 170.
Bouchiat M. A. *et al.* (1984), *Phys. Lett.* **134B**, 463.
Bourquin M. *et al.* (1984), *Nucl. Phys.* **B241**, 1.
Brennan E. C. (1985), *Proc. 13th SLAC Summer Inst. on Particle Physics*, SLAC Report No 296.
Brodsky S. J. & Chertok B. T. (1976), *Phys. Rev.* **D14**, 3003.
Brosi A. R. (1962), *Nucl. Phys.* **33**, 353.
Buchmüller W. *et al.* (1986), *Physics at LEP*, *CERN Report No* 86–02.
Buras A. J. (1980), *Rev. Mod. Phys.* **52**, 199.
Buras A. J. (1985), *Proc. HEP85*, ed. L. Nitti & G. Preparata, EPS, Geneva.
Burkard H. *et al.* (1985), *Phys. Lett.* **150B**, 242.
Cabibbo N. (1963), *Phys. Rev. Lett.* **10**, 531.
Caldwell D. O. *et al.* (1987), *Phys. Rev. Lett.* **59**, 419.
Callan C. G. & Gross D. (1969), *Phys. Rev. Lett.* **22**, 156.
Carlitz R. & Kaur J. (1976), *Phys. Rev. Lett.* **38**, 673.
Carr J. *et al.* (1983), *Phys. Rev. Lett.* **51**, 627.

Cashmore R. *et al.* (1986), *Z. Phys.* **C30**, 125.
Cavaignac J. F. *et al.* (1984), *Phys. Lett.* **148B**, 387.
Chau L.–L. (1983), *Phys. Reports* **95**, 1.
Cheng T.–P. & Li L.–F. (1984), *Gauge Theory of Elementary Particle Physics*, Clarendon, Oxford.
Chikashige Y. *et al.* (1981), *Phys. Lett.* **98B**, 265.
Chounet L. M. *et al.* (1972), *Phys. Reports* **4**, 199.
Christensen C. J. *et al.* (1970), *Phys. Lett.* **B28**, 411.
Christenson J. H. *et al.* (1964), *Phys. Rev. Lett.* **13**, 138.
Close F. E. (1979), *An Introduction to Quarks and Partons*, Academic, London.
Close F. E. *et al.* (1987), *Rutherford and Appleton Laboratory Report No* RAL-87-027.
Cnops A. M. *et al.* (1978), *Phys. Rev. Lett.* **40**, 144.
Coleman S. & Weinberg E. (1973), *Phys. Rev.* **D7**, 1888.
Collins P. D. B. & Martin A. D. (1984), *Hadron Interactions*, Adam Hilger, Bristol.
Combley F. & Picasso E. (1974), *Phys. Reports* **14**, 3
Combridge B. L. *et al.* (1977), *Phys. Lett.* **70B**, 234.
Commins E. D. & Bucksbaum P. H. (1983), *Weak Interactions of Leptons and Quarks*, Cambridge University Press, Cambridge, UK.
Corriveau F. *et al.* (1981), *Phys. Rev.* **D24**, 2004.
Dally E. B. *et al.* (1980), *Phys. Rev. Lett.* **45**, 232.
Danby G. *et al.* (1962), *Phys. Rev. Lett.* **9**, 36.
Daniel H. (1968), *Rev. Mod. Phys.* **40**, 659.
Derrick M. *et al.* [HRS] (1987), *Phys. Lett.* **189B**, 260.
Dokshitzer Y. L. *et al.* (1980), *Phys. Reports* **58**, 269.
Duff M. (1985), *Proc. HEP85*, ed. L. Nitti & G. Preparata, EPS, Geneva.
Duke D. W. & Owens J. F. (1980), *Phys. Rev.* **D22**, 2280.
Duke D. W. & Roberts R. G. (1985), *Phys. Reports* **120**, 276.
Duncan M. J. *et al.* (1985), *Phys. Lett.* **153B**, 165.
Eichten E. *et al.* (1984), *Rev. Mod. Phys.* **56**, 579.
Egger J. (1980), *Nucl. Phys.* **A335**, 87.
Elliot S. R. *et al.* (1987), *Phys. Rev. Lett.* **59**, 2020.
Ellis J. & Jaffe R. L. (1974), *Phys. Rev.* **D9**, 1444.
Ellis J. *et al.* (1976a), *Nucl. Phys.* **B106**, 292.
Ellis J. *et al.* (1976b), *Nucl. Phys.* **B109**, 213.
Engelbrecht C. A. (1986), Univ. of Manchester, Preprint No M/C TH 86/28.
Engfer R. & Walter H. K. (1986), *Ann. Rev. Nucl. Sci.* **36**, 327.
Erozolimskii B. G. *et al.* (1976), *Sov. Phys. JEPT* **23**, 663.
Farhi E. & Susskind L. (1981), *Phys. Rep.* **74**, 277.
Feinberg G. & Weinberg S. (1961), *Phys. Rev. Lett.* **6**, 381.
Feltesse J. (1985), *Proc. HEP85*, ed. L. Nitti & G. Preparata, EPS, Geneva.
Fermi E. (1934), *Z. Phys.* **88**, 161.
Fernandez E. *et al.* [MAC] (1985), *Phys. Rev. Lett.* **54**, 1620.
Fernandez E. *et al.* [MAC] (1987), *Phys. Rev.* **D35**, 10.
Fetscher W. (1984), *Phys. Lett.* **140B**, 117.
Feynman R. P. (1949), *Phys. Rev.* **76**, 749 & 769.
Feynman R. P. (1961), *Quantum Electrodynamics*, Benjamin, Reading, Mass.

Feynman R. P. (1972), *Photon–Hadron Interactions*, Benjamin, Reading, Mass.
Feynman R. P. & Gell-Mann M. (1958), *Phys. Rev.* **109**, 193.
Fierz M. (1937), *Z. Phys.* **104**, 553.
Filippone B. W. (1981), *Proc. Int. Conf. on Neutrino Physics and Astrophysics*, ed. R. Cence, University of Hawaii.
Fiorini E. (1986), *Proc. '86 Massive Neutrinos in Astrophysics and in Particle Physics*, ed. D. Fackler & J. Tran Thanh Van, Kim Hup Lee, Singapore.
Fischbach E. *et al.* (1986), *Phys. Rev. Lett.* **56**, 3.
Flamm D. & Schöberl F. (1982), *Introduction to the Quark Model of Elementary Particles*, Gordon and Breach, New York.
Fogli G. L. (1982), *Nucl. Phys.* **B207**, 322.
Ford W. T. *et al.* [MAC] (1987), *Phys. Rev.* **D36**, 1971.
Friedman J. I. & Kendall H. W. (1972), *Ann. Rev. Nuc. Sci.* **22**, 203.
Fritschi M. *et al.* (1986), *Phys. Lett.* **173B**, 485.
Fritze P. *et al.* (1980), *Phys. Lett.* **96B**, 427.
Fryberger D. (1968), *Phys. Rev.* **166**, 1379.
Gaemers K. J. F. & Gounaris G. J. (1979), *Z. Phys.* **C1**, 259.
Gaillard M. K. (1979), *CERN Report No* 79–01.
Gaillard M. K. & Lee B. W. (1974), *Phys. Rev.* **D10**, 897.
Gaillard J. M. & Sauvage G. (1984), *Ann. Rev. Nucl. Sci.* **34**, 351.
Gan K. K. *et al.* [MARK II] (1987), *Phys. Lett.* **197B**, 561.
Gasiorowicz S. (1966), *Elementary Particle Physics*, Wiley, New York.
Gasiorowicz S. (1974), *Quantum Physics*, Wiley, New York.
Gasser J. & Leutwyler H. (1982), *Phys. Reports* **87**, 77.
Gastmans R. (1975), *Weak and Electromagnetic Interactions at High Energies*, ed. M. Levy *et al.*, Plenum, New York & London.
Gell-Mann M. (1962), *Phys. Rev.* **125**, 1067.
Gell-Mann M. (1964), *Phys. Lett.* **8**, 214.
Gell-Mann M. & Levy M. (1960), *Nuovo Cimento* **16**, 705.
Georgi H. & Glashow S. L. (1974), *Phys. Rev. Lett.* **33**, 438.
Geweniger C. *et al.* (1974), *Phys. Lett.* **48B**, 487.
Gibson W. M. & Pollard B. R. (1976), *Symmetry Principles in Elementary Particle Physics*, Cambridge University Press, Cambridge, UK.
Gilman F. J. (1967), *Phys. Rev.* **167**, 1365.
Gilman F. J. (1986), *Seventh Vanderbilt Conference on High Energy Physics*, *SLAC Publication No* 4002.
Gilman F. J. & Hagelin J. S. (1983), *Phys. Lett.* **133B**, 443.
Gilman F. J. & Rhie S. H. (1985), *Phys. Rev.* **D31**, 1066.
Gjesdal S. *et al.* (1974), *Phys. Lett.* **52B**, 113.
Glashow S. L. (1961), *Nucl. Phys.* **22**, 579.
Glashow S. L., Iliopoulos J. & Maiani L. (1970), *Phys. Rev.* **D2**, 1285.
Gluck M. & Reya E. (1983), *Phys. Rev.* **D28**, 2749.
Goldhaber M. *et al.* (1958), *Phys. Rev.* **109**, 1015.
Goldhaber G. *et al.* (1977), *Phys. Lett.* **69B**, 503.
Green M. B. & Schwarz J. H. (1984), *Phys. Lett.* **149B**, 117.
Green M. B., Schwarz J. H. & Witten E. (1987), *Superstring Theory*, Cambridge University Press, Cambridge, UK.
Gross D. J. & Llewellyn Smith C. H. (1969), *Nucl. Phys.* **B14**, 337.

Haber H. E. & Kane G. L. (1985), *Phys. Reports* **117**, 75.
Halzen F. & Martin A. D. (1984), *Quarks and Leptons*, Wiley, New York.
Hand L. N. (1963), *Phys. Rev.* **120**, 1834.
Harari H. & Nir Y. (1987), *Phys. Lett.* **195B**, 586.
Hasert F. J. *et al.* [Gargamelle] (1973a), *Phys. Lett.* **46B**, 121.
Hasert F. J. *et al.* [Gargamelle] (1973b), *Phys. Lett.* **46B**, 138.
Haxton W. C. & Stephenson G. C. (1985), *Prog. in Particle and Nuclear Physics*, **12**, 409.
Herb S. W. *et al.* (1977), *Phys. Rev. Lett.* **39**, 252.
Higgs P. W. (1964), *Phys. Lett.* **12**, 132.
Higgs P. W. (1966), *Phys. Rev.* **145**, 1156.
Hirata K. *et al.* [Kamiokande] (1987), *Phys. Rev. Lett.* **58**, 1490.
Hofstadter R. (1956), *Rev. Mod. Phys.* **28**, 214.
Holder M. *et al.* [CDHS] (1978), *Phys. Lett.* **74B**, 277.
Hollweg J. V. (1974), *Phys. Rev. Lett.* **32**, 961.
Hughes V. W. & Kuti J. (1983), *Ann. Rev. Nucl. Sci.* **33**, 611.
Hulth P. O. (1984), *Proc. 11th Int. Conf. on Neutrino Physics and Astrophysics,* World Scientific, Singapore.
Hung P. Q. & Sakurai J. J. (1981), *Ann. Rev. Nucl. Sci.* **31**, 375.
Itzykson C. & Zuber J. (1980), *Quantum Field Theory*, McGraw-Hill, New York.
Jackson J. D. *et al.* (1957), *Phys. Rev.* **106**, 517.
Jacob M. (1958), *Nuovo Cimento* **9**, 826.
Jaffe, R. L. *et al.* (1984), *Phys. Lett.* **134B**, 449.
Jaffe R. L. (1987a), *Phys. Lett.* **193B**, 101.
Jaffe R. L. (1987b), *Proc. XI Int. Conf. on Particles and Nuclei* (PANIC). University of Kyoto, Japan.
Jarlskog C. (1979), *New Phenomena in Lepton–Hadron Physics*, ed. D. E. C. Fries & J. Wess, Plenum, New York & London.
Jarlskog C. (1983), *Proc. of HEP83, Brighton,* ed. J. Guy & C. Costain, Rutherford and Appleton Laboratory, Didcot, UK.
Jaros J. A. (1983), *Proc. 3rd Int. Conf. on Physics in Collisions, Como, Italy, SLAC Publication No.* 3248.
Jeckelmann B. *et al.* (1986), *Phys. Rev. Lett.* **56**, 1444.
Jegerlehner F. (1986), *Z. Phys.* **C32**, 425.
Johnson C. H. *et al.* (1963), *Phys. Rev.* **132**, 1149.
Jones, G. T. *et al.* (1986), *Phys. Lett.* **178B**, 329.
Jonker M. *et al.* [CHARM] (1981), *Phys. Lett.* **102B**, 67.
Jonker M. *et al.* [CHARM] (1983), *Z. Phys.* **C17**, 211.
Källen G. (1964), *Elementary Particle Physics*, Addison–Wesley, Reading, Mass.
Kinoshita T. (1959), *Phys. Rev. Lett.* **2**, 477.
Kinoshita T. *et al.* (1984), *Phys. Rev. Lett.* **52**, 717.
Kinoshita T. & Sirlin A. (1957), *Phys. Rev.* **108**, 844.
Kirsten T. *et al.* (1983), *Phys. Rev. Lett.* **50**, 474.
Klanner R. (1984), *Proc. 22nd Int. Conf. on High Energy Physics,*, Academie der Wissenschaften, Leipzig, DDR.
Kleinknecht K. (1986), *Detectors for Particle Radiation*, Cambridge University Press, Cambridge, UK.
Kluttig H. *et al.* (1977), *Phys. Lett.* **71B**, 446.

Kobayashi M. & Maskawa K. (1973), *Prog. Theor. Phys.* **49**, 652.
Kodaïra J. *et al.* (1979), *Phys. Rev.* **D20**, 627.
Koks F. & Van Klinken J. (1976), *Nucl. Phys.* **A272**, 61.
Kolb E. W. *et al.* (1987), *Phys. Rev.* **D35**, 3598.
Koltick D. S. (1986), *Proc. '86 Massive Neutrinos in Astrophysics and Particle Physics*, ed. U. Fackler & J. Tran Thanh Van, Kim Hup Lee, Singapore.
Kremer M. *et al.* (1983), *Johannes-Gutenberg University, Mainz, Report No* MZ-TH/83-12.
Lai C. H. (1981), *Gauge Theory of Weak and Electromagnetic Interactions*, ed. C. H. Lai, World Scientific, Singapore.
Langacker P. (1981), *Phys. Rep.* **72**, 185.
Langacker P. (1985), *Proc. Int. Symp. on Lepton and Photon Interactions at High Energies*, ed. M. Konuma & K. Takahashi, University of Kyoto, Japan.
Lattes C. M. G. *et al.* (1947), *Nature* **159**, 694.
Leader, E. & Predazzi E. (1982), *Introduction to Gauge Theories and the 'New Physics'*, Cambridge University Press, Cambridge, UK.
Lee T. D. & Yang C. N. (1956), *Phys. Rev.* **104**, 254.
Lee Y. *et al.* (1963), *Phys. Rev. Lett.* **10**, 253.
Leutwyler H. & Roos M. (1984), *Z. Phys.* **C25**, 91.
Llewellyn Smith C. H. (1977), *Proc. Roy. Soc., London* **Ser. A355**, 585.
Llewellyn Smith C. H. (1983), *Nucl Phys.* **B228**, 205.
Llewellyn Smith C. H. (1985), *Nucl. Phys.* **A434**, 35C.
Locci E. (1986), *Proc. 8th European Symp. on Nucleon–Antinucleon Interactions, Thessaloniki, CERN Report No* EP–86–159.
Louis W. C. *et al.* (1986), *Phys. Rev. Lett.* **56**, 1027.
Maiani L. (1977), *Proc. 1977 Int. Symp. on Lepton and Photon Interactions at High Energies*, DESY, Hamburg.
Majorana E. (1937), *Nuovo Cimento* **14**, 171.
Maki Z. *et al.* (1962), *Prog. Theor. Phys.* **28**, 870.
Mandl F. & Shaw G. (1984), *Quantum Field Theory*, Wiley, New York.
Mannelli I. (1987), *CERN Report No* EP/87–177.
Marciano W. J. & Parsa Z. (1986), *Ann. Rev. Nucl. Sci.* **36**, 171.
Marciano W. J. & Senjanović G. (1982), *Phys. Rev.* **D25**, 3092.
Marciano W. J. & Sirlin A. (1981), *Phys. Rev. Lett.* **46**, 163.
Marciano W. J. & Sirlin A. (1984), *Phys. Rev.* **D29**, 945.
Marciano W. J. & Sirlin A. (1986), *Phys. Rev. Lett.* **56**, 22.
Marshall R. (1985), *Proc. XVI Int. Symp. on Multiparticle Dynamics*, Kirvat Anavim, Israel, ed. J. Grunhaus, Kim Hup Lee, Singapore.
Mestayer M. D. *et al.* (1983), *Phys. Rev.* **D27**, 285.
Mikheyev S. P. & Smirnov A. Y. (1986), *Proc. '86 Massive Neutrinos in Astrophysics and in Particle Physics*, ed. D. Fackler & J. Tran Thanh Van, Kim Hup Lee, Singapore.
Mohapatra R. N. & Senjanovic G. (1981), *Phys. Rev.* **D23**, 165.
Mohapatra R. N. (1985), *Quarks, Leptons and Beyond*, ed. H. Fritzsch *et al.*, Plenum, New York & London.
Muirhead H. (1965), *The Physics of Elementary Particles*, Pergamon, Oxford.
Nash T. (1983) *Proc. Int. Symp. on Lepton and Photon Interactions at High Energies*, ed. D. G. Cassel & D. L. Kreinick, Cornell University, USA.

Neddermeyer S. H. & Anderson C. D. (1937), *Phys. Rev.* **51**, 884; **54**, 88.

Nicolaev N. N. & Zakarov V. I. (1975), *Phys. Lett.* **55B**, 397.

Nilles H. P. (1984), *Phys. Reports* **110**, 1.

Novikov V. A. *et al.* (1978), *Phys. Reports* **41**, 1.

Olive D. *et al.* (1979), *Riv. Nuovo Cimento* **2**, No 8.

Pais A. & Trieman S. B. (1975), *Phys. Rev.* **D12**, 2744.

Pakvasa S. *et al.* (1979), *Phys. Rev.* **D20**, 2862.

Panofsky W. K. H. & Phillips M. (1969), *Classical Electricity and Magnetism*, Addison–Wesley, Reading, Mass.

Particle Data Group (1986), *Phys. Lett.* **170B**.

Paschos F. & Wolfenstein L. (1973), *Phys. Rev.* **70**, 91.

Paul H. (1970), *Nucl. Phys.* **A154**, 160.

Pauli W. (1933), *Handbuch der Physik* **24**, 1; 233.

Pauli W. & Weisskopf V. F. (1934), *Helv. Phys. Acta* **7**, 709.

Pauli W. & Villars F. (1949), *Rev. Mod. Phys.* **21**, 434

Peccei R. D. & Quinn H. R. (1977), *Phys. Rev. Lett.* **38**, 1440.

Peccei R. D. (1986), *Proc. XXIII Int. Conf. on High Energy Physics, Berkeley, California*, World Scientific, Singapore.

Pennington M. R. (1983), *Rep. Prog. Phys.* **46**, 393.

Peoples J. (1966), Preprint Nevis-147, unpublished.

Perl M. L. (1980), *Ann. Rev. Nucl. Sci.* **30**, 299.

Perl M. L. *et al.* (1975), *Phys. Rev. Lett.* **35**, 1489.

Peterson C. *et al.* (1983), *Phys. Rev.* **D27**, 105.

Pham X. Y. (1982), *Proc. 17th Recontre de Moriond*, ed. J. Tran Thanh Van, Editions Frontières, Gif sur Yyette, France.

Poggio E. C. & Schnitzer H. J. (1979), *Phys. Rev.* **D20**, 1175; **D21**, 2034.

Politzer, H. D. (1974), *Phys. Reports* **14**, 129.

Pontecorvo B. (1968a), *Sov. Phys. JEPT* **26**, 984.

Pontecorvo B. (1968b), *Phys. Lett.* **26B**, 630.

Prescott C. Y. *et al.* (1979), *Phys. Lett.* **84B**, 524 (see also *Phys. Lett.* **77B**, 347).

Prescott C. Y. (1983), *SLAC Publications No* 3120.

Qiu J. (1987), *Nucl. Phys.* **B291**, 746.

Reines F. & Cowan C. L. (1953), *Phys. Rev.* **92**, 830.

Reines F. & Cowan C. L. (1959), *Phys. Rev.* **113**, 273.

Reines F. *et al.* (1976), *Phys. Rev. Lett.* **37**, 315.

Renard F. M. (1979), *Z. Phys.* **C1**, 225.

Rochester G. D. & Butler C. C. (1947), *Nature* **160**, 855.

Ross G. (1985), *Grand Unified Theories*, Benjamin/Cummings, California.

Sachs A. M. & Sirlin A. (1975), *Muon Physics Vol. II*, ed. V. W. Hughes, Academic Press, New York.

Sakurai J. J. (1967), *Advanced Quantum Mechanics*, Addison–Wesley, Reading, Mass.

Salam A. (1968), *Elementary Particle Theory*, ed. N. Svartholm, Stockholm, Almquist & Wiksells.

Sass J. (1985), *Physics in Collisons* **5**, ed. B. Aubert & L. Montanet, Kim Hup Lee, Singapore.

Scharf G. & Twerenbold D. (1987), *Phys. Lett.* **198B**, 389.

Scheck F. (1983), *Leptons, Hadrons and Nuclei*, North-Holland, Netherlands.

Schiff L. I. (1954), *Prog. Theor. Phys.* **11**, 288.
Schiff L. I. (1955), *Quantum Mechanics*, McGraw-Hill, New York.
Schindler R. H. (1985), *SLAC Publication No* 3799.
Schopper H. F. (1966), *Weak Interactions and Beta Decay*, North-Holland, Netherlands.
Schwartz D. M. (1967), *Phys. Rev.* **162**, 1306.
Schweppe J. *et al.* (1983), *Phys. Rev. Lett.* **51**, 2261.
Sciulli F. (1985), *Proc. Int. Symp. on Lepton and Photon Interactions at High Energies*, ed. M. Konuma & K. Takahashi, University of Kyoto, Japan.
Sehgal L. M. (1977), *Phys. Lett.* **71B**, 99.
Sehgal L. M. (1985), *Electromagnetic Effects at High Energies*, ed. H. Newman, Plenum, New York.
Shifman M. A. *et al.* (1977), *Nucl. Phys.* **B120**, 316.
Simon G. G. *et al.* (1980), *Nucl. Phys.* **A333**, 381.
Simpson J. J. (1985), *Phys. Rev. Lett.* **54**, 1891.
Sirlin A. (1975), *Nucl. Phys.* **B100**, 291.
Sirlin A. (1980), *Phys. Rev.* **D22**, 971.
Sirlin A. (1984), *Phys. Rev.* **D29**, 89.
Sirlin A. & Zucchini R. (1986), *Phys. Rev. Lett.* **57**, 1994.
Steinberg R. I. *et al.* (1976), *Phys. Rev.* **D13**, 2469.
Stone S. (1983), *Proc. Int. Symp. on Lepton and Photon Interactions at High Energies, Cornell, USA*, ed. D. G. Cassel and D. L. Kreinick, F. R. Newman Laboratory of Nuclear Studies, New York.
Street J. C. & Stevenson E. C. (1937), *Phys. Rev.* **52**, 1003.
Tanner N. W. & Dalitz R. H. (1986), *Ann. Phys.* **171**, 463.
Taylor J. C. (1978), *Gauge Theories of Weak Interactions*, Cambridge University Press, Cambridge, UK.
't Hooft G. (1971), *Nucl. Phys.* **B33**, 173; B**35**, 167.
't Hooft G. & Veltman M. (1972), *Nucl. Phys.* **B44**, 189.
Thorndike E. H. (1985), *Proc. Int. Symp on Lepton and Photon Interactions at High Energies*, ed. M. Konuma & K. Takahashi, University of Kyoto, Japan.
Totsuka Y. (1985), *Proc. Int. Symp. on Lepton and Photon Interactions at High Energies*, ed. M. Konuma & K. Takahashi, University of Kyoto, Japan.
Tsai Y.–S. (1971), *Phys. Rev.* **D4**, 2821.
Van Royen R. & Weisskopf V. F. (1967), *Nuovo Cimento* **50A**, 617; E **51A**, 583.
Veltman M. (1977), *Nucl. Phys.* **B123**, 89.
Vergados J. D. (1986), *Phys. Reports* **133**, 1.
Wachsmuth H. (1984), *Proc. 22nd Int. Conf. on High Energy Physics*, Academie der Wissenschaften, Leipzig, DDR.
Webber B. R. (1983), *Nucl. Phys.* **B238**, 492.
Weinberg S. (1958), *Phys. Rev.* **112**, 1375.
Weinberg S. (1967), *Phys. Rev. Lett.* **19**, 1264.
Wheater J. F. & Llewellyn Smith C. H. (1982), *Nucl. Phys.* **B208**, 27; E **B226**, 547.
Williams E. & Olsen P. (1979), *Phys. Rev. Lett.* **42**, 1575.
Williams H. H. (1986), *Ann Rev. Nucl. Sci.* **36**, 361.
Willis S. E. *et al.* (1980), *Phys. Rev. Lett.* **42**, 522; E **45**, 1370.
Wilson K. G. (1969), *Phys. Rev.* **179**, 1499.

Winter K. (1983), *Proc. Int. Symp. on Leptons and Photon Interactions at High Energies, Cornell, USA*, ed. D. G. Cassel & D. L. Kreinick, F. R. Newman Laboratory of Nuclear Studies, New York.
Winter K. (1986), *Proc. 2nd ESO/CERN Symp. on Cosmology, Astronomy and Fundamental Physics, CERN Report No* EP/86–61.
Witten E. (1977), *Nucl. Phys.* **B120**, 189.
Wolfenstein L. (1964), *Phys. Rev. Lett.* **13**, 562.
Wolfenstein L. (1978), *Phys. Rev.* **D17**, 2369.
Wolfenstein L. (1983), *Phys. Rev. Lett.* **51**, 1945.
Wolfenstein L. (1986), *Ann. Rev. Nucl. Sci.* **36**, 137.
Wu C. S. *et al.* (1957), *Phys. Rev.* **105**, 1413.
Yamanaka T. *et al.* (1986), *Phys. Rev.* **D34**, 85.
Yang C. N. & Mills R. L. (1954), *Phys. Rev.* **96**, 191.
Yukawa H. (1935), *Proc. Phys. Math. Soc. Japan* **17**, 48.
Zacek V. *et al.* (1985), *Phys. Lett.* **164B**, 193.
Zweig G. (1964), *CERN Report No* 8419/Th 412.

Index